Advanced Statistics from an Elementary Point of View

Advanced Statistics from an Elementary Point of View

Michael J. Panik

University of Hartford

ELSEVIER
ACADEMIC
PRESS

AMSTERDAM • BOSTON • HEIDELBERG • LONDON • NEW YORK • OXFORD
PARIS • SAN DIEGO SAN FRANCISCO • SINGAPORE • SYDNEY • TOKYO

Acquisition Editor: Tom Singer
Project Manager: Sarah Hajduk
Marketing Manager: Linda Beattie
Cover Design: Eric DeCicco
Cover Image: Getty Images
Composition: Cepha Imaging Pvt Ltd
Cover Printer: Phoenix Color
Interior Printer: The Maple-Vail Book Manufacturing Group

Elsevier Academic Press
30 Corporate Drive, Suite 400, Burlington, MA 01803, USA
525 B Street, Suite 1900, San Diego, California 92101-4495, USA
84 Theobald's Road, London WC1X 8RR, UK

This book is printed on acid-free paper. ∞

Library of Congress Cataloging-in-Publication Data

Panik, Michael J.
   Advanced statistics from an elementary point of view / Michael Panik.
      p. cm.
   Includes bibliographical references and index.
   ISBN 0-12-088494-1 (acid-free paper)
  1.   Mathematical statistics—Textbooks.   I. Title.
   QA276.P224 2005
   519.5—dc22   2005009834

British Library Cataloguing in Publication Data
A catalogue record for this book is available from the British Library

ISBN 13: 978-0-12-088494-0
ISBN 10: 0-12-088494-1

For all information on all Elsevier Academic Press Publications
visit our Web site at www.books.elsevier.com

Printed and bound by CPI Group (UK) Ltd, Croydon, CR0 4YY

Transferred to Digital Print 2011

Working together to grow
libraries in developing countries

www.elsevier.com | www.bookaid.org | www.sabre.org

ELSEVIER   BOOK AID
International   Sabre Foundation

To the Memory of

Frank C. Grella

Friend and Colleague

Contents

12 Tests of Parametric Statistical Hypotheses 483

16 Bivariate Linear Regression and Correlation 669

Appendix A 717

Preface

This book is intended as an introduction to probability and statistical inference for junior- or senior-level students in one- or two-semester courses offered by departments of mathematics or statistics. It can also serve as the foundation text for first-year graduate students in disciplines such as engineering, economics, and the physical and life sciences, where considerable use of statistics is the norm.

No previous study of probability or statistical inference is assumed. The only prerequisite is the standard introductory course in the calculus. Review sections dealing with set algebra, functions, and basic combinatories are included for your convenience.

A strength of this book is that it is highly readable. Great care has been taken to fully develop statistical concepts and definitions. Detailed explanations of theorems, tests, and results are offered without compromising the rigor and integrity of the subject matter. An objective of this work is to get students to concentrate on the statistics without being overwhelmed by the calculations. Students who have used this book should be well on their way to thinking like a statistician when it comes to problem solving- and decision- making.

An important feature of this text is the considerable attention given to sampling distributions, point and interval estimation, parametric and distributional hypothesis testing, and linear regression and correlation. These topics typically constitute the heart and soul of most statistics courses, and this book has been written with this notion in mind.

This book can be used at a variety of levels. If theorem content but not theorem proof is important, then the general flow of the various chapters can be followed and the task-oriented/applications exercises found at the end of each chapter can be selectively chosen. However, if proofs and derivations are an integral part of the course, then the exercises that address the same can be attempted. The motivation underlying the execution of each proof as well as step-by-step details necessary for its completion are offered in the *Instructor's Manual*. So although the main text is certainly not devoid of proofs (it engages the reader in proofs that are more or less constructive or that reinforce the conceptual notions and definitions at hand), the more complex and mathematically challenging proofs are available as standalone items and presented without impeding the continuity of presentation of the basic material.

After a review in Chapter 2 of some basic descriptive concepts, Chapter 3 develops the rudiments of probability theory. The latter is a key chapter since it sets the stage for the study of a broad range of inferential statistical techniques that follow. Chapter 4 treats general univariate probability distributions in considerable detail, and Chapter 5 does the same for general bivariate probability distributions. Chapters 6 and 7 introduce you to a variety of important specific discrete and continuous probability distributions, respectively. The bivariate normal distribution is also introduced in Chapter 7.

Chapter 8 exposes you to the concept of random sampling and the sampling distribution of a statistic (including those of the mean, proportion, and variance). Laws of large numbers and a Central Limit Theorem are carefully developed and explained. Chapter 9 deals with a set of derived distributions (chisquare, t, and F) and revisits the sampling distribution of the variance under the normality assumption.

Point estimation is the topic of Chapter 10. Small-sample as well as largesample properties of point estimators are covered along with a variety of techniques for finding good point estimators. This is a critical chapter in that you are introduced to the Cramér-Rao lower bound, the Fisher-Neyman Factorization Theorem, and the theorems of Lechmann-Scheffé and Rao-Blackwell. The methods of least squares, maximum likelihood, and best linear unbiased estimation are fully explored. The method of moments technique is also addressed in the chapter exercises.

Chapter 11 introduces you to the construction of a variety of single-sample and two-sample confidence intervals. Independent populations as well as paired comparisons are considered. In addition, the joint estimation of a family of population parameters is conducted using the Bonferroni method.

Parametric statistical hypothesis testing is the topic of Chapter 12. Great care is taken to develop the preliminaries. That is, issues such as statistical hypothesis formulation, the research question, varieties of decision outcomes, errors in testing, devising tests, types of tests, and so on, are treated in detail before any actual testing is undertaken. Determining the best test for a statistical hypothesis, the power of a test, and generalized likelihood ratio tests are included to complete the hypothesis testing methodology. Various one-sample and two-sample hypothesis tests are conducted, where the latter involve both independent and dependent populations. Hypothesis tests for Spearman's rank correlation coefficient round out the chapter. Throughout all of the presentation, the appropriate reporting of hypothesis testing results is emphasized.

Chapter 13 involves a collection of nonparametric hypothesis tests. Here, both single-sample and two-sample tests are executed. Comparisons between parametric and non-parametric tests are frequently made as successive tests are developed.

Testing goodness of fit is the thrust of Chapters 14 and 15. Chapter 14 treats distributional hypotheses (via chi-square Kolmogorov-Smirnov, Lilliefors, and

Shapiro-Wilk procedures) and Chapter 15 employs contingency tables to test for independence, homogeneity, and uniformity among a set of proportions. Issues concerning strength of association are also explored.

Chapter 16 offers an extremely detailed discussion of bivariate regression and correlation. Topics treated include the assumptions underlying the strong vs. weak classical linear regression models, the Gauss-Markov Theorem, least squares estimation, hypothesis test for the population parameters, confidence bands, prediction, decomposition of the sample variation in the dependent variable, the correlation model, and inferences about the population correlation coefficient. Embellishments of the basic regression model (e.g., dummy variables, nonlinearities, etc.) are found in the chapter exercises.

Many individuals have helped to make this work possible. A debt of gratitude is owed to each of them. First and foremost is my wife, Paula, whose support, patience, and encouragement helped sustain me throughout all the writing and rewriting. I would also like to thank the Department of Economics, Finance and Insurance, the Barney School, and the University's Coffin Grant Committee for financial assistance. Additionally, I am grateful to Alice Schoenrock and to a whole host of graduate assistants, particularly Amlan Datta, who helped with many aspects of the preparation of the basic manuscript and accompanying tables. Special accolades go to Marilyn Baleshiski who expertly typed the final draft of the entire manuscript.

A special note of thanks is extended to the Editorial and Production Departments at Elsevier. Acquisition Editors Barbara Holland and Tom Singer, along with the Project Manager, Sarah Hajduk, made the entire publication process quite painless. Their kindness and professionalism are deeply appreciated. Furthermore, the following reviewers generously offered many valuable comments about the manuscript: John Travis, Mississippi College; Laura McSweeney, Fairfield University; Eric Slud, University of Maryland at College Park; Tom Short, Indiana University of Pennsylvania; Cristopher Mechlin, Murray State University; Pierre Grillet, Tulane University; and Mohammad Shakil, Miami Dade University. I appreciate their insight and constructive suggestions.

Introduction

1.1 Statistics Defined

Broadly defined, statistics encompasses the theory and methods of collecting, organizing, presenting, analyzing, and interpreting data sets so as to determine their essential characteristics. Although the collection, organization, and presentation of data will be addressed frequently throughout the text, primary emphasis will be placed upon the analysis of data and the interpretation of the results. Underlying the analysis of data is the vast mathematical apparatus of abstract concepts, theorems, formulae, algorithms, and so on that constitute the statistical tools that we will employ to study a data set. In this regard, our goal is to develop *a kit of tools* with which to analyze and interpret data. These tools will then enable us to build a framework for good decision making.

1.2 Types of Statistics

There are essentially two major categories of statistics: descriptive and inductive. *Descriptive statistics* includes any treatment of data designed to *summarize* their essential features. Here we are interested in arranging data in a readable form; for example, we may construct tables, charts, and graphs; and we can compute percents, averages, rates of change, and so on. In this regard, we do not go beyond the data at hand.

With *inductive statistics*, we are concerned with making estimates, predictions or forecasts, and generalizations. Since induction is the process of reasoning from the specific to the general, the essential characteristic of inductive statistics is termed *statistical inference*—the process of inferring something about *the whole* from an examination of only *the part*. Specifically, this process is carried out through *sampling*; that is, a representative subgroup of items is subject to study (the part) and the conclusions derived therefrom are assumed to characterize the entire group (the whole). Moreover, since an exhaustive *census* (or complete enumeration) of the whole is not being undertaken, so that our conclusions are hostage to the characteristics of those items actually comprising the part, some level of error will most assuredly taint our conclusions about the whole. Hence we must accompany any generalization about the whole by a measure of the uncertainty of the inference made. Such measures can be calculated once the rudiments of probability theory are covered.

For instance, suppose we want to gain some insight into the relative popularity of a collection of candidates for the presidency of the United States. Can we realistically poll each and every individual of voting age (the whole)? Certainly not. But we can, using scientific polling techniques, elicit the preferences of only a small segment of all potential voters (the part). Hence, we may possibly conclude from this exercise that candidate A is the choice of 64% of *all* eligible voters. But this conclusion is not couched in absolute certainty. Some margin of error emerges since only a sample of individuals was taken. Hence we would accompany the 64% figure with a statement reading something like: the degree of precision of our estimate is ±3% with a 95% reliability level. The notions of *precision* and *reliability* will play a key role in the development of our inferential techniques.

In sum, if we only want to summarize or present data or just catalog facts, then we will use descriptive statistics. But if we want to make inferences based on sample data or make decisions in the face of uncertainty, then we must rely on inductive statistical methods.

1.3   Levels of Discourse: Sample vs. Population

Let us further elaborate on the concepts of *the whole* and *the part*. To set the stage for this discussion let us define an *experiment* as any process of observation. It may involve something like quality control inspection of electric motors or simply monitoring the flow of various types of motor vehicles along a particular street on a given day and over a given time interval. Next, a *variable* (denoted as X) can be thought of as any observable or measurable magnitude. Here X may depict an individual's set of exam scores, height (in inches) of sixth graders in a particular school system, weight (in pounds) of dressed turkeys, elapsed time (in minutes) taken to perform a certain task, and so on.

In this regard, let us define the sample (the part) as those things or events that both happened and were observed. It is drawn from the population (the whole), which involves things or events that happened but were not necessarily observed. Equivalently, we may think of the *population* as representing all conceivable observations on some variable X, whereas a *sample* is simply a subset of the population, that is, a given collection of observations taken from the population (see Figure 1.1). By convention, for a finite population, let us depict the population size by N and the sample size by n, where obviously $n \leq N$. Moreover, if X depicts a population variable, then the items within the population can be listed as $X : X_1, X_2, X_3, \ldots, X_N$ or as $X : X_i, i = 1, \ldots, N$. And if X represents a sample variable, then the collection of sample items can be written as $X : X_1, X_2, \ldots, X_n$ or $X : X_i, i = 1, \ldots, n$.

A sample is of finite size, but a population can be finite or infinite. That is, the collection of all individuals who have read part or all of this book can be thought

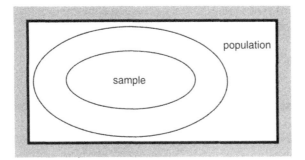

Figure 1.1   A sample as a subset of a population.

of as constituting a finite population, whereas an infinite population can easily be generated by sampling with replacement. For instance, if a population consists of N items and we *sample without replacement*, then the item obtained on any draw is set aside before the next item is chosen. But if we *sample with replacement* (on each draw we select an item from the population and then replace it before the next one is chosen), then clearly we can effectively sample in this fashion forever; that is, we can operate *as if* the population is infinite even though in reality it is not. So under sampling with replacement, a member of the population can appear more than once in the sample; in sampling without replacement, we disregard any item that has already been chosen for inclusion in the sample.

A few additional points merit our attention. First, it is important to mention that oftentimes inductive statistics is described as *decision making under uncertainty*. For our immediate purposes, note that an important source of uncertainty is the concept of *randomness*. That is, if the outcome of an experiment is not predictable, then it is said to occur in a random fashion. Next, we stated earlier that in sampling, a representative group of items is desired. In this regard, we may deem a sample as representative if it is *typical*, that is, it adequately reflects the attributes of the population. Moreover, we will frequently engage in the process of random sampling. Specifically, a sample is *random* if each and every item in the population has an equal (and known) chance of being included in the sample.

It is important to remember that the process of random sampling does not guarantee that a representative sample will be obtained. Randomization is likely but by no means certain to generate a representative sample. This is because randomization gives the same chance of selection to every sample of a given size—representative ones as well as nonrepresentative ones.

We end this section by noting that the term *parameter* (denoted θ) is used to represent any descriptive measure of a population, whereas a *statistic* (denoted $\hat{\theta}$) is any descriptive measure of a sample. Here $\hat{\theta}$ serves as an *estimator* of (the unknown) θ; that is, $\hat{\theta}$ is some function of the sample values used to discern the level of θ. The $\hat{\theta}$ actually obtained by the estimation process will be called an *estimate* of θ. If $\hat{\theta}$ represents a single numerical value, then it is termed a *point estimator* of θ. An *interval estimator* of θ enables us to state just how confident we are, in terms of probability, that θ lies within some range of values. We shall return to these notions in Chapters 10 and 11.

1.4 Levels of Discourse: Target vs. Sampled Population

The *target population* is defined as the population to be studied; it is the population about which information is desired. This is in contrast to the *sampled population*—the population from which the sample is actually obtained. This latter population concept is alternatively called the *sampling frame*, or just *frame* for short. Based upon these two population notions, consequently we can describe a sample (or study for that matter) as being *valid* if the target and sampled populations have similar characteristics. (Note that some items in the target population may not be a member of the frame.)

Continuing in this vein, an *elementary sampling unit* is an item in the frame and an *observation* is a piece of information possessed by an elementary sampling unit. A *sample*, then, is that portion of the sampled population actually studied. (Note also that a sample may be representative—it adequately reflects the attributes of the frame—but not valid.) So under random sampling, each item in the frame has the same chance of being chosen.

For example, suppose we want to develop a profile of the membership of the local country club. This is the target population. The sampled population or frame is the club's membership list. If the list is up to date, then the target and sampled populations coincide. However, if the list has not been updated recently, then the target and sampled populations may be widely disparate and thus the question of validity comes to the fore. Elementary sampling units are the individual members, whereas an observation consists of a particular data point or value of some characteristic of interest. In what follows we shall depict each characteristic by a separate variable. Hence an observation is the value of the variable representing a characteristic of an elementary sampling unit.

For instance, assume that a country club has 3000 ($= N$) members and that we want a sample of size 100 ($= n$). If the membership list is arranged alphabetically (which we can assume to be a random arrangement), then we can easily engage in a sampling process called *systematic random sampling*, with a *sampling cycle* of 30 = 3000/100 ($= N/n$). That is, we start at the beginning of the list, select one member at random from the first 30 members listed, and then pick from that item on every thirtieth number for inclusion in the sample.

Which characteristic of a member (our elementary sampling unit) might we be interested in? We can possibly list them as sex, number of years as a member, number of years service as a board member, type of membership (e.g., founding, charter, regular), annual gift amount, major activity preferred (e.g., golf, tennis, swimming, bridge), and so on. As indicated earlier, each characteristic is represented by a variable and the value of the variable is an observation of the characteristic, as listed in Table 1.1.

Table 1.1

Country Club Membership Characteristics		
Characteristic	**Variable**	**Observation Values**
Sex	X_1	Male (1); Female (2)
Years	X_2	Number of years
Board	X_3	Number of years
Type	X_4	Founding (3); Charter (2); Regular (1)
Gift	X_5	$
Activity	X_6	Golf (1); Tennis (2); Swimming (3); Bridge (4)

1.5   Measurement Scales

What types or varieties of data have we defined in the preceding example on country club membership characteristics? We may generally refer to *data* as a collection of facts, values, observations, or measurements. So if our data consists of observations that can be *measured* (i.e., classified, ordered, or quantified), then at what level does the measurement take place? Here we are interested in the *forms* in which data is found or the *scales* on which data is measured. These scales, stated in terms of increasing information content, are classified as nominal, ordinal, interval, and ratio.

Let us first consider the *nominal* scale. Here nominal should be associated with the word *name* since this scale identifies categories. Observations on a nominal scale possess neither numerical values nor order. However, observations on this type of scale can be given numerical codes such as "0 or 1" or "$1, 2, 3, \ldots$." Variables X_1 and X_6 in Table 1.1 are nominal in nature and are termed *qualitative* or *categorical* variables. Note that when dealing with a nominal scale, the categories defined must be *mutually exclusive* (each item falls into one and only one category) and *collectively exhaustive* (the list of categories is complete in that each item can be classified). Since X_1 has only two categories, it is called a binary or *dichotomous* variable. Note also that a number code has been used to classify the members as either 1 (male) or 2 (female). These numbers serve only as *identifiers*; the magnitude of the differences between these numerical values is meaningless. The only valid operations for variables represented by a nominal scale are the determination of "=" or "$\neq$."

The *ordinal* scale (think of the word *order*) includes all properties of the nominal scale with the additional property that the observations can be ranked from the smallest to the largest or from the least important to the most important. (Note that nominal measurements cannot be ordered—all items are treated equally.) If the country club mentioned earlier is a hierarchical organization wherein founding members are more important or ranked higher (in terms of privileges) than charter members, which are in turn ranked above regular members, then X_4 is an ordinal (and thus qualitative) variable. (Although chartered is in some sense *better* than regular, the ranking does not indicate *how much better*.) That is, since the numerical values assigned to X_4 are 3, 2, and 1, these numbers serve only to indicate a *pecking order* of levels of the variable—the differences between these numerical values are meaningless (we could just as effectively use 100, 57, and 10). In this regard, the only valid operations for ordinally scaled variables are "=, ≠, <, >."

Both the nominal and ordinal scales are termed *nonmetric* scales since differences among their values are of no consequence.

Next comes *interval* data. This scale includes all the properties of the ordinal scale with the additional property that distance between observations is meaningful. Here the numbers assigned to the observations indicate order and possess the property that the difference between any two consecutive values is the same as the difference between any other two consecutive values (the difference $10 - 9 = 1$ has the same meaning as $3 - 2 = 1$). It is important to note that while an interval scale has a zero point, its location may be arbitrary. Hence ratios of interval scale values have no meaning. For example, at the country club golf course hole number five is a par three hole. Member A completes the hole in four strokes while member B does so in five strokes. The club pro takes only two strokes. How do we describe the skill level of member A relative to member B? The answer depends on our point of reference or zero point. If the zero point is taken to be the performance of the club pro, then A is +2 above and B is +3 above. Can we then say that member A is one and one-half times as good as member B? What if the zero point is instead taken to be par itself? Now player A is one over par while B is two over par. Do we now conclude that member A is twice as good a golfer as B?

Other examples abound. For instance, since 0°C (zero degrees centigrade) does not imply the absence of heat (it is simply the temperature at which water freezes), 40°C is not twice as cold as 20°C. Likewise, a score of zero on a standardized national test does not imply lack of knowledge. In this regard, since the zero point is again arbitrary, a student with a score of 700 is not twice as smart as one with a score of 350.

The operations for handling variables measured on an interval scale are "=, ≠, >, <, +, −."

Our final level of measurement is the *ratio* scale. It includes all the properties of the interval scale with the added property that ratios of observations are meaningful. This is because *absolute zero is uniquely defined*. Clearly variable X_5 is a ratio variable in that $0 measures the absence of any gift and a gift of $2000 is twice as large as a gift of $1000 (the ratio is $2/1 = 2$). Valid operations for variables measured on a ratio scale are "=, ≠, >, <, +, −, ×, ÷." Both the interval and the ratio scales are said to be *metric* scales (since differences between values measured on these scales are meaningful), and variables measured on these scales are said to be *quantitative* variables.

It should be evident from the preceding discussion that any variable measured on one scale automatically satisfies all the properties of a *less informative* scale.

In addition to variables being qualitative or quantitative in nature, we may also identify variables as being discrete or continuous. A *discrete* variable takes on only a countable number of values (e.g., sex, makes of automobiles, residential housing units in a particular area of a city, etc.), whereas a *continuous* variable can take on any value over some range—that is, fractional values are admitted (e.g., time, liquid measures, interest rates, etc. are all continuous variables).

1.6   Sampling and Sampling Errors

If we undertake a complete census of some population, then we can readily describe, in varying degrees of detail, many of its salient features or characteristics. However, if we only sample from the population, then it should be intuitively clear that the different possible samples obtained will exhibit different sets of sample values (although some duplication in the sample values can occur). Hence any conclusion reached about the population from any one sample may differ slightly from that reached on the basis of examining some other sample. Again, this is because different samples typically possess different sample values. Given that none of the samples will look exactly like the population at large, we will most assuredly find a difference between some true population parameter (θ) and the statistic ($\hat{\theta}$) used to estimate it. Hence the degree of *sampling error* is $\hat{\theta} - \theta$. It should be obvious the sampling error is inescapable—it reflects the inherent natural variability among a parameter and a statistic used to estimate it. For some samples the amount of sampling error will be large; for others it may be quite small. (If the true θ is known, then the difference between $\hat{\theta}$ and θ reflects the degree of *accuracy* of our estimate of θ. Since θ is usually unknown, the best we can do is talk about a long-term degree of *precision* of our estimate of θ; i.e., we can determine an *error bound* on $\hat{\theta}$ as an estimate of θ.)

This is in contrast to the notion of *nonsampling error*, which is due essentially to unsound sampling or experimental techniques, and which can be controlled. Here human or mechanical factors distort the observed values, thus contributing to the difference between $\hat{\theta}$ and θ. Nonsampling error emerges when you have a *biased sample*. For instance, some items in the population may be more likely to be included in the sample than others or errors of observation or measurement result in a systematic accumulation of inaccuracies in a single direction. When this occurs $\hat{\theta}$ may be larger on the average (a positive bias) or smaller on the average (a negative bias) than θ. If on the average $\hat{\theta}$ equals θ, then $\hat{\theta}$ is said to be *unbiased*; that is, the long-run average of the sampling error is zero.

1.7   Exercises

1-1. For the following variables, determine whether the data are best characterized as categorical (nominal or ordinal) or numerical (continuous or discrete).

(a) X: red, white, blue

(b) X: poor, fair, good, excellent

(c) X: below average, average, above average

(d) X: African–American, Caucasian, Hispanic, Native American, Other

(e) X: birth weight (in ounces)

(f) X: private, private first class, ..., four star general

(g) X: monthly motor vehicle accidents in Boston, MA

(h) X: number of visits to the dentist per year

1-2. For the following variables, determine whether the data are measured on an interval or ratio scale.

(a) X: attendance at last night's concert

(b) X: cost of parking your car at the local civic center

(c) X: capacity of your vehicle's gas tank

(d) X: melting point (in degrees Celsius) of a lead ingot

(e) X: national unemployment rate

(f) X: IQ scores of your age cohort

(g) X: LSAT scores for college seniors

(h) X: interest rate on new car loans at the local credit union

Elementary Descriptive Statistical Techniques

2.1 Summarizing Sets of Data Measured on a Ratio or Interval Scale

In what follows we shall consider a variety of descriptive statistical techniques that may be employed to conveniently summarize, for the most part, a set of ratio- or interval-scale data. Unless otherwise stated, this data set will be assumed to represent the entire population. Two basic approaches will be advanced. First, we may summarize the data in a tabular (and, eventually, graphical) fashion. Second, rather than work with the entire mass of data in, say, tabular form, we may derive from it a set of concise quantitative summary characteristics (either computed or positional values arrived at largely from a set of formulas), which themselves describe the salient features of the data set. Once these summary measures have been obtained, they may be used to assess the current status of a particular population, measure differences between two or more populations, or consider changes in a given population over time. These approaches are contrasted in Figure 2.1.

With respect to our second approach regarding various sets of quantitative/positional summary figures, we note briefly that:

(a) *Measures of Central Location*, broadly construed, are essentially *averages* that describe the location of the *center* of the data set.

(b) *Measures of Dispersion* (typically taken about some point of central location) describe the spread or scatter of the observations along the horizontal axis.

(c) *Measures of Skewness* indicate the degree of departure of the data distribution from symmetry and thus serve as measures of *asymmetry*.

(d) *Measures of Kurtosis* indicate how flat or rounded the peak of the data distribution happens to be.

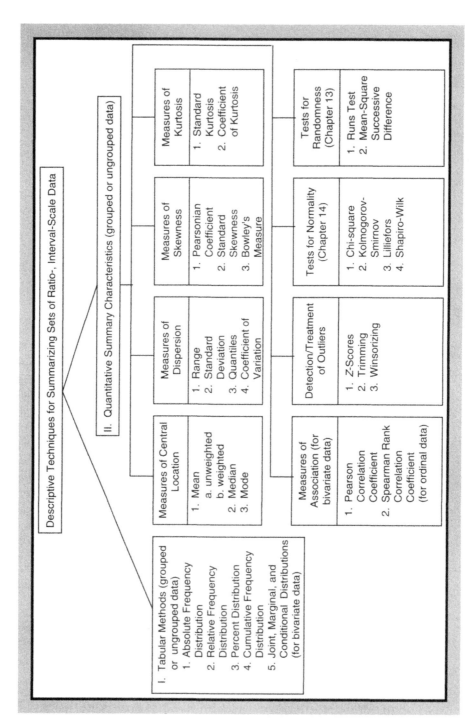

Figure 2.1   Two approaches for summarizing data.

(e) *Tests of Randomness* determine if the data set (typically a sample) can be thought of as generated in a random or nonsystematic fashion.

(f) *Tests of Normality* are used to determine if the data set closely approximates a bell-shaped symmetrical distribution.

(g) The *Determination of Outliers* or abnormally large or small extreme values is important if we do not want the values of some of our other descriptive measures to be distorted.

(h) *Measures of Association* are used to determine the strength of the (linear) relationship between two variables or between two sets of ranked values.

Except for tests of normality and tests of randomness, all of the items indicated in Figure 2.1 will be examined in this chapter. Moreover, most of these items will be subject to further study when we make the transition from descriptive to inferential statistics.

2.2   Tabular Methods

In this section we seek to determine at a glance a global picture of the entire population data set so as to detect some underlying pattern of behavior; that is, to make a quick and rough assessment of the characteristics of the population. So, given an amorphous collection of data, how may we arrange it in a systematic and easily readable form?

Let us assume that we have a set of N observations on a population variable $X : X_1, X_2, \ldots, X_N$. For instance, such a data set appears in Table 2.1. We first construct the *absolute frequency distribution*, which shows the absolute frequencies with which the various values of a variable X occur in a set of data.

Here *absolute frequency* represents the number of times a particular value of X is recorded. This distribution is illustrated in Table 2.2. In the left-hand column we have the *different* values of X listed (there is a total of 50 observations

Table 2.1

X: Time Required to Perform a Given Production Task (N = 50 Observations on Worker A) (Minutes)				
10	9	13	9	13
8	13	12	15	14
13	19	17	18	17
13	14	13	16	13
10	12	15	14	12
15	13	11	12	11
13	12	11	11	16
16	17	14	13	6
20	12	13	7	12
15	16	19	18	14

Table 2.2

Absolute Frequency Distribution of Performance Time for Worker A (Minutes)	
X	Absolute Frequency
6	1
7	1
8	1
9	2
10	2
11	4
12	7
13	11
14	5
15	4
16	4
17	3
18	2
19	2
20	1
	$\overline{50}$

on X but only 15 different values presented) and in the right-hand column we have their absolute frequencies recorded. Note that the sum of the absolute frequencies must equal the total number of observations $N = 50$. The graphical description of the absolute frequency distribution is the *absolute frequency function* (see Figure 2.2).

Next, we may transform an absolute frequency distribution into a *relative frequency distribution* by expressing the absolute frequencies as a fraction of the total number of observations N (as Table 2.3 reveals, relative frequency = (absolute frequency)/N). For instance, $X = 14$ occurs $\frac{1}{10}$ of the time. Note that the sum of the relative frequencies must be 1. The graphical description of the relative frequency distribution is the *relative frequency function* (see Figure 2.3).

Furthermore, the preceding relative frequency distribution may be transformed into a *percent distribution* by multiplying the relative frequencies in Table 2.3 by 100 so as to convert them to percentages (see Table 2.4). Here $X = 11$ occurs 8% of the time. Clearly the percent column must sum to 100%. The accompanying percent distribution function appears as Figure 2.4.

Finally, you can construct a less than or equal to cumulative frequency distribution. To do so, cumulate the absolute frequencies as X increases from its lowest to its highest value (see Table 2.5). That is, one observation is less than or equal to 6; two observations are less than or equal to 7; three observations are less than or equal to 8; and so on.

The purpose of this distribution is to answer questions such as: out of the 50 observations presented, how many times did worker A take no more than

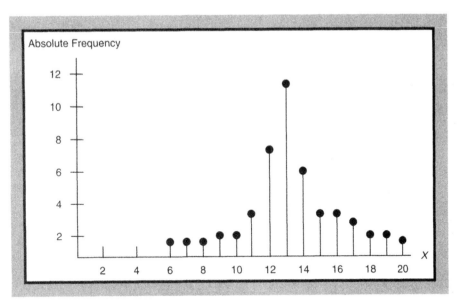

Figure 2.2   Absolute frequency function for performance time for worker A (minutes).

Table 2.3

Relative Frequency Distribution of Performance Time for Worker A (Minutes)	
X	Relative Frequency
6	1/50
7	1/50
8	1/50
9	2/50
10	2/50
11	4/50
12	7/50
13	11/50
14	5/50
15	4/50
16	4/50
17	3/50
18	2/50
19	2/50
20	1/50
	1

16 minutes to perform the given task? (Answer: 42 times) The graphical description of the less than or equal to cumulative distribution appears in Figure 2.5. Note that since X is made up of a set of discrete values, the graph is a *step function* (e.g., at $X = 6$, the less than or equal to cumulative frequency is 1; and it remains at

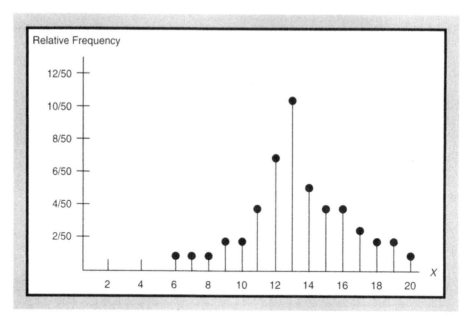

Figure 2.3   Relative frequency function for performance time for worker A (minutes).

Table 2.4

Percent Distribution of Performance Time for Worker A (Minutes)	
X	Percent of N
6	2
7	2
8	2
9	4
10	4
11	8
12	14
13	22
14	10
15	8
16	8
17	6
18	4
19	4
20	2
	100

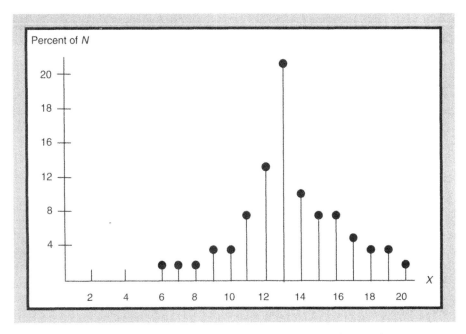

Figure 2.4 Percent distribution function for performance time for worker A (minutes).

Table 2.5

	Cumulative Frequency Distributions of Performance Time for Worker A (Minutes)		
X	Absolute Frequency	Less Than or Equal to Cumulative Frequency	Greater Than Cumulative Frequency
6	1	1	$50 - 1 = 49$
7	1	2	$50 - 2 = 48$
8	1	3	$50 - 3 = 47$
9	2	5	$50 - 5 = 45$
10	2	7	$50 - 7 = 43$
11	4	11	$50 - 11 = 39$
12	7	18	$50 - 18 = 32$
13	11	29	$50 - 29 = 21$
14	5	34	$50 - 34 = 16$
15	4	38	$50 - 38 = 12$
16	4	42	$50 - 42 = 8$
17	3	45	$50 - 45 = 5$
18	2	47	$50 - 47 = 3$
19	2	49	$50 - 49 = 1$
20	1	50	$50 - 50 = 0$
	$\overline{50}$		

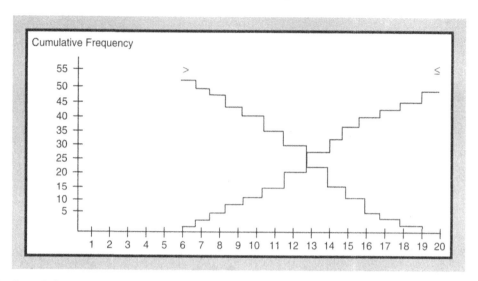

Figure 2.5   Cumulative frequency functions for performance time for worker A (minutes).

1 until X increases to 7, at which point the cumulative frequency takes an abrupt jump to 2).

2.3 Quantitative Summary Characteristics

Our goal is now to describe the characteristics of a set of data by an assortment of numerical measures. These measures then serve as an alternative to simply obtaining a global picture of the population data set in the form of, say, an absolute frequency distribution. So again, given an amorphous collection of data, what sort of quantitative summary measures best describe the properties of the set of observations? As we shall now see, these characteristics of interest may be classified as measures of central location, measures of dispersion, measures of skewness, measures of kurtosis, and detectors of outliers.

2.3.1 Measures of Central Location

These descriptive measures (which are either of a computed or positional nature) serve to locate the *center* of X's absolute frequency distribution on the horizontal axis.

Arithmetic Mean.   Given a variable $X : X_1, \ldots, X_N$, the *arithmetic mean* of X (or of X's absolute frequency distribution) is

$$\mu = \frac{1}{N} \sum_{i=1}^{N} X_i. \tag{2.1}$$

For instance, if X assumes the values 1, 4, 7, 8, 10, the arithmetic mean of X is, from (2.1), 30/5 = 6. Geometrically, the mean of X is its center of gravity, that is, X's frequency distribution will balance at the mean (see Figures 2.6a, b). In addition to this physical interpretation of the mean, we may view the word *balance* in a purely statistical sense. That is to say, if we define the ith deviation from the mean as $X_i - \mu$, then clearly the sum of the positive and negative deviations from the mean must be zero (see Table 2.6).

As far as the properties of the mean are concerned, the mean always exists, is unique, is relatively reliable (i.e., it does not vary considerably under repeated sampling from the same population), and is affected by extreme values.

Weighted Mean.   In (2.1), each observation X_i had the same relative importance. But what if some observations are in some sense more important than others?   Given that the observed X_i values have a different relative

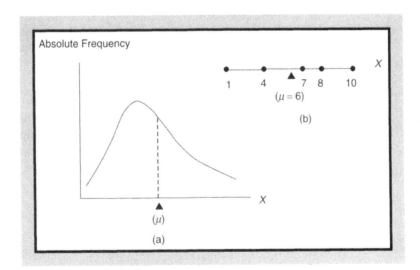

Figure 2.6 X's frequency distribution balances at the mean.

Table 2.6

X_i	$X_i - \mu$
1	−5
4	−2
7	1
8	2
10	4
	0

importance, we must form their average by using a *weighted mean*

$$\mu_w = \frac{\sum\limits_{i=1}^{N} w_i X_i}{\sum\limits_{i=1}^{N} w_i},$$ (2.2)

where w_i represents the weight attached to $X_i, i = 1, \ldots, N$.

Example 2.3.1.1   Let us assume that an individual earns the following set of scores on three separate exams: $X_1 = 90, X_2 = 70$, and $X_3 = 65$. Clearly the average score is $\mu = 225/3 = 75$. However, what if the exams were timed: the first took 15 minutes to complete; the second took one hour to complete; and the third took an hour and a half to complete. Since it is intuitively clear that those tests that took one hour and one and one-half hours to finish should be relatively more important in calculating the average score than the exam that took only 15 minutes to finish, we may assign relative importance according to time length. If we set $w_1 = 1$, then, since the second exam took four times longer to finish than the first one, $w_2 = 4$; and since the third exam took six times longer to finish than the first one, $w_3 = 6$. Then from (2.2) and Table 2.7, $\mu_w = 760/11 = 69.1$.

Here $\mu_w < \mu$ since the exams with the lower scores had relatively more weight than the exam with the higher score. ■

Arithmetic Mean of a Frequency Distribution.   When a set of data is arranged in the form of an absolute frequency distribution, we may compute the mean of the same by relying upon the concept of a weighted mean, where the weights of the *different* values of X are their absolute frequencies f_j; that is,

$$\mu = \frac{\sum\limits_{j=1}^{k} f_j X_j}{\sum\limits_{j=1}^{k} f_j} = \frac{1}{N} \sum\limits_{j=1}^{k} f_j X_j,$$ (2.3)

where k represents the number of different (net of any duplication) values of X.

Table 2.7

X_i	w_i	$X_i w_i$
90	1	90
70	4	280
65	6	390
	11	760

Table 2.8

X_j	f_j	$f_j X_j$
10	3	30
15	2	30
17	1	17
20	3	60
	9	137

Example 2.3.1.2   Given X: 10, 15, 20, 10, 17, 20, 10, 15, 10, Table 2.8 and (2.3) provide us with the value $\mu = 137/9 = 15.22$. (Note that although $N = 9$, there are only $k = 4$ *different* values of X.) ■

Median.   The *median* of a variable X (or of X's absolute frequency distribution) is the value that divides the observations on X (or the total area under the absolute frequency distribution of X) into two equal parts, given that the said observations are arranged in order of increasing magnitude. In this regard, the median is a positional value and not a computed value for a set of data. Looking to Figure 2.7, panel (a) shows the absolute frequency distribution of X; immediately below in panel (b) we have X's less than or equal to cumulative frequency distribution. If we find the 50% point (C) on the less than or equal to cumulative frequency distribution in panel (b) and project it downward to the X-axis, the resulting X value represents the median—half of the observations on X are above the median and half of them are below the median. In addition, as mentioned earlier, the median also divides the total area under X's absolute frequency distribution into two equal parts (area $A =$ area B).

To actually determine the median from a set of observations on X, let us mirror the salient features of this graphical location of the median. The steps involved in arriving at the median are as follows:

(1) Arrange the observations on X in an increasing sequence.

(2) (a) For an odd number of observations, there is always a middle term whose value is the median. In this instance the median corresponds to the value of X that occupies the $\frac{N+1}{2}^{nd}$ position; that is, the median $= X_{((N+1)/2)}$.

   (b) For an even number of observations, there is never a specific middle term. In this regard, interpolate the median as the average of the two middle terms; that is, it is midway between the $\frac{N}{2}^{nd}$ and the $(\frac{N}{2} + 1)^{nd}$ terms or the

$$\text{median} = \frac{X_{N/2} + X_{(N/2)+1}}{2}.$$

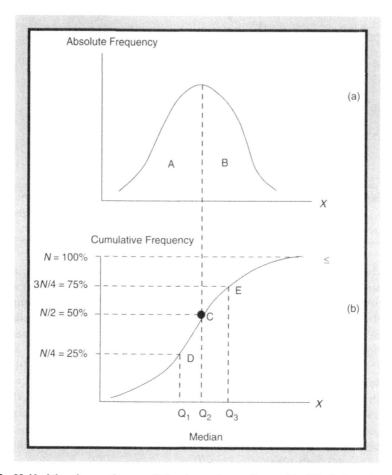

Figure 2.7 Half of the observations on X lie above the median and half lie below the median.

Example 2.3.1.3 Find the median for each of the following sets of observations:

(a) X: 8, 7, 11, 7, 2, 4, 3, 10, 8

(b) X: 4, 9, 1, 6, 7, 2

From (a), let us arrange the observations on X in an increasing sequence or 2, 3, 4, 7, 7, 8, 8, 10, 11. Since $N = 9$, the median is the $\frac{N+1}{2} = 5^{th}$ term or median $= 7$. And from (b), when the $N = 6$ observations are rearranged to yield 1, 2, 4, 6, 7, 9, we know that the median lies between the $\frac{N}{2} = 3^{rd}$ and $\frac{N}{2} + 1 = 4^{th}$ terms or median $= \frac{4+6}{2} = 5$. ∎

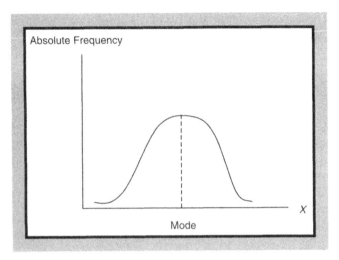

Figure 2.8   The mode occurs at the peak of the absolute frequency distribution.

As far as the properties of the median are concerned, the median always exists, is unique, may or may not equal the mean, and is not affected by extreme values.

Mode.   The *mode* of a variable X (or of X's absolute frequency distribution) is the value that occurs with the highest absolute frequency; that is, it is the most common or most probable value in a set of data. As Figure 2.8 reveals, the mode occurs at the peak of the absolute frequency distribution.

Example 2.3.1.4   Given X: 1, 2, 4, 4, 5, 6, the mode is 4 (it has the highest frequency of occurrence, namely 2). Given X: 3, 3, 4, 5, 5, 6, there are two modes, namely 3 and 5. And for X: 1, 2, 3, 6, 9, there is no mode (since each value of X occurs with an absolute frequency of 1).   ■

Turning to the properties of the mode, the mode may not exist, may not be unique, may or may not equal the mean and median, is not affected by extreme values, and always corresponds to one of the actual observations (unlike the mean and median).

2.3.2   Measures of Dispersion

These descriptive measures serve to characterize the extent of the scatter or spread of the X values along the horizontal axis. Here we typically determine the dispersion of the data around some measure of central location, where the latter usually is taken to be the arithmetic mean.

Range.   The *range* of a variable X (or of X's absolute frequency distribution) is the largest value of X less its smallest value or

$$\text{range} = \max X_i - \min X_i. \tag{2.4}$$

Clearly the range is a positional measure of the total spread of the data. For instance, given X: 3, 10, 18, 20, 30, the range $= 30 - 3 = 27$.

Turning to the properties of the range, note that the range is obviously sensitive to the extreme values in the data set. It indicates nothing about the behavior of the X values within the bounds max X_i and min X_i; for example, are the X values evenly distributed between max X_i and min X_i or are they all concentrated at one end of the data set? Since this latter property of the range is a definite drawback in most statistical applications, we shall now turn to a measure of dispersion that reflects variability about the mean.

Standard Deviation.   The *standard deviation* of the variable X (or of X's absolute frequency distribution) is the positive square root of the *variance* of X, where the latter is defined as the average of the squared deviations from the mean. That is, if the variance of X is denoted as

$$\sigma^2 = \frac{1}{N} \sum_{i=1}^{N} (X_i - \mu)^2, \tag{2.5}$$

then the standard deviation is

$$\sigma = \sqrt{\frac{1}{N} \sum_{i=1}^{N} (X_i - \mu)^2} = \sqrt{\frac{1}{N} \sum_{i=1}^{N} X_i^2 - \mu^2}. \tag{2.6}$$

(Note that σ^2 is measured in units of X squared and σ is expressed in the same units as X.)

Example 2.3.2.1   The standard deviation of the data set presented in Table 2.9 is, from (2.6), $\sigma = \sqrt{\frac{230}{5} - 36} = \sqrt{10} = 3.16$; that is, on the average, the individual values of X are approximately 3.16 units away from the mean, lying either above or below the mean.   ∎

As far as the properties of the standard deviation are concerned, the standard deviation:

(1) Is a measure of *average spread* or dispersion about the mean.

(2) Serves as an index of distance from the mean; that is, the distance between any observation X_i and μ may be expressed in terms of standard deviation units. For example, if for some variable X we find that $\mu = 30$ and $\sigma = 10$, then if $X_1 = 25$ and $X_2 = 50$ are two individual observations on X, it is evident

Table 2.9

X_i	X_i^2
1	1
4	16
7	49
8	64
10	100
30	230

that X_1 is 5 units or $\frac{1}{2}$ a standard deviation (denoted $\frac{\sigma}{2}$) below μ and X_2 lies 20 units or 2 standard deviations (denoted 2σ) above μ.

(3) (*Chebyshev's Theorem*) for any variable X and any constant $k > 1$, at least $1 - \frac{1}{k^2}$ of the N observations on X lie within k standard deviations of the mean (see Figure 2.9). For instance, if $\mu = 10$, $\sigma = 12.5$, and $k = 2$, at least $\frac{3}{4}$ of all the observations on X lie within the interval $\mu \pm 2\sigma$ or within 10 ± 5. Hence the interval is $(5, 15)$. (Note: as a rule of thumb, about two-thirds of all X values lie within one standard deviation of the mean.) Let us now get a bit more specific.

(4) (*Empirical Rule*) if the variable X follows a *normal distribution* (i.e., a continuous bell-shaped distribution that is symmetrical about its mean and whose tails are asymptotic to the horizontal axis),[1] the interval:

(a) $\mu \pm \sigma$ contains 68.26% of the observations on X

(b) $\mu \pm 2\sigma$ contains 95.46% of the observations on X

(c) $\mu \pm 3\sigma$ contains 99.73% of the observations on X (see Figure 2.10)

Conversely:

(a′) 50% of the observations on X fall between $\mu \pm 0.674\sigma$

(b′) 95% of the observations on X fall between $\mu \pm 1.960\sigma$

(c′) 99% of the observations on X fall between $\mu \pm 2.576\sigma$

In view of 4c, it is evident that for a normal distribution, we may approximate σ as $\frac{\text{range}}{6}$.

Standard Deviation of a Frequency Distribution.   When a set of data is arranged in the form of an absolute frequency distribution, we may compute

---

[1] A distribution is *symmetrical* if, when *creased* at its mean and folded in half, the right-hand side fits exactly over the left-hand side and conversely; i.e., the right-hand side is the *mirror image* of the left-hand side. The tails of a distribution are *asymptotic* to the horizontal axis if, as they extend away from the center of the distribution to $-\infty$ and $+\infty$, they gradually approach, but never touch, the horizontal axis.

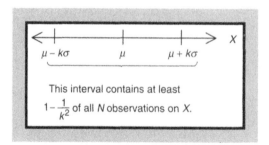

Figure 2.9 Chebyshev's theorem.

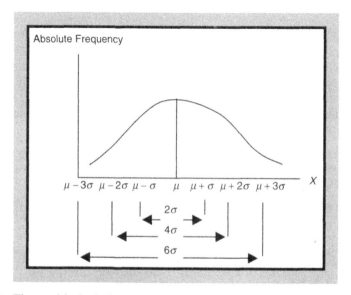

Figure 2.10 The empirical rule for a normal distribution.

the standard deviation of the same by weighting the squares of the deviations of the *different* values of X from the mean by their absolute frequencies; that is,

$$\sigma = \sqrt{\sum_{j=1}^{k} (X_j - \mu)^2 f_j / N} = \sqrt{\frac{\sum_{j=1}^{k} f_j X_j^2}{N} - \left(\frac{\sum_{j=1}^{k} f_j X_j}{N}\right)^2}, \qquad (2.7)$$

where k represents the number of different (net of duplication) values of X.

Table 2.10

X_j	f_j	X_j^2	$f_j X_j$	$f_j X_j^2$
1	1	1	1	1
4	2	16	8	32
6	5	36	30	180
9	3	81	27	243
10	1	100	10	100
	12			556

Example 2.3.2.2   Given X: 1, 6, 6, 4, 6, 4, 9, 9, 6, 9, 10, 6, Table 2.10 and (2.7), with $k = 5$, provides us with the value $\sigma = \sqrt{\frac{556}{12} - \left(\frac{76}{12}\right)^2} = 2.65$.   ∎

Quantiles.   The *quantiles* of a variable X (or of X's absolute frequency distribution) are positional measures that divide the observations on X (or the total area under the absolute frequency distribution of X) into a number of equal portions, given that the said observations have been ordered in an increasing sequence. Such measures are:

- The median—splits an absolute frequency distribution into two equal parts

- Quartiles—split an absolute frequency distribution into four equal parts

- Deciles—split an absolute frequency distribution into 10 equal parts

- Percentiles—split an absolute frequency distribution into 100 equal parts

The three *quartiles* Q_1, Q_2, and Q_3 divide the absolute frequency distribution into four equal parts. In particular, the first quartile Q_1 corresponds to the point below which 25% of the observations on X lie; the second quartile Q_2 is the median, the point below which 50% of the observations on X are found; and the third quartile Q_3 is the point below which 75% of the values of X lie.

In this regard, two additional (positional) measures of dispersion may be introduced. First, if we are interested in the location of the middle 50% of the observations on X, then we may determine the *interquartile range* $(I.Q.) = Q_3 - Q_1$. (Here the interquartile range is a *limited* type of range in that it excludes extreme values from its computation.) Then the *quartile deviation* $Q.D. = \frac{Q_3 - Q_1}{2}$ measures half the distance between the first and third quartiles. For a normal distribution, since the quartile deviation considers half the range of the middle 50% of the values of X, it serves as a measure of dispersion about the median $= \frac{Q_1 + Q_3}{2}$. Given that the quartile deviation describes dispersion in the central half of the distribution of X values, it clearly has the advantage of being able to be used with an absolute frequency distribution possessing outliers. And since it is not affected by any observation on X lying below Q_1 and above Q_3 (i.e., lying in the tails of the absolute frequency distribution), it suffers from a defect opposite to that of the range, where the latter is determined only by values below Q_1 and above Q_3.

Next, nine *deciles* $D_1, \ldots, D_9$ divided the absolute frequency distribution into 10 equal parts. Thus the first decile D_1 corresponds to the point below which 10% of the observations on X lie; D_2 is the point below which 20% of the observations on X are found, and so on. In general, the decile D_j, $j = 1, \ldots, 9$, corresponds to the data point below which $10j\%$ of the observations on X lie and above which $(100 - 10j\%)$ of the data points are found.

Finally, 99 *percentiles* $P_1, \ldots, P_{99}$ divide the absolute frequency distribution into 100 equal parts. Here the first percentile P_1 corresponds to the point below which 1% of the observations on X lie, the second percentile P_2 is the value below which 2% of the observations on X lie, and so on. In general, the percentile P_j, $j = 1, \ldots, 99$, corresponds to the point below which $j\%$ of the observations on X are found and above which $(100 - j)\%$ of the data points lie.

If the values of X are arranged in an increasing sequence, then by an appropriate choice of j, we may easily specify a given percentile, quartile, or decile. That is:

(a) The positional locations of the various quartiles are: Q_1 at $\frac{N+1}{4}$, Q_2 at $\frac{N+1}{2}$, and Q_3 at $\frac{3(N+1)}{4}$

(b) The positional locations of the deciles are: D_j at $\frac{j(N+1)}{10}$, $j = 1, \ldots, 9$

(c) The positional locations of the percentiles are: P_j at $\frac{j(N+1)}{100}$, $j = 1, \ldots, 99$

Example 2.3.2.3   Let us find P_{20} given X: 3, 8, 8, 6, 7, 9, 5, 10, 11, 20, 19, 15, 11, 16. Upon rearranging the $N = 14$ observations in an increasing sequence we have

$$3, 5, 6, 7, 8, 8, 9, 10, 11, 11, 15, 16, 19, 20.$$

Since 20% of the observations on X lie at or below P_{20}, we may determine $\frac{j(N+1)}{100} = \frac{20(15)}{100} = 3$; that is, the twentieth percentile of X is found at the third observation in the previous increasing sequence or $P_{20} = 6$. Similarly, Q_3 is located at a position that is one fourth the distance between the eleventh and twelfth observation in the increasing sequence (i.e., $\frac{3(N+1)}{4} = 11.25$) or $Q_3 = 15.25$, and D_6 is located at the ninth observation in the same (since $\frac{6(N+1)}{10} = 9$) or $D_6 = 11$.   ∎

2.3.3   Standardized Variables

Let the variable $X : X_1, \ldots, X_N$ have a mean μ and a standard deviation $\sigma \neq 0$. If we form a new variable $Z = \frac{X - \mu}{\sigma}$ with values

$$Z_1 = \frac{X_1 - \mu}{\sigma}, \ldots, Z_N = \frac{X_N - \mu}{\sigma}, \tag{2.8}$$

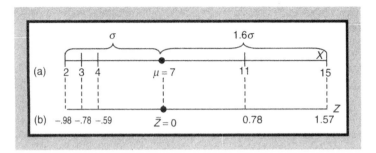

Figure 2.11   Distance from the mean expressed in standard deviation units.

then Z has a mean of zero and a standard deviation of one. The set of data depicted in (2.8) is termed a *unit distribution* or set of *Z-scores*.

Example 2.3.3.1   Given $X: 2, 3, 4, 11, 15$, it follows that $\mu = 7, \sigma = \sqrt{26} = 5.1$. To derive (2.8) we first shift the origin to the mean by subtracting μ from each X_i to obtain the sequence of values $-5, -4, -3, 4, 8$. Each of these adjusted values is then divided by σ to yield the unit distribution

$$Z : -\frac{5}{5.1}, -\frac{4}{5.1}, -\frac{3}{5.1}, \frac{4}{5.1}, \frac{8}{5.1},$$

or $-0.98, -0.78, -0.59, 0.78, 1.57$.

How may we interpret these Z-scores? First, a positive (respectively, negative) Z_i value indicates that the original X_i value lies above (respectively, below) the mean. Second, the size of Z_i indicates how many standard deviations separate X_i and μ (i.e., distance from the mean is put in terms of standard deviations). In this regard, from panels a and b of Figure 2.11, $Z_1 = -0.98$ tells us that $X_1 = 2$ lies approximately 1 standard deviation below the mean, and $Z_5 = 1.57$ informs us that $X_5 = 15$ lies about 1.6 standard deviations above the mean. ■

For large N, the distribution of Z values in (2.8) typically will range in size from -3 to 3. In view of this observation, we can posit a simple test for the detection of outliers (or inordinately large or small extreme values) in a data set: any X_i with a $|Z_i| > 3$ is termed an *outlier*.

We may note further that:

(1) Z-scores are dimensionless (independent of units).

(2)   If $Z_i > 0$, X_i lies above μ; if $Z_i < 0$, X_i lies below μ; and if $Z_i = 0$, $X_i = \mu$.

(3)   The size of the Z-score indicates the number of standard deviations separating X_i and μ.

(4) Z-scores are used to *place* one observation relative to others in a data set.

(5) The mean and standard deviation of a set of Z-scores are zero and one, respectively.

(6) For a normal distribution (bell-shaped and symmetrical), it can be shown (see Chapter 7) that the Z-scores corresponding to the first and third quartiles are $\frac{(Q_1 - \mu)}{\sigma} = -0.67$ and $\frac{(Q_3 - \mu)}{\sigma} = 0.67$, respectively. Since the standard deviation of a set of Z-scores is 1, it follows that the *standardized interquartile range* $= I.Q./\sigma = (0.67 - (-0.67))/1 = 1.34$ can be used to test for the normality of a data set; that is, if $I.Q./\sigma \approx 1.34$, then the data are approximately normal.

For N small, the range of Z-scores typically will be smaller than that for large data sets.

To account for the sensitivity of the arithmetic mean μ to outliers in a data set, we can determine the α-*trimmed mean*, $0 < \alpha < \frac{1}{2}$, and use it in place of μ. In this regard, for a set of data points $X_1, \ldots, X_N$ arranged in an increasing sequence, the α-trimmed mean μ_N^α is computed by deleting the αN smallest observations and the αN largest observations, and then taking the arithmetic mean of the remaining data points.

Example 2.3.3.2   Suppose we have a variable X with the $N = 30$ values:

$$
\begin{array}{cccccc}
2 & 25 & 26 & 28 & 29 & 30 \\
15 & 25 & 26 & 28 & 29 & 30 \\
21 & 25 & 26 & 28 & 29 & 30 \\
25 & 25 & 27 & 28 & 29 & 39 \\
25 & 25 & 28 & 28 & 30 & 55
\end{array}
$$

Here the extreme values for this data set are 2 and 55. But are they outliers? Let us first find

$$
\mu = \frac{1}{N} \sum_{i=1}^{N} X_i = \frac{816}{30} = 27.2
$$

and

$$
\sigma = \left[\left(\frac{1}{N} \sum_{i=1}^{N} X_i^2 \right) - \mu^2 \right]^{1/2} = \left[\left(\frac{24{,}018}{30} \right) - 739.84 \right]^{1/2} = 7.79.
$$

Then the Z-scores for 2 and 55 are, respectively,

$$
\frac{2 - 27.2}{7.79} = -3.23, \quad \frac{55 - 27.2}{7.79} = 3.56.
$$

Since each of these Z-scores exceeds 3 (in absolute value), both of these values of X are deemed outliers. Since the presence of outliers might distort the value of the mean μ, let us replace μ by, say, $\mu_{30}^{0.10}$, the 10% trimmed mean. Hence we must delete the $\alpha N = 3$ smallest and largest data values (which are 2,15,21; and 30,39,55, respectively) and use the remaining 24 data points to obtain $\mu_{30}^{0.10} = 27.3$. Clearly this process has not appreciably changed the average obtained for this data set. However, for other data sets, the trimming process might yield a marked change in the average.   ■

A second technique for handling outliers is called *Winsorization*. Here, too, we seek to minimize the effect of extraneous values in a data set on the value of the mean. In this regard, for *k-level Winsorization*, the k smallest and largest observations are given the values of their nearest neighbors; in first-level Winsorization, the smallest and largest data points are given the values of their nearest neighbors; in second-level Winsorization, the two smallest and largest observations are replaced by their nearest neighbors, and so on. In general, the *k-level Winsorized mean* (determined under k-level Winsorization) will be denoted as $\mu_{w,k}$.

Example 2.3.3.3   Given the preceding data set involving $N = 30$ observations, first-level Winsorization replaces 2 and 55 by 15 and 39, respectively. Second-level Winsorization replaces 2 and 15 by 21 and 39 and 55 by 30. It is easily shown that the second-level Winsorized mean is $\mu_{w,2} = \frac{807}{30} = 26.9$ and the third-level Winsorized mean is $\mu_{w,3} = \frac{819}{30} = 27.3$.   ■

2.3.4   Moments

The hth *moment about the origin* is represented as

$$m_h = \frac{\sum\limits_{i=1}^{N} X_i^h}{N}, \quad h = 1, 2, 3, \dots . \tag{2.9}$$

Moreover, the hth *moment about the mean or the hth central moment* is

$$v_h = \frac{\sum\limits_{i=1}^{N} (X_i - \mu)^h}{N}, \quad h = 1, 2, 3, \dots . \tag{2.10}$$

And if $Z_i = (X_i - \mu)/\sigma, i = 1, \dots, N$, then the hth *standard moment* is

$$a_h = \frac{\sum\limits_{i=1}^{N} Z_i^h}{N} = \frac{\sum\limits_{i=1}^{N} \left[\frac{X_i - \mu}{\sigma} \right]^h}{N} = \frac{v_h}{\sigma^h}, \quad h = 1, 2, 3, \dots . \tag{2.11}$$

For computational expediency, we may write the first few moments about the mean in terms of moments about the origin as

$$v_1 = m_1 - m_1 = 0,$$

$$v_2 = m_2 - m_1^2,$$

$$v_3 = m_3 - 3m_1 m_2 + 2m_1^3, \tag{2.12}$$

$$v_4 = m_4 - 4m_1 m_3 + 6m_1^2 m_2 - 3m_1^4.$$

It is evident that the first moment about the origin is the mean $\mu = m_1 = \sum_{i=1}^{N} \frac{X_i}{N}$ and the second moment about the mean is the variance $\sigma^2 = v_2 = \sum_{i=1}^{N} \frac{(X_i - \mu)^2}{N}$, and thus the standard deviation is $\sigma = \sqrt{v_2}$.

Example 2.3.4.1   Given X: 2, 3, 5, 7, 8, 11, find a_3 and a_4. From Table 2.11, $m_1 = \frac{36}{6} = 6$, $m_2 = \frac{272}{6} = 45.33$, $m_3 = \frac{2,346}{6} = 391$, and $m_4 = \frac{21,860}{6} = 3,643.33$. Then from (2.12),

$$v_2 = 45.33 - 36 = 9.33,$$

$$v_3 = 391 - 3(6)(45.33) + 2(216) = 7.06,$$

$$v_4 = 3,643.33 - 4(6)(391) + 6(36)(45.33) - 3(1,296) = 162.61.$$

Since $\sigma = \sqrt{v_2} = 3.055$, it follows that $\sigma^3 = 28.512$, $\sigma^4 = 87.105$. Then $a_3 = \frac{v_3}{\sigma^3} = 0.2476$, $a_4 = \frac{v_4}{\sigma^4} = 1.8668$. ∎

If the data set happens to be arranged in the form of an absolute frequency distribution, then the hth *central moment* is

$$m_h = \frac{\sum_{j=1}^{k} f_j X_j^h}{N}, \quad h = 1, 2, 3, \ldots; \tag{2.9.1}$$

Table 2.11

X_i	X_i^2	X_i^3	X_i^4
2	4	8	16
3	9	27	81
5	25	125	625
7	49	343	2,401
8	64	512	4,096
11	121	1,331	14,641
36	272	2,346	21,860

and the hth *moment about the mean* is

$$v_h = \frac{\sum_{j=1}^{k} f_j(X_j - \mu)^h}{N}, \quad h = 1, 2, 3\ldots, \tag{2.10.1}$$

where k represents the number of different (net of duplication) values of X. The hth *standard moment* is again $\frac{v_h}{\sigma^h}$, $h = 1, 2, 3, \ldots$.

2.3.5   Skewness and Kurtosis

An absolute frequency distribution is said to be *skewed* if it lacks symmetry; *kurtosis* refers to how flat or sharp the peak of the absolute frequency distribution happens to be. In particular, measures of skewness and kurtosis are shape parameters and essentially reflect departures from normality. Looking at Figure 2.12, panel (a) indicates that X's absolute frequency distribution is symmetrical (about the mean) and panel (b) (respectively, (c)) indicates that the said distribution is skewed to the right (respectively, left). Alternatively, the absolute frequency distribution in panel (b) (respectively, (c)) of Figure 2.12 is said to be positively (respectively, negatively) skewed. For positive (respectively, negative) skewness, there are some relatively large (respectively, small) values of X for which there are no comparable or offsetting small (respectively, large) ones.

As far as the measurement of skewness is concerned, we may utilize the *Pearsonian coefficient of skewness*

$$SK_p = \frac{\mu - mode}{\sigma}. \tag{2.13}$$

Note that this quantity is a pure number and thus expressed as a percent (i.e., it is dimensionless or independent of units). If the mode cannot readily be determined,

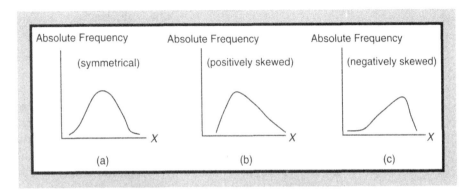

Figure 2.12   Symmetrical versus skewed absolute frequency functions.

then we may employ a relationship among the mean, median, and mode that holds approximately for moderately skewed unimodal distributions, namely

$$\mu - \text{mode} \approx 3(\mu - \text{median}), \qquad (2.14)$$

to obtain

$$SK_p \approx \frac{3(\mu - \text{median})}{\sigma}. \qquad (2.13.1)$$

Another option is to use as our measure of skewness the third standard moment $a_3 = \frac{v_3}{\sigma^3}$ (called *standard skewness*). Here, too, we have a dimensionless figure expressed as a percent. Clearly all three of these measures of skewness are zero when X's absolute frequency distribution is symmetrical, positive when the distribution is skewed to the right, and negative when the distribution is skewed to the left. Moreover, the magnitude of each indicates the degree of skewness so that each is a relative measure of skewness. By convention, if $|a_3| > 0.5$ or, if, using (2.13.1), $|SK_p| > 1$, an absolute frequency distribution is said to possess a substantial amount of skewness.

Sometimes the quartiles of an absolute frequency distribution are used to depict skewness. As a point of reference, if the distance between Q_1 and Q_2 is the same as the distance between Q_2 and Q_3—$Q_2 - Q_1 = Q_3 - Q_2$—then the distribution is symmetrical. (Looked at in another fashion, for a symmetrical absolute frequency distribution, if the quartile deviation is subtracted from and added to the median (Q_2), we obtain the first and third quartiles or $Q_1 = Q_2 - Q.D.$ and $Q_3 = Q_2 + Q.D.$) However, if $Q_2 - Q_1 > Q_3 - Q_2$, the absolute frequency distribution is positively skewed or skewed to the right; if $Q_2 - Q_1 < Q_3 - Q_2$, the said distribution is skewed to the left or negatively skewed. In view of these observations we may also introduce *Bowley's coefficient of skewness*

$$SK_b = \frac{(Q_3 - Q_2)(Q_2 - Q_1)}{Q_3 - Q_1} \qquad (2.15)$$

Clearly this expression is positive (respectively, negative) if X's absolute frequency distribution is skewed to the right (respectively, left).

Looking next to a measure of kurtosis, we may employ the fourth standard moment $a_4 = \frac{v_4}{\sigma^4}$ (termed *standard kurtosis*). This quantity is independent of units (or expressed as a percent). As a benchmark, let us employ the notion that for a normal distribution $a_4 = 3$. Then our estimate of the degree of kurtosis is $a_4 - 3$ (called *the excess*). So if $a_4 - 3 > 0$ (respectively, <0), an absolute frequency distribution has a peak that is sharper (respectively, flatter) than that of a normal distribution and thus is said to exhibit positive (respectively, negative) kurtosis. These cases are depicted in Figure 2.13.

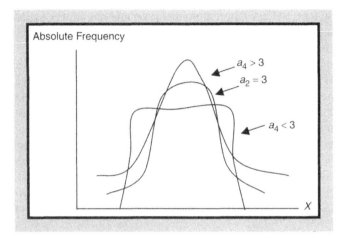

Figure 2.13   Standard kurtosis as a measure of kurtosis.

Another measure of kurtosis is the *coefficient of kurtosis*, which consists of the quartile deviation divided by the range between the 10th and 90th percentiles or

$$K = \frac{Q.D.}{P_{90} - P_{10}}. \tag{2.16}$$

For a sharply peaked curve, $K \approx 0.5$ (since $Q_3 - Q_1 \approx P_{90} - P_{10}$); for a very flat curve, $K \approx 0$ (since $P_{90} - P_{10}$ exceeds $Q_3 - Q_1$ considerably); and for a normal curve, $K = 0.263$.

2.3.6   Relative Variation

How may we compare the dispersions of two (or more) absolute frequency distributions? For convenience sake, let us limit ourselves to the discussion of distributions A and B. Our approach is the following:

(1) If they have the same (or nearly the same) mean and are expressed in the same units, then we can simply compare their standard deviations. (Here the standard deviation serves as a measure of *absolute variation or dispersion*.)

(2) If they have different means or are expressed in different units, then we need to convert the standard deviation of each distribution to a measure of *relative variation*.

Why is the conversion mentioned in step two necessary? The difficulty of comparing items measured in different units should be obvious (e.g., if the data for distribution A is in miles per hour and that for distribution B is in metric tons, then clearly no reasonable comparison of absolute dispersion can be made). What about the *different means* issue? Even though units may not be a problem, it is the case that a set of very large numbers invariably has a large standard deviation relative to a set of very small numbers, even though there may be more variability associated with the values within the latter.

For instance, we may address the question: Is there more variation among weights of mice relative to that of elephants? It should be clear that the standard deviation computed from the elephants (converted to grams) will always be larger than the comparable figure for mice (also in grams) even though there may be more uniformity among the weights of the elephants.

To carry out the procedure hinted at in step two, earlier, consider the following: to achieve comparability, the measures of absolute variation must be changed to relative forms; that is, express the standard deviation as a fraction of the average about which the deviations are taken (remember that within σ we have the term $\Sigma_{i=1}^{N}(X_i - \mu)^2$). This leads us to define the *coefficient of variation* as

$$V = \frac{\sigma}{\mu} \times 100. \tag{2.17}$$

Here V is a pure number (independent of units) expressed as a percent. Once V has been obtained, we may conclude that *the standard deviation is V% of the mean*. For example, if $V = 0.25 \times 100$, we may say that "the standard deviation is 25% of the mean."

So, given distributions A and B, we compute V for each. Then the distribution with the smaller V value has a greater degree of uniformity or homogeneity among its values (or the distribution with the larger V has a greater degree of variability or heterogeneity among its values).

Example 2.3.6.1   Is there more variation among incomes received by nurses affiliated with a certain large urban hospital (distribution A) or a group of private practice nurses (distribution B)? For each group we compute

$$A : \mu_A = \$45,000, \qquad \sigma_A = \$2,500;$$
$$B : \mu_B = \$39,000, \qquad \sigma_B = \$2,000.$$

Since $V_A = \frac{\sigma_A}{\mu_A} \times 100 = 0.05 \times 100$ and $V_B = \frac{\sigma_B}{\mu_B} \times 100 = 0.06 \times 100$, we conclude that the standard deviation of the incomes received by nurses in distribution B are 6% of the mean, whereas the standard deviation of the incomes received by nurses in distribution A are 5% of the mean. Hence there is slightly more variability in distribution B incomes relative to those in distribution A. ■

2.3.7   Comparison of the Mean, Median, and Mode

When an absolute frequency distribution is *normal* or bell-shaped and symmetrical, its mean = median = mode (see Figure 2.14a). For a distribution that is skewed to the right (respectively, left), mean > median > mode (respectively, mean < median < mode) (see Figures 2.14b, c).

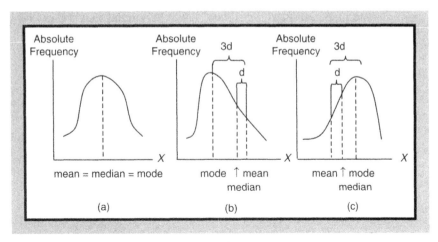

Figure 2.14   The relative positions of the mean, median, and mode.

In fact, for a moderately skewed unimodal absolute frequency distribution, (2.14) indicates that the distance between the mean and the mode (3d) is three times the distance between the mean and the median (d).

2.3.8   The Sample Variance and Standard Deviation

Up to this point in our data analysis we have assumed that the data set under discussion represents a population of size N. But if the set of observations is actually a sample of size n extracted from some population, then we must modify our calculation of the variance and standard deviation slightly to take into account what may be called a *correction for degrees of freedom*. Given a variable $X : X_1, X_2, \ldots, X_n$, we may think of degrees of freedom as the number of independent observations in the sample; that is, it is calculated as the sample size less the number of prior estimates made. For instance, to compute the *sample variance S^2* we correct $\sum_{i=1}^{n} (X_i - \overline{X})^2$ for degrees of freedom by dividing this expression by $n - 1$ to obtain

$$S^2 = \frac{\sum\limits_{i=1}^{n} (X_i - \overline{X})^2}{n - 1}, \tag{2.18}$$

where the *sample mean*

$$\overline{X} = \frac{\sum\limits_{i=1}^{n} X_i}{n} \tag{2.19}$$

is the *single prior estimate made*. Then the sample standard deviation appears as

$$S = \sqrt{\frac{\sum_{i=1}^{n} (X_i - \overline{X})^2}{n - 1}}. \tag{2.20}$$

Looked at in another fashion, we may view degrees of freedom as a property of a sum of squares under the linear restriction $\sum_{i=1}^{n} (X_i - \overline{X}) = 0$. That is, if $\sum_{i=1}^{n} (X_i - \overline{X}) = 0$ must hold and $\overline{X}$ is given, then once we select any $n - 1$ of the observations $X_i, i = 1, \ldots, n$, the nth is uniquely determined. For instance, if

$$\left(X_1 - \overline{X}\right) + \left(X_2 - \overline{X}\right) + \left(X_3 - \overline{X}\right) = 0$$

and $X_1 = 2$, $X_2 = 7$, and $\overline{X} = 5$, then we are not free to choose the value of X_3—X_3 must equal 6. And if $\overline{X}$ is changed to 9, then we are free to arbitrarily select the values of only two of these data points. Once any two of them are chosen, the third is uniquely determined by virtue of the requirement $(X_1 - 9) + (X_2 - 9) + (X_3 - 9) = 0$.

There is another compelling reason to divide by $n - 1$ in (2.18) rather than by n. As will be discussed in greater detail later, dividing $\sum_{i=1}^{n} (X_i - \overline{X})^2$ by $n - 1$ produces an unbiased estimate of the population variance σ^2.

For a sample of size n, the equivalent of (2.3) is

$$\overline{X} = \frac{\sum_{j=1}^{k} f_j X_j}{n} \tag{2.19.1}$$

and the computational equivalents of (2.6) and (2.7) are, respectively,

$$S = \sqrt{\frac{\sum_{i=1}^{n} X_i^2}{n - 1} - \frac{\left(\sum_{i=1}^{n} X_i\right)^2}{n (n - 1)}}; \tag{2.21}$$

$$S = \sqrt{\frac{\sum_{j=1}^{k} f_j X_j^2}{n - 1} - \frac{\left(\sum_{j=1}^{k} f_j X_j\right)^2}{n (n - 1)}}, \tag{2.22}$$

where k is the number of different (net of duplication) values of a variable X and the f_j's are their absolute frequencies.

Example 2.3.8.1 If the data in Table 2.9 depicts a sample of size $n = 5$, then (2.21) yields

$$S = \sqrt{\frac{230}{4} - \frac{(30)^2}{5(4)}} = 3.535.$$

And if the data in Table 2.10 represents a sample of size $n = 12$, then (2.22) renders

$$S = \sqrt{\frac{556}{11} - \frac{(76)^2}{12(11)}} = 2.61.$$

We close this section by offering formulae useful for updating the sample mean and variance when additional sample information becomes available. To this end we first examine a recursion formula for the sample mean of a variable X. This expression allows us to efficiently determine the mean of $n + 1$ data points from the average of the first n of them (2.19). That is, suppose we know $\overline{X}_n$ (the average of the values $X_1, \ldots, X_n$) and another observation X_{n+1} becomes available. What is the average of all $n + 1$ observations? From the definition of the mean $\overline{X}_n$ it is easily demonstrated that

$$\overline{X}_{n+1} = \frac{X_{n+1} + n\overline{X}_n}{n+1} \tag{2.19.2}$$

Based upon this notation, we immediately can see how a recursion relationship works: the result for the $(n + 1)^{st}$ term is obtained from the value of the nth term.

Example 2.3.8.2 Suppose that the mean of a set of $n = 50$ values of some variable X is $\overline{X}_n = \overline{X}_{50} = 63.76$. Let us now assume that a fifty-first observation $X_{n+1} = X_{51} = 22$ becomes available. What is the average of all 51 data points? From (2.19.1) it is readily found that

$$\overline{X}_{n+1} = \overline{X}_{51} = \frac{22 + 50(63.76)}{51} = 62.94. \quad \blacksquare$$

Next, to establish a recursion formula for the sample variance of a variable X (we seek to find the variance of $n + 1$ data points from the variance of the first n of them), let us suppose we know S_n^2 (the variance of the values $X_1, \ldots, X_n$) and another observation X_{n+1} becomes available. What is the variance of all $n + 1$ observations? Given $S_n^2 = \sum_{i=1}^{n} (X_i - \overline{X}_n)^2/(n-1)$ and (2.19.1), we can find

$$S_{n+1}^2 = \left(\frac{n-1}{n}\right) S_n^2 + \frac{1}{n}(X_{n+1} - \overline{X}_n)^2. \tag{2.18.1}$$

Example 2.3.8.3   Suppose that the mean and variance of a set of 25 values of some variable X are $\overline{X}_n = \overline{X}_{25} = 10$ and $S_n^2 = S_{25}^2 = 8.72$, respectively. Let us now assume that a twenty-sixth observation $X_{n+1} = X_{26} = 14$ is obtained. What is the variance of all 26 data points? Using (2.18.1) we have

$$S_{n+1}^2 = S_{26}^2 = \left(\frac{24}{25}\right)(8.72) + \frac{1}{25}(14 - 10)^2 = 9.0112. \quad \blacksquare$$

2.4   Correlation between Variables X and Y

For a set of N ordered pairs of observations (X_i, Y_i), $i = 1, \ldots, N$, a measure that depicts the joint variation of X_i and Y_i is the *covariance of X and Y*,

$$COV(X, Y) = \sigma_{XY} = \frac{\sum\limits_{i=1}^{N}(X_i - \mu_X)(Y_i - \mu_Y)}{N}, \quad (2.23)$$

or, as it is alternatively represented,

$$\sigma_{XY} = \frac{\sum\limits_{i=1}^{N} X_i Y_i}{N} - \mu_X \mu_Y \quad (2.23.1)$$

In this regard:

(a) If higher values of X are associated with higher values of Y, then $\sigma_{XY} > 0$

(b) If higher values of X are associated with lower values of Y, then $\sigma_{XY} < 0$

(c) If X is neither higher nor lower for higher values of Y, then $\sigma_{XY} = 0$

There are a couple of pitfalls connected with using the covariance as a measure of association: (1) its magnitude can be arbitrarily increased by increasing the number of data points; and (2) it is arbitrarily influenced by the units in which X and Y are measured. However, these two defects can be corrected by simply standardizing the product of the deviations from the mean in (2.23), that is, we divide σ_{XY} by the product of the standardized deviations of the X and Y variables to obtain the Pearson product moment correlation coefficient

$$\rho_{XY} = \frac{COV(X, Y)}{\sigma_X \sigma_Y} = \frac{\sigma_{XY}}{\sigma_X \sigma_Y}. \quad (2.24)$$

Here ρ_{XY} serves as a measure of *linear association* between X and Y, is independent of units, and varies between ± 1, inclusively. That is, $-1 \le \rho_{XY} < 1$, with the sign of ρ_{XY} determined by the sign of σ_{XY}. Specifically:

(1) If $\sigma_{XY} \ne 0$, then $\rho_{XY} \ne 0$ and X and Y are said to be linearly related or correlated.

   (a) If $\sigma_{XY} > 0$, then $0 < \rho_{XY} < 1$ and X and Y are said to be directly related or positively correlated (panel a of Figure 2.15).

   (b) If $\sigma_{XY} < 0$, then $-1 < \rho_{XY} < 0$ and X and Y are said to be inversely related or negatively correlated (panel b of Figure 2.15).

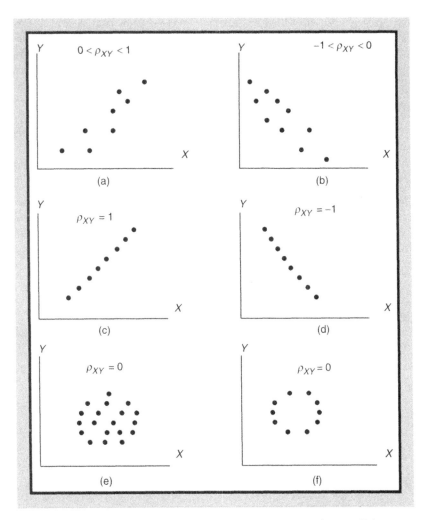

Figure 2.15   The range of values of the Pearson product moment correlation coefficient.

(c) If $\sigma_{XY} = \sigma_X\sigma_Y$, then $\rho_{XY} = 1$ and we have the case of perfect positive association between X and Y (panel c of Figure 2.15).

(d) If $\sigma_{XY} = -\sigma_X\sigma_Y$, then $\rho_{XY} = -1$ and we have the case of perfect negative association between X and Y (panel d of Figure 2.15).

(2) If $\sigma_{XY} = 0$, then $\rho_{XY} = 0$ and we say that there is no linear relationship between X and Y or that X and Y are uncorrelated (panel e of Figure 2.15).

(3) The correlation coefficient ρ_{XY} is invariant under a linear transformation. That is, if we define new variables $U = a + bX$ and $V = c + dY$, then $\rho_{UV} = \pm\rho_{XY}$, where the sign attached to ρ_{XY} depends upon the signs of b and d.

It is important to remember that if X and Y are found to be uncorrelated ($\rho_{XY} = 0$), then it does not necessarily follow that X and Y are independent. This is because the correlation coefficient only detects linear association between these variables—the true underlying relationship may actually be highly nonlinear (panel f of Figure 2.15). So if X and Y are independent, then there is no relationship of any sort (linear or otherwise) between these variables so that X and Y must also be uncorrelated. But if X and Y are uncorrelated, we cannot generally conclude that they are also independent.

It is also important to note that *correlation does not imply causation*. That is, finding that $\rho_{XY} \neq 0$ does not imply that movement in X causes movement in Y or vice versa. In fact, some other unobserved variable Z may be causing movements in both X and Y.

Example 2.4.1   Given the $N = 10$ observations presented in Table 2.12, determine if the variables X and Y are positively correlated. We first find $\mu_X = 5$ and $\mu_Y = 4$ and thus $\sigma_{XY} = \frac{42}{10} = 4.2$, $\sigma_X = \sqrt{\frac{54}{10}} = 2.324$, and $\sigma_Y = \sqrt{\frac{44}{10}} = 2.097$.

Table 2.12

X	Y	$X_i - \mu_X$	$Y_i - \mu_Y$	$(X_i - \mu_X)(Y_i - \mu_Y)$	$(X_i - \mu_X)^2$	$(Y_i - \mu_Y)^2$
4	2	−1	−2	2	1	4
5	5	0	1	0	0	1
2	3	−3	−1	3	9	1
4	3	−1	−1	1	1	1
7	4	2	0	0	4	0
2	2	−3	−2	6	9	4
9	8	4	4	16	16	16
6	6	1	2	2	1	4
8	6	3	2	6	9	4
3	1	−2	−3	6	4	9
50	40	0	0	42	54	44

Then from (2.24), $\rho_{XY} = \frac{\sigma_{XY}}{\sigma_X \sigma_Y} = \frac{4.2}{(2.324)(2.097)} = 0.86$. We may thus conclude that there is a fairly strong degree of positive linear association between X and Y. ■

A broader view of the covariance concept emerges when we are interested in determining the variance of a variable Y that is defined as the sum of a collection of p other variables $X_1, \ldots, X_p$ or $Y = \sum_{j=1}^{p} X_j$. As we shall now see, the variance of Y not only depends upon the variances of the individual X_j's, $j = 1, \ldots, p$, it also depends upon the covariances among all pairs of variables X_j and X_k (with $j < k$) as both $j, k = 1, \ldots, p$. Let us express the ith observation on Y as $Y_i = \sum_{j=1}^{p} X_{ji}$, $i = 1, \ldots, N$. Then

$$\sigma_Y^2 = \sum_{j=1}^{p} \sigma_j^2 + 2 \sum_{j<k}^{p} \sigma_{jk}, \tag{2.25}$$

where σ_j^2 is the variance of X_j and σ_{jk} denotes the covariance between X_j and X_k. For instance, if $Y = X_1 + X_2 + X_3$, then, according to (2.25),

$$\sigma_Y^2 = \sigma_1^2 + \sigma_2^2 + \sigma_3^2 + 2\sigma_{12} + 2\sigma_{13} + 2\sigma_{23}.$$

It is instructive to derive (2.25) for the case where $Y_i = X_{1i} + X_{2i}, i = 1, \ldots, N$. Since

$$\sum_{i=1}^{N} Y_i = \sum_{i=1}^{N} X_{1i} + \sum_{i=1}^{N} X_{2i},$$

it follows, after dividing both sides of this expression by N, that $\mu_Y = \mu_1 + \mu_2$. Then

$$Y_i - \mu_Y = (X_{1i} + X_{2i}) - (\mu_1 + \mu_2) = (X_{1i} - \mu_1) + (X_{2i} - \mu_2)$$

and thus

$$\sum_{i=1}^{n} (Y_i - \mu_Y)^2 = \sum_{i=1}^{N} (X_{1i} - \mu_1)^2 + \sum_{i=1}^{N} (X_{2i} - \mu_2)^2 + 2 \sum_{i=1}^{N} (X_{1i} - \mu_1)(X_{2i} - \mu_2).$$

Upon dividing by N we have

$$\sigma_Y^2 = \sigma_1^2 + \sigma_2^2 + 2\sigma_{12}. \tag{2.26}$$

You should now be able to easily demonstrate that:

(1) If $Y_i = X_{1i} - X_{2i}, i = 1, \ldots, N$, then $\mu_Y = \mu_1 - \mu_2$ and

$$\sigma_Y^2 = \sigma_1^2 + \sigma_2^2 - 2\sigma_{12}. \tag{2.27}$$

(2) If $Y_i = X_{1i}X_{2i}, i = 1,\ldots,N$, then $\mu_Y = \mu_1\mu_2 + \sigma_{12}$ and

$$\sigma_Y^2 = \frac{\sum\limits_{i=1}^{N}(X_{1i}X_{2i})^2}{N} - \mu_Y^2. \tag{2.28}$$

Note that if X_1 and X_2 are independent variables, then $\sigma_{12} = 0$ and, for $Y_i = X_{1i} \pm X_{2i}, i = 1,\ldots,N$, it follows that $\sigma_Y^2 = \sigma_1^2 + \sigma_2^2$; and for $Y_i = X_{1i}X_{2i}$, $i = 1,\ldots,N$, we have $\mu_Y = \mu_1\mu_2$ and

$$\sigma_Y^2 = \frac{\sum\limits_{i=1}^{N}(X_{1i}X_{2i})^2}{N} - (\mu_1\mu_2)^2. \tag{2.29}$$

2.5 Rank Correlation between Variables *X* and *Y*

We noted in the preceding section that the Pearson correlation coefficient ρ_{XY} serves as an index of linear association between two variables X and Y; that is, it measures the strength of the linear relationship between X and Y. Although the observations on X and Y used to determine ρ_{XY} are numerical scores measured on an interval or ratio scale, it may be the case that the X and Y values are ordinal in nature and thus represent numerical values that depict *ranks*. In this latter instance the Pearson correlation coefficient can be transformed into what will be termed a rank-order correlation coefficient, namely the *Spearman rank correlation coefficient* ρ_S, where $-1 \le \rho_S \le 1$.

The magnitude of the Spearman coefficient suggests a tendency for the X and Y variables to relate in a *monotone* fashion. That is to say, if ρ_S is positive and near 1, then the relationship between X and Y is *monotone-increasing* (there exists a direct relationship between X and Y); and if ρ_S is negative and near -1, then the relationship between X and Y is *monotone-decreasing* (X and Y are inversely related). If $|\rho_S|$ happens to be small, then this is an indication that the relationship between X and Y is nonmonotone. Finally, if $\rho_S = 0$, then clearly these variables exhibit no relationship at all.

Since the Spearman coefficient only measures the degree of monotone association between X and Y over rank values, its characterization of the relationship between X and Y is not as narrowly focused as that of ρ_{XY}; that is, ρ_S measures the tendency toward monotonicity and ρ_{XY} measures the tendency toward linearity, with the tendency toward linearity emerging as a special case of the tendency toward monotonicity. (However, ρ_S can be thought of as an index of the strength of the linear relationship between the rank of X and the rank of Y.) Since the X and Y values are simply ranks, it should be evident that any outliers present in a data set will distort the value of ρ_{XY} but will not affect the magnitude of ρ_S since the latter measure treats all observations equally.

Let us assume that we have a data set involving N ordered pairs (X_i, Y_i), $i = 1,\ldots,N$, where X_i is the ith observation on the variable X and Y_i is the

i^{th} observation on the variable Y. Suppose that X_i is an ordinal value that depicts the *rank* associated with the ith observation on some variable W (i.e., $X_i = \text{rank } (W_i)$, $i = 1,\dots,N$), where the ranks are specified from low to high or from 1 through N. Additionally, let the Y_i's constitute a second ranking of the W_i's, $i = 1,\dots,N$. The question that now emerges is: How much *agreement* is there between the two sets of rankings depicted by the X and Y variables? It should be intuitively clear that if the rank orders agree, then the ranks appearing in the sets of X and Y values should be positively associated or positively correlated. And if the rank orders disagree, then the said ranks should be negatively correlated. Note that if no association exists between the ranks described by the X and Y variables, then a zero correlation should emerge.

As previously indicated, a descriptive measure of the association or agreement between ranks is the Spearman rank correlation coefficient that applies the ordinary or Pearson correlation coefficient for interval or ratio scale data (2.24) to compute the correlation over ranks. To obtain this rank correlation coefficient, let us rewrite (2.24) as

$$\rho_S = \frac{\sigma_{XY}}{\sigma_X \sigma_Y} = \frac{\dfrac{\sum\limits_{i=1}^{N} X_i Y_i}{N} - \mu_X \mu_Y}{\sqrt{\dfrac{\sum\limits_{i=1}^{N} X_i^2}{N} - \mu_X^2} \; \sqrt{\dfrac{\sum\limits_{i=1}^{N} Y_i^2}{N} - \mu_Y^2}}. \qquad (2.24.1)$$

Given that we are dealing with sets of rank values $\{X_1,\dots,X_N\}$ and $\{Y_1,\dots,Y_N\}$, it can be shown, in the absence of any ties in ranks, that

$$\sum_{i=1}^{N} X_i = \sum_{i=1}^{N} Y_i = \frac{N(N+1)}{2}, \qquad (2.30)$$

$$\sum_{i=1}^{N} X_i^2 = \sum_{i=1}^{N} Y_i^2 = \frac{N(N+1)(2N+1)}{6}. \qquad (2.31)$$

If we let $D_i = X_i - Y_i$ represent the difference between the ranks associated with the ordered pair $(X_i, Y_i), i = 1,\dots,N$, then

$$\sum_{i=1}^{N} D_i^2 = \sum_{i=1}^{N} X_i^2 + \sum_{i=1}^{N} Y_i^2 - 2 \sum_{i=1}^{N} X_i Y_i$$

and thus

$$\sum_{i=1}^{N} X_i Y_i = \frac{\sum_{i=1}^{N} X_i^2 + \sum_{i=1}^{N} Y_i^2 - \sum_{i=1}^{N} D_i^2}{2} \tag{2.32}$$

Upon substituting (2.30)–(2.32) into (2.24.1) and simplifying ultimately yields

$$\rho_S = 1 - \left[\frac{6 \sum_{i=1}^{N} D_i^2}{N(N^2 - 1)} \right], \tag{2.33}$$

Spearman's index of correlation over ranks.

Equation (2.33) cannot be used if there are ties in either or both rankings. In the presence of ties, convention dictates that we assign mean ranks to sets of tied observations. Once this is done, (2.33) is then used to determine the rank correlation over a set of ranks corrected for ties.

A few additional points concerning (2.33) are in order. First, as was the case for ρ_{XY}, ρ_S is dimensionless with $|\rho_S| \le 1$. Second, since $D = \sum_{i=1}^{N} D_i^2$ measures deviation from complete agreement, it follows that if the rank orders are in perfect or complete agreement, $D = 0$ (each $D_i = 0, i = 1, \ldots, N$) so that $\rho_S = 1$. And if the ranks assigned to one set of observations are exactly the opposite of the ranks assigned to the other set (the two rankings are in complete disagreement), then $D = N \frac{N^2-1}{3}$ and thus $\rho_S = -1$. (In this latter case of perfect disagreement, let $X_i = i$. Then $Y_i = N - i + 1$ so that $D_i = 2i - (N + 1)$.)

Example 2.5.1   To see exactly how (2.33) is utilized, let us assume that two well-known celebrities were asked to rate $N = 10$ different brands of bottled sparkling water on the basis of overall product quality. Their rankings are presented in Table 2.13. What is the extent of their agreement based upon their individual rankings of the various brands of sparkling water? From (2.33),

$$\rho_S = 1 - \left[\frac{6(12)}{10(99)} \right] = 1 - 0.07 = 0.93.$$

Clearly there is a considerable amount of agreement in their rankings.

Table 2.13 indicates that the celebrities placed the various brands in rank order, with 10 being the highest rank assigned and 1 being the lowest. However, let us now assume that they were asked to rate each brand on a scale from 0 to 100, with 100 being the highest rating (see Table 2.14). Clearly there are ties present in this alternative ranking scheme. As previously indicated, ties are handled by assigning mean ranks to the tied observations. That is, we assign to each tied value the average of the ranks that would have been assigned had there been no ties.

Table 2.13

Brand	Celebrity1 Rankings X_i	Celebrity2 Rankings Y_i	$D_i = X_i - Y_i$	D_i^2
	Rankings of Bottled Sparkling Water			
1	4	5	−1	1
2	3	4	−1	1
3	8	8	0	0
4	7	6	1	1
5	10	9	1	1
6	5	3	2	4
7	9	10	−1	−1
8	6	7	−1	1
9	1	2	−1	1
10	2	1	1	1
				10

Table 2.14

Brand	Celebrity 1 Scores	Celebrity 2 Scores	Celebrity 1 Rankings X_i	Celebrity 2 Rankings Y_i	$D_i = X_i - Y_i$	D_i^2
	Rankings of Bottled Sparkling Water					
1	85	88	3	3	0	0
2	86	88	4	3	1	1
3	90	93	8.5	7.5	1	1
4	88	91	7	5.5	1.5	2.25
5	95	95	10	9	1	1
6	87	88	5.5	3	2.5	6.25
7	90	98	8.5	10	−1.5	2.25
8	87	93	5.5	7.5	2	4
9	81	91	2	5.5	−3.5	12.25
10	80	85	1	1	0	0
						30

In this regard, a second application of (2.33) yields

$$\rho_S = 1 - \left[\frac{6(30)}{10(99)}\right] = 1 - 0.18 = 0.82.$$

Based upon this revised rating system, there still exists a substantial amount of agreement between the two rankings.   ■

The Pearson and Spearman coefficients of correlation have been presented as purely descriptive measures of association/agreement; tests of their statistical significance will be offered in later chapters, where these indexes will be treated as statistics computed from sample data.

2.6 Exercises

2-1. Given the following set of $n = 35$ observations:

$$
\begin{array}{ccccccc}
10 & 19 & 18 & 21 & 20 & 20 & 18 \\
18 & 18 & 16 & 11 & 12 & 16 & 12 \\
18 & 19 & 19 & 13 & 15 & 17 & 17 \\
16 & 17 & 14 & 15 & 14 & 18 & 19 \\
16 & 14 & 15 & 19 & 17 & 17 & 18
\end{array}
$$

construct an absolute frequency distribution, the relative frequency distribution, and the less than or equal to cumulative frequency distribution.

2-2. For each of the following sample data sets, find the mean, median, mode, and standard deviation:

(A)   7, 8, 10, 7, 3, 11, 13, 10, 4, 14

(B)   6, 2, 1, 3, 0, 10, 5, 12, 10

2-3. Which data set in Exercise 2-2 has more variability associated with it?

2-4. Use Chebyshev's theorem to determine the amount of data lying within 2.5 standard deviations of the mean. What is implied interval for data set (A) in Exercise 2-2?

2-5. For each of the data sets appearing in Exercise 2-2, find $Q_1, Q_2, P_{10}, P_{20}, D_4$, and D_8. What is the quartile deviation for each of these data sets?

2-6. Transform the following collection of sample observations into a set of Z-scores: 2, 11, 22, 25, 26, 22, 25, 27, 32, 40. Are any of the data values outliers? Find the 20% trimmed mean. What are the second and third level Winsorized means?

2-7. For data set (B) in Exercise 2-2, find v_3 and v_4. What are the third and fourth standard moments? Interpret their values.

2-8. Given the following $N = 10$ data points of the form (X_i, Y_i):

$$(1,1), (2,3), (3,1), (4,3), (6,3), (6,5), (6,6), (8,6), (8,8), (9,10),$$

find ρ_{XY}.

2-9. If $Y = X/S_X$, where S_X is the standard deviation of X, prove that $\overline{Y} = \overline{X}/S_X$ and $S_Y = 1$. Also verify that for $Z = (X - \overline{X})/S_X, \overline{Z} = 0$ and $S_Z = 1$.

2-10. Two judges were asked to rate $N = 10$ different perfumes on a scale from 1 (poor) to 10 (excellent). Given the following rating outcomes, determine if there exists a reasonably strong degree of agreement between the two judges.

Perfume	Judge 1 Rankings	Judge 2 Rankings
1	3	5
2	4	2
3	7	8
4	6	9
5	9	10
6	10	7
7	5	1
8	2	6
9	8	4
10	1	3

2-11. Two middle-school teachers were asked to rate $N = 10$ different essays on the basis of clarity of presentation. They were asked to assign grades of A, B, C, D, or F to each essay, with A being the highest grade and F considered failing. Is there a substantial amount of agreement between the grading standards of these teachers?

Essay	Judge 1 Grades X_i	Judge 2 Grades Y_i
1	D	C
2	D	D
3	F	D
4	D	F
5	A	B
6	B	A
7	A	B
8	C	B
9	B	C
10	C	C

2-12. Absolute/Relative Frequency Distribution (Grouped Data)

An *absolute frequency distribution* shows the absolute frequencies with which the various values of a variable X are distributed among chosen classes. Here the absolute frequency of class j, f_j, is the number of items falling into the jth class. For instance, an example of a typical absolute frequency distribution is:

Classes of X	f_j	$\frac{f_j}{N}$	m_j	Class Boundaries
0–9	2	0.058	4.5	−0.5–9.5
10–19	7	0.205	14.5	9.5–19.5
20–29	10	0.294	24.5	19.5–29.5
30–39	8	0.235	34.5	29.5–39.5
40–49	6	0.176	44.5	39.5–49.5
50–59	1	0.029	54.5	49.5–59.5
	$\overline{34}$	$\overline{1.000}$		

The absolute frequency distribution consists of columns one and two, whereas a relative frequency distribution involves columns one and three, where the relative frequency of class j is $\frac{f_j}{N}$. For k classes, $\sum_{j=1}^{k} f_j = N$ and $\sum_{j=1}^{k} \frac{f_j}{N} = 1$. (If the data set is a sample of size n, then N is replaced by n in the preceding discussion.) In addition, the *class mark* of the jth class, m_j, is the midpoint of the jth class and thus serves as a representative or typical value from the class; it is determined as the average of the class limits. Next, *class boundaries* (sometimes called *real class limits*) are "impossible" values and serve to avoid gaps between the classes. Finally, the *class interval* (denoted as c) is the length of a class and is computed as the difference between the upper and lower boundaries of the class. (Note that if the classes are of equal length, then the class interval can be determined by taking the difference between two successive class marks.) Thus the common class interval for this distribution is, using the second class, $19.5-9.5 = 10 = c$ (which also equals $m_2 - m_1 = 14.5 - 4.5 = 10$). Given the $n = 25$ observations that follow, determine an absolute frequency distribution using the classes: $30-39, 40-49, \ldots, 90-99$. What are the relative frequencies, class marks, class boundaries, and the class interval?

$$33 \quad 87 \quad 39 \quad 40 \quad 44 \quad 44 \quad 97 \quad 48 \quad 53 \quad 50$$

$$67 \quad 37 \quad 51 \quad 66 \quad 94 \quad 55 \quad 55 \quad 68 \quad 71 \quad 76$$

$$83 \quad 87 \quad 70 \quad 89 \quad 90$$

2-13. Descriptive Statistical Measures (Grouped Data)

Given an absolute frequency distribution involving k classes (as defined in the preceding exercise), we may define the following descriptive statistical measures:

a. Mean

$$\mu \approx \frac{\sum_{j=1}^{k} m_j f_j}{N} \; ; \quad \overline{X} \approx \frac{\sum_{j=1}^{k} m_j f_j}{n} . \qquad (2.E.1)$$
$$\text{(population)} \qquad \text{(sample)}$$

b. Standard Deviation

$$\sigma \approx \sqrt{\frac{\sum_{j=1}^{k} f_j (m_j - \mu)^2}{N}} \; ; \quad S \approx \sqrt{\frac{\sum_{j=1}^{k} f_j (m_j - \overline{X})^2}{n-1}} . \qquad (2.E.2)$$
$$\text{(population)} \qquad \text{(sample)}$$

c. Median

$$\text{median} \approx L + \left(\frac{\frac{N}{2} - CF}{f_{med}}\right) c; \quad \text{median} \approx L + \left(\frac{\frac{n}{2} - CF}{f_{med}}\right) c, \quad (2.E.3)$$
(population)   (sample)

where L is the lower boundary of the class containing the median, CF is the cumulative frequency of the preceding classes, f_{med} is the frequency of the class containing the median, and c is the length of the class containing the median.

d. Mode

$$\text{mode} \approx L + \left(\frac{\Delta_1}{\Delta_1 + \Delta_2}\right) c, \quad (2.E.4)$$
(population or sample)

where L is the lower boundary of the class containing the mode, Δ_1 is the difference between the frequency of the modal classes and the premodal class, Δ_2 is the difference between the frequency of the modal classes and the postmodal class, and c is the length of the modal class. Given the data set appearing in Exercise 2-12, find $\overline{X}, S$, the median, and the mode.

2-14. Given the sample values $X_1, \ldots, X_n$, let us define:

a. Geometric Mean (G.M.)

$$\text{G.M.} = \sqrt[n]{X_1 X_2, \cdots, X_n}. \quad (2.E.5)$$

(Note that $\log(\text{G.M.}) = \frac{1}{n} \sum_{i=1}^{n} \log X_i$.)

b. Harmonic Mean (H.M.)

$$\text{H.M.} = \frac{n}{\sum_{i=1}^{n} \frac{1}{X_i}}. \quad (2.E.6)$$

Given the following $n = 5$ sample values, demonstrate that H.M. $\leq$ G.M. $\leq \overline{X}$. $X: 1, 4, 5, 7, 10$. Under what circumstance does strict equality hold?

2-15. Measures of dispersion, stated in either absolute or relative terms (e.g., the standard deviation or coefficient of variation, respectively) are standard fare when it comes to analyzing interval scale data. However, if a variable is measured on, say, a nominal or categorical scale, then these measures cannot be employed to assess variation in a data set. In this instance a so-called *"measure of diversity"* is used to analyze the distribution of observations over categories. That is to say, observations evenly distributed over a set of mutually exclusive and collectively exhaustive categories are said to exhibit "*high diversity*," and observations distributed in a fashion such that most of the data points occur in only a few of the said categories reflect "*low diversity*."

In what follows we shall employ an information-theoretic measure of diversity.

To this end, suppose we have a finite population of size N containing k distinct and nonoverlapping groups with N_j elements belonging to the jth group, $j = 1, \ldots, k$, where $N = \sum_{j=1}^{k} N_j$. Then an appropriate measure of diversity (see Brillouin (1962) and Pielou (1966)) in this circumstance is

$$H = \frac{1}{N} \ln \frac{N!}{N_1! N_2! \ldots N_k!} = \frac{1}{N} \left(\ln N! - \sum_{j=1}^{k} \ln N_j! \right), \qquad (2.E.7)$$

where, via Stirling's formula,

$$\ln N! \approx N(\ln N - 1) + \frac{1}{2} \ln(2\pi N). \qquad (2.E.8)$$

Equation (2.E.7) is based upon the use of information content as a measure of diversity and thus is related to the notion of uncertainty. That is, in a population exhibiting low diversity, we can be relatively certain of the identity of an item chosen at random. But in a high diversity population, it is much more difficult to predict the identity of an item selected at random. Hence, from an information theoretic viewpoint, low diversity is associated with low uncertainty and high diversity is associated with high uncertainty.

A moment's reflection about the nature of (2.E.7) reveals that this index takes into account both group *richness* (the number of distinct groups) and the *evenness* of the distribution of items among the k groups. Then for a given N and k, the notion of evenness considers how closely a set of observed group frequencies align with those from a collection of groups having maximum possible diversity. To obtain an index of evenness or relative diversity, let us first note that the maximum possible value of H for a set of N observations distributed among k groups is when $N_j = \frac{N}{k}$ or

$$H_{\max} = \frac{1}{N} [\ln N! - (k - d)c! - d \ln(c + 1)!], \qquad (2.E.9)$$

where c is the integer portion of $\frac{N}{k}$ and d is the remainder. Then the evenness of the distribution of N observations among the k categories is expressible as

$$E = \frac{H}{H_{\max}} \qquad (2.E.10)$$

while

$$J = I - E \qquad (2.\text{E}.11)$$

serves as an index of *dominance* or *heterogeneity*.

Suppose we have three populations A, B, and C, with each population made up of $n = 100$ items distributed over $k = 10$ categories, where N_j is the number of items in category j. Compare these populations:

	Populations		
Category Abundance	**A**	**B**	**C**
N_1	10	1	90
N_2	10	2	2
N_3	10	3	1
N_4	10	5	1
N_5	10	7	1
N_6	10	9	1
N_7	10	12	1
N_8	10	16	1
N_9	10	22	1
N_{10}	10	23	1
	100	100	100

with respect to their indices of diversity, evenness, and dominance.

Probability Theory

In the introductory chapter, we mentioned that the application of the process of statistical inference involves determining one or more characteristics of a population by examining only a small subset of the same called a sample. In addition, an important component of this process has us offer a measure of the uncertainty or reliability of the inference made. To measure the magnitude of this uncertainty, we need to apply the basics of probability theory.

As we shall soon see, only *events* have probabilities associated with them. And since events are best described using set notation, the following section reviews the rudiments of set algebra. If you feel comfortable with set notation and operations (including functions), proceed directly to Section 3.2.

3.1 Mathematical Foundations: Sets, Set Relations, and Functions

In this chapter we shall see exactly how probabilities are assigned to events. Since events are basically outcomes of either real or conceptual experiments, they may occur individually or in some collective fashion. That is, once an experiment is performed:

- Event A may occur by itself
- Event A and event B may occur together
- Event A or event B or both events A and B may result
- Event A does not occur when event B does occur
- And so on

If we are to describe outcomes such as these in a concise and efficient fashion, then we need to develop the appropriate logical apparatus for structuring events. In this regard, as will be exhibited later on, we may conveniently represent events as sets. Thus this foundation section is designed to introduce you to the essentials of set algebra. Once this material is reviewed and the chapter proper is started, it should become clear that set notation is the appropriate descriptive vehicle to be employed in the specification of events and the calculation of their probabilities.

Consider a *set* as a well-defined collection or grouping of objects called *elements*. Sets will be denoted by uppercase letters and elements typically will be written as lowercase letters. The set A is said to be *well-defined* if there is a rule that determines whether or not a particular element is in A. When x is an element of A we write $x \in A$ (read "x is in A"); if x is not a member of A, then the element inclusion symbol $\in$ is negated or $x \notin A$ (read "x is not in A").

Sets may be defined by listing their elements or by indicating some unique feature that their elements possess. For the latter, we shall use the notation

$$X = \{x|\text{ a rule for specifying } x \in X \text{ is given}\},$$

where X is the set under discussion, x is a representative element, and the vertical bar is read "such that." For instance, one way of denoting the set N of all odd numbers is $N = \{n|n + 1 \text{ is even}\}$.

When set A is a member of set B we write $A \subseteq B$ (read "A is a subset of B"); if A is not a member of B, then the subset inclusion symbol $\subseteq$ is negated or $A \nsubseteq B$ (read "A is not a subset of B"). We note further that if $A \subseteq B$, then either

1. $A \subset B$ (read "A is a *proper subset* of B").

2. $A = B$ (both A and B possess the same elements).

Here $A \subset B$ means that B contains the same elements that A does along with some additional elements that are not in A, that is, B is a larger set.

The *universal set* U is the set containing all elements under a particular discussion and the *null set* ϕ is the empty set. Also, the *complement of set* A, $\overline{A}$, is the collection of elements within U that lie outside of A. The relationship between U, A and $\overline{A}$ can very conveniently be exhibited by the *Venn diagram* in Figure 3.1 Note that $\overline{U} = \phi$, $\overline{\phi} = U$, and $\overline{\overline{A}} = A$.

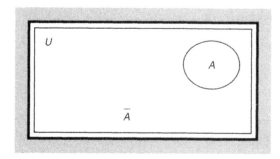

Figure 3.1   Set A and the complement of A.

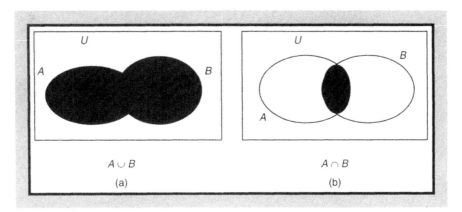

Figure 3.2   (a) The union of sets A and B; (b) The intersection of sets A and B.

Next, the *union* of sets A and B, denoted $A \cup B$, is the collection of elements in A or in B or in both A and B (see Figure 3.2a); the *intersection* of sets A and B, denoted $A \cap B$, is the collection of elements common to both A and B (see Figure 3.2b).

Example 3.1.1   If $A = \{1, 3, 6, 8, 9, 15\}$ and $B = \{1, 3, 7, 10\}$, then

$$A \cup B = \{1, 3, 6, 7, 8, 9, 10, 15\}; A \cap B = \{1, 3\}. \quad \blacksquare$$

Note that

$$A \cup \phi = A \qquad A \cap \overline{A} = \phi$$
$$A \cap \phi = \phi \qquad A \cup A = A$$
$$A \cup U = U \qquad A \cap A = A$$
$$A \cap U = A \qquad \left. \begin{array}{l} \overline{A \cup B} = \overline{A} \cap \overline{B} \\ \overline{A \cap B} = \overline{A} \cup \overline{B} \end{array} \right\} \text{ De Morgan's Laws}$$
$$A \cup \overline{A} = U$$

The *difference* of sets A and B, $A - B$, is the set containing those elements of A which are not in B—that is, $A - B = A \cap \overline{B}$ (see Figure 3.3). Moreover, sets A and B are *mutually exclusive* or *disjoint* if they have no elements in common; that is, $A \cap B = \phi$.

The sets $A_1, \ldots, A_n$ form a *partition* of U if they are pairwise disjoint (intersecting any two of them yields the null set) and their union is U; that is, $A_i \cap A_j = \phi$ for $i \neq j$ and $U = A_1 \cup \cdots \cup A_n$.

For example, the following Venn diagram (see Figure 3.4) indicates that U has been partitioned by sets $A_1, \ldots, A_4$.

How might a diverse assortment of facts and figures be organized into a concise *information set,* which subsequently can serve as a valuable aid in decision making?

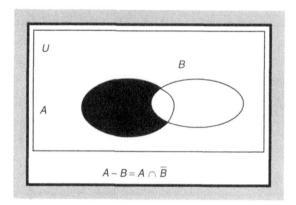

Figure 3.3   The difference of sets A and B.

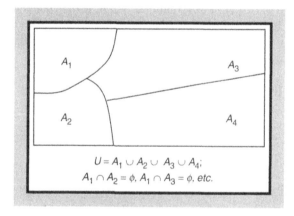

Figure 3.4   Sets $A_1, \ldots, A_4$ form a partition of the universal set.

Two examples are offered to illustrate the versatility of set notation. The approach that follows also serves as a general guide to the structuring of events so that their probabilities can easily be determined.

Example 3.1.2   Consider a group of 100 managers having the following work habits: 50 work overtime during the regular work week, 30 work weekends, and 10 of those who work weekends also put in overtime during the week. If we define sets

A: managers working overtime

B: managers working weekends

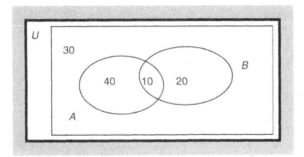

Figure 3.5   Distribution of work habits for 100 managers.

then Figure 3.5 summarizes this information. Let us describe the following statements (a)–(f) in terms of sets and then determine the number of managers appearing in each of the resulting sets.

(a) Managers who work overtime and on weekends—$A \cap B = \{10\}$

(b) Managers who work overtime or on weekends—$A \cup B = \{40+20+10\} = \{70\}$

(c) Managers who do work overtime but do not work on weekends—$A - B = \{40\}$

(d) Managers who do not work overtime—$\overline{A} = \{30 + 20\} = \{50\}$

(e) Managers who neither work overtime nor work weekends—$\overline{A} \cap \overline{B} = \overline{A \cup B} = \{30\}$

(f) Managers who do not work weekends or do not work overtime—$\overline{A} \cup \overline{B} = \overline{A \cap B} = \{30 + 40 + 20\} = \{90\}$ ∎

Example 3.1.3   Out of a group of 30 people traveling together on a bus tour, 10 speak English, 12 speak French, and 16 speak Italian. Five of those who speak English also speak French; six of those who speak French know Italian; six of those fluent in Italian are also fluent in English; and three members of the group speak all three languages. Let us illustrate the distribution of language proficiency by constructing a Venn diagram (see Figure 3.6) using sets

A: individuals who speak English

B: individuals who speak French

C: individuals who speak Italian

Let us again describe the following statements in terms of sets and then count the number of persons in each. (Hint: Avoid the trap of counting certain elements

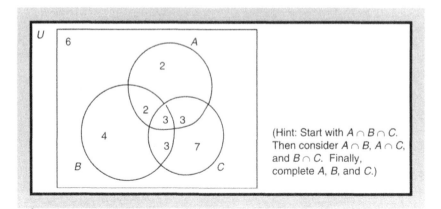

Figure 3.6   Distribution of language proficiency for 30 travelers.

more than once or eliminating elements completely.) To this end we have:

(a) Individuals who speak English or French or Italian—$A \cup B \cup C = \{2 + 2 + 3 + 3 + 4 + 3 + 7\} = \{24\}$

(b) Individuals who speak English or Italian—$A \cup C = \{2+2+3+3+3+7\} = \{20\}$

(c) Individuals who do not speak any of the three languages mentioned— $\overline{A \cup B \cup C} = \{6\}$

(d) Individuals who speak English or Italian but not French—$A \cup C - (A \cap B) - (B \cap C) + (A \cap B \cap C) = \{20 - 5 - 6 + 3\} = \{12\}$

(e) Individuals who speak at least two languages—$(A \cap B) \cup (A \cap C) \cup (B \cap C) - 2(A \cap B \cap C) = \{5 + 6 + 6 - 2(3)\} = \{11\}$

(f) Individuals who speak only French—$B - (A \cap B) - (B \cap C) + (A \cap B \cap C) = \{12 - 5 - 6 + 3\} = \{4\}$. ∎

We end this section with the definition of some important mathematical concepts that will prove useful in conceptualizing the meaning of a random variable and the probability of an event. Random variables and event probabilities are defined in subsequent sections of this chapter.

First, let X and Y be two nonempty sets of real numbers. Then a single-valued function or point-to-point mapping $f : X \rightarrow Y$ is a rule or law of correspondence that associates with each element $x \in X$ a unique element $y \in Y$. Here $y = f(x)$, the *image of x under rule f*, is termed the *dependent variable* and x is the *independent variable*. Set X is called the *domain of f* (see Figure 3.7a) and denoted D_f; the collection of those y's in Y that are the image of at least one $x \in X$ is called the *range of f* and denoted R_f. Clearly R_f is a subset of Y. If $R_f = Y$, then f is termed an *onto mapping*; if $R_f \subset Y$, then f is called an *into mapping*. Also, f is termed *one-to-one* if no $y \in Y$ is the image of more than one $x \in X$.

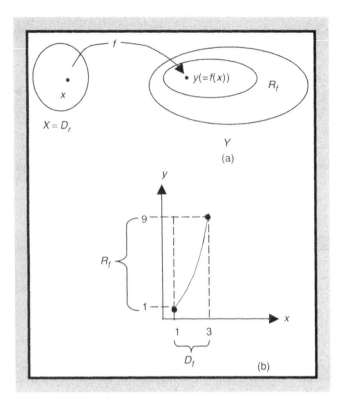

Figure 3.7   (a) f is a point-to-point mapping; (b) The point-to-point mapping $y = f(x) = x^2$.

Example 3.1.4   Let $y = f(x) = x^2, x \in X = D_f = [1,3]$ with $Y = [0,14]$ (see Figure 3.7b). Clearly $R_f = [1,9] \subset Y$. What is the *rule* that tells us how to get a y value from an x value?   ∎

Next, let X be a class or collection of sets and Y a set of *nonnegative* real numbers. Then a measure (set) function or set-to-point mapping $M : X \rightarrow Y$ is a rule or law of correspondence that associates with each set $A \subseteq X$ a unique element $0 \leq y \in Y$. This time y is the image of A under rule M (see Figure 3.8).

Examples of measure or set functions are *length, area, volume,* and so on. That is, as Figure 3.9a reveals, length of $A = M(A) = b - a \geq 0$. Here the *rule* is $b - a$. From Figure 3.9b, area of $A = M(A) = b \times h \geq 0$.

3.2   The Random Experiment, Events, Sample Space, and the Random Variable

We may think of a *random experiment* as a class of occurrences that can happen repeatedly, for an unlimited number of times, under essentially unchanged conditions. Here we require that, in the long run, the circumstances under which

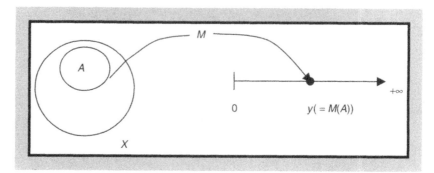

Figure 3.8 M is a set-to-point mapping.

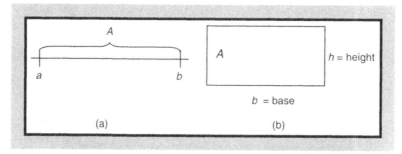

Figure 3.9 (a) Length is a measure function; (b) Area is a measure function.

the various trials of a random experiment are performed are virtually the same. Relative to this definition, we have the concept of a *random phenomenon*—a phenomenon such that, on any given trial of a random experiment, the outcome is unpredictable. It is important to note that all random experiments generate a collection of *potential* outcomes. If the *actual* or *realized* outcome on any given trial cannot be ascertained with complete certainty, then it is characterized as a random phenomenon. That is, as the trials of the random experiment progress, it is assumed that the results of the same do not exhibit any systematic favoring of one outcome relative to any other possible outcome. Here the outcome of a random experiment will be termed an *event*. In particular, an event may be *simple* (denoted E_i) or *compound* (denoted $A, B, C, \ldots$). In the former case, we have a possible outcome of a random experiment that cannot be decomposed into a combination of other events; in the latter, we have an event that can be decomposed into a collection of simple events.

Next, we may view the *sample space* (denoted S) as the collection or record of all possible outcomes of a random experiment. Here S serves as the *universal set of outcomes*. As Figure 3.10 indicates, S is a collection of n points or simple events; that is, there exists a one-to-one correspondence between the simple events E_i and the points of the sample space S. In fact, $S = \cup_{i=1}^{n} E_i$. A compound

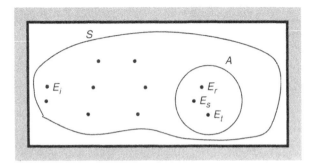

Figure 3.10   A compound event is the union of simple events.

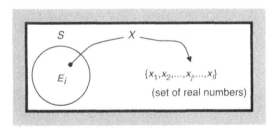

Figure 3.11   A random variable X is a point-to-point mapping.

event A is thus the union of its constituent simple events in S or, as Figure 3.10 reveals, $A = E_r \cup E_s \cup E_t \subset S$. Thus simple events serve as the *basic building blocks* of compound events. In general, when we speak of an event (simple or compound) we are speaking of some subset of S. How are the possible outcomes of a random experiment defined? The answer involves the specification of a *random variable* (denoted X). Specifically, X is a chance variable or a variable whose values have probabilities associated with them. More formally, it is a real-valued function defined on the elements of a sample space; that is, it is a rule or law of correspondence that associates with each $E_i \in S$ some (unique) real number $X(E_i) = x_j, \ j = 1, \ldots, l$.

What is the connection between $E_i, S,$ and X? Looking to Figure 3.11, the random variable X takes a simple event $E_i \in S$ and associates with it a real number x_j.

Example 3.2.1   Let us define as our random experiment the rolling of a pair of six-sided dice. If the pair of dice is *fair*, each possible outcome is a random phenomenon. For the sample space (see Figure 3.12), $S = \cup_{i=1}^{36} E_i$. How may we specify the random variable X? Since its role is to define the possible outcomes of the random experiment, it may be specified by the experimenter. In this regard, let us define X as *the sum of the faces showing*. We could have just as easily defined X as the product of the faces showing or, if one die is, say, red and the other blue, we

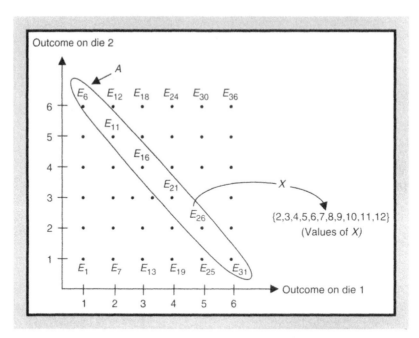

Figure 3.12   The sample space $S = \bigcup_{i=1}^{36} E_i$.

could let X be the outcome showing on the red die divided by the outcome showing on the blue one. The point being made is that, within the confines of the physical structure of the random experiment, the random variable may be whatever we want it to be. (However, it is important to keep in mind that for some random experiments, the specification of the random variable may be uniquely dictated by the physical structure of the experiment itself. For instance, if the random experiment involves the tossing of a single die, the random variable X is typically the face showing.) With X specified as the sum of the faces showing, X has 11 values: $x_1 = 2, x_2 = 3, \ldots, x_{11} = 12$. Then X takes a point such as $E_{26} \in S$ and associates with it the real number 7. To form the compound event $A = \{$getting a sum of 7$\}$, we simply form the union $A = E_6 \cup E_{11} \cup E_{16} \cup E_{21} \cup E_{26} \cup E_{31}$. It is evident from Figure 3.12 that there are more simple events in S than there are values of X; that is, $i = 1, \ldots, 36$ and $j = 1, \ldots, 11$. This is because there is more than one way of getting the sum of, say, 6. However, this need not always be true. Looking back to the case where the random experiment involves the tossing of a single die, it is clear that S contains six simple events and the random variable assumes exactly six values, namely 1, 2, 3, 4, 5, 6. ∎

3.3 Axiomatic Development of Probability Theory

What general properties must the probability of an event possess? Once these properties are stated for any idealized event, we can then employ them as an aid

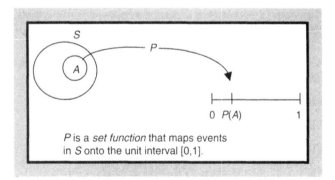

Figure 3.13   A probability measure P on S is a set-to-point mapping.

in assigning probabilities to simple events and ultimately to the specification of probabilities of compound events.

Let us begin with the concept of a *probability measure* on S. Specifically, a probability measure is a special type of set-to-point function. That is, it is a set function P from the sample space to the unit interval $[0, 1]$. As Figure 3.13 reveals, P is a rule that ascribes to each event $A \subseteq S$ a real number between 0 and 1 inclusive, called its *probability of occurrence*. In this regard, if event A never occurs, $P(A) = 0$; if event A is the *certain event* (i.e., it always occurs), $P(A) = 1$. Note that $P(A)$ is larger in value the greater the likelihood of the occurrence of event A.

We are now ready to state the usual minimal set of *probability axioms* (an *axiom* is an accepted or established principle—it is simply taken as given and needs no formal proof of its legitimacy). In this regard, let A be any event in S. If P is a probability measure on S, then:

AXIOM 1. $P(A)$ is defined for every $A \subseteq S$.

AXIOM 2. $P(A) \geq 0$ for all $A \subseteq S$.

AXIOM 3. $P(S) = 1$.

AXIOM 4. If $A_1, \ldots, A_k$ is a sequence of mutually exclusive events in $S(A_i \cap A_j = \phi, i \neq j)$, then

$$P\left(\cup_{i=1}^{k} A_i\right) = \sum_{i=1}^{k} P(A_i),$$

that is, the probability of the union of a set of mutually exclusive events is the sum of their individual probabilities.

Note that by virtue of axiom 3, S is the *certain event*; that is, an outcome in S *must* occur when a random experiment is performed. Next come the corollaries to the axiom system (a *corollary* is a consequence—it is something that naturally follows). Then for any events $A, B \subseteq S$ and P a probability measure on S:

COROLLARY 1. The probability that event A does not occur is one minus the probability that it does occur or $P(\overline{A}) = 1 - P(A)$.

COROLLARY 2. $0 \leq P(A) \leq 1$.

COROLLARY 3. $P(\phi) = 0$.

COROLLARY 4. If $A \subset B$, then $P(A) < P(B)$.

COROLLARY 5. If $A \subset B$, then $P(B - A) = P(B) - P(A)$.

COROLLARY 6. $P(A) = P(A \cap B) + P(A \cap \overline{B})$.

COROLLARY 7. $P(A - B) = P(A \cap \overline{B}) = P(A) - P(A \cap B)$.

Clearly these axioms and corollaries are applicable to the simple events $E_i, i = 1, \ldots, n$, of S. In the next section we shall see that they can actually be used to readily determine the probability of any compound event $A \subseteq S$.

3.4   The Occurrence and Probability of an Event

When can we say that some compound event A has occurred? Since A is a subset of points or simple events in S, event A occurs if any one of the simple events comprising A occurs when a random experiment is conducted. That is, in Figure 3.10, event A occurs if on any given trial of a random experiment, either E_r or E_s or E_t occurs. Although we shall leave to a subsequent section the *sources* or *alternative interpretations* of probabilities, we shall specify, within the context of the preceding example problem involving the rolling of a pair of dice, the probability of an event A as the sum of the probabilities of the individual simple events comprising A. Since $S = \cup_{i=1}^{36} E_i$ contains $n = 36$ simple events, the probability of any E_i is $P(E_i) = \frac{1}{n} = \frac{1}{36}$. Then by axiom 3,

$$
\begin{aligned}
P(A) &= P(\text{getting a sum of 7}) \\
&= P(E_6 \cup E_{11} \cup E_{16} \cup E_{21} \cup E_{26} \cup E_{31}) \\
&= P(E_6) + P(E_{11}) + P(E_{16}) + P(E_{21}) + P(E_{26}) + P(E_{31}) \\
&= \frac{1}{36} + \frac{1}{36} + \frac{1}{36} + \frac{1}{36} + \frac{1}{36} + \frac{1}{36} = \frac{6}{36}.
\end{aligned}
$$

In general, if a random experiment has n equiprobable outcomes (i.e., $P(E_i) = \frac{1}{n}$ for all i) and if n_A of these outcomes constitute event A, then $P(A) = \frac{n_A}{N}$. Here $\frac{n_A}{n}$ is termed the relative frequency of event A on S. So with A made up of $n_A = 6$ simple events and with S containing $n = 36$ simple events, $P(A) = \frac{6}{36} = \frac{1}{6}$. (Note that this relative frequency concept circumvents the direct application of axiom 3.)

3.5   **General Addition Rule for Probabilities**

For events $A, B \subseteq S, P(A \cup B) \leq P(A) + P(B)$ (*Boole's inequality*). That is, if $A \cap B = \phi$, axiom 3 informs us that $P(A \cup B) = P(A) + P(B)$; and if $A \cap B \neq \phi$, $P(A \cup B) < P(A) + P(B)$. In this latter instance we must determine *how much less than*? Since $A \cap B$ (the shaded portion of Figure 3.14) is contained in A as well as in B, the sum $P(A) + P(B)$ counts the probability of the intersection twice. To avoid double counting this portion of $A \cup B, P(A \cap B)$ is netted out once.

Thus the *general addition rule* appears as

$$P(A \cup B) = P(A) + P(B) - P(A \cap B). \tag{3.1}$$

For sets $A, B, C \subseteq S$,

$$P(A \cup B \cup C) = P(A) + P(B) + P(C) - P(A \cap B) - P(A \cap C)$$
$$- P(B \cap C) + P(A \cap B \cap C).$$

In general, for events $A_1, A_2, \ldots, A_n$ within S,

$$P(A_1 \cup A_2 \cup \cdots \cup A_n) = \sum_{i=1}^{n} P(A_i) - \sum\sum_{i<j} P(A_i \cap A_j)$$
$$+ \sum\sum\sum_{i<j<k} P(A_i \cap A_j \cap A_k)$$
$$- \cdots + (-1)^{n+1} P(A_1 \cap A_2 \cap \cdots \cap A_n). \tag{3.2}$$

The total number of items in each individual summation (double summation, etc.) term may be computed from the combinational expression $\frac{n!}{r!(n-r)!}$ as $r = 1, 2, \ldots, n$.

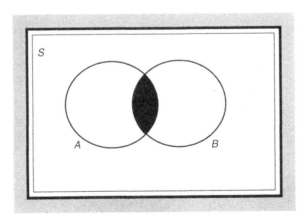

Figure 3.14 $A \cap B \neq \phi$.

Example 3.5.1   To see exactly how (3.2) is applied, it is easily verified that

$$P(A_1 \cup A_2 \cup A_3 \cup A_4) = P(A_1) + P(A_2) + P(A_3) + P(A_4)$$
$$- [P(A_1 \cap A_2) + P(A_1 \cap A_3) + P(A_1 \cap A_4)$$
$$+ P(A_2 \cap A_3) + P(A_2 \cap A_4) + P(A_3 \cap A_4)]$$
$$+ [P(A_1 \cap A_2 \cap A_3) + P(A_1 \cap A_2 \cap A_4)$$
$$+ P(A_1 \cap A_3 \cap A_4) + P(A_2 \cap A_3 \cap A_4)]$$
$$- P(A_1 \cap A_2 \cap A_3 \cap A_4).$$

Clearly the number of terms of the form $P(A_i)$ is $\frac{4!}{1!3!} = 4$; there are $\frac{4!}{2!2!} = 6$ terms of the form $P(A_i \cap A_j)$; we have $\frac{4!}{3!1!} = 4$ terms of the form $P(A_i \cap A_j \cap A_k)$; and there is $\frac{4!}{4!0!} = 1$ term involving the probability of the intersection of all four sets.   ∎

3.6   Joint, Marginal, and Conditional Probability

Assume that the sample space S consists of the n simple events E_i, each with probability $P(E_i) = \frac{1}{n}, i = 1, \ldots, n$. Let the events $A_1, A_2, \ldots, A_r$ represent a *partition* of S (here $A_j \subseteq S, A_p \cap A_q = \phi, p \neq q$, and $\cup_{j=1}^{r} A_j = S$). In addition, let a second collection of events $B_1, B_2, \ldots, B_s$ form a second partition of $S (B_k \subseteq S, B_p \cap B_q = \phi, p \neq q$, and $\cup_{k=1}^{s} B_k = S$). Hence the n simple events of S may be classified in a two-way table, Table 3.1. Here n_{jk} depicts the number of points or simple events in S having attributes A_j and B_k.

Based upon this partitioning of S, we may first define the notion of *joint probability* as the probability of the simultaneous occurrence of two (or more)

Table 3.1

	Partition B					
S	B_1	B_2	$\cdots$	B_s		
A_1	n_{11}	n_{12}	$\cdots$	n_{1s}	$n_{1.}$	
A_2	n_{21}	n_{22}	$\cdots$	n_{2s}	$n_{2.}$	Row Totals
$\vdots$	$\vdots$	$\vdots$	$\cdots$	$\vdots$	$\vdots$	
A_r	n_{r1}	n_{r2}	$\cdots$	n_{rs}	$n_{r.}$	
	$n_{.1}$	$n_{.2}$	$\cdots$	$n_{.s}$		

Partition A

$$\text{Column Totals} \qquad n = \sum_{j=1}^{r} n_{j.} = \sum_{k=1}^{s} n_{.k}$$

events. For instance, the probability that events A_2 and B_1 occur together is $P(A_2 \cap B_1) = \frac{n_{21}}{n}$. In general,

$$P(A_j \cap B_k) = \frac{n_{jk}}{n}. \tag{3.3}$$

Next, a *marginal probability* is computed whenever one or more criteria of classification are ignored. For instance, we may be interested only in the probability of a single event A_1 occurring (here we ignore the B partition) or in the probability of the event B_2 occurring alone (we now ignore the A partition). In the former case $P(A_1) = \frac{n_{1.}}{n}$; in the latter, $P(B_2) = \frac{n_{.2}}{n}$. Generally,

$$P(A_j) = \sum_{k=1}^{s} P(A_j \cap B_k) = \frac{n_{j.}}{n}; P(B_k) = \sum_{j=1}^{r} P(A_j \cap B_k) = \frac{n_{.k}}{n}. \tag{3.4}$$

Finally, we have the concept of a *conditional probability*, where we are interested in the probability of occurrence of one event given that another event has definitely occurred. For example, if the event B_1 has occurred, what is the probability that A_2 has also occurred? To answer this we need to consider the concept of the *effective sample space*. That is, since B_1 has definitely occurred, we can simply ignore the last $s - 1$ columns in Table 3.1 and take as our effective sample space only the first column of this table. Then the probability of A_2 given B_1 (denoted as $P(A_2|B_1)$ and read "the probability of A_2 given B_1") is the number of elements common to both A_2 and B_1 divided by the number of elements in B_1 (the effective sample space) or $\frac{n_{21}}{n_{.1}}$. In general,

$$P(A_j|B_k) = \frac{P(A_j \cap B_k)}{P(B_k)} = \frac{n_{jk}/n}{n_{.k}/n} = \frac{n_{jk}}{n_{.k}};$$

$$P(B_k|A_j) = \frac{P(A_j \cap B_k)}{P(A_j)} = \frac{n_{jk}/n}{n_{j.}/n} = \frac{n_{jk}}{n_{j.}}. \tag{3.5}$$

Since the notion of a conditional probability is extremely important in its own right, let us redefine this concept for any general sample space S. To this end, let events $A, B \subseteq S$ with $P(B) \neq 0$. Then

$$P(A|B) = \frac{P(A \cap B)}{P(B)}. \tag{3.6}$$

And if $P(A) \neq 0$,

$$P(B|A) = \frac{P(A \cap B)}{P(A)}. \tag{3.7}$$

Since $P(A \cap B)$ appears in the numerator of each of these conditional probabilities, it follows that

$$P(A \cap B) = P(A/B) \cdot P(B) = P(B/A) \cdot P(A). \tag{3.8}$$

Here (3.8) is often referred to as the *multiplication law for probabilities.*

We noted earlier that, by virtue of equation (3.4), $P(A_j) = \sum_{k=1}^{s} P(A_j \cap B_k)$; that is, a marginal probability can be calculated as the sum of a set of joint probabilities. Moreover, since each of these joint probabilities can be written, via the multiplication law (3.8), as the product between a conditional probability and an appropriate marginal probability ($P(A_j \cap B_k) = P(B_k) \cdot P(A_j/B_k)$), it then follows that the preceding expression for $P(A_j)$ can be rewritten as

$$P(A_j) = \sum_{k=1}^{s} P(B_k) \cdot P(A_j/B_k). \tag{3.4.1}$$

Equation (3.4.1) is termed the *law of total probability* (of event A_j on the partition $B_1, \ldots, B_s$ of S). In general, for events $B_1, \ldots, B_4 \subseteq S$ given in Figure 3.15 with $S = B_1 \cup B_2 \cup B_3 \cup B_4$, and $B_1 \cap B_2 \cap B_3 \cap B_4 = \phi$, we have, for event $A \subseteq S$,

$$P(A) = \sum_{i=1}^{4} P(A \cap B_i) = \sum_{i=1}^{4} P(B_i) \cdot P(A/B_i). \tag{3.4.2}$$

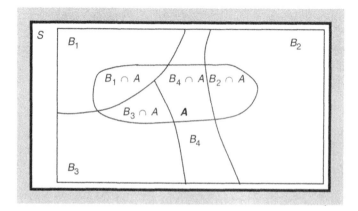

Figure 3.15   Law of total probability of A on the partition $B_1, \ldots, B_4$ of S: $P(A) = \sum_{i=1}^{4} P(A \cap B_i)$.

For nonempty events $A, B, C \subseteq S$, the following conditional probability rules hold under the assumption that the indicated conditional probabilities exist:

$$P(A/A) = 1;$$

$$P(\phi/A) = 0 \left(\text{Note: } P(A/\phi) \text{ is undefined.}\right);$$

$$\text{if } A \subset B, \text{then } P(A/C) < P(B/C);$$

$$P(\overline{A}/B) = 1 - P(A/B);$$

$$P(A/\overline{B}) = \frac{P(A) - P(A \cap B)}{1 - P(B)};$$

$$P(A \cup B/C) = P(A/C) + P(B/C) - P(A \cap B/C);$$

$$\text{if } A \cap B = \phi, \text{then } P(A/B) = P(B/A) = 0;$$

$$P(A/S) = P(A).$$

Moreover, for events $A_1, A_2, \ldots, A_n$ within $S, P\left(\cup_{i=1}^{n} A_i/B\right) \leq \sum_{i=1}^{n} P\left(A_i/B\right)$. If this sequence of events is mutually exclusive, then the expression holds as a strict equality.

Example 3.6.1   Given the partition of S in Table 3.2, find $P(A_2 \cap B_3), P(A_1 \cup B_2), P(B_1), P(\overline{A}_2)$, and $P(A_1/B_3)$. Based upon the preceding discussion:

$$P(A_2 \cap B_3) = \frac{8}{20},$$

$$P(A_1 \cup B_2) = P(A_1) + P(B_2) - P(A_1 \cap B_2) = \frac{8}{20} + \frac{3}{20} - \frac{1}{20} = \frac{10}{20},$$

$$P(B_1) = \frac{5}{20},$$

$$P(\overline{A}_2) = 1 - P(A_2) = 1 - \frac{12}{20} = \frac{8}{20},$$

$$P(A_1/B_3) = \frac{P(A_1 \cap B_3)}{P(B_3)} = \frac{4/20}{12/20} = \frac{4}{12}. \quad \blacksquare$$

If the events $C_1, C_2, \ldots, C_t$ constitute a third partition of $S(C_l \subseteq S, C_p \cap C_q = \phi, p \neq q, \text{and } \cup_{l=1}^{t} C_l = S)$, then n_{jkl} represents the number of simple events within

Table 3.2

	B_1	B_2	B_3	
A_1	3	1	4	8
A_2	2	2	8	12
	5	3	12	20

S possessing the attributes A_j, B_k, and C_l. Hence the joint probability that events A_j, B_k, and C_l occur together is

$$P(A_j \cap B_k \cap C_l) = \frac{n_{jkl}}{n}. \qquad (3.9)$$

The associated set of marginal probabilities for any pair of events is

$$P(A_j \cap B_k) = \sum_{l=1}^{t} P(A_j \cap B_k \cap C_l) = \frac{n_{jk.}}{n},$$

$$P(A_j \cap C_l) = \sum_{k=1}^{s} P(A_j \cap B_k \cap C_l) = \frac{n_{j.l}}{n}, \qquad (3.10)$$

$$P(B_k \cap C_l) = \sum_{j=1}^{r} P(A_j \cap B_k \cap C_l) = \frac{n_{.kl}}{n},$$

and the set of marginal probabilities involving each event individually is, by virtue of (3.10),

$$P(A_j) = \sum_{k=1}^{s} \sum_{l=1}^{t} P(A_j \cap B_k \cap C_l) = \sum_{k=1}^{s} P(A_j \cap B_k) = \frac{n_{j..}}{n},$$

$$P(B_k) = \sum_{j=1}^{r} \sum_{l=1}^{t} P(A_j \cap B_k \cap C_l) = \sum_{j=1}^{r} P(A_j \cap B_k) = \frac{n_{.k.}}{n}, \qquad (3.11)$$

$$P(C_l) = \sum_{j=1}^{r} \sum_{k=1}^{s} P(A_j \cap B_k \cap C_l) = \sum_{j=1}^{r} P(A_j \cap C_l) = \frac{n_{..l}}{n}.$$

If we combine (3.5) and (3.11), then we may alternatively express these marginal probabilities of individual events in terms of conditional probabilities as

$$P(A_j) = \sum_{k=1}^{s} P(A_j \cap B_k) = \sum_{k=1}^{s} P(A_j/B_k)P(B_k),$$

$$P(B_k) = \sum_{j=1}^{r} P(A_j \cap B_k) = \sum_{j=1}^{r} P(B_k/A_j)P(A_j), \qquad (3.11.1)$$

$$P(C_l) = \sum_{j=1}^{r} P(A_j \cap C_l) = \sum_{j=1}^{r} P(C_l/A_j)P(A_j).$$

As far as conditional probabilities are concerned, we may, for instance, start with

$$P(A_j/B_k \cap C_l) = \frac{P(A_j \cap B_k \cap C_l)}{P(B_k \cap C_l)}; \; P(A_j \cap B_k/C_l) = \frac{P(A_j \cap B_k \cap C_l)}{P(C_l)}. \quad (3.12)$$

Clearly other conditional probabilities may be determined by reordering the events $A_j, B_k,$ and C_l. In addition, (3.12) may be used to determine a set of equivalent representations for $P(A_j \cap B_k \cap C_l)$, for example,

$$P(A_j \cap B_k \cap C_l) = P(A_j/B_k \cap C_l) \cdot P(B_k \cap C_l)$$
$$= P(A_j/B_k \cap C_l) \cdot P(B_k/C_l) \cdot P(C_l) \quad (3.13)$$

by an application of (3.8).

Since the use of expressions such as (3.8) and (3.13) are quite common in practice (especially when repeated sampling from the same population is undertaken or when decisions are sequential in nature), the generalization of these expressions appears as follows: if $n(\geq 2)$ is any integer and $A_1, A_2, \ldots, A_n$ represents a sequence of events in S for which $P(A_1 \cap A_2 \cap \cdots \cap A_{n-1}) \neq 0$, then

$$P(A_1 \cap A_2 \cap \cdots \cap A_n) = P(A_1) \cdot (A_2/A_1) \cdot P(A_3/A_1 \cap A_2) \cdots$$
$$P(A_n/A_1 \cap A_2 \cap \cdots \cap A_{n-1}). \quad (3.14)$$

Example 3.6.2 Suppose a jar contains 10 black and six white marbles. What is the probability of selecting four black ones under repeated drawings without replacement. If B_j denotes *getting a black marble on the jth draw*, then from (3.14),

$$P(B_1 \cap B_2 \cap B_3 \cap B_4) = P(B_1) \cdot P(B_2/B_1) \cdot P(B_3/B_1 \cap B_2) \cdot P(B_4/B_1 \cap B_2 \cap B_3)$$
$$= \frac{10}{16} \cdot \frac{9}{15} \cdot \frac{8}{14} \cdot \frac{7}{13} = 0.1153.$$

Letting W_i denote *getting a white marble on the jth draw*, the *decision tree* implied by this experiment appears as Figure 3.16. You can readily demonstrate that:

$$P(B_1 \cap W_2 \cap W_3 \cap B_4) = \frac{10}{16} \cdot \frac{6}{15} \cdot \frac{5}{14} \cdot \frac{9}{13} = 0.0618,$$

$$P(W_1 \cap W_2 \cap B_3 \cap W_4) = \frac{6}{16} \cdot \frac{5}{15} \cdot \frac{10}{14} \cdot \frac{4}{13} = 0.0275,$$

and so on. ■

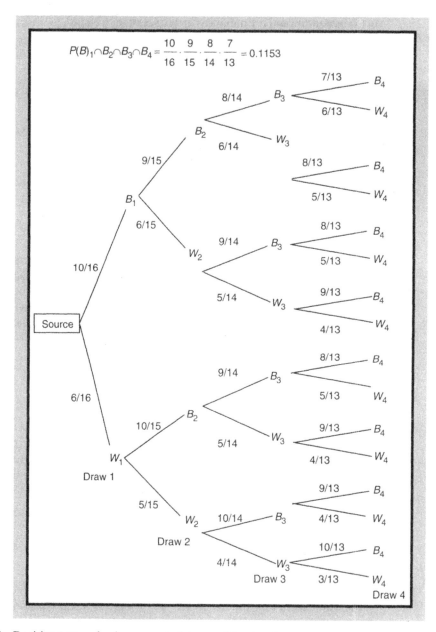

$$P(B)_1 \cap B_2 \cap B_3 \cap B_4 = \frac{10}{16} \cdot \frac{9}{15} \cdot \frac{8}{14} \cdot \frac{7}{13} = 0.1153$$

Figure 3.16   Decision tree under four repeated draws without replacement.

3.7   **Classification of Events**

First, events $A, B \subseteq S$ are *mutually exclusive* if, on any given trial of a random experiment, they cannot occur together. Here $A \cap B = \phi$ so that $P(A \cap B) = 0$. For instance, if our random experiment involves the daily observation of the price of a given stock and we define events $A = \{$the price of the stock increases$\}$, $B = \{$the price of the stock decreases$\}$, then clearly these events cannot occur simultaneously on any given day so that they are indeed mutually exclusive. We may next classify events as *collectively exhaustive* if, on any given trial of a random experiment, at least one of them must occur. The events A and B just defined are not collectively exhaustive since, in structuring these events, we have not accounted for all possible outcomes concerning the stock's behavior. However, if we introduce a third event $C = \{$the price is unchanged$\}$, then clearly the events $A, B,$ and C taken together are collectively exhaustive.

It is evident from the preceding discussion that events may be mutually exclusive but not collectively exhaustive. Likewise, events may be collectively exhaustive but not mutually exclusive. For instance, if we decide to observe the daily prices of two stocks, then we may specify two additional events: $D = \{$the price of at least one increases$\}$, $E = \{$the price of at least one decreases$\}$. Clearly all possible outcomes are subsumed by these events and they can occur together. Finally, it may be the case that events are simultaneously *mutually exclusive and collectively exhaustive*, in which case exactly one of them occurs; for example, A, B, C defined earlier satisfy this criterion. (See Figure 3.17 for a summary of these characterizations.)

An additional classification scheme for a collection of events is that of independence. In this regard, events $A, B \subseteq S$ are *independent* if, on any given trial of random experiment, the occurrence of one of them in no way affects the probability of occurrence of the other. That is, *A is independent of B* if and only if

$$P(A/B) = P(A), \text{ provided } P(B) > 0; \qquad (3.15)$$

and *B is independent of A* if and only if

$$P(B/A) = P(B), \text{ provided } P(A) > 0. \qquad (3.16)$$

It is evident that if A is independent of B, then B is independent of A. In this light, the notion of independence is sometimes referred to as *mutual independence*. From (3.8), we may form the *multiplication rule for probabilities*

$$P(A \cap B) = P(B/A) \cdot P(A) = P(A/B) \cdot P(B).$$

Under independence, this expression becomes, using (3.15) and (3.16),

$$P(A \cap B) = P(B) \cdot P(A) = P(A) \cdot P(B).$$

In short, events A and B are independent if and only if

$$P(A \cap B) = P(A) \cdot P(B). \qquad (3.17)$$

If inequality holds in (3.17), then A and B are said to be *dependent* events. In general, events $A, B \subseteq S$ are independent if and only if any one of the

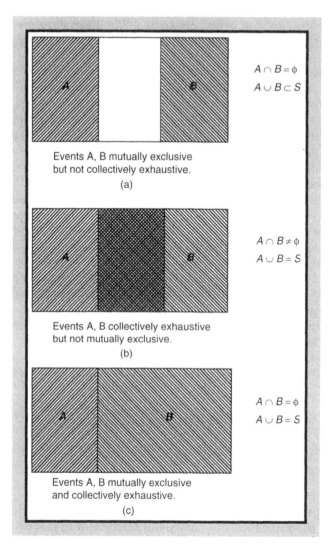

$A \cap B = \phi$
$A \cup B \subset S$

Events A, B mutually exclusive
but not collectively exhaustive.
(a)

$A \cap B \neq \phi$
$A \cup B = S$

Events A, B collectively exhaustive
but not mutually exclusive.
(b)

$A \cap B = \phi$
$A \cup B = S$

Events A, B mutually exclusive
and collectively exhaustive.
(c)

Figure 3.17   (a) Events A, B mutually exclusive but not collectively exhaustive; (b) Events A, B collectively exhaustive but not mutually exclusive; (c) Events A, B mutually exclusive and collectively exhaustive.

expressions (3.15) through (3.17) holds. We may note further that if A and B are independent, then so are the pairs of events $A, \overline{B}$; $B, \overline{A}$; and $\overline{A}, \overline{B}$.

Can events be simultaneously mutually exclusive and independent? The answer is, typically, no. If $A \cap B = \phi$ (i.e., A and B are mutually exclusive), the occurrence of A on a particular trial of a random experiment precludes the occurrence of B (and conversely) so that the independence of these events should be anticipated. In a sense, mutually exclusive events are actually *dependent* upon each other; that is, the occurrence of one depends on the other not occurring.

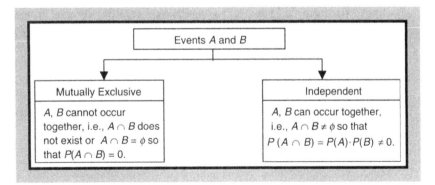

Figure 3.18   Mutually exclusive versus independent events.

In general, $P(A \cap B) = 0 \neq P(A) \cdot P(B)$ unless one or both $P(A)$ and $P(B)$ are zero. Thus two mutually exclusive events are independent if and only if either $P(A) = 0$ or $P(B) = 0$ or both $P(A)$ and $P(B)$ equal zero. So if both $P(A)$ and $P(B)$ are different from zero, then A and B mutually exclusive implies that they are not independent; and the dependence of A and B implies that they are not mutually exclusive. These remarks are summarized in Figure 3.18.

Looking to some special cases, if $A \subset B$, then B occurs if A does so that A and B are dependent; that is, $P(A \cap B) = P(A) \neq P(A) \cdot P(B)$ unless $B = S$ or $P(A) = 0$. In addition, A and S are independent for any $A \subset S$ (since $P(A \cap S) = P(A) = P(A) \cdot P(S)$) as are A and ϕ (since $P(A \cap \phi) = P(\phi) = 0 = P(A) \cdot P(\phi)$).

The events $A, B, C \subseteq S$ are independent if and only if

(a) A, B, and C are pairwise independent; that is,

$$P(A \cap B) = P(A) \cdot P(B),$$
$$P(A \cap C) = P(A) \cdot P(C),$$
$$P(B \cap C) = P(B) \cdot P(C), \text{ and}$$

(b) $P(A \cap B \cap C) = P(A) \cdot P(B) \cdot P(C)$

More generally, the events $A_1, A_2, \ldots, A_n$ within S are independent if and only if

$$P(A_i \cap A_j) = P(A_i) \cdot P(A_j), i \neq j,$$
$$P(A_i \cap A_j \cap A_k) = P(A_i) \cdot P(A_j) \cdot P(A_k), i \neq j, i \neq k, j \neq k, \ldots, \qquad (3.18)$$
$$P(\cap_{i=1}^{n} A_i) = \prod_{i=1}^{n} P(A_i).$$

Let us now consider two particular types of dependence that may exist between events $A, B \subseteq S$. These events are termed *supplements* if event B is more likely to occur if event A has occurred; that is, A and B are supplements when $P(B/A) > P(B)$. Since from (3.8) $P(A \cap B) = P(B/A) \cdot P(A)$, it follows that A and B are supplements if $P(A \cap B) > P(B) \cdot P(A)$. Next, events A and B are *substitutes* if event B is less likely to occur when event A has occurred; that is, $P(B/A) < P(B)$. Thus, by an argument similar to that used earlier, A and B are .substitutes if $P(A \cap B) < P(B) \cdot P(A)$. If A and B are mutually exclusive and if both $P(A), P(B) \neq 0$, then $P(A) \cdot P(A) > 0$ so that mutually exclusive events are by nature substitutes.

Example 3.7.1 Let us define a random experiment as the drawing of a card from an ordinary deck of 52 playing cards. Based upon this simple experiment, let us define the following set of events:

$A = \{$a black card is drawn$\}$

$B = \{$a club is drawn$\}$

$C = \{$a red card is drawn$\}$

It is easily seen that A and C are mutually exclusive but A and B as well as B and C are not. Moreover, events A and C are collectively exhaustive but A and B and B and C are not. Furthermore, since $P(A) = \frac{26}{52}$, $P(B) = \frac{4}{52}$, $P(C) = \frac{26}{52}$, $P(A \cap B) = \frac{2}{52}$, $P(B \cap C) = \frac{2}{52}$, and $P(A \cap C) = 0$, it follows that

$$\underset{(1/26)}{P(A \cap B)} = \underset{(1/26)}{P(A) \cdot P(B)},$$

$$\underset{(0)}{P(A \cap C)} \neq \underset{(1/4)}{P(A) \cdot P(C)},$$

$$\underset{(1/26)}{P(B \cap C)} = \underset{(1/26)}{P(B) \cdot P(C)}.$$

Thus the pairs of events A and B and B and C are independent, and A and C are substitutes (and thus dependent). Looking at these results from another perspective, since $P(A/B) = \frac{P(A \cap B)}{P(B)} = \frac{2}{4} = P(A)$ and $P(B/C) = \frac{P(B \cap C)}{P(C)} = \frac{2}{26} = P(B)$, the pairs of events A and B and B and C are neither supplements nor substitutes (i.e., not dependent); in fact, they are independent as verified earlier. Also, $P(A/C) = \frac{P(A \cap C)}{P(C)} = 0 < P(A)$ so that, as verified, A and C are substitutes. ∎

Finally, let us combine the notion of conditional probability with that of independence of events in order to develop the concept of conditional independence. Specifically, events $A, B \subseteq S$ are *conditionally independent* given event $C \subseteq S$ if $P(A \cap B \cap C) = P(A/C) \cdot P(B/C)$. Given the presence of C, the occurrence of, say, B does not affect the probability of occurrence of A. However, in the absence of C, the occurrence of B may or may not influence the probability of A. In this regard, events can be conditionally independent without being independent $\left(P(A \cap B) \neq P(A) \cdot P(B)\right)$ and vice versa.

Example 3.7.2   Suppose a student is selected at random from a statistics class and let the following events be defined on S:

A: the student scored a 95 on the latest statistics exam

B: the student completes all of the assigned statistics homework problems

C: the student drinks root beer

Although we can safely assume that events A and B are probably not independent or that $P(A \cap B) \neq P(A) \cdot P(B)$, it may be true that $P(A \cap B/C) = P(A/C) \cdot P(B/C)$; that is, studying has no effect on the student's test score given that he or she likes root beer.   ■

3.8   Sources of Probabilities

In this section we shall consider what may be called the sources or alternative interpretations of probabilities. That is, how are probabilities determined?

First, we have what is termed the *classical* or *a priori* interpretation of probability. Here we compute a *theoretical probability* without having to actually perform a random experiment. In this regard, the classical approach exploits the physical structure of an implied experiment. The argument goes something like this. If a random experiment has n mutually exclusive and equiprobable outcomes, and if n_A of them possess attribute A, then the probability of event A is its relative frequency n_A/n. For instance, we may ask: "What is the probability of obtaining an odd number on the roll of a single die?" Obviously there are six possible outcomes. Moreover, they are mutually exclusive and, if the die is properly balanced, the six outcomes are equiprobable; that is, for the random variable $X : X_1 = 1, X_2 = 2, \ldots, X_6 = 6$, and $P(X = X_i) = \frac{1}{6}$, $i = 1, \ldots, 6$. Thus we can anticipate that each face would have the same relative frequency if this random experiment were to be repeated a large number of times. In this regard, the probability of obtaining an odd number on the roll of a die is $\frac{n_A}{n} = \frac{3}{6}$. Note that to calculate this probability we did not have to actually perform a random experiment. If we are not familiar with the physical properties of the die, we could argue via the principle of insufficient reason that we have no reason or evidence to suppose that the die is biased in a particular direction. Thus, given our ignorance about the die, we may impute equal probabilities to the outcomes given that the outcomes are still mutually exclusive. These probabilities can then be modified through experimentation if we collect empirical evidence to the contrary.

The second major source of probabilities is that of the *subjectivist* (or *intuitivist*) approach. Here we are relying upon a *best guess* probability—a *quantified degree of belief*. Such probabilities often relate to experiments that cannot be repeated (e.g., the portfolio manager thinks that there is a 30% chance that the price of a given stock will increase by 6% tomorrow).

Finally, the third interpretation of probability involves the *empirical-frequency* (or *a posteriori*) approach. Here we determine an objective or empirical probability. The typical approach employed in determining empirical probabilities is to compute the relative frequency of an event by actually performing a random experiment. (Hence we do not obtain a theoretical relative frequency but an empirical one.) To this end, if A is an event for a random experiment and if that experiment is repeated n times and A is observed in n_A of the n trials, then the relative frequency of an event A is $\frac{n_A}{n}$. Since this ratio varies from trial to trial, what number is assigned to A as its probability? The answer lies in the *frequency limit principle*: if in a large number n of repetitions of a random experiment the relative frequencies $\frac{n_A}{n}$ of event A approach some number $P(A)$, then this number is ascribed to A as its probability,

$$\lim_{n \to \infty} \frac{n_A}{n} = P(A), \tag{3.19}$$

where $P(A)$ is to be interpreted as the *expected limit* of the relative frequencies. Thus $P(A)$ is some *long-run stable value* to which the relative frequencies converge (see Figure 3.19).

Are these various interpretations of probability mutually exclusive? The answer is, no. As Figure 3.20 reveals, there is considerable overlapping among these three approaches to the determination of probabilities.

As a practical matter, there is really no need to strictly or formally interpret the concept of probability at all. Each of these interpretative approaches to probability theory can be viewed as a special subcase of a noninterpretative or axiomatic approach to the calculation of probabilities. In fact, this is the approach that we actually have been employing throughout this chapter. As we noted earlier, all we need in order to determine the probability of some event is a probability measure (set-to-point function) defined on S. The probability axioms 1–4 then specify the general properties that the measure possesses. If we now consider this axiomatic

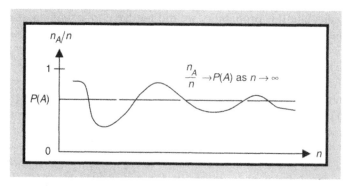

Figure 3.19   The frequency limit principle: $P(A)$ is the expected limit of n_A/n as $n \to \infty$.

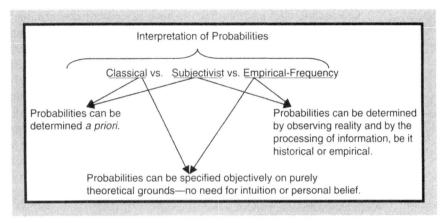

Figure 3.20 Sources of probabilities.

approach as a fourth interpretation of probability, then it is evident that all four approaches have the following in common:

(a) They enumerate events; that is, each possible outcome of some process under study must be identified.

(b) This identification in (a) leads to the specification of S.

(c) Probability is simply the relative frequency of A in S.

3.9 Bayes' Rule

Let the events $A_1, A_2, \ldots, A_n$ form a partition of the sample space $S(S = \cup_{i=1}^{n} A_i, A_i \cap A_j = \phi$ for $i \neq j$ and $i, j = 1, \ldots, n)$ with $P(A_i) > 0$ for all i. Let event $B \subset S$ with $P(B) > 0$ and suppose that B can occur only if one of the events $A_i, i = 1, \ldots, n$, occurs (see Figure 3.21).

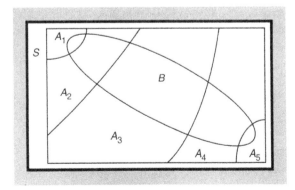

Figure 3.21 Events $A_1, \ldots, A_5$ partition S.

The set of relationships that we have described can now be viewed in the following light. Suppose we have certain *factors* (causes, inducements, stimuli, or attributes) $A_1, A_2, \ldots, A_n$ with *prior probabilities* $P(A_1), P(A_2), \ldots, P(A_n)$, respectively, that can each produce an *outcome* (effect, response, or consequence) B with *likelihood probabilities* $P(B/A_i), i = 1, \ldots, n$. The probabilities $P(A_i), i = 1, \ldots, n$, are established *prior* to any sampling or experimentation or gathering of empirical evidence; the probabilities $P(B/A_i), i = 1, \ldots, n$, are termed *likelihoods* in that they indicate how likely the outcome B is given each of the possible factors $A_i, i = 1, \ldots, n$. Furthermore, we are interested in determining the *posterior probabilities* $P(A_i/B), i = 1, \ldots, n$, of the factors once the outcome B has been observed. In general, a posterior probability is essentially a *revised* prior probability that is based upon additional (sample or empirical) information; it is calculated after some experimental outcome is observed.

In sum, for a fixed i, given the prior probability $P(A_i)$ and the likelihood $P(B/A_i)$, we seek to determine the posterior probability $P(A_i/B), i = 1, \ldots, n$. Hence we want the probability of factor A_i given that we have observed the experimental result of B. So after a trial of some random experiment is performed, the prior probabilities are adjusted to reflect the experimental response and thus subsequently replaced by the posterior probabilities. As we shall now see, the posterior probabilities can be expressed in terms of prior probabilities and the likelihoods of the sample outcome B given the factors $A_i, i = 1, \ldots, n$. To demonstrate this assertion we have *Bayes' Theorem*.

BAYES' THEOREM 3.9.1.   Let the events $A_1, A_2, \ldots, A_n$ form a partition of the sample space S with $P(A_i) > 0, i = 1, \ldots, n$. For any event $B \subset S$ with $P(B) > 0$,

$$P(A_i/B) = \frac{P(A_i)P(B/A_i)}{\sum_{i=1}^{n} P(A_i)P(B/A_i)}, \quad i = 1, \ldots, n. \tag{3.20}$$

Expression (3.20) is known as *Bayes' Rule*. We may rationalize this result as follows. Given $B \subset S = \cup_{i=1}^{n} A_i$, it follows that $B = \cup_{i=1}^{n}(B \cap A_i)$. If the A_i's are mutually exclusive, then so are the events $B \cap A_i, i = 1, \ldots, n$, and thus, from axiom 4,

$$P(B) = P\left(\cup_{i=1}^{n}(B \cap A_i)\right) = \sum_{i=1}^{n} P(B \cap A_i). \tag{3.21}$$

From the multiplication rule (3.8),

$$P(B \cap A_i) = P(A_i)P(B/A_i), \quad i = 1, \ldots, n. \tag{3.22}$$

Then from (3.21) and (3.22),

$$P(A_i/B) = \frac{P(B \cap A_i)}{P(B)} = \frac{P(B \cap A_i)}{\sum_{i=1}^{n} P(B \cap A_i)} = \frac{P(A_i)P(B/A_i)}{\sum_{i=1}^{n} P(A_i)P(B/A_i)}, \quad i = 1,\dots,n.$$

$$(3.23)$$

Example 3.9.1   Given events $A_1 = \{$an attribute is present$\}$ and $A_2 = \{$an attribute is not present$\}$, it is readily seen that $S = A_1 \cup A_2$ and $A_1 \cap A_2 = \phi$. Suppose that within some population there is a 5% chance that the attribute is present and a 95% chance that it is not present. The implied prior probability distribution is thus provided by Table 3.3a.

Let event B represent the occurrence of a positive outcome on a test performed to detect the presence of the attribute. Experience dictates that: (a) if the attribute is present, the test gives a positive result 90% of the time; (b) if the attribute is not present, the test gives a false positive result 3% of the time. Suppose we now randomly select an item from the population and test for the presence of the attribute. Given a positive test result, what is the probability that the item actually selected has the attribute? Here we are interested in the posterior probability (see(3.20))

$$P(A_1/B) = \frac{P(A_1)P(B/A_1)}{P(A_1)P(B/A_1) + P(A_2)P(B/A_2)}$$

$$= \frac{0.05(0.90)}{0.05(0.90) + 0.95(0.03)} = 0.6122.$$

Hence there is about a 61% chance that the item selected actually has the attribute given that it tested positively for it. Similarly,

$$P(A_2/B) = \frac{P(A_2)P(B/A_2)}{P(A_1)P(B/A_1) + P(A_2)P(B/A_2)}$$

$$= \frac{0.95(0.03)}{0.05(0.90) + 0.95(0.03)} = 0.3878.$$

Table 3.3

a.		b.		c.	
A_i	$P(A_i)$	B/A_i	$P(B/A_i)$	A_i/B	$P(A_i/B)$
A_1	$P(A_1) = 0.05$	B/A_1	$P(B/A_1) = 0.90$	A_1/B	$P(A_1/B) = 0.6122$
A_2	$P(A_2) = \dfrac{0.95}{1}$	B/A_2	$P(B/A_2) = 0.03$	A_2/B	$P(A_2/B) = \dfrac{0.3878}{1}$

That is, there is about a 39% chance that the item selected does not have the attribute even though it tested positively for it. The complete posterior probability distribution is given in Table 3.3c. ■

Example 3.9.2 Suppose that three vendors (call them A_1, A_2 , and A_3) each supply the XYZ Corp. with a miniature electronic component for its new DVD player. Records indicate that vendor A_1 supplies 45% of the gross shipment, vendor A_2 supplies 20% of the total shipment, and vendor A_3 supplies the remaining 35%, with each vendor's shipment to XYZ arriving at 7:00 A.M. each Monday. Furthermore, it has been determined that, on average, 8% of vendor A's components are defective, and the percents defective for vendors A_2 and A_3 are 5% and 3%, respectively. If one component is chosen at random from the week's combined shipment and it is found to be defective (event D), what is the probability that it was supplied by vendor A_2 ? Given the previous information sets, the prior and likelihood distributions are provided by Tables 3.4a and b. Then from (3.20), the posterior probability

$$P(A_2/D) = \frac{P(A_2)P(D/A_2)}{P(A_1)P(D/A_1) + P(A_2)P(D/A_2) + P(A_3)P(D/A_3)}$$

$$= \frac{0.20(0.05)}{0.45(0.08) + 0.20(0.05) + 0.35(0.03)} = 0.1754.$$

Table 3.4

	a.		**b.**		**c.**
A_i	$P(A_i)$	D/A_i	$P(D/A_i)$	A_i/D	$P(A_i/D)$
A_1	$P(A_1) = 0.45$	D/A_1	$P(D/A_1) = 0.08$	A_1/D	$P(A_1/D) = 0.6316$
A_2	$P(A_2) = 0.20$	D/A_2	$P(D/A_2) = 0.05$	A_2/D	$P(A_2/D) = 0.1754$
A_3	$P(A_3) = \dfrac{0.35}{1}$	D/A_3	$P(D/A_3) = 0.03$	A_3/D	$P(A_3/D) = \dfrac{0.1842}{1}$

So given that a defective item was randomly obtained, there is about an 18% chance that it was supplied by vendor A_2. The remaining posterior probabilities are given in Table 3.4c. ■

3.10 Exercises

3-1. The nonmanagerial employees of ABC Corp. are classified into the following mutually exclusive groups: production, development, sales, and handling. The results of a recent union poll pertaining to the issue of expanded health benefits in lieu of wage increases are presented in the accompanying table. Given that an employee is to be chosen at random, determine each of the following probabilities.

	Yes (Y)	No (N)	Indifferent (I)
Production (P)	57	49	5
Development (D)	22	10	1
Sales (S)	37	55	2
Handling (H)	14	38	5

(Hint: First obtain the marginal totals and then convert this table into a relative frequency table.)

(a) $P(D \cap N)$

(b) $P(D \cup N)$

(c) $P(Y \cap I)$

(d) $P(H \cap \overline{Y})$

(e) $P(S \cup \overline{N})$

(f) $P(D|Y)$

(g) $P(S|\overline{Y})$

(h) Are *employment status* and *poll result* statistically independent?

3-2. For events A and B within a sample space S, let $P(A) = 0.4$, $P(B) = 0.3$, and $P(A \cap B) = 0.1$. Are events A and B mutually exclusive? Are they collectively exhaustive? Are they independent events? Also, find:

(a) $P(\overline{A})$

(b) $P(A \cup B)$

(c) $P(A|B)$

(d) $P(B|A)$

(e) $P(B|\overline{A})$

(f) $P(\overline{A} \cap \overline{B})$

(g) $P(B|A \cup B)$

(h) $P(B|A \cap B)$

(i) $P(A \cap B|A \cup B)$

3-3. Verify that for S a sample space and $A \subseteq S$, $P(A) \neq 0$:

(a) $P(S|A) = 1$

(b) $P(\phi|A) = 0$

3-4. A fair pair of (six-sided) dice is tossed and the values of the faces showing are observed. List the points or simple events with coordinates (I, II) in the sample space S, where "I" denotes the outcomes on the first die and "II" denotes the outcome on the second die. Let us define four subsets of S as:

A: the second die shows odd

B: the first die shows even

C: at least one number in the pair is odd

D: the sum of the faces is at least 8

Then list the simple events in

(a) $\overline{B}$

(b) $A \cup C$

(c) $B \cup D$

(d) $A \cap C$

(e) $B \cap D$

(f) $\overline{A} \cup C$

(g) $\overline{B} \cup D$

(h) $\overline{A} \cap \overline{B}$

Are B and $\overline{C}$ mutually exclusive?

3-5. At a particular university students have either undergraduate (U) or graduate (G) status and either live in a dorm (D) or live off campus (O). One hundred students were chosen at random: 77 were undergraduates, 80 lived in the dorms, and 70 undergraduates lived in the dorms. Find:

(a) $P(U \cup D)$

(b) $P(U \cap \overline{D})$

(c) $P(G|D)$

(d) $P(D|U)$

(e) $P(D|\overline{U})$

3-6. Two ointments (call them A and B) for a particular type of skin disorder are being tested. Ten percent of patients did not show any improvement after one week. Half of them used ointment A. If (under randomization) it was determined that 40% of the patients were assigned ointment A, find:

(a) The probability that a patient will not improve if he or she is assigned ointment A.

(b) The probability that the patient will improve, given that he/she was treated with ointment A.

3-7. For events $A, B \subseteq S$, the set difference $A - B$ is the set of elements in A that are not in B. Does $A - B = A \cap \overline{B}$? Verify that $P(A - B) = P(A \cap \overline{B}) = P(A) - P(A \cap B)$.

3-8. If A and B are independent events with $P(A) = 0.4$ and $P(B) = 0.3$, find:

(a) $P(A \cup B)$

(b) $P(\overline{A} \cap \overline{B})$

(c) $P(\overline{A} \cup \overline{B})$

(d) $P(A|B)$

(e) $P(B|\overline{A})$

3-9. In a lot of 200 manufactured items, 35 are defective. Two are drawn at random without replacement. Find the probability that both are defective.

3-10. Given the following partitioning of the sample space S, find:

S	A	$\overline{A}$
B	3	4
C	1	5
D	1	1
E	2	3

(a) $P(\overline{A} \cap B)$

(b) $P(A \cup C)$

(c) $P(B \cup D)$

(d) $P(C \cap E)$

(e) $P(A|D)$

(f) $P(D|C)$

(g) $P(\overline{A}|C)$

(h) $P(A|\overline{B})$

(i) $P(E|\overline{C})$

(j) $P(\overline{E}|A)$

3-11. A vessel contains six blue and four red marbles. Three random draws are to be made from the vessel in the following fashion: for each draw a marble is selected and its color is recorded. It is then returned to the vessel along with an additional marble of the same color. Determine the probability that a red marble is selected on each of the three draws.

3-12. Forty percent of the subscribers to the Dear Hearts dating service are men (M) and 60% are women (W). Among the men, 70% indicate that they enjoy travel (T) and 80% of the women state that travel is one of their favorite activities. If a member is selected at random, what is the probability that he or she enjoys travel? (Hint: $T = (M \cap T) \cup (W \cap T)$.)

3-13. For events $A, B \subseteq S$, prove that $P(A \cup B) = P(A) + P(B) - P(A \cap B)$.

3-14. A jar contains six red marbles and four blue ones. Three are drawn at random without replacement. What is the probability that all are blue? What is the probability of a blue on the first draw and red on the second and third draws?

3-15. Suppose A and B are events within the sample space S and that $P(A) = P(B) = \frac{1}{2}$.
Answer the following:

(a) If $P(B|A) = \frac{1}{2}$, are A and B independent? Are A and B mutually exclusive?

(b) If A and B are independent, what is the probability that "A occurs and B does not occur or B occurs and A does not occur"?

(c) If event $C \subset S$ and $P(C|A \cap B) = \frac{1}{2}$, find $P(A \cap B \cap C)$.

(d) If A and B are independent and $P(B|A) = \frac{1}{2}$, find $P(A \cup B)$.

3-16. A fair coin is tossed twice in succession. Find:

(a) The probability of two heads given a head on the first toss.

(b) The probability of two heads given at least one head.

3-17. Suppose A, B, and C are events within the sample space S. Demonstrate that $P(A \cap B \cap C) = P(A) \cdot P(B|A) \cdot P(C|A \cap B)$, given that $P(A \cap B) \neq 0$.

3-18. A red and a blue die (each fair) are tossed simultaneously. Given the events:

$A = \{$the sum of the faces is odd$\}$

$B = \{$the blue die shows 1$\}$

$C = \{$the sum of the faces is 7$\}$

determine if:

(a) A and B are independent

(b) A and C are independent

(c) B and C are independent

3-19. For events $A, B \subseteq S$, demonstrate that:

(a) if $B \subseteq A$, then $P(B) \leq P(A)$

(b) $P(A \cap B) \geq 1 - P(\overline{A}) - P(\overline{B})$

3-20. A red and a blue die (both fair) are tossed simultaneously. Given the events:

$A_1 = \{$the red die shows odd$\}$

$A_2 = \{$the blue die shows odd$\}$

$A_3 = \{$the sum of the faces is odd$\}$

determine if A_1, A_2, and A_3 are pairwise independent. Are A_1, A_2, and A_3 all taken together independent?

3-21. Suppose we toss a fair six-sided blue die and a fair six-sided red die simultaneously. What is the probability that:

(a) The sum of the faces showing is nine, given that the outcome on the red die is five?

(b) A five occurs on the red die given that the sum of the faces showing is nine?

3-22. A fair single six-sided die is tossed. For the events $A_1 = \{$the face shows even$\}$; $A_2 = \{$the face shows odd$\}$; $A_3 = \{$3 or 4 shows$\}$; $A_4 = \{$at least a 3 shows$\}$; $A_5 = \{$5 or 6 shows$\}$; and $A_6 = \{$at most a 4 shows$\}$, determine which events are mutually exclusive, collectively exhaustive, and simultaneously mutually exclusive and collectively exhaustive.

3-23. Let the proportion of the women in the population with at least one college degree be w_c and the proportion of men in the population with at least one college degree be m_c. If the proportion of women in the population is w, what is the probability that an individual chosen at random from the population at large holds at least one college degree?

3-24. Two cards are to be randomly selected from an ordinary deck of 52 playing cards. Determine the probability of drawing two aces if: (a) the first card is not replaced; and (b) the first card is replaced before the second is drawn.

3-25. Suppose we toss a fair coin twice in succession (or we toss two fair coins simultaneously). Define H_1 as "heads on the first toss" and H_2 as "heads on the second toss." Verify that H_1 and H_2 are independent events.

3-26. A coin and a die (each fair) are tossed simultaneously. Determine the simple events within the sample space. Find:

(a) P(at least a 4 on the die)

(b) P(the coin shows heads)

(c) P(not more than 3 on the die and the coin shows tails)

3-27. If a theorem is specified in terms of n and involves a statement that some relationship holds when n is any positive integer, then a proof of the theorem by *mathematical induction* proceeds as follows:

1. Verify the theorem for $n = 1$ (usually).
2. Assume that the theorem holds for $n = p$.
3. Prove that the theorem holds for $n = p + 1$.

Use mathematical induction to prove that for the finite sequence of events $A_1, A_2, \ldots, A_n$ within S, $P(\cup_{i=1}^{n} A_i) \leq \sum_{i=1}^{n} P(A_i)$. (Hint: Start with $n = 2$.)

3-28. A jar contains three white and seven black marbles. Two are drawn at random without replacement. What is the probability that they are both white?

What is the probability that they are of a different color? Represent the outcomes of this experiment by a decision tree and determine all terminal node probabilities.

3-29. Demonstrate that for a finite sequence of mutually exclusive events $A_1, A_2, \ldots, A_n$ within S, $P(\cup_{i=1}^{n} A_i \,|B) = \sum_{i=1}^{n} P(A_i \,|B)$.

3-30. Demonstrate that if $A, B \subseteq S$ are independent events, then so are:

(a) A and $\overline{B}$

(b) $\overline{A}$ and B

(c) $\overline{A}$ and $\overline{B}$

3-31. Four aces are removed from an ordinary deck of 52 playing cards. The remaining 48 cards are put aside and the four aces are shuffled and laid face down on a table. Let us define the events:

C: the ace of clubs;

D: the ace of diamonds;

H: the ace of hearts; and

S: the ace of spades.

One card is to be selected at random. Determine if the events $R = \{C, D\}$, $Q = \{C, H\}$, and $T = \{C, S\}$ are independent.

3-32. Let $A_1, \ldots, A_n$ be a set of events defined on S. Suppose $S = \cup_{i=1}^{n} A_i$, $P(A_i) > 0$ for all i and $A_i \cap A_j = \phi, i \neq j$. Then for any event B, verify that $P(B) = \sum_{i=1}^{n} P(B\,|A_i) \cdot P(A_i)$. (Hint: Use the distributive law $A \cap (B \cup C) = (A \cap B) \cup (A \cap C)$.)

3-33. Suppose A and B are independent events in S. Does $P(\overline{A} \cap \overline{B}) = P(\overline{A}) \cdot P(\overline{B})$? (Hint: Apply the general addition rule, corollary 1, and DeMorgan's law.)

3-34. A manufacturer of automatic garage door openers buys circuit boards from three different suppliers: 40% of the total supply comes from Vendor A, 10% comes from Vendor B, and 50% comes from Vendor C. Since the quality control standards of the vendors vary somewhat, experience indicates that 2% of Vendor A's shipments are defective, Vendor B ships 3% defective, and 1% of Vendor C's shipments have proven to be defective. The manufacturer wants to determine the proportion of circuit boards in the latest shipments that are defective.

3-35. Suppose we toss a single fair six-sided die and we observe the face showing. Given the following events:

A: an even number occurs;

B: an odd number occurs; and

C: a 3 or 4 occurs,

determine if:

(a) A and B are independent

(b) A and C are independent

(c) B and C are independent

(d) A and B are mutually exclusive

(e) B and C are mutually exclusive

3-36. Applicants at the ABC employment agency were asked if they had graduated high school. The following table displays their responses to this
question.

High School	Gender	
Graduate (HSG)	Male (M)	Female (F)
Yes (Y)	40	30
No (N)	20	10

What is the probability that an individual chosen at random will be:

(a) Male and a HSG

(b) Either female or a HSG

(c) A HSG

(d) Male, given that the person chosen is a HSG

(e) Not a HSG, given that the person chosen is female

Are the categories HSG and Gender independent?

3-37. Twenty percent of the students in the freshman class of a certain college
have decided on a major and 80% have not. Among those who have already
chosen a major, 20% are planning to attend graduate school and only 2%
of those not having chosen a major are planning to go to graduate school. If
a freshman student is chosen at random, what is the probability that he or
she will be planning to attend graduate school? (As a solution aid, construct
a Venn diagram for this problem.)

3-38. Experience dictates that at a particular clinic, about 80% of those attempting to quit smoking by undergoing hypnosis actually have stopped smoking
for at least six months. If those patients are given this treatment in an independent fashion, what is the probability that at least one will be successful
and stop smoking for at least six months?

3-39. Suppose we select two balls at random without replacement from a vessel
containing four red and six blue balls. Find the probability that:

(a) Both are red

(b) The second ball is red

3-40. Prove that for events A, B, and C within S,

$$P(A \cup B | C) = P(A | C) + P(B | C) - P(A \cap B | C).$$

(Hint: Use the distributive law $A \cap (B \cup C) = (A \cap B) \cup (A \cap C)$.)

3-41. Prove that for events $A, B \subseteq S, P(\overline{A} \cap B) = P(B) - P(A \cap B)$.
(Hint: Let $B = (\overline{A} \cap B) \cup (A \cap B)$.)

3-42. Bayes' law has often been characterized as involving a *reverse* process of reasoning; that is, *reasoning from effect to cause.* Comment.

3-43. Suppose events $A_1, \ldots, A_n$ are contained within the sample space S. For $r = 3$ and 4, verify that $2^r - r - 1$ equalities must hold for these events to be independent.

3-44. Suppose A, B, and C are events in the sample space S. If events A and B are independent, they might not be independent given C. Comment.

3-45. Sunflower seeds purchased from vendor A_1 have an 85% germination rate. Those purchased from vendor A_2 have a 90% germination rate, and such seeds purchased from vendor A_3 have a 75% germination rate. A garden center obtains 60% of its sunflower seeds from vendor A_1, 10% from vendor A_2, and 30% from vendor A_3 and mixes them together to form what it calls its Sunshine Blend. Find:

   (a) The probability that a seed selected at random from the blend will germinate. Given that a seed has germinated, find:

   (b) The probability that the seed was supplied by vendor A_1

   (c) By vendor A_2

   (d) By vendor A_3

3-46. Passenger cars account for about 25% of all vehicles on the road. If a passenger car is in an accident, there is about a 10% chance of a fatality; the chance of a fatality is only 3% if an accident does not involve a passenger car. Suppose an accident involving a fatality has occurred. What is the probability that a passenger car was involved?

3-47. ACE Metal Products Corp. produces steel pins of a specific diameter. It uses three machines (denoted as A, B, and C) in the initial grinding phase of production. Machines A, B, and C have a history of producing 3%, 1%, and 2.5% defective pins, respectively. Machine A produces 45% of total production and machines B and C contribute 35% and 20%, respectively, to total production. If a pin is randomly chosen from the total amount produced for a given grinding phase production run, what is the probability that it was produced by machine A, given that the pin was found to conform to specifications?

3-48. A&E Opticians provide free adult screening for glaucoma with the purchase of a pair of glasses. Let G be the event that the customer actually exhibits the early stages of glaucoma and let event T occur when the test detects glaucoma. Suppose $P(G) = 0.001$. Determine the probability that a customer actually has glaucoma given that the test result is positive.

3-49. Suppose a gem dealer has a 90% chance of correctly discriminating between natural versus color-enhanced gemstones. If the dealer does appraisals and 75% of the gems that he or she examines are actually natural, what is the probability that a gemstone is really color-enhanced, given that the dealer classifies it as such?

3-50. Vessel A_1 contains four white (W) marbles, vessel A_2 contains two red (R) and two white marbles, and vessel A_3 contains four red marbles. Suppose $P(A_1) = 0.5, P(A_2) = 0.4$, and $P(A_3) = 0.1$. If a vessel is selected using these probabilities and a marble is drawn at random, find:

(a) The probability of drawing a white marble $P(W)$

(b) The probability of selecting a red marble $P(R)$

(c) $P(A_1|W)$

(d) $P(A_2|R)$

(e) $P(A_1|R)$

Random Variables and Probability Distributions

4.1 Random Variables

It was mentioned in the preceding chapter that a *random variable X* is a real-valued function defined on a sample space *S*. In this regard, its role is to specify the possible outcomes of a random experiment by relating the result of any such experiment to a single numerical value. Furthermore, random variables may be termed discrete or continuous.

A random variable is *discrete* if the number of values it can assume forms a *countable set*; that is, this set has either a finite number of elements or its elements are *countably infinite* in that they can be put into a one-to-one correspondence with the positive integers. For instance, if the random variable X represents the number of points obtained in the roll of a single six-sided die, then X takes on a finite set of possible outcomes {1, 2, 3, 4, 5, 6}. But if the random variable X is defined according to the rule "roll a single six-sided die repeatedly until a 4 appears for the first time," then this could happen on the first roll ($X = 1$), on the second roll ($X = 2$), on the third roll ($X = 3$), and so on. Clearly there are infinitely many possibilities and thus X assumes the countably infinite set of values {1, 2, 3, ...}.

A random variable X is *continuous* if it can assume an infinite or uncountable number of values over some interval. For instance, if on a line segment of length L two points A and B are chosen at random, then a random variable can be defined as $X = |A - B|$ or the distance between A and B. Clearly X assumes an infinite number of values on the interval $0 \leq X \leq L$. In fact, variables measured in terms of temperature, ounces, pints, and so on, can take on essentially any real value over some appropriate interval.

4.2 **Discrete Probability Distributions**

Let a sample space S consist of a finite or countable collection of simple events $E_i, i = 1, 2, \ldots$, each with a known probability $P(E_i), i = 1, 2, \ldots$. In addition, for X a discrete random variable defined on the simple events in S, let the range of X be the set values $R = \{X_i, i = 1, 2, \ldots\}$. Let f be a function that associates with each $X_i \in R$ a number $f(X_i) \in [0, 1]$ representing the probability that $X = X_i$; that is, $P(X = X_i) = f(X_i), i = 1, 2, \ldots$. Here $f(X_i)$ is termed the *probability mass* at X_i. In general, a function $f(X)$, which assigns a probability $f(X_i)$ to each number X_i within the range of a discrete random variable X, is called a probability mass function if:

$$\text{(a)}\ f(X_i) \geq 0 \text{ for all } i;$$

$$\text{(b)}\ \sum_i f(X_i) = 1; \text{ and} \tag{4.1}$$

$$\text{(c)}\ f(X_i) = 0 \text{ for all } X_i \notin R.$$

Note that the value of the probability mass function is zero at all points other than the X_i's in R. The connection between S, X, and f is illustrated in Figure 4.1. Typically f is represented by a table, a graph, or a mathematical equation.

To construct a discrete probability distribution we need two specific pieces of information: (1) we need to specify the random experiment; and (2) we need to define the random variable. Once this is done, the discrete probability distribution will be completely specified once the sequence of probabilities $f(X), i = 1, 2, \ldots$, is given; that is, the discrete probability distribution is completely described by the probability mass function taken over all values of the discrete random variable X. Hence it can be represented as the set of all possible pairs $(X_i, f(X_i)), i = 1, 2, \ldots$. Once a discrete probability distribution is determined, we may compute, for

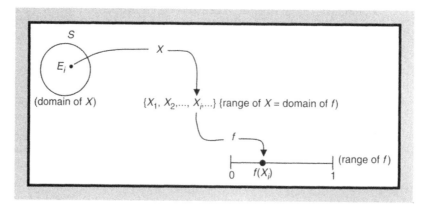

Figure 4.1   The probability mass function f assigns a probability to each value of a discrete random variable X.

instance, probabilities such as:

$$\text{(a) } P(a \leq X \leq b) = \sum_{a \leq X_i \leq b} f(X_i),$$

$$\text{(b) } P(X \geq a) = \sum_{X_i \geq a} f(X_i), \tag{4.2}$$

$$\text{(c) } P(X < a) = 1 - P(X \geq a) = \sum_{X_i < a} f(X_i).$$

Example 4.2.1 Let a random experiment involve the rolling of a pair of fair six-sided dice and let the random variable X be defined as the sum of the faces showing. The sample space S (the domain of X) and the range of X are depicted in Figure 3.12. If we place the sequence of X_i's in the first column of Table 4.1, we can easily determine the $f(X_i)$'s, $i = 1, \ldots, 11$, given that $P(E_j) = \frac{1}{36}$ for all $j = 1, \ldots, 36$. The probabilities associated with the X_i's are given in column two of Table 4.1. Hence the discrete probability distribution consists of the X_i's and their associated probabilities $P(X = X_i) = f(X_i), i = 1, \ldots, 11$. In addition, the graph of the discrete probability distribution or the probability mass function $f(X)$ appears in Figure 4.2. ∎

If in the preceding example our random experiment had consisted of rolling a single fair six-sided die, then the resulting discrete probability distribution would have $f(X_i) = \frac{1}{6}, i = 1, \ldots, 6$. Any discrete probability distribution that has all probability masses equal to $\frac{1}{n}$ for each $X_i, i = 1, \ldots, n$, is called a *uniform probability distribution*.

It should be apparent that a discrete probability distribution is essentially a *theoretical frequency distribution* in that it describes how the outcomes $X_i, i = 1, 2, \ldots$, are *expected* to vary (or how we would *expect* our random experiment to

Table 4.1

X	$f(X)$	$F(X)$
$X_1 = 2$	$f(X_1) = 1/36$	$F(X_1) = 1/36$
$X_2 = 3$	$f(X_2) = 2/36$	$F(X_2) = 3/36$
$X_3 = 4$	$f(X_3) = 3/36$	$F(X_3) = 6/36$
$X_4 = 5$	$f(X_4) = 4/36$	$F(X_4) = 10/36$
$X_5 = 6$	$f(X_5) = 5/36$	$F(X_5) = 15/36$
$X_6 = 7$	$f(X_6) = 6/36$	$F(X_6) = 21/36$
$X_7 = 8$	$f(X_7) = 5/36$	$F(X_7) = 26/36$
$X_8 = 9$	$f(X_8) = 4/36$	$F(X_8) = 30/36$
$X_9 = 10$	$f(X_9) = 3/36$	$F(X_9) = 33/36$
$X_{10} = 11$	$f(X_{10}) = 2/36$	$F(X_{10}) = 35/36$
$X_{11} = 12$	$f(X_{11}) = \underline{1/36}$	$F(X_{11}) = 1$
	1	

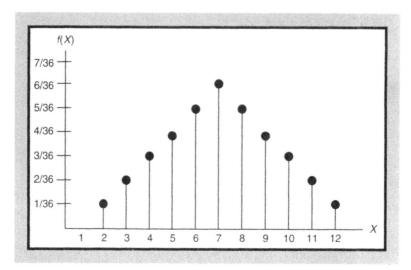

Figure 4.2 The probability mass function for X equal to the sum of faces showing on the roll of a pair of dice.

perform) under repeated sampling or trials in the long run. What is the connection between a discrete probability distribution and the concept of a relative frequency distribution introduced in Chapter 2? A relative frequency distribution for a variable X is a discrete probability distribution if the total number of observations or sum of the absolute frequencies corresponds to the population size N and the observations are chosen at random.

Example 4.2.2   For the discrete probability distribution exhibited in Table 4.1, it is easily demonstrated, using (4.2), that:

$$P(4 \leq X \leq 7) = f(X_3) + f(X_4) + f(X_5) + f(X_6) = \frac{18}{36},$$

$$P(4 \leq X < 7) = f(X_3) + f(X_4) + f(X_5) = \frac{12}{36},$$

$$P(X \geq 9) = f(X_8) + f(X_9) + f(X_{10}) + f(X_{11}) = \frac{10}{36},$$

$$P(X > 9) = f(X_9) + f(X_{10}) + f(X_{11}) = \frac{6}{36},$$

$$P(X < 9) = 1 - P(X \geq 9) = \frac{26}{36},$$

$$P(X < 4) = f(X_1) + f(X_2) = \frac{3}{36}. \quad \blacksquare$$

At times we may find that a second random variable Y is defined on the same sample space S and, in this circumstance, we may inquire as to whether or not X and Y follow the same distribution. To determine this we need only note that the random variables X and Y are *identically distributed* if for every event $A \subseteq S$, $P(X \in A) = P(Y \in A)$. Here it is not necessarily the case that $X = Y$.

Although $f(X_i)$ gives the probability that the discrete random variable X equals X_i exactly, we may consider a related function (derived from the f values) that renders the probability that X assumes a value less than or equal to X_i. In this regard, the *cumulative distribution function* (also called the *cumulative mass function*) of a discrete random variable X, $F(X_i)$, gives the probability that $X \leq X_i$ or

$$F(X_i) = P(X \leq X_i) = \sum_{j \leq i} f(X_j), \qquad (4.3)$$

with the X_i's placed in increasing order.

Here the cumulative distribution function serves as an alternative representation of a discrete probability distribution.

Example 4.2.3   For the discrete random variable X depicted in Table 4.1, its cumulative distribution function is displayed in column three of the same. Note that, as required by (4.3),

$$F(X_1) = \sum_{j \leq 1} f(X_j) = f(X_1) = \frac{1}{36},$$

$$F(X_2) = \sum_{j \leq 2} f(X_j) = f(X_1) + f(X_2) = \frac{3}{36},$$

$$F(X_3) = \sum_{j \leq 3} f(X_j) = f(X_1) + f(X_2) + f(X_3) = \frac{6}{36},$$

$$\vdots$$

$$F(X_{11}) = \sum_{j \leq 11} f(X_j) = f(X_1) + \cdots + f(X_{11}) = 1. \quad \blacksquare$$

As far as the properties of $F(X_i)$ are concerned:

(a) $0 \leq F(X_i) \leq 1$ for any X_i

(b) $F(X_s) \leq F(X_t)$ when $X_s \leq X_t$ (F is a monotone nondecreasing function of X_i)

(c) $P(X > X_i) = 1 - P(X \leq X_i) = 1 - F(X_i)$

(d) $P(X = X_i) = F(X_i) - F(X_i - 1)$ (X_i an integer)

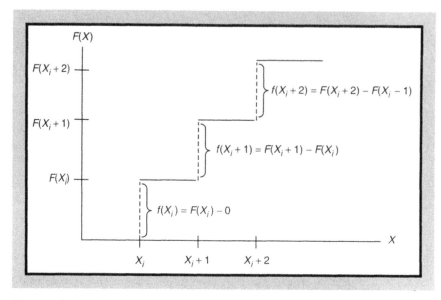

Figure 4.3   The cumulative distribution function for a discrete random variable X.

(e) $P(X_s \leq X \leq X_t) = F(X_t) - F(X_s - 1)$ (X integer valued)

(f) $P(X_s < X \leq X_t) = F(X_t) - F(X_s)$

(g) $F(-\infty) = \lim\limits_{X_i \to -\infty} F(X_i) = 0, F(+\infty) = \lim\limits_{X_i \to +\infty} F(X_i) = 1$ (4.4)

(h) F is continuous from the right, that is, $\lim\limits_{\substack{h \to 0 \\ h > 0}} F(X_i + h) = F(X_i)$

Two additional characteristics of (4.3) merit our attention. First, the domain of the cumulative distribution function F is the set of *all* real values of X, not just those within the range of $X(R)$. Second, based upon (4.4h), the graph of F is a step function that assumes the upper value at each jump; that is, F has a finite step or discontinuity at any point X_i such that $f(X_i) > 0$, where the step size is $f(X_i)$ itself. In fact, if X assumes r distinct values and each has a positive probability, then F displays r distinct steps (see Figure 4.3). In this regard, if we know X's probability mass function f, then we can easily find X's cumulative distribution function F via (4.3). Conversely, if we are given X's cumulative distribution function, we can just as easily find the associated probability mass function since each f value is the finite step size occurring at each point within the range of X.

Example 4.2.4   Let us define a random experiment as the tossing of a fair coin three times in succession and the associated random variable X as the number of heads obtained in the three tosses. Then each outcome point or simple event in S

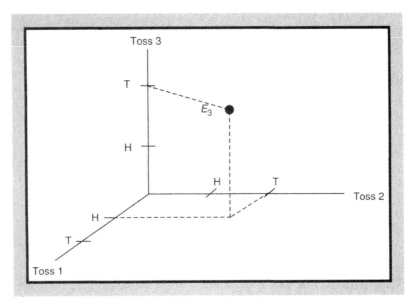

Figure 4.4 Sample space when tossing a fair coin three times in succession.

is a triple of the form

(outcome on toss 1, outcome on toss 2, outcome on toss 3).

Hence the collection of all possible outcomes involves the eight simple events:

$$E_1 = (H, H, H), E_2 = (H, H, T), E_3 = (H, T, T), E_4 = (T, T, T),$$
$$E_5 = (T, H, H), E_6 = (T, T, H), E_7 = (H, T, H), E_8 = (T, H, T).$$

Each is a point in a three-dimensional sample space. (For instance, $E_3 = (H, T, T)$ is plotted in S in Figure 4.4.)

Based upon the structure of S, the derived discrete probability distribution, as characterized by its probability mass function, is presented in Table 4.2 along with its cumulative distribution function. Alternatively, these two functions may

Table 4.2

X	$P(X = X_i) = f(X_i)$	$P(X \leq X_i) = F(X_i)$
0	1/8	1/8
1	3/8	4/8
2	3/8	7/8
3	1/8	1
	1	

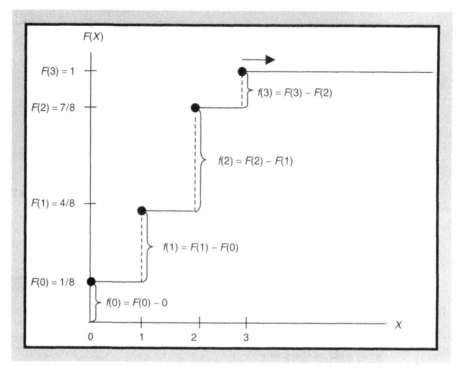

Figure 4.5   Cumulative distribution function for X equal to the number of heads obtained in three tosses of a fair coin.

be expressed mathematically as

$$f(X) = \begin{cases} \frac{1}{8} & \text{for} \quad X = 0, 3; \\ \frac{3}{8} & \text{for} \quad X = 1, 2 \end{cases}$$

and

$$F(X) = \begin{cases} 0 & \text{for} \quad X < 0; \\ \frac{1}{8} & \text{for} \quad 0 \le X < 1; \\ \frac{4}{8} & \text{for} \quad 1 \le X < 2; \\ \frac{7}{8} & \text{for} \quad 2 \le X < 3; \\ 1 & \text{for} \quad X \ge 3 \end{cases}$$

respectively. Finally, Figure 4.5 illustrates the cumulative distribution function and exhibits its connection to the probability mass function. Note that by virtue of (4.4d), the size of each vertical jump at X_i amounts to the probability mass there or $f(X_i) = P(X = X_i) = F(X_i) - F(X_i - 1)$. ∎

4.3   **Continuous Probability Distributions**

Let a sample space S consist of a class of events representable by all open and closed intervals (as well as by all half-open intervals and the rays $[a, +\infty)$, $(-\infty, b]$). Hence any random variable X defined on S must be continuous since it can assume any value within some event $A \subseteq S$.[1] In addition, let P be a *probability measure* or set function that associates with each event $A \subseteq S$ a number $P(A)$ representing the probability that event A occurs. In general, a function $f(x)$, that defines the probability measure

$$P(A) = P(X \in A) = \int_A f(x)dx \text{ for every } A \subseteq S \qquad (4.5)$$

is called a probability density function if

(a) $f(x) \geq 0$ for all real $x \in (-\infty, +\infty)$ and $f(x) > 0$ for $x \in A$

(b) $\displaystyle\int_A f(x)dx$ exists for every $A \subseteq S$ (e.g., $f(x)$ has at most a finite

      number of discontinuities over $A \subseteq S$)

$$\qquad (4.6)$$

(c) $\displaystyle\int_S f(x)dx = \int_{-\infty}^{+\infty} f(x)dx = 1$

So if the collection of all open and closed intervals constitutes the sample space for a random variable X and if $P(A)$ is determined by a probability density function $f(x)$ as in (4.5), then X has a *continuous probability distribution*. In fact, a continuous probability distribution is completely determined once the probability density function $f(x)$ is given.

If X is a continuous random variable with probability density $f(x)$, then for $A = \{x | a \leq x \leq b\}$,

$$P(A) = P(a \leq X \leq b) = \int_a^b f(x)dx, \qquad (4.7)$$

that is, the probability that a continuous random variable X assumes a value between a and b (inclusive) is the area under X's probability density function $f(x)$ from a to b (see Figure 4.6a). Note that for $X = a$,

$$P(X = a) = \int_a^a f(x)dx = 0 \qquad (4.7.1)$$

---

[1] If event A is an *open interval*, then $A = \{x | a < x < b\}$; if A is a *closed interval*, then $A = \{x | a \leq x \leq b\}$.

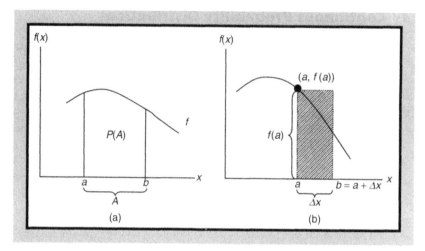

Figure 4.6 (a) Probability as area under the probability density function; (b) $f(a)$ is the probability density of X at point $x = a$.

since the area under $f(x)$ at a single point is zero. (The probability assigned to any finite set of points is also zero.) Hence it is immaterial whether A is specified as an open or closed interval; that is,

$$P(a \leq X \leq b) = P(X = a) + P(a < X < b) + P(X = b) = P(a < X < b).$$

In view of (4.7.1) it is important to remember that for a continuous random variable $X, f(a) \neq P(X = a)$. For $X = a, f(a)$ is termed the *probability density of X at a*. How shall we interpret $f(a)$? If we consider only a small section of f and the rectangle formed under a slight increase in the value of x from a to b, then, as Figure 4.6b indicates,

$$P(a \leq X \leq b) = P(a \leq x \leq a + \Delta x) \approx f(a)\Delta x$$

for Δx small. Then for a fixed,

$$\lim_{\Delta x \to 0} \frac{P(a \leq x \leq a + \Delta x)}{\Delta x} = f(a), \tag{4.8}$$

the probability density of X at a. In general, the probability density at any fixed $x = a$ is the limiting value of the incremental change in the probability that $a \leq x \leq a + \Delta x$ per unit change in x as $\Delta x \to 0$. It is not $P(X = a)$.

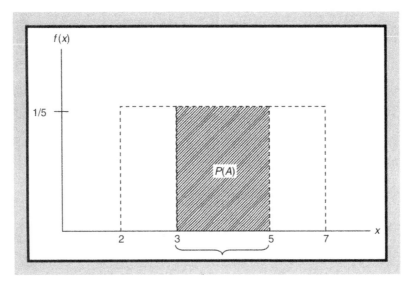

Figure 4.7   The probability of event A given a uniform probability density function.

Example 4.3.1   Let $f(x)$ be a uniform probability density function defined by

$$f(x) = \begin{cases} \frac{1}{\beta - \alpha}, & \alpha < x < \beta; \\ 0 & \text{elsewhere.} \end{cases}$$

For $\alpha = 2$ and $\beta = 7$,

$$f(x) = \begin{cases} 0, & x \le 2; \\ \frac{1}{5}, & 2 < x < 7; \\ 0, & x \ge 7 \end{cases}$$

(see Figure 4.7). Find $P(A) = P(3 \le X \le 5)$. From (4.7),

$$P(A) = \frac{1}{5} \int_3^5 dx = \frac{1}{5} x \Big]_3^5 = \frac{2}{5}. \quad \blacksquare$$

Under what circumstances can a function $g(x)$ be used to obtain a probability density function for a continuous random variable X? In general, any function $g(x)$ that satisfies:

$$\text{(a) } g(x) \ge 0, x \in (-\infty, +\infty) \tag{4.6.a}$$

(b) $\displaystyle\int_{-\infty}^{+\infty} g(x)dx$ exists   (4.6.b)

(c) $\displaystyle\int_{-\infty}^{+\infty} g(x)dx = K \neq 0$ (4.6.c)

may be used to obtain a probability density function of the form

$$f(x) = kg(x),$$

where $k = \frac{1}{K}$ is termed the *normalizing constant*, which renders

$$\int_{-\infty}^{+\infty} f(x)dx = k\int_{-\infty}^{+\infty} g(x)dx = \frac{1}{K}\int_{-\infty}^{+\infty} g(x)dx = 1$$

as required by (4.6c).

Example 4.3.2 Can

$$g(x) = \begin{cases} e^{-2x}, & x > 0; \\ 0, & x \le 0 \end{cases}$$

be used to obtain a probability density function? Clearly $g(x) \ge 0$ while

$$K = \int_{-\infty}^{+\infty} e^{-2x}dx = \lim_{a\to\infty}\int_{0}^{a} e^{-2x}dx = \lim_{a\to\infty}\left(-\frac{1}{2}\, e^{-2x}\Big]_{0}^{a}\right) = \frac{1}{2}.$$

Then $k = \frac{1}{K} = 2$ and thus

$$f(x) = kg(x) = \begin{cases} 2e^{-2x}, & x > 0; \\ 0, & x < 0. \end{cases}$$

If we choose $A = \{1 \le X \le 4\}$, then

$$P(A) = \int_{1}^{4} f(x)dx = 2\int_{1}^{4} e^{-2x}dx = -\, e^{-2x}\Big]_{1}^{4} = 0.13496. \quad \blacksquare$$

Although (4.7) gives the probability that a continuous random variable X with probability density function $f(x)$ assumes a value within the interval $\{x|a \le x \le b\}$, we may define a related function (derived from f) that

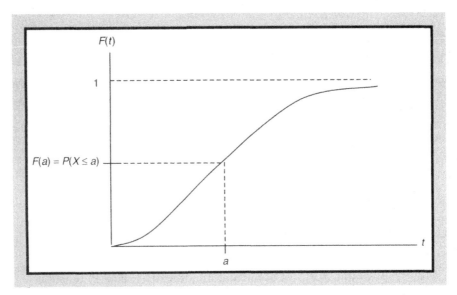

Figure 4.8   Cumulative distribution function for a continuous random variable X.

yields the probability that X takes on a value less than or equal to some real number t. This is the *cumulative distribution function* for a continuous random variable X, or

$$F(t) = P(X \leq t) = \int_{-\infty}^{t} f(x)dx \left(= \lim_{a \to +\infty} \int_{-a}^{t} f(x)dx \right) \tag{4.9}$$

(see Figure 4.8). $F(t)$ is a continuous function of t, its derivatives exist at every point of continuity of $f(x)$, and, at each such point, $dF(t)/dt = f(t)$. (Hence the integrand $f(x)$ in (4.9) must be the value of the derivative of F at x.) Clearly the cumulative distribution function serves as an alternative representation of a continuous probability distribution. In fact, if we know X's probability density function $f(x)$, then we can determine its cumulative distribution function using (4.9). Conversely, if the cumulative distribution function $F(t)$ is given, then we can determine the probability density $f(t)$ at each of its points of continuity by $dF(t)/dt = f(t)$.

Additional properties of the cumulative distribution function are:

(a) $0 \leq F(t) \leq 1$

(b) $F(a) \leq F(b)$ when $a < b$ (F is a nondecreasing function of t)

(c) $F(-\infty) = \lim_{t \to -\infty} F(t) = 0$ and $F(+\infty) = \lim_{t \to +\infty} F(t) = 1$

(d) F is everywhere continuous from the right at each t

(e) for $a < b, P(a \leq X \leq b) = P(a < X \leq b)$

$$= F(b) - F(a) = \int_a^b f(x)dx \text{ (or (4.7))}$$

(f) if F has a point of discontinuity at t, then $P(X = t)$ is the size of the jump which F exhibits at t; and if F is continuous at t, then $P(X = t) = 0$
(4.10)

By virtue of (4.10e), we may rewrite (4.8) as

$$\lim_{\Delta x \to 0} \frac{F(b) - F(a)}{\Delta x} = \lim_{\Delta x \to 0} \frac{\Delta F}{\Delta x} = F' = f(a)$$

as required at each point of continuity of f.

Example 4.3.3   For the uniform probability density function

$$f(x) = \begin{cases} 0, & x \leq 2; \\ \frac{1}{5}, & 2 < x < 7; \\ 0, & x \geq 7 \end{cases}$$

we may obtain, from (4.9),

$$F(t) = \int_{-\infty}^2 (0)dx + \frac{1}{5}\int_2^t dx = \frac{1}{5} x]_2^t = \frac{t - 2}{5}.$$

Then the cumulative distribution function may be expressed as

$$F(t) = \begin{cases} 0, & t < 2; \\ \frac{t-2}{5}, & 2 \leq t \leq 7; \\ 1, & t > 7 \end{cases}$$

(see Figure 4.9). Since $F(3) = \frac{1}{5}$ and $F(5) = \frac{3}{5}, P(A) = P(3 \leq X \leq 5) = F(5) - F(3) = \frac{2}{5}$ as determined earlier (see Figure 4.9). ∎

4.4   Mean and Variance of a Random Variable

Given a discrete probability distribution $(X_i, f(X_i)), i = 1, 2, \ldots$, let us define the *mean of the discrete random variable* X (or of X's discrete probability

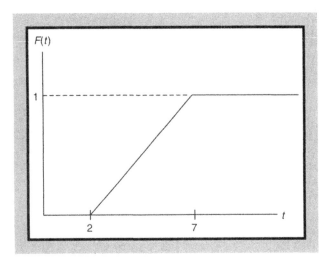

Figure 4.9   Cumulative distribution function for a uniform probability density function.

distribution) as

$$E(X) = \sum_i X_i f(X_i) (= \mu).^2 \tag{4.11}$$

Here $E(X)$ denotes what is termed the *mathematical expectation* or simply the *expected value* of X. Why does the mean of a discrete random variable have this form? It should be intuitively clear that those values of X with relatively high probabilities should carry relatively more weight in the calculation of the mean of X. Hence (4.11) is actually a weighted mean of X, where the weights (or probabilities) sum to one.

We may also note that $E(X)$ is actually the long-run average value of the discrete random variable X. That is, if we repeat (either actually or conceptually) a random experiment over and over for an indefinitely large number of trials, then $E(X)$ is the value that we would *expect* to observe in the long run. On any given trial the outcome $E(X)$ may not occur and, in fact, $E(X)$ might not even be an actual value of X. In this regard, $E(X)$ measures central tendency in that it is the quantity around which the values of X tend to cluster. For instance, given the following discrete probability distribution (see Table 4.3), it is evident that

$$E(X) = 30(06.) + 45(0.4) = 36.$$

---

[2] $E(X)$ exists if the sum $\sum_i X_i f(X_i)$ is absolutely convergent; that is, if $\sum_i |X_i| f(X_i)$ converges and is finite. For the vast majority of random variables absolute convergence of the sum in (4.11) occurs and thus $\sum_i X_i f(X_i)$ itself converges or is finite as well. In what follows then we shall assume that all of the given expectations exist.

Table 4.3

X	$f(X)$
30	0.60
45	0.40
	1

Hence in the long run, since 30 occurs 60% of the time and 45 occurs 40% of the time, we expect the average value of X to be 36.

In general, if $g(X)$ is a single-valued real function of the discrete random variable X, then the expected value of $g(X)$ is

$$E[g(X)] = \sum_i g(X_i)f(X_i).^3 \tag{4.12}$$

Some important properties of the expectation (linear) operator are:

(a) $E(a) = a$ (a is constant)

(b) $E(a \pm bX) = a \pm bE(X)$ (a, b constants)

(c) $E[(a + bX)^n] = \sum_{i=0}^{n} \binom{n}{i} a^i b^{n-i} E(X^{n-i})$ (a, b constants)

(d) $E\left(\sum_{i=1}^{n} X_i\right) = \sum_{i=1}^{n} E(X_i)$

(e) $E[g(X) \pm h(X)] = E[g(X)] \pm E[h(X)]$ (for g, h functions of X)

(f) in general, for $E(X) \neq 0$, $E\left(\dfrac{1}{X}\right) \neq \dfrac{1}{E(X)}$

$$\tag{4.13}$$

The *variance* of a discrete random variable X (or of X's discrete probability distribution) is defined as the expected value of the squared deviations from the expectation or

$$V(X) = E\left[X - E(X)^2\right] = \sum_i (X_i - E(X))^2 f(X_i)(= \sigma^2). \tag{4.14}$$

Example 4.4.1   Given the discrete probability distribution presented in Table 4.4, it is easily demonstrated that

$$E(X) = 1(0.10) + 3(0.60) + 8(0.30) = 4.3$$

---

[3] In general, if the discrete sum $\sum_i |g(X_i)| f(X_i)$ is finite, then $E[g(X)]$ exists.

and thus, from (4.14),

$$V(X) = 10.89(0.10) + 1.69(0.60) + 13.69(0.30) = 6.21. \quad \blacksquare$$

If we perform the indicated squaring in (4.14) and employ the properties of expectations given in (4.13) (remembering that $E(X)$ is a constant), then we can derive an alternative (computational) formula for calculating the variance of a discrete random variable X,

$$V(X) = E(X^2) - E(X)^2 = \sum_i X_i^2 f(X_i) - E(X)^2. \qquad (4.14.1)$$

Employing the preceding probability distribution (see Table 4.4) we have, from (4.14.1),

$$V(X) = 1(0.10) + 9(0.60) + 64(0.30) - (4.3)^2 = 6.21.$$

If we take the positive square root of $V(X)$, then we obtain the *standard deviation* of our discrete random variable X or $S(X) = \sqrt{V(X)}\,(= \sigma)$. Here $S(X)$ serves as a measure of dispersion of the X_i's around the mean or expectation of X.
    A few of the key properties of the variance of a discrete random variable X are:

(a) $V(a) = 0$ (a is constant)

(b) $V(aX) = a^2 V(X)$

(c) the variance of a random variable need not always exist   (4.15)

(d) if $E(X^2)$ exists, then $E(X)$ exists and thus $V(X)$ exists. Hence the

     existence of $V(X)$ implies that $E(X)$ exists

Given a discrete random variable X, we may express distance from the mean of X in terms of standard deviation units by forming the *standardized variable* $Z = \frac{X-\mu}{\sigma}$, where $E(Z) = 0, S(Z) = 1$. For any value X_i of X, $Z_i = \frac{X_i-\mu}{\sigma}$ tells us how many standard deviations X_i is from the mean μ; that is, for any X_i, Z_i describes the relative location of X_i in the discrete probability distribution.

Table 4.4

X	$f(X)$	$X - E(X)$	$\left(X - E(X)\right)^2$	X^2
1	0.10	−3.3	10.89	1
3	0.60	−1.3	1.69	9
8	0.30	3.7	13.69	64
	1.00			

If X represents a continuous random variable with probability density function $f(x)$, then the *mean of X* or the *expectation* of a continuous probability distribution is written as

$$E(X) = \int_{-\infty}^{+\infty} xf(x)dx \qquad (4.16)$$

(provided, of course, that the integral in (4.16) converges to a finite value). In general, if $g(X)$ is a single-valued real function of a continuous random variable X, then

$$E[g(X)] = \int_{-\infty}^{+\infty} g(x)f(x)dx \qquad (4.17)$$

provided $\int_{-\infty}^{+\infty} |g(x)| f(x)dx$ exists.

It is interesting to note that if X is a continuous random variable with probability density function

$$f(x) \begin{cases} > 0 & \text{for } 0 < x < a < +\infty; \\ = 0 & \text{elsewhere} \end{cases}$$

and cumulative distribution function $F(x)$, then

$$E(X) = \int_0^a [1 - F(x)]\, dx. \qquad (4.16.1)$$

Example 4.4.2   For instance, let

$$f(x) = \begin{cases} 0, & x \le 2; \\ \frac{1}{5}, & 2 < x < 7; \\ 0, & x \ge 7 \end{cases}$$

with

$$F(x) = \begin{cases} 0, & x < 2; \\ \frac{x-2}{5}, & 2 \le x \le 7; \\ 1, & x > 7. \end{cases}$$

Then from (4.11.1),

$$E(X) = \int_0^2 dx + \int_2^7 \left(1 - \frac{x-2}{5}\right)dx = x\big]_0^2 + \int_2^7 \left(\frac{7}{5} - \frac{x}{5}\right)dx$$

$$= 2 + \frac{7}{5}x\bigg]_2^7 - \frac{1}{10}x^2\bigg]_2^7 = 4.5 \quad \blacksquare$$

To obtain the variance of a continuous probability distribution we may calculate

$$V(X) = \int_{-\infty}^{+\infty} ((x - E(X))^2 f(x) dx \qquad (4.18)$$

or

$$V(X) = E(X^2) - E(X)^2 = \int_{-\infty}^{+\infty} x^2 f(x) dx - E(X)^2. \qquad (4.18.1)$$

Note that properties (4.13) and (4.15), as stated previously in terms of a discrete random variable, carry over to continuous random variables as well.

4.5   Chebyshev's Theorem for Random Variables

For X a discrete or continuous random variable with a finite expectation $(E(X) = \mu)$ and variance $(V(X) = \sigma^2)$, *Chebyshev's Theorem* provides us with limits, which are independent of the form of the distribution of X, to the probabilities associated with events described in terms of a random variable and its mean and variance (or standard deviation). Specifically,

> **CHEBYSHEV'S THEOREM.**   The probability that a random variable X will differ absolutely from μ by at least ε is always less than or equal to $\frac{\sigma^2}{\varepsilon^2}$; that is,

$$P(|X - \mu| \geq \varepsilon) \leq \frac{\sigma^2}{\varepsilon^2}. \qquad (4.19)$$

Hence any deviation from μ by ε or more units can be no more probable than $\frac{\sigma^2}{\varepsilon^2}$.

If $\varepsilon = k\sigma$, then (4.19) can be rewritten as

$$P(|X - \mu| \geq k\sigma) \leq \frac{1}{k^2}. \qquad (4.19.1)$$

or, alternatively,

$$P(|X - \mu| < k\sigma) \geq 1 - \frac{1}{k^2}. \qquad (4.19.2)$$

A slightly more transparent way to view (4.19.2) is to rewrite it as

$$(\mu - k\sigma < X < \mu + k\sigma) \geq 1 - \frac{1}{k^2}, \qquad (4.19.3)$$

that is, the probability that X lies within k standard deviations of μ is at least $1 - \frac{1}{k^2}$.

Let us rewrite (4.19.1) as

$$P\left(\left|\frac{X - \mu}{\sigma}\right| \geq k\right) = P(|Z| \geq k) \leq \frac{1}{k^2} \tag{4.20}$$

or

$$P(|Z| < k) \geq 1 - \frac{1}{k^2}. \tag{4.20.1}$$

But this latter expression is equivalent to

$$P(-k < Z < k) \geq 1 - \frac{1}{k^2}. \tag{4.20.2}$$

Hence the probability that the standardized random variable Z assumes a value within k standard deviations of its mean is at least $1 - \frac{1}{k^2}$. (Remember that the mean of Z is zero and its standard deviation is one.) By virtue of (4.20), the probability of obtaining an X whose associated Z is at least 2 is at most $\frac{1}{4}$; and from (4.20.1), the probability of obtaining an X with a corresponding Z less than 2 is at least $\frac{3}{4}$.

Example 4.5.1   Suppose that a random variable X has a mean of 50 and a standard deviation of 10. According to (4.19.1), the probability that an X value will be at least 2.5 standard deviations from the mean or at least $k\sigma = 25$ units from the mean is no greater than $1/(2.5)^2 = 0.16$. But from (4.20), this is equivalent to the probability bound that Z will lie at least 2.5 units away from zero (in either direction). Next, what is the probability limit on X lying within ± 15 units of the mean or within the interval $35 < X < 65$? Setting $X = X_0 = 65$, $Z_0 = \frac{65-50}{10} = 1.5$. Then from (4.20.1), $P(|Z| < Z_0) \geq 1 - 1/Z_0^2$ or $P(|Z| < 1.5) \geq 1 - 0.444 = 0.556$. Hence the probability is at least 0.556.   ■

Although the probabilities determined by Chebyshev's Theorem are *distribution free*, the bounds offered by these inequalities can be strengthened somewhat if we assume that X's probability distribution is both symmetrical and unimodal. For instance, under these restrictions, (4.20) becomes

$$P(|Z| \geq k) \leq \frac{4}{9}\left(\frac{1}{k^2}\right). \tag{4.20.3}$$

Clearly this inequality provides a tighter bound on the indicated probability than (4.20) itself.

A generalization of Chebyshev's Theorem can be stated as follows: if $g(X)$ is a single-valued nonnegative real function of a random variable X and $E[g(X)]$ exists, then, for any positive constant ε,

$$P[g(X) \geq \varepsilon] \leq \frac{E\left[g(X)\right]}{\varepsilon}.$$

For $g(X) = (X - \mu)^2$ and $\varepsilon = k^2\sigma^2$, we obtain (4.19.1).

In addition, we may examine an *asymmetric* variant of Chebyshev's Theorem, which applies to intervals containing $E(X) = \mu$, but not necessarily as their midpoint; that is, for a random variable X with $E(X) = \mu < +\infty$ and $\sqrt{V(X)} = \sigma < +\infty$ and for real constants $t_1, t_2 > 0$,

$$P(\mu - t_1\sigma < X < \mu + t_2\sigma) \geq 1 - \frac{4 + (t_2 - t_1)^2}{(t_2 + t_1)^2}. \tag{4.19.4}$$

4.6   Moments of a Random Variable

A *moment* of a random variable X is defined as the expected value of some particular function of X. In general, the moments of a probability distribution amount to a collection of descriptive measures that can be used to characterize the location and shape of the distribution. Hence a probability distribution can be completely specified in terms of its moments. As we shall now see, moments of a random variable typically are defined in terms of having either zero or the expectation of X as the reference point.

For X a discrete random variable, the rth *moment about zero* is

$$\mu'_r = E(X^r) = \sum_i X_i^r f(X_i) \tag{4.21}$$

(note that the first moment about zero is the mean of X or $\mu'_1 = E(X) = \mu$) and the rth *central moment of X* or the rth moment about the mean of X is

$$\mu_r = E\left[(X - \mu)^r\right] = \sum_i (X_i - \mu)^r f(X_i). \tag{4.22}$$

If X is a continuous random variable with probability density $f(x)$, then, provided the following integrals exist, we may correspondingly define

$$\mu'_r = E(X^r) = \int_{-\infty}^{+\infty} x^r f(x)dx; \tag{4.23}$$

and

$$\mu_r = E\left[(X - \mu)^r\right] = \int_{-\infty}^{+\infty} (x - \mu)^r f(x)dx. \tag{4.24}$$

It is easily verified that:

(a) the *zero^{th} central moment of X* is one ($\mu_0 = E(X - \mu)^0 = E(1) = 1$)

(b) the *first central moment of X* is zero ($\mu_1 = E(X - \mu) = E(X) - \mu = 0$)

(c) the *second central moment of X* is the variance of X or

$$\mu_2 = [E(X - \mu)^2] = V(X) \tag{4.25}$$

The first moment about zero locates the mean or measures central tendency of a probability distribution and the second moment about the mean describes its shape in terms of variation or dispersion about the mean. Additional information about the shape of a probability distribution, as characterized by measures of skewness and kurtosis, are provided by the third and fourth central moments of X, respectively. In particular, we shall develop standardized (independent of units and taken relative to σ) measures of skewness and kurtosis.

In this regard, the *third central moment of X* is

$$\mu_3 = E[(X - \mu)^3] \tag{4.26}$$

and the standardized third moment or the *coefficient of skewness* is

$$\alpha_3 = \frac{\mu_3}{\sigma^3} = \frac{\mu_3}{(\mu_2)^{3/2}}, \tag{4.27}$$

where the sign of α_3 is determined by that of μ_3; that is, for unimodal probability distributions:

(a) If $\mu_3 > 0$, then $\alpha_3 > 0$ and thus X's probability distribution is *positively skewed* or *skewed to the right*.

(b) If $\mu_3 < 0$, then $\alpha_3 < 0$ and thus X's probability distribution is *negatively skewed* or *skewed to the left*.

(c) If $\mu_3 = 0$, then $\alpha_3 = 0$ and thus X's probability distribution is *symmetrical* (about the mean).

Next, the *fourth central moment of X* is

$$\mu_4 = E[(X - \mu)^4]$$

and the standardized fourth moment or the *coefficient of kurtosis* is

$$\alpha_4 = \frac{\mu_4}{\sigma^4} = \frac{\mu_4}{(\mu_2)^2}. \tag{4.29}$$

If the peak of X's probability distribution mirrors that of a normal distribution, then $\alpha_4 = 3$. If $\alpha_4 > 3$ (respectively, < 3), then the peak of the probability distribution is sharper (respectively, flatter) than that of a normal distribution.

For purposes of computational expedience, we may express any central moment of a random variable X in terms of its moments about zero. Specifically, since

$$(X - \mu)^r = \sum_{j=0}^{r} (-1)^j \binom{r}{j} \mu^j X^{r-j},$$

it follows that,

$$\mu_r = E\left[(X - \mu)^r\right] = \sum_{j=0}^{r} (-1)^j \binom{r}{j} \mu^j E(X^{r-j}) = \sum_{j=0}^{r} (-1)^j \binom{r}{j} \mu^j \mu'_{r-j} \quad (4.30)$$

by virtue of (4.21) and the properties of the expectation operator depicted in (4.13). So according to (4.30), we can readily demonstrate that:

$$\mu_2 = \mu'_2 - \mu^2; \qquad (4.25.1)$$

$$\mu_3 = \mu'_3 - 3\mu\mu'_2 + 2\mu^3; \qquad (4.26.1)$$

$$\mu_4 = \mu'_4 - 4\mu\mu'_3 + 6\mu^2\mu'_2 - 3\mu^4. \qquad (4.28.1)$$

If we standardize the random variable X to obtain $Z = \frac{X-\mu}{\sigma}$, then, since $E(Z) = 0$, the rth central moment of Z can be expressed in terms of the rth central moment of X as

$$\mu_r(Z) = E(Z^r) = E\left[\left(\frac{X-\mu}{\sigma}\right)^r\right] = \frac{1}{\sigma^r}\left[E(X-\mu)^r = \frac{\mu_r(X)}{\sigma^r}\right] = \frac{\mu_r(X)}{\left(\mu_2(X)\right)^{r/2}}.$$
$$(4.31)$$

Also, $V(Z) = \mu_2(Z) = 1$ and $\alpha_3(Z) = \alpha_3(X)$ and $\alpha_4(Z) = \alpha_4(X)$. Hence standardizing a random variable X affects its mean and variance but not its standardized third and fourth moments.

Table 4.5

X	$f(X)$
1	0.2
2	0.3
3	0.4
5	0.1
	1.0

Example 4.6.1   Given the discrete probability distribution in Table 4.5, determine and interpret its standardized third and fourth moments or the coefficients of skewness and kurtosis. From (4.21):

$$\mu = E(X) = \sum X_i f(X_i) = 1(0.2) + 2(0.3) + 3(0.4) + 5(0.1) = 2.50,$$

$$\mu_2' = E(X^2) = \sum X_i^2 f(X_i) = 1(0.2) + 4(0.3) + 9(0.4) + 25(0.1) = 7.50,$$

$$\mu_3' = E(X^3) = \sum X_i^3 f(X_i) = 1(0.2) + 8(0.3) + 27(0.4) + 125(0.1) = 25.90,$$

$$\mu_4' = E(X^4) = \sum X_i^4 f(X_i) = 1(0.2) + 16(0.3) + 81(0.4) + 625(0.1) = 99.9.$$

Then from (4.25.1), (4.26.1), and (4.28.1), respectively,

$$\mu_2 = V(X) = \mu_2' - \mu^2 = 7.5 - (2.5)^2 = 1.25,$$

$$\mu_3 = \mu_3' - 3\mu\mu_2' + 2\mu^3 = 0.90,$$

$$\mu_4 = \mu_4' - 4\mu\mu_3' + 6\mu^2\mu_2' - 3\mu^4 = 4.96.$$

Based upon this latter set of values we have, from (4.27) and (4.29), respectively,

$$\alpha_3 = \frac{\mu_3}{\mu_2^{3/2}} = 0.64, \quad \alpha_4 = \frac{\mu_4}{\mu_2^2} = 3.17.$$

Since $\alpha_3 > 0$, this discrete probability distribution is slightly skewed to the right. And with $\alpha_4 > 3$, the distribution has a peak that is slightly sharper than that of a normal distribution. ■

Example 4.6.2   Let the probability density function for a continuous random variable X be

$$f(x) = \begin{cases} 2x, & 0 < x < 1; \\ 0 & \text{elsewhere.} \end{cases}$$

Then from (4.23):

$$\mu = E(X) = \int_{-\infty}^{+\infty} xf(x)dx = 2\int_0^1 x^2 dx = \frac{2}{3} x^3\Big]_0^1 = 0.666,$$

$$\mu_2' = E(X^2) = \int_{-\infty}^{+\infty} x^2f(x)dx = 2\int_0^1 x^3 dx = \frac{1}{2} x^4\Big]_0^1 = 0.500,$$

$$\mu_3' = E(X^3) = \int_{-\infty}^{+\infty} x^3f(x)dx = 2\int_0^1 x^4 dx = \frac{2}{5} x^5\Big]_0^1 = 0.400,$$

$$\mu_4' = E(X^4) = \int_{-\infty}^{+\infty} x^4f(x)dx = 2\int_0^1 x^5 dx = \frac{1}{3} x^6\Big]_0^1 = 0.333$$

and thus, from (4.25.1), (4.26.1), and (4.28.1), respectively,

$$\mu_2 = V(X) = \mu_2' - \mu^2 = 0.055,$$

$$\mu_3 = \mu_3' - 3\mu\mu_2' + 2\mu^3 = -0.007,$$

$$\mu_4 = \mu_4' - 4\mu\mu_3' + 6\mu^2\mu_2' - 3\mu^4 = 0.007.$$

Finally, (4.27) and (4.29), respectively, yield

$$\alpha_3 = \frac{\mu_3}{\mu_2^{3/2}} = -0.534, \quad \alpha_4 = \frac{\mu_4}{\mu_2^2} = 2.333.$$

With $\alpha_3 < 0$ we see that this continuous probability distribution is moderately skewed to the left, and $a_4 < 3$ indicates that its peak is a bit flatter than that of a normal distribution. ∎

4.7   Quantiles of a Probability Distribution

The *quantiles* of a probability distribution (or of a random variable X) are positional values that divide the probability distribution (or the total area under X's probability density function) into a number of equal portions. In this regard, the *quantile of order p*, $0 < p < 1$, of a discrete probability distribution is a value γ_p such that $P(X < \gamma_p) \leq p$ and $P(X \leq \gamma_p) \geq p$. If X is a continuous random variable, then γ_p is the value such that $P(X \leq \gamma_p) = p$. Note that if $p = 0.04$, then we are interested in finding the fourth percentile; if $p = 0.3$, we seek the thirtieth percentile or third decile; and if $p = 0.5$, we are looking for the median (equivalently, the fiftieth percentile or fifth decile) of X's probability distribution.

So for $p = 0.5$, the median of a discrete random variable X is the value $\gamma_{0.5}$ such that $P(X < \gamma_{0.5}) \leq 0.5$ and $P(X \leq \gamma_{0.5}) \geq 0.5$. If X is a continuous random variable, then the median is the value $\gamma_{0.5}$ such that $P(X \leq \gamma_{0.5}) = 0.5$. Alternatively, in the continuous case, $\gamma_{0.5}$ is the value that satisfies

$$\int_{-\infty}^{\gamma_{0.5}} f(x)dx = \int_{\gamma_{0.5}}^{+\infty} f(x)dx = 0.5.$$

Hence $\gamma_{0.5}$ divides the probability distribution (or the area under the probability density function) into two equal parts. Although $\gamma_{0.5}$ is unique for a continuous random variable, this may not be true for the discrete case.

Quantiles of a probability distribution can also be determined by directly employing the cumulative distribution function of a random variable X. In this regard, for the discrete case, the quantile of order p, γ_p, is the smallest number γ satisfying $F(\gamma) \geq p$. For a continuous random variable, γ_p is the smallest number γ satisfying $F(\gamma) = p$.

Example 4.7.1   For the discrete probability distribution in Table 4.6, find the fifteenth percentile or $\gamma_{0.15}$ and the median or $\gamma_{0.5}$. Since $\gamma_{0.15}$ must satisfy $P(X < \gamma_{0.15}) \leq 0.15$ and $P(X \leq \gamma_{0.15}) \geq 0.15$, it is readily seen from Table 4.6 that $P(X < 4) = 0.0834 < 0.15$ and $P(X \leq 4) \geq 0.1668 > 0.15$. Hence $\gamma_{0.15} = 4$. Note also that $X = 4$ is the smallest number satisfying $F(4) \geq 0.15$. To find the median of this discrete probability distribution, let us consider the requirement that $P(X < \gamma_{0.5}) \leq 0.5$ and $P(X \leq \gamma_{0.5}) \geq 0.5$. Since $P(X < 7) = 0.4170 < 0.5$ and $P(X \leq 7) = 0.5838 > 0.5$, we shall take $\gamma_{0.5} = 7$. Additionally, $X = 7$ is the smallest value for which $F(7) \geq 0.5$.   ■

Table 4.6

X	$f(X)$	$F(X)$
2	0.0278	0.0278
3	0.0556	0.0834
4	0.0834	0.1668
5	0.1112	0.2780
6	0.1390	0.4170
7	0.1668	0.5838
8	0.1390	0.7228
9	0.1112	0.8340
10	0.0834	0.9174
11	0.0556	0.9730
12	0.0278	1
	1	

Example 4.7.2   Given that the probability density function for a continuous random variable X appears as

$$f(x) = \begin{cases} 2x, & 0 < x < 1; \\ 0 & \text{elsewhere} \end{cases}$$

let us find the fortieth percentile of the associated continuous probability distribution. Since generally $P(X \le \gamma_p) = p$, we have

$$P(X \le \gamma_{0.4}) = \int_{-\infty}^{\gamma_{0.4}} f(x)dx = 2\int_0^{\gamma_{0.4}} x\,dx = x^2\big]_0^{\gamma_{0.4}} = \gamma_{0.4}^2.$$

Setting $\gamma_{0.4}^2 = 0.4$ renders $\gamma_{0.4} = +\sqrt{0.4} = 0.16$. Hence 40% of the area beneath $f(x)$ lies below $\gamma_{0.4} = 0.16$. Alternatively, since X's cumulative distribution function is

$$F(t) = \int_{-\infty}^{t} f(x)dx \quad \text{or} \quad F(t) = \begin{cases} 0, & t < 0; \\ t^2, & 0 \le t \le 1; \\ 1, & t > 1 \end{cases}$$

it is easily demonstrated that for $F(\gamma) = 0.4$ or $\gamma^2 = 0.4$, we again get $\gamma = \gamma_{0.4} = +\sqrt{0.4} = 0.16$. To also find the median of this probability distribution, let us calculate

$$\int_{\gamma_{0.5}}^{+\infty} f(x)dx = 2\int_{\gamma_{0.5}}^{1} xdx = x^2\big]_{\gamma_{0.5}}^{1} = 1 - \gamma_{0.5}^2.$$

Then, since we require that $1 - \gamma_{0.5}^2 = 0.5, \gamma_{0.5}^2 = 0.5$ or $\gamma_{0.5} = +\sqrt{0.5} = 0.7071$. And since $F(\gamma) = 0.5$ or $\gamma^2 = 0.5$, we again obtain $\gamma = \gamma_{0.5} = 0.7071$. ■

4.8   Moment-Generating Function

In this section we shall consider a convenient alternative to the direct computation of moments of a random variable X. Specifically, we shall examine what is called the *moment-generating function* of X. Although this function is a useful device for determining moments about zero as well as central moments of a random variable, it can also be used to easily derive the distribution of a sum of independent random variables. Moreover, once the moment-generating function of a random variable X has been determined, it can be compared to the moment-generating functions of some well-known random variables. If it turns out that the moment-generating function of X is identical to that of a well-known random variable Y, then X and Y must have identical probability distributions.

Let X be a discrete or continuous random variable. The *moment-generating function of X*, denoted $m_X(t)$, is defined as $E(e^{tX})$ provided that the expectation exists for every t satisfying $|t| < t_0, t_0 > 0$. That is, $m_X(t)$ exists if there exists a positive constant t_0 such that $E(e^{tX})$ is finite for all $-t_0 < t < t_0$. Hence $E(e^{tX})$ must exist near or within some suitably restricted neighborhood of $t = 0$. In this regard, either

$$m_X(t) = E(e^{tX}) = \sum_i e^{tX_i} f(X_i) \; [X \text{ discrete}] \tag{4.32}$$

or

$$m_X(t) = E(e^{tX}) = \int_{-\infty}^{+\infty} e^{tx} f(x) dx. \; [X \text{ continuous}] \tag{4.33}$$

(If $t = 0$, then $m_X(0) = E(e^0) = 1$.)

We note briefly that:

(a) $m_X(t)$ is a function of the dummy real variable t; that is, the parameter t has no real meaning save that it is a purely mathematical device which, as we shall soon see, serves as the *moment generator*.

(b) If $m_X(t)$ exists, then it is unique and completely specifies the probability distribution of X.

(c) If two random variables have the same moment-generating function, then they have the same probability distribution.

(d) If $m_X(t)$ exists, then it is continuously differentiable in some neighborhood of the origin; that is, the derivatives of all orders of $m_X(t)$ will exist at $t = 0$. Hence $m_X(t)$ will generate all moments of X about zero under suitable differentiation. In this regard, if $m_X(t)$ exists, then for any positive integer r,

$$m_X^{(r)}(0) = \frac{d^r}{dt^r} m_X(t) \mid_{t=0} = E(X^r) = \mu'_r. \tag{4.34}$$

If μ'_r exists, then all moments μ'_k exist for $k \le r$.

(Note that if X is a continuous random variable, then we are implicitly assuming that we can differentiate under the integral sign.)

To see exactly how (4.34) is applied, let us assume that the derivative operator distributes over the expectation operator. In fact, interchanging the operations of differentiation and expectation is permissible if the sum or integral defining the moment-generating function converges uniformly. Then

$$m_X^{(1)}(t) = \frac{d}{dt} E(e^{tX}) = E\left(\frac{d}{dt} e^{tX}\right) = E(Xe^{tX})$$

so that

$$m_X^{(1)}(0) = E(X) = \mu_1';$$

$$m_X^{(2)}(t) = \frac{d^2}{dt^2} E(e^{tX}) = E\left(\frac{d^2}{dt^2} e^{tX}\right) = E\left(\frac{d}{dt} X e^{tX}\right) = E(X^2 e^{tX})$$

and thus

$$m_X^{(2)}(0) = E(X^2) = \mu_2'.$$

In general,

$$m_X^{(r)}(t) = \frac{d^r}{dt^r} E(e^{tX}) = E\left(\frac{d^r}{dt^r} e^{tX}\right) = E(X^r e^{tX})$$

with

$$m_X^{(r)}(0) = E(X^r) = \mu_r'.$$

It is instructive to obtain (4.34) in an alternative fashion by replacing e^{tX} by its Maclaurin's series expansion

$$e^{tX} = 1 + tX + \frac{(tX)^2}{2!} + \frac{(tX)^3}{3!} + \cdots . \tag{4.35}$$

(In general, $e^{g(t)} = 1 + g(t) + g(t)^2/2! + g(t)^3/3! + \cdots .$) Under this representation $m_X(t)$ is expressible as a function of all moments about zero. That is, for X discrete and μ_i' finite, $i = 1, 2, 3, \ldots,$

$$m_X(t) = E(e^{tX}) = \sum_i e^{tX_i} f(X_i)$$

$$= \sum_i \left[1 + tX_i + \frac{(tX_i)^2}{2!} + \frac{(tX_i)^3}{3!} + \cdots \right] f(X_i)$$

$$= \sum_i f(X_i) + t \sum_i X_i f(X_i) \tag{4.36}$$

$$+ \frac{t^2}{2!} \sum_i X_i^2 f(X_i) + \frac{t^3}{3!} \sum_i X_i^3 f(X_i) + \cdots$$

$$= 1 + t\mu_1' + \frac{t^2}{2!} \mu_2' + \frac{t^3}{3!} \mu_3' + \frac{t^4}{4!} \mu_4' + \cdots .$$

Thus the coefficient on $\frac{t^k}{k!}$ is $\mu_k' = E(X^k), k = 1, 2, 3, \ldots .$

Given (4.36), the rth derivative of $m_X(t)$ with respect to t evaluated at $t = 0$ yields μ'_r. That is, for, say,

$$m_X^{(1)}(t) = \mu'_1 + \mu'_2 t + \mu'_3 \frac{t^2}{2!} + \mu'_4 \frac{t^3}{3!} + \cdots ,$$

$$m_X^{(2)}(t) = \mu'_2 + \mu'_3 t + \mu'_4 \frac{t^2}{2!} + \cdots , \text{ and}$$

$$m_X^{(3)}(t) = \mu'_3 + \mu'_4 t + \cdots ,$$

it follows that $m_X^{(1)}(0) = \mu'_1$, $m_X^{(2)}(0) = \mu'_2$, and $m_X^{(3)}(0) = \mu'_3$ as required by (4.34). So if we can determine $E(e^{tX})$, then we can find any of the moments about zero for X.

In a similar fashion we can determine that if X is continuous and $m_X(t)$ exists, then, via (4.35),

$$m_X(t) = E(e^{tX}) = \int_{-\infty}^{+\infty} e^{tx} f(x) dx$$

$$= \int_{-\infty}^{+\infty} \left(1 + tx + \frac{(tx)^2}{2!} + \frac{(tx)^3}{3!} + \cdots \right) f(x) dx$$

$$= \int_{-\infty}^{+\infty} f(x) dx + t \int_{-\infty}^{+\infty} x f(x) dx \qquad (4.37)$$

$$+ \frac{t^2}{2!} \int_{-\infty}^{+\infty} x^2 f(x) dx + \frac{t^3}{3!} \int_{-\infty}^{+\infty} x^3 f(x) dx + \cdots$$

$$= 1 + t\mu'_1 + \frac{t^2}{2!} \mu'_2 + \frac{t^3}{3!} \mu'_3 + \frac{t^4}{4!} \mu'_4 + \cdots$$

for μ'_i finite, $i = 1, 2, 3, \ldots$. Here too it is easily demonstrated via the successive differentiation of (4.37) that (4.34) holds.

Example 4.8.1   A fair coin is tossed three times in succession, with a success defined as getting heads on any individual toss. If the random variable X depicts the number of heads obtained, then X's probability distribution appears in Table 4.7.

From (4.32), the moment-generating function for X is

$$m_X(t) = E(e^{tX}) = \sum_i e^{tX_i} f(X_i)$$

$$= \frac{1}{8} e^{t(0)} + \frac{3}{8} e^{t(1)} + \frac{3}{8} e^{t(2)} + \frac{1}{8} e^{t(3)}$$

$$= \frac{1}{8} + \frac{3}{8} e^t + \frac{3}{8} e^{2t} + \frac{1}{8} e^{3t}.$$

Table 4.7

X	$P(X = X_i) = f(X_i)$
0	1/8
1	3/8
2	3/8
3	1/8
	1

Then

$$m_X^{(1)} = \frac{3}{8}e^t + \frac{6}{8}e^{2t} + \frac{3}{8}e^{3t}, \quad m_X^{(2)} = \frac{3}{8}e^t + \frac{12}{8}e^{2t} + \frac{9}{8}e^{3t},$$

and thus

$$m_X^{(1)}(0) = \mu_1' = 1.5, \quad m_X^{(2)}(0) = \mu_2' = 3.$$

Hence for this discrete probability distribution $\mu = 1.5$ and $\sigma^2 = \mu_2' - (\mu_1')^2 = 0.75.$ ∎

Example 4.8.2 Let the (exponential) probability density function for the continuous random variable X appear as

$$f(x) = \begin{cases} \theta e^{-\theta x}, & x > 0, \theta > 0; \\ 0 & \text{elsewhere.} \end{cases}$$

From (4.33) and for $t < \theta$,

$$m_X(t) = E(e^{tX}) = \theta \int_{-\infty}^{+\infty} e^{tx} e^{-\theta x} dx$$

$$= \theta \int_0^{+\infty} e^{(t-\theta)x} dx = \theta \lim_{a \to \infty} \int_0^a e^{(t-\theta)x} dx$$

$$= \frac{\theta}{t-\theta} \lim_{a \to \infty} e^{(t-\theta)x} \Big]_0^a = \frac{\theta}{\theta - t}.$$

Then

$$m_X^{(1)}(t) = \frac{\theta}{(\theta - t)^2}, \quad m_X^{(2)}(t) = \frac{2\theta}{(\theta - t)^3}$$

and thus

$$m_X^{(1)}(0) = \mu_1' = \frac{1}{\theta}, \quad m_X^{(2)}(0) = \mu_2' = \frac{2}{\theta^2}.$$

Again $\mu = \frac{1}{\theta}$ while $\sigma^2 = \mu_2' - (\mu_1')^2 = \frac{1}{\theta^2}$. ■

Example 4.8.3   Suppose that a continuous random variable X exhibits a uniform distribution with probability density function

$$f(x) = \begin{cases} 1, & 0 < x < 1; \\ 0 & \text{elsewhere.} \end{cases}$$

From (4.34),

$$\mu_k' = \int_0^1 x^k dx = \frac{1}{k+1} x^{k+1} \Big]_0^1 = \frac{1}{k+1}.$$

And from (4.33), for $t \neq 0$,

$$m_X(t) = E(e^{tX}) = \int_0^1 e^{tx} dx = \frac{1}{t} e^{tx} \Big]_0^1 = \frac{1}{t}(e^t - 1).$$

To get μ_k' from $m_X(t)$, let us employ (4.35) so as to obtain

$$m_x(t) = \frac{1}{t}\left(1 + t + \frac{t^2}{2!} + \frac{t^3}{3!} + \cdots + \frac{t^{k+1}}{(k+1)!} + \cdots - 1\right)$$

$$= 1 + \frac{t}{2!} + \frac{t^2}{3!} + \cdots + \frac{t^k}{(k+1)k!} + \cdots.$$

Since μ_k' is the coefficient on $\frac{t^k}{k!}$, it follows from the preceding expression that $\mu_k' = \frac{1}{k+1}$. ■

Next, let $g(X)$ be a single-valued function of the random variable X. The moment-generating function of $g(X)$, denoted $m_{g(X)}(t)$, is defined as $E(e^{g(X)t})$ provided that the expectation exists (is finite) for every t satisfying $|t| < t_0, t_0 > 0$. In this regard, either

$$m_{g(X)}(t) = E(e^{g(X)t}) = \sum_i e^{tg(X_i)} f(X_i) \; [X \text{ discrete}] \tag{4.38}$$

or

$$m_{g(X)}(t) = E(e^{g(X)t}) = \int_{-\infty}^{+\infty} e^{tg(x)} f(x) dx. \; [X \text{ continuous}] \tag{4.39}$$

To generate central moments of a random variable X, let us set $g(X) = X - \mu$ in either (4.38) or (4.39). Then the moment-generating function of X-μ or the *central moment-generating function*, denoted $m_{X-\mu}(t)$, is defined as $E(e^{t(X-\mu)})$ provided that the expectation exists for every t satisfying $|t| < t_0, t_0 > 0$. Hence either

$$m_{X-\mu}(t) = E(e^{t(X-\mu)}) = \sum_i e^{t(X_i-\mu)} f(X_i) \quad [X \text{ discrete}] \qquad (4.38.1)$$

or

$$m_{X-\mu}(t) = E(e^{t(X-\mu)}) = \int_{-\infty}^{+\infty} e^{t(x-\mu)} f(x) dx. \quad [X \text{ continuous}] \qquad (4.39.1)$$

If $m_{X-\mu}(t)$ exists, then it is continuously differentiable in some neighborhood of $t = 0$. Hence successively differentiating $m_{X-\mu}(t)$ with respect to t and evaluating the resulting derivatives at $t = 0$ enables us to obtain the central moments of X (provided that the derivative operator again distributes over the expectation operator). That is, if $m_{X-\mu}(t)$ is finite, then for any positive integer r,

$$m_{X-\mu}^{(r)}(0) = \frac{d^r}{dt^r} m_{X-\mu}(t)\bigg|_{t=0} = E[(X-\mu)^r] = \mu_r. \qquad (4.40)$$

To see exactly how the central moments of X are derived using the central moment-generating function, let us find

$$m_{X-\mu}^{(1)}(t) = \frac{d}{dt} E(e^{t(X-\mu)}) = E\left(\frac{d}{dt} e^{t(X-\mu)}\right) = E\left[(X-\mu)e^{t(X-\mu)}\right]$$

so that

$$m_{X-\mu}^{(1)}(0) = E(X - \mu) = \mu_1 = 0;$$

$$m_{X-\mu}^{(2)}(t) = \frac{d^2}{dt^2} E(e^{t(X-\mu)}) = E\left(\frac{d^2}{dt^2} e^{t(X-\mu)}\right)$$

$$= E\left[\frac{d}{dt}(X-\mu)e^{t(X-\mu)}\right] = E\left[(X-\mu)^2 e^{t(X-\mu)}\right]$$

and thus

$$m_{X-\mu}^{(2)}(0) = E\left[(X-\mu)^2\right] = \mu_2.$$

In general,

$$m_{X-\mu}^{(r)}(t) = \frac{d^r}{dt^r} E\left(e^{t(X-\mu)}\right) = E\left(\frac{d^r}{dt^r} e^{t(X-\mu)}\right) = E\left[(X-\mu)^r e^{t(X-\mu)}\right]$$

with

$$m_{X-\mu}^{(r)}(0) = E[(X - \mu)^r] = \mu_r.$$

We may obtain the central moments of X in an alternative fashion by replacing $e^{t(X-\mu)}$ by its Maclaurin's series expansion

$$e^{t(X-\mu)} = 1 + t(X - \mu) + \frac{t^2(X - \mu)^2}{2!} + \frac{t^3(X - \mu)^3}{3!} + \cdots. \tag{4.41}$$

Then for X discrete or continuous and for μ_i finite, $i = 1, 2, 3, \ldots,$

$$m_{X-\mu}(t) = E\left(e^{t(X-\mu)}\right) = 1 + t\mu_1 + \frac{t^2}{2!}\mu_2 + \frac{t^3}{3!}\mu_3 + \frac{t^4}{4!}\mu_4 + \cdots. \tag{4.42}$$

Thus the coefficient on $\frac{t^k}{k!}$ is $E[(X - \mu)^k] = \mu^k, k = 1, 2, 3, \ldots.$ If we now take the derivatives of (4.42) with respect to t and evaluate the same at $t = 0$, then (4.40) holds for positive integer values of r. So if we can determine $E\left(e^{t(X-\mu)}\right)$, then we can find any of the central moments of X.

Example 4.8.4   Given the probability density function appearing in example 4.8.3 above, let us directly determine the central moment-generating function of X for $t \neq 0$ as

$$m_{X-\mu}(t) = E\left(e^{t(X-\mu)}\right) = \int_0^1 e^{t(x-\mu)}dx = e^{-t\mu}\int_0^1 e^{tx}dx = \frac{1}{t}e^{-t\mu}(e^t - 1).$$

Then

$$m_{X-\mu}(t) = \frac{1}{t}e^{-\mu t}(e^t - 1), t \neq 0.$$

Using (4.35) we may rewrite $m_{X-\mu}(t), t \neq 0$, as

$$m_{X-\mu}(t) = e^{-t\mu}\left(1 + \frac{t}{2!} + \frac{t^2}{3!} + \cdots + \frac{t^k}{(k+1)!} + \cdots\right).$$

You should verify that (4.40) holds for $r = 1, 2$; that is, $m_{X-\mu}^{(1)}(0) = 0$ and $m_{X-\mu}^{(2)}(0) = \frac{1}{12}$. ∎

What is the connection between the moment-generating function of X and the moment-generating function of $X - \mu$? It is readily seen that

$$m_{X-\mu}(t) = E\left(e^{t(X-\mu)}\right) = e^{-t\mu}E(e^{tX}) = e^{-t\mu}m_X(t). \tag{4.43}$$

So if $m_{X-\mu}(t)$ is known and we desire to find moments of X about zero, (4.43) renders

$$m_X(t) = e^{t\mu}m_{X-\mu}(t); \tag{4.43.1}$$

and if $m_X(t)$ is known, central moments of X directly follow from (4.43). (Note that the connection between $m_X(t)$ and $m_{X-\mu}(t)$ as given by (4.43) was anticipated from the form of the solution offered in the preceding example problem.)

In what follows we shall assume that for each of the functional forms specified for $g(X)$, $E(e^{tg(X)})$ exists in some neighborhood of $t = 0$. Then for X discrete or continuous and $b > 0$:

(a) if $g(X) = a + bX$, then $m_{g(X)}(t) = e^{ta}m_X(tb), -\dfrac{t_0}{b} < t < \dfrac{t_0}{b}$;

(b) if $g(X) = a + bh(X)$, then $m_{g(X)}(t) = e^{ta}m_{h(X)}(tb), -\dfrac{t_0}{b} < t < \dfrac{t_0}{b}$; and

$$\tag{4.44}$$

(c) if $g(X) = \dfrac{X+a}{b}$, then $m_{g(X)}(t) = e^{ta/b}m_X\left(\dfrac{t}{b}\right), -bt_0 < t < bt_0$.

(What adjustment in the neighborhood of t must be made if $b < 0$?)

Example 4.8.5   Suppose X is a continuous random variable with mean μ and variance σ^2. Let $Z = (X - \mu)/\sigma$. Clearly the first and second moments of Z about zero are $E(Z) = 0$ and $E(Z^2) = 1$ respectively and, from (4.44.c),

$$m_Z(t) = E(e^{tZ}) = e^{-t\mu/\sigma}m_X\left(\frac{t}{\sigma}\right), -\sigma t_0 < t < \sigma t_0. \quad \blacksquare$$

4.9   Probability-Generating Function

In what follows we shall consider a convenient computational device for generating: (1) the probabilities associated with a certain class of discrete probability distributions; and (2) the factorial moments of the same. Specifically, suppose that a discrete random variable X serves to index a *count* of items and thus assumes only nonnegative integer values; that is, $X = 0, 1, 2, \ldots$. (Examples of such distributions are the uniform, Bernoulli, binomial, negative binomial, hypergeometric, geometric, and Poisson, among others.)

Our usual notation for depicting the probability mass function for a discrete random variable X is to list the various values of X as $X_1, X_2, \ldots, X_n$ and their respective probabilities as $f(X_1), f(X_2), \ldots, f(X_n)$, where $f(X_i) = P(X = X_i)$, $i = 1, \ldots, n$. However, since X is a count random variable, let us set $X_i = i$ with $f(i) = P(X = i), i = 0, 1, 2, \ldots$. Armed with these considerations, the *probability-generating (factorial-generating) function of X*, denoted $\phi_X(t)$, is defined as $E(t^X)$, with t serving as a dummy parameter, provided that the expectation exists; that is, $\phi_X(t)$ is finite over some range of values containing both $t = 0$ and $t = 1$. (As we shall now see, t has a dual role—factorial moments may be obtained from the derivatives of $E(t^X)$ as the *factorial generator* $t \to 1$, and probabilities may be obtained from the same as the *probability generator* $t \to 0$.) In this regard,

$$\phi_X(t) = E(t^X) = \sum_i t^i f(i). \tag{4.45}$$

(If $t = 0, \phi_X(0) = f(0)$; if $t = 1, \phi_X(1) = 1$.)

We note briefly that:

(a) $\phi_X(t)$, if it exists, is unique and completely determines the probability distribution of the discrete random variable X.

(b) If two discrete random variables have the same probability-generating function, then they have the same probability distribution.

(c) If $\phi_X(t)$ exists, then it is continuously differentiable near $t = 0$ and $t = 1$. Hence $\phi_X(t)$ will generate all factorial moments of X near $t = 1$ and all probabilities of X near $t = 0$ under suitable differentiation; that is, if $\phi_X(t)$ exists, then for a positive integer r, the rth *factorial moment of X* is

$$\phi_X^{(r)} = \frac{d^r}{dt^r} \phi_X(t)\Big|_{t=1} = E[X(X-1)\cdots(X-r+1)] = f_r \tag{4.46}$$

(notice the factorial structure to (4.46)) while the probability that $X = r$ is

$$\phi_X^{(r)}(0) = \frac{1}{r!} \frac{d^r}{dt^r} \phi_X(t)\Big|_{t=0} = f(r). \tag{4.47}$$

To generate the factorial moments of X, let us first successively differentiate (4.45) with respect to t so as to obtain:

$$\phi_X^{(1)}(t) = \frac{d}{dt}E(t^X) = E\left(\frac{d}{dt}t^X\right) = E\left(Xt^{X-1}\right);$$

$$\phi_X^{(2)}(t) = \frac{d^2}{dt^2}E(t^X) = E\left(\frac{d}{dt}Xt^{X-1}\right) = E\left[X(X-1)t^{X-2}\right]; \text{ and}$$

$$\vdots$$

$$\phi_X^{(r)}(t) = \frac{d^r}{dt^r}E(t^X) = E\left(\frac{d^r}{dt^r}t^X\right) = E\left[X(X-1)\cdots(X-r+1)t^{X-r}\right],$$

where r is a positive integer. (Here we are implicitly assuming that the derivative operator distributes over the expectation operator.) Then at $t = 1$:

$$\phi_X^{(1)}(1) = E(X) = f_1;$$

$$\phi_X^{(2)}(1) = E\left[X(X-1)\right] = E(X^2) - E(X) = f_2; \text{ and}$$

$$\vdots$$

$$\phi_X^{(r)}(1) = E\left[X(X-1)\cdots(X-r+1)\right] = f_r.$$

Note that $f_1 = \mu$ while $\sigma^2 = f_2 + f_1 - f_1^2 = f_2 + \mu - \mu^2$.

Example 4.9.1   For $r = 1, 2, 3, 4$, let us express (4.46) as a finite sum when $i = 1, 2, 3, 4$. To this end:

$$\phi_X^{(1)}(t) = \sum_{i=1}^{4} i t^{i-1} f(i) = f(1) + 2tf(2) + 3t^2 f(3) + 4t^3 f(4);$$

$$\phi_X^{(2)}(t) = \sum_{i=2}^{4} i(i-1)t^{i-2} f(i) = 2f(2) + 6tf(3) + 12t^2 f(4);$$

$$\phi_X^{(3)}(t) = \sum_{i=3}^{4} i(i-1)(i-2)t^{i-3} f(i) = 6f(3) + 24tf(4); \text{ and}$$

$$\phi_X^{(4)}(t) = \sum_{i=4}^{4} i(i-1)(i-2)(i-3)t^{i-4} f(i) = 24f(4).$$

Then for $t = 1$:

$$\phi_X^{(1)}(1) = f(1) + 2f(2) + 3f(3) + 4f(4);$$

$$\phi_X^{(2)}(1) = 2f(2) + 6f(3) + 12f(4);$$

$$\phi_X^{(3)}(1) = 6f(3) + 24f(4); \text{ and}$$

$$\phi_X^{(4)}(1) = 24f(4).$$

In general, for $X = 0, 1, \ldots, n$ and $1 \leq r \leq n$,

$$\phi_X^{(r)}(t) = \sum_{i=r}^{n} i(i-1)(i-2) \cdots (i-r+1)t^{i-r}f(i). \quad \blacksquare \qquad (4.48)$$

What is the connection between the moment-generating function $m_X(t)$ and the factorial-generating function $\phi_X(t)$? A moments reflection reveals that knowledge of $f_k, k = 1, 2, \ldots$, is equivalent to knowledge of the moments of a discrete random variable X about zero and conversely. So if we are able to determine the factorial moments of X, then we should be able to determine X's moments about zero. More precisely, since $t^X = e^{X \log_e t}$, it follows that

$$\phi_X(t) = m_X(\log_e t) = E\left(e^{X \log_e t}\right) \qquad (4.49)$$

so that

$$m_X(t) = \phi_X(e^t) = E(e^{tX}). \qquad (4.50)$$

Next, to generate probabilities associated with the values of a discrete random variable X, let us again assume that X is a count variable and that the range of X is the set of integers $X = 0, 1, \ldots$. Then from (4.45),

$$\phi_X(t) = \sum_{i=0}^{n} t^i f(i) = f(0) + tf(1) + t^2 f(2) + t^3 f(3) + \cdots + t^n f(n).$$

(Note again that $\phi_X(0) = f(0)$.) Hence for, say $i = 0, 1, \ldots, n$,

$$\phi_X^{(1)}(t) = f(1) + 2tf(2) + 3t^2 f(3) + \cdots + nt^{n-1}f(n);$$

$$\phi_X^{(2)}(t) = 2f(2) + 6tf(3) + \cdots + n(n-1)t^{n-2}f(n);$$

$$\vdots$$

$$\phi_X^{(n-1)}(t) = (n-1)!f(n-1) + n!\, tf(n); \text{ and}$$

$$\phi_X^{(n)}(t) = n!f(n).$$

At $t = 0$: $\phi_X^{(1)}(0) = f(1), \phi_X^{(2)}(0) = 2!f(2),\ldots,\phi_X^{(n-1)}(0) = (n-1)!f(n-1)$, and $\phi_X^{(n)}(0) = n!f(n)$. Hence, as anticipated from (4.47): $f(1) = \phi_X^{(1)}(0)$,

$$f(2) = \frac{1}{2!}\phi_X^{(2)}(0),\ldots,f(n-1) = \frac{1}{(n-1)!}\phi_X^{(n-1)}(0), \text{ and } f(n) = \frac{1}{n!}\phi_X^{(n)}(0).$$

Example 4.9.2   Given the discrete probability distribution appearing in Table 4.7, determine the probability-generating function as well as the factorial moments f_1 and f_2. What is the value of σ^2? Also, we shall demonstrate that $\phi_X(t)$ does indeed generate the requisite probabilities. From (4.45),

$$\phi_X(t) = E(t^X) = \sum_{i=0}^{3} t^i f(i) = \frac{1}{8} + \frac{3}{8}t + \frac{3}{8}t^2 + \frac{1}{8}t^3$$

with

$$\phi_X^{(1)}(t) = \frac{3}{8} + \frac{6}{8}t + \frac{3}{8}t^2, \phi_X^{(2)}(t) = \frac{6}{8} + \frac{6}{8}t, \text{ and } \phi_X^{(3)}(t) = \frac{6}{8}.$$

Then $f_1 = \phi_X^{(1)}(1) = 1.5 = \mu$, $f_2 = \phi_X^{(2)}(1) = 1.5 = E(X^2) - 1.5$, and thus $\sigma^2 = f_2 + \mu - \mu^2 = 0.75$. Also, $\phi_X(0) = \frac{1}{8} = f(0), \phi_X^{(1)}(0) = \frac{3}{8} = f(1)$, $\phi_X^{(2)}(0) = \frac{6}{8} = 2!f(2)$ (so that $f(2) = 3/8$), and $\phi_X^{(3)}(0) = \frac{6}{8} = 3!f(3)$ (and thus $f(3) = \frac{1}{8}$).   ∎

Example 4.9.3   Suppose that a discrete random variable X has a probability mass function of the form $f(X;k) = 2^{-k}, k = 1,2,3,\ldots$. Then the probability generating function for X appears as

$$\phi_X(t) = \sum_{i=1}^{\infty} t^i 2^{-i} = \sum_{i=1}^{\infty}\left(\frac{t}{2}\right)^i = \frac{1}{1-\left(\frac{t}{2}\right)} - 1 = \frac{t}{2-t}, |t| < 2.$$

Also, since

$$\phi_X^{(1)}(t) = \frac{2}{(2-t)^2}, \phi_X^{(2)}(t) = \frac{4}{(2-t)^3}, \text{ and } \phi_X^{(3)}(t) = \frac{12}{(2-t)^4},$$

it follows that $f_1 = \phi_X^{(1)}(1) = 2$, $f_2 = \phi_X^{(2)}(1) = 4$, and $f_3 = \phi_X^{(3)}(1) = 12$. (For this discrete probability distribution $\mu = f_1 = 2$ and $\sigma^2 = f_2 + f_1 - f_1^2 = 2$.) In addition, $f(1) = \phi_X^{(1)}(0) = \frac{1}{2}, f(2) = \frac{1}{2!}\phi_X^{(2)}(0) = \frac{1}{4}$, and $f(3) = \frac{1}{3!}\phi_X^{(3)}(0) = \frac{1}{8}$. In general, $f(k) = \frac{1}{k!}\phi_X^{(k)}(0) = 2^{-k}, k = 1,2,3,\ldots$.   ∎

4.10 Exercises

4-1. A random variable X has a probability density function of the form

$$f(x) = \begin{cases} ke^{-x/3} & x > 0; \\ 0 & \text{elsewhere} \end{cases}$$

for a specific constant k. Determine:

(a) the value of k

(b) the cumulative distribution function

(c) $P(X \leq 6)$

(d) $P(1 \leq X \leq 4)$

4-2. For a discrete random variable X, determine the constant k such that $f(X) = k/X$, $X = 1, 2, 3, 4, 5$, is a probability mass function. Then determine:

(a) The cumulative distribution function

(b) $P(X \leq 3)$

(c) $P(X < 3)$

(d) $P(1 \leq X \leq 4)$

4-3. Determine the constant k such that

$$f(x) = \begin{cases} kx^2, & -2 \leq x \leq 2; \\ 0 & \text{elsewhere} \end{cases}$$

is a probability density function. Then find:

(a) The cumulative distribution function

(b) $P(X > 1/2)$

(c) $P(X \leq 1)$

(d) $P(-1 \leq X \leq 1)$

4-4. Let the cumulative distribution function for a random variable X appear as

$$F(t) = \begin{cases} 0, & t < 0; \\ 2t - t^2 & 0 \leq t \leq 1; \\ 1, & t > 0. \end{cases}$$

Find:

(a) $P(X \leq 1/4)$

(b) $P(X \geq 1/4)$

(c) $P(1/3 \leq X \leq 3/4)$

(d) The probability density function $f(x)$

4-5. Determine the cumulative distribution function $F(X)$ associated with the following probability mass function.

X	f(X)
0	0.215
1	0.433
2	0.288
3	0.064
	1.000

4-6. Determine the cumulative distribution function $F(X)$ given the following probability mass function. Use $F(X)$ to determine:

(a) $P(0 < X \leq 2)$

(b) $P(X \leq 1)$

(c) $P(-1 < X \leq 2)$

(d) $P(-1 < X \leq 0)$

X	f(X)
−1	0.5787
1	0.3472
2	0.0694
3	0.0046
	1.0000

4-7. Given the following cumulative distribution function, find:

(a) $f(0)$

(b) $f(2)$

(c) $f(3)$

(Hint: $f(X_i) = F(X_i) - F(X_i - 1)$.)

$$F(X) = \begin{cases} 0, & X < 0; \\ \frac{1}{2}, & 0 \leq X \leq 2; \\ \frac{5}{6}, & 2 \leq X \leq 3; \\ 1, & X \geq 3. \end{cases}$$

4-8. Does $F(t)$ given below satisfy the properties of a cumulative distribution function? Find the associated probability density function and use it to find $P(0.30 < X < 0.80)$.

$$F(t) = \begin{cases} 0, & t < 0; \\ t, & 0 \le t \le 1; \\ 1, & t > 1. \end{cases}$$

4-9. Is the expression $F(t) = (1 + e^{-t})^{-1}$ a legitimate (continuous) cumulative distribution function?

4-10. For $0 < p < 1$, is the expression $F(X) = 1 - (1 - p)^X$, $X = 1, 2, \ldots$, a legitimate (discrete) cumulative distribution function?

4-11. Suppose $f(x) = \begin{cases} k(3 - x), & 0 \le x \le 3; \\ 0 & \text{elsewhere.} \end{cases}$

What value of k makes $f(x)$ a legitimate probability density function? Find the cumulative distribution function $F(t)$. Using $f(x)$, find $P(1 \le X \le 2)$. Using $F(t)$, also find $P(1 \le X \le 2)$.

4-12. Let a random variable X be defined as the number of heads obtained in two flips of a fair coin. Determine the sample space S and the associated (discrete) probability distribution. Verify that the sequence of probabilities can be determined from the probability mass function

$$f(X) = \begin{cases} \dfrac{\binom{2}{X}}{4}, & X = 0, 1, 2; \\ 0 & \text{elsewhere.} \end{cases}$$

Given $f(X)$ find the cumulative distribution function.

4-13. Given the cumulative distribution function

$$F(X) = \begin{cases} 0, & X < 0; \\ \frac{1}{2}, & 0 \le X < 1; \\ \frac{3}{4}, & 1 \le X < 4; \\ 1, & X \ge 4, \end{cases}$$

determine the associated probability mass function. (Hint: Find the points of discontinuity. At such points $f(x) > 0$.)

4-14. Let the random variable X have the probability density function

$$f(x) = \begin{cases} x^{-2}, & 1 < x < +\infty; \\ 0 & \text{elsewhere.} \end{cases}$$

For events $A = \{x|1 < x < 2\}$ and $B = \{x|4 < x < 5\}$, find the probability that A or B occurs.

4-15. Can the expression

$$F(t) = \begin{cases} 0, & t < 0; \\ 1 - e^{-t}, & t \geq 0 \end{cases}$$

serve as a cumulative distribution function for a continuous random variable X? Find the associated probability density function and use it to determine:

(a) $P(2 < X < 5)$

(b) $P(X < 3)$

(c) $P(X > 6)$

4-16. Comment on the following statement: The probability mass function of a discrete random variable X has 1 as an upper bound but the probability density function of a continuous random variable X need not be bounded. Is the function

$$f(x) = \begin{cases} \frac{3}{2}x^{1/2}, & 0 < x < 1; \\ 0 & \text{elsewhere} \end{cases}$$

bounded over its domain? Is it a probability density function?

4-17. Comment on the following statement: If X is a continuous random variable, then its probability density function need not be continuous; however, its cumulative distribution function will be continuous. Let

$$f(x) = \begin{cases} \frac{1}{3}, & x \in \{0 < x < 1\} \cup \{4 < x < 6\}; \\ 0 & \text{elsewhere.} \end{cases}$$

Is f discontinuous over its domain? If so, where? Can f serve as a probability density function?
(Hint: $\int_{x \varepsilon A \cup B} f(x)dx = \int_{x \varepsilon A} f(x)dx + \int_{x \varepsilon B} f(x)dx$.)

4-18. Can the expression

$$f(x) = \begin{cases} \frac{6 - |X - 7|}{36}, & x = 2, 3, \ldots, 12; \\ 0 & \text{elsewhere} \end{cases}$$

serve as a probability mass function? How is the random variable X defined?

4-19. Verify that the cumulative distribution function given by (4.3):

   (a) has values restricted to $[0, 1]$

   (b) is monotone nondecreasing

   (c) is continuous from the right

   (d) defines $P(X_s < X \le X_t) = F(X_t) - F(X_s)$

   (e) requires at points of discontinuity $X = X_i, i = 1, 2, \ldots,$ that $F(X_i + 0) - F(X_i - 0) = F(X_i) - F(X_i - 0) = f(X_i)$

   (f) requires $\lim_{X_i \to -\infty} F(X_i) = 0$ and $\lim_{X_i \to +\infty} F(X_i) = 1$

4-20. Verify that the cumulative distribution function given by (4.9):

   (a) has values restricted to $[0,1]$

   (b) is monotone nondecreasing and continuous in t, with its derivatives existing at every point of continuity of f, and at each such point $dF(t)/dt = f(t)$

   (c) is continuous from the right

   (d) for every pair of real numbers $a \le b$, defines $P(a < X \le b) = F(b) - F(a)$

   (e) requires $\lim_{t \to -\infty} F(t) = 0$ and $\lim_{t \to +\infty} F(t) = 1$

4-21. Do you agree or disagree with statements (a)–(f)? Given that the indicated limits exist:

   (a) $\int_a^{+\infty} f(x)dx = \lim_{b \to +\infty} \int_a^b f(x)dx$

   (b) $\int_{-a}^b f(x)dx = \lim_{a \to +\infty} \int_{-a}^b f(x)dx$

   (c) $\int_{-\infty}^{+\infty} f(x)dx = \lim_{a \to +\infty} \int_{-a}^a f(x)dx$

   (d) $\int_{-\infty}^{+\infty} f(x)dx = \int_{-\infty}^0 f(x)dx + \int_0^{+\infty} f(x)dx$ (provided that the integrals on the right-hand side of this equation exist).

     Moreover, if $f(x)$ is unbounded on $[a, b]$ but bounded on $[a+\varepsilon, b]$, $\varepsilon > 0$ and small, then, provided that the indicated limit exists,

   (e) $\int_a^b f(x)dx = \lim_{\substack{\varepsilon \to 0 \\ \varepsilon > 0}} \int_{a+\varepsilon}^b f(x)dx$

     If $f(x)$ is unbounded on $[a, b]$ though bounded on $[a, b - \varepsilon]$, $\varepsilon > 0$ and small, then, provided that the indicated limit exists,

   (f) $\int_a^b f(x)dx; = \lim_{\substack{\varepsilon \to 0 \\ \varepsilon > 0}} \int_a^{b-\varepsilon} f(x)dx$

     In addition, find:

   (g) $\int_0^2 x^{-1/2}dx$

   (h) $\int_0^{+\infty} e^{-1/2x}dx$

   (i) $\int_{-\infty}^{+\infty} e^{-|x|}dx$

4-22. Suppose a probability density function has the form

$$f(x) = \begin{cases} \frac{40+x}{1600}, & -40 < x < 0; \\ \frac{40-x}{1600} & 0 \le x < 40; \\ 0 & \text{elsewhere.} \end{cases}$$

For $A = \{-10 \le X \le 10\}$, find $P(A)$. (Hint: Write $P(A)$ as the sum of two separate integrals for $A = \{x \mid -10 \le X \le 0\} \cup \{0 \le X \le 10\}$.)

4-23. For the probability density function

$$f(x) = \begin{cases} \frac{1}{200}e^{-x/200}, & x > 0; \\ 0 & \text{elsewhere.} \end{cases}$$

find the associated cumulative distribution function. Use the latter to determine:

(a) $P(X > 80)$

(b) $P(50 < X \le 90)$

4-24. Suppose a continuous random variable has a probability density function of the form

$$f(x) = \begin{cases} e^{-x}, & x > 0; \\ 0 & \text{elsewhere.} \end{cases}$$

Find $P(X^2 < \theta), \theta > 0$.

4-25. Given the following discrete probability distribution, find:

(a) $E(X)$

(b) $V(X)$

X	1	5	7	9
$f(X)$	1/6	2/6	2/6	1/6

4-26. Given the probability density function

$$f(x) = \begin{cases} 2(1-x), & 0 \le x \le 1; \\ 0 & \text{elsewhere} \end{cases}$$

find:

(a) $E(X)$

(b) $V(X)$

4-27. A fair coin is tossed three times in succession (or three fair coins are tossed once). List the points or simple events in the sample space S. Determine the associated probability distribution. Find its mean and variance.

4-28. For the probability density function $f(x) = \lambda^{-1} e^{-x/\lambda}$, $\lambda > 0$, $0 \le x < \infty$, demonstrate that $E(X) = \lambda$ and $V(X) = \lambda^2$.

4-29. Suppose $f(x) = \begin{cases} x^{-2}, & 1 \le x < +\infty; \\ 0 & \text{elsewhere.} \end{cases}$

Does $E(x)$ exist?

4-30. Comment on the following statement: For X a continuous random variable with probability density function $f(x)$, the existence of $E(X)$ implies that

$$\int_{-\infty}^{0} xf(x)dx, \int_{0}^{+\infty} xf(x)dx, \int_{-\infty}^{+\infty} |x|f(x)dx$$

all converge.

4-31. Evaluate the expression $E\left[(aX + b)^n\right] = \sum_{i=0}^{n} \binom{n}{i} a^{n-i} b^i E(X^{n-i})$ for $n = 1, 2$.

4-32. The cumulative distribution function for a continuous random variable X is

$$F(t) = \begin{cases} 1 - e^{-t/\theta}, & \theta > 0, t \ge 0; \\ 0 & t < 0. \end{cases}$$

Demonstrate that θ is the mean of X.

4-33. Suppose that the lifetime (in hours) of a particular piece of electrical equipment can be characterized by the probability density function

$$f(x) = \begin{cases} \frac{1}{200} e^{-x/200}, & x > 0; \\ 0 & \text{elsewhere.} \end{cases}$$

Verify that the average life of this electrical component is 200 hours. Suppose that an event A is defined as "the component fails in 60 or fewer hours." Find $P(A)$.

4-34. Given the probability mass function

X	$f(X)$
1	1/4
3	1/2
9	1/4
	1

let $g(X) = X^2 - 1$. Find $E\left(g(X)\right)$.

4-35. The probability density function for a continuous random variable X is

$$f(x) = \begin{cases} \frac{1}{5} & 5 < x < 10; \\ 0 & \text{elsewhere.} \end{cases}$$

Find $E(X)$ and $E(X^2)$. What is $V(X)$?

4-36. Verify that if X is a discrete or continuous random variable whose expectation exists, then, for a and b constants:

(a) $E(a) = a$

(b) $E(a \pm bX) = a \pm bE(X)$

For finite k, demonstrate that:

(c) $E\left(\sum_{j=1}^{k} g_j(X)\right) = \sum_{j=1}^{k} E(g_j(X))$, where $g_j(X)$ is a function of the random variable X.

4-37. Verify that if X is a discrete or continuous random variable whose variance exists, then, for a and b constants:

(a) $V(a) = 0$

(b) $v(a + X) = V(X)$

(c) $V(a + bX) = b^2 V(X)$

(d) $V(X) = E(X^2) - E(X)^2$

4-38. Let X be a random variable such that $P(X < 0) = 0$. If $E(X) = \alpha \geq 0$ exists, then, for $t \geq 1, P(X < \alpha t) \geq 1 - \frac{1}{t}$. Establish this result for:

(a) X discrete

(b) X continuous

4-39. Use the preceding result to derive the (4.19.2) form of Chebyshev's Theorem.

4-40. Let X be a continuous random variable whose mean $E(X) = \mu$ and variance $V(X) = \sigma^2$ exist. For any $\varepsilon > 0$ and small, verify that the (4.19) version of Chebyshev's Theorem holds.

(Hint: start with $V(X) = \int_{-\infty}^{+\infty} (x - \mu)^2 f(x)dx$ and let

$\{X \mid -\infty < X < +\infty\} = \{X \mid -\infty < X < \mu - \varepsilon\} \cup \{X \mid \mu + \varepsilon < X < +\infty\}$.)

4-41. Over the long run the daily price/barrel (denoted X) of a certain grade of crude oil has averaged about $17.23 with a standard deviation of $1.26. Given that the probability distribution of X is unknown, determine a lower bound estimate of the probability that X will fall between $14.23 and $20.23.

4-42. If X is a random variable with $E(X) = 3$ and $V(X) = 4$, use Chebyshev's Theorem to determine a lower-bound for $P(-2 < X < 8)$.

4-43. Given that a continuous random variable X has a cumulative distribution function of the form

$$F(t) = \begin{cases} 0 & t < 0; \\ t^{0.5} & 0 \le t \le 1; \\ 1 & t > 1, \end{cases}$$

find the quartiles of X. That is, find:

(a) $\gamma_{0.25}$

(b) $\gamma_{0.50}$

(c) $\gamma_{0.75}$

What is the quartile deviation of X?

4-44. Suppose that the probability density function for a continuous random variable X is

$$f(x) = \begin{cases} \frac{1}{2}x^{-1/2} & 0 < x < 1; \\ 0 & \text{elsewhere.} \end{cases}$$

Does X have a symmetrical probability density function? (Hint: Compare $E(X)$ and $\gamma_{0.5}$, with the latter determined from $P(X \le \gamma_{0.5}) = \int_{-\infty}^{\gamma_{0.5}} f(x)dx = 0.5$.)

4-45. Let the probability mass function for a discrete random variable X be

$$f(X) = \begin{cases} 0.3 & \text{for } X = 0; \\ 0.7 & \text{for } X = 1; \\ 0 & \text{elsewhere.} \end{cases}$$

Find $\mu'_r, r = 1, 2, \ldots$.

4-46. For the discrete probability distribution appearing in Exercise 4-25, find:

(a) μ'_3

(b) μ_3

(c) α_3

(d) α_4

(e) $\gamma_{0.5}$

(f) $\gamma_{0.3}$

4-47. For the probability density function

$$f(x) = \begin{cases} \frac{1}{10}e^{-x/10}, & x > 0; \\ 0 & \text{elsewhere} \end{cases}$$

find:

(a) $E(X)$

(b) $V(X)$

(c) μ'_3

(d) μ_3

(e) α_3

(f) α_4

(g) $\gamma_{0.50}$

(h) $\gamma_{0.75}$

4-48. Given that the random variable X has the probability density function

$$f(x) = \begin{cases} \frac{1}{5}, & 5 < x < 10; \\ 0 & \text{elsewhere} \end{cases}$$

find:

(a) $F(t)$

(b) $E(X)$

(c) $V(X)$

(d) the median

(e) the interquartile range

4-49. Given the probability density function

$$f(x) = \begin{cases} \frac{1}{2}x^2 e^{-x}, & 0 < x < +\infty; \\ 0 & \text{elsewhere,} \end{cases}$$

for the random variable X, find the mode of X.

4-50. If a random variable X has the probability density function

$$f(x) = \begin{cases} 4x^3, & 0 < x < 1; \\ 0 & \text{elsewhere,} \end{cases}$$

find its median. Also determine $\gamma_{0.3}$.

4-51. Suppose a continuous random variable X has a probability density function of the form

$$f(x) = \begin{cases} \frac{1}{5} & 20 < x < 25; \\ 0 & \text{elsewhere.} \end{cases}$$

Find μ'_r. Then determine μ'_1, μ'_2, μ'_3, and μ'_4. Also find μ_2, μ_3, and μ_4. Finally, determine α_3 and α_4.

4-52. Let X be a continuous random variable with probability mass function $f(x)$. If a random variable $Y = a + bX, b \neq 0$, then the probability density function of Y is

$$g(y) = \frac{1}{|b|} f\left(\frac{y-a}{b}\right).$$

Verify this result (two separate ways) when

$$f(x) = \begin{cases} 3(1-x)^2, & 0 < \infty < 1; \\ 0 & \text{elsewhere} \end{cases}$$

and $Y = 4 + 2X$. What if $Y = 4 - 2X$? (Hint: One approach is to use this expression for $g(y)$ directly. Another is to determine the cumulative distribution function $G(f)$ and then use $\frac{dG}{dt} = g(f)$.)

4-53. For X a continuous random variable, use the transformation employed in the preceding problem to verify that $E(cX) = cE(X)$, c a constant. (Hint: Let $f(y)$ be the probability density function for $Y = cX$.)

4-54. Let X be a discrete random variable with probability mass function $f(X)$. If a random variable $Y = a + bX, b \neq 0$, then the probability mass function of Y is

$$g(y) = f\left(\frac{y-a}{b}\right).$$

Verify this result when

$$f(x) = \begin{cases} p^X(1-p)^{1-X}, & X = 0 \text{ or } 1, 0 \leq p \leq 1; \\ 0 & \text{elsewhere} \end{cases}$$

and $Y = 1 + 3X$.

4-55. Suppose X is a continuous random variable with probability density function

$$f(x) = \begin{cases} \frac{1}{2}x, & 0 < x < 2; \\ 0 & \text{elsewhere.} \end{cases}$$

Find $m_X(t), \mu_1', \mu_2'$, and σ^2.

4-56. Let the continuous random variable X have a probability density function of the form

$$f(x) = \begin{cases} 100e^{-100x}, & x > 0; \\ 0 & \text{elsewhere.} \end{cases}$$

Find $m_X(t)$. What is the restriction on t? Also determine $m_X^{(1)}(0), m_X^{(2)}(0)$, and σ^2. Next, use (4.43) to determine $m_{X-\mu}(t)$ along with $m_{X-\mu}^{(1)}(0)$, $m_{X-\mu}^{(2)}(1)$, and σ.

4-57. Suppose the continuous random variable X has a probability density function of the exponential variety or

$$f(x) = \begin{cases} \frac{1}{\lambda}e^{-x/\lambda}, & x \geq 0, \lambda > 0; \\ 0 & \text{elsewhere.} \end{cases}$$

Verify that the moment-generating function of X is $m_X(t) = (1 - \lambda t)^{-1}$, $t < \lambda^{-1}$.

4-58. Let the continuous random variable X be uniformly distributed with probability density function

$$f(x) = \begin{cases} \frac{1}{\beta - \alpha}x, & \alpha < x < \beta; \\ 0 & \text{elsewhere.} \end{cases}$$

Show that the moment-generating function for X is

$$m_X(t) = \frac{e^{t\beta} - e^{t\alpha}}{t(\beta - \alpha)}, t \neq 0.$$

4-59. Let X be a discrete random variable with probability mass function

$$f(X) = \begin{cases} p(1 - p)^{X-1}, & X = 1, 2, \ldots; \\ 0 & \text{elsewhere.} \end{cases}$$

Determine $m_X(t)$.

4-60. For X a continuous random variable with probability density function

$$f(x) = \begin{cases} \frac{1}{3}, & -1 < x < 2; \\ 0 & \text{elsewhere,} \end{cases}$$

demonstrate that

$$m_X(t) = \frac{e^{2t} - e^{-t}}{3t}, \quad t \neq 0.$$

4-61. Suppose a fair coin is tossed twice. Let the random variable X depict the number of heads obtained. Determine the probability mass function for X. Also, find the moment-generating function for X and use it to obtain μ and σ.

4-62. For the discrete random variable X defined in the preceding exercise, find X's probability generating function. Then determine $\phi_X(0), \phi_X^{(1)}(0), \phi_X^{(2)}(0)$ along with $\phi_X^{(1)}(1)$ and $\phi_X^{(2)}(1)$.

4-63. For a continuous random variable X with probability density function

$$f(x) = \begin{cases} 10e^{-10x}, & x > 0; \\ 0 & \text{elsewhere,} \end{cases}$$

find X's moment generating function. Verify that $\mu = \sigma = \frac{1}{10}$. Then find the moment-generating function of $X - \mu$. Demonstrate that $m_{X-\mu}^{(1)}(0) = \mu_1 = 0, m_{X-\mu}^{(2)}(0) = \mu_2$.

4-64. For each of the following moment-generating functions, find the associated probability density function:

(a) $m_X(t) = \frac{e^{7t} - e^{3t}}{4t}, t \neq 0$

(b) $m_X(t) = (1 - 0.8t)^{-1}, t \leq 1.25$

(c) $m_X(t) = \frac{2}{2-t}, t < 2$

4-65. For the probability mass function utilized in Example 4.8.1, determine $\phi_X(e^t)$. What is its interpretation? What restriction is placed on t? Use $\phi_X(e^t)$ to determine the mean of X.

4-66. Suppose X is a continuous random variable with mean μ and standard deviation σ. Let $Z = (X - \mu)/\sigma$. Demonstrate that:

(a) $E(Z) = 0$

(b) $V(Z) = 1$

4-67. Consider the problem of obtaining the probability distribution of a random variable Y from information about the probability distribution of a random variable X, where $y = g(x)$ is a functional relationship between the values X and Y. To perform the indicated *change of variable* or transformation we shall utilize the following

> **THEOREM.** Let the probability density function of the random variable X be given by $f(x)$ and let the function $y = g(x)$ define a one-to-one transformation between X and Y. (A one-to-one transformation implies that g is either an increasing or decreasing function for all admissible x.) In addition, let the unique inverse transformation of g be denoted as $x = w(y)$ and let $dx/dy = w'(y)$ be continuous and not vanish for all admissible y's. Then the probability density function of Y is given by
>
> $$h(y) = |dw(y)/dy| f(w(y)), \quad dw(y)/dy \neq 0.$$

Here $|dw(y)/dy|$ denotes the absolute value of $dw(y)/dy$. If $f(x) = e^{-x}$, $x \geq 0$, and $y = g(x) = 3x$, find $h(y)$.

Bivariate Probability Distributions

5.1 Bivariate Random Variables

The previous chapter focused on the determination of probabilities involving a single discrete or continuous random variable X. Such models reflect only a single characteristic of some random phenomenon. However, if a second attribute of a random phenomenon is also of interest and is observed along with values of X, then it can be represented as a separate set of values of an additional random variable Y. Hence the two random variables X and Y are jointly measured when defining the outcomes of a random experiment. In this regard, we form what is termed a *bivariate probability distribution*. Such distributions can be discrete or continuous. Moreover, they can be depicted as a listing or table, a graph, or written as a specific function or equation.

5.2 Discrete Bivariate Probability Distributions

Suppose (X, Y) is a pair of real-valued functions defined on a sample space S. The pair (X, Y) is a *bivariate random variable* if both X and Y map elements in S into real numbers. Hence (X, Y) defines the possible outcomes of some random experiment. In addition, if the range of (X, Y) is a discrete ordered set of points $R_{x,y} = \{(X_i, Y_j), \ i = 1, \ldots, n; \ j = 1, \ldots, m\}$ in the Cartesian plane, then (X, Y) is termed a *discrete bivariate random variable*. Clearly X and Y must be discrete random variables when considered individually. (Note that although a discrete random variable can actually assume a countable infinite set of values, we have assumed, for convenience, that both X and Y take on only a finite number of values.)

Given a random variable $X : X_1, X_2, \ldots, X_n$ (with $X_1 < X_2 < \cdots < X_n$) and a second random variable $Y : Y_1, Y_2, \ldots, Y_m$ (with $Y_1 < Y_2 < \cdots < Y_m$), let us

147

denote the joint probability that $X = X_i$ and $Y = Y_j$ as $P(X = X_i, Y = Y_j) = f(X_i, Y_j), i = 1, \ldots, n; j = 1, \ldots, m$. Here $f(X_i, Y_j)$ is termed the probability mass at the point (X_i, Y_j). In general, a function $f(X, Y)$, which assigns a probability $f(X_i, Y_j)$ within the ranges $R_x = \{X_i, i = 1, \ldots, n\}$ and $R_y = \{Y_j, j = 1, \ldots, m\}$ of the discrete random variables X and Y, respectively, is called a *bivariate probability mass function* if:

$$\text{(a) } f(X_i, Y_j) \geq 0 \quad \text{for all } i, j; \text{ and}$$

$$\text{(b) } \sum_i \sum_j f(X_i, Y_j) = 1. \tag{5.1}$$

To completely specify a discrete joint or bivariate probability distribution it is sufficient to list the set of probabilities $f(X_i, Y_j), i = 1, \ldots, n; j = 1, \ldots, m$. Hence the set of nm events (X_i, Y_j) together with their associated probabilities $f(X_i, Y_j), i = 1, \ldots, n; j = 1, \ldots, m$, constitutes the *discrete bivariate probability distribution* of the random variables X and Y.

Example 5.2.1   Let us assume that a total of 2405 students currently are enrolled at a local college. For each student, two attributes immediately present themselves: gender and year or class. Let us represent *class* by a random variable X and *gender* by the random variable Y; that is,

X (Class)	Y (Gender)
X_1—freshman	Y_1—male
X_2—sophomore	Y_2—female
X_3—junior	
X_4—senior	
X_5—other	

The joint classification of the values of these random variables, expressed in terms of *number of students*, is depicted in the two-way table, Table 5.1.

Table 5.1

Classification of Students by Class and Gender

X \\ Y	Y_1	Y_2	
X_1	300	200	500
X_2	280	215	495
X_3	275	215	490
X_4	250	230	480
X_5	200	240	440
	1305	1100	2405

To convert Table 5.1 into a discrete bivariate probability distribution, let us divide each item therein by 2405 (assuming, of course, that the individual outcomes are equiprobable). Hence we ultimately obtain Table 5.2.

Table 5.2

Bivariate Probability Mass Function $f(X, Y)$		
X \ Y	Y_1	Y_2	
X_1	0.125	0.083	0.208
X_2	0.116	0.089	0.205
X_3	0.115	0.089	0.204
X_4	0.104	0.095	0.200
X_5	0.083	0.100	0.183
	0.543	0.456	1.000

As this table reveals, the probability that a randomly selected student is a *male and a junior* is $P(X = X_3, Y = Y_1) = f(X_3, Y_1) = 0.115$. Other combinations of events for X and Y and their joint probabilities are interpreted in a similar fashion. ∎

Given a bivariate probability mass function $f(X, Y)$, the joint probability that $X \leq X_r$ and $Y \leq Y_s$ is provided by the *bivariate cumulative distribution function*

$$F(X_r, Y_s) = P(X \leq X_r, Y \leq Y_s) = \sum_{i \leq r} \sum_{j \leq s} f(X_i, Y_j). \tag{5.2}$$

Example 5.2.2 Given the bivariate probability mass function provided by Table 5.3, we may easily determine that

$$F(X_3, Y_2) = P(X \leq X_3, Y \leq Y_2)$$

$$= \sum_{i \leq 3} \sum_{j \leq 2} f(X_i, Y_j)$$

$$= \sum_{i \leq 3} [f(X_i, Y_1) + f(X_i, Y_2)]$$

$$= f(X_1, Y_1) + f(X_1, Y_2) + f(X_2, Y_1)$$

$$+ f(X_2, Y_2) + f(X_3, Y_1) + f(X_3, Y_2)$$

$$= 0.07 + 0.11 + 0.05 + 0.09 + 0.10 + 0.01 = 0.43. \blacksquare$$

Bivariate probability distributions tell us something about the joint behavior of the random variables X and Y. However, if we want information about each of these two random variables taken individually, then we need to examine

Table 5.3

Bivariate Probability Mass Function $f(X,Y)$			
X ＼ Y	Y_1	Y_2	Y_3	$g(X_i)$
---	---	---	---	---
X_1	0.07	0.11	0.03	0.21
X_2	0.05	0.09	0.16	0.30
X_3	0.10	0.01	0.02	0.13
X_4	0.11	0.18	0.07	0.36
$h(Y_j)$	0.33	0.39	0.28	1.00

their marginal probabilities. In this regard, for X and Y discrete random variables, the summation of the bivariate probability mass function $f(X,Y)$ over all $Y \in R_y$ yields the univariate probability mass function $g(X)$ called the marginal probability mass function of X

$$g(X) = \sum_j f(X, Y_j). \qquad (5.3)$$

Similarly, taking the summation of $f(X,Y)$ over all $X \in R_x$ produces the univariate *marginal probability mass function of Y*

$$h(Y) = \sum_i f(X_i, Y). \qquad (5.4)$$

Using (5.3), we may readily determine the (marginal) probability that $X = X_r$ irrespective of the value of Y as

$$P(X = X_r) = g(X_r) = \sum_j f(X_r, Y_j). \qquad (5.5)$$

And from (5.4), the marginal probability that $Y = Y_s$ irrespective of the value of X is

$$P(Y = Y_s) = h(Y_s) = \sum_i f(X_i, Y_s). \qquad (5.6)$$

Moreover, once the marginal probability mass functions (5.3) and (5.4) are available, we can also find probabilities such as

$$P(X_a \leq X \leq X_b) = \sum_{a \leq i \leq b} g(X_i); \qquad (5.7)$$

$$P(Y_c \leq Y \leq Y_d) = \sum_{c \leq j \leq d} h(Y_j). \qquad (5.8)$$

Example 5.2.3 Using the bivariate probability mass function depicted in Table 5.3 we can easily find, via (5.5),

$$P(X = X_2) = g(X_2) = \sum_j f(X_2, Y_j)$$

$$= f(X_2, Y_1) + f(X_2, Y_2) + f(X_2, Y_3)$$

$$= 0.05 + 0.09 + 0.16 = 0.30;$$

from (5.6),

$$P(Y = Y_1) = h(Y_1) = \sum_i f(X_i, Y_1)$$

$$= f(X_1, Y_1) + f(X_2, Y_1) + f(X_3, Y_1) + f(X_4, Y_1)$$

$$= 0.07 + 0.05 + 0.10 + 0.11 = 0.33;$$

from (5.7),

$$P(X_2 \leq X \leq X_4) = \sum_{2 \leq i \leq 4} g(X_i) = g(X_2) + g(X_3) + g(X_4)$$

$$= 0.30 + 0.13 + 0.36 = 0.79;$$

and from (5.8),

$$P(Y_2 \leq Y \leq Y_3) = \sum_{2 \leq j \leq 3} h(Y_j) = h(Y_2) + h(Y_3) = 0.39 + 0.28 = 0.67.$$

Based upon these calculations it should be evident that the marginal probabilities of the X_j's are the row totals and the marginal probabilities of the Y_j's are the column totals in Table 5.3. Hence we may label the set of row totals as $g(X_i)$ and the set of column totals as $h(Y_j)$. ■

Using (5.3) (or (5.5)) we can define the set of all X_i's together with their marginal probabilities $g(X_i)$, $i = 1, \ldots, n$, as the *marginal probability distribution of X*. This distribution depicts the probability distribution of a single discrete random variable X when the levels of the random variable Y are ignored. Moreover, we require that: (a) $g(X_i) \geq 0$ for all i; and (b) $\Sigma_i\ g(X_i) = 1$. (Table 5.4a depicts X's marginal probability distribution derived from Table 5.3.) Similarly, from (5.4) (or (5.6)), the set of all Y_j's together with their set of marginal probabilities $h(Y_j)$, $j = 1, \ldots, m$, is the *marginal probability distribution of Y*—the probability distribution of the single random variable Y when the levels of the random variable X are ignored. Here, too: (a) $h(Y_j) \geq 0$ for all j; and (b) $\Sigma_j\ h(Y_j) = 1$. (Table 5.4b portrays Y's marginal probability distribution calculated from Table 5.3.)

Table 5.4

a.		b.	
Marginal Probability Distribution of X		**Marginal Probability Distribution of Y**	
X	$g(X)$	Y	$h(Y)$
X_1	$g(X_1) = 0.21$	Y_1	$h(Y_1) = 0.33$
X_2	$g(X_2) = 0.30$	Y_2	$h(Y_2) = 0.39$
X_3	$g(X_3) = 0.13$	Y_3	$h(Y_3) = \underline{0.28}$
X_4	$g(X_4) = \underline{0.36}$		1.00
	1.00		

Next, let X and Y be discrete random variables with bivariate probability mass function $f(X, Y)$ and marginal probability mass functions $g(X)$ and $h(Y)$, respectively. Given these relationships, we may now pose the following problem. What if, on some trial of a random experiment, we specify the level of Y but we allow the values of X to be determined by chance? Then we may determine, for instance, the probability that $X = X_i$ given $Y = Y_j$ for fixed j. In like fashion, we may be interested in finding the probability that $Y = Y_j$ given that $X = X_i$ when i is held fixed. In this regard, if (X, Y) is any point at which $h(Y) > 0$, then the *conditional probability mass function of X given Y* is

$$g(X|Y) = \frac{f(X, Y)}{h(Y)}. \tag{5.9}$$

Here $g(X|Y)$ is a function of X alone and, for each X_i given $Y = Y_s$, the probability that $X = X_i$ given $Y = Y_s$ is

$$P(X = X_i|Y = Y_s) = g(X_i|Y_s) = \frac{f(X_i, Y_s)}{h(Y_s)}, \ h(Y_s) > 0, \ i = 1,\ldots,n. \tag{5.10}$$

Similarly, at any point (X, Y) at which $g(X) > 0$, the *conditional probability mass function of Y given X* (a function of Y alone) is

$$h(Y|X) = \frac{f(X, Y)}{g(X)}. \tag{5.11}$$

So for any Y_j given $X = X_r$, the probability $Y = Y_j$ given $X = X_r$ is

$$P(Y = Y_j|X = X_r) = h(Y_j|X_r) = \frac{f(X_r, Y_j)}{g(X_r)}, \ g(X_r) > 0, \ j = 1,\ldots,m. \tag{5.12}$$

Example 5.2.4   Using equations (5.10) and (5.12) we may determine, from Table 5.3, that

$$P(X = X_3 | Y = Y_1) = g(X_3 | Y_1) = \frac{f(X_3, Y_1)}{h(Y_1)} = \frac{0.10}{0.33} = 0.303;$$

$$P(Y = Y_2 | X = X_1) = h(Y_2 | X_1) = \frac{f(X_1, Y_2)}{g(X_1)} = \frac{0.11}{0.21} = 0.523,$$

and so on.   ■

Note that if we solve for $f(X, Y)$ from both (5.9) and (5.11), then we may state the *multiplication theorem for probability mass functions*:

$$f(X, Y) = g(X) \cdot h(Y | X) = g(X | Y) \cdot h(Y). \tag{5.13}$$

From (5.10) we can define the set of all X_i's together with their conditional probabilities $g(X_i | Y_j)$, $i = 1, \dots, n$, as the *conditional probability distribution of X given $Y = Y_j$*. This distribution depicts the probability distribution of a single discrete random variable X when the level of Y is fixed at Y_j. For this distribution it must be true that: (a) $g(X_i | Y_j) \geq 0$ for all i; and (b) $\sum_i g(X_i | Y_j) = 1$. And from (5.12), the set of all Y_j's along with their conditional probabilities $h(Y_j | X_i), j = 1, \dots, m$, represents the *conditional probability distribution of Y given $X = X_i$*. It portrays the univariate probability distribution of the random variable Y when the level of X is set at X_i. It is also the case that: (a) $h(Y_j | X_i) \geq 0$ for all j; and (b) $\sum_j h(Y_j | X_i) = 1$.

Example 5.2.5   Using the bivariate probability mass function given in Table 5.3, we may easily derive the conditional probability distribution of X given $Y = Y_2$ (see Table 5.5a) as well as the conditional probability distribution of Y given $X = X_4$ (see Table 5.5b).   ■

Let X and Y be discrete random variables with bivariate probability mass function $f(X, Y)$ and marginal probability mass functions $g(X)$ and $h(Y)$, respectively. Then the *random variable X is independent of the random variable Y* if

$$g(X | Y) = g(X) \tag{5.14}$$

for all values of X and Y for which both of these functions are defined. Similarly, the *random variable Y is independent of the random variable X* if

$$h(Y | X) = h(Y) \tag{5.15}$$

for all X and Y at which these functions exist. It should be evident that if X is independent of Y, then Y is independent of X; that is, X and Y are *mutually*

Table 5.5

a.	b.
Conditional Probability Distribution of X Given $Y = Y_2$ $\left(g(X_i|Y_2) = \frac{f(X_i,Y_2)}{h(Y_2)}, \ i = 1,2,3,4\right)$	**Conditional Probability Distribution of Y Given** $X = X_4$ $\left(h(Y_j|X_4) = \frac{f(X_4,Y_j)}{g(X_4)}, \ j = 1,2,3\right)$

X	$g(X|Y_2)$	Y	$h(Y|X_4)$
X_1	$g(X_1|Y_2) = 0.282$	Y_1	$h(Y_1|X_4) = 0.306$
X_2	$g(X_2|Y_2) = 0.231$	Y_2	$h(Y_2|X_4) = 0.500$
X_3	$g(X_3|Y_2) = 0.025$	Y_3	$h(Y_3|X_4) = \underline{0.194}$
X_4	$g(X_4|Y_2) = \underline{0.462}$		1.000
	1.000		

independent random variables. Let (5.14) (or (5.15)) hold. Then the multiplication theorem for probability mass functions (5.13) renders

$$f(X, Y) = g(X) \cdot h(Y). \tag{5.16}$$

As this expression reveals, X and Y are independent random variables if and only if their joint probability mass function $f(X, Y)$ can be written as the product of their individual marginal probability mass functions $g(X)$ and $h(Y)$, respectively. Hence, under independence,

$$P(X = X_i, Y = Y_j) = P(X = X_i) \cdot P(Y = Y_j) \tag{5.16.1}$$

for all points (X,Y). In fact, the equalities in (5.16) and (5.16.1) must hold for all possible pairs of X and Y values. If inequality obtains for at least one point (X, Y), then the random variables X and Y are said to be *dependent.*

5.3 Continuous Bivariate Probability Distributions

Let the two-dimensional sample space S consist of a class of events representable by all open and closed rectangles. Hence the random variables X and Y defined on S are continuous since they can assume any values within some event $A = \{(X, Y)|a \leq X \leq b, \ c \leq Y \leq d\} \subseteq S$. In addition, let P be a (joint) *probability measure* or set function that associates with each $A \subseteq S$ a number $P(A)$. In general, a function $f(X, Y)$, which defines the probability measure

$$P(A) = P(a \leq X \leq b, \ c \leq Y \leq d) = \int_a^b \int_c^d f(x,y) \, dy \, dx, \tag{5.17}$$

is called a *bivariate probability density function* if

(a) $f(x, y) \geq 0$ for all real x and y such that $-\infty < x, y < +\infty$
and $f(x, y) > 0$ for $(x, y) \in A$; and

(5.18)

(b) $\displaystyle\int_{-\infty}^{+\infty} \int_{-\infty}^{+\infty} f(x, y) \, dy \, dx = 1.$

So if the collection of all open and closed rectangles in two dimensions constitutes the sample space S and if $P(A)$ is determined by the bivariate probability density $f(x, y)$ as in (5.17), then the random variables X and Y follow a *continuous bivariate probability distribution*.

Example 5.3.1 Given the probability density function

$$f(x, y) = \begin{cases} x + y, & 0 < x < 1 \text{ and } 0 < y < 1; \\ 0 & \text{elsewhere}, \end{cases}$$

find $P(\frac{1}{3} < X < \frac{1}{2}, \frac{1}{2} < Y < 1)$. From (5.17),

$$P\left(\frac{1}{3} < X < \frac{1}{2}, \frac{1}{2} < Y < 1\right) = \int_{1/3}^{1/2} \int_{1/2}^{1} (x + y) \, dy \, dx$$

$$= \int_{1/3}^{1/2} \left\{ \left(xy + \frac{1}{2}y^2 \right) \right]_{1/2}^{1} \right\} dx = \int_{1/3}^{1/2} \left(\frac{1}{2}x + \frac{3}{8} \right) dx$$

$$= \left(\frac{1}{4}x^2 + \frac{3}{8}x \right) \Big]_{1/3}^{1/2} = \frac{7}{72}.$$

Here $\frac{7}{72}$ is the amount of volume between the three-dimensional surface $f(x, y)$ and the rectangle $A = \{(X, Y) | \frac{1}{3} < X < \frac{1}{2}, \frac{1}{2} < Y < 1\}$ within the XY-plane. ∎

It must be remembered that event A involves a rectangle defined by letting both X and Y vary over intervals. Probabilities involving events containing a single point are always zero; for example, $P(X = a, Y = c) = 0$.

Let us now consider the circumstances under which a function $g(x, y)$ can be used to obtain a bivariate probability density function for the continuous random variables X and Y. Specifically, any function $g(x, y)$ for which:

(a) $g(x, y) \geq 0, \quad -\infty < x, y < +\infty$;

(b) $\displaystyle\int_{-\infty}^{+\infty} \int_{-\infty}^{+\infty} g(x, y) \, dy \, dx$ exists; and

(c) $\displaystyle\int_{-\infty}^{+\infty} \int_{-\infty}^{+\infty} g(x, y) \, dy \, dx = K \neq 0$

may be used to determine a bivariate probability density function of the form

$$f(x, y) = kg(x, y),$$

where $k = \frac{1}{K}$ is the *normalizing constant*, which ensures that

$$\int_{-\infty}^{+\infty} \int_{-\infty}^{+\infty} f(x, y) \, dy \, dx = k \int_{-\infty}^{+\infty} \int_{-\infty}^{+\infty} g(x, y) \, dy \, dx$$

$$= \frac{1}{K} \int_{-\infty}^{+\infty} \int_{-\infty}^{+\infty} g(x, y) \, dy \, dx = 1$$

per (5.18b).

Example 5.3.2 Can the function

$$g(x, y) = \begin{cases} 12xy, & 0 < x < 1 \text{ and } 0 < y < 1; \\ 0 & \text{elsewhere} \end{cases}$$

be used to generate a bivariate probability density function? It is easily demonstrated that $g(x, y) \geq 0$ while

$$\int_{-\infty}^{+\infty} \int_{-\infty}^{+\infty} g(x, y) \, dy \, dx = 12 \int_0^1 \int_0^1 xy \, dy \, dx$$

$$= 12 \int_0^1 \left\{ \frac{1}{2}xy^2 \Big]_0^1 \right\} dx = 6 \int_0^1 x \, dx = 3x^2 \Big]_0^1 = 3 = K.$$

Then $k = 1/K = 1/3$ and thus

$$f(x, y) = kg(x, y) = \begin{cases} 4xy, & 0 < x < 1 \text{ and } 0 < y < 1; \\ 0 & \text{elsewhere.} \end{cases}$$

Next, does

$$f(x, y) = \begin{cases} e^{-x-y}, & 0 < x, y < +\infty; \\ 0 & \text{elsewhere} \end{cases}$$

represent a probability density function? Since $f(x, y) \geq 0$ and

$$\int_{-\infty}^{+\infty} \int_{-\infty}^{+\infty} f(x, y)\, dy\, dx = \int_{0}^{+\infty} \int_{0}^{+\infty} e^{-x-y} dy\, dx$$

$$= \int_{0}^{+\infty} \left\{ \lim_{a \to +\infty} \int_{0}^{a} e^{-x-y}\, dy \right\} dx$$

$$= \int_{0}^{+\infty} \left\{ e^{-x} \right\} dx = \lim_{a \to +\infty} \int_{0}^{a} e^{-x} dx = 1,$$

the answer is yes.   ∎

Whereas (5.17) defines the probability that the continuous random variables X and Y with bivariate probability density $f(x, y)$ assume a value within the rectangle $A = \{(X, Y)|a \leq X \leq b, \ c \leq Y \leq d\}$, we may introduce a related function (derived from f) that yields the (joint) probability that the random variable X assumes a value less than or equal to some number t and the random variable Y takes on a value less than or equal to the number s. This is the *bivariate cumulative distribution function* for continuous random variables X and Y or

$$F(t, s) = P(X \leq t, Y \leq s) = \int_{-\infty}^{t} \int_{-\infty}^{s} f(x, y)\, dy\, dx. \tag{5.19}$$

$F(t, s)$ is a continuous function of t and s and, at every point of continuity of $f(x, y)$, $\frac{\partial^2 F(t,s)}{\partial s \partial t} = f(t, s)$. Hence knowing the bivariate probability density function $f(x, y)$ enables us to determine its bivariate cumulative distribution function $F(t, s)$ via (5.19). Conversely, if $F(t, s)$ is known, we can determine the probability density $f(t, s)$ at each of its points of continuity by finding $\frac{\partial^2 F}{\partial s \partial t}$.

Example 5.3.3   From the bivariate probability density function

$$f(x, y) = \begin{cases} 4xy, & 0 < x < 1 \text{ and } 0 < y < 1; \\ 0 & \text{elsewhere} \end{cases}$$

we have

$$F(t, s) = 4 \int_{-\infty}^{t} \int_{-\infty}^{s} xy\, dy\, dx = 4 \int_{0}^{t} \int_{0}^{s} xy\, dy\, dx$$

$$= 4 \int_{0}^{t} \left\{ \frac{1}{2} xy^2 \right\}_{0}^{s} dx = 2s^2 \int_{0}^{t} x\, dx = s^2 t^2$$

or

$$F(t,s) = \begin{cases} 0, & t < 0 & \text{and } s < 0; \\ s^2 t^2, & 0 \le t \le 1 & \text{and } 0 \le s \le 1; \\ 1, & t > 1 & \text{and } s > 1. \end{cases}$$

And, as required, $\frac{\partial^2 F}{\partial s \partial t} = 4st$. ■

Example 5.3.4 Given a bivariate cumulative distribution function $F(t,s)$ we can readily determine the joint probability $P(A) = P(a < X \le b, \, c < Y \le d)$ as

$$P(A) = F(b,d) - F(a,d) - F(b,c) + F(a,c) \tag{5.20}$$

See Figure 5.1a.

For instance, using the preceding bivariate probability density function $f(x,y)$ and its associated cumulative distribution function $F(t,s)$, let us find $P(A)$, where $A = \{(X,Y)|\frac{1}{2} < X \le 1, 0 < Y \le \frac{1}{3}\}$. Calculating $P(A)$ directly using the probability density function yields

$$P(A) = 4 \int_{1/2}^{1} \int_{0}^{1/3} xy \, dy \, dx = \frac{1}{12}.$$

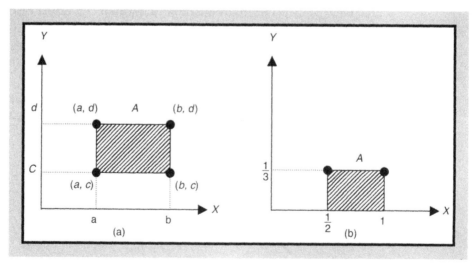

Figure 5.1 (a) Determining the joint probability $P(A) = P(a < X \le b, c < Y \le d)$; (b) Determining the joint probability $P(A) = P(\frac{1}{2} < X \le 1, 0 < Y \le \frac{1}{3})$.

However, according to (5.20),

$$P(A) = F\left(1, \frac{1}{3}\right) - F\left(\frac{1}{2}, \frac{1}{3}\right) - F(1, 0) + F\left(\frac{1}{2}, 0\right)$$

$$= (1)\left(\frac{1}{9}\right) - \left(\frac{1}{4}\right)\left(\frac{1}{9}\right) - 0 + 0 = \frac{1}{12}$$

as required (see Figure 5.1b).   ■

Given the continuous random variables X and Y with bivariate probability density $f(x, y)$, the *marginal probability density function of X* is

$$g(x) = \int_{-\infty}^{+\infty} f(x, y) \, dy \qquad (5.21)$$

and is a function of x alone with: (a) $g(x) \geq 0$; and (b) $\int_{-\infty}^{+\infty} g(x) \, dx = 1$. In addition, the *marginal probability density function of Y* is

$$h(y) = \int_{-\infty}^{+\infty} f(x, y) \, dx \qquad (5.22)$$

and is a function of y alone with: (a) $h(y) \geq 0$; and (b) $\int_{-\infty}^{+\infty} h(y) \, dy = 1$.

Given the joint cumulative distribution function $F(t, s)$ from (5.19), the *marginal cumulative distribution functions of X and Y* are

$$P(X \leq t) = F(t) = \int_{-\infty}^{t} \int_{-\infty}^{+\infty} f(x, y) \, dy \, dx = \int_{-\infty}^{t} g(x) \, dx$$

and

$$P(Y \leq s) = F(s) = \int_{-\infty}^{s} \int_{-\infty}^{+\infty} f(x, y) \, dx \, dy = \int_{-\infty}^{s} h(y) \, dy$$

respectively, where $g(x)$ is the marginal probability density of X and $h(y)$ is the marginal probability density of Y.

If $f(x, y)$ is the bivariate probability density function for the random variables X and Y and the marginal probability density functions $g(x)$ and $h(y)$ are known, then we may define, for Y fixed at y, the *conditional probability density function for X given Y* as

$$g(x|y) = \frac{f(x, y)}{h(y)}, h(y) > 0; \qquad (5.23)$$

and for X fixed at x, the *conditional probability density function for Y given X* is

$$h(y|x) = \frac{f(x,y)}{g(x)}, g(x) > 0. \tag{5.24}$$

Here $g(x|y)$ is a function of x alone with properties: $g(x|y) > 0$; and (b) $\int_{-\infty}^{+\infty} g(x|y)\, dx = 1$. Similarly, $h(y|x)$ depends upon y alone and: (a) $h(y|x) \geq 0$; and (b) $\int_{-\infty}^{+\infty} h(y|x)\, dy = 1$.

How may we interpret (5.23) and (5.24)? If y is the observed level of Y but the value of X is not known, then the function $g(x|y)$ gives the probability density of X along a fixed line $Y = y$ in the XY-plane. And if X is observed at the level x but the Y value is unknown, then the function $h(y|x)$ yields the probability density of Y along a fixed line $X = x$ in the XY-plane.

Once the conditional probabilities $g(x|y)$ and $h(y|x)$ are determined, we may calculate probabilities such as

$$P(a \leq X \leq b|y) = \int_a^b g(x|y)\, dx \tag{5.25}$$

and

$$P(c \leq Y \leq d|x) = \int_c^d h(y|x)\, dy. \tag{5.26}$$

In this regard, (5.25) addresses the question: given that Y is set at y, what is the probability that X varies between a and b inclusive? Equation (5.26) is interpreted in a similar fashion.

Example 5.3.5   For the bivariate probability density function

$$f(x,y) = \begin{cases} x + y, & 0 < x < 1 \text{ and } 0 < y < 1; \\ 0 & \text{elsewhere} \end{cases}$$

we have, from (5.21) and (5.22), respectively,

$$g(x) = \int_{-\infty}^{+\infty} (x+y)\, dy = \int_0^1 (x+y)\, dy = \left(xy + \frac{1}{2}y^2\right)\Big]_0^1 = x + \frac{1}{2};$$

$$h(y) = \int_{-\infty}^{+\infty} (x+y)\, dx = \int_0^1 (x+y)\, dx = \left(\frac{1}{2}x^2 + xy\right)\Big]_0^1 = \frac{1}{2} + y.$$

And from (5.23) and (5.24), respectively,

$$g(x|y) = \frac{x+y}{\frac{1}{2}+y}; \quad h(y|x) = \frac{x+y}{x+\frac{1}{2}}.$$

Then for, say, $X = x = 2$, $h(y|2) = \frac{4}{5} + \frac{2}{5}y$ and thus

$$P\left(\frac{1}{3} \leq Y \leq \frac{1}{2} \middle| x = 2\right) = \int_{1/3}^{1/2} \left(\frac{4}{5} + \frac{2}{5}y\right) dy = \left(\frac{4}{5}y + \frac{1}{5}y^2\right)\Bigg]_{1/3}^{1/2} = \frac{29}{180}. \quad \blacksquare$$

Let X and Y be continuous random variables with bivariate probability density function $f(x,y)$ and marginal probability densities $g(x)$ and $h(y)$, respectively. Then the *random variable X is independent of the random variable Y* if

$$g(x|y) = g(x) \tag{5.27}$$

for all values of X and Y for which both of these functions exist. In like fashion we may state that the *random variable Y is independent of the random variable X* if

$$h(y|x) = h(y) \tag{5.28}$$

for all X and Y for which both of these functions are defined. Using either (5.27) or (5.28) we have

$$f(x,y) = g(x) \cdot h(y); \tag{5.29}$$

that is, *X and Y are independent random variables* if and only if their joint probability density function $f(x,y)$ can be written as the product of their individual marginal probability densities $g(x)$ and $h(y)$, respectively. So under independence,

$$P(a \leq X \leq b, c \leq Y \leq d) = \int_a^b \int_c^d f(x,y)\, dy\, dx$$

$$= \int_a^b \int_c^d g(x)h(y)\, dy\, dx$$

$$= \left(\int_a^b g(x)dx\right)\left(\int_c^d h(y)dy\right)$$

$$= P(a \leq X \leq b) \cdot P(c \leq Y \leq d). \tag{5.29.1}$$

Example 5.3.6 The random variables X and Y with bivariate probability density function

$$f(x,y) = \begin{cases} x+y, & 0 < x < 1 \text{ and } 0 < y < 1; \\ 0 & \text{elsewhere} \end{cases}$$

are not (mutually) independent since the product of their marginal probability densities $g(x) = x + \frac{1}{2}$ and $h(y) = \frac{1}{2} + y$ does not equal $f(x, y)$. However, if

$$f(x, y) = \begin{cases} e^{-x-y}, & 0 < x < +\infty \text{ and } 0 < y < +\infty; \\ 0 & \text{elsewhere} \end{cases}$$

then clearly the random variables X and Y are independent since (5.29) holds. In fact, $f(x, y) = e^{-x-y} = (e^{-x})(e^{-y}) = g(x) \cdot h(y)$ as required. ■

5.4 Expectations and Moments of Bivariate Probability Distributions

In what follows it is assumed that all relevant expectations exist. For X and Y discrete random variables with bivariate probability mass function $f(X, Y)$, the *expectation of a function of X and Y, $\varphi(X, Y)$,* is

$$E[\varphi(X, Y)] = \sum_i \sum_j \varphi(X_i, Y_j) f(X_i, Y_j). \tag{5.30}$$

In view of this specification, we may define the rth *moment of the random variable X about zero* as

$$E(X^r) = \sum_i \sum_j X_i^r f(X_i, Y_j) = \sum_i X_i^r g(X_i) \tag{5.31}$$

and the sth *moment of the random variable Y about zero* is

$$E(Y^s) = \sum_i \sum_j Y_j^s f(X_i, Y_j) = \sum_j Y_j^s h(Y_j). \tag{5.32}$$

Furthermore, the rth *and s^{th} product or joint moment of X and Y about the origin* (0,0) is

$$E(X^r Y^s) = \sum_i \sum_j X_i^r Y_j^s f(X_i, Y_j) \tag{5.33}$$

and, for r and s nonnegative integers, the rth *and s^{th} product or joint moment of X and Y about the mean* is

$$E\left[(X - \mu_X)^r (Y - \mu_Y)^s\right] = \sum_i \sum_j (X_i - \mu_X)^r (Y_j - \mu_Y)^s f(X_i, Y_j), \tag{5.34}$$

where $\mu_X = E(X)$ and $\mu_Y = E(Y)$ are the means of the random variables X and Y, respectively.

If X and Y are continuous random variables with joint probability density function $f(x, y)$, then the *expectation of a function of X and Y, $\varphi(X, Y)$*, is

$$E\left[\varphi(X, Y)\right] = \int_{-\infty}^{+\infty} \int_{-\infty}^{+\infty} \varphi(x, y) f(x, y) \, dy \, dx \tag{5.35}$$

and thus the rth *moment of X about zero* is

$$E(X^r) = \int_{-\infty}^{+\infty} \int_{-\infty}^{+\infty} x^r f(x, y) \, dy \, dx = \int_{-\infty}^{+\infty} x^r g(x) \, dx \tag{5.36}$$

and the sth *moment of Y about zero* is

$$E(Y^s) = \int_{-\infty}^{+\infty} \int_{-\infty}^{+\infty} y^s f(x, y) \, dy \, dx = \int_{-\infty}^{+\infty} y^s h(y) \, dy. \tag{5.37}$$

In addition, the rth *and s^{th} product or joint moment of X and Y about the origin* is

$$E(X^r Y^s) = \int_{-\infty}^{+\infty} \int_{-\infty}^{+\infty} x^r y^s f(x, y) dy \, dx \tag{5.38}$$

and, for r and s nonnegative integers, the rth *and s^{th} product or joint moment of X and Y about the mean* is

$$E\left[(X - \mu_X)^r (Y - \mu_Y)^s\right] = \int_{-\infty}^{+\infty} \int_{-\infty}^{+\infty} (x - \mu_X)^r (y - \mu_Y)^s f(x, y) \, dy \, dx. \tag{5.39}$$

Example 5.4.1 If $\varphi(X, Y)$ is a linear function of the random variables X and Y or $\varphi = aX \pm bY$, where a and b are constants, then, for either the discrete or continuous case,

$$E(aX \pm bY) = aE(X) \pm bE(Y). \tag{5.40}$$

In general, if $X_i, i = 1, \ldots, k$, are discrete or continuous random variables and φ is expressible as a linear combination of the X_i or $\varphi = \sum_{i=1}^{k} a_i X_i, a_i$ constant for all i, then, since E is a linear operator,

$$E\left(\sum_{i=1}^{k} a_i X_i\right) = \sum_{i=1}^{k} a_i E(X_i). \quad \blacksquare \tag{5.40.1}$$

In (5.34) and (5.39) let $r = s = 1$. Then the expression

$$E\left[(X - \mu_X)(Y - \mu_Y)\right] = E(XY) - \mu_X\mu_Y = COV(X, Y) = \sigma_{XY} \qquad (5.41)$$

is called the *covariance* of the random variables X and Y. Note that

$$COV(aX, bY) = ab\ \sigma_{XY}. \qquad (5.42)$$

As mentioned in Chapter 2, σ_{XY} depicts the joint variability of the random variables X and Y. That is:

(a) If the probability is high that large values of $X - \mu_X$ are associated with large values of $Y - \mu_Y$ and small values of $X - \mu_X$ are associated with small values of $Y - \mu_Y$, then X and Y are positively related and $\sigma_{XY} > 0$.

(b) If the probability is high that large values of $X - \mu_X$ are associated with small values of $Y - \mu_Y$ and small values of $X - \mu_X$ are associated with large values of $Y - \mu_Y$, then X and Y are negatively related and $\sigma_{XY} < 0$.

Whereas the sign of σ_{XY} indicates the *direction* of the relationship between the discrete or continuous random variables X and Y, its magnitude depends upon the units in which X and Y are measured. To correct for the scaling of X and Y, let us divide σ_{XY} by $\sigma_X\sigma_Y$, where σ_X (respectively, σ_Y) is the standard deviation of X (respectively, Y). This yields the *coefficient of correlation* between the random variables X and Y or

$$\rho_{XY} = \frac{\sigma_{XY}}{\sigma_X\sigma_Y}, \quad -1 \leq \rho_{XY} \leq 1 \qquad (5.43)$$

or

$$|\sigma_{XY}| \quad \leq \quad \sigma_X\ \sigma_Y. \qquad (5.43.1)$$

As also indicated in Chapter 2, ρ_{XY} measures the *strength* as well as the *direction* of the *linear* relationship between X and Y. At the extremes, when $\rho_{XY} = 1$, we have perfect positive association between X and Y. And when $\rho_{XY} = -1$, the case of perfect negative association emerges.

Example 5.4.2   If we standardize the discrete or continuous random variables X and Y so as to obtain

$$Z_X = \frac{X - \mu_X}{\sigma_X} \quad \text{and} \quad Z_Y = \frac{Y - \mu_Y}{\sigma_Y},$$

respectively (where σ_X is the standard deviation of X and σ_Y is the standard deviation of Y), then

$$\rho_{Z_X Z_Y} = \frac{\sigma_{Z_X Z_Y}}{\sigma_{Z_X}\sigma_{Z_Y}} = \sigma_{Z_X Z_Y};$$

that is, since $\sigma_{Z_X} = \sigma_{Z_Y} = 1$, the correlation coefficient of two standardized random variables Z_X and Z_Y is just their covariance itself. ∎

If X and Y are independent discrete or continuous random variables, then $\sigma_{XY} = 0$ and thus:

(a) from (5.41), $E(XY) = \mu_X \mu_Y$

(b) X and Y are uncorrelated or $\rho_{XY} = 0$ (but if $\sigma_{XY} = 0$, it does not necessarily follow that X and Y are independent random variables)

If φ is a real-valued function of the discrete or continuous random variables X and Y, then the *variance of a function of X and Y, $\varphi(X, Y)$*, is

$$V[\varphi(X, Y)] = E\left\{[\varphi(X, Y) - E(\varphi(X, Y))]^2\right\} = \sigma_\varphi^2. \tag{5.44}$$

Based upon this definition it can be shown that:

(a) If $\varphi(X, Y) = aX + bY$, then

$$V(aX + bY) = a^2\sigma_X^2 + b^2\sigma_Y^2 + 2ab\sigma_{XY}, \tag{5.45}$$

where a and b are constants and $\sigma_X^2 = V(X)$ and $\sigma_Y^2 = V(Y)$ are the variances of the random variables X and Y, respectively.

(b) If $\varphi(X, Y) = aX - bY$, then

$$V(aX - bY) = a^2\sigma_X^2 + b^2\sigma_Y^2 - 2ab\sigma_{XY}, \tag{5.46}$$

with a and b constants.

(c) If $\varphi(X, Y) = aX \pm bY$ and X and Y are independent random variables, then

$$V(aX \pm bY) = a^2\sigma_X^2 + b^2\sigma_Y^2, \tag{5.47}$$

with a and b constants.

(d) In general, if $X_i, i = 1, \ldots, k$, are random variables and φ is a linear combination of the X_i or $\varphi = \sum_{i=1}^k a_i X_i$, with a_i constant for all $i = 1, \ldots, k$, then

$$V\left(\sum_{i=1}^k a_i X_i\right) = \sum_{i=1}^k a_i^2 \sigma_i^2 + 2\sum_{\substack{i=1 \\ i<j}}^k \sum_{j=1}^k a_i a_j \sigma_{ij}, \tag{5.48}$$

where $\sigma_i^2 = V(X_i)$ and $\sigma_{ij} = COV(X_i, X_j), i \neq j$.

If the $X_i, i = 1, \ldots, k$, are independent random variables, then

$$V\left(\sum_{i=1}^{k} a_i X_i\right) = \sum_{i=1}^{k} a_i^2 \sigma_i^2. \tag{5.48.1}$$

(e) For random variables X and Y,

$$E(XY) = \mu_X \mu_Y + \sigma_{XY}. \tag{5.49}$$

If X and Y are independent,

$$E(XY) = \mu_X \mu_Y \tag{5.49.1}$$

as indicated earlier.

(f) If $\varphi(X, Y) = XY$, then, from (e),

$$
\begin{aligned}
V(XY) = E\left\{[XY - E(XY)]^2\right\} &= E\left\{[XY - (\mu_X \mu_Y + \sigma_{XY})]^2\right\} \\
&= \mu_Y^2 \sigma_X^2 + \mu_X^2 \sigma_Y^2 + 2\mu_X \mu_Y \sigma_{XY} - \sigma_{XY}^2 \\
&\quad + E\left[(X - \mu_X)^2 (Y - \mu_Y)^2\right] \\
&\quad + 2\mu_Y E\left[(X - \mu_X)^2 (Y - \mu_Y)\right] \\
&\quad + 2\mu_X E\left[(X - \mu_X)(Y - \mu_Y)^2\right]. \tag{5.50}
\end{aligned}
$$

If X and Y are independent random variables, then

$$V(XY) = \mu_Y^2 \sigma_X^2 + \mu_X^2 \sigma_Y^2 + \sigma_X^2 \sigma_Y^2. \tag{5.50.1}$$

(g) In general, if $X_i, i = 1, \ldots, k$, are random variables and both φ_1 and φ_2 are linear combinations of the X_i or $\varphi_1 = \sum_{i=1}^{k} a_i X_i$ and $\varphi_2 = \sum_{i=1}^{k} b_i X_i$, where a_i and b_i are constants, $i = 1, \ldots, k$, then

$$\sigma_{\varphi_1 \varphi_2} = COV(\varphi_1, \varphi_2) = \sum_{i=1}^{k} a_i b_i \sigma_i^2 + \sum_{\substack{i=1 \\ i<j}}^{k} \sum_{j=1}^{k} (a_i b_j + a_j b_i) \sigma_{ij}. \tag{5.51}$$

If the $X_i, i = 1, \ldots, k$, are independent random variables, then

$$\sigma_{\varphi_1 \varphi_2} = COV(\varphi_1, \varphi_2) = \sum_{i=1}^{k} a_i b_i \sigma_i^2. \tag{5.51.1}$$

(h) For random variables X and Y,

$$E\left(\frac{X}{Y}\right) \approx \left(\frac{\mu_X}{\mu_Y}\right) - \left(\frac{\sigma_{XY}}{\mu_Y^2}\right) + \left(\frac{\mu_X \sigma_Y^2}{\mu_Y^3}\right); \tag{5.52}$$

$$V\left(\frac{X}{Y}\right) \approx \left(\frac{\mu_X}{\mu_Y}\right)^2 \left[\left(\frac{\sigma_X^2}{\mu_X^2}\right) + \left(\frac{\sigma_y^2}{\mu_y^2}\right) - \left(\frac{2\sigma_{xy}}{\mu_x \mu_y}\right)\right]. \tag{5.53}$$

If X and Y are independent random variables,

$$E\left(\frac{X}{Y}\right) \approx \left(\frac{\mu_X}{\mu_Y}\right) + \left(\frac{\mu_X \sigma_Y^2}{\mu_Y^3}\right); \tag{5.52.1}$$

$$V\left(\frac{X}{Y}\right) \approx \left(\frac{\mu_X}{\mu_Y}\right)^2 \left[\left(\frac{\sigma_X^2}{\mu_X^2}\right) + \left(\frac{\sigma_y^2}{\mu_y^2}\right)\right]. \tag{5.53.1}$$

Let X and Y be discrete random variables with $g(X|Y)$ the conditional probability mass function of X given Y and $h(Y|X)$ the conditional probability mass function of Y given X. Then the *conditional expectation (mean) of X given $Y = Y_s$* is defined as

$$E(X|Y_s) = \sum_i X_i g(X_i|Y_s); \tag{5.54}$$

and the *conditional variance of X given $Y = Y_s$* is

$$V(X|Y_s) = E\left\{[X - E(X|Y_s)]^2 |Y_s\right\} = E(X^2|Y_s) - E(X|Y_s)^2, \tag{5.55}$$

where

$$E(X^2|Y_s) = \sum_i X_i^2 g(X_i|Y_s). \tag{5.56}$$

In general, both $E(X|Y_s)$ and $V(X|Y_s)$ are functions of the chosen Y value. Similarly, the *conditional expectation (mean) of Y given $X = X_r$* is

$$E(Y|X_r) = \sum_j Y_j h(Y_j|X_r); \tag{5.57}$$

and the *conditional variance of Y given $X = X_r$* is

$$V(Y|X_r) = E\left\{[Y - E(Y|X_r)]^2 |X_r\right\} = E(Y^2|X_r) - E(Y|X_r)^2, \tag{5.58}$$

with

$$E(Y^2|X_r) = \sum_j Y_j^2 h(Y_j|X_r). \tag{5.59}$$

Here both $E(Y|X)$ and $V(Y|X)$ generally depend upon the given value of X.

If X and Y are continuous random variables with $g(x|y)$ and $h(y|x)$ representing conditional probability density functions for X given Y and Y given X, respectively, then the *conditional expectation (mean) of X given $Y = y$* is

$$E(X|y) = \int_{-\infty}^{+\infty} x\, g(x|y)\, dx; \tag{5.60}$$

and the *conditional variance of X given $Y = y$* is

$$V(X|y) = E\left\{[X - E(X|y)]^2\,|y\right\} = E(X^2|y) - E(X|y)^2, \tag{5.61}$$

where

$$E(X^2|y) = \int_{-\infty}^{+\infty} x^2\, g(x|y)\, dx. \tag{5.62}$$

Both $E(X|y)$ and $V(X|y)$ are functions of y and give the mean and variance of the random variable X, respectively, along the line $Y = y$ in the XY-plane. In like fashion we may define the *conditional expectation (mean) of Y given $X = x$* as

$$E(Y|x) = \int_{-\infty}^{+\infty} y\, h(y|x)\, dy; \tag{5.63}$$

and the *conditional variance of Y given $X = x$* is

$$V(Y|x) = E\left\{[Y - E(Y|x)]^2\,|x\right\} = E(Y^2|x) - E(Y|x)^2, \tag{5.64}$$

where

$$E(Y^2|x) = \int_{-\infty}^{+\infty} y^2\, h(y|x)\, dy. \tag{5.65}$$

Here $E(Y|x)$ and $V(Y|x)$ depend upon x and give the mean and variance of the random variable Y, respectively, along the line $X = x$ in the XY-plane.

If X and Y are independent discrete or continuous random variables, then the preceding (four) sets of conditional means and variances equal their unconditional counterparts. For instance, from (5.60) and (5.61), under independence, $E(X|y) = E(X)$ and $V(X|y) = V(X)$.

Finally, for X and Y any two random variables (discrete or continuous), it can be demonstrated that:

$$\text{(a)}\quad E(X) = E[E(X|Y)]$$
$$\text{(b)}\quad V(X) = E[V(X|Y)] + V[E(X|Y)] \tag{5.66}$$

given that the indicated expectations exist.

5.5   Chebyshev's Theorem for Bivariate Probability Distributions

Let X and Y be jointly distributed discrete or continuous random variables with individual means and variances: $E(X) = \mu_X, V(X) = \sigma_X^2$; and $E(Y) = \mu_Y$, $V(Y) = \sigma_Y^2$, respectively. Let us define the events:

$$A_X = \left\{X|\ |X - \mu_X| < \sqrt{2}\,k\sigma_X\right\}; \quad A_Y = \left\{Y|\ |Y - \mu_Y| < \sqrt{2}\,k\sigma_Y\right\},$$

where $k > 0$. Then

$$P(A_X \cap A_Y) \geq 1 - \frac{1}{k^2};$$

that is, the probability that X lies within $\sqrt{2}\,k$ standard deviations of μ_X *and* Y lies within $\sqrt{2}\,k$ standard deviations of μ_Y is at least $1 - \frac{1}{k^2}$.

5.6   Joint Moment–Generating Function

Let X and Y be discrete or continuous random variables. The *joint moment-generating function of X and Y*, denoted $m_{X,Y}(t_1, t_2)$, is defined as $E(e^{t_1 X + t_2 Y})$ provided that the expectation exists for values of the *moment-generators* t_1 and t_2 such that $-t_0 < t_i < t_0$, $i = 1, 2$. In this regard, either

$$m_{X,Y}(t_1, t_2) = E(e^{t_1 X + t_2 Y})$$
$$= \sum_i \sum_j e^{t_1 X_i + t_2 Y_j} f(X_i, Y_j)\ [X, Y \text{ discrete}] \tag{5.67}$$

or

$$m_{X,Y}(t_1, t_2) = E(e^{t_1 X + t_2 Y})$$
$$= \int_{-\infty}^{+\infty} \int_{-\infty}^{+\infty} e^{t_1 x + t_2 y} f(x, y)\, dy\, dx.\ [X, Y \text{ continuous}] \tag{5.68}$$

(If $t_1 = t_2 = 0$, then $m_{X,Y}(0, 0) = E(e^0) = 1$.)

Since either (5.67) or (5.68) completely specifies the joint distribution of X and Y, the marginal distributions of X and Y can be determined from the *marginal moment-generating functions* as $m_X(t_1) = m_{X,Y}(t_1, 0) = E(e^{t_1 X})$, $m_Y(t_2) = m_{X,Y}(0, t_2) = E(e^{t_2 Y})$.

If $m_{X,Y}(t_1, t_2)$ exists, then it is continuously differentiable in some neighborhood of the origin $(0,0)$. Hence $m_{X,Y}(t_1, t_2)$ will generate all moments of X and Y under suitable differentiation. That is, for X and Y discrete random variables, we may employ (5.67) to obtain

$$\frac{\partial^{r+s} m_{X,Y}(t_1, t_2)}{\partial t_{t_1}^r \partial t_{t_2}^s} = \sum_i \sum_j X_i^r Y_j^s e^{t_1 X_i + t_2 Y_j} f(X_i, Y_j)$$

(provided a uniform convergence criterion holds for infinite sums) so that the rth and s^{th} joint moment of X and Y about the origin $(0,0)$ is

$$E(X^r Y^s) = \left. \frac{\partial^{r+s} m_{X,Y}(t_1, t_2)}{\partial t_1^r \partial t_2^s} \right|_{t_1 = t_2 = 0} = \sum_i \sum_j X_i^r Y_j^s f(X_i, Y_j) \qquad (5.69)$$

(given that r and s are not simultaneously zero).

And for X and Y continuous random variables, we may use (5.68) to find

$$\frac{\partial^{r+s} m_{X,Y}(t_1, t_2)}{\partial t_{t_1}^r \partial t_{t_2}^s} = \int_{-\infty}^{+\infty} \int_{-\infty}^{+\infty} x^r y^s e^{t_1 x + t_2 y} f(x, y) \, dy \, dx$$

(provided that the operations of integration and differentiation can be interchanged). Then the rth and s^{th} joint moment of X and Y about the origin $(0,0)$ (with r and s not simultaneously zero) is

$$E(X^r Y^s) = \left. \frac{\partial^{r+s} m_{X,Y}(t_1, t_2)}{\partial t_1^r \partial t_2^s} \right|_{t_1 = t_2 = 0} = \int_{-\infty}^{+\infty} \int_{-\infty}^{+\infty} x^r y^s f(x, y) \, dy \, dx. \qquad (5.70)$$

Given either (5.69) or (5.70), it should be obvious that:

$$\frac{\partial m_{X,Y}(0,0)}{\partial t_1} = E(X) = \mu_X, \quad \frac{\partial m_{X,Y}(0,0)}{\partial t_2} = E(Y) = \mu_Y,$$

$$\sigma_X^2 = E(X^2) - \mu_X^2 = \frac{\partial^2 m_{X,Y}(0,0)}{\partial t_1^2} - \mu_X^2,$$

$$\sigma_Y^2 = E(Y^2) - \mu_Y^2 = \frac{\partial^2 m_{X,Y}(0,0)}{\partial t_2^2} - \mu_Y^2, \text{ and}$$

$$\sigma_{XY} = E(XY) - \mu_X \mu_Y = \frac{\partial^2 m_{X,Y}(0,0)}{\partial t_1 \partial t_2} - \mu_X \mu_Y.$$

Let us replace $e^{t_1X+t_2Y}$ by its Maclaurin's series expansion

$$e^{t_1X+t_2Y} = 1 + t_1X + t_2Y + \frac{1}{2!}(t_1^2X^2 + 2t_1t_2XY + t_2^2Y^2) + \cdots. \qquad (5.71)$$

Then if X and Y are, say, continuous random variables and $m_{X,Y}(t_1,t_2)$ exists, (5.68) can be expressed as

$$m_{X,Y}(t_1,t_2) = E(e^{t_1X+t_2Y})$$

$$= \int_{-\infty}^{+\infty}\int_{-\infty}^{+\infty}\left(1 + t_1x + t_2y + \frac{1}{2!}t_1^2x^2 + t_1t_2xy + \frac{1}{2!}t_2^2y^2 + \cdots\right)f(x,y)\,dy\,dx$$

$$= 1 + t_1\mu_X + t_2\mu_Y + \frac{t_1^2}{2!}E(X^2) + t_1t_2E(XY) + \frac{t_2^2}{2!}E(Y^2) + \cdots. \qquad (5.72)$$

So given (5.72), the partial derivatives of various orders of $m_{X,Y}(t_1,t_2)$ with respect to the generators t_1 and t_2, evaluated at $t_1 = t_2 = 0$, yield the various moments of X and Y about the origin. That this is indeed the case is left to you as an exercise.

Example 5.6.1   Suppose X and Y are continuous random variables with bivariate probability density function

$$f(x,y) = \begin{cases} e^{-x-y}, & x > 0, y > 0; \\ 0 & \text{elsewhere.} \end{cases}$$

From (5.68), the joint moment-generating function of X and Y is

$$m_{X,Y}(t_1,t_2) = E(e^{t_1X+t_2Y})$$

$$= \int_{-\infty}^{+\infty}\int_{-\infty}^{+\infty} e^{t_1x+t_2y}e^{-x-y}\,dy\,dx$$

$$= \int_0^{+\infty} e^{-x(1-t_1)}\,dx \int_0^{+\infty} e^{-y(1-t_2)}\,dy = (1-t_1)^{-1}(1-t_2)^{-1}.$$

Then by direct computation we see, via (5.70), that

$$\frac{\partial m_{X,Y}(t_1, t_2)}{\partial t_1} = (1 - t_1)^{-2}(1 - t_2)^{-1}, \quad \frac{\partial m_{X,Y}(t_1, t_2)}{\partial t_2} = (1 - t_1)^{-1}(1 - t_2)^{-2},$$

$$\frac{\partial^2 m_{X,Y}(t_1, t_2)}{\partial t_1^2} = 2(1 - t_1)^{-3}(1 - t_2)^{-1},$$

$$\frac{\partial^2 m_{X,Y}(t_1, t_2)}{\partial t_2^2} = 2(1 - t_1)^{-1}(1 - t_2)^{-3}, \text{ and}$$

$$\frac{\partial^2 m_{X,Y}(t_1, t_2)}{\partial t_1 \partial t_2} = (1 - t_1)^{-2}(1 - t_2)^{-2}.$$

Then

$$\frac{\partial m_{X,Y}(0,0)}{\partial t_1} = \frac{\partial m_{X,Y}(0,0)}{\partial t_2} = \frac{\partial^2 m_{X,Y}(0,0)}{\partial t_1 \partial t_2} = 1,$$

$$\frac{\partial^2 m_{X,Y}(0,0)}{\partial t_1^2} = \frac{\partial^2 m_{X,Y}(0,0)}{\partial t_2^2} = 2,$$

and thus $\mu_X = \mu_Y = 1, \sigma_X^2 = \sigma_Y^2 = 1$, and $\sigma_{XY} = 0$.
Looking at (5.72), let us write

$$m_{X,Y}(t_1, t_2) = \int_0^{+\infty} \int_0^{+\infty} \left(1 + t_1 x + t_2 y + \frac{1}{2!} t_1^2 x^2 + t_1 t_2 xy + \frac{1}{2!} t_2^2 y^2 + \cdots \right) e^{-x-y} \, dy \, dx$$

$$= 1 + t_1 + t_2 + t_1^2 + t_1 t_2 + t_2^2 + \cdots.$$

Then

$$\frac{\partial m_{X,Y}(t_1, t_2)}{\partial t_1} = 1 + 2t_1 + t_2 + \cdots,$$

$$\frac{\partial m_{X,Y}(t_1, t_2)}{\partial t_2} = 1 + t_1 + 2t_2 + \cdots,$$

$$\frac{\partial^2 m_{X,Y}(t_1, t_2)}{\partial t_1^2} = 2 + \text{higher order terms involving } t_1 \text{ and } t_2,$$

$$\frac{\partial^2 m_{X,Y}(t_1, t_2)}{\partial t_2^2} = 2 + \text{higher order terms involving } t_1 \text{ and } t_2,$$

$$\frac{\partial^2 m_{X,Y}(t_1, t_2)}{\partial t_1 \partial t_2} = 1 + \text{higher order terms involving } t_1 \text{ and } t_2.$$

Upon evaluating all these partial derivatives at (0,0) renders the exact same set of moments about the origin that we just determined above. ∎

Example 5.6.2 Let X and Y be discrete random variables with the bivariate probability mass function depicted in Table 5.6. Then the joint moment-generating function of X and Y is, from (5.67),

$$m_{X,Y}^{(t_1,t_2)} = e^{t_1 X_1 + t_2 Y_1} f(X_1, Y_1) + e^{t_1 X_2 + t_2 Y_1} f(X_2, Y_1)$$
$$+ e^{t_1 X_1 + t_2 Y_2} f(X_1, Y_2) + e^{t_1 X_2 + t_2 Y_2} f(X_2, Y_2)$$
$$= 0.25 + 0.25 e^{t_1} + 0.25 e^{t_2} + 0.25 e^{t_1 + t_2}.$$

Table 5.6

X \\ Y	$Y_1 = 0$	$Y_2 = 1$
$X_1 = 0$	0.25	0.25
$X_2 = 1$	0.25	0.25

Then from (5.69),

$$\frac{\partial m_{X,Y}(0,0)}{\partial t_1} = \frac{\partial m_{X,Y}(0,0)}{\partial t_2} = 0.50,$$

$$\frac{\partial^2 m_{X,Y}(0,0)}{\partial t_1^2} = \frac{\partial^2 m_{X,Y}(0,0)}{\partial t_2^2} = 0.50, \text{ and}$$

$$\frac{\partial^2 m_{X,Y}(0,0)}{\partial t_1 \partial t_2} = 0.25. \quad ■$$

Next, suppose that X and Y are independent discrete or continuous random variables and let $U = X + Y$. Then the moment-generating function for U is

$$m_U(t) = E(e^{tU}) = E(e^{t(X+Y)})$$
$$= E(e^{tX} e^{tY}) = E(e^{tX}) \cdot E(e^{tY}) = m_X(t) \cdot m_Y(t) \quad (5.73)$$

under independence (provided that the individual expectations exist). Obviously this result is readily generalized for the case where we have n mutually independent random variables $X_1, \ldots, X_n$ with moment-generating functions $m_{X_1}(t), \ldots, m_{X_n}(t)$, respectively. If $U_1 = \sum_{i=1}^{n} X_i$, then

$$m_{U_1}(t) = \prod_{i=1}^{n} m_{X_i}(t). \quad (5.73.1)$$

And if $U_2 = \sum_{i=1}^{n} a_i X_i$, then, under the mutual independence of the $X_i, i = 1, \ldots, n,$

$$m_{U_2}(t) = \prod_{i=1}^{n} m_{X_i}(a_i t). \tag{5.73.2}$$

Finally, if, in addition to being mutually independent random variables, the X_i have the same probability density function $f(x)$ for all $i = 1, \ldots, n$, more then (5.73.1) and (5.73.2) become, respectively,

$$m_{U_1}(t) = [m_X(t)]^n, \tag{5.73.3}$$

$$m_{U_2}(t) = [m_X(a_i t)]^n. \tag{5.73.4}$$

Note that if in (5.73.4) we set $a_i = \frac{1}{n}$ for all i, then the resulting expression depicts the moment-generating function for the sample mean $\overline{X}$ or

$$m_{\overline{X}}(t) = \left[m_X \left(\frac{t}{n} \right) \right]^n. \tag{5.73.5}$$

5.7 Exercises

5-1. Given the following bivariate probability distribution, determine the sets of marginal probabilities for the X and Y random variables. Find:

X \ Y	0	1	2	3
0	$\frac{3}{87}$	$\frac{3}{87}$	$\frac{5}{87}$	$\frac{5}{87}$
1	$\frac{4}{87}$	$\frac{6}{87}$	$\frac{8}{87}$	$\frac{10}{87}$
2	$\frac{7}{87}$	$\frac{9}{87}$	$\frac{12}{87}$	$\frac{15}{87}$

(a) $F(1,2)$

(b) $P(1 \leq Y \leq 3)$

(c) The marginal probability distributions of X and Y

(d) $P(X = 1 | Y = 1)$

(e) $P(Y = 2 | X = 0)$

(f) The conditional distribution of X given $Y = 3$

(g) The conditional distribution of Y given $X = 2$. Are X and Y independent?

5-2. Let a random experiment consist of tossing a fair coin twice. Define the random variable X to be the number of heads obtained in the two tosses and let the random variable Y be the opposite face outcome. Determine the points within the sample space S and the values of X and Y on S. Next, specify the bivariate probability distribution between X and Y. Find:

(a) $P(X = 2 | Y = 1)$

(b) The marginal distribution of Y

(c) The conditional distribution of X given $Y = 0$. Are X and Y independent?

5-3. Let a random experiment consist of rolling a fair pair of six-sided dice. Let the random variable X (respectively, Y) be defined as the face value of die 1 (respectively, die 2). Determine the sample space S and find:

(a) $P(3 \le X \le 5, 2 \le Y \le 4)$

(b) $P(3, 3)$

(c) The marginal probability distributions of X and of Y

(d) The conditional probability distribution of X given $Y = Y_4$

5-4. Given the following bivariate probability distribution, find:

X \ Y	-1	2	3
1	$\dfrac{1}{14}$	$\dfrac{1}{14}$	$\dfrac{2}{14}$
2	$\dfrac{2}{14}$	$\dfrac{3}{14}$	$\dfrac{1}{14}$
3	$\dfrac{1}{14}$	0	$\dfrac{3}{14}$

(a) $E(X)$ and $E(Y)$

(b) $V(X)$ and $V(Y)$

(c) ρ_{XY}

(d) $E(X | Y = 2)$ and $E(Y | X = 3)$

(e) $V(X | Y = -1)$ and $V(Y | X = 3)$

5-5. Let a random experiment consist of rolling a fair pair of six-sided dice. Let the random variable $X = $ |difference of the faces| and let the random variable $Y = $ sum of the faces. Determine the points in S and find:

(a) $P(4 \le X \le 6, 0 \le Y \le 4)$

(b) $P(X \ge 4, Y = 3)$

(c) The marginal probability distribution of Y

(d) The conditional probability distribution of X given $Y = 3$

(e) $E(Y)$ and $V(Y)$

(f) $E(X|Y = 1)$

(g) $V(Y|X = 4)$

5-6. Let the bivariate probability mass function of the discrete random variables X and Y be given as

$$f(X, Y) = \begin{cases} \frac{X+Y}{36}, & X = 1, 2, 3 \text{ and } Y = 1, 2, 3; \\ 0 & \text{elsewhere.} \end{cases}$$

Find:

(a) $P(X = 2, Y \le 2)$

(b) $F(2, 2)$

(c) The marginal probability mass functions of X and Y

(d) The marginal probability distribution of X

(e) The conditional probability distribution of Y given $X = 1$

Are the X and Y random variables independent?

5-7. Let the bivariate probability mass function for the discrete random variables X and Y appear as

$$f(X, Y) = \begin{cases} \frac{1}{9}, & X = 1, 2, 3 \text{ and } Y = 1, 2, 3; \\ 0 & \text{elsewhere.} \end{cases}$$

Are X and Y independent?

5-8. A jar contains three balls numbered 1,2, and 3, respectively. Two balls are drawn at random with replacement. Let X be the number of the ball on the first draw and let Y be the number on the second ball drawn. Determine the resulting bivariate probability distribution between X and Y. What would the said distribution look like if the draws were made without replacement? For which of these two bivariate distributions is $E(X)$ highest? Are the random variables for either/both of the two bivariate distributions independent?

5-9. For X, Y discrete random variables with bivariate probability mass function $f(X, Y)$, demonstrate that the marginal probability mass functions have the form:

$$g(X) = \sum_j f(X, Y_j); \quad h(Y) = \sum_i f(X_i, Y).$$

5-10. Given the following bivariate probability mass function for the discrete random variables X and Y, find $E(Z)$, where $Z = X - XY + 2Y$.

X \ Y	0	1	2	3
0	0.05	0.30	0.20	0.05
1	0.05	0.10	0.20	0.05

5-11. Suppose a fair pair of six-sided dice is tossed and (X, Y) is a bivariate random variable defined on the sample space $S = \{E_i, i = 1, \ldots, 36\}$. Here each simple event E_i is a point $(X = j, Y = j), j = 1, \ldots, 6$, where X is the outcome on the first die and Y is the outcome on the second die. If the random variable $Z = X + Y$ is the sum of the faces showing, find the probability that $X = 3$ given $Z = 7$. What is $E(X \mid Z = 7)$?

5-12. Comment on the following statement: Knowledge of the marginal probability mass functions $g(X)$ and $h(Y)$ is generally not equivalent to knowledge of the bivariate probability mass function $f(X, Y)$. Under what circumstance can $f(X, Y)$ be obtained from knowledge about $g(X)$ and $h(Y)$?

5-13. Given the bivariate probability density function

$$f(x, y) = \begin{cases} 3x(1 - xy), & 0 < x < 1 \text{ and } 0 < y < 1; \\ 0 & \text{elsewhere,} \end{cases}$$

find:

(a) $P\left(X \leq \frac{1}{2}, Y \leq \frac{1}{2}\right)$

(b) the marginal densities of X and Y

(c) $P\left(\frac{1}{4} < X < \frac{1}{3}, \frac{1}{3} < Y < \frac{1}{2}\right)$

(d) The conditional density function of X given Y

(e) The conditional density function of Y given X

Are the random variables X and Y independent?

5-14. For random variables X and Y, find the value of k that makes

$$f(x, y) = \begin{cases} kxy^2, & 0 \leq x \leq 1 \text{ and } 0 \leq y \leq 1; \\ 0 & \text{elsewhere,} \end{cases}$$

a bivariate probability density function. Then find:

(a) The cumulative distributive function

(b) $P\left(\frac{1}{2} \leq X \leq \frac{3}{4}, \frac{1}{4} \leq Y \leq \frac{1}{2}\right)$

(c) $F\left(\frac{1}{2}, \frac{1}{3}\right)$

(d) The marginal density functions of X and Y

(e) The conditional density function for Y given X

5-15. Given the bivariate probability density function

$$f(x,y) = \begin{cases} 2y, & 0 < x \le 1 \text{ and } 0 < y \le 1; \\ 0 & \text{elsewhere,} \end{cases}$$

find the marginal density functions for X and Y. Are X and Y independent random variables?

5-16. Suppose a bivariate random variable (X,Y) has a probability density function of the form

$$f(x,y) = \begin{cases} \theta^2 e^{-\theta(x+y)}, & \theta > 0, x \ge 0, y \ge 0; \\ 0 & \text{elsewhere.} \end{cases}$$

Find $P(0 \le X \le 100, 0 \le Y \le 100)$.

5-17. For the bivariate probability density function appearing in Exercise 5–13, find:

(a) $E(X)$ and $E(Y)$

(b) $E(XY)$

(c) $V(X)$ and $V(Y)$

(d) $COV(X,Y)$

5-18. Given the bivariate probability density function for the continuous random variables X and Y,

$$f(x,y) = \begin{cases} \lambda, & 0 < x < 1 \text{ and } 0 < y < 2 \\ 0 & \text{elsewhere.} \end{cases}$$

Find:

(a) λ

(b) $P\left(0 < X < \frac{1}{2}, \frac{1}{2} < Y < \frac{3}{4}\right)$

5-19. Given the bivariate probability density function specified in the preceding exercise, define event

$$A = \{(X,Y)\,|\,X + Y \le 1\}.$$

Find $P(A)$.

5-20. Let X_1, X_2, and X_3 be random variables, where:

$$E(X_1) = 1 \quad V(X_1) = 1 \quad COV(X_1, X_2) = 2$$
$$E(X_2) = -1 \quad V(X_2) = 5 \quad COV(X_1, X_3) = -1$$
$$E(X_3) = 3 \quad V(X_3) = 2 \quad COV(X_2, X_3) = 0.5$$

For $U = X_1 - 2X_2 + 3X_3$, find:

(a) $E(U)$

(b) $V(U)$

(c) $E\left(\frac{U}{X_1}\right)$

(d) $V\left(\frac{U}{X_2}\right)$

(e) $E(U \cdot X_3)$

5-21. Let X, Y be independent random variables with respective probability density functions

$$f(x) = \begin{cases} \frac{1}{2}e^{-x/10} & x > 0; \\ 0 & \text{elsewhere} \end{cases} \qquad g(y) = \begin{cases} \frac{1}{2}e^{-y/10} & y > 0; \\ 0 & \text{elsewhere.} \end{cases}$$

If $Z = X + Y$, find $h(z)$, the probability density function of Z, from Z's cumulative distribution function.

5-22. Given the joint probability density function

$$f(x, y) = \begin{cases} k(x + 2y), & 0 < x \le 2 \text{ and } 0 < y \le 1; \\ 0 & \text{elsewhere,} \end{cases}$$

find:

(a) The value of k

(b) The marginal distribution of X

(c) The cumulative distribution function

(d) $P\left(1 \le X \le 2, \frac{1}{4} \le Y \le \frac{1}{2}\right)$

(e) The conditional density function of Y given X

5-23. Find the value of k that makes

$$f(x, y) = \begin{cases} k\left(\frac{x}{y}\right), & 0 < x < 2 \text{ and } 1 < y < 3; \\ 0 & \text{elsewhere} \end{cases}$$

a bivariate probability density function.

5-24. If $F(t,s) = ts(t+s)/2$, $0 \leq x \leq 1$ and $0 \leq y \leq 1$, find the joint probability density function $f(t,s)$.

5-25. Given the bivariate cumulative distribution function $F(t,s)$, the joint probability $P(a < X \leq b, \, Y \leq d) = F(b,d) - F(a,d)$. Using this expression, find $P\left(\frac{1}{3} < X \leq 1, Y \leq \frac{2}{3}\right)$ when

$$f(x,y) = \begin{cases} 4xy, & 0 < x < 1 \text{ and } 0 < y < 1; \\ 0 & \text{elsewhere.} \end{cases}$$

5-26. Can the expression

$$f(x,y) = \begin{cases} xe^{-x-xy}, & x > 0 \text{ and } y > 0; \\ 0 & \text{elsewhere} \end{cases}$$

serve as a joint probability density function?

5-27. Suppose $f(x,y)$ is a bivariate probability density function and we want to find

$$P(a < X < b) = P(a < X < b, -\infty < Y < +\infty) = \int_a^b \left[\int_{-\infty}^{+\infty} f(x,y)dy \right] dx.$$

How is the term within the square brackets interpreted?

5-28. Suppose that the random variables X and Y have the following joint probability density function

$$f(x,y) = \begin{cases} 2, & 0 < x < y < 1; \\ 0 & \text{elsewhere.} \end{cases}$$

Find:

(a) The marginal distributions of X and Y

(b) The cumulative distribution function

(c) The conditional density function of Y given X

(d) $E(X)$

(e) $V(Y)$

(f) $E(XY)$

(g) $E(X/Y)$

(h) $V(Y/X)$

5-29. Suppose the bivariate cumulative distribution function for the continuous random variables X and Y appears as

$$F(t,s) = \begin{cases} 0, & t < 0 \text{ and } s < 0; \\ 1 - e^{-2t} - e^{-2s} + e^{-2(t+s)}, & t \geq 0 \text{ and } s \geq 0. \end{cases}$$

What is the associated probability density function?

5-30. Are the random variables X and Y with probability density function

$$f(x,y) = \begin{cases} 12(xy - xy^2), & 0 < x < 1 \text{ and } 1 < y < 1; \\ 0 & \text{elsewhere} \end{cases}$$

independent?

5-31. For the joint probability density function

$$f(x,y) = \begin{cases} x + y, & 0 < x < 1 \text{ and } 0 < y < 1; \\ 0 & \text{elsewhere} \end{cases}$$

find ρ_{xy}.

5-32. What is the value of the k that will make

$$f(x,y) = \begin{cases} kxy^{-1}, & 0 < x < 1 \text{ and } 1 < y < 2; \\ 0 & \text{elsewhere} \end{cases}$$

a valid probability density function?

5-33. Let X, Y be independent random variables with probability mass functions $f(X), g(Y)$, respectively (or probability density functions $f(x), g(y)$, respectively). Let $Z = X + Y$ with probability density function $h(z)$. Then:

(a) If X, Y are discrete, $h(Z) = \sum_X f(X)g(Z - X)$

(b) If X, Y are continuous, $h(z) = \int_{-\infty}^{+\infty} f(x)g(z - x)\, dx$

Suppose we have the following two independent probability density functions for the continuous random variables X and Y, respectively;

$$f(x) = \begin{cases} e^{-x}, & x > 0; \\ 0 & \text{elsewhere;} \end{cases} \qquad g(y) = \begin{cases} e^{-y}, & y > 0; \\ 0 & \text{elsewhere.} \end{cases}$$

Let $Z = X + Y$. Use part (b) to determine $h(z)$.

5-34. Suppose $\alpha \leq X \leq \beta$ and $\alpha \leq Y \leq \beta$. What is the form of the uniform probability density function for the bivariate random variable (X, Y)?

5-35. Suppose a bivariate random variable (X, Y) has a probability density function of the form

$$f(x, y) = \begin{cases} e^{-x-y}, & x > 0 \text{ and } y > 0; \\ 0 & \text{elsewhere.} \end{cases}$$

Find $P(X + Y > 10)$.

5-36. Suppose that (X, Y) is a bivariate random variable with probability density function

$$f(x, y) = \begin{cases} k(x + y), & 0 \le x \le 1 \text{ and } 0 \le y \le 1; \\ 0 & \text{elsewhere.} \end{cases}$$

Find:

(a) The constant k

(b) The bivariate cumulative distribution function $F(t, s)$

(c) Demonstrate that $F\left(\frac{1}{2}, \frac{1}{2}\right) = P\left(0 \le X \le \frac{1}{2}, 0 \le Y \le \frac{1}{2}\right)$

5-37. Suppose a bivariate probability density function has the form

$$f(x, y) = \begin{cases} 2 - x - y, & 0 < x < 1 \text{ and } 0 < y < 1; \\ 0 & \text{elsewhere.} \end{cases}$$

Determine:

(a) The marginal probability density functions of the continuous random variables X and Y

(b) The marginal cumulative distribution functions for X and Y

(c) The conditional probability density functions for X given Y (respectively, for Y given X)

5-38. What is the justification for using equation (5.20) for finding $P(a < X \le b, c < Y \le d)$? Can (5.20) simply be replaced by $F(b, d) - F(a, c)$?

5-39. Suppose (X, Y) is a bivariate random variable and X and Y are discrete (or continuous). If X and Y are independent random variables, which of the following is a valid equality?

(a) $E\left(\frac{X}{Y}\right) = E(X) \cdot E\left(\frac{1}{Y}\right)$

(b) $E\left(\frac{X}{Y}\right) = \frac{E(X)}{E(Y)}$

Verify your choice via the following random experiment. Vessel A contains three identical chips numbered 1, 2, 3; vessel B also contains three identical chips numbered 1, 2, 3. A chip is drawn at random from each vessel simultaneously. Construct the sample space S (there are nine simple events in S) by plotting the outcome obtained from vessel A (the random variable X) on the horizontal axis and the outcome obtained from vessel B (the random variable Y) on the vertical axis. Construct the discrete probability distribution for $X, Y, 1/Y$, and X/Y. Find $E(X), E(Y), E(1/Y)$, and $E(X/Y)$.

5-40. For the bivariate random variable (X, Y), demonstrate that $\rho_{XY} = 0$ when the probability density function is of the form

$$f(x,y) = \begin{cases} e^{-x-y}, & x > 0 \text{ and } y > 0; \\ 0 & \text{elsewhere.} \end{cases}$$

5-41. Suppose the bivariate random variable (X, Y) has a probability density function of the form

$$f(x,y) = \begin{cases} 2, & 0 < x < 1 \text{ and } x < y < 1; \\ 0 & \text{elsewhere.} \end{cases}$$

Find:

(a) $E(X)$

(b) $E(Y)$

(c) $V(X)$

(d) $V(Y)$

5-42. Verify that for X, Y independent random variables, $V(X + Y) = V(X) + V(Y)$. (Hint: Use $V(X + Y) = E\left[(X + Y)^2\right] - \left[(E(X + Y))\right]^2$, with $E(X + Y) = E(X) + E(Y)$ and $E(XY) = E(X) \cdot E(Y)$.)

5-43. Verify that for the bivariate random variable (X, Y), $E\left[\alpha(X) + \beta(Y)\right] = E\left(\alpha(X)\right) + E\left(\beta(Y)\right)$. Consider both the discrete and continuous cases.

5-44. For (X, Y) a bivariate random variable, suppose that $V(X)$ and $V(Y)$ exist. Then $V(aX + bY)$ exists and $V(aX + bY) = a^2V(X) + b^2V(Y) + 2abCOV(X, Y)$. Verify this result for X, Y continuous random variables.

5-45. For (X, Y) a bivariate random variable, suppose X and Y are independent random variables whose individual expectations exist. Then $E(XY)$ exists and $E(XY) = E(X) \cdot E(Y)$. Verify this result for X,Y continuous. (Hint: For X, Y independent, $f(x, y) = g(x) \cdot h(y)$.)

5-46. For (X, Y) a bivariate random variable, suppose that $V(X)$ and $V(Y)$ both exist. Then $COV(X, Y)$ exists. Comment on this assertion.

5-47. For the bivariate random variable (X, Y) having the probability density function

$$f(x, y) = \begin{cases} x^{-1}, & 0 < x < 1 \text{ and } 0 < y < x; \\ 0 & \text{elsewhere,} \end{cases}$$

find the marginal probability density functions $g(x)$ and $h(y)$.

5-48. For (X, Y) a bivariate random variable, suppose X and Y are independent random variables whose individual expectations exist. Then $COV(X, Y) = 0$. Verify this result.

5-49. Verify that for (X, Y) a bivariate random variable, $V(Y) = E[V(Y|X)] + V[E(Y|X)]$. (Hint: Begin with the expression $E[V(Y|X)] = E\{E(Y^2|X) - [E(Y|X)]^2\}$.)

5-50. Verify that for (X, Y) a bivariate random variable, $COV(X, Y) = E(XY) - E(X)E(Y)$.

5-51. Demonstrate that if (X, Y) is a bivariate random variable with σ_X and σ_Y finite, then $-1 \le \rho_{XY} \le 1$. (Hint: (a) For X, Y continuous, apply Schwarz's Inequality for two variables; and (b) for X, Y discrete, apply Cauchy's Inequality for a finite double sum.)
Note:

(1) For real-valued functions $\alpha(x, y)$ and $\beta(x, y)$ such that the integrals

$$\int_a^b \int_c^d [\alpha(x,y)]^2 \, dy \, dx \text{ and } \int_a^b \int_c^d [\beta(x,y)]^2 \, dy \, dx$$

exist $(-\infty \le a < b \le +\infty, -\infty \le c < d \le +\infty)$, it follows that

$$\left(\int_a^b \int_c^d \alpha(x,y)\beta(x,y) \, dy \, dx \right)^2 \le \left(\int_a^b \int_c^d [\alpha(x,y)]^2 \, dy \, dx \right) \left(\int_a^b \int_c^d [\beta(x,y)]^2 \, dy \, dx \right).$$

[*Schwarz's Inequality*]

(2) For finite double sums $\sum_i \sum_j a_{ij}^2$ and $\sum_i \sum_j b_{ij}^2$ (having the same number of terms in each subscript), it follows that

$$\left(\sum_i \sum_j a_{ij} b_{ij} \right)^2 \le \left(\sum_i \sum_j a_{ij}^2 \right) \left(\sum_i \sum_j b_{ij}^2 \right).$$

[*Cauchy's Inequality*]

5-52. Given the following bivariate probability density functions, find the joint moment-generating functions for the random variables X and Y:

(a) $f(x,y) = \begin{cases} 4xy, & 0 < x < 1, \ 0 < y < 1 \\ 0 & \text{elsewhere} \end{cases}$

(b) $f(x,y) = \begin{cases} x+y, & 0 < x < 1, \ 0 < y < 1 \\ 0 & \text{elsewhere} \end{cases}$

5-53. Suppose that X and Y are independent continuous random variables with bivariate probability density function $f(x,y)$. Verify that for $U = X + Y$, $m_U(t) = m_X(t) \cdot m_Y(t)$. (Hint: Use equation (5.29).)

5-54. Suppose that $X_1, \ldots, X_n$ are independent continuous random variables, where the probability density function of each X_i appears as

$$f^i(x) = \begin{cases} \frac{1}{\lambda} e^{-x_i/\lambda}, & x_i > 0, \ \lambda > 0, \ i = 1, \ldots, n; \\ 0 & \text{elsewhere.} \end{cases}$$

For $U = \sum_{i=1}^{n} X_i$, verify that the moment-generating function for U has the form

$$m_U(t) = (1 - \lambda t)^{-n}, \ t < \lambda^{-1}.$$

5-55. Suppose that (X, Y) is a bivariate random variable and that X and Y are independent continuous random variables with moment-generating functions $m_X(t)$ and $m_Y(t)$, respectively. Verify that for real numbers a, b, and c, $m_{aX+bY+c}(t) = e^{ct} m_X(at) \cdot m_Y(bt)$.

5-56. (*Bayes' Rule for discrete random variables*) Using (5.4) and (5.13), express (5.9) as

$$g(X|Y) = \frac{g(X)h(Y|X)}{\sum_i g(X)h(Y|X)}.$$

Similarly, show that (5.11) can be written, via (5.3) and (5.13), as

$$h(Y|X) = \frac{h(Y)g(X|Y)}{\sum_i h(Y)g(X|Y)}.$$

Let vessel A_1 contain one red (R) marble and two white (W) marbles and let vessel A_2 contain three red marbles and one white marble.

Suppose a marble is randomly drawn from vessel A_1 and placed into vessel A_2. Call its color the value of the random variable Y. Then a marble is randomly selected from vessel A_2. Call its color the value of random variable X. If $X = W$, what is the probability that $Y = R$? (Hint: Find $h(R|W)$.)

5-57. (*Bayes' Rule for continuous random variables*) We may posit the *multipli-cation theorem for probability density functions* as $f(x,y) = g(x)h(y|x) = h(y)g(x|y)$. Then using this expression, along with (5.22), demonstrate that (5.23) can be written as

$$g(x|y) = \frac{g(x)h(y|x)}{\int_{-\infty}^{+\infty} g(x)h(y|x)dx}.$$

Similarly, using the preceding multiplication theorem along with (5.21), verify that (5.24) can be expressed as

$$h(y|x) = \frac{h(y)g(x|y)}{\int_{-\infty}^{+\infty} h(y)g(x|y)dy}.$$

Given the probability density function

$$f(x,y) = \begin{cases} x+y, & 0 < x < 1 \text{ and } 0 < y < 1; \\ 0 & \text{elsewhere}, \end{cases}$$

use Bayes' Rule to find $g(x|y)$ and $h(y|x)$.

Discrete Parametric Probability Distributions

6.1 Introduction

In this chapter we shall examine a variety of important discrete (univariate) parametric probability distributions that have been found useful for modeling many types of real-life random phenomena often encountered in the actual practice of statistics. These are not *empirical distributions* arising from the repeated performance of some random experiment, but rather, *theoretical distributions* derived mathematically under a particular set of assumptions concerning a hypothetical sampling process; for example, assumptions about sampling with or without replacement, the types of events that may occur, the nature of the random variable involved, and so on.

These theoretical distributions are expressed formally by a set of rules that describe a family of probability mass functions in terms of certain parameters. In this regard:

1. The members of a *parametric family* of discrete theoretical distributions share the same generic rule for determining a probability mass function.

2. A *parameter* is an unspecified mathematical constant that enters into the rule that determines the theoretical probability mass function. It can assume any value within some range of possible values.

So by assigning specific values to the parameters in a theoretical probability mass function, we obtain a particular probability mass function that is a member of some parametric family of probability mass functions, and as we vary the value of the parameter, we generate a whole collection of different probability mass functions, each one corresponding to an admissible parameter value. For instance, given a random variable X and a single parameter α, let $f(X; \alpha)$ represent a rule that depicts some discrete theoretical probability mass function. If α is deemed a positive integer, then it serves to index a whole collection of different probability mass functions: $f(X;1), f(X;2), f(X;3), \ldots$. So as α ranges over the positive integers, we obtain the various members of the parametric family of discrete distributions $F = \{f(X;\alpha) | \alpha$ is a positive integer$\}$.

What types of parameters are useful in characterizing a family of discrete probability mass functions for a random variable X? The following list typifies the types of parameters used to structure probability mass functions:

count parameter—records the number of times a particular outcome occurs in a random experiment

proportion parameter—records the number of times a particular outcome occurs relative to the total number of trials of a random experiment

location parameter—specifies the position (relative to the origin) of the probability mass function on the X-axis

scale parameter—specifies the physical units in which a random variable is measured, thus influencing the dispersion of a probability mass function by compressing or elongating its graph

shape parameter—influences the form of a probability mass function (e.g., symmetry)

rate parameter—specifies the intensity of occurrence of the outcomes of some random process over time, space, or volume (e.g., the number of occurrences of an event in a given time period)

6.2 Counting Rules

This section is designed to develop a set of mathematical tools (generally called *combinatorial formulas*) that are useful for solving so-called *problems in counting*. For instance, these types of problems arise when we ask questions such as:

- Among a given set of alternatives, how many ways can a choice be made? (e.g., how many committees of three people can be chosen from a group of eight people?)

- How many ways are there of doing something? (e.g., how many ways are there of labeling four items with four different labels?)

- How many different ways can some event occur? (e.g., how many ways are there of getting three *tails* in ten flips of a fair coin?)
  And so on

Our first general counting technique is called the *Multiplication Principle*:

If sets $A_1, A_2, \ldots, A_r$ have, respectively, $n_1, n_2, \ldots, n_r$ elements, then the total number of different ways in which one can sequentially choose an element from A_1, then an element from $A_2, \ldots$, and finally an element from A_r is

$$n_1 \cdot n_2 \cdots n_r = \prod_{i=1}^{r} n_i. \qquad (6.1)$$

(Stated alternatively, if a process has steps $S_1, S_2, \ldots, S_r$ of which the first can be performed in any of n_1 ways, the second in any of n_2 ways, $\ldots$, and finally the rth in any of n_r ways, then all r steps can be performed sequentially in $n_1 \cdot n_2 \cdots n_r$ different ways.)

Intuitively, we can establish the validity of the multiplication principle by visualizing a *tree diagram* representing all the different ways in which the sequence of steps can be undertaken. There would be n_1 branches emanating from the starting node. From the nodes at the ends of these n_1 branches there would be n_2 new branches, each ending at a node from which n_3 new branches now emanate, and so on. The total number of different paths through the tree diagram would correspond to the product in (6.1).

Example 6.2.1 If A_1 has $n_1 = 2$ elements $\{a_{11}, a_{12}\}$, A_2 has $n_2 = 3$ elements $\{a_{21}, a_{22}, a_{23}\}$, and A_3 has $n_3 = 2$ elements $\{a_{31}, a_{32}\}$, then Figure 6.1 illustrates the implied tree diagram, which exhibits a total of $n_1 \cdot n_2 \cdot n_3 = 2 \cdot 3 \cdot 2 = 12$ different possible paths. ■

Let us now apply this multiplication principle to the situation in which repeated selections are made from the same set of elements and the order in which the elements are selected is important. Specifically, let r items be selected from a set of n distinct items ($r \leq n$) and arranged in a definite order. In general, any particular arrangement of these r items is called a *permutation*. (If any two of the r items have their respective positions interchanged, then we obtain a completely different permutation.) How many different permutations of n different items taken r at a time are there? The answer is provided by Theorem 6.1.

THEOREM 6.1. The number of different permutations of r items selected from a set of n distinct items ($r \leq n$) is

$$_nP_r = n(n-1)(n-2) \cdots (n-r+1). \qquad (6.2)$$

The symbol $_nP_r$ is read "the number of permutations of n distinct items taken r at a time."

The rationalization of (6.2) is straightforward. That is, there are n ways to make the first selection, $n-1$ ways to make the second selection, $n-2$ ways to make the third selection, and so on. Finally, there are $n - (r-1) = n - r + 1$ items to choose from to fill the rth position after the first $r-1$ selections have been made. Then by virtue of (6.1), (6.2) immediately follows given that $n_1 = n$, $n_2 = n-1$, $n_3 = n-2, \ldots, n_r = n - r + 1$.

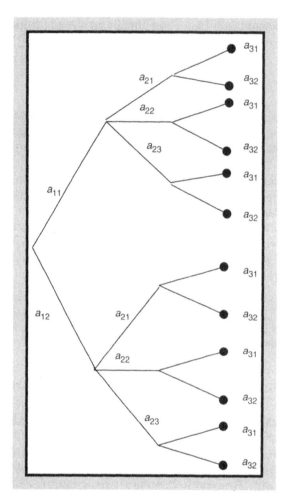

Figure 6.1   Determining the total number of paths through a tree diagram when $n_1 = 2$, $n_2 = 3$, and $n_3 = 2$.

Let

$$n! = n(n-1)(n-2)\cdots 2\cdot 1 \tag{6.3}$$

(the symbol $n!$ is read "n factorial"), where $0! \equiv 1$.[1] Then if (6.2) is multiplied by $\frac{(n-r)!}{(n-r)!}$, we obtain an alternative formula for $_nP_r$ or

$$_nP_r = \frac{n!}{(n-r)!}, \tag{6.2.1}$$

---

[1] When n is large, $n!$ can be approximated by *Stirling's formula*:

$$n! \approx (n/e)^n\sqrt{2\pi n},$$

where $e \approx 2.718282$ and $\pi = 3.1416$. As n increases in value, the error of this approximation approaches zero.

where n is a positive integer and $r = 0, 1, \ldots, n$. If $n = r$, then we seek to determine the total number of permutations of n distinct items taken all together or

$$_nP_n = n! \qquad (6.2.2)$$

Example 6.2.2   Given the set of letters {a,b,c,d,e}, the number of permutations of these five letters taken three at a time is $_5P_3 = \frac{5!}{(5-3)!} = 60$. And the number of permutations of these five letters taken all together is $_5P_5 = 5! = 120$.   ∎

Up to this point in our discussion of permutations we have been assuming that the n items are all distinct. However, it may be the case that a number of items may display certain similarities. In this regard, let us assume that within the collection of n items there are k distinct types of items, of which there are r_1 of the first type, r_2 of the second type,..., and finally r_k of the kth type, where $r_1 + r_2 + \ldots + r_k = n$. We then have Theorem 6.2:

The total number of permutations of n items of which r_1 are alike, r_2 are alike, ..., and r_k are alike is

$$\binom{n}{r_1, r_2, \ldots, r_k} = \frac{n!}{r_1! r_2! \ldots r_k!} = \frac{n!}{\prod\limits_{i=1}^{k} r_i!}, \qquad (6.4)$$

where the $r_i, i = 1, \ldots, k$, are nonnegative integers and $\sum_{i=1}^{k} r_i = n$.

Example 6.2.3   Let us determine the total number of permutations of the letters in the word *potato*. Here there is one *p*, one *a*, two *o*'s, and two *t*'s. Setting $r_1 = 1$, $r_2 = 1$, $r_3 = 2$, and $r_4 = 2$ in (6.4) yields

$$\binom{6}{1,1,2,2} = \frac{6!}{1!\,1!\,2!\,2!} = 180. \quad ∎$$

A permutation is an ordered arrangement of items, whereas a *combination* is an arrangement of objects without regard to order; that is, we are interested in *what items are selected* irrespective of the order in which their actual selection occurs. (So although *abc* and *bac* are different permutations of the three letters *a*, *b*, and *c*, they represent the same combination of these three letters.) Based upon this definition we may now address the question, how many combinations of n distinct items taken r at a time are there? To answer this we look to Theorem 6.3.

THEOREM 6.3.   The number of different combinations of r items selected from a set of n distinct items $(r \leq n)$ is

$$\binom{n}{r} = \frac{n!}{r!(n-r)!},\tag{6.5}$$

where n is a positive integer and $r = 0, 1, \ldots, n$.

The symbol $\binom{n}{r}$ is read "the number of combinations of n distinct items taken r at a time." If $n = r$, then we are interested in the total number of combinations of n items taken all together or $\binom{n}{n} = 1$. And since $\binom{n}{r} = \frac{nP_r}{r!}$, it is evident that the number of combinations of n distinct items taken r at a time is smaller than the number of permutations of the n items taken r at a time.

To verify (6.5) we need to remember only that a permutation is obtained by first selecting r of the n distinct objects and then arranging them in a given order, and a combination is obtained by performing only the first step of this process. Hence the total number of permutations is obtained by taking every possible combination and arranging them in all possible ways. Since the total number of arrangements of r items taken all together is $r!$, it follows that $_nP_r = \binom{n}{r} r!$, from which (6.5) immediately obtains.

Example 6.2.4   How many committees of $r = 5$ people each can be selected from a group of $n = 10$ people? Since the order of selection for membership on a committee is irrelevant, our answer is provided by

$$\binom{10}{5} = \frac{10!}{5! \, 5!} = 252. \quad \blacksquare$$

It is instructive to view (6.5) in a slightly different light. Since a combination of r items selected from n distinct items is just a subset of r elements (since the elements within a set or subset represent an amorphous collection of items and they do not display any particular structure or order), we may note that the number of subsets of size r that a set of n distinct items has is just the number of combinations of n things taken r at a time and thus given by (6.5).

As a practical consideration in solving certain types of counting problems, we may wonder whether the number of permutations or combinations should be computed for a given n and r. The choice is clear: If order is important, compute permutations; if order is not important, find combinations.

A few additional comments pertaining to these counting issues merit our attention. First, the combinatorial expression $\binom{n}{r}$ often is termed a *binomial coefficient* since it is the coefficient on the term $x^r y^{n-r}$ in the *binomial expansion* of $(x+y)^n$; that is, for x and y any two real numbers and n a positive integer, it can be shown that

$$(x + y)^n = \sum_{r=0}^{n} \binom{n}{r} x^r y^{n-r}. \tag{6.6}$$

To expedite the calculation of these binomial coefficients we may employ the *recursion formula*

$$\binom{n}{r+1} = \frac{n-r}{r+1} \binom{n}{r}, \quad r = 0, 1, \ldots, n. \tag{6.7}$$

Example 6.2.5   To see how (6.7) is applied, let us select $n = 5$ with $r = 0, 1, \ldots, 5$. Since $\binom{5}{0} = 1$, we have:

$$\binom{5}{1} = \frac{5-0}{0+1} \binom{5}{0} = 5 \cdot 1 = 5;$$

$$\binom{5}{2} = \frac{5-1}{1+1} \binom{5}{1} = 2 \cdot 5 = 10;$$

$$\binom{5}{3} = \frac{5-2}{2+1} \binom{5}{2} = 1 \cdot 10 = 10;$$

$$\binom{5}{4} = \frac{5-3}{3+1} \binom{5}{3} = \frac{1}{2} \cdot 10 = 5;$$

$$\binom{5}{5} = \frac{5-4}{4+1} \binom{5}{4} = \frac{1}{5} \cdot 5 = 1. \quad \blacksquare$$

As an alternative to (6.7) we may construct *Pascal's triangle* (see Figure 6.2) to determine binomial coefficients. Here we first write 1's down along the sides of the triangle. Then any number in the triangle is simply the sum of the two adjacent numbers in the row immediately above. So to find the binomial coefficients $\binom{5}{r}$, $r = 0, 1, \ldots, 5$, we look in the row corresponding to $n = 5$ and determine where the diagonal line corresponding to $r = 0$, then to $r = 1, \ldots$, and finally to $r = 5$ intersects this row.

Finally, the combinatorial expression (6.4) is oftentimes termed a *multinomial coefficient* since it is the coefficient on the term $x_1^{r_1} x_2^{r_2} \cdots x_k^{r_k}$, where the r_i's, $i = 1, \ldots, k$, are nonnegative integers that sum to n in the *multinomial expansion* of $(x_1 + x_2 + \cdots + x_k)^n$; that is, for the x_j's, $j = 1, \ldots, k$, real numbers and n a positive integer, it can be shown that

$$\left(\sum_{j=1}^{k} x_j \right)^n = \sum_{r_1 + r_2 + \cdots + r_k = n} \binom{n}{r_1, r_2, \ldots, r_k} \prod_{i=1}^{k} x_i^{r_i}. \tag{6.8}$$

Here (6.8) serves as a generalization of the binomial expansion (6.6).

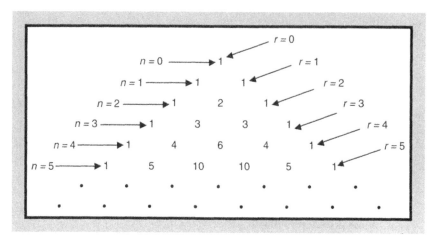

Figure 6.2 Binomial coefficients determined via Pascal's triangle.

Armed with the preceding set of counting rules, we may now examine an assortment of important discrete parametric probability distributions. By far the simplest such distribution is the discrete uniform distribution.

6.3 Discrete Uniform Distribution

Let a discrete random variable X take on k different values $X = 1, 2, \ldots, k$, with each value of X having the same probability $\frac{1}{k}$. Then the probability mass function for the *discrete uniform distribution* appears as

$$f(X; k) = \begin{cases} 1/k, & X = 1, 2, \ldots, k; \\ 0 & \text{elsewhere} \end{cases} \qquad (6.9)$$

where the parameter k ranges over the positive integers (see Figure 6.3). And for each different value of k, we obtain a whole family of individual discrete uniform distributions. For instance, if $k = 6$, the resulting distribution is provided by Table 6.1.

For the discrete uniform distribution the mean and standard deviation are, respectively,

$$\mu = E(X) = \frac{k+1}{2}, \qquad \sigma = \sqrt{V(X)} = \left(\frac{k^2 - 1}{12}\right)^{1/2}. \qquad (6.10)$$

At times a discrete uniform distribution may be represented in a slightly different but equivalent fashion. To this end let a and b be positive or negative integers

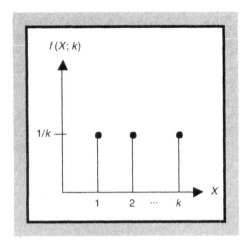

Figure 6.3   Probability mass function for a discrete uniform distribution.

Table 6.1

X	$f(X;6)$
1	1/6
2	1/6
3	1/6
4	1/6
5	1/6
6	1/6
	1

such that $a \leq X \leq b$. Then an alternative version of the probability mass function (6.9) is

$$f(X;k) = \begin{cases} \frac{1}{b-a+1}, & X = a, a+1, \ldots, b-1, b; \\ 0 & \text{elsewhere} \end{cases} \qquad (6.9.1)$$

where now both a and b serve as parameters. Given (6.9.1) we now have

$$\mu = E(X) = \frac{a+b}{2}, \quad \sigma = \sqrt{V(X)} = \left[\frac{(b-a+1)^2 - 1}{12} \right]^{1/2}. \qquad (6.10.1)$$

6.4   The Bernoulli Distribution

Let us define a *simple alternative experiment* (or a *Bernoulli experiment*) as one that admits only two possible outcomes, success or failure; for example, some event either occurs (we record a success) or does not occur (we record a failure).

Here the sample space S consists of only the two simple events: $E_1 = $ success; $E_2 = $ failure. Then the associated discrete random variable X (called a *Bernoulli variable*) takes on only two possible values: If a success occurs, we set $X = 1$; if a failure occurs, we set $X = 0$. In addition, let the probability of a success (or *Bernoulli probability*) be denoted as p. Then, with only two possible simple events in S, the probability of a failure must be $1 - p$.

Based upon these considerations, let a random experiment consist of a single trial of a simple alternative (called a *Bernoulli trial*). Then the random variable X follows a *Bernoulli probability distribution* with probability mass function

$$b\,(X;p) = \begin{cases} p^X(1-p)^{1-X}, & X = 0 \text{ or } 1, \quad 0 \le p \le 1; \\ 0 & \text{elsewhere} \end{cases} \qquad (6.11)$$

where p serves as a parameter that defines a family of Bernoulli probability distributions. An alternative representation of a Bernoulli distribution is provided by Table 6.2; its accompanying graph (for $p > \frac{1}{2}$) is illustrated by Figure 6.4.

Table 6.2

X	$b(X;p)$
0	$1 - p$
1	p
	1

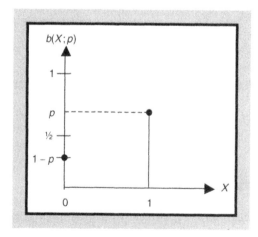

Figure 6.4   Bernoulli probability mass function, $p > \frac{1}{2}$.

For X a Bernoulli random variable, its mean and standard deviation are, respectively,

$$\mu = E(X) = p, \quad \sigma = \sqrt{V(X)} = \sqrt{p(1-p)}; \quad\quad (6.12)$$

and its coefficients of skewness and kurtosis are

$$\alpha_3 = \frac{1-2p}{\sqrt{p(1-p)}}, \quad \alpha_4 = \frac{1-3p(1-p)}{p(1-p)}, \quad\quad (6.13)$$

respectively.

It is easily verified that:

(a) for $p = \frac{1}{2}$, $\alpha_3 = 0$ and $\alpha_4 = 1$ (the minimum value of α_4)

(b) for $p < \frac{1}{2}$, $\alpha_3 > 0$ and $\alpha_4 > 1$

(c) for $p > \frac{1}{2}$, $\alpha_3 < 0$ and $\alpha_4 > 1$

6.5   The Binomial Distribution

As indicated in the preceding section, the Bernoulli probability distribution arises from a single trial of a Bernoulli or simple alternative experiment. Let us generalize this experiment to n separate Bernoulli trials while sampling with replacement. What results then is a *Bernoulli process*—a series of repeated or identical independent Bernoulli trials. If the trials are identical, then the probability of a success p must be constant from trial to trial. In this regard, we may think of the binomial probability distribution as arising from a Bernoulli process with p constant from trial to trial. To summarize, the essential characteristics of a binomial experiment are:

• X is a discrete random variable

• We have a simple alternative experiment

• The trials are identical and independent

• p is constant from trial to trial

Let the random variable X depict the number of successes obtained in n independent trials of a Bernoulli experiment. In particular, we are interested in the probability of getting $X = k(\leq n)$ successes in the n trials. To simplify our analysis, let us assume that we get all k successes on the first k trials. Then we must have $n - k$ failures on the last $n - k$ trials. As was the case for the Benoulli experiment, when a success occurs we set $X = 1$ with $P(\text{success}) = p$; when a failure occurs,

$X = 0$ and $P(\text{failure}) = 1 - p$. Hence, under the independence of these events,

$P(k$ successes on the first k trials $\cap\ n - k$ failures on the last $n - k$ trials$)$

$$= P(k \text{ 1's}) \cdot P(n - k \text{ 0's})$$

$$= \underbrace{(pp \cdots p)}_{\text{taken } k \text{ times}} \cdot \underbrace{\left((1-p)(1-p)\cdots(1-p)\right)}_{\text{taken } n-k \text{ times}} = p^k(1-p)^{n-k}. \qquad (6.14)$$

But this is only one way to get k successes in n independent trials. (In fact, the probability of any given sequence of outcomes in n independent Bernoulli trials depends only on the number of successes k and p. Hence the probability of k successes and $n - k$ failures in any other order is exactly the same as that provided by (6.14).) Since we are interested in getting k successes irrespective of the order in which they occur, the total number of ways or orders in which k successes can occur is $\binom{n}{k}$. Hence the probability of k successes and $n - k$ failures in any order in n independent trials is

$$b(X = k;\ n,p) = \binom{n}{k} p^k(1-p)^{n-k}. \qquad (6.14.1)$$

In light of this discussion, let us define the random variable X as the number of successes obtained in the n outcomes of a Bernoulli process. Then X follows a *binomial probability distribution* with probability mass function

$$b(X;n,p) = \begin{cases} \binom{n}{X} p^X(1-p)^{n-X}, & X = 0,1,\ldots,n, \\ & 0 \le p \le 1,\ n \text{ is a positive integer;} \\ 0 & \text{elsewhere} \end{cases} \qquad (6.15)$$

where the parameters n and p serve to define an entire family of binomial probability distributions. Based upon our rationalization of the form of a binomial probability distribution in (6.15), it should be evident that

$$b(X;n,p) = b(n - X;n,\ 1 - p). \qquad (6.15.1)$$

(Note that this transformation has us essentially interchange the labels *success* and *failure*.) An alternative representation of a binomial distribution is provided by Table 6.3.

The binomial probability distribution is so named because the probabilities given in (6.15) are the successive terms in the binomial expansion (6.6)

$$(p + (1-p))^n = \sum_{X=0}^{n} \binom{n}{X} p^X(1-p)^{n-X} = \sum_{X=0}^{n} b(X;n,p) = 1.$$

A few additional comments pertaining to the binomial distribution are in order. First, note that for $n = 1$, the binomial probability mass function (6.15) reduces

Table 6.3

X	$b(X; n, p)$
0	$b(0; n, p)$
1	$b(1; n, p)$
$\vdots$	$\vdots$
n	$\underline{b(n; n, p)}$
	1

to (6.11), the Bernoulli probability mass function. Also, to expedite the calculation of successive binomial probabilities, we may employ the *binomial recursion formula*

$$b(X + 1; \, n, p) = \frac{(n - X)p}{(X + 1)(1 - p)} \, b(X; \, n, p), \; X = 0, 1, \ldots, n - 1. \qquad (6.16)$$

Next, the probability that a binomial random variable X assumes a value less than or equal to some given integer level x, $0 \leq x \leq n$, is provided by the *binomial cumulative distribution function*

$$P(X \leq x) = B(x; n, p) = \sum_{i \leq x} \binom{n}{i} \, p^i (1 - p)^{n - i} = \sum_{i \leq x} b(i; n, p). \qquad (6.17)$$

Given this expression, we can also find

$$P(X > x) = P(X \geq x + 1) = 1 - P(X \leq x) = 1 - B(x; n, p). \qquad (6.18)$$

Finally, with X integer valued, we may employ (6.17) to calculate the binomial probabilities given by (6.15); that is,

$$b(X = x; \, n, p) = B(x; \, n, p) - B(x - 1; n, p). \qquad (6.15.1)$$

Example 6.5.1   Let a random experiment consist of rolling a pair of fair six-sided dice five times in succession and let a success be defined as getting an eight as the sum of the faces showing. What is the probability of obtaining two eights in the five rolls? Since P(the sum is eight)$= 5/36$, we have, from (6.15),

$$b\left(2; \, 5, \, \frac{5}{36}\right) = \binom{5}{2}\left(\frac{5}{36}\right)^2\left(\frac{31}{36}\right)^3 = 0.1232.$$

Table 6.4

X	b(X; 5, 5/36)
0	0.4735
1	0.3818
2	0.1232
3	0.0199
4	0.0016
5	0.0000
	1.0000

To determine the entire binomial probability distribution associated with this experiment let us calculate

$$b\left(X; 5, \frac{5}{36}\right) = \binom{5}{X}\left(\frac{5}{36}\right)^X \left(\frac{31}{36}\right)^{5-X}, \quad X = 0, 1, \ldots, 5.$$

Our results are depicted in Table 6.4. For instance, when $X = 0$ we have

$$b\left(0; 5, \frac{5}{36}\right) = \binom{5}{0}\left(\frac{5}{36}\right)^0 \left(\frac{31}{36}\right)^5 = 0.4735;$$

and for $X = 1$ we obtain

$$b\left(1; 5, \frac{5}{36}\right) = \binom{5}{1}\left(\frac{5}{36}\right)^1 \left(\frac{31}{36}\right)^4 = 0.3818.$$

To determine the binomial probabilities when $X = 3$, 4, and 5, let us use the recursion formula (6.16). Hence

$$b\left(3; 5, \frac{5}{36}\right) = \frac{(5-2)\left(\frac{5}{36}\right)}{(2+1)\left(\frac{31}{36}\right)} b\left(2; 5, \frac{5}{36}\right) = 0.0199;$$

$$b\left(4; 5, \frac{5}{36}\right) = \frac{(5-3)\left(\frac{5}{36}\right)}{(3+1)\left(\frac{31}{36}\right)} b\left(3; 5, \frac{5}{36}\right) = 0.0016;$$

$$b\left(5; 5, \frac{5}{36}\right) = \frac{(5-4)\left(\frac{5}{36}\right)}{(4+1)\left(\frac{31}{36}\right)} b\left(4; 5, \frac{5}{36}\right) = 0.0000.$$

What is the probability of no more than two eights in the five rolls? Here we want $P(X \leq 2)$ or, from (6.17),

$$B\left(2; 5, \frac{5}{36}\right) = \sum_{i \leq 2} b\left(i; 5, \frac{5}{36}\right)$$

$$= b\left(0; 5, \frac{5}{36}\right) + b\left(1; 5, \frac{5}{36}\right) + b\left(2; 5, \frac{5}{36}\right) = 0.9785.$$

Moreover, the probability of at least three eights in the five rolls is, from (6.18),

$$P(X > 2) = P(X \geq 3) = 1 - P(X \leq 2) = 1 - B\left(2; 5, \frac{5}{36}\right) = 0.0215.$$

Finally, from (6.15.2), we can easily determine, for instance, that

$$b\left(3; 5, \frac{5}{36}\right) = B\left(3; 5, \frac{5}{36}\right) - B\left(2; 5, \frac{5}{36}\right) = 0.9984 - 0.9785 = 0.0199$$

as expected.   ■

To offer expressions for the mean and standard deviation of a binomial random variable X we need only note that since a binomial variable can be represented as a sum of identical and independent Bernoulli random variables X_i, $i = 1, \ldots, n$, each with mean p and variance $p(1 - p)$, it follows that for $X = \sum_{i=1}^{n} X_i$,

$$\text{(a)} \quad \mu = E(X) = \sum_{i=1}^{n} E(X_i) = np;$$

$$\text{(b)} \quad \sigma^2 = V(X) = \sum_{i=1}^{n} V(X_i) = np(1 - p)$$

(6.19)

and thus

$$\sigma = \sqrt{V(X)} = \sqrt{np(1 - p)}.$$

(6.20)

(For $n = 1$ in (6.19) and (6.20), the Bernoulli subcases obtain as in (6.12).) Moreover, the coefficients of skewness and kurtosis for a binomial distribution are, respectively,

$$\alpha_3 = \frac{1 - 2p}{\left(np(1 - p)\right)^{1/2}},$$

(6.21)

$$\alpha_4 = 3 + \frac{(1 - 6p(1 - p))}{np(1 - p)}.$$

(6.22)

Based upon the preceding discussion we can now examine some of the important properties of the binomial probability distribution:

1. For a binomial random variable X, $E(X) > V(X)$.

2. A binomial distribution is symmetrical ($\alpha_3 = 0$) if $p = \frac{1}{2}$; otherwise it is skewed. In this regard, it is:

   (a) Positively skewed ($\alpha_3 > 0$) if $p < \frac{1}{2}$

   (b) Negatively skewed ($\alpha_3 < 0$) if $p > \frac{1}{2}$

3. The binomial distribution is relatively flat-topped when $p = \frac{1}{2}$ since α_4 attains its minimum value at $p = \frac{1}{2}$. For any other value at p, the binomial distribution has a relatively high peak. And as $n \to \infty$, $\alpha_4 \to 3$ (the benchmark value of α_4 for the normal distribution) for any p.

4. As the binomial random variable X takes on values from 0 to n, the terms $b(X;n,p)$ first increase monotonically if $X < (n+1)p$; they then decrease monotonically if $X > (n+1)p$. The binomial terms $b(X;n,p)$ attain their maximum value when $X = m = (n+1)p$. (In this instance m is termed the "most probable number of successes.") If m turns out to be an integer, then $b(m;n,p) = b(m-1;n,p)$ and the largest binomial probability is not unique. In fact, there exists exactly one such integer m for which $(n+1)p - 1 \le m \le (n+1)p$.

5. For $p = \frac{1}{2}$, the binomial probability mass function has a unique maximum at $X = \frac{n}{2}$ when n is even; when n is odd, this function has maxima at $X = \frac{(n-1)}{2}$ and $X = \frac{(n+1)}{2}$.

Example 6.5.2   Turning back to the binomial probability distribution provided by Table 6.4, it is readily verified that

$$\mu = 5\left(\frac{5}{36}\right) = 0.6944, \quad \sigma = \left(5\left(\frac{5}{36}\right)\left(\frac{31}{36}\right)\right)^{1/2} = 0.7733,$$

$$\alpha_3 = \frac{1 - 2\left(\frac{5}{36}\right)}{0.7733} = 0.9339, \text{ and}$$

$$\alpha_4 = 3 + \frac{\left(1 - 6\left(\frac{5}{36}\right)\left(\frac{31}{36}\right)\right)}{5\left(\frac{5}{36}\right)\left(\frac{31}{36}\right)} = 3.4722.$$

Hence this particular binomial distribution is positively skewed and its peak is a bit sharper than that of a normal distribution. And since $X = 0 < (n+1)p = 6\left(\frac{5}{36}\right) = 0.8333$ and $X > (n+1)p = 0.8333$ for $X = 1,\ldots,5$, it is evident that the sequence of binomial values $b(X;\ 5,\ \frac{5}{36})$ is monotonically decreasing for $X = 0,1,\ldots,5$.   ∎

For many binomial experiments the determination of binomial probabilities $b(k;\ n,p)$ for various values of n and p can easily be determined by appealing to Table A.6 of the Appendix. For instance, from the first page of this table we can easily find $b(3;\ 6,\ 0.20) = 0.0819$. Note that the largest value of p in this table is 0.50. If $p > 0.50$, we may use equation (6.15.1) (or $b(X;\ n,p) = b(n - X;\ n,\ 1 - p)$) to readily calculate the desired binomial probability. For instance, to find $b(k;\ n,p) = (3;\ 8,\ 0.70)$ we simply find instead $b(n - k;\ n,p) = b(5;\ 8,\ 0.30) = 0.0467$.

6.6   **The Multinomial Distribution**

This particular probability distribution is a generalization of the binomial distribution to the case where the n identical and independent trials of a random experiment give rise to more than two possible outcomes whose associated probabilities are constant from trial to trial (we again sample with replacement or from an infinite population). In this regard, let each of the n trials represent a *k-fold alternative*; that is, each trial results in any one of k mutually exclusive and collectively exhaustive outcomes $E_1, \ldots, E_k$ with respective probabilities $P(E_1) = p_1, \ldots, P(E_k) = p_k$ and $\sum_{i=1}^{k} p_i = 1$. In addition, let the random variable X_i, $i = 1, \ldots, k$, depict the number of times the ith type of outcome (call it the ith success) E_i occurs in the n trials. To summarize, the essential characteristics of a multinomial experiment are:

- Each trial gives rise to a k-fold alternative

- The random variables X_i, $i = 1, \ldots, k$, associated with the k outcomes are discrete

- The trials are identical and independent

- The probabilities p_i, $i = 1, \ldots, k$, associated with the k outcomes are constant from trial to trial

Based upon these considerations, we now seek to answer the following question. What is the probability of getting $X_1 = x_1$ outcomes of the first type, $X_2 = x_2$ outcomes of the second type,..., and $X_k = x_k$ outcomes of the kth type in the n trials? Here the x_i, $i = 1, \ldots, k$, are arbitary nonnegative integers such that $\sum_{i=1}^{k} x_k = n$. To offer an answer let us assume that we get x_1 outcomes of type one on the first x_1 trials, x_2 outcomes of type two on the next x_2 trials,..., and finally x_k outcomes of type k on the last x_k trials. Then by virtue of an argument similar to the one underlying the derivation of (6.14), we can readily see that for this *specific assortment of outcomes*,

$$P(x_1 \text{ outcomes of type } 1 \cap x_2 \text{ outcomes of type }$$

$$2 \cap \cdots \cap x_k \text{ outcomes of type } k)$$

$$= \prod_{i=1}^{k} P(x_i \text{ outcomes of type } i)$$

$$= \prod_{i=1}^{k} \underbrace{(p_i \, p_i \cdots p_i)}_{\text{taken } x_i \text{ times}} = p_1^{x_1} p_2^{x_2} \cdots p_k^{x_k}. \qquad (6.23)$$

But this expression, in general, applies to *any particular sequence of n trials* yielding x_1 outcomes of the first kind, x_2 outcomes of the second kind,..., and x_k

outcomes of the kth kind. Since ultimately we are interested in getting this assortment of outcomes irrespective of the order in which they occur, the total number of ways or orders in which this collection of outcomes can occur is given by the multinomial coefficient $\frac{n!}{x_1!x_2!\cdots x_k!}$. Hence the desired probability is the product of this coefficient and (6.23) or

$$m(X_1 = x_1,\ldots, X_k = x_k; n, p_1,\ldots, p_k) = \binom{n}{x_1, x_2, \ldots, x_k} p_1^{x_1} p_2^{x_2} \cdots p_k^{x_k}.$$
(6.23.1)

On the basis of the preceding discussion let us define the random variable X_i, $i = 1,\ldots,k$, as the number of times the ith outcome type occurs in n identical and independent trials of a k-fold alternative experiment. Then the X_i follow a *multinomial probability distribution* with probability mass function

$$m(X_1,\ldots,X_k; n, p_1,\ldots,p_k) = \begin{cases} \binom{n}{X_1, X_2, \ldots, X_k} p_1^{X_1} p_2^{X_2} \cdots p_k^{X_k}, \\ \quad X_i = 0, 1,\ldots,n, \sum_{i=1}^{k} X_i = n, \\ \quad 0 \le p_i \le 1 \text{ for each } i, \\ \quad n \text{ a positive integer}; \\ 0 \quad \text{elsewhere.} \end{cases}$$
(6.24)

Here the parameters $n, p_1,\ldots,p_k$ define an entire family of multinomial probability distributions. This distribution gets its name from the fact that the probabilities provided by (6.24) are the successive terms of the multinomial expansion (6.8)

$$\left(\sum_{i=1}^{k} p_i\right)^n = \sum_{X_1+X_2+\cdots+X_k=n} \binom{n}{X_1, X_2, \ldots, X_k} \prod_{i=1}^{k} p_i^{X_i}$$

$$= \sum_{X_1+X_2+\cdots+X_k=n} m(X_1,\ldots,X_k; n, p_1,\ldots,p_k) = 1.$$
(6.25)

Note also that the binomial probability distribution (6.15) is a special case of (6.24) for $k = 2$, $X_1 = X$, $X_2 = n - X$, $p_1 = p$, and $p_2 = 1 - p$.

Example 6.6.1 Let us assume that in a particular city 50% of the registered voters are Democrats, 30% are Republicans, 15% are Independents, and 5% are Libertarians. Given a random sample of 20 individuals taken as an exit poll during the last local election, what is the probability that of those polled, eight are Democrats, six are Republicans, three are Independents, and three are Libertarians? Here $n = 20$ and

$$
\begin{array}{ll}
x_1 = 8 & p_1 = 0.50 \\
x_2 = 6 & p_2 = 0.30 \\
x_3 = 3 & p_3 = 0.15 \\
x_4 = \dfrac{3}{20} & p_4 = \dfrac{0.05}{1.00}
\end{array}
$$

Then from (6.23.1),

$$
m(x_1 = 8, x_2 = 6, x_3 = 3, x_4 = 3; n = 20, p_1 = 0.50, p_2 = 0.30, p_3 = 0.15, p_4 = 0.05)
$$

$$
= \frac{20!}{8!\,6!\,3!\,3!}(0.50)^8(0.30)^6(0.15)^3(0.05)^3 = 0.0028. \quad \blacksquare
$$

At times, we may be interested only in the occurrence of some proper subset of the k possible outcomes of a multinomial experiment. For instance, let us assume that we are interested in, say, only the first $j < k$ outcomes of the original k-fold alternative experiment. That is, we choose to focus only on the probability of occurrence of $X_1 = x_1$ outcomes of type 1, $X_2 = x_2$ outcomes of type 2,..., and $X_j = x_j$ outcomes of type j. Since the last $k - j$ possible outcomes are of no interest to us, we can formulate a new $(j + 1)$-fold alternative experiment with outcomes $E_1, E_2, \ldots, E_j, E'_{j+1} = \{E_{j+1}, E_{j+2}, \ldots, E_k\}$, where E'_{j+1} collectively contains the last $k - j$ outcomes or events of the original multinomial experiment. In this regard, we seek to find the probability of occurrence of $X_1 = x_1$ outcomes of type 1, $X_2 = x_2$ outcomes of type 2,..., $X_j = x_j$ outcomes of type j and $X'_{j+1} = x'_{j+1} = n - \sum_{r=1}^{j} x_j$ outcomes of type $j + 1$.

Clearly the probabilities associated with these $j + 1$ mutually exclusive and collectively exhaustive events are $P(E_1) = p_1, P(E_2) = p_2, \ldots, P(E_j) = p_j$, and $P(E'_{j+1}) = p'_{j+1} = \sum_{r=j+1}^{k} p_r = p_{j+1} + \cdots + p_k$ with $\sum_{r=1}^{j} p_r + p'_{j+1} = 1$. Then by virtue of (6.23.1), the desired probability is

$$
m\left(x_1, \ldots, x'_{j+1}; n, p_1, \ldots, p_j, p'_{j+1}\right) = \binom{n}{x_1, \ldots, x_j, x'_{j+1}} p_1^{x_1} \cdots p_j^{x_j} \left(p'_{j+1}\right)^{x'_{j+1}}.
$$

$$(6.23.2)$$

Example 6.6.2   What if in the preceding example problem we are now interested in the probability that, of those polled, five are registered Republicans and one is a registered Independent.

For $n = 20$ we now have

$$X_2 = 5 \qquad\qquad p_2 = 0.30$$
$$X_3 = 1 \qquad\qquad p_3 = 0.15$$
$$X' = 20 - 6 = 14 \qquad p' = p_1 + p_4 = 0.55$$

Then from (6.23.2),

$$m(X_2 = 5,\; X_3 = 1,\; X' = 14;\; n = 20,\; p_2 = 0.30,\; p_3 = 0.15,\; p' = 0.55)$$

$$= \frac{20!}{5!1!4!}(0.30)^5(0.15)(0.55)^{14} = 1.88E - 11. \quad\blacksquare$$

6.7 The Geometric Distribution

As was the case for the binomial probability distribution, let the outcomes of a random experiment be generated by a Bernoulli process; that is, by a series of repeated independent trials of a Bernoulli (simple alternative) experiment with p (the probability of a success) constant from trial to trial. But unlike the binomial case in which the random variable X depicts the number of successes obtained in a fixed number (n) of independent trials, let us now consider the random variable X to be the number of the trial on which the *first success* is obtained. In this instance the trials, although not fixed in number, are still independent and thus the probability of a success p is again constant from trial to trial. To summarize, the basic characteristics of this so-called geometric experiment are:

- The number of the trial X on which the first success is observed is a discrete random variable
- We have a simple alternative experiment
- The trials are identical and independent
- p is constant from trial to trial

Now, the probability of $x - 1$ failures in a row is $(1 - p)^{x-1}$ and the probability of actually obtaining a success on trial $X = x$ is p. Hence the probability of getting the first success on the xth trial after $x - 1$ consecutive failures is, under independence,

$$P(\text{a success on trial } x \cap x - 1 \text{ failures on the first } x - 1 \text{ trials})$$

$$= g(X = x;\; p) = p(1 - p)^{x-1}. \tag{6.26}$$

Based upon these considerations, let us define the random variable X as the trial on which the first success is observed in a Bernoulli process. Then X follows a *geometric probability distribution* with probability mass function

$$g(X;p) = \begin{cases} p(1-p)^{X-1}, & X = 1,2,3,\dots, \ 0 \le p \le 1; \\ 0 & \text{elsewhere.} \end{cases} \tag{6.27}$$

And as the parameter p is varied, we generate an entire family of geometric probability distributions. Note that the calculation of geometric probabilities can be expedited by employing the identity

$$g(X;p) = \frac{1}{X}b(1;X,p), \tag{6.28}$$

where $b(1;X,p)$ is the (tabular) value of the binomial probability mass function. Here the usual binomial random variable (the number of successes) is set at one and X varies over binomial trials.

For the geometric probability distribution, we may define the *geometric cumulative distribution function* as

$$P(X \le x) = G(i;p) = \sum_{i \le x} g(i;p) = \sum_{i \le x} p(1-p)^{i-1}, \ x = 1,2,3,\dots, \tag{6.29}$$

or

$$P(X \le x) = 1 - (1-p)^x, \ x = 1,2,3,\dots. \tag{6.30}$$

In addition, it can be shown that $\sum_{i=1}^{\infty} g(i;p) = 1$ as required.

Expressions for the mean and standard deviation of a geometric random variable are, respectively,

$$\mu = E(X) = \frac{1}{p}; \ \sigma = \sqrt{V(X)} = \frac{\sqrt{1-p}}{p}. \tag{6.31}$$

Moreover, the coefficients of skewness and kurtosis are, respectively,

$$\alpha_3 = \frac{(2-p)}{\sqrt{1-p}}; \tag{6.32}$$

$$\alpha_4 = 3 + \frac{p^2 + 6(1-p)}{1-p}. \tag{6.33}$$

Some of the important properties of the geometric probability distribution are:

1. The geometric distribution is always skewed to the right ($\alpha_3 > 0$).

2. There is a considerable amount of variability associated with the geometric distribution when p is small.

3. The mode always occurs at $X = 1$ since $g(1; p) = p$ and for $X = 2, 3, \ldots$, the geometric probabilities decrease monotonically since $(1-p)^{X-1}$ does likewise.

4. $E(X) < V(X)$.

5. The peak of a geometric distribution is flatter than that of a normal distribution ($\alpha_4 < 3$) and, as p increases in value, the distribution becomes flatter.

6. The geometric distribution is so named because, for successive values of X, the probabilities constitute a geometric progression (i.e., $p, p(1-p), p(1-p)^2, \ldots$).

7. Unlike the binomial probability distribution, the geometric distribution does not specify a fixed sample size n since sampling continues until the first success occurs on the Xth trial.

8. Since sampling continues until the desired outcome occurs for the first time, there is no effective upper limit to the number of trials performed and thus we may view the sample values as being extracted from an infinite population.

Example 6.7.1   Suppose our random experiment consists of rolling a fair pair of six-sided dice until a success occurs, where the latter is taken to be the sum of the faces equaling eight. Here $p = \frac{5}{36}$. What is the probability that it will take $X = 5$ rolls for the first eight to occur? From (6.26) we have

$$g\left(X = 5; \frac{5}{36}\right) = \left(\frac{5}{36}\right)\left(1 - \frac{5}{36}\right)^{5-1} = 0.0763.$$

And from (6.30), the probability that it will take at most $x = 5$ rolls to get the first eight is

$$P(X \leq 5) = 1 - \left(1 - \frac{5}{36}\right)^5 = 0.5296. \quad \blacksquare$$

6.8 The Negative Binomial Distribution

The sampling (Bernoulli) process underlying the determination of binomial probabilities admits a sequence of a specific number (n) of independent trials of a simple alternative experiment with p constant from trial to trial. Instead of fixing n and counting the number of successes $X(= 0, 1, 2, \ldots, n)$, suppose we count the trials until a fixed number of successes (k) is obtained. Now the random variable is the number of trials $Y(= k, k+1, k+2, \ldots)$ necessary to observe exactly k successes. In this regard, for the binomial case, we seek the probability of exactly $X = k$ successes in n trials; in the negative binomial case, since sampling continues until a fixed number (k) of successes is obtained, we seek the probability of getting the kth success on the Yth trial, where $Y = k, k+1, k+2, \ldots$. Clearly the number of trials in a negative binomial experiment may become infinite.

In sum, the salient features of the negative binomial experiment are:

1. The number of the trial on which the kth success occurs is a discrete random variable.

2. We have a simple alternative experiment.

3. The trials are identical and independent.

4. p is constant from trial to trial.

Let the random variable Y associated with the Bernoulli process depict the trial on which the kth success occurs. The event $Y = y$ can occur only if we observe $k - 1$ successes in the first $y - 1$ trials, and a success on the yth trial. From the binomial probability mass function (6.15), the probability of $k - 1$ successes in $y - 1$ trials is

$$\binom{y-1}{k-1} p^{k-1} (1 - p)^{y-k} .$$

Now, the probability of a success on the yth trial is simply p. Hence, under independence,

$P(k - 1 \text{ successes in } y - 1 \text{ trials} \cap \text{a successes on trial } y)$

$$= n(Y = y; k, p) = \left[\binom{y-1}{k-1} p^{k-1} (1-p)^{y-k} \right] \cdot p$$

$$= \binom{y-1}{k-1} p^k (1-p)^{y-k}, \ k \le y. \quad (6.34)$$

Let us view this negative binomial process in a slightly different light. Specifically, let us define a new random variable X as the number of failures observed before exactly k successes occur. Then for the yth trial:

- If $Y = k$, no failures are observed on the first k trials and thus $X = 0$

- If $Y = k + 1$, exactly $X = 1$ failure occurs before k successes are obtained

- If $Y = k + 2$, exactly $X = 2$ failures are observed before k successes occur, and so on

Hence we may write $Y = k + X$, where $X = 0, 1, 2, \ldots$.

Now, consider again this Bernoulli process, but this time let the random variable $X = x$ depict the total number of failures in this process before the kth success occurs. Then $k + X$ is the number of trials necessary to render exactly k successes. And since the last trial must result in a success, we must have $k - 1$ successes and

x failures among the first $k + x - 1$ trials. Then by an argument similar to the one supporting the derivation of (6.26), we may respecify the same as

$$n\,(X = x; k, p) = \binom{k + x - 1}{x} p^k (1 - p)^x, \; x \geq 0. \qquad (6.34.1)$$

Based upon the preceding developments, let us define the random variable X as the number of failures observed before exactly k successes occur in a Bernoulli process and let the number of trials required to achieve exactly k successes be $k + X$. Then X follows a *negative binomial distribution* with probability mass function

$$n(X; k, p) = \begin{cases} \binom{k + X - 1}{X} p^k (1 - p)^X, & X = 0, 1, 2, \ldots, \\ k \text{ is a positive integer}, & 0 \leq p \leq 1; \\ 0 & \text{elsewhere}. \end{cases} \qquad (6.35)$$

And as the parameters k and p are varied, we obtain an entire family of negative binomial distributions.

In addition, the *negative binomial cumulative distribution function* appears as

$$P(X \leq x) = N(X; k, p) = \sum_{i \leq x} n(i; k, p) = \sum_{i \leq x} \binom{k + i - 1}{i} p^k (1 - p)^i \qquad (6.36)$$

with $\sum_{i=0}^{\infty} n(i; k, p) = 1$ (a verification of this latter equality is provided in [Feller (1968), pp. 155–156]).

A few important features of the negative binomial distribution merit our consideration. First, the negative binomial distribution can be viewed as an extension of the geometric probability distribution to $k > 1$ successes; that is, we are interested in the kth success rather than the first. In fact, if in the geometric distribution we define the random variable X as the number of failures observed before the first success occurs, then (6.27) may be rewritten as

$$g(X; p) = \begin{cases} p(1 - p)^X, & X = 0, 1, 2, \ldots, 0 \leq p \leq 1; \\ 0 & \text{elsewhere}. \end{cases} \qquad (6.27.1)$$

Then when $k = 1$ in (6.35), we obtain this version of the geometric distribution as a special case.

Second, since the binomial coefficients with negative integers may be defined as

$$\binom{-k}{X} = (-1)^X \binom{k + X - 1}{X}, \; k > 0, \qquad (6.37)$$

(on all this see [Feller (1968), pp. 61,155]) we can easily rewrite (6.35) as

$$n(X;k,p) = (-1)^X \binom{-k}{X} p^k (1-p)^X$$

$$= \binom{-k}{X} p^k \left[-(1-p) \right]^X, \quad X = 0,1,2,\ldots. \tag{6.35.1}$$

Hence the negative binomial distribution gets its name from (6.37), and the sequence of probabilities (6.35.1) corresponds to the successive terms of the binomial expansion of $p^k \left[1 - (1-p) \right]^{-k}$.

Next, we may conveniently represent negative binomial probabilities in terms of binomial probabilities by utilizing the relationship

$$n(X; k,p) = \frac{k}{k+X} b(k; k+X, p), \quad X = 0,1,2,\ldots, \tag{6.38}$$

where obviously the number of Bernoulli trials is $n = k + X$. Alternatively, a negative binomial probability can also be expressed in terms of successive values of its cumulative distribution function as

$$n(X; k,p) = N(X; k,p) - N(X-1:k,p). \tag{6.39}$$

But since the probability of at most x failures in $k + x$ trials equals the probability of at least k successes in $k + x$ trials or

$$N(X; k,p) = 1 - B(k-1; k+X, p), \tag{6.40}$$

we also have

$$n(X; k,p) = B(k-1; k+X-1, p) - B(k-1; k+X, p). \tag{6.41}$$

Example 6.8.1   Let us use (6.35) to calculate the probability that $k + X = 10$ flips of a fair coin will need to be performed to obtain exactly $k = 4$ heads, where the fourth head is obtained on the tenth flip. Then

$$n\left(X = 6;\ 4, \frac{1}{2} \right) = \binom{9}{6} \left(\frac{1}{2} \right)^4 \left(\frac{1}{2} \right)^6 = 0.0820.$$

Alternatively, using (6.38), we also have

$$n\left(X = 6;\ 4, \frac{1}{2} \right) = \frac{4}{10} b\left(4;\ 10, \frac{1}{2} \right) = \frac{4}{10} \binom{10}{4} \left(\frac{1}{2} \right)^4 \left(\frac{1}{2} \right)^6 = 0.0820. \quad \blacksquare$$

For the negative binomial random variable X, we may express its mean and variance as

$$\mu = E(X) = \frac{k(1-p)}{p} \tag{6.42}$$

and

$$\sigma^2 = V(X) = \frac{k(1-p)}{p^2} = \mu + \frac{1}{k}\mu^2 \tag{6.43}$$

(or $\sigma = \sqrt{V(X)} = \sqrt{k(1-p)}/p$) respectively. In addition, its coefficients of skewness and kurtosis are respectively,

$$\alpha_3 = \frac{2-p}{[k(1-p)]^{1/2}}; \tag{6.44}$$

$$\alpha_4 = 3 + \frac{p^2 - 6p + 6}{k(1-p)}. \tag{6.45}$$

Some key properties of the negative binomial distribution are:

1. $E(X) < V(X)$.

2. It is positively skewed ($\alpha_3 > 0$).

3. Its peak is sharper than that of a normal distribution ($\alpha_4 > 3$) and $\alpha_4 \rightarrow 3$ as $k \rightarrow \infty$.

4. As $p \rightarrow 0$ (it takes more failures to observe k successes) the dispersion of this distribution increases and its peak becomes flatter.

6.9   The Poisson Distribution

For some random experiments we are not able to actually perform a finite number of trials or observe a finite sequence of outcomes. What happens instead is that observations occur at random over a continuum (such as time, length, space, or volume).[2] Suppose we are interested in the number of occurrences (k) of some event over an interval of unit length. If we divide this interval into n equal subintervals, each of length $\frac{1}{n}$, and n is large enough to preclude multiple occurrences in any single subinterval (or at least the probability of multiple occurrences in any

---

[2] Examples of such occurrences are the number of vehicles per minute arriving at a toll booth, the number of deaths per year attributed to a specific disease, the number of customers per minute arriving at a store checkout counter, the number of defects on a length of cable, the number of machines that break down per day on a production line, the number of typographic errors per page in a volume of printed material, and so on.

single subinterval is approximately equal to zero), then these subintervals can be thought of as *n identical and independent trials of a Bernoulli process.*

Now, either a particular subinterval is empty, or it contains at least one random occurrence of the event. We shall refer to these two possibilities as the outcomes of a Bernoulli or simple alternative experiment; that is, the former depicts *failure* and the latter represents *success.* Since the subintervals are all of the same length, the probability of success (p) is the same for all n subintervals. Then the assumed independence of these disjoint subintervals implies that we have n Bernoulli trials and the probability of exactly k successes is given by the binomial probability $b(k; n,p)$.

To summarize, the essential characteristics of a Poisson experiment are:

- The discrete random variable X represents the number of occurrences of independent events that take place at a constant rate over a continuous interval.

- The number (n) of independent trials is very large.

- Only one outcome occurs on each trial, or the probability of more than one occurrence per trial is negligible.

- The probability of occurrence of the event on each trial is small and is proportional to the length of the subinterval.

To approximate the previously mentioned binomial probability, let us assume that the probability of two or more random occurrences within any subinterval is, in the limit, approximately zero so that we experience a set of isolated random outcomes. Then the probability of finding exactly k such outcomes in the unit interval is the limiting value of $b(k; n,p)$ as $n \to \infty$. If $np \to \lambda = $ constant, then $b(k; n,p) \approx b(k; n, \frac{\lambda}{n}) \to p(k; \lambda)$, where

$$p(k; \lambda) = e^{-\lambda} \left(\frac{\lambda^k}{k!} \right) \tag{6.46}$$

denotes a *Poisson probability.*[3]

A couple of observations pertaining to the limiting process underlying (6.46) are in order. First, (6.46) is the limiting form of a binomial probability when $n \to \infty$ and $p = \frac{\lambda}{n}$ for some constant $\lambda > 0$. Then

$$\lim_{\substack{n \to \infty \\ p \to 0}} b\left(k; n, \frac{\lambda}{n}\right) = \frac{e^{-\lambda} \lambda^k}{k!}.$$

---

[3] By definition, $e = \lim_{n \to \infty} \left(1 + \frac{1}{n}\right)^n \approx 2.71828$, or, expressed as an exponential series, $e = 1 + \frac{1}{1!} + \frac{1}{2!} + \frac{1}{3!} + \cdots$; and $e^x = 1 + x + \frac{x^2}{2!} + \frac{x^3}{3!} + \cdots$.

Stated alternatively, (6.46) is the limiting form of a binomial probability when $n \to \infty$ and $p \to 0$ in a fashion such that $np = $ constant. Then

$$\lim_{\substack{n \to \infty \\ p \to 0 \\ np=\text{constant}}} b(k;n,p) = \frac{e^{-np}(np)^k}{k!}.$$

Second, if this unit interval is replaced by an arbitrary interval of length T, then this interval may be divided into nT subintervals, each of length $\frac{1}{n}$. If these sub-intervals are viewed as nT identical and independent trials of a Bernoulli process, then the probability of k successes in the interval is given by the limit of $b(k; nT, p)$ as $n \to \infty$. And if $nTp \to \lambda T$, then, as we pass to the said limit, (6.46) is replaced by

$$p(k; \lambda T) = e^{-\lambda T} \frac{(\lambda T)^k}{k!}. \tag{6.46.1}$$

Based upon the preceding discussion, let us define the random variable X as the number of independent random occurrences of some event that takes place at a constant rate over a given continuous interval. Then X follows a *Poisson probability distribution* with probability mass function

$$p(X; \lambda) = \begin{cases} \frac{e^{-\lambda}\lambda^X}{X!}, & X = 0, 1, 2, \ldots, \ \lambda > 0; \\ 0 & \text{elsewhere.} \end{cases} \tag{6.47}$$

And as the parameter λ is varied over a set of positive values, we generate a whole family of Poisson probability distributions. Here λ represents the average number of occurrences (or expected rate of occurrence) of some random event over a unit interval. In this regard, λ can be thought of as the *intensity* of the random process per standardized unit (or λT is the *intensity* per period of arbitrary length T); it determines the *density* of the outcomes on the unit interval.

A few additional comments pertaining to the Poisson probability distribution are in order. First, to expedite the calculation of Poisson probabilities we may use the *Poisson recursion formula*

$$p(X + 1; \lambda) = \frac{\lambda}{X + 1} p(X; \lambda). \tag{6.48}$$

Next, the probability that a Poisson random variable X assumes a value less than or equal to some given nonnegative integer level x is provided by the *Poisson cumulative distribution function*

$$P(X \le x) = P(x; \lambda) = \sum_{i \le x} \frac{e^{-\lambda}\lambda^i}{i!} = \sum_{i \le x} p(i; \lambda). \tag{6.49}$$

Based upon this expression, individual Poisson probabilities may be calculated as

$$p(x; \lambda) = P(x; \lambda) - P(x - 1; \lambda). \tag{6.50}$$

Moreover,

$$P(X < x) = P(X \leq x - 1) = P(x - 1; \lambda) \tag{6.51}$$

and

$$P(X \geq x) = 1 - P(X \leq x - 1) = 1 - P(x - 1; \lambda). \tag{6.52}$$

And given (6.47), it must be the case that $\sum_{X=0}^{\infty} p(X; \lambda) = 1$ while $p(X; \lambda) \to 0$ as $X \to \infty$.

Example 6.9.1   To show exactly how Poisson probabilities are calculated, let us consider how events can occur over, say, time. For concreteness, suppose a secretary is monitoring incoming telephone calls at a clinic. Here the *number* of calls occurring in a fixed interval of time T (clearly T can be a minute, five minutes, an hour, day, etc.) is a Poisson random variable X. In this regard, although the time between calls is a continuous random variable, the number of calls over T assumes only integer values ($X = 0, 1, 2, \ldots$) and thus represents a discrete random variable.

Now, suppose we choose a time interval of 60 minutes, which is then subdivided into 12 subintervals of $T = 5$ minutes each. Assume that over this one hour period the clinic receives 20 telephone calls. Since there are 20 occurrences in the 12 subperiods, the average number of calls per subperiod of $T = 5$ minutes is $\lambda T = \frac{20}{12} \approx 1.66$ calls per 5 minutes. (If the initial time interval of one hour were divided into 60 subperiods of $T = 1$ minute each, then $\lambda T = \lambda = \frac{20}{60} \approx 0.33$ calls per minute; and if we had six subperiods of $T = 10$ minutes each, then $\lambda T = \frac{20}{6} \approx 3.33$ calls per 10 minutes.)

Under the assumptions of randomness and the independence of the occurrence of calls in these subintervals, these occurrences depict a Poisson process and, from (6.46.1), are distributed as

$$p(X; 1.66) = e^{-1.66} \frac{(1.66)^X}{X!}, \quad X = 0, 1, 2, \ldots. \tag{6.53}$$

What is the probability that the clinic gets exactly two telephone calls in a five-minute period? From the preceding equation,

$$p(X = 2;\ 1.66) = e^{-1.66} \frac{(1.66)^2}{2!} = 0.26197.$$

To complete the Poisson probability distribution associated with this experiment, let us vary X in (6.53) over the set of nonnegative integer values. Our results are

Table 6.5

X	$B(X; 5, 5/36)$
0	0.4735
1	0.3818
2	0.1232
3	0.0199
4	0.0016
5	0.0000
	1.0000

depicted in Table 6.5. For $X = 0$,

$$p(0; \ 1.66) = e^{-1.66} = 0.19014;$$

and for $X = 1$,

$$p(1; \ 1.66) = e^{-1.66}1.66 = 0.31563.$$

To determine the Poisson probabilities when $X = 3, 4, \ldots$, let us employ the recursion formula

$$p(X + 1; \lambda T) = \frac{\lambda T}{X + 1} p(X; \lambda T). \tag{6.48.1}$$

Thus

$$p(3; 1.66) = \frac{1.66}{3} p(2; 1.66) = 0.14496;$$

$$p(4; 1.66) = \frac{1.66}{4} p(3; 1.66) = 0.06015;$$

$$p(5; 1.66) = \frac{1.66}{5} p(4; 1.66) = 0.01997,$$

and so on. In addition, it is easily demonstrated that the probability of no more than three telephone calls per five minutes is, from Table 6.5 and (6.49),

$$P(X = 2; 1.66) = \sum_{i \leq 2} p(i; 1.66)$$

$$= p(0; 1.66) + p(1; 1.66) + p(2; 1.66) = 0.7677;$$

and the probability of at least five calls per five minutes is

$$P(X \geq 5) = 1 - P(X \leq 4) = 1 - P(4; 1.66) = 1 - 0.97285 = 0.0272.$$

Finally, from (6.50), it is readily verified that, for instance,

$$p(3; 1.66) = P(3; 1.66) - P(2; 1.66) = 0.91270 - 0.76774 = 0.14496$$

as expected. ∎

As was the case for binomial probabilities, there are certain parameter values for which it is easier to read Poisson probabilities from a table than to calculate them directly. A table of Poisson probabilities is provided in the Appendix (see Table A.8). Here the table renders Poisson probabilities for selected λT values and appropriate nonnegative integer values of $X = k$. For instance, we can easily find the Poisson probability $p(9; 3.2) = 0.0040$.

For a Poisson random variable X, it can be shown that the mean and variance are equal; that is

$$\mu = E(X) = \lambda; \quad \sigma^2 = V(X) = \lambda \tag{6.54}$$

and thus

$$\sigma = \sqrt{V(X)} = \sqrt{\lambda}. \tag{6.55}$$

Also, the coefficients of skewness and kurtosis are, respectively,

$$\alpha_3 = \frac{1}{\sqrt{\lambda}}; \tag{6.56}$$

$$\alpha_4 = 3 + \frac{1}{\lambda}, \tag{6.57}$$

where $\lambda > 0$.

Some of the key aspects/properties of the Poisson probability distribution are:

1. The Poisson distribution is used to calculate the probabilities of independent random events that occur at a constant average rate over some continuous standardized unit or interval. The distribution is completely determined by its mean λ, which changes proportionately whenever the standardized unit changes. So while λ is the intensity of the random process per unit interval, λT is the intensity in an interval of length T. Then for this latter case, $E(X) = V(X) = \lambda T$.

2. The Poisson distribution is positively skewed for any $\lambda > 0$ (i.e., $\alpha_3 > 0$) with the degree of skewness inversely related to λ.

3. The Poisson distribution has a peak that is sharper than that of a normal distribution ($\alpha_4 > 3$); this distribution approaches normality as λ increases without bound.

4. The Poisson distribution is bimodal if λ (or λT) is an integer: the two modes are located at $k = \lambda - 1$ and $k = \lambda$; if λ is not an integer, there is a unique mode which corresponds to the integer value between $\lambda - 1$ and λ.

5. When n is large and p is small, binomial probabilities are extremely tedious to calculate. But since the limiting form of the binomial distribution when $n \to \infty$, $p = \frac{\lambda}{n} \to 0$, and $np = \lambda = $ constant is the Poisson distribution, we can use the latter to approximate binomial probabilities. The approximation is appropriate when $n \geq 20$ and $np \leq 1$. Moreover, if $n \geq 50$ and $np \leq 5$, or if $n \geq 100$ and $np \leq 10$, then the Poisson approximation to binomial probabilities is excellent.

6. By virtue of the preceding comment, it must be the case that the limiting forms of, for instance, the mean and variance of the binomial distribution must be the mean and variance of the Poisson distribution, respectively; that is,

$$\lim_{p \to 0} np = \lim_{p \to 0} \lambda = \lambda;$$

$$\lim_{p \to 0} np(1 - p) = \lim_{p \to 0}(np - np^2) = \lim_{p \to 0}(\lambda - \lambda p) = \lambda.$$

Example 6.9.2 For the Poisson distribution given in Table 6.5, it is easily verified that $\mu = 1.66$, $\sigma^2 = 1.66$, $\alpha_3 = \frac{1}{\sqrt{1.66}} = 0.7762$, and $\alpha_4 = 3 + \frac{1}{\sqrt{1.66}} = 3.602$.

Clearly the peak of this distribution is sharper than that of a normal distribution since $\alpha_4 > 3$ and, since $\lambda T = 1.66$ is not an integer, the distribution has a unique mode at the integer value between $\lambda T - 1 = 0.66$ and $\lambda T = 1.66$ or at $k = 1$. ∎

As mentioned in item 5, earlier, the Poisson distribution is the limiting form of the binomial distribution if $n \to \infty$, $p \to 0$, and $np = $ constant. So when p is close to zero and n is large, a Poisson probability with $\lambda T = np$ should provide a fairly good approximation to a binomial probability.

Example 6.9.3 Let $p = 0.05$ and $n = 50$. Then the binomial probability that $k = 5$ is

$$b(5; 50, 0.05) = \binom{50}{5}(0.05)^5 (0.95)^{45} = 0.0656.$$

However, it is easily verified from Table A.8 that, for $\lambda T = np = 2.5$ and $k = 5$, $p(5; 2.5) = 0.0668$. Clearly this is an excellent approximation to the preceding binomial probability value. ∎

6.10 The Hypergeometric Distribution

For a finite population of size N, let us assume (as in binomial case) that each element belongs to only one of two mutually exclusive and collectively exhaustive classes—success or failure, where k of the N items are labeled success and $N - k$ items are labeled failure. Suppose we extract a random sample of size n without replacement (or, equivalently, we select all n items at the same time). Unlike the binomial experiment involving sampling with replacement (or from an infinite population), the probability of a success does not remain constant over successive draws since, in this instance, the outcomes (draws) are not independent; that is, the probability of a success or failure on, say, the rth draw is conditional on the number of successes or failures obtained on the preceding $r - 1$ draws. Given this sampling scheme, we seek to determine the probability of obtaining exactly x successes and $n - x$ failures in a random sample of size n, where $x \leq k$ and $n - x \leq N - k$ (see Figure 6.5).

To summarize, the essential characteristics of a hypergeometric experiment are:

- The number of successes obtained (X) is a discrete random variable.

- Each draw results in a simple alternative (success or failure).

- The n draws from a finite population of size $N(n \leq N)$ are not independent under sampling without replacement.

- The actual number of successes in the population is known.

Based upon the preceding discussion, it is evident that the x successes can be chosen in a total of $\binom{k}{x}$ different ways and the $n - x$ failures can be selected in a total of $\binom{N-k}{n-x}$ different ways. Then the total number of ways to draw x successes

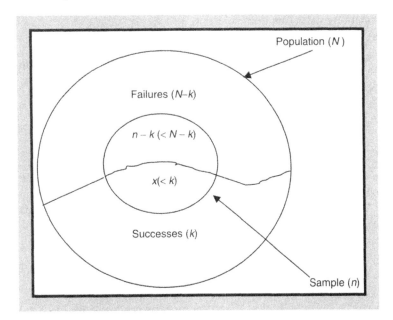

Figure 6.5   Sampling scheme for a hypergeometric experiment.

and $n - x$ failures is

$$\binom{k}{x}\binom{N-k}{n-x} \tag{6.58}$$

Now, the total number of distinct ways in which a sample of n items can be selected from a population of N items is $\binom{N}{n}$. Hence the probability of choosing any particular sample of size n is $\dfrac{1}{\binom{N}{n}}$ and thus the probability of observing exactly x successes and $n - x$ failures in a sample size n is, from (6.58),

$$P(x \text{ successes} \cap n - x \text{ failures}) = \frac{\left(\begin{array}{c}\text{total number of samples exhibiting } x \\ \text{successes} \cap n - x \text{ failures}\end{array}\right)}{(\text{total number of samples})}$$

or

$$h(X = x; N, n, k) = \frac{\binom{k}{x}\binom{N-k}{n-x}}{\binom{N}{n}}. \tag{6.59}$$

Let us generalize this development by defining a random variable X as the number of successes obtained in a sample of size n that is drawn without replacement from a finite simple alternative population of size N in which the number of successes k is known. Then X follows a *hypergeometric probability distribution* with probability mass function

$$h(X; N, n, k) = \begin{cases} \dfrac{\binom{K}{X}\binom{N-k}{n-X}}{\binom{N}{n}}, & \begin{array}{l} X = 0, 1, \ldots, n, \ N \text{ is a positive integer,} \\ k(\leq N) \text{ is a nonnegative integer,} \\ n(\leq N) \text{ is a positive integer,} \\ X \leq k, \text{ and } n - X \leq N - k; \end{array} \\ 0 \text{ elsewhere} \end{cases} \tag{6.60}$$

where the parameters N, n, and k define an entire family of hypergeometric distributions. (Note that if $k < n$, then $X = 0, 1, \ldots, k$ in (6.60). Hence we ultimately require $X = 0, 1, \ldots, \min\{n, k\}$.)

To streamline somewhat the calculation of hypergeometric probabilities, we may employ the *hypergeometric recursion formula*

$$h(X + 1; N, n, k) = \frac{(n - X)(k - X)}{(X + 1)(N - k - n + X + 1)} h(X; N, n, k). \tag{6.61}$$

Next, the probability that a hypergeometric random variable X takes on a value less than or equal to some given integer level x, $0 \leq x \leq n$, is provided by the *hypergeometric cumulative distribution function*

$$P(X \leq x) = H(x; N, n, k) = \sum_{i \leq x} h(i; N, n, k) = \sum_{i \leq x} \frac{\binom{k}{i}\binom{N-k}{n-i}}{\binom{N}{n}}. \tag{6.62}$$

Then from this expression we can also find

$$P(X > x) = 1 - H(x; N, n, k). \tag{6.63}$$

Example 6.10.1   To calculate a set of hypergeometric probabilities, let us consider a vessel containing $N = 20$ balls of which eight are white and 12 are black. If we extract a sample of size $n = 5$ without replacement, what is the probability of obtaining $x = 3$ white balls? Here the acquisition of a white ball is considered to be a success (see Figure 6.6). From (6.59) it is readily seen that

$$h(3; 20, 5, 8) = \frac{\binom{8}{3}\binom{12}{2}}{\binom{20}{5}} = 0.2379.$$

To derive the complete hypergeometric probability distribution associated with this experiment, let us determine

$$h(X; 20, 5, 8) = \frac{\binom{8}{X}\binom{12}{5-X}}{\binom{20}{5}}, \quad X = 0, 1, \dots, 5.$$

The implied distribution appears in Table 6.6. In this regard, for $X = 0$ we have

$$h(0; 20, 5, 8) = \frac{\binom{8}{0}\binom{12}{5}}{\binom{20}{5}} = 0.0511.$$

Table 6.6

X	$h(X; 20, 5, 8)$
0	0.0511
1	0.2555
2	0.3974
3	0.2379
4	0.0541
5	0.0037
	≈ 1.0000

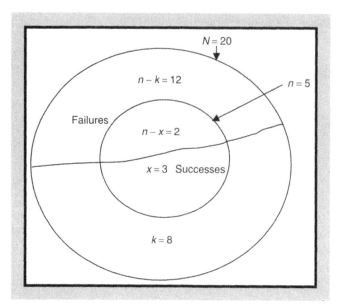

Figure 6.6   Sampling scheme for a hypergeometric experiment involving $x = 3$ successes and $5 - x = 2$ failures.

From recursion formula (6.61) we may calculate the remaining hypergeometric probabilities as:

$$h(1; 20, 5, 8) = \frac{(5-0)(8-0)}{(0+1)(12-5+0+1)} h(0; 20, 5, 8) = 0.2555;$$

$$h(2; 20, 5, 8) = \frac{(5-1)(8-1)}{(1+1)(12-5+1+1)} h(1; 20, 5, 8) = 0.3974;$$

$$h(4; 20, 5, 8) = \frac{(5-3)(8-3)}{(0+3)(12-5+3+1)} h(3; 20, 5, 8) = 0.0541;$$

$$h(5; 20, 5, 8) = \frac{(5-4)(8-4)}{(4+1)(12-5+4+1)} h(4; 20, 5, 8) = 0.0037.$$

Finally, an application of equation (6.62) reveals that the probability of selecting no more than three white balls or $P(X \leq 3)$ is

$$H(3; 20, 5, 8) = \sum_{i \leq 3} h(i;\ 20, 5, 8)$$

$$= h(0; 20, 5, 8) + h(1; 20, 5, 8) + h(2; 20, 5, 8) + h(3; 20, 5, 8) = 0.9419. \quad \blacksquare$$

We next turn to the specification of the mean and variance of a hypergeometric random variable X. Here

$$\mu = E(X) = \frac{nk}{N} \tag{6.64}$$

and

$$V(X) = \frac{nk(N-k)}{N^2}\left(\frac{N-n}{N-1}\right). \tag{6.65}$$

Moreover, the coefficients of skewness and kurtosis for a hypergeometric distribution are respectively

$$\alpha_3 = \frac{(N-2k)(N-2n)\sqrt{N-1}}{(N-2)\sqrt{nk(N-k)(N-n)}} \tag{6.66}$$

and

$$\alpha_4 = \frac{(N-1)\{N^3(N+1)-6nN^2(N-n)+3k(N-k)[N^2(n-2)-Nn^2+6n(N-n)]\}}{nk(N-n)(N-2)(N-3)(N-k)}. \tag{6.67}$$

Some of the important properties of the hypergeometric distribution are:

1. If we represent the proportion of successes in the finite population as $p=k/N$, then we may rewrite (6.60) as

$$h(X;N,n,p) = \frac{\binom{Np}{X}\binom{N(1-p)}{n-X}}{\binom{N}{n}},\ 0 \le p \le 1.$$

If n and p are held fixed and N increases without limit, then this hypergeometric distribution with parameters N, n, and p approaches the binomial distribution with parameters n and p; that is,

$$\lim_{N\to\infty} h(X;N,n,p) = b(X;n,p).$$

So when $\frac{n}{N}$ is small (say $\frac{n}{N} < 0.05$), the binomial probability mass function provides an excellent approximation to the hypergeometric probability mass function. Hence the distinction between the binomial and the hypergeometric probability distributions is important only when the sample of units chosen without replacement represents a sizeable proportion of the finite population; that is, when n is large relative to N the binomial distribution does not provide us with a satisfactory approximation to the true distribution of successes; in this instance there is no substitute for the hypergeometric distribution.

2. If we again express the proportion of successes in the finite population as $p = \frac{k}{N}$, then the hypergeometric mean (6.64) becomes $E(X) = np$ and thus corresponds to the mean of the binomial distribution.

3. If we set $p = \frac{k}{N}$ and $1 - p = \frac{N-k}{N}$, then the hypergeometric variance (6.65) becomes

$$V(X) = np(1-p)\left(\frac{N-n}{N-1}\right);\qquad(6.55.1)$$

that is, the variance of the hypergeometric distribution equals the variance of the binomial distribution times the finite population correction factor $\frac{N-n}{N-1}$. If N is large relative to n, then the correction factor is approximately one and thus the hypergeometric and binomial distributions have virtually the same variance. Otherwise the variance of the hypergeometric distribution is smaller than that of the binomial distribution.

4. Looking to (6.66), it is evident that, for $N > 2$, the hypergeometric distribution is symmetrical ($\alpha_3 = 0$) if $N = 2k$ or $N = 2n$; it is positively skewed ($\alpha_3 > 0$) if $N > 2k$ and $N > 2n$; and it is negatively skewed ($\alpha_3 < 0$) if $N < 2k$ or $N < 2n$.

Example 6.10.2 If we consider the hypergeometric distribution derived earlier (see Table 6.6) for $N = 20$, $n = 5$, and $k = 8$, then it is easily seen that:

$$\mu = E(X) = \frac{5(8)}{20} = 2;$$

$$V(X) = \frac{(5)(8)(12)}{400}\left(\frac{20-5}{20-1}\right) = 0.947$$

and thus

$$\sigma = \sqrt{V(X)} = 0.973;$$

$$\alpha_3 = \frac{(4)(10)\sqrt{19}}{18\sqrt{40(12)(15)}} = 0.114;$$

and

$$\alpha_4 = \frac{19\{8000(21) - 30(400)(15) + (24)(12)[400(3) - 500 + 30(15)]\}}{40(15)(18)(17)(12)} = 2.753.$$

Hence this hypergeometric distribution is positively skewed (since we have $N = 20 > 2k = 16$ and $N = 20 > 2n = 10$) and, with $\alpha_4 < 3$, its peak is somewhat flatter than that of a normal distribution. ∎

6.11   **The Generalized Hypergeometric Distribution**

The probability distribution considered in this section is essentially a generalization of the hypergeometric distribution to more than two categories of outcomes. In fact, it can also be viewed as a variant of the multinomial probability distribution introduced earlier. In this regard, let us assume that we sample without replacement from a population containing a finite number of elements N, the totality of which are divided into r mutually exclusive and collectively exhaustive categories of types of events $E_1,\ldots,E_r$. Additionally, we posit that there are k_1 items of first type, k_2 items of second type, $\ldots$, and k_r items of the rth type within the population, where $\sum_{i=1}^{r} k_i = N$.

To summarize, the salient features of this generalized hypergeometric experiment are:

- Each trial gives rise to an r-fold alternative.

- The random variables X_i, $i = 1,\ldots,r$, associated with the r outcomes are discrete.

- The n draws from a finite population of size $N(n \leq N)$ are not independent.

- The actual number of items of each type k_i, $i = 1,\ldots,r$, within the population is known.

Suppose a sample of n items is randomly selected from this finite population. Since sampling is done without replacement, the n trials are not independent and thus the probabilities associated with the events E_i, $i = 1,\ldots,r$, are not constant from trial to trial. Additionally, let us assume that the sample contains x_1 items of type one, x_2 items of type two, $\ldots$, and x_r items of type r, where $\sum_{i=1}^{r} x_i = n$. Then by an argument similar to the one underlying the derivation of (6.59), the probability of obtaining this collection of outcomes is

$$g(X_1 = x_1,\ldots,X_r = x_r; N,n,k_1,\ldots,k_r) = \frac{\binom{k_1}{x_1}\binom{k_2}{x_2}\cdots\binom{k_r}{x_r}}{\binom{N}{n}}, \qquad (6.67)$$

where $N = \sum_{i=1}^{r} k_i$ and $n = \sum_{i=1}^{r} x_i$.

On the basis of the preceding arguments, let us define the random variables $X_i, i = 1,\ldots,r$, as the number of items of type i obtained in a random sample of size n that is drawn without replacement from an r-fold alternative population of size N in which the number of items of type i or $k_i, i = 1,\ldots,r$, is known and $\sum_{i=1}^{r} k_i = N$. Then the X_i follow a *generalized hypergeometric distribution* with probability mass function

$$g(X_1,\ldots,X_r; N,\ n,k_1,\ldots,k_r) = \begin{cases} \dfrac{\binom{k_1}{X_1}\binom{k_2}{X_2}\cdots\binom{k_r}{X_r}}{\binom{N}{n}}, & X_i = 0,1,\ldots,n, \sum_{i=1}^{r} X_i = n \text{ is a positive integer,} \\ & N \text{ is a positive integer, } k_i(\leq N) \text{ is a} \\ & \text{nonnegative integer, and } X_i \leq k_i \text{ for all } i; \\ 0 & \text{elsewhere.} \end{cases}$$

$$(6.68)$$

Here the parameters N, n, and k_i, $i = 1, \ldots, r$, define a whole family of generalized hypergeometric distributions.

Example 6.11.1 To apply these notions, let us assume that we have a population containing $N = 30$ items that can be partitioned into four distinct groups: I, II, III, and IV. Suppose group I contains $k_1 = 10$ items, group II contains $k_2 = 5$ items, group III has $k_3 = 7$ items, and group IV has $k_4 = 8$ items (all labeled accordingly), with $\sum_{i=1}^{4} k_i = 30$. What is the probability that a sample of size $n = 8$ will contain two items from each group? Here $x_i = 2$ for all i and $\sum_{i=1}^{4} x_i = 8$. From (6.67) we seek to find

$$g(X_1 = 2, \; X_2 = 2, \; X_3 = 2, \; X_4 = 2; 30, 8, \; k_1 = 10, k_2 = 5, \; k_3 = 7, \; k_4 = 8)$$

$$= \frac{\binom{10}{2} \binom{5}{2} \binom{7}{2} \binom{8}{2}}{\binom{30}{8}} = 0.452. \quad \blacksquare$$

It was mentioned earlier that the generalized hypergeometric distribution is a variant of the multinomial distribution. The distinction between these two distributions is important only when the population size is small. If the population is large, sampling without replacement has a negligible effect on the probabilities of the various outcomes for successive samplings. So in this latter instance of large N, the generalized hypergeometric probabilities are closely approximated by multinomial probabilities and the distinction between the generalized hypergeometric and multinomial distribution blurs.

6.12 Exercises

Uniform Distribution

6-1. For a particular variety of electronic bulletin board the number of components having unacceptable reliability coefficients was evenly distributed between 7 and 30. Determine the probability of finding at least 15 unacceptable components. What is the average number of unacceptable components? What is the standard deviation?

6-2. Let a random variable X follow a discrete uniform distribution with probability mass function given by (6.9). Determine X's probability-generating function.

6-3. Let a random variable X follow a discrete uniform distribution with probability mass function given by (6.9). Find X's moment-generating function.

6-4. Verify that if a random variable X follows a discrete uniform distribution with probability mass function given by (6.9), then $E(X) = \frac{k+1}{2}$, $V(X) = \frac{k^2-1}{12}$.

Binomial Distribution

6-5. Suppose a vessel contains 10 identical marbles of which four are red and six are blue. For each of four random draws from the vessel a marble's color is recorded and the marble is returned to the vessel. If the random variable X is the number of red marbles drawn, find the probability that $X = 3$. What is the probability of drawing no more than two red marbles?

6-6. Suppose that for a Bernoulli process $p = 0.15$. Also, suppose that $n = 10$ items are drawn from this process and we find that the first, third, and eighth items have the same particular characteristics of interest. What is the probability of obtaining this specific sequence of items? What is the probability that no items in the sequence have the characteristic? What is the probability that the characteristic in question appears on the first five items?

6-7. Suppose a dart player has a probability of 8/9 of hitting the bullseye and that his throws are independent. If the player is given three darts, what is the probability that:

(a) he hits the bullseye all three times

(b) he misses the bullseye all three times

6-8. We toss a fair coin $n = 10$ times. What is the probability of observing four heads followed by six tails?

6-9. A student takes a multiple choice exam that contains 15 questions, each with four possible answers. If a student guesses at the answer to each question, what is the probability that he or she passes the exam, given that 10 or more correct answers are required to pass?

6-10. A fair coin is tossed five times. Using the binomial formula, what is the probability of X successes in the five trials? Determine the resulting probability distribution. Find $E(X)$ and $V(X)$. Then find:

(a) The probability of at least two heads

(b) The probability of at most one head

(c) $P(2 < X \leq 4)$

6-11. Let the *proportion of successes* in a sample of size n be $Y = X/n$. Since $Y = y$ and $X = nY = ny$ are equivalent events (there exists a one-to-one correspondence between X and Y), it follows that $P(Y = y) = P(X = ny)$. Hence the distribution of Y can be determined from the (binomial) distribution of X. So for $n = 5$ trials of the random experiment in Exercise 6-10, determine the probability distribution of Y. Then find $E(Y)$ and $V(Y)$ in terms of $E(X)$ and $V(X)$.

6-12. A true or false exam has 20 questions. If a student guesses at the answer to each question, what is the probability that he or she guesses correctly on more than half of the questions?

6-13. A fair pair of dice is rolled $n = 4$ times. Define a success as *the sum of the faces is nine*. Determine the implied probability distribution. What is the probability of at least two nines? What is the probability of at most one nine? Find $E(X)$ and $V(X)$.

6-14. Suppose a fair coin is tossed 10 times in succession. What is the probability of obtaining at least seven heads?

6-15. Suppose that 30% of all persons taking a particular driver's education course fail the written exam for a driver's license. If the written exam is given to five randomly chosen individuals, what is the probability that three of them will fail? What is the implied probability distribution?

6-16. Experience dictates that a certain basketball player makes 80% of his free throws and that those free throws are independent. If he gets eight free throws in a particular game, what is the probability that he misses them all? What is the probability that he makes them all? What is the probability that he makes more than half of them? What is his expected number of free throws for this game?

6-17. Let the random variable X be binomially distributed with probability mass function $b(X; 3, p) = \binom{3}{X} p^X (1 - p)^{3-X}$, $X = 0, 1, 2, 3$. Determine the cumulative distribution function $P(X \le t) = B(t; 3, p)$.

6-18. Verify that the probability-generating function for the Bernoulli random variable X is $\phi_X(t) = (1 - p) + pt$. Use it to find $E(X)$, $V(X)$.

6-19. Verify that (6.15) is a legitimate probability mass function.

6-20. Verify that the probability generating function for a binomial random variable X is $\phi_X(t) = [(1 - p) + pt]^n$. Use it to determine the mean and variance of X.

6-21. Let X be a binomial random variable with probability mass function provided by equation (6.15). Determine the moment-generating function for X.

6-22. For X a binomial random variable with probability mass function given by equation (6.15), find $E(X)$, $V(X)$ using equation (6.15).

6-23. Suppose that a random variable X follows a binomial distribution with $E(X) = np$, $V(X) = np(1-p)$. Here X is the *number of successes* obtained in n independent trials of a simple alternative experiment. Now, consider the random variable X/n. How is this variable defined? What are its values? Find $E(X/n)$ and $V(X/n)$.

Multinomial Distribution

6-24. Suppose that in a certain locale 30% of the households purchase Brand A soap powder, 50% purchase Brand B soap powder, and 20% of the households use Brand C soap powder. For a random sample of $n = 10$ households selected with replacement, what is the probability that six use Brand A, three use Brand B, and one uses Brand C?

6-25. Suppose that in a particular game a play renders 1, 2, 3 or 4 points with associated probabilities $p_1 = 0.60$, $p_2 = 0.30$, $p_3 = 0.08$, and $p_4 = 0.02$. Under the assumption that these probabilities are the same for all plays and that the plays are independent, determine the probability of having four plays yielding 1 point, four plays resulting in 2 points, two plays yielding 3 points, and two plays resulting in 4 points in a sequence of 12 plays.

6-26. Twelve six-sided fair dice are tossed simultaneously. What is the probability that each odd number appears exactly two times and that each even number appears exactly two times?

6-27. Customers at the Big Deal Outlet Store pay for their purchases by check (10%), cash (5%), major credit card (70%), or store credit card (15%). What is the probability that among the next eight customers passing through the checkout line, exactly two will pay by each mode of payment?

Geometric Distribution

6-28. Suppose an individual is going to roll a bowling ball until he or she get a strike. If, under independent rolls, this person has a probability of 0.20 of making a strike and if X is the number of rolls it takes to make the first strike, determine the probability that fewer than five rolls will be needed to make the first strike. What is the average number of rolls that the person needs to make to get the first strike?

6-29. Suppose an individual stands at the free throw line on a basketball court and shoots until she makes a basket. Let the free throws be independent. If experience dictates that this person has an 80% chance of making each free throw, find the probability that it takes five shots to make the first basket. What is the probability that it takes fewer than five throws? What is the probability that it takes at least six shots? What is the average number of shots made?

6-30. Suppose that 1 in 1000 bottle cups of Zesty Cola contains the letter Z printed on its underside. This special bottle cap is redeemable at any store selling this cola for a special prize. If X is the number of bottles of Zesty you need to purchase to obtain a Z-cap, determine the probability that fewer than 10 bottles will need to be purchased. What is the average number of bottles needed to be purchased to get a Z-cap?

6-31. A major candy company markets 35 different products. Past experience dictates that there is a 2% chance that in any given year a dissatisfied customer will file a complaint with his or her consumer protection agency against the company concerning any one of its products. A new product is to be distributed to the company's retail outlets. What is the probability of the first complaint in the second year of sales? In the fifth year? What is the mean and variance for the random variable X (the year in which the first complaint is filed)?

6-32. The caps on soda bottles are examined with a scanning device in order to determine if they are properly set. Experience dictates the probability of detecting an improperly set cap is 0.01. What is the probability that the first improperly set cap will be detected on the tenth bottle? On the two hundredth bottle? Find the mean and the variance of the number of bottles examined until the next improperly set cap is found.

6-33. Given that a random variable X follows a geometric probability distribution, verify that (6.27) is a legitimate probability mass function.

6-34. Given that a random variable X follows a geometric probability distribution, find X's probability-generating function. Use it to obtain $E(X)$, $V(X)$.

6-35. Given that a random variable X follows a geometric probability distribution, find X's moment-generating function.

Negative Binomial Distribution

6-36. Suppose that for a large number of red and blue marbles within a jar the probability of obtaining a red marble is $\frac{1}{4}$ and the probability of getting a blue one is $\frac{3}{4}$ on any given draw. Marbles are selected from the jar until the fourth red marble is observed. What is the probability of observing the fourth red marble on the tenth draw? What is the probability that at most nine marbles have to be drawn in order to find the fourth red one?

6-37. Suppose that a basketball player has a 75% of chance of making a free-throw shot. The player shoots free throws until a total of 10 are made. What is the probability that 12 shots will be required in order to sink 10 of them? What is the probability that it will take 15 shots in order to sink 10 of them? If X is the minimum number of free throws needed to sink 10 of them, find $E(X)$ and $V(X)$.

6-38. A fair coin is tossed a larger number of times. What is the probability that the seventh head is obtained on the tenth toss? What is the probability that it is obtained on the fifteenth toss?

6-39. A girl scout is selling American flags door to door and has a quota of 10 flags to be sold. Let the probability of a sale be 0.25. What is the probability that her quota will be made at the fourteenth house?

6-40. Given that a random variable X follows a negative binomial probability distribution, determine X's moment-generating function.

6-41. Suppose a random variable X follows a negative binomial distribution. Find $E(X)$, $V(X)$, using the moment-generating function.

Poisson Distribution

6-42. Suppose that the number of moving violations captured by a traffic camera at a given traffic light is Poisson distributed with three moving violations recorded per day. Let the random variable X be the number of moving violations captured by the camera during a five-day work week. What is the probability of capturing exactly 10 moving violations over the five-day period? What is the probability of detecting at least six moving violations over this period?

6-43. Suppose that the intensity of the process generating incoming calls at a switching station is two per minute. What is the probability that the station receives exactly 12 calls in a 5-minute period? What is the probability of exactly 13 calls over this period?

6-44. Suppose that the 911 emergency facility of a particular metropolitan area experiences 300 calls per hour. Let X be the number of calls that arrive in a one-minute period. Then X is Poisson distributed with $\lambda = 5$. What is the probability of more than four calls but no more than 10 calls during a one-minute period?

6-45. Suppose the number of suicides in a certain locale is Poisson distributed with parameter $\lambda = 2$ per day, where X is the number of suicides per week. What is the probability of exactly 10 suicides during this period? What is the probability of no more than eight suicides during this period? Find the probability of at least 10 suicides during this week. What is the probability of less than five suicides within this period? What is the expected number of suicides in the one-week period? What is the variance of this Poisson random variable?

6-46. Suppose that the average number of incoming telephone requests for emergency service received by staff members of the regional auto club office is 30 per hour. What is the probability that at least five calls will arrive in a five-minute period? What is the probability of no calls arriving in a five-minute period?

6-47. Suppose that in a wooded area the airborne particles of a certain type of pollen occur at an average rate of six per cubic foot of air and that the number of particles X found in a cubic foot sample of air is Poisson distributed with parameter $\lambda = 6$. Determine the Poisson probability mass function. What is the probability of finding four particles of pollen in our air sample? What is the probability of finding none?

6-48. A night watchman (based on his past experience) estimates that there is only about a 1% chance of completing his rounds early and having to wait to punch in his code on the hour at a security station. Suppose that being early on any given patrol of the facility does not affect his being early or not on any other patrol. Let X be the number of times he will be early on his next 100 arrivals at the security station. Then X is binomially distributed with $p = 0.01$ and $n = 100$. Compare the binomial and Poisson probabilities for $X = 0, 1$ and 2; that is, how would you rate the Poisson approximation to the binomial probabilities?

6-49. Verify that equation (6.47) is a legitimate probability mass function.

6-50. Given that a random variable X is Poisson distributed, find X's probability-generating function. What is $E(X)$, $V(X)$?

6-51. Given that a random variable X follows a Poisson probability distribution, determine X's moment-generating function.

6-52. For a Poisson random variable X with probability mass function given by equation (6.47), find $E(X)$, $V(X)$ using equation (6.47).

Hypergeometric Distribution

6-53. Suppose that out of 50 property owners from a given city it is known that 30 support a bond issue for the addition of a new wing to the public library but 20 do not. If five property owners are selected at random at a town meeting, what is the probability that more than one of them will favor the bond issue?

6-54. Suppose that out of 120 applicants for a particular job only 80 of them have prior experience at operating a certain piece of milling machinery. If 10 applicants are randomly chosen for an interview, and an inventory of their skills is made, what is the probability that exactly four of the 10 will list experience on the milling machine as a skill?

6-55. A vessel contains 10 chips of which six are red and four are blue. Three are drawn at random without replacement. If X represents the number of red chips drawn, find $E(X)$, $V(X)$.

6-56. Seated at a table are five individuals, four of which are registered Democrats. If two individuals are selected at random without replacement, what is the probability that the non-Democrat will be one of those selected? (Let the random variable X be the number of non-Democrats selected.) What is the probability that two Democrats will be selected?

6-57. Suppose that in a group of 20 consumers, five prefer brand A espresso, six prefer brand B espresso, and six prefer brand C espresso. A sample of 10 consumers is selected at random without replacement. What is the probability that six of them will prefer brand A, two will prefer brand B, and two will prefer brand C?

6-58. Verify that (6.60) is a legitimate probability mass function.

6-59. For each of the following moment-generating functions, find the associated probability mass function:

(a) $m_X(t) = 0.0156(1 - 0.75e^t)^{-3}$, $t < 0.2877$

(b) $m_X(t) = 0.45e^t(1 - 0.55e^t)^{-1}$, $-\infty < t < +\infty$

(c) $m_X(t) = [0.60e^t + 0.04]^{10}$, $-\infty < t < +\infty$

(d) $m_X(t) = e^{-2}e^{2e^t}$, $-\infty < t < +\infty$

(e) $m_X(t) = 0.70e^t + 0.30$, $-\infty < t < +\infty$

(f) $m_X(t) = e^t(1 - e^{4t})[4(1 - e^t)]^{-1}$, $t \neq 0$

Continuous Parametric Probability Distributions

7.1 Introduction

This chapter focuses on a set of important continuous parametric probability distributions that are commonly employed in many areas of theoretical as well as applied statistics. Such distributions are essentially idealized models or mathematical constructs that can closely approximate the patterns of behavior of random phenomena observed in many diverse settings. Their forms are specified by a function or rule that tells us how to obtain probability density values from the values assumed by a continuous random variable X. Moreover, these probability density functions are described in terms of certain parameters that allow us to generate or index a whole family of density functions as we change the value(s) of the parameter(s) over some admissible range.

For the most part, the parameters of interest herein fall within the two broad categories of location and scale. In this regard, for a random variable X, let the associated probability density function appear as $f(x)$. If the probability density function $f(x; \alpha) = f(x - \alpha), -\infty < \alpha < +\infty, \alpha$ a parameter, has the same shape as $f(x)$ but is shifted or translated a distance α along the x-axis, then α is termed a *location parameter*. If $\alpha > 0 (< 0)$, f shifts a distance α to the left (right). So if the original origin is $x = 0$, then the new origin is $x = \alpha$ since now $x - \alpha = \alpha - \alpha = 0$.

Next, let the probability density function for a random variable X again appear as $f(x)$. If the probability density function $f(x; \beta) = \frac{1}{\beta} f(\frac{x}{\beta}), \beta > 0, \beta$ a parameter, has the same location and general shape as $f(x)$ but its graph is either stretched ($\beta > 1$) or contracted ($\beta < 1$) a bit, then β is termed a *scale parameter*.

If $f(x)$ is a probability density function for a random variable X and a location and/or scale parameter is introduced into f, then the resulting expression is also a legitimate probability density function. That is, if $f(x)$ is a probability

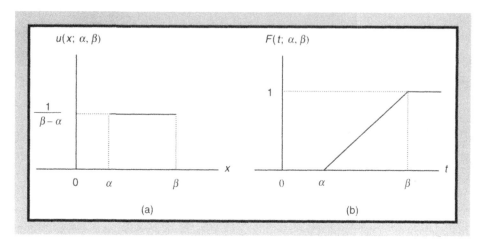

Figure 7.1   (a) A uniform probability density function; (b) Cumulative distribution function for X uniformly distributed.

density function and $\alpha(-\infty < \alpha < +\infty)$ and $\beta\ (>0)$ are location and scale parameters, respectively, then the function $f(x;\alpha,\beta) = \frac{1}{\beta} f\left(\frac{x-\alpha}{\beta}\right)$ is a probability density function.

7.2   The Uniform Distribution

The *uniform* (or *rectangular*) *probability distribution* is one whose probability densities are constant over some subinterval of the real line. That is, the random variable X is said to be uniformly distributed over the interval (α, β) if its probability density function is given by

$$u(x;\alpha,\beta) = \begin{cases} \frac{1}{\beta-\alpha}, & \alpha < x < \beta, \\ & \alpha \text{ and } \beta \text{ real numbers;} \\ 0 & \text{elsewhere} \end{cases} \qquad (7.1)$$

(see Figure 7.1a). Here the parameters α and β may be varied so as to generate a whole family of uniform distributions. Moreover, X's *cumulative distribution function* appears as

$$P(X \le t) = F(t;\alpha,\beta) = \frac{1}{\beta-\alpha}\int_{-\infty}^{t} dx = \begin{cases} 0, & t < \alpha; \\ \frac{t-\alpha}{\beta-\alpha}, & \alpha \le t \le \beta; \\ 1, & t > \beta \end{cases} \qquad (7.2)$$

(see Figure 7.1b).

To find $P(a \leq X \leq b)$, where $\alpha \leq a \leq b \leq \beta$, we may either integrate (7.1) directly between the limits a and b or use the relation

$$P(a \leq X \leq b) = F(b) - F(a) = \frac{b - a}{\beta - \alpha}. \tag{7.3}$$

For a uniformly distributed random variable X:

$$\mu = E(X) = \frac{\alpha + \beta}{2}; \tag{7.4}$$

$$\sigma = \sqrt{V(X)} = \frac{\beta - \alpha}{\sqrt{12}}; \tag{7.5}$$

the coefficient of skewness is $\alpha_3 = 0$; and the coefficient of kurtosis is $\alpha_4 = \frac{9}{5}$.

In addition, the quantile γ_p, the value for which $P(X \leq \gamma_p) = p$, is determined by setting $F(\gamma_p) = p$. Then from (7.2),

$$\gamma_p = \alpha + (\beta - \alpha)p. \tag{7.6}$$

As far as its essential properties are concerned, the uniform probability distribution:

- Is symmetrical (since $\alpha_3 = 0$)

- Has no mode

- Has a peak that is flatter than that of a normal distribution (since $\alpha_4 < 3$)

- Has its median equal to its mean (since $\gamma_{0.50} = \alpha + (\beta - \alpha)0.5 = 0.5\alpha + 0.5\beta = \mu$)

Example 7.2.1   Let the random variable X be uniformly distributed over the interval $(3, 9)$. Then

$$u(x; 3, 9) = \begin{cases} \frac{1}{6}, & 3 < x < 9; \\ 0 & \text{elsewhere.} \end{cases}$$

Also,

$$F(t; 3, 9) = \begin{cases} 0, & t < 3; \\ \frac{t-3}{6}, & 3 \leq t \leq 9; \\ 1, & t > 9. \end{cases}$$

What is the probability that the uniform random variable X takes on a value between 4 and 6? Using (7.3),

$$P(4 \leq X \leq 6) = \frac{6-4}{6} = \frac{1}{3};$$

and

$$P(X > 6) = 1 - P(X \leq 6) = 1 - F(6) = 1 - \frac{6-3}{6} = \frac{1}{2}.$$

In addition, it is easily verified that: $E(X) = 6$, $\sqrt{V(X)} = \frac{6}{\sqrt{12}}$; and the fortieth percentile is $\gamma_{0.40} = 3 + 6(0.4) = 5.40$. ∎

7.3   The Normal Distribution

7.3.1   Introduction to Normality

The *normal* (or *Gaussian*) *distribution* is a continuous bell-shaped distribution that is symmetrical about its mean and asymptotic to the horizontal axis. In this regard, a random variable X has a normal probability density function if

$$f(x; \mu, \sigma) = \frac{1}{\sigma\sqrt{2\pi}} e^{-\frac{1}{2}\left(\frac{x-\mu}{\sigma}\right)^2}, \quad -\infty < x, \ \mu < +\infty, \ \sigma > 0, \tag{7.7}$$

where μ and σ are, respectively, the mean and standard deviation of X and e denotes the base of the natural logarithm. Here (7.7) depicts the equation of the so-called *normal curve*. It can be shown that $f(x; \mu, \sigma) > 0$ for all x and

$$\int_{-\infty}^{+\infty} f(x; \mu, \sigma)dx = 1.$$

In general, if X follows a normal distribution with mean μ and standard deviation σ, then we shall describe this situation by simply saying that "X is $N(\mu, \sigma)$." Clearly this notation underscores the fact that (7.7) can admit a whole family of probability density functions because the parameters μ and σ are varied. Here μ is termed a location parameter and σ a scale parameter; that is, as we vary μ with σ held constant (see Figure 7.2a), we obtain a translation of the graph of f along the x-axis; and as we vary σ with μ held constant (see Figure 7.2b), the dispersion of f changes in a fashion such that the total area under the graph of f remains equal to one.

Some of the key properties of the normal distribution are as follows.

1. The normal probability density function (7.7) attains its (global) maximum or modal value of $\frac{1}{\sigma\sqrt{2\pi}}$ at $x = \mu$; and since the normal distribution is symmetrical, it follows that $\mu = \text{median} = \text{mode}$.

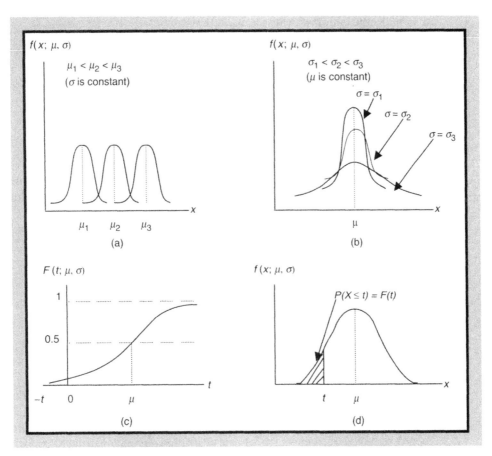

Figure 7.2 (a) Varying μ with σ constant; (b) Varying σ with μ constant; (c) The normal cumulative distribution function $F(t)$; (d) The value of F at t is the area under f from $-\infty$ to t.

2. f has (nonhorizontal) points of inflection at $\mu \pm \sigma$.

3. The mean and variance of the normal random variable are defined as

$$\mu = E(X) = \int_{-\infty}^{+\infty} x f(x; \mu, \sigma) dx \tag{7.8}$$

and

$$\sigma^2 = V(X) = \int_{-\infty}^{+\infty} (x - \mu)^2 f(x; \mu, \sigma) dx$$

$$\left(= \int_{-\infty}^{+\infty} x^2 f dx - 2\mu \int_{-\infty}^{+\infty} x f dx + \mu^2 \int_{-\infty}^{+\infty} f dx \right.$$

$$\left. = E(X^2) - 2\mu(\mu) + \mu^2(1) = E(X^2) - \mu^2 \right), \tag{7.9}$$

respectively.

4. The coefficient of skewness is $\alpha_3 = 0$.

5. The coefficient of kurtosis is $\alpha_4 = 3$.

If X is $N(\mu, \sigma)$, the probability that X assumes a value less than or equal to some number t is given by X's *cumulative distribution function*

$$P(X \le t) = F(t; \mu, \sigma) = \frac{1}{\sigma\sqrt{2\pi}} \int_{-\infty}^{t} e^{-\frac{1}{2}\left(\frac{x-\mu}{\sigma}\right)^2} dx \qquad (7.10)$$

(see Figure 7.2c). Clearly F is nondecreasing as required and exhibits a (nonhorizontal) point of inflection at $t = \mu$. In addition, Figure 7.2d relates the value of F at t to the area under f from $-\infty$ to t.

Since the total area under the normal probability density function is one, (7.7) and (7.10) may be used to determine probabilities such as

$$P(a \le X \le b) = \int_{a}^{b} f(x; \mu, \sigma) dx = \int_{-\infty}^{b} f dx - \int_{-\infty}^{a} f dx = F(b) - F(a) \qquad (7.11)$$

as illustrated in Figures 7.3a, b, respectively.

7.3.2 The *Z* Transformation

How may we calculate (7.11)? Direct integration of (7.7) between the (numerical) limits a and b is extremely cumbersome and (7.10) does not even have a closed form. Fortunately, tables of areas under (7.7) and cumulative probabilities up to and including various values of t in (7.10) have been constructed to expedite the determination of probabilities such as (7.11). But since we get a different normal probability density and cumulative distribution function for each possible

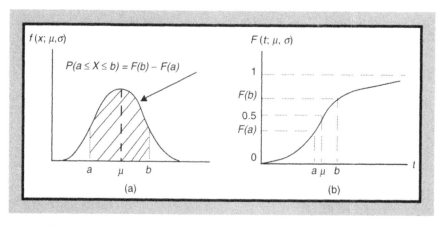

Figure 7.3   (a), (b) $\left(P(a \le X \le b) = P(X \le b) - P(X \le a) = F(b) - F(a)\right)$.

combination of μ and σ values, we might be tempted to surmise that a rather large (indeed infinite) number of tables would be needed. In fact, only one table is required, and it is called the Table of Cumulative Distribution Function Values for the Standard Normal Distribution (Table A.2 of the Appendix). To use this table, let us employ the following:

Transformation of Variable: If the random variable X is $N(\mu, \sigma)$, then the random variable

$$Z = \frac{X - \mu}{\sigma} \tag{7.12}$$

is $N(0, 1)$; that is, Z is a normally distributed random variable with a mean of zero and a standard deviation of one and will be termed a *standard normal random variable.*

Hence any normal X, irrespective of the values of its mean μ and standard deviation $\sigma (\neq 0)$, can be transformed into a standard normal or $N(0, 1)$ random variable Z.

Applying this transformation to (7.7) enables us to express the *standard normal probability density function* for the $N(0, 1)$ random variable Z as

$$f(z; 0, 1) = \frac{1}{\sqrt{2\pi}} e^{-\frac{1}{2}z^2} \left(\approx 0.3989(0.6065)^{z^2} \right) \tag{7.13}$$

(see Figure 7.4a). Here, too, $f(z; 0, 1) > 0$ for all z and

$$\int_{-\infty}^{+\infty} f(z; 0, 1) dz = 1.$$

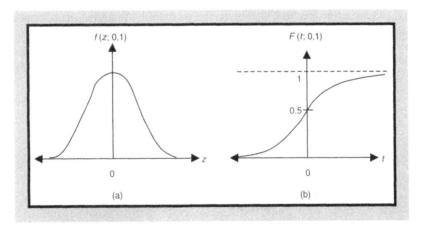

Figure 7.4 (a) The standard normal probability density function; (b) The standard normal cumulative distribution function.

Note that this density function depends only on $|z|$, which is conveniently measured in standard deviation units.

A second application of the Z-transformation, this time to (7.10), renders (for $dx = \sigma \, dz$) the standard normal cumulative distribution function

$$P(X \leq t) = P\left(Z \leq \frac{t-\mu}{\sigma}\right) = F\left(\frac{t-\mu}{\sigma};0,1\right) = \frac{1}{\sqrt{2\pi}} \int_{-\infty}^{\frac{t-\mu}{\sigma}} e^{-\frac{1}{2}z^2} dz \quad (7.14)$$

(see Figure 7.4b). So under this transformation, the cumulative normal probabilities are unchanged and equal the cumulative standard normal probabilities for each t; that is,

$$F(t;\mu,\sigma) = F\left(\frac{t-\mu}{\sigma};0,1\right).$$

Based upon this discussion, we see that to calculate probabilities involving a $N(\mu,\sigma)$ random variable X, we need only use the Z-transformation to calculate equivalent probabilities involving a $N(0,1)$ random variable Z. In this regard, if X is $N(\mu,\sigma)$, then

$$P(a \leq X \leq b) = \int_{a}^{b} f(x;\mu,\sigma)dx$$
$$= \int_{(a-\mu)/\sigma}^{(b-\mu)/\sigma} f(z;0,1)dz = P\left(\frac{a-\mu}{\sigma} \leq Z \leq \frac{b-\mu}{\sigma}\right), \quad (7.15)$$

where again Z is $N(0,1)$. Hence the probability that the normal random variable X assumes a value between a and b is equal to the probability that the standard normal random variable Z takes on a value between $\frac{(a-\mu)}{\sigma}$ and $\frac{(b-\mu)}{\sigma}$. The equivalence of these two probability calculations is illustrated in Figure 7.5.

In terms of (7.14), (7.15) can be rewritten as

$$P(a \leq X \leq b) = F\left(\frac{b-\mu}{\sigma};0,1\right) - F\left(\frac{a-\mu}{\sigma};0,1\right) \quad (7.15.1)$$

Additionally,

$$P(X \leq a) = P\left(Z \leq \frac{a-\mu}{\sigma}\right) = F\left(\frac{a-\mu}{\sigma};0,1\right); \quad (7.16)$$

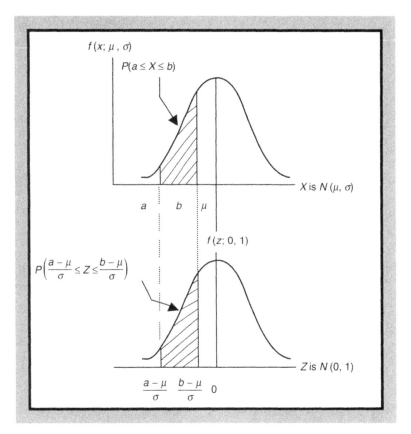

Figure 7.5   If X is $N(\mu, \sigma)$ and Z is $N(0, 1)$, then $P(a \leq X \leq b) = P\left(\frac{a-\mu}{\sigma} \leq Z \leq \frac{b-\mu}{\sigma}\right)$.

and

$$P(X > a) = 1 - P(X \leq a)$$

$$= P\left(Z > \frac{a-\mu}{\sigma}\right) = 1 - P\left(Z \leq \frac{a-\mu}{\sigma}\right) = 1 - F\left(\frac{a-\mu}{\sigma}; 0, 1\right).$$

$$(7.17)$$

Armed with these considerations, let us now turn to the table of $N(0, 1)$ cumulative distribution function values (see Table A.2). As the accompanying figure indicates (replicated here as Figure 7.6a), the table gives values of the standard normal cumulative distribution function F from $-\infty$ to any point a on the nonnegative z-axis or $P(Z \leq a) = F(a)$. For $-a < 0$, we may use the expression

$$P(Z \leq -a) = F(-a) = 1 - F(a) \qquad (7.18)$$

since the $N(0, 1)$ probability density is symmetrical about zero (see Figure 7.6b).

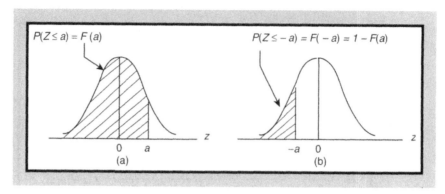

Figure 7.6   (a), (b) Probability determined via the standard normal cumulative distribution function.

Example 7.3.1   It is easily seen that:

1. $P(Z \leq 0) = F(0) = 0.5$.

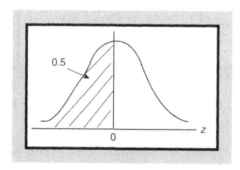

2. $P(Z \leq 1.2) = F(1.2) = 0.8849$.

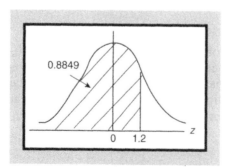

3. $P(Z \leq -0.34) = F(-0.34) = 1 - F(0.34) = 1 - 0.6331 = 0.3669.$

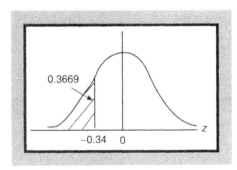

4. $P(0.42 \leq Z \leq 1.35) = F(1.35) - F(0.42) = 0.9115 - 0.6628 = 0.2487.$

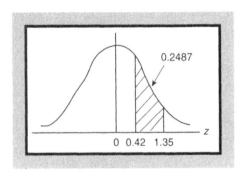

5. $P(-0.30 \leq Z \leq 2.16) = F(2.16) - F(-0.30) = F(2.16) - \left(1 - F(0.30)\right)$
$$= 0.9846 - (1 - 0.6179) = 0.6025. \quad \blacksquare$$

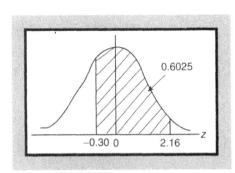

Now that we have Table A.2 at our disposal, the probability calculations implied by (7.15) and (7.15.1) are straightforward.

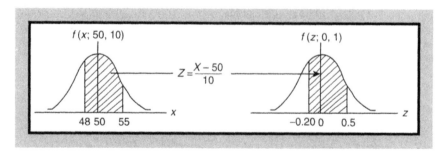

Figure 7.7 $P(48 \le X \le 55)$ and $P(-0.2 \le Z \le 0.5)$ are equivalent probability statements.

Example 7.3.2   Given that the random variable X is $N(50, 10)$, find $P(48 \le X \le 55)$. Proceeding as just shown, let us transform this given probability statement into an equivalent statement involving the $N(0, 1)$ random variable Z as follows (see Figure 7.7):

$$P(48 \le X \le 55) = P\left(\frac{48 - 50}{10} \le \frac{X - \mu}{\sigma} \le \frac{55 - 50}{10}\right)$$

$$= P(-0.2 \le Z \le 0.5)$$

$$= F(0.5) - F(-0.2)$$

$$= F(0.5) - \left(1 - F(0.2)\right) = 0.6914 - (1 - 0.5793) = 0.2707.$$

Also, it is easily shown that

$$P(X \le 45) = P\left(\frac{X - \mu}{\sigma} \le \frac{45 - 50}{10}\right) = P(Z \le -0.5) = F(-0.5) = 1 - F(0.5)$$

$$= 1 - 0.6914 = 0.3086;$$

and

$$P(X > 60) = 1 - P(X \le 60) = 1 - P\left(\frac{X - \mu}{\sigma} \le \frac{60 - 50}{10}\right)$$

$$= 1 - P(Z \le 1) = 1 - F(1) = 1 - 0.8413 = 0.1587. \quad \blacksquare$$

To reinforce our understanding of the use of the Z-transformation, let us examine a set of important (and equivalent) probability statements connecting the $N(\mu, \sigma)$ random variable X and the $N(0,1)$ random variable Z. In this regard,

$$P(X \le a) = P\left(Z \le \frac{a - \mu}{\sigma}\right);$$ (7.19)

and

$$P(Z \le z) = P(X \le \mu + z\sigma).$$ (7.20)

Additionally,

$$P(a \le X \le b) = P\left(\frac{a - \mu}{\sigma} \le Z \le \frac{b - \mu}{\sigma}\right);$$ (7.21)

$$P(c \le Z \le d) = P(\mu + c\sigma \le X \le \mu + d\sigma);$$ (7.22)

$$P(|X| \le a) = P(-a \le X \le a) = P\left(\frac{-a - \mu}{\sigma} \le Z \le \frac{a - \mu}{\sigma}\right);^{1}$$ (7.23)

$$P(|Z| \le z) = P(-z \le Z \le z) = P(\mu - z\sigma \le X \le \mu + z\sigma);$$ (7.24)

$$P(|X| > a) = P(X < -a) + P(X > a)$$
$$= P\left(Z < \frac{-a - \mu}{\sigma}\right) + P\left(Z > \frac{a - \mu}{\sigma}\right);$$ (7.25)

and

$$P(|Z| > z) = P(Z < -z) + P(Z > z)$$
$$= P(X < \mu - z\sigma) + P(X > \mu + z\sigma).$$ (7.26)

If we take $z = 1, 2,$ and 3 in (7.24), then we may use this expression to directly compare the X and Z scales (see Figure 7.8). So for $Z = \frac{X - \mu}{\sigma}$:

if $X = \mu$, then $Z = 0$;

if $X = \mu + \sigma$, then $Z = 1$;

if $X = \mu - 2\sigma$, then $Z = -2$; and so on.

---

[1] Here the *absolute value function* is defined as $|x| = \begin{cases} x, & x \ge 0; \\ -x, & x < 0. \end{cases}$

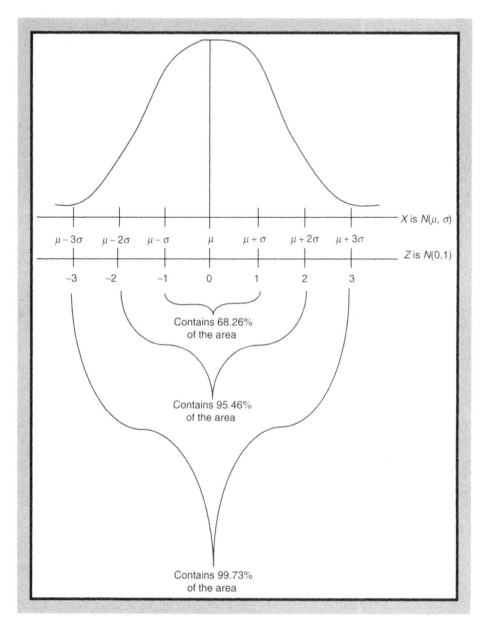

Figure 7.8 Comparing the X and Z scales illustrates the Empirical Rule.

Moreover, this discussion highlights the notion that although the tails of the normal (or standard normal) probability density function extend from $-\infty$ to $+\infty$, there is a negligible amount of area under this function once we depart more than three standard deviations from the mean μ of X (or from the mean 0 of Z). To verify this let us again consider (7.24) for $z = 1, 2,$ and 3. Then the $N(0, 1)$

cumulative distribution function values table (see Table A.2) allows us to easily determine the following probabilities:

1.

$$P(\mu - \sigma \leq X \leq \mu + \sigma) = P(-1 \leq Z \leq 1)$$
$$= F(1) - F(-1) = F(1) - \left(1 - F(1)\right)$$
$$= 2F(1) - 1 = 0.6826;$$

(68.26% of the area under $N(\mu, \sigma)$ lies within the interval $\mu \pm \sigma$ or within one standard deviation of the mean)

2.

$$P(\mu - 2\sigma \leq X \leq \mu + 2\sigma) = P(-2 \leq Z \leq 2)$$
$$= F(2) - F(-2) = F(2) - \left(1 - F(2)\right)$$
$$= 2F(2) - 1 = 0.9546;$$

(95.46% of the area under $N(\mu, \sigma)$ lies within the interval $\mu \pm 2\sigma$ or within two standard deviations of the mean)

3.

$$P(\mu - 3\sigma \leq X \leq \mu + 3\sigma) = P(-3 \leq Z \leq 3)$$
$$= F(3) - F(-3) = F(3) - \left(1 - F(3)\right)$$
$$= 2F(3) - 1 = 0.9973.$$

(99.73% of the area under $N(\mu, \sigma)$ lies within the interval $\mu \pm 3\sigma$ or within three standard deviations of the mean)

The preceding set of three probability statements (typically referred to as *one-sigma*, *two-sigma*, and *three-sigma* probabilities) constitute the so-called *Empirical Rule* introduced earlier in Chapter 2 when we first encountered the concept of dispersion.

7.3.3   Moments, Quantiles, and Percentage Points

A few additional items pertaining to the use of the normal distribution are in order. First, if $Z = \frac{X - \mu}{\sigma}$ is standard normal or $N(0, 1)$, then the rth *order moments of* Z are

$$\mu_r = \mu'_r = E(Z^r) = \frac{1}{\sqrt{2\pi}} \int_{-\infty}^{+\infty} z^r e^{-\frac{1}{2}z^2} dz.$$

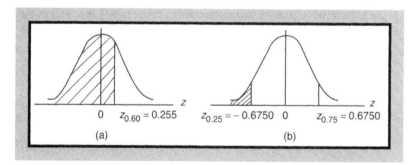

Figure 7.9 (a) The 60th percentile $\gamma_{0.60}$ satisfies $F(z_{0.60}; 0, 1) = 0.60$; (b) The 25th percentile $\gamma_{0.25}$ satisfies $F(z_{0.25}; 0, 1) = 0.25$.

If r is even, it can be shown that $\mu_r = (r-1)(r-3)\cdots 3 \cdot 1$; if r is odd, $\mu_r = 0$. (See Appendix 7.A for details pertaining to the moment-generating functions for $N(0, 1)$ and $N(\mu, \sigma)$ random variables.)

Second, if X is $N(\mu, \sigma)$, then we may readily determine a *quantile of* $N(\mu, \sigma)$ by remembering (from Chapter 4) that a quantile of order $p(0 < p < 1)$ is a value γ_p such that $P(X \leq \gamma_p) = p$ or, for our purposes, a value that satisfies

$$F(\gamma_p; \mu, \sigma) = \int_{-\infty}^{\gamma_p} f(x; \mu, \sigma)\,dx = p, \tag{7.27}$$

where $f(x; \mu, \sigma)$ is a normal probability density function. So if X is $N(\mu, \sigma)$, then

$$P(X \leq \gamma_p) = P\left(Z \leq \frac{\gamma_p - \mu}{\sigma} = z_p\right)$$

$$= F(z_p; 0, 1) = \int_{-\infty}^{z_p} f(z; 0, 1)\,dz = p. \tag{7.27.1}$$

Example 7.3.3 If X is $N(30, 5)$, then the sixtieth percentile of this distribution is the value $\gamma_{0.60}$, which satisfies $F(z_{0.60}; 0, 1) = 0.60$ or

$$z_{0.60} = \frac{\gamma_{0.60} - 30}{5} = 0.255$$

(note that we start with an F value and then read Table A.2 in reverse so as to obtain a z value). From this equality we obtain $\gamma_{0.60} = 30 + 5(0.255) = 31.275$.

Hence we conclude that 60% of the area under $N(30, 5)$ lies below 31.275 (see Figure 7.9a). Similarly, the twenty-fifth percentile is the value $\gamma_{0.25}$, which satisfies $F(z_{0.25}; 0, 1) = 0.25$ or

$$z_{0.25} = \frac{\gamma_{0.25} - 30}{5}.$$

Since $N(0,1)$ is symmetric about zero, it must be true that $z_{0.25} = -z_{0.75} = -0.675$ (see Figure 7.9b). Hence

$$z_{0.25} = \frac{\gamma_{0.25} - 30}{5} = -0.675$$

or $\gamma_{0.25} = 30 + 5(-0.675) = 26.625$. Thus 25% of the area under $N(30,5)$ lies below 26.625. In general, since $|z_p| = |z_{1-p}|$, take

$$z_p = -z_{1-p}, \, p < 0.50. \quad \blacksquare \qquad (7.28)$$

Third, for Z a $N(0,1)$ random variable, the quantity z_α, defined by the expression

$$P(Z > z_\alpha) = \alpha, \qquad (7.29)$$

is termed an *upper α–percentage point* of the standard normal distribution. It is the point on the (positive) z-axis such that the probability of a larger value is α (see Figure 7.10a).

Similarly, the quantity $-z_\alpha$, defined by the relationship

$$P(Z < -z_\alpha) = \alpha, \qquad (7.30)$$

is called a *lower α-percentage point* of the standard normal distribution. Now it is the point on the (negative) z-axis such that the probability of a smaller value is α (see Figure 7.10b). And if α is divided equally between both tails of the $N(0,1)$

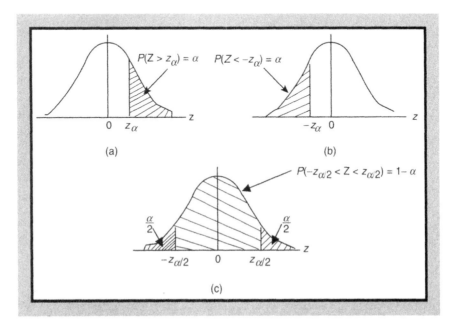

Figure 7.10   Locating α and $\alpha/2$ percentage points for the $N(0,1)$ distribution.

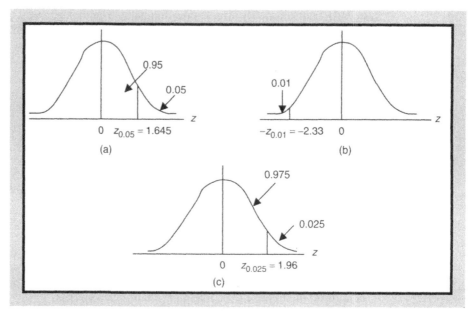

Figure 7.11   (a) Finding the upper $\alpha = 0.05$ percentage point; (b) Finding the lower $\alpha = 0.01$ percentage point; (c) Finding the upper $\alpha/2 = 0.025$ percentage point.

distribution (thus $\alpha/2$ is the area in each tail), then the remaining area of $1 - \alpha$ is located under the $N(0, 1)$ curve between $-z_{\alpha/2}$ and $z_{\alpha/2}$; that is,

$$P(|Z| < z_{\alpha/2}) = 1 - \alpha$$

or

$$P(-z_{\alpha/2} < Z < z_{\alpha/2}) = 1 - \alpha.$$

This latter area appears in Figure 7.10c. Hence the *lower and upper $\alpha/2$-percentage points* of the standard normal distribution are, respectively, $-z_{\alpha/2}$ and $z_{\alpha/2}$.

Example 7.3.4   To find these $N(0, 1)$ percentage points in actual practice, let us note that α is always chosen in advance; that is, any given α implies a z_{α} or $z_{\alpha/2}$ value. If $\alpha = 0.05$, then $z_{\alpha} = z_{0.05} = 1.645$ since $F(1.645; 0, 1) = 0.95$ via Table A.2 (see Figure 7.11a). If $\alpha = 0.01$, $-z_{\alpha} = -z_{0.01} = -2.33$ since $F(2.33; 0, 1) = 0.99$ and $|z_{\alpha}| = |z_{1-\alpha}|$ by virtue of the symmetry (about zero) of the standard normal distribution (see Figure 7.11b). And if $\alpha = 0.05$, then $z_{\alpha/2} = z_{0.025} = 1.96$ since $F(1.96; 0, 1) = 0.975$ (see Figure 7.11c). ∎

Next, if X_1 and X_2 are independent normally distributed random variables, then:

- Any linear function of X_1 and X_2 is also normally distributed
- If $E(X_1) = E(X_2) = 0$, then $X_1 X_2 (X_1^2 + X_2^2)^{-1/2}$ is normally distributed

- If $E(X_1) = E(X_2) = 0$ and $V(X_1) = V(X_2)$, then $(X_1^2 - X_2^2)(X_1^2 + X_2^2)^{-1}$ is normally distributed
- If X_1 and X_2 are each $N(\mu, \sigma)$, then $\overline{X} = \frac{1}{2}(X_1 + X_2)$ is $N(\mu, \frac{\sigma}{\sqrt{2}})$

7.3.4   The Normal Curve of Error

Let us address the question of the source of the normal distribution: How may we rationalize the form of the normal distribution? What we find more often than not is that the type of experiment that produces a random variable X that is approximately normally distributed is one in which the various values of X are actual *measurements* that tend to cluster symmetrically about some central value, or *measurement errors* that tend to cluster symmetrically about zero. For instance, the distribution of traits such as height, weight, IQ, and so on tend to more or less follow a normal distribution.

Let us consider the so-called *normal curve of error* in greater detail. Whereas the notion of *error* generally involves any deviation from systematic behavior, we shall get a bit more specific and consider a situation in which a large number of individuals are asked to measure the length of an object. It should be intuitively clear that each measurement reported may be subject to an error that depicts the resolution of essentially an infinite number of diverse (albeit minute) factors that operate on one individual independently of other individuals. Moreover, we shall assume that for each individual these factors produce small errors or departures from the true length and that positive and negative deviations are equally likely and no one factor can be expected to predominate. Thus we expect that the actual measurements are normally distributed about the true length or the errors are normally distributed about zero. That is, small errors are likely to occur more frequently than large ones and very large errors are highly unlikely.

So although we cannot predict the exact type of measurement error that a single individual will make, we can predict, with a reasonable degree of certainty, how the errors resulting from measurements made by a large number of individuals will behave: more people will make small errors rather than large ones; the larger the error, the fewer the number of people making it; approximately the same number of individuals will overestimate as will underestimate the true length; and the average error, taken over all individuals, will be zero. Hence this random error, which results from many like or repeated measurements actually made, may be viewed as a sample drawn from an infinitely large normal population.

7.4   **The Normal Approximation to Binomial Probabilities**

In this section we shall see exactly how and under what circumstances normal probabilities can serve as useful approximations to binomial probabilities. The reason why such approximations are legitimate is that the normal distribution is the limiting form of the binomial distribution when n is very large and p, the probability of a success on any given trial of a simple alternative or Bernoulli experiment, is not too close to either 0 or 1.

Let the binomial random variable S represent the number of successes in n independent trials of a Bernoulli experiment. Then from the binomial probability mass function (6.15), the probability that $S = s$ is

$$b(s;n,p) = \frac{n!}{s!(n-s)!}p^s(1-p)^{n-s}.$$

Moreover, $E(S) = np$ and $\sigma = \sqrt{V(S)} = \sqrt{np(1-p)}$. Now, the binomial variable S is a discrete random variable and the normal random variable X is continuous. Hence $b(s;n,p) \neq 0$ but $P(X = s) = 0$. However, we may approximate $b(s;n,p)$ as follows. Let us standardize S as $X = \frac{S-np}{\sqrt{np(1-p)}}$ with $x = \frac{s-np}{\sqrt{np(1-p)}}$. Then

$$b(s;n,p) \approx \frac{1}{\sqrt{2\pi}}\frac{1}{\sqrt{np(1-p)}}e^{-1/2x^2} = \frac{1}{\sqrt{2\pi}}\frac{1}{\sqrt{np(1-p)}}e^{-\frac{1}{2}\frac{(s-np)^2}{np(1-p)}}, \qquad (7.32)$$

where s is assumed close to np but not too close to either 0 or n.

The justification for this approximation (7.32) is Theorem 7.4.1, the *DeMoivre-Laplace-Gauss Limit Theorem* (1718, 1812).

THEOREM 7.4.1.   Let S depict a binomial random variable with mean np and standard deviation $\sqrt{np(1-p)}$. Then for fixed p, as the number of independent Bernoulli trials increases without bound, the standardized random variable

$$X = \frac{S - np}{\sqrt{np(1-p)}} \quad \rightarrow \quad Z \sim N(0,1)$$

(where "$\sim$" means "*is distributed as*").

Stated alternatively: given p, the cumulative distribution function of $X = \frac{S-np}{\sqrt{np(1-p)}}$ approaches the standard normal cumulative distribution function $F(z;0,1)$ asymptotically as $n \rightarrow \infty$ or

$$\lim_{\substack{n \to \infty \\ p=\text{constant}}} F(x) = \int_{-\infty}^{x} f(z;0,1)\,dz = F(x;0,1).$$

So given p, the limiting distribution as $n \rightarrow \infty$ of the standardized binomial variable S is the standard normal distribution (or the limiting distribution as $n \rightarrow \infty$ of the binomial variable S is, for fixed p, a normal distribution with mean np and standard deviation $\sqrt{np(1-p)}$. This approximation works well provided $np > 5$ when $p \leq 0.5$; or $n(1-p) > 5$ when $p > 0.5$.

If S is again taken to be a binomial random variable with mean np and standard deviation $\sqrt{np(1-p)}$, then for a and b integers, the preceding limit theorem is also the basis for the probability calculation

$$
P(a \leq S \leq b) \rightarrow \frac{1}{\sqrt{2\pi}} \int_{\frac{a-np}{\sqrt{np(1-p)}}}^{\frac{b-np}{\sqrt{np(1-p)}}} e^{-\frac{1}{2}z^2} dz
$$

$$
= P\left(\frac{a - np}{\sqrt{np\,(1-p)}} \leq Z \leq \frac{b - np}{\sqrt{np\,(1-p)}} \right)
$$

$$
= F\left(\frac{b - np}{\sqrt{np\,(1-p)}}; 0, 1 \right) - F\left(\frac{a - np}{\sqrt{np\,(1-p)}}; 0, 1 \right) \qquad (7.33)
$$

as $n \rightarrow \infty$; or

$$
P\left(np + a\sqrt{np(1-p)} \leq S \leq np + b\sqrt{np(1-p)} \right) \rightarrow \frac{1}{\sqrt{2\pi}} \int_a^b e^{-\frac{1}{2}z^2} dz
$$

$$
= P(a \leq Z \leq b) = F(b; 0, 1) - F(a; 0, 1) \qquad (7.34)
$$

as $n \rightarrow \infty$.

To improve the approximations provided by (7.32) to (7.34), let us introduce what is commonly called the *continuity correction* (since we are using a continuous distribution to approximate a discrete one). Consider integers a, b, and c, where $a < b < c$. Since the binomial random variable S is discrete, it has no probability mass between a and b and between b and c. In contrast to this situation, the standard normal variable Z has nonzero probability only over intervals. For $s = a$, the binomial probability is $b(a; n, p)$. But from the standard normal cumulative distribution function, the probability that Z takes on a value less than or equal to a is $F(a; 0, 1)$. Clearly the binomial probability does not equal $F(a; 0, 1)$. But if we employ the continuity correction and calculate the standard normal probability over the interval $a \pm 0.5$, we get $P(a - 0.5 \leq Z \leq a + 0.5) = F(a + 0.5; 0, 1) - F(a - 0.5; 0, 1) \approx b(a; n, p)$. Similarly, the binomial probability $P(a \leq S \leq c) = b(a; n, p) + b(b; n, p) + b(c; n, p)$. In terms of the standard normal cumulative distribution function, $P(a \leq Z \leq c) = F(c; 0, 1) - F(a; 0, 1)$, which understates the preceding binomial probability. But if we again employ the continuity correction and calculate instead $P(a - 0.5 \leq Z \leq c + 0.5) = F(c + 0.5; 0, 1) - F(c - 0.5; 0, 1)$, then an improved approximation to $P(a \leq S \leq c)$ obtains.

In sum, if we assume that the binomial probability $b(s; n, p)$ is the probability associated with an interval of length one and centered on s, then, via the continuity correction,

$$b(s; n, p) \approx P(s - 0.5 \leq S \leq s + 0.5)$$

$$= P\left(\frac{s - 0.5 - np}{\sqrt{np(1-p)}} \leq Z \leq \frac{s + 0.5 - np}{\sqrt{np(1-p)}} \right) \qquad (7.32.1)$$

$$= F\left(\frac{s + 0.5 - np}{\sqrt{np(1-p)}}; 0, 1 \right) - F\left(\frac{s - 0.5 - np}{\sqrt{np(1-p)}}; 0, 1 \right).$$

Thus adding and subtracting 0.5 (the continuity correction) accounts for the gaps between integer values where the binomial probability mass function is undefined.

Next, incorporating the continuity correction in the calculation of (7.33) yields

$$P(a \leq S \leq b) \approx P\left(\frac{a - 0.5 - np}{\sqrt{np(1-p)}} \leq Z \leq \frac{b + 0.5 - np}{\sqrt{np(1-p)}} \right)$$

$$= F\left(\frac{b + 0.5 - np}{\sqrt{np(1-p)}}; 0, 1 \right) - F\left(\frac{a - 0.5 - np}{\sqrt{np(1-p)}}; 0, 1 \right). \qquad (7.33.1)$$

We note briefly that for any finite n, the standard normal probability approximations to binomial probabilities will be very accurate when $p = 0.5$ (the binomial distribution is symmetrical). When $p \neq 0.5$ (the binomial distribution is skewed), the standard normal probability approximations to binomial probabilities will not be as accurate as in the preceding case. In this instance, the greater the difference between p and 0.5, the larger n must be in order to ensure an adequate approximation. For n sufficiently large, the standard normal probabilities will always provide a satisfactory approximation to binomial probabilities for any p not too close to 0 or 1. In general, the standard normal approximation to binomial probabilities is warranted if n is large enough to satisfy $\min \{np, n(1-p)\} \geq 5$.

Example 7.4.1   To see exactly how these approximations are made, let a random experiment consist of tossing a fair pair of dice 1000 times and let the random variable X represent the number of times the event *the sum of the faces showing equals 7* occurs. Clearly X is binomially distributed with $n = 1,000, p = \frac{1}{6}, E(x) = np = 166.66$, and $\sigma = \sqrt{np(1-p)} = 11.78$. Find $P(100 \leq X \leq 150)$. Under the direct calculation of binomial probabilities we have

$$P(100 \leq X \leq 150) = \sum_{i=100}^{150} \binom{1000}{i} \left(\frac{1}{6}\right)^i \left(\frac{5}{6}\right)^{1000-i}.$$

Since the amount of effort required to execute these numerical computations is prohibitive, let us approximate this probability using (7.33.1) (and Table A.2) or

$$P(100 \leq X \leq 150) \approx F\left(\frac{150 + 0.5 - 166.66}{11.78}; 0, 1\right)$$
$$- F\left(\frac{100 - 0.5 - 166.66}{11.78}; 0, 1\right)$$
$$= F(-1.37) - F(-5.70) = 1 - F(1.37) - \left(1 - F(5.70)\right)$$
$$= 1 - F(1.37) = 0.0854.$$

In addition, from (7.32.1),

$$P(X = 180) \approx F\left(\frac{180 + 0.5 - 166.66}{11.78}; 0, 1\right) - F\left(\frac{180 - 0.5 - 166.66}{11.78}; 0, 1\right)$$
$$= F(1.17) - F(1.09) = 0.0169. \quad \blacksquare$$

7.5   The Normal Approximation to Poisson Probabilities

The purpose of this section is to briefly mention that the De Moivre-Laplace-Gauss Limit Theorem (Section 7.4) applied to the approximation of binomial probabilities also applies to the approximation of Poisson probabilities. That is, there exists an asymptotic relationship between the discrete Poisson probability distribution and the continuous standard normal probability distribution. Specifically, if X is a Poisson random variable with parameter $\lambda = \mu = \sigma^2$, then the cumulative distribution function of the standardized variable $Y = \frac{X - \lambda}{\sqrt{\lambda}}$ approaches the standard normal cumulative distribution function as λ increases without bound; that is,

$$\lim_{\lambda \to \infty} F(y) = \int_{-\infty}^{y} f(z; 0, 1)dz = F(y; 0, 1).$$

This result may be utilized to find probabilities such as:

$$P(a \leq X \leq b) \to F\left(\frac{b - \lambda}{\sqrt{\lambda}}; 0, 1\right) - F\left(\frac{a - \lambda}{\sqrt{\lambda}}; 0, 1\right) \tag{7.35}$$

as $\lambda \to \infty$; or

$$P(\lambda + a\sqrt{\lambda} \leq X \leq \lambda + b\sqrt{\lambda}) \to F(b; 0, 1) - F(a; 0, 1) \tag{7.36}$$

as $\lambda \to \infty$.

7.6 The Exponential Distribution

7.6.1 Source of the Exponential Distribution

The exponential probability distribution has its origins in the Poisson probability model. To see this let us return briefly to Chapter 6 wherein we found that a (discrete) Poisson random variable R can be considered as the *number* of independent random occurrences of some event A, which takes place at a constant (average) rate over a given continuous interval of length T. For convenience, let T depict a fixed interval of time. Then R's probability mass function appears as $p(R = r; \lambda T) = e^{-\lambda T} \frac{(\lambda T)^r}{r!}, r = 1, 2, \dots$. If the parameter λ depicts the intensity or constant rate of occurrence of the Poisson random process per standard unit or specified time period, then the parameter λT is the intensity of this process over T time periods. Moreover, we also found that $E(R) = \lambda T = V(R)$.

Let the occurrence of an event A follow a Poisson probability distribution. If the underlying Poisson process operates at a rate of λ occurrences per unit of time, then, as we shall now see, the length of time we have to wait before the first occurrence of this event follows an exponential probability distribution.

For a Poisson random variable R, the probability of no occurrence of A prior to T or over the time interval $[0,T)$ is $p(R = 0; \lambda T) = e^{-\lambda T}$. If we define a *new* random variable X as the time taken for the *first* occurrence of event A (clearly X is a continuous random variable), then the probability of the first occurrence of A at or after T is $P(X > T) = p(0; \lambda T) = e^{-\lambda T}$. (Note that since the Poisson process describing R is independent, the random variable X, the time until the first occurrence of A, may be defined equivalently as the time *between* successive occurrences of A).

So for this new random variable X, its cumulative distribution function can be written as

$$F(T) = P(X \leq T) = 1 - P(X > T) = 1 - e^{-\lambda T}, T > 0. \qquad (7.37)$$

Hence its associated probability density function is

$$f(T; \lambda) = F'(T) = \begin{cases} \lambda e^{-\lambda T}, & \lambda > 0, T > 0; \\ 0 & \text{elsewhere.} \end{cases} \qquad (7.38)$$

In general, if the random variable R, depicting the number of occurrences of some event A, follows a discrete Poisson probability distribution with parameter λ, then the random variable X, representing the length of time between successive occurrences of A, follows a continuous *exponential probability distribution* with parameter λ and probability density function

$$f(x; \lambda) = \begin{cases} \lambda e^{-\lambda x}, & \lambda > 0, x \geq 0; \\ 0 & \text{elsewhere.} \end{cases} \qquad (7.39)$$

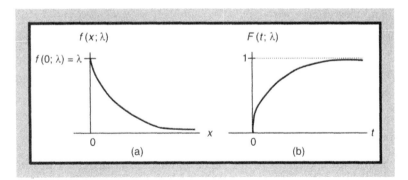

Figure 7.12   (a) Exponential probability density function; (b) Exponential cumulative distribution function.

(see Figure 7.12a). And as we vary the parameter λ, we obtain an entire family of exponential probability distributions and, as required, it is easily demonstrated that for each such distribution

$$\int_{-\infty}^{+\infty} f(x;\lambda)dx = \int_{0}^{+\infty} f(x;\lambda)dx = 1.$$

Note that since (7.39) is negatively sloped for all $x \geq 0$ (for this reason it is sometimes called the *negative exponential distribution*), it follows that the probability of experiencing a long time interval until the first occurrence of A should be less than the probability of a short time interval. So if, for instance, the number of telephone calls arriving at a switchboard is Poisson distributed, then the time between each arrival is exponentially distributed (since time starts over at zero after each arrival and thus the rate of arrival is constant).

Given that the instantaneous rate of occurrence of some event A is a constant λ, the probability density function (7.39) gives the instantaneous probability of occurrence of A at any given (time) $x > 0$. Moreover, the *exponential cumulative distribution function* yields the probability of occurrence of A prior to t or

$$P(X \leq t) = F(t;\lambda) = \begin{cases} 1 - e^{-\lambda t}, & t > 0; \\ 0, & t \leq 0 \end{cases} \tag{7.40}$$

(see Figure 7.12b).

From (7.39) and (7.40):

$$P(X \leq a) = 1 - e^{-\lambda a};$$

$$P(X > a) = 1 - P(X \leq a) = e^{-\lambda a};$$

and for $a < b$,

$$P(a \leq X \leq b) = P(X > a) - P(X > b) = e^{-\lambda a} - e^{-\lambda b}.$$

The mean and variance of the exponential probability distribution can be shown to be

$$E(X) = \frac{1}{\lambda}, \quad V(X) = \frac{1}{\lambda^2},$$

respectively. Since λ is the mean number of occurrences per unit time of the Poisson process, the exponential mean or reciprocal of λ may be thought of as the mean length of time between successive Poisson occurrences of event A or the expected time until the first occurrence of a Poisson event.

7.6.2 Features/Uses of the Exponential Distribution

An assortment of key properties of the exponential probability distribution are:

1. As Figure 7.12a indicates, the exponential distribution is highly skewed to the right with its median equal to $\frac{1}{\lambda} \ln 2$ and its mode located at zero.

2. The coefficients of skewness and kurtosis are, respectively, $\alpha_3 = 2$ and $\alpha_4 = 9$.

3. The pth quantile γ_p, the value of x which satisfies $P(X \leq \gamma_p) = p$, is $\gamma_p = \frac{1}{\lambda} \ln \left(\frac{1}{1-p} \right)$.

4. If a nonzero location parameter θ is introduced into the exponential distribution, then (7.39) is replaced by

$$f(x; \lambda, \theta) = \begin{cases} \lambda e^{-\lambda(x-\theta)}, & x > \theta \geq 0, \lambda > 0; \\ 0 & \text{elsewhere.} \end{cases} \tag{7.39.1}$$

Here the graph of f is translated a distance θ along the positive x-axis. The median is now $\theta + \frac{1}{\lambda} \ln 2$, and the mode is at the lowest value of x or at θ. Additionally, the pth quantile is, in this instance, $\gamma_p = \theta + \frac{1}{\lambda} \ln \left(\frac{1}{1-p} \right)$.

5. The exponential probability distribution is applicable whenever a random variable depicts the time until the next occurrence of an event, or the time between successive occurrences of two independent Poisson events. In this regard, the relationship between the Poisson and exponential distributions is analogous to that between the binomial and geometric distributions; that is, the Poisson and binomial distributions are concerned with the number of occurrences of some event; the continuous exponential and discrete geometric distributions deal with the amount of time and the number of trials, respectively, until the first (next) occurrence of an event. Specifically, the geometric random variable is the number of trials until the first success in a Bernoulli

process with parameter p; the exponential random variable is the length of time until the first occurrence of an event in a Poisson process with parameter λ. The mean of the geometric random variable is $\frac{1}{p}$; that of the exponential random variable is $\frac{1}{\lambda}$.

6. Given that the exponential random variable X typically is taken as the time until the first occurrence of some Poisson event (such as an arrival per time unit or the failure per time unit of an item), and if, say, the instantaneous failure rate of an item over its entire lifetime is a constant λ, then the probability of no failure in time T is termed an item's *reliability* and will be denoted $r(T; \lambda) = p(R = 0; \lambda T) = e^{-\lambda t}$. This is so because $F(T) = 1 - e^{-\lambda T} = 1 - r(T; \lambda)$ gives the probability of failure *prior to* T and thus reliability is the probability of no failure for *at least* time T or $r(T; \lambda) = 1 - F(T)$. In this context then, (7.38) serves as the *failure probability density function* and its mean $\frac{1}{\lambda}$ is termed the *average life* of an item or the mean time between failures.

7. The exponential probability distribution exhibits the so-called *memoryless property* in that the probability of occurrence of present or future events does not depend upon events that have already happened in the past. That is, if X is an exponential random variable, then

$$P(X > a + b | X > a) = P(X > b) \quad \text{for } a, b > 0. \tag{7.41}$$

Thus the fact that X has already attained some level "a" has no effect on the probability of achieving a larger level "$a + b$" (since the events in a Poisson process are independent). For instance, suppose the time to failure of a particular item (e.g., a component in a stereo system) is exponentially distributed. If this component has already been used 1000 hours without failure, then the probability that it will last another 500 hours equals the probability of its lasting 500 hours when it was new. Thus the chance of failure in a given time T will be the same regardless of when it occurs in an item's life.

   To summarize, if events occur according to a Poisson process, then the time between successive occurrences of these events is exponentially distributed. That is:

Poisson random variable X—the number of occurrences in a finite interval T; exponential random variable X—the length of time between Poisson occurrences or the time until the first Poisson occurrence.

Furthermore:

$\lambda =$ mean number of occurrences per time unit T of the Poisson process. (If T is in hours, then λ is stated in occurrences/hr.)

$\frac{1}{\lambda} =$ mean time interval between successive Poisson occurrences or the expected time until the first Poisson occurrence

(If events follow a Poisson process at a rate of $\lambda = 10$/min., then the average time between these Poisson occurrences is $\frac{1}{\lambda} = 0.10$/min.) So if the number of occurrences in a fixed time interval T is Poisson distributed with parameter λT and if some event A has just occurred, then the length of time we have to wait for the next occurrence of A is exponentially distributed with parameter λT.

Example 7.6.1   Let us define an exponential random variable X as the length of life of a particular component within a piece of precision machinery. Given that the average life of this component is $\frac{1}{\lambda} = 5000$ hours (and thus the expected number of failures in one hour is $\lambda = 0.0002$), the failure probability density function appears as

$$f(T;\lambda) = 0.0002e^{-0.0002T}, \ T > 0.$$

Hence the cumulative distribution function is

$$F(T;\lambda) = 1 - e^{-0.0002T}$$

and thus the reliability or probability of no failure for at least T hours is

$$r(T;\lambda) = e^{-0.0002T}.$$

Then the probability that the component will fail before 3000 hours is

$$P(T \leq 3000) = F(3000) = 1 - e^{-0.0002(3000)} = 1 - 0.5488 = 0.4512.$$

The probability that it will last 6000 hours or more is

$$r(6000) = e^{-0.0002(6000)} = 0.3012.$$

The probability that the component will last at least 4000 hours but not more than 7000 hours is

$$P(4000 \leq X \leq 7000) = F(7000) - F(4000)$$
$$= P(X > 4000) - P(X > 7000)$$
$$= e^{-0.0002(4000)} - e^{-0.0002(7000)}$$
$$= 0.4493 - 0.2466 = 0.2027.$$

And the probability that the component will last the average life or less is

$$P(X \leq 5000) = F(5000) = 1 - e^{-0.0002(5000)} = 1 - e^{-1} = 0.6322. \quad \blacksquare$$

Example 7.6.2   Let us assume that the number of incoming calls at an answering service during a 15-minute period follows a Poisson distribution with $\lambda = 3$ (thus the average number of calls per 15-minute period is three). Hence the corresponding exponential random variable X is thus the time between calls with $\frac{1}{\lambda} = 0.333$ so that the average time between calls is approximately five minutes. Then:

$$f(T;\lambda) = 3e^{-3T}, T > 0;$$

and

$$F(T;\lambda) = 1 - e^{-3T},$$

where T is expressed in 15-minute intervals. For instance, the probability of no calls in a 15-minute period is equivalent to the probability that the time between calls is greater than one 15-minute period or

$$P(T > 1) = e^{-3(1)} = 0.04898.$$

The probability that it takes no more than three minutes to get the next call is

$$P\left(T \le \frac{1}{5}\right) = 1 - e^{-3(1/5)} = 1 - 0.5488 = 0.4512;$$

and the probability that the time between calls is at least 10 minutes is

$$P\left(T > \frac{2}{3}\right) = e^{-3(2/3)} = 0.1353.$$

An alternative approach to this problem is to use time intervals of one minute. In this regard, based upon the information just given, the average number of calls per minute is $\lambda = \frac{3}{15} = 0.20$ and thus the average duration between calls is, as expected, $\frac{1}{\lambda} = \frac{1}{0.20} = 5$ minutes. So for intervals of one minute duration, we have

$$f(T;\lambda) = 0.20e^{-0.20T}, \quad T > 0;$$

and

$$F(T;\lambda) = 1 - e^{-0.20T}.$$

Then the probability that the time between calls is greater than 15 minutes is

$$P(T > 15) = e^{-0.20(15)} = e^{-3} = 0.0498;$$

the probability that the next call arrives in three minutes or less is

$$P(T \le 3) = 1 - e^{-0.20(3t)} = 1 - e^{-0.6} = 0.4512;$$

and the probability that the time between calls is at least 10 minutes is

$$P(T > 10) = e^{-0.20(10)} = e^{-2} = 0.1353. \quad \blacksquare$$

7.7 Gamma and Beta Functions

For any real number $\alpha > 0$, the *gamma function* (of α) is the improper integral[2]

$$\Gamma(\alpha) = \int_0^\infty y^{\alpha-1} e^{-y} dy. \tag{7.42}$$

The gamma function is well defined and continuous for $\alpha > 0$ and convergent if and only if $\alpha > 0$. So if α is positive, $\Gamma(\alpha)$ exists and is a positive number.
   It can be shown that:

(a) $\Gamma(1) = 1$

(b) $\Gamma\left(\dfrac{1}{2}\right) = \sqrt{\pi}$

(c) (7.42) satisfies the recursion formula $\Gamma(\alpha) = (\alpha - 1)\Gamma(\alpha - 1), \alpha > 0$

(d) $\Gamma(\alpha) = (\alpha - 1)!, \alpha$ a positive integer

(e) $\Gamma(\alpha + 1) = \alpha!, \alpha$ a nonnegative integer

(f) $\dbinom{n + \alpha - 1}{n} = \dfrac{\Gamma(n + \alpha)}{\Gamma(n)\Gamma(\alpha)}$ $\qquad\qquad$ (7.43)

(g) $\Gamma(\alpha) \to +\infty$ as $\alpha \to 0+$

(h) $\Gamma(\alpha)$ has a unique minimum for $1 < \alpha < 2$

(i) $\Gamma(\alpha)$ is concave upward for all $\alpha > 0$

(j) $\Gamma(\alpha) \approx \sqrt{2\pi/\alpha}\ \alpha^\alpha e^{-\alpha}$ for large α. (If $\alpha = n + 1, n$ a positive integer, then, for large $n, n! \approx \sqrt{z\pi} n^{n+\frac{1}{z}} e^{-n}$ (Stirling's formula).)

---

[2] In general, an integral is deemed *improper* if it is an integral of an unbounded function or if it is an integral over an unbounded interval. The improper integral $\int_a^\infty f(x)dx$ is an integral in which $f(x)$ is: (a) defined for $x \ge a$; and (b) integrable over every finite closed interval $[a, b]$. This integral is then defined as the limit $\lim\limits_{b \to \infty} \int_a^b f(x)dx$. If this limit exists, then the integral is said to be *convergent*.

The function of the variables α and t,

$$\Gamma_t(\alpha) = \int_0^t y^{\alpha-1} e^{-y} \, dy, \qquad (7.44)$$

is called the *incomplete gamma function*. It is defined for all $\alpha > 0$ and $t \geq 0$.

For real numbers $\alpha, \beta > 0$, the *beta function* (of α and β) is the integral

$$B(\alpha, \beta) = \int_0^1 y^{\alpha-1}(1-y)^{\beta-1} \, dy. \qquad (7.45)$$

The beta function is well defined for $\alpha, \beta > 0$. If $\alpha \geq 1$ and $\beta \geq 1$, the integral is proper (an integral is *proper* if the integrand is bounded or the interval of integration is bounded). And if $\alpha, \beta > 0$ and either $\alpha < 1$ or $\beta < 1$ (or both), then this integral is improper though convergent.

Some key properties of (7.45) are:

(a) $B(\alpha, \beta) = B(\beta, \alpha)$

(b) $B(\alpha, 1) = \frac{1}{\alpha}$

(c) $B\left(\frac{1}{2}, \frac{1}{2}\right) = \pi$

(d) if $\alpha, \beta > 0$, then the relationship between the gamma and beta functions is expressed as

$$B(\alpha, \beta) = \frac{\Gamma(\alpha)\Gamma(\beta)}{\Gamma(\alpha + \beta)} \qquad (7.46)$$

(e) $\beta B(\alpha + 1, \beta) = \alpha B(\alpha, \beta + 1)$

(f) $B(\alpha + 1, \beta) = \frac{\alpha}{\alpha+\beta} B(\alpha, \beta)$

(g) $B(\alpha + 1, \beta) + B(\alpha, \beta + 1) = B(\alpha, \beta)$

(h) if $\alpha + 1$ and $\beta + 1$ are positive integers, then

$$B(\alpha + 1, \beta + 1) = \frac{\alpha! \beta!}{(\alpha + \beta + 1)!}.$$

The function of the variables α, β, and t,

$$B_t(\alpha, \beta) = \int_0^t y^{\alpha-1}(1-y)^{\beta-1} \, dy, \qquad (7.47)$$

is termed the *incomplete beta function*. It is defined for all real $\alpha > 0, \beta > 0$, and $0 \leq t \leq 1$.

7.8 The Gamma Distribution

Within the gamma function (7.42)

$$\Gamma(\alpha) = \int_0^\infty y^{\alpha-1} e^{-y} \, dy$$

let us introduce a new variable x by writing $y = \frac{x-\tau}{\theta}, \theta > 0, x > \tau$. Then since $dy = \frac{dx}{\theta}$,

$$\Gamma(\alpha) = \int_0^\infty \left(\frac{x-\tau}{\theta}\right)^{\alpha-1} e^{-\frac{x-\tau}{\theta}} \left(\frac{1}{\theta}\right) dx$$

or

$$1 = \int_0^\infty \frac{1}{\Gamma(\alpha)\theta^\alpha}(x-\tau)^{\alpha-1} e^{-\frac{x-\tau}{\theta}} \, dx.$$

And since the integrand in this expression is nonnegative (for $\alpha, \beta, \Gamma(\alpha)$, and $x(> \tau)$ all positive), it can serve as a probability density function. Consistent with most applications of the gamma distribution, we shall restrict the (location) parameter τ to zero.

In this regard, a continuous random variable X is said to follow a *gamma probability distribution* if its probability density function is

$$f(x;\alpha,\theta) = \begin{cases} \frac{1}{\Gamma(\alpha)\theta^\alpha} x^{\alpha-1} e^{-x/\theta}, & 0 \leq x < +\infty, \text{ and } \alpha, \theta > 0; \\ 0 & \text{elsewhere.} \end{cases} \tag{7.48}$$

Clearly $f(x;\alpha,\theta) \geq 0$ and it is easily verified that

$$\int_{-\infty}^{+\infty} f(x;\alpha,\theta)dx = \int_0^\infty f(x;\alpha,\theta)dx = 1.$$

Here the parameter α determines the shape of the gamma distribution and θ serves as scale parameter. And as we vary these parameters, we generate a whole family of gamma distributions. Specifically, as θ varies, the basic shape of the gamma distribution is unchanged while:

(a) For small values of $\alpha(\alpha < 1)$, the gamma distribution is reverse J-shaped with an elongated right-hand tail.

(b) For $0 < \alpha < 1$, the gamma probability density function has no mode. For $\alpha > 1$, this function has a unique mode occurring at $x = \theta(\alpha - 1)$.

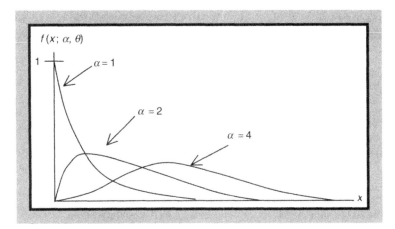

Figure 7.13   Gamma probability density functions for selected values of θ and α.

For instance, a specific collection of gamma probability density functions for $\theta = 1$ and $\alpha = 1, 2,$ and 4 is illustrated in Figure 7.13.

If the random variable X is gamma distributed, then the *gamma cumulative distribution function* appears as

$$P(X \le t) = F(t; \alpha, \theta) = \frac{1}{\Gamma(\alpha)\theta^\alpha} \int_0^t x^{\alpha-1} e^{-\frac{x}{\theta}} dx, x > 0. \qquad (7.49)$$

Alternatively, if the variable of integration is changed to $y = \frac{x}{\theta}$, then $dx = \theta dy$ and (7.49) becomes

$$P(X \le t) = F(t; \alpha, \theta) = \frac{1}{\Gamma(\alpha)\theta^\alpha} \int_0^{\frac{t}{\theta}} (\theta y)^{\alpha-1} e^{-y} \theta dy$$

$$= \frac{1}{\Gamma(\alpha)} \int_0^{\frac{t}{\theta}} y^{\alpha-1} e^{-y} dy$$

$$= \frac{\Gamma_{t/\theta}(\alpha)}{\Gamma(\alpha)} = I_t(\alpha). \qquad (7.49.1)$$

Hence the gamma cumulative distribution function is expressible as the ratio between the incomplete gamma function (see (7.44)) and the (complete) gamma function (7.42). For given values of α and $\frac{t}{\theta}$, (7.49.1) is termed the *incomplete gamma function ratio*. Although (7.49.1) does not admit a closed-form representation if α is not an integer, tables of gamma function integrals are available (see Pearson (1956)). For instance, such tables would be needed to determine

$$P(a \le X \le b) = F(b; \alpha, \theta) - F(a; \alpha, \theta) = \int_a^b \frac{x^{\alpha-1} e^{-\frac{x}{\theta}}}{\Gamma(\alpha)\theta^\alpha} dx.$$

If α is a positive integer, then it can be shown that (7.49) can be rewritten as the closed form expression

$$F(t;\alpha,\theta) = 1 - \left[1 + \frac{t}{\theta} + \frac{1}{2!}\left(\frac{t}{\theta}\right)^2 + \cdots + \frac{1}{(\alpha-1)!}\left(\frac{t}{\theta}\right)^{\alpha-1}\right]e^{-\frac{t}{\theta}}. \qquad (7.49.2)$$

The gamma probability distribution has the following properties:

(a) Its mean and standard deviation are, respectively,

$$\mu = E(X) = \alpha\theta; \qquad (7.50)$$

$$\sigma = \sqrt{V(X)} = \theta\sqrt{\alpha}. \qquad (7.51)$$

(b) The gamma distribution is positively skewed since the coefficient of skewness is

$$\alpha_3 = \frac{2}{\sqrt{\alpha}} > 0. \qquad (7.52)$$

And since $\lim_{\alpha\to\infty} \alpha_3 = 0$, this distribution approaches a symmetrical distribution for very large α.

(c) The coefficient of kurtosis is

$$\alpha_4 = 3 + \frac{6}{\alpha}. \qquad (7.53)$$

Hence the peak of a gamma probability density function is sharper than that of a normal probability density function. But since $\lim_{\alpha\to\infty} \alpha_4 = 3$, the degree of sharpness of the gamma probability density function approaches that of a normal probability density function for very large α.

(d) If a random variable X is gamma distributed with mean and standard deviation given by (7.50) and (7.51), respectively, then, for a fixed θ, the variable $Z = (X - \alpha\theta)/\theta\sqrt{\alpha} \to N(0,1)$ as $\alpha \to \infty$.

(e) When $\alpha = 1$ and $\frac{1}{\theta} = \lambda$ in (7.48), the gamma probability density function reduces to the exponential probability density function. Hence the gamma probability distribution can be viewed as a generalization of the exponential distribution (7.39). Remember that the exponential random variable can be regarded as the length of time until the occurrence of the *first* Poisson event (or the time between successive occurrences of two independent Poisson events). In general, when α in (7.48) is taken to be any positive integer (not necessarily unity), the gamma distribution specializes to what is called the *Erlang probability distribution* (see (7.54)). As indicated earlier, a Poisson distribution with intensity parameter $\lambda T = \frac{t}{\theta}$ is used to model the *number* of independent random occurrences of an event in a given time interval t. For a particular integer α, the *length of time* until the occurrence of the αth Poisson event (or the time between the occurrence of the αth Poisson event and the next α consecutive Poisson events) is modeled by an Erlang probability distribution. Hence the Poisson random variable X is the number of independent events taking place at the constant rate $\frac{1}{\theta}$ over t whereas the Erlang random variable T is the waiting time or the time interval t until the αth Poisson event obtains. Specifically, let t denote the length of time until the αth occurrence of some event in a Poisson process with intensity parameter $\lambda T = \frac{t}{\theta}$. Then the total time for α Poisson events is expressible as $T = \sum_{i=1}^{\alpha} X_i$ (it is the sum of α independent and identically distributed exponential random variables X_i) and follows an *Erlang probability distribution* with probability density function

$$f(t;\alpha,\theta) = \begin{cases} \frac{t^{\alpha-1}e^{-t/\theta}}{(\alpha-1)!\theta^\alpha}, & t \geq 0, \alpha \text{ an integer;} \\ 0 & \text{elsewhere.} \end{cases} \tag{7.54}$$

Then as we might anticipate, $E(T) = \alpha\theta = \frac{\alpha}{\lambda}, V(T) = \alpha\theta^2 = \frac{\alpha}{\lambda^2}$. Clearly the Erlang distribution is simply a gamma distribution with α integer valued.

(f)   Let X be a Poisson random variable with parameter $\lambda T = \frac{t}{\theta}$ and let the random variable Y be Erlang distributed. Then Erlang probabilities can be determined by using the Poisson distribution (table); that is, for α integer valued and any real number $t > 0$,

$$P(Y \leq t) = F(t;\alpha,\theta)$$

$$= \begin{cases} 1 - P(X \leq \alpha - 1) = P(X \geq \alpha), & t > 0, \alpha \text{ an integer;} \\ 0, & t \leq 0. \end{cases} \tag{7.55}$$

Thus the probability that the length of time until the αth Poisson event occurs does not exceed t is the same as the probability that the number of Poisson events occurring over t is at least α. This expression is valid because incomplete gamma functions with integer α's can be written in terms of sums of Poisson probabilities as

$$P(Y > t) = \int_t^\infty \frac{y^{\alpha-1}}{\Gamma(\alpha)\theta^\alpha} e^{-\frac{y}{\theta}} dy$$

$$= \sum_{j=0}^{\alpha-1} \frac{(\frac{t}{\theta})^j}{j!} e^{-\frac{t}{\theta}} = P(X \leq \alpha - 1) = F(\alpha - 1; \frac{t}{\theta}). \tag{7.55.1}$$

Here the probability that the length of time until the αth Poisson event occurs exceeds t equals the probability that the number of Poisson events occurring over t does not exceed $\alpha - 1$.

(g) The negative binomial probability distribution is the discrete analog of the gamma probability distribution; that is, the discrete negative binomial random variable is the number of failures observed before exactly k successes occur in a simple alternative experiment, and the continuous gamma random variable is the length of time until the rth Poisson event occurs.

(h) The gamma distribution has the *reproductive property*: If X_1 and X_2 are independent random variables each distributed as (7.48) with common θ's but possibly with different α's (call them α_1 and α_2, respectively), then $X_1 + X_2$ also has a distribution of this form with the same θ and with an $\alpha = \alpha_1 + \alpha_2$.

7.9 The Beta Distribution

Starting with the beta function ((7.45))

$$B(\alpha, \beta) = \int_0^1 x^{\alpha-1}(1-x)^{\beta-1}dx$$

let us form (using (7.46d))

$$1 = \int_0^1 \frac{\Gamma(\alpha + \beta)}{\Gamma(\alpha)\Gamma(\beta)}x^{\alpha-1}(1-x)^{\beta-1}dx.$$

Since the integrand in this specification is nonnegative, it can be viewed as a probability density function. In this regard, a continuous random variable $X \in (0,1)$ is said to follow a *beta probability distribution* if its probability density function is

$$f(x; \alpha, \beta) = \begin{cases} \frac{\Gamma(\alpha+\beta)}{\Gamma(\alpha)\Gamma(\beta)}x^{\alpha-1}(1-x)^{\beta-1}, & 0 < x < 1, \text{ with } \alpha, \beta > 0; \\ 0 & \text{elsewhere.} \end{cases} \tag{7.56}$$

As required, $f(x; \alpha, \beta) \geq 0$ and

$$\int_{-\infty}^{+\infty} f(x; \alpha, \beta)dx = \int_0^1 f(x; \alpha, \beta)dx = 1.$$

Moreover, the beta distribution satisfies a symmetry condition if in (7.56) we substitute $1 - x$ for x or

$$f(x; \alpha, \beta) = f(1 - x; \beta, \alpha). \tag{7.57}$$

An alternative version of (7.56) results if α and β are taken to be positive integers and we set $n = \alpha + \beta$.

Then from (7.43d), (7.56) becomes

$$f(x;n,\alpha) = \begin{cases} \frac{(n-1)!}{(\alpha-1)!(n-\alpha-1)!}x^{\alpha-1}(1-x)^{n-\alpha-1}, & 0 < x < 1, \\ & \text{with } n, \alpha > 0; \\ 0 & \text{elsewhere.} \end{cases} \qquad (7.56.1)$$

The parameters α and β determine the shape of the beta distribution. In this regard:

(a) If both $\alpha, \beta < 1$, the beta distribution is U-shaped

(b) If $\alpha \geq 1$ and $\beta < 1$, the beta distribution is J-shaped with an elongated left–hand tail

(c) If $\alpha < 1$ and $\beta \geq 1$, the beta distribution is reverse J-shaped with its right–hand tail elongated

(d) If $a > 2$, the beta distribution is tangent to the horizontal axis at $x = 0$; if $\beta > 2$, it is tangent to the horizontal axis at $x = 1$

(e) For $\alpha < 0$ and $\beta < 1$, the beta density $f(x; \alpha, \beta) \to \infty$ as $x \to 0$ or 1

(f) When both $\alpha, \beta > 1$, the beta probability density function has a unique mode occurring at $x = \frac{(\alpha-1)}{(\alpha+\beta-2)}$

(g) The beta distribution is symmetrical when $\alpha = \beta$

(h) If $\alpha = \beta = 1, f(x; 1,1) = \Gamma(2)$ and thus the beta random variable X is uniformly distributed over $(0,1)$

(i) For $\alpha, \beta > 0$, the beta distribution has points of inflection at

$$\frac{\alpha-1}{\alpha+\beta-2} \pm \frac{1}{\alpha+\beta-2}\sqrt{\frac{(\alpha-1)(\beta-1)}{\alpha+\beta-3}}.$$

(Note that these values must be real and are restricted to the interval $(0,1)$.)

A limited selection of beta probability density functions for specific values of α and β is illustrated in Figure 7.14.

Given that the random variable X follows a beta distribution, the *beta cumulative distribution function* is

$$P(X \leq t) = F(t; \alpha, \beta)$$

$$= \begin{cases} 0, & x \leq 0; \\ \frac{\Gamma(\alpha+\beta)}{\Gamma(\alpha)\Gamma(\beta)} \int_0^t x^{\alpha-1}(1-x)^{\beta-1}dx, & 0 < x < 1 \\ 1, & x \geq 1. \end{cases} \qquad (7.58)$$

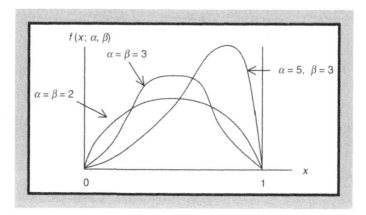

Figure 7.14 Beta probability density functions for selected values of α and β.

Since $P(X \leq t) = 1 - P(1 - X \leq 1 - t)$, the symmetry condition (7.57) allows as to write

$$F(t; \alpha, \beta) = 1 - F(1 - t; \beta, \alpha). \tag{7.59}$$

In addition, from (7.46d) and (7.47), we may rewrite (7.58) as

$$P(X \leq t) = F(t; \alpha, \beta)$$

$$= \frac{\Gamma(\alpha + \beta)}{\Gamma(\alpha)\Gamma(\beta)} B_t(\alpha, \beta) = \frac{B_t(\alpha, \beta)}{B(\alpha, \beta)} = I_t(\alpha, \beta). \tag{7.58.1}$$

Thus the beta cumulative distribution function can be expressed as the ratio between the incomplete beta function (7.47) and the (complete) beta function (7.45). For given values of α and β, (7.58.1) is termed the *incomplete beta function ratio*. As was the case for the incomplete gamma function ratio, (7.58.1) does not admit a closed form representation if α and β are not both integers. However, beta function integral tables are available (Pearson (1956)). Note that if we again invoke the symmetry condition we have, from (7.59) and (7.58.1),

$$I_t(\alpha, \beta) = 1 - I_{1-t}(\alpha, \beta). \tag{7.59.1}$$

Some key properties of the beta probability distribution are:

(a) Its mean and variance are, respectively,

$$\mu = E(x) = \frac{\alpha}{\alpha + \beta}; \tag{7.60}$$

$$\sigma^2 = V(X) = \frac{\alpha\beta}{(\alpha + \beta)^2(\alpha + \beta + 1)}. \tag{7.61}$$

It is interesting to note that (7.60) and (7.61) may be used to readily demonstrate that the parameters α and β can be written in terms of μ and σ^2 as

$$\alpha = \mu \left[\frac{\mu(1-\mu)}{\sigma^2} - 1 \right], \quad \beta = (1-\mu)\left[\frac{\mu(1-\mu)}{\sigma^2} - 1 \right].$$

(b) The coefficient of skewness is

$$\alpha_3 = \frac{2(\beta-\alpha)}{\alpha+\beta+2}\sqrt{\frac{\alpha+\beta+1}{\alpha\beta}}. \tag{7.62}$$

As indicated earlier, the beta distribution is symmetrical when $\alpha = \beta$ so that, as expected, $\alpha_3 = 0$. (Note that when $\alpha = \beta$, (7.60) simplifies to $\mu = 0.5$. So when the mean of the beta distribution is close to 0.5, the distribution itself is nearly symmetrical.) Also, if $\alpha < \beta$, then $\alpha_3 > 0$ and the beta distribution is skewed to the right. But if $\alpha > \beta$, then $\alpha_3 < 0$ and the distribution is skewed to the left.

(c) The coefficient of kurtosis is

$$\alpha_4 = \frac{3(\alpha+\beta+1)\left[2(\alpha^2+\beta^2)+\alpha\beta(\alpha+\beta-6)\right]}{\alpha\beta(\alpha+\beta+2)(\alpha+\beta+3)}. \tag{7.63}$$

(d) If X_1 and X_2 are independent gamma random variables, then the random variable $Y = \frac{X_1}{X_1+X_2}$ is beta distributed with probability density function $f(y;\alpha,\beta)$, where $y = \frac{x_1}{x_1+x_2}$.

The beta distribution generally is used to model the behavior of random variables whose values are restricted to intervals of finite length. In fact, up to this point in our discussion of this distribution, we have taken $X \in (0,1)$. Although the unit interval is used most commonly for defining the beta distribution, (7.56) can be

---

[3] If the beta probability density is formulated as (7.56.1), then (7.60) through (7.63) appear as

$$\mu = \alpha/n; \tag{7.60.1}$$

$$\sigma^2 = \alpha(n-\alpha)/n^2(n+1); \tag{7.61.1}$$

$$\alpha_3 = \frac{2(n-2\alpha)}{n+2}\sqrt{\frac{n+1}{\alpha(n-\alpha)}} \tag{7.62.1}$$

(here $\alpha_3 = 0$ when $\alpha = \frac{n}{2}$ so that the distribution is symmetrical; if $\alpha < \frac{n}{2}$, then $\alpha_3 < 0$ and the distribution is positively skewed; and if $\alpha > \frac{n}{2}$, then $\alpha_3 < 0$ and the distribution is negatively skewed);

$$\alpha_4 = \frac{3(n+1)\left[2n^2+\alpha(n-\alpha)(n-6)\right]}{\alpha(n-\alpha)(n+2)(n+3)} \tag{7.63.1}$$

generalized to the case where a continuous random variable X takes on values within the interval $(a, b), 0 < a < b < +\infty$.

For instance, if $X \in (a, b)$, then, under a suitable translation and change of scale, the beta distribution can be applied to the case $a < X < b$ by forming a new random variable $Y = \frac{(X-a)}{(b-a)}$. Then $0 < Y < 1$ as required by (7.56) and (7.58). As an alternative, we may rewrite (7.56) and (7.58) explicitly in terms of a and b. To this end, $X \in (a, b)$ follows a *beta probability distribution* if its probability density function is

$$
f(x; \alpha, \beta) = \begin{cases} 0, & x \le a; \\ \frac{(b-a)^{1-\alpha-\beta}}{B(\alpha,\beta)}(x - a)^{\alpha-1}(b - x)^{\beta-1}, & a < x < b, \text{ with } \alpha, \beta > 0 \quad (7.64) \\ 0, & x \ge b. \end{cases}
$$

Here too $f(x; \alpha, \beta) \ge 0$ and $\int_{-\infty}^{+\infty} f(x; \alpha, \beta) = 1$.

Also, for $X \in (a, b)$, the *beta cumulative distribution function* is

$$
P(X \le t) = F(t; \alpha, \beta) = \begin{cases} 0, & x \le a; \\ \frac{(b-a)^{1-\alpha-\beta}}{B(\alpha,\beta)} \int_a^t (x - a)^{\alpha-1}(b - x)^{\beta-1}\, dx, & a < x < b; \\ 0, & x \ge b. \end{cases}
$$

$$(7.65)$$

This expression may be written alternatively as

$$
P(X \le t) = F(t; \alpha, \beta) = \frac{1}{B(\alpha, \beta)} \int_0^{\frac{t-a}{b-a}} x^{\alpha-1}(1 - x)^{\beta-1} dx
$$

$$
= B_{\frac{t-a}{b-a}}(\alpha, \beta)/B(\alpha, \beta).[4] \quad (7.65.1)
$$

For a given finite interval (a, b) and specific values of α and β, (7.65.1) is referred to as the *incomplete beta function ratio*.

A glance back at (7.56.1) reveals that the beta probability density function resembles the probability mass function (6.15) of the binomial distribution. Since p, the probability of a success on any given trial of a Bernoulli experiment, can assume an infinite number of values between 0 and 1, it follows that we can actually use the continuous beta distribution to represent the distribution of p;

---

[4] Equation (7.65.1) has been derived from (7.65) by using the following integral transformations. Specifically, for real numbers $a, b,$ and c:

$$
(1) \int_a^b f(x)dx = \int_{(a+c)}^{(b+c)} f(x - c)dx;
$$

$$
(2) \int_a^b f(x)dx = \frac{1}{c} \int_{ac}^{bc} f(\tfrac{x}{c})dx. \quad (7.n4)
$$

that is, if p is beta distributed with parameters n and α, where n and α are integer valued and $n > \alpha > 0$, then the probability density function of p is

$$f(p;n,\alpha) = \begin{cases} \frac{(n-1)!}{(\alpha-1)!(n-\alpha-1)!} p^{\alpha-1}(1-p)^{n-\alpha-1}, & 0 < p < 1, \\ & \text{with } n > \alpha > 0; \qquad (7.65.2) \\ 0 & \text{elsewhere.} \end{cases}$$

Here the parameters n and α may be viewed as the number of trials and the number of successes, respectively, in a Bernoulli process with random variable Y. And since n and α are integers, this beta density value can be found readily in tables for the binomial probability distribution. (If n and α are not integers, the factorial terms must be replaced by gamma function values; i.e., see (7.43d).) To actually calculate (7.56.2) using binomial probabilities, let us rewrite this expression as

$$f_{beta}(p;n,\alpha) = (n-1)\frac{(n-2)!}{(\alpha-1)!(n-\alpha-1)!}p^{\alpha-1}(1-p)^{n-\alpha-1}$$

$$= (n-1)P_{binom}(\alpha-1;n-2,p). \qquad (7.56.3)$$

Example 7.9.1 If $n = 6, \alpha = 3$, and $p = 0.40$, the beta probability density value is, via Table A.6,

$$f_{beta}(0.40;6,3) = 5P_{binom}(2;4,0.40) = 5(0.3456) = 1.728.$$

Armed with (7.56.2), we can now determine, for various values of n and α, probabilities involving p; for example;

$$P(a \le p \le b) = \int_a^b f(p;n,\alpha)dp, \qquad (7.66)$$

where obviously $0 \le a \le b \le 1$. (Keep in mind, however, that beta distribution integral tables must generally be employed in order to find this probability.)

Finally, if α and β are integers and we set $t = p, 0 < p < 1$, then it is easily seen (using (7.58)) that the beta cumulative distribution function is related to the binomial cumulative distribution function as

$$P(X \le p) = F(p;\alpha,\beta) = I_p(\alpha,\beta) = \sum_{i=\alpha}^{n-1} \binom{n-1}{i} p^i(1-p)^{n-1-i}, \qquad (7.67)$$

where $n = \alpha + \beta$. Alternatively, this same result may be obtained from (7.56.1) by finding

$$P(X \le p) = F(p;n,\alpha) = I_p(n,\alpha) = \int_0^p f(x;n,\alpha)dx.$$

In fact, we may operationalize the process of determining cumulative beta probabilities from tables for the binomial distribution (again n and α are deemed integers) by writing the beta cumulative distribution function as

$$F_{beta}(p;n,\alpha) = \begin{cases} P_{binom}(Y \geq \alpha; n-1, p) = 1 - P_{binom}(Y \leq \alpha-1; n-1, p), p \leq 0.50; \\ P_{binom}(Y < n-\alpha; n-1, 1-p), p \geq 0.50. \quad \blacksquare \end{cases}$$

Example 7.9.2   In this regard, if we again take $n = 6, \alpha = 3$, and $p = 0.40$, then the beta cumulative distribution function value is

$$F_{beta}(0.40; 6, 3) = P_{binom}(Y \geq 3; 5, 0.40)$$

$$= 1 - P_{binom}(Y \leq 2; 5, 0.40) = 1 - B(2; 5, 0.40) = 0.3174. \quad \blacksquare$$

7.10   Other Useful Continuous Distributions

Although the preceding sections of this chapter have offered detailed descriptions of many important families of continuous parametric probability distributions, many other families of such distributions exist, although they may not have the same degree of notoriety and applicability as in the former case. In this regard, only an abbreviated sketch will be offered of the remaining (by no means exhaustive) collection of probability distributions.

7.10.1   The Lognormal Distribution

Let the random variable Y be defined as $Y = \ln X$, where X is a positive random variable. If Y is normally distributed, then $X = e^Y$ has a *lognormal probability distribution* with probability density function

$$f(x; \mu, \sigma) = \frac{1}{\sigma x \sqrt{2\pi}} e^{-\frac{1}{2}\left(\frac{\ln x - \mu}{\sigma}\right)^2}, \quad x > 0, -\infty < \mu < +\infty, \sigma > 0. \qquad (7.68)$$

(Note that (7.68) *does not* describe the probability density function of the logarithm of a normal random variable; it depicts the distribution of an exponential function of a normal random variable.) For a lognormal random variable X, it can be shown that

$$E(X) = e^{\mu + \frac{1}{2}\sigma^2}; \qquad (7.69)$$

$$V(X) = e^{2\mu}\left(e^{2\sigma^2} - e^{\sigma^2}\right). \qquad (7.70)$$

Hence μ and σ are *not* location and scale parameters, respectively, for the lognormal random variable X. In fact, μ and σ are the mean and standard deviation, respectively, of the normally distributed random variable Y, that is, $E(\ln X) = \mu$, $V(\ln X) = \sigma^2$. The *lognormal cumulative distribution function* is

$$P(X \le t) = F(t; \mu, \sigma) = \frac{1}{\sigma \sqrt{2\pi}} \int_0^t \frac{1}{x} e^{-\frac{1}{2}(\frac{\ln x - \mu}{\sigma})^2} \, dx. \qquad (7.71)$$

Additional characteristics of a lognormal random variable X are:

(a) The median is equal to e^μ and the (unique) mode is $e^{(\mu - \sigma^2)}$.

(b) The lognormal distribution is skewed to the right since its coefficient of skewness

$$\alpha_3 = (e^{\sigma^2} - 1)^{1/2}(e^{\sigma^2} + 2) \qquad (7.72)$$

is positive. And since its coefficient of kurtosis

$$\alpha_4 = (e^{\sigma^2})^4 + 2(e^{\sigma^2})^3 + 3(e^{\sigma^2})^2 - 3 \qquad (7.73)$$

is also positive, the peak of the lognormal probability density function is sharper than that of a normal distribution.

(c) The lognormal probability density function has two points of inflection occurring at

$$x = e^{\left[\mu - \frac{3}{2}\sigma^2 \pm \sigma(1 + \frac{1}{4}\sigma^2)^{1/2} \right]}.$$

(d) As $\sigma \to 0$, the standardized lognormal probability density function approaches a standard normal probability density function. And as $\sigma \to +\infty$, the lognormal probability density function quickly becomes highly nonnormal in shape.

(e) Values of the lognormal probability density function and cumulative distribution function can be obtained from the standard normal tables (Tables A.2 and A.3 of the Appendix) once x, μ, and σ are given. In this regard, the probability density values for the lognormal (LN) and normal (N) distributions satisfy

$$f_{LN}(x; \mu, \sigma) = \frac{1}{x} f_N(\ln x; \mu, \sigma) = \frac{1}{\sigma x} f_N\left(z = \frac{\ln x - \mu}{\sigma}; 0, 1\right); \qquad (7.74)$$

and the cumulative distribution function values for these distributions satisfy

$$F_{LN}(x; \mu, \sigma) = F_N(\ln x; \mu, \sigma) = F_N\left(z = \frac{\ln x - \mu}{\sigma}; 0, 1\right). \qquad (7.75)$$

Example 7.10.1.1   Suppose we choose to find the probability density function values for a lognormal variable X with $\mu = 0.2$ and $\sigma = 1.2$ at $x = 0.8$. From (7.74), $z = \frac{\ln 0.8 - 0.2}{1.2} = -0.3526$ and

$$f_{LN}(0.8; 0.2, 1.2) = \frac{1}{1.2(0.8)} f_N(-0.3526; 0, 1) = \frac{0.375}{0.96} = 0.3906$$

and from (7.75)

$$F_{LN}(0.8; 0.2, 1.2) = FN(-0.3526; 0, 1) = 0.363. \quad \blacksquare$$

(f) The quantile γ_p for the lognormal (LN) distribution can be obtained from the quantile $\ln \gamma_p$ for the normal (N) distribution since

$$F_{LN}(\gamma_p; \mu, \sigma) = F_N(\ln \gamma_p; \mu, \sigma) = F_N\left(z_p = \frac{\ln \gamma_p - \mu}{\sigma}; 0, 1\right) = p.$$

That is, if the random variable X is lognormal and the random variable Y is normal, then

$$P(X \le \gamma_p) = \int_0^{\gamma_p} f_{LN}(x; \mu, \sigma) dx$$

$$= \int_{-\infty}^{\ln \gamma_p} f_N(y; \mu, \sigma) dy$$

$$= \int_{-\infty}^{\frac{\ln \gamma_p - \mu}{\sigma}} f_N(z; 0, 1) dz = \int_{-\infty}^{z_p} f_N(z; 0, 1) dz = p.$$

Hence

$$z_p = \frac{\ln \gamma_p - \mu}{\sigma}$$

and thus

$$\gamma_p = e^{(\mu + \sigma z_p)}. \tag{7.76}$$

Hence the quantile γ_p of the lognormal distribution (satisfying $F_{LN}(\gamma_p; \mu, \sigma) = p$) is obtained by finding the quantile $z_p = \frac{\ln \gamma_p - \mu}{\sigma}$ of the standard normal distribution (which satisfies $F_N(z_p; 0, 1) = p$).

Example 7.10.1.2   For instance, for X a lognormal random variable with $\mu = 2$ and $\sigma = 0.6$, find $\gamma_{0.30}$. Since $z_{0.30} = -0.52$, (7.76) yields

$$\gamma_{0.30} = e^{[2 + 0.6(-0.52)]} = 5.4087. \quad \blacksquare$$

(g) If X_1 and X_2 are independent random variables, each distributed as (7.68), then so is $X_1 X_2$. Also, if X is distributed as (7.68), then so are X^α and αX, $\alpha \neq 0$.

It may be the case that at least one x value is negative. In this circumstance we must redefine the origin of X so as to make all quantities positive; that is, let $y = \log(x - \alpha)$, where $\alpha < \min\{x_i\}$ is termed the *threshold value* of the lognormal model. Then the probability density function for our new lognormal random variable Y is

$$f(y; \mu, \sigma) = \frac{1}{\sigma(x - \alpha)\sqrt{2\pi}} e^{-\frac{1}{2}\left[\frac{\ln(x-\alpha)-\mu}{\sigma}\right]^2}. \tag{7.68.1}$$

7.10.2 The Logistic Distribution

For convenience sake, discussions pertaining to a logistic probability distribution typically start with its associated cumulative distribution function. In this regard, for a logistic random variable X, the *logistic cumulative distribution function* has the form

$$F(x; \mu, \beta) = \frac{1}{1 + e^{-\frac{x-\mu}{\beta}}}, \quad -\infty < x < +\infty, -\infty < \mu < +\infty, \beta > 0; \tag{7.77}$$

or, as it is equivalently expressed,

$$F(x; \mu, \beta) = \frac{e^{\frac{x-\mu}{\beta}}}{1 + e^{\frac{x-\mu}{\beta}}}, \quad -\infty < x < +\infty, -\infty < \mu < +\infty, \beta > 0. \tag{7.77.1}$$

Here μ is the location parameter and β serves as a scale parameter.

For a logistic random variable X, it can be shown that:

$$E(X) = \mu; V(X) = \frac{(\beta\pi)^2}{3}.$$

Moreover, the *logistic probability density function* $f(x; \mu, \beta)$ can be obtained from (7.77) by differentiation; that is,

$$f(x; \mu, \beta) = F'(x; \mu, \beta) = \frac{1}{\beta} \frac{e^{-\frac{x-\mu}{\beta}}}{\left(1 + e^{-\frac{x-\mu}{\beta}}\right)^2}, \quad -\infty < x, \mu < +\infty, \beta > 0. \tag{7.78}$$

And since $F(\mu - a; \mu, \beta) = 1 - F(\mu + a; \mu, \beta)$, this probability density function is symmetrical about its mean μ (the coefficient of skewness is $\alpha_3 = 0$) with its shape mirroring that of a normal probability distribution (the coefficient of kurtosis is

$\alpha_4 = 4.2$ so that its peak is a bit sharper than that of a normal distribution). In addition, it is readily seen that

$$f(x; \mu, \beta) = \frac{1}{\beta} F(x; \mu, \beta) [1 - F(x; \mu, \beta)];$$

and

$$x = \mu + \beta \ln \left\{ F(x; \mu, \beta) / [1 - F(x; \mu, \beta)] \right\}.$$

Appendix 7.A   Moment-Generating Function for the Normal Distribution

For the $N(0,1)$ random variable $Z = \frac{X - \mu}{\sigma}$, it is easily demonstrated that its moment-generating function is of the form

$$m_Z(t) = e^{\frac{t^2}{2}}, \quad -\infty < t < +\infty. \tag{7.A.1}$$

To this end set

$$m_Z(t) = E(e^{tZ}) = \frac{1}{\sqrt{2\pi}} \int_{-\infty}^{+\infty} e^{tz} e^{-\frac{z^2}{2}} dz = \frac{1}{\sqrt{2\pi}} e^{\frac{t^2}{2}} \int_{-\infty}^{+\infty} e^{-\frac{(z-t)^2}{2}} dz.$$

Then for $u = z - t$,

$$m_Z(t) = e^{\frac{t^2}{2}} \left(\frac{1}{\sqrt{2\pi}} \int_{-\infty}^{+\infty} e^{-\frac{u^2}{2}} du \right) = e^{\frac{t^2}{2}}.$$

Replacing the right-hand side of $m_Z(t)$ by its Maclaurin's series expansion yields

$$m_Z(t) = 1 + \frac{t^2}{2} + \frac{\left(\frac{t^2}{2}\right)^2}{2!} + \frac{\left(\frac{t^2}{2}\right)^3}{3!} + \cdots . \tag{7.A.2}$$

Then we may generate all moments of Z about zero by finding $m_Z^{(r)}(0) = \mu_r'$, $r = 1, 2, \ldots$. That is, using (7.A.2), you are asked to verify that $m_Z^{(1)}(0) = \mu_1' = 0$, $m_Z^{(2)}(0) = \mu_2' = 1$, and $m_Z^{(3)}(0) = \mu_3' = 0$. In fact, all odd-order moments of Z about zero are equal to zero.

If X is $N(\mu, \sigma)$, what is the form of the moment-generating function for X? Since $Z = \frac{X - \mu}{\sigma}$ is $N(0, 1)$ if and only if X is $N(\mu, \sigma)$, let us work with $X = \mu + \sigma Z$ and obtain, via (4.44.a) and (7.A.1),

$$m_X(t) = e^{t\mu} m_Z(\sigma t) = e^{t\mu} e^{t^2 \sigma^2 / 2} = e^{t\mu + t^2 \sigma^2 / 2}, \quad -\infty < t < +\infty. \tag{7.A.3}$$

In addition, if X is $N(\mu, \sigma)$, then the moment-generating function of $X - \mu$ is, from (4.43),

$$m_{X-\mu}(t) = e^{-t\mu} m_X(t) = e^{\frac{t^2\sigma^2}{2}}, \quad -\infty < t < +\infty, \tag{7.A.4}$$

or, in terms of its Maclaurin's series expansion,

$$m_{X-\mu}(t) = 1 + \frac{t^2\sigma^2}{2} + \frac{\left(\frac{t^2\sigma^2}{2}\right)^2}{2!} + \frac{\left(\frac{t^2\sigma^2}{2}\right)^3}{3!} + \cdots. \tag{7.A.5}$$

Then the rth central moment of X is determined as $m^{(r)}_{X-\mu}(0) = \mu_r, r = 1, 2, \ldots$. (Verify that $m^{(1)}_{X-\mu}(0) = 0$, $m^{(2)}_{X-\mu}(0) = \sigma^2$, and $m^{(3)}_{X-\mu}(0) = 0$.) In fact, all odd-order central moments of X are zero.

Our final result in this section addresses the issue of the form of the distribution of a sum of independent normal random variables. Specifically, Theorem 7.A.1:

THEOREM 7.A.1.   The sum of a finite number of independent normally distributed random variables is itself normally distributed.

To verify this assertion, let $U = \sum_{i=1}^{n} X_i$, where each X_i is $N(\mu_i, \sigma_i)$, $i = 1, \ldots, n$. Then from (5.73.1) and (7.A.3),

$$m_U(t) = \prod_{i=1}^{n} m_{X_i}(t) = \prod_{i=1}^{n} e^{t\mu_i + \frac{t^2\sigma_i^2}{2}} = e^{\sum_{i=1}^{n}\left(t\mu_i + \frac{t^2\sigma_i^2}{2}\right)} = e^{t\mu + \frac{t^2\sigma^2}{2}}, \tag{7.A.6}$$

where $\mu = \sum_{i=1}^{n} \mu_i$ and $\sigma^2 = \sum_{i=1}^{n} \sigma_i^2$. Hence U is $N(\mu, \sigma)$.

If we replace the sum $U = \sum_{i=1}^{n} X_i$ by the linear combination $Y = \sum_{i=1}^{n} a_i X_i$, then (7.A.6) becomes

$$m_Y(t) = e^{t\mu_Y + \frac{t^2\sigma_Y^2}{2}}, \tag{7.A.7}$$

where $\mu_Y = \sum_{i=1}^{n} a_i\mu_i$ and $\sigma_Y^2 = \sum_{i=1}^{n} a_i^2\sigma_i^2$. Thus Y is $N(\mu_Y, \sigma_Y)$.

Appendix 7.B   The Bivariate Normal Probability Distribution

In this section we seek to generalize the univariate normal distribution involving the random variable X to the bivariate case involving the random variables X and Y. Specifically, let the two-dimensional random variable (X, Y) have the joint probability density function

$$f(x, y) = \frac{1}{2\pi\sigma_X\sigma_Y(1-\rho^2)^{1/2}} e^{-\frac{1}{2}Q}, \quad -\infty < x, y < +\infty, \tag{7.B.1}$$

where

$$Q = \frac{1}{1-\rho^2}\left[\left(\frac{x-\mu_X}{\sigma_X}\right)^2 - 2\rho\left(\frac{x-\mu_X}{\sigma_X}\right)\left(\frac{y-\mu_Y}{\sigma_Y}\right) + \left(\frac{y-\mu_Y}{\sigma_Y}\right)^2\right], \quad (7.B.2)$$

and $\mu_X, \mu_Y, \sigma_X, \sigma_Y$, and ρ are all parameters with, $-\infty < \mu_X, \mu_Y < +\infty$, $\sigma_X > 0, \sigma_Y > 0$, and $-1 < \rho < 1$ ($f(x,y)$ is undefined for $\rho = \pm 1$). Here $\mu_X = E(X), \mu_Y = E(Y), \sigma_X^2 = V(X), \sigma_Y^2 = V(Y)$, and ρ is the coefficient of (linear) correlation between X and Y; that is, it can be demonstrated that

$$\rho_{XY} = \frac{COV(X,Y)}{\sigma_X \sigma_Y}$$

$$= E\left[\left(\frac{X-\mu_X}{\sigma_X}\right)\left(\frac{Y-\mu_Y}{\sigma_Y}\right)\right]$$

$$= \int_{-\infty}^{+\infty}\int_{-\infty}^{+\infty}\left(\frac{x-\mu_X}{\sigma_X}\right)\left(\frac{y-\mu_Y}{\sigma_Y}\right)f(x,y)dydx = \rho.$$

If (7.B.1) and (7.B.2) hold, then X and Y are said to follow a *bivariate normal distribution* with joint probability density function (7.B.1). In this regard, any random variable (X,Y) having a probability density function given by (7.B.1) is said to be bivariate $N(\mu_X, \mu_Y, \sigma_X, \sigma_Y, \rho)$.

Looking to the properties of (7.B.1), $f(x,y)$ depicts a bell-shaped surface that is centered around the point (μ_X, μ_Y) in the x-, y-plane (so that any plane perpendicular to the x-, y-plane will intersect the surface in a curve that has the univariate normal form) with:

(a) $f(x,y) > 0$

(b) The probability that a point (X, Y) will lie within a region $\mathcal{A}$ of the x, y-plane is

$$P[(X,Y) \in \mathcal{A}] = \iint_{\mathcal{A}} f(x,y)dydx$$

(c) As required of any legitimate joint probability density function,

$$\int_{-\infty}^{+\infty}\int_{-\infty}^{+\infty} f(x,y)dydx = 1$$

Note that if in (7.B.1) we set

$$u = \frac{x-\mu_X}{\sigma_X} \text{ and } v = \frac{y-\mu_Y}{\sigma_Y}, \quad (7.B.3)$$

then $f(x,y)$ can be expressed in terms of a single parameter ρ.

Suppose that the random variable (X, Y) is bivariate $N(\mu_X, \mu_Y, \sigma_X, \sigma_Y, \rho)$. Then it can be shown that the marginal distribution of X is $N(\mu_X, \sigma_X)$ and the marginal distribution of Y is $N(\mu_Y, \sigma_Y)$; that is, the individual marginal distributions of X and Y are univariate normal. Here the *marginal distribution of X* is obtained by integrating out y in (7.B.1) or

$$g(x) = \int_{-\infty}^{+\infty} f(x, y) dy = \frac{1}{\sqrt{2\pi}\sigma_X} e^{-\frac{1}{2}\left(\frac{x-\mu_X}{\sigma_X}\right)^2}, -\infty < x < +\infty; \qquad (7.B.4)$$

and the *marginal distribution of Y* is obtained by integrating out x in (7.B.1) or

$$h(y) = \int_{-\infty}^{+\infty} f(x, y) dx = \frac{1}{\sqrt{2\pi}\sigma_Y} e^{-\frac{1}{2}\left(\frac{y-\mu_Y}{\sigma_Y}\right)^2}, -\infty < y < +\infty. \qquad (7.B.5)$$

(It is important to note that if the marginal distributions of the random variables X and Y are each univariate normal, this does not imply that (X, Y) is bivariate normal.)

We noted earlier (in Chapter 5) that if the random variables X and Y are independent, then they must also be uncorrelated or $\rho_{XY} = 0$. However, the converse of this statement does not generally follow. Interestingly enough, *it does apply* if X and Y follow a bivariate normal distribution, as Theorem 7.B.1 attests:

THEOREM 7.B.1.   Let the random variables X and Y follow a bivariate normal distribution with probability density function given by (7.B.1). Then X and Y are independent random variables if and only if $\rho_{XY} = \rho = 0$.

Hence, by virtue of this theorem, $\rho = 0$ is a necessary and sufficient condition for independence under the assumption of bivariate normality. Given that $\rho = 0$, (7.B.1) factors as $f(x, y) = g(x)h(y)$, where $g(x)$ and $h(y)$ are the univariate normal marginal distributions appearing in (7.B.4) and (7.B.5), respectively.

Next, if the random variable (X, Y) is bivariate $N(\mu_X, \mu_Y, \sigma_X, \sigma_Y, \rho)$, then the conditional distribution of X given $Y = y$ is univariate $N(\mu_X + \rho\frac{\sigma_X}{\sigma_Y}(y - \mu_Y), \sigma_X\sqrt{1 - \rho^2})$. Here the conditional mean of X given $Y = y$ is $E(X|Y = y) = \mu_X + \rho\left(\frac{\sigma_X}{\sigma_Y}\right)(y - \mu_Y)$ and the conditional variance of X given $Y = y$ is $V(X|Y = y) = \sigma_X^2(1-\rho^2)$. (Note that $V(X|Y = y) \leq V(X)$. Also, as $\rho \to \pm 1$, the conditional variance $\sigma_X^2(1 - \rho^2) \to 0$, whereas if $\rho = 0$, then $E(X|Y = y) = \mu_X$ and $V(X|Y = y) = \sigma_X^2$.) The conditional distribution of X given $Y = y$ is obtained from the joint and marginal distributions (7.B.1) and (7.B.5), respectively, for $h(y) \neq 0$, as

$$g(x|y) = \frac{f(x, y)}{h(y)} = \frac{1}{\sqrt{2\pi}\sqrt{V(X|Y = y)}} e^{-\frac{(x-E(X|Y=y))^2}{2V(X|Y=y)}}$$

$$= \frac{1}{\sqrt{2\pi}\sigma_X\sqrt{1 - \rho^2}} e^{-\frac{\left(x-\mu_X-\rho(\sigma_X/\sigma_Y)(y-\mu_Y)\right)^2}{2\sigma_X^2(1-\rho^2)}}, -\infty < x < +\infty.$$

$$(7.B.6)$$

Similarly, the conditional distribution of Y given $X = x$ is univariate $N\left(\mu_Y + \rho(\frac{\sigma_Y}{\sigma_X})(x - \mu_X), \sigma_Y\sqrt{1-\rho^2}\right)$. Thus the conditional mean of Y given $X = x$ is $E(Y|X = x) = \mu_Y + \rho\left(\frac{\sigma_Y}{\sigma_X}\right)(x - \mu_X)$ and the conditional variance of Y given $X = x$ is $V(Y|X = x) = \sigma_Y^2(1 - \rho^2)$. Hence the *conditional distribution of Y given $X = x$* is obtained from the joint and marginal distributions (7.B.1) and (7.B.4), respectively, given that $g(x) \neq 0$, as

$$h(y|x) = \frac{f(x,y)}{g(x)} = \frac{1}{\sqrt{2\pi}\sqrt{V(Y|X = x)}}e^{-\frac{(y-E(Y|X=x))^2}{2V(Y|X=x)}}$$

$$= \frac{1}{\sqrt{2\pi}\sigma_Y\sqrt{1-\rho^2}}e^{-\frac{(y-\mu_Y-\rho(\sigma_Y/\sigma_X)(x-\mu_X))^2}{2\sigma_Y^2(1-\rho^2)}}, \quad -\infty < y < +\infty.$$

$$(7.B.7)$$

It is well known that the mean of a random variable in a conditional distribution is termed a regression curve when expressed as a function of the fixed variable in the said distribution. For instance, the *regression of Y on $X = x$* is the mean of Y in the conditional density of Y given $X = x$ and is written as a function $y = j(x)$, where

$$j(x) = E(Y|X = x) = \int_{-\infty}^{+\infty} yh(y|x)dy = \int_{-\infty}^{+\infty} y\frac{f(x,y)}{g(x)}dy. \qquad (7.B.8)$$

The expression $y = j(x)$, the locus of the means of the conditional distribution $E(Y|X = x)$ when plotted in the x-, y-plane, gives the *regression curve of y on x* for Y given $X = x$. That is to say, it is a curve that specifies the location of the means of Y for various values of X in the conditional distribution of Y given $X = x$. (The regression curve of x on y is defined in a similar fashion.)

Example 7.B.1 For example, to obtain the curve of regression of y on x, let

$$f(x,y) = \begin{cases} \frac{2}{3} - \frac{1}{3}xy, & 0 < x < 2, 0 < y < 1; \\ 0 & \text{elsewhere.} \end{cases}$$

Then

$$g(x) = \int_0^1 f(x,y)dy = \frac{2}{3} - \frac{1}{6}x$$

and thus

$$h(y|x) = \frac{f(x,y)}{g(x)} = \frac{\frac{2}{3} - \frac{1}{3}xy}{\frac{2}{3} - \frac{1}{6}x}.$$

Hence, from (7.B.8),

$$j(x) = E(Y|X = x) = \int_0^1 yh(y|x)dy = \frac{1 - \frac{1}{3}x}{2 - \frac{1}{2}x}. \quad \blacksquare$$

Although the regression function defined in (7.B.8) need not be a straight line (as the preceding example reveals), it will be linear if the random variable (X, Y) is bivariate $N(\mu_X, \mu_Y, \sigma_X, \sigma_Y, \rho)$; that is, the regression curve of y on x for X and Y jointly normally distributed has the linear form $y = j(x)$ or

$$y = E(Y|X = x) = \mu_Y + \rho\frac{\sigma_Y}{\sigma_X}(x - \mu_X)$$

$$= \left(\mu_Y - \rho\left(\frac{\sigma_Y}{\sigma_X}\right)\mu_X\right) + \rho\left(\frac{\sigma_Y}{\sigma_X}\right)x = \beta_0 + \beta_1 x, \quad (7.B.9)$$

where $\beta_0 = \mu_Y - \rho\left(\frac{\sigma_Y}{\sigma_X}\right)\mu_X$ is the vertical intercept and $\beta_1 = \rho\left(\frac{\sigma_Y}{\sigma_X}\right)$ is the slope of the linear regression curve of y on x. So if the random variable (X, Y) follows a bivariate $N(\mu_X, \mu_Y, \sigma_X, \sigma_Y, \rho)$ distribution, then a linear regression model (such as (7.B.9)) is the correct specification to be employed when studying the behavior of Y conditioned on X.

Finally, we note briefly that if X and Y follow a bivariate normal distribution, then the joint moment-generating function of X and Y is

$$m_{X,Y}(t_1, t_2) = E(e^{t_1 X + t_2 Y}) = e^{\left[t_1\mu_X + t_2\mu_Y + \frac{1}{2}(t_1^2\sigma_X^2 + 2\rho t_1 t_2\sigma_X\sigma_Y + t_2^2\sigma_Y^2)\right]}. \quad (7.B.10)$$

And if (7.B.3) holds, then

$$m_{U,V}(t_1, t_2) = e^{\frac{1}{2}(t_1^2 + 2\rho t_1 t_2 + t_2^2)}. \quad (7.B.11)$$

7.11   Exercises

Uniform Distribution

7-1. If a continuous random variable X is uniformly distributed over the interval $(3,9)$, find its probability density function. What is its cumulative distribution function? What is $P(3.5 \leq X < 7)$? Find the mean and standard deviation of X. What is the value of $\gamma_{0.75}$?

7-2. The life X of a new miniature battery is uniformly distributed between 48 and 55 hours of continuous use. Law enforcement use requires that it last at least 51 hours. What is the probability that X exceeds the law enforcement standard? What is the probability that X is within $\pm\sigma$ of the 51 hour requirement?

7-3. If X is uniformly distributed on (α, β), find a value of k such that

$$f(x) = \begin{cases} k, & \alpha < x < \beta; \\ 0 & \text{elsewhere} \end{cases}$$

is a legitimate probability density function.

7-4. Given that a random variable X is uniformly distributed with probability density function (7.1), find $E(X), V(X)$.

7-5. Given that a random variable X has a uniform probability distribution with probability density function (7.1), find X's moment-generating function. Use it to determine $E(X), V(X)$.

Normal Distribution

7-6. X is $N(50, 10)$. Find:

(a) $P(X \leq 60)$

(b) $P(X \geq 55)$

(c) $P(40 \leq X \leq 49)$

(d) $\gamma_{0.70}$

7-7. X is $N(10, 8)$. Find:

(a) $P(|X| \leq 5)$

(b) $P(|X| > 8)$

7-8. Chebyshev's Theorem informs us that the probability that X will deviate from μ by not more than 2σ is at least 0.75. Is this statement valid if X is $N(40, 2)$? Verify your answer.

7-9. If X is $N(50, 10)$, find a number b such that $P(-b \leq X \leq b) = 0.90$.

7-10. Find the 20th percentile of the distribution that is $N(60, 20)$.

7-11. If X is $N(\mu, 10)$ and $P(X \leq 90) = 0.95$, find μ.

7-12. Given that X is $N(19, \sigma)$ and $P(X < 22) = 0.65$, find σ.

7-13. The semester stipend given to college interns at Big Bucks Investments Inc. is normally distributed with a mean of $4,000 and a standard deviation of $600. If the bottom 5% of interns is to receive an increase in their stipend, what is the level below which an intern will receive an increase?

7-14. Suppose that the amount of cereal X (measured in ounces) placed in a box is $N(16, 0.06)$. Let a denote the amount of cereal such that 95% of all boxes contains at least a ounces. Find a.

7-15. Verify that if a random variable X is $N(\mu, \sigma)$, then (7.7) is a legitimate probability density function.

7-16. A real-valued function $y = f(x)$ is said to be an *odd function* if $f(-x) = -f(x)$. For any such function $\int_{-a}^{a} f(x)dx = 0$. Using this concept, demonstrate that $\int_{-\infty}^{+\infty} xe^{-\frac{1}{2}x^2} dx = 0$.

7-17. Suppose a random variable X is $N(\mu, \sigma)$. Find $E(X), V(X)$ using (7.7).

7-18. Suppose that a random variable Z is $N(0, 1)$. (1) Demonstrate that $E(Z) = 0, V(Z) = 1$ using (7.13). (2) Suppose Z is not $N(0, 1)$ but, instead, is a continuous random variable with $\sigma > 0$. For $Z = \frac{X-\mu}{\sigma}$, does $E(Z) = 0, V(Z) = 1$ still hold? Verify your answer.

7-19. Suppose a random variable X is $N(\mu, \sigma)$. Which (if any) of the following statements are true?

(a) Since the normal probability density function $f(x; \mu, \sigma)$ is symmetrical about $x = \mu$ and the maximum of f occurs at $x = \mu$, it follows that $\mu = $ median $=$ mode.

(b) The normal probability density function $f(x; \mu, \sigma)$ is symmetrical since (7.A.2) does not include any odd powers of t.

(c) Only the presence of even-order central moments of X in (7.A.5) indicates that the normal probability density function $f(x; \mu, \sigma)$ is symmetrical about $x = \mu$.

7-20. Rewrite (7.A.5) in terms of even powers of σt. Then demonstrate that $m_{X-\mu}^{(2)}(0) = \sigma^2$.

7-21. Suppose a random variable Z is $N(0, 1)$. Rewrite (7.A.2) in terms of even powers of t.

(1) Verify that $m_Z(t) = \sum_{n=0}^{\infty} a_n t^n$, where

$$a_n = \begin{cases} 0 & \text{if } n \text{ is odd;} \\ \frac{1}{2^{n/2}(\frac{n}{2})!} & \text{if } n \text{ is even.} \end{cases}$$

(2) Using (4.36), demonstrate that

$$E(Z^n) = \mu'_n = \begin{cases} 0 & \text{if } n \text{ is odd;} \\ \frac{n!}{2^{n/2}(\frac{n}{2})!} & \text{if } n \text{ is even.} \end{cases}$$

7-22. Given that a random variable X is $N(\mu, \sigma)$ with probability density function (7.7), use (7.7) to determine X's moment-generating function. Once this function is obtained, use it to determine the form of the associated probability density functions given that:

(a) $m_X(t) = e^{4t+3t^2}, -\infty < t < +\infty$

(b) $m_X(t) = e^{t^2/2}, -\infty < t < +\infty$

7-23. Given that a random variable X is $N(\mu, \sigma)$ with probability density function (7.7), use (7.7) to determine X's central moment-generating function.

Normal Approximation to Binomial

7-24. Suppose that X is binomially distributed $b(X; 15, 0.5)$. Use the normal approximation to the calculation of binomial probabilities to determine $P(X = 8)$ and $P(5 \leq X \leq 10)$.

7-25. About 10% of all purchasers of the Zapper microwave oven fill out a marketing survey card that accompanies the warranty statement. What is the approximate probability that at least 20 cards will be returned out of the next 250 ovens sold? What is the approximate probability that between 25 and 30 cards will be returned?

7-26. Let X be distributed as $b(X; 20, 0.3)$. Use the normal distribution to find the approximate probabilities:

(a) $P(X \leq 5)$

(b) $P(X \geq 10)$

(c) $P(3 \leq X \leq 7)$

7-27. Suppose that 60% of the customers of a local department store that receive a circular in the mail announcing a special sale actually respond to the announcement. If 1000 circulars are mailed, what is the approximate probability that between 400 and 600 individuals will attend the special sale? What is the approximate probability that at least half will attend?

Normal Approximation to Poisson

7-28. Suppose X is a Poisson random variable with parameter $\lambda = 45$. Use (7.35) to find $P(40 \leq X \leq 50)$. Recalculate this probability by employing the continuity correction $P(a \leq X \leq b) \rightarrow F\left(\frac{b+0.5-\lambda}{\sqrt{\lambda}}; 0, 1\right) - F\left(\frac{a-0.5-\lambda}{\sqrt{\lambda}}; 0, 1\right)$. Also use the continuity correction to find $P(X = a) \rightarrow F\left(\frac{a+0.5-\lambda}{\sqrt{\lambda}}; 0, 1\right) - F\left(\frac{a-0.5-\lambda}{\sqrt{\lambda}}; 0, 1\right)$ for $a = 48$.

7-29. Traffic safety records indicate that the number of accidents occurring along a particular stretch of road is Poisson distributed with a mean of two accidents per week. Determine the approximate probability that X varies between 100 and 110 over the course of a year. What is the approximate probability that $X \leq 100$ over a year?

7-30. Let X be Poisson random variable with mean $\lambda = 20$. Approximate the following probabilities using the normal distribution:

    (a) $P(16 \le X \le 20)$

    (b) $P(16 < X \le 20)$

    (c) $P(16 < X < 20)$

Exponential Distribution

7-31. Suppose a random variable X depicts the life of a circuit board component with probability density function

$$f(x; 0.0005) = \begin{cases} 0.0005e^{-0.0005x}, & x > 0; \\ 0 & \text{elsewhere.} \end{cases}$$

What is the probability that the component will not last more than 1000 hours?

7-32. Suppose that four calls per minute arrive at a switchboard and that the arrival of calls follows a Poisson process. What is the probability that the next call will arrive within seven minutes? Suppose that a malfunction disables the switchboard for two minutes. What is the probability that no calls arrive during this period?

7-33. System failures at an industrial facility are approximately Poisson distributed with $\lambda = 0.25$ per hour (the system experiences a failure every four hours). If system startup begins at 7 A.M. on a typical work day and T is the time until the first failure occurs, find

    (a) The probability that it is at least two hours until the first failure occurs

    (b) The probability that it is no more than six hours until the first failure occurs

    (c) The average time to the first failure

    (d) The probability that the time to the next failure exceeds the average time to failure

7-34. Imperfections in a certain grade of sail canvas occur randomly with a mean of one flaw per 70 square feet of canvas. What is the probability that a piece of sail canvas with dimensions 30 feet by 9 feet will have at least one flaw?

7-35. The length of time required to wash and dry a car at ACE Hand Wash is exponentially distributed with a mean of 13 minutes. What percentage of people will have their car ready within 10 minutes? Within 15 minutes? Determine the length of time X_0 such that the probability that the service time will take more than X_0 minutes is 0.70.

7-36. The distribution of length of operating life before the first malfunction of a certain electrical component is exponential with $E(X) = 350$ hours. What is the form of X's probability density function and cumulative distribution function? What is the probability that the component will operate effectively less than 300 hours? More than 400 hours? Between 310 hours and 375 hours?

7-37. Let X follow an exponential distribution with parameter $\lambda = 3$. Find:

(a) $P(X > 1)$

(b) $P(2 < X \leq 5)$

(c) $E(X)$ and $V(X)$

7-38. The amount of time required for a customer at Big Bank to cash a check is exponentially distributed with a mean time of 1.35 minutes. Find:

(a) λ

(b) The probability that a customer will cash a check in under 1.25 minutes

(c) The probability that it will take a customer more than two minutes to cash a check

7-39. Verify that if a random variable X follows an exponential probability distribution, then (7.39) is a legitimate probability density function. (Note: Readers not familiar with the gamma function (7.42) can return to this exercise when coverage of this function is attained.)

7-40. Suppose that X is an exponential random variable with probability density function (7.39). Demonstrate that $E(X) = \frac{1}{\lambda}$; $V(X) = \frac{1}{\lambda^2}$.

7-41. For a random variable X with exponential probability density function (7.39), verify that (7.41) holds.

7-42. Let a random variable X have a probability density function given by (7.39). Determine X's moment-generating function and use it to determine $E(X), V(X)$.

Gamma Distribution

7-43. Suppose a random variable X is gamma distributed with parameters $\alpha = 2$, $\theta = 10$. Find:

(a) X's probability density function

(b) $E(X)$ and $V(X)$

(c) X's cumulative distribution function

(d) $P(15 < X < 25)$ using (7.49.2)

Recalculate the probability in (d) by using the Poisson approximation provided by (7.55).

7-44. Suppose that 20 customers per hour arrive at a gas station, where arrivals are assumed to follow a Poisson process. If a minute is one unit, find λ. What is the probability that the station attendant will wait more than five minutes before the first two customers arrive? (Let X represent the waiting time in minutes until the second customer arrives.)

7-45. If X is gamma distributed with $\theta = 4$ and $\alpha = 2$, find $P(X < 8)$.

7-46. Given that a random variable X is gamma distributed with probability density function (7.48), verify that (7.48) is a legitimate density function.

7-47. Let a random variable X be gamma distributed with probability density function given by (7.48). What is the role of θ? Demonstrate that if X is gamma distributed with parameters α and θ, then θX is gamma distributed $\alpha, 1$.

7-48. Given the gamma function (7.42), use integration by parts to demonstrate that:

(a) $\Gamma(\alpha) = (\alpha - 1)\Gamma(\alpha - 1)$
(b) If α is a positive integer, $\Gamma(\alpha) = (\alpha - 1)(\alpha - 2)\ldots 2 \cdot 1\Gamma(1)$
(c) $\Gamma(1) = 1$
(d) $\Gamma(\alpha) = (\alpha - 1)!$

7-49. Cars arrive at a toll booth at an average rate of 10 cars every 20 minutes via a Poisson process. Determine the probability that the toll booth operator will have to wait longer than 30 minutes before the ninth toll.

7-50. Let X be gamma distributed with $\alpha = 2$ and $\theta = 50$. Find:

(a) The probability that X assumes a value within two standard deviations of the mean
(b) The probability that X assumes a value below its mean

7-51. Suppose a random variable X is gamma distributed with probability density function given by (7.48). Demonstrate that $E(X) = \alpha\theta$, $V(X) = \alpha\theta^2$.

7-52. Suppose a random variable X follows a gamma distribution with probability density function given by (7.48). Determine X's moment-generating function.

7-53. Suppose a random variable X is gamma distributed with probability density function (7.48). Demonstrate that the rth moment about zero, μ'_r, can be written as $\mu'_r = \theta^r\Gamma(\alpha + r)/\Gamma(\alpha)$. Then use this expression to determine $E(X), V(X)$.

7-54. Suppose a random variable X follows a gamma distribution with cumulative distribution function given by (7.49). Demonstrate that (7.49.2) can be obtained from (7.49) by successive integration by parts. (Obtaining the first few terms will obviously suffice.)

Beta Distribution

7-55. Suppose a random variable X is beta distributed with parameters $\alpha = \beta = 3$. Find X's probability density function. Then determine $P(X \geq 0.75)$.

7-56. Suppose that for a beta distribution $n = 12, \alpha = 6$ and $p = 0.30$. Using (7.68), find the probability that X is less than or equal to p. What is the probability that X is less than or equal to 0.7?

7-57. Bill and Alice's Country Store has propane tanks that are filled every Wednesday. Experience indicates that the proportion of gas used between fills is beta distributed with $\alpha = 3$ and $\beta = 2$. Determine the probability that at least 85% of the gas will be used between fills.

7-58. The proportion of defective parts produced by a metal stamping machine follows a beta distribution with $\alpha = 1$ and $\beta = 10$. Find:

(a) The probability that the proportion of defective units exceeds 5%

(b) The probability that the proportion defective is more than one standard deviation above the mean

7-59. Let a random variable X be beta distributed with $\alpha = 2$ and $\beta = 3$. Find:

(a) $P(X < 0.1)$

(b) The probability that X will assume a value within one standard deviation of the mean

7-60. Verify that if a random variable X is beta distributed with probability density function (7.56), then (7.56) is a legitimate probability density.

7-61. Suppose a random variable X is beta distributed with probability density function given by (7.56). Use this density function to obtain μ'_r, the rth moment about zero. Use the resulting expression to obtain $E(X), V(X)$.

Sampling and the Sampling Distribution of a Statistic

In the introduction (Chapter 1) we rather loosely explored an assortment of concepts such as *statistical inference, population, random sample, statistic*, and so on. In this chapter we shall firm up these notions by redefining them in a much more rigorous and technically correct fashion. This added degree of formality will then enable us to fully develop the concept of random sampling and, in turn, the sampling distribution of a statistic.

8.1 The Purpose of Random Sampling

Statistical or inductive inference generally involves extracting a sample of a given size from an unknown population distribution in order to discern or infer something about the characteristics or behavior of that distribution. Given that conclusions or generalizations about a population are made from sample data and that (typically) only a small portion of the population is being examined, we essentially have *incomplete information* about the population. Hence an element of *uncertainty* enters into our analysis so that any conclusions or assertions about the characteristics of the population must be accompanied by a quantitative measure of the risk or degree of uncertainty of the inference made. As was mentioned in Chapter 1, this process of inductively reaching conclusions in the face of uncertainty about the characteristics of a population (or some phenomenon), as well as quantitatively measuring the risk of the same, is called *statistical inference*. As we shall now see, the act of making such inferences is carried out using the technique of random sampling.

Inductive inference involves random sampling so that the rules of probability theory can be applied in evaluating the magnitude of the risk inherent in this process. Hence uncertain inferences can be made, and the degree of uncertainty can be measured, if the sampling experiment is performed in accordance with certain fundamental principles that prescribe (limit) the way in which a sample is obtained.

In this regard, inferences will be made under a process of random sampling and, in particular, by taking *simple random samples*; that is, samples for which the sampling process gives each item in a population of size N an equal probability of being included in the sample, and gives each sample so obtained an equal probability of being drawn. In what follows we shall consider both random sampling with and without replacement. In the former case (which is equivalent to sampling from an infinite population) the same member of a population can be observed more than once; in the latter, the same element of a population cannot be chosen more than once.

The *target population* is the population about which information is desired, whereas a random sample is actually selected from what is called the *sampled population*, or from a population with probability density function (or probability mass function) $f(x; \theta)$. (Hence a random sample is defined for a population that has a density (mass) function—a distribution of numerical values that describe the population.) Obviously the sampled population may or may not coincide with the target population. In this regard, inferences about the sampled population can be made using random samples obtained from the same, but inferences about the target population cannot be made unless the target and sampled populations are one and the same. By virtue of this discussion, we shall restrict our attention to selecting a sample of size n from the sampled population $f(x; \theta)$, and on the basis of this sample information, make inferences or probability statements about $f(x; \theta)$.

8.2 Sampling Scenarios

Two general sampling situations present themselves. The first involves the infinite population case.

8.2.1 Data Generating Process or Infinite Population

Here an experiment is performed, which results in an infinite number of possible values of a measurable or observable characteristic of some phenomenon. Hence the occurrence of the phenomenon can be thought of as an information generating process. For instance, a physical measurement such as weight in fluid ounces may be taken on the output (fill amount) of an automatic bottling machine.

If we choose to investigate the occurrence of any given phenomenon, then we must utilize the information that it generates. This information set is simply called *data*—a collection of measurements or observations indicating how some particular characteristic of the phenomenon manifests itself. In order to eventually construct a statistical model of a *data generating process*, we need to assume that the said process is *probabilistic* or *random* in that the *inherent natural variation* in the data values precludes us from predicting their level or magnitude.

The *generated data* or output of a random phenomenon serves as input into a *measurement process*. We may think of this process as a means of communicating data. It may involve specialized instrumentation or equipment (needed to measure, for instance, the brightness of a light bulb), or simply a count via visual observation (as in the case of an inspector recording the number of defects of a particular type occurring on the surface of a manufactured product). The output of the measurement process is simply a set of *observed data* values that is to be used by the experimenter or analyst for the study at hand.

But the observed quantities rendered by the measurement process may themselves be subject to error. That is, there may exist *systematic error* (the values obtained from the data generating process are modified in some consistent or predictable fashion) or *random error* (possibly due to imprecise or outdated measurement techniques). If we ignore outright the presence of any systematic error in the measurement process and we assume that this process imparts only a negligible random error, then we may assume that the (predominant) source of randomness in the observed data is due to the data generating process itself. In this regard, we need not worry about any *experimental or nonsampling error* that results when the generated data is different from the observed data, where any such differences are introduced, either systematically or randomly, by the measurement process itself.

In sum, the researcher is faced with deciphering the output of a *sequential two-stage process*, which consists of a data generating phase and a measurement phase (see Figure 8.1). And if the measurement process does not taint the generated data with any systematic or random error, then the observed data mirrors the properties of the generated data and we may now set out to construct a statistical or probability model describing the behavior of this random phenomenon or process.

Most data generating processes, at least in principle, are capable of producing an infinite number of data points on a multitude of measurable characteristics $X, Y, Z, \ldots$ of some random phenomenon. Hence the typical observed data set is infinitely large. Yet, in practice, we deal with finite data sets. Clearly the distinction we are making here is that between a population and a sample. To fully develop the difference(s) between these two levels of abstraction, let us restrict our attention to a single measurable characteristic (X) of some random phenomenon. In this regard, for any such phenomenon, the *population* can be thought of as the state of nature or collection of all possible values of a measurable characteristic X. It can be finite or infinite. The *sample* is the finite collection of actual measurements on a characteristic X obtained from the population in order to acquire representative information about the same. It is generated by performing repeated trials of a sampling experiment under conditions that are identical in terms of all controllable environmental factors; that is, all sample outcomes are assumed to be *drawn from the same source and obtained under the same conditions*. Hence the various trials of the random experiment are *independent* (the outcome obtained on any one trial does not affect nor is it affected by the outcome obtained on any other trial) and thus may be viewed as constituting a process of *random sampling with replacement* or from an infinite population.

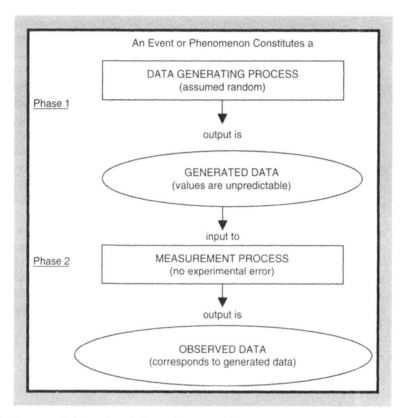

Figure 8.1   A sequential two-stage information generating process.

In this regard, to obtain (sequential) readings on a measurable characteristic X, let us repeat a random experiment n times in succession under identical conditions. Then trial 1 of our sampling experiment generates the outcome X_1; trial 2 generates the outcome $X_2;\dots$; and trial n generates the outcome X_n. Hence we have obtained n generated data values on the characteristic $X : X_1, X_2, \dots, X_n$. Once X_i has been subject to the measurement process, its observed (numerical) level may be denoted as x_i; that is, x_i is termed a *realization* of the random variable $X_i, i = 1, \dots, n$.

If $f(x; \theta)$ depicts the probability density function[1] of the population distribution for the measurable characteristic X, then, by virtue of the preceding sampling process, we may view the generated outcomes $X_i, i = 1, \dots, n$, as *sample random variables*. And if the X_i have the same probability density function as that of the population distribution and their joint probability density function can be expressed as the product of their individual marginal probability densities $f(x_i; \theta)$,

---

[1] Although the discussion that immediately follows is framed in terms of the probability density function for a continuous random variable, a parallel treatment can be given using the probability mass function for a discrete random variable.

then the X_i represent n mutually independent[2] and identically distributed random variables that constitute a random sample from the underlying (infinite) population. To reiterate, the random variables X_i should be viewed as a *random sample*, and their realizations x_i form the observed data set.

Let us write the joint probability density function of the sample random variables $X_1, \ldots, X_n$ as

$$\ell(x_1, x_2, \ldots, x_n; \theta, n) = \prod_{i=1}^{n} f(x_i; \theta), \tag{8.1}$$

where θ and n are parameters and

$$f(x_i; \theta) = f(x; \theta), \; i = 1, \ldots, n. \tag{8.2}$$

So if (8.1) and (8.2) hold, then our sampling routine is deemed random and thus the sample is connected in probability to the parent population $f(x; \theta)$.

Example 8.2.1.1   Let $X_1, \ldots, X_n$ depict a random sample of n mutually independent and identically distributed random variables obtained from a population that follows a normal distribution with probability density function

$$f(x; \mu, \sigma) = \frac{1}{\sqrt{2\pi}\sigma} e^{-\frac{1}{2}\left(\frac{x-\mu}{\sigma}\right)^2}, \; -\infty < x < +\infty.$$

When the sample random variable X_1 is generated and its realization x_1 is observed, the marginal probability density of X_1 mirrors that of the parent population and thus

$$f(x_1; \mu, \sigma) = \frac{1}{\sqrt{2\pi}\sigma} e^{-\frac{1}{2}\left(\frac{x_1-\mu}{\sigma}\right)^2}, \; -\infty < x_1 < +\infty.$$

Next, let the sample random variable X_2 admit the realization x_2. Since X_1, X_2 are independent random variables with the same marginal densities, it follows that

$$f(x_2; \mu, \sigma) = \frac{1}{\sqrt{2\pi}\sigma} e^{-\frac{1}{2}\left(\frac{x_2-\mu}{\sigma}\right)^2}, \; -\infty < x_2 < +\infty.$$

Then the joint probability density function of X_1 and X_2 is the bivariate function

$$\ell(x_1, x_2; \mu, \sigma, n = 2) = f(x_1; \mu, \sigma) \cdot f(x_2; \mu, \sigma)$$

$$= \left(\frac{1}{\sqrt{2\pi}\sigma}\right)^2 e^{-\frac{1}{2}\left[\left(\frac{x_1-\mu}{\sigma}\right)^2 + \left(\frac{x_2-\mu}{\sigma}\right)^2\right]}, \; -\infty < x_1, x_2 < +\infty.$$

---

[2] Here *mutual independence* of the n random variables $X_1, \ldots, X_n$ implies that they are *pairwise independent* (X_i, X_j are independent for any $i \neq j$ with their bivariate probability density function $f(x_i, x_j; \theta) = f(x_i; \theta) \cdot f(x_j; \theta)$) *but not conversely*.

In general, for a random sample of size n, the joint probability density function of the n mutually independent and identically distributed random variables $X_1, \ldots, X_n$ is

$$\ell(x_1, \ldots, x_n; \mu, \sigma, n) = \left(\frac{1}{\sqrt{2\pi}\sigma} \right)^n e^{-\frac{1}{2} \sum\limits_{i=1}^{n} \left(\frac{x_i - \mu}{\sigma} \right)^2} \quad -\infty < x_i < +\infty, i = 1, \ldots, n.$$

(8.3)

So $f(x; \mu, \sigma)$ is the distribution of the population, and the joint density (8.3) is the *distribution of the (random) sample*. Given that the joint probability density function (8.3) factors as (8.1) (the random variables X_i are mutually independent and have a common probability density function by virtue of (8.2)), we may conclude that $X_1, \ldots, X_n$ is a random sample of size n from an infinite population with probability density $f(x; \mu, \sigma)$. (Note again that the *common distribution* from which we are sampling is actually the *population*.) ■

Example 8.2.1.2   A second depiction of a random sample, this time taken from a finite (discrete) population, involves tossing a fair coin n times in succession (the *data generating process*) and counting the number of successes obtained in the n trials, where a success is defined as *getting heads* on any toss. And since we can flip the coin, at least conceptually, an unlimited number of times under identical conditions, these flips are independent of each other and thus can also be viewed as involving a process of *sampling with replacement*. Here the discrete population random variable X has two values: $x = 1$ when heads occurs; $x = 0$ when tails occurs. And since the sample space S admits but two simple events (heads or tails), $P(X = 1) = p = \frac{1}{2}$; $P(X = 0) = 1 - p = \frac{1}{2}$. In general, the probability mass function for this discrete population random variable X is

$$P(X = x) = f(x) = \begin{cases} p^x (1 - p)^{1-x}, & x = 0 \text{ or } 1; \\ 0 & \text{otherwise.} \end{cases}$$

(8.4)

Hence the population is the finite set $\{0, 1\}$ and the distribution of the population is the distribution of X. So on each toss of the coin, we are selecting either one or the other of these two population values as the observed outcome.

In this regard, if the value of the random variable X_i is determined on the ith toss of our coin, then each X_i has two possible realizations: either $x_i = 1$ or $0, i = 1, \ldots, n$. Clearly these discrete sample random variables X_i are mutually independent (the population is unaffected from toss to toss) and each has the same distribution as X; that is, $f(x_i) = p^{x_i}(1-p)^{1-x_i}$, $x_i = 0$ or 1, so that $f(x_i = 1) = p$ and $f(x_i = 0) = 1 - p$. Hence a random sample consists of the n random variables $X_1, \ldots, X_n$ with realizations $x_1, \ldots, x_n$, respectively. Thus the n tosses of the coin will result in an *observed data set* consisting of the realizations $x_1, \ldots, x_n$ with each x_i being either 0 or 1, $i = 1, \ldots, n$.

The probability of the event $\{X_1 = x_1, \ldots, X_n = x_n\}$ is given by the joint probability mass function of the discrete random variables $X_1, \ldots, X_n$ and, under the mutual independence of the X_i, appears as

$$P(X_1 = x_1, \ldots, X_n = x_n) = \ell(x_1, \ldots, x_n; n) = \prod_{i=1}^{n} f(x_i), \tag{8.5}$$

where n is a parameter and

$$f(x_i) = f(x), \quad i = 1, \ldots, n. \tag{8.6}$$

So if, via (8.6), each X_i is distributed as (8.4), then (8.5) becomes

$$\ell(x_1, \ldots, x_n; n) = \begin{cases} p^{\sum_{i=1}^{n} x_i} (1-p)^{n - \sum_{i=1}^{n} x_i}, & x_i = 0 \text{ or } 1; \\ 0 & \text{otherwise,} \end{cases} \tag{8.7}$$

where $\sum_{i=1}^{n} x_i$ = number of heads obtained in n tosses of a coin (it is the number of 1's in the observed data set $x_1, \ldots, x_n$). Hence for each observed data set at which $\sum_{i=1}^{n} x_i = v$,

$$\ell(x_1, \ldots, x_n; n) = p^v (1-p)^{n-v}. \tag{8.7.1}$$

So given that (8.7) factors as (8.5) (the discrete random variables X_i are mutually independent and, from (8.6), have a common probability mass function), we may conclude that the X_i, $i = 1, \ldots, n$, constitute a random sample of size n from a finite or discrete population $\{0, 1\}$ with probability mass function $f(x)$ (see (8.4)). ■

8.2.2 Drawings from a Finite Population

The preceding two examples of a random sample involved observing the data obtained from an experimental or data generating process that, in effect, constituted an infinite population. In contrast, we may be interested in examining a measurable characteristic (X) of some finite collection of actual or material objects, where the said characteristic may be a quantitative measure (such as the individual lengths, in inches, of an assortment of steel rods) or possibly the presence or absence of an attribute (a motor vehicle either passes or fails a safety inspection).

We have two ways of extracting a sample of size n from this type of population (assumed to be of size N). The first involves *sampling with replacement* and is essentially a special case of the preceding sampling or data generating process. Given the N items in the population, the first is randomly drawn and the measurable characteristic X is observed and recorded as $X_1 = x_1$. (It is assumed that the measurement process does not exhibit any nonsampling error.) This first item is then put back into the population, the N items are randomized in some fashion, and another item is selected. This yields the second observation on X that is recorded as $X_2 = x_2$. Once the second item selected is returned to the population and the latter is again randomized, a third item is drawn, and so on. This process continues until all n items have been chosen, with the item obtained on any one trial replaced in the population so that it is available for selection on the next trial. Hence on each trial each of the N items is equally likely to be chosen. We ultimately obtain a sample of n observations $X_1, \ldots, X_n$ on the measurable characteristic X, where each $X_i, i = 1, \ldots, n$, is a discrete random variable that takes on each of the N possible realizations $x_1, \ldots, x_N$ with equal probability $\frac{1}{N}$. So under the scheme of sampling with replacement from a finite population, the sample random variables $X_1, \ldots, X_n$ are mutually independent and identically distributed (the probability distribution of each $X_i, i = 1, \ldots, n$, is the same as that of the population) and thus constitutes what is commonly called a *simple random sample*.

A second method of sampling from a finite population of size N is *sampling without replacement*. If the items in the population are randomized before selection begins, then we simply extract n items in succession, one after the other, without replacing the objects drawn on any one trial before the next trial begins. This process also renders a collection of n sample random variables X_i whose realizations $x_i, i = 1, \ldots, n$, are the elements of an observed data set. However, these random variables *do not constitute a random sample* as previously defined since they are *not mutually independent*, although they are identically distributed since the marginal distribution of each $X_i, i = 1, \ldots, n$, is the same as that of the parent population. Under this sampling scheme, once an item is selected, it is unavailable for selection on any subsequent trial. Hence N items are available for the first draw and thus the probability of selecting any one item is $\frac{1}{N}$. Without replacement, there are only $N - 1$ items available for the second draw and thus the probability of obtaining any one of these $N - 1$ population values is $\frac{1}{N-1}$ and so on for successive draws.

To verify that the random variables X_i are not mutually independent but are identically distributed under a regime of sampling without replacement, let u and v (with $u \neq v$) represent distinct elements of the population. Then on draw one, $P(X_1 = u) = \frac{1}{N}$ and, under sampling without replacement, for draw two, $P(X_2 = u \,|X_1 = u) = 0$ (u is not available for selection on draw two) and $P(X_2 = v \,|X_1 = u) = \frac{1}{N-1}$ (since there is now one fewer element in the population). Hence the (conditional) probability distribution of X_2 depends on the realization of X_1 so that X_1 and X_2 are not independent random variables. But, as noted earlier, X_1 and X_2 are identically distributed; that is, the marginal distribution of each $X_i, i = 1, \ldots, n$, is the same as the population distribution. In this regard, for $i = 1$, the marginal distribution of X_1 is $P(X_1 = u) = \frac{1}{N}$ for any u in the population. And for $i = 2$, the marginal distribution of X_2 is, via (3.4.2),

$$P(X_2 = v) = \sum_{i=1}^{N} P(X_2 = v \,|X_1 = x_i) \cdot P(X_1 = x_i). \qquad (8.8)$$

For $i = h$, $v = x_h$ and $P(X_2 = x_h|X_1 = x_h) = 0$. For all other $j \neq h$, $P(X_2 = v|X_1 = x_j) = \frac{1}{N-1}$. Hence (8.8) reduces to

$$P(X_2 = v) = \sum_{j \neq h} P(X_2 = v|X_1 = x_j) \cdot P(X_1 = x_j) = (N-1)\left(\frac{1}{N-1} \cdot \frac{1}{N}\right) = \frac{1}{N}.$$

$$(8.8.1)$$

Hence each X_i, $i = 1, \ldots, n$, has the same marginal distribution.

So under sampling without replacement from a finite population, the withdrawal of an item from the population obviously affects the probability distribution of the same. Hence we must modify the preceding definition of simple random sampling to accommodate this change. The usual modification is to introduce randomness in sampling in a fashion such that *all possible sequences of n trials (or all possible samples of size n) have the same chance of occurring*. To achieve this, the first item must be drawn in a way that ensures that all N items in the population have the same probability $\frac{1}{N}$ of being chosen. Having selected the first item, the second sample element is drawn from the $N-1$ remaining elements in a way that ensures that each has the same probability $\frac{1}{N-1}$ of being drawn. On the third trial, the remaining $N-2$ elements of the population each have the same probability $\frac{1}{N-2}$ of being chosen. This process continues until, finally, on the nth draw, all $N-(n-1)$ remaining items have the same probability $\frac{1}{N-(n-1)}$ of being selected. Hence each sample of size n taken without replacement from a finite population of size N has $\frac{1}{N} \cdot \frac{1}{N-1} \cdot \frac{1}{N-2} \cdots \frac{1}{N-(n-1)}$ as its probability of being drawn. So if the sampling process is executed in this fashion, we again obtain a simple random sample. In general, if the finite population is extremely large relative to n, then sampling without replacement is, for all intents and purposes, equivalent to sampling with replacement; in practice we can treat the sample random variables $X_1, \ldots, X_n$ as if they were independent and identically distributed since, for extremely large N, the probabilities $\frac{1}{N}, \frac{1}{N-1}, \frac{1}{N-2}, \ldots$ will be approximately equal. In this instance the random variables $X_i, i = 1, \ldots, n$, may be termed *nearly independent* in that, for large N, the conditional distribution of X_i given $X_1, \ldots, X_{i-1}$ is approximated closely by the marginal distribution of X_i alone if $i(\leq n)$ is small relative to N. Hence, for distinct population elements v, s, and t, it is assumed that $P(X_i = t|X_1 = r, \ldots, X_{i-1} = s) = \frac{1}{N-(i-1)} \approx \frac{1}{N}$.

If the population size N is relatively small to begin with, then random sampling without replacement may be the preferred sampling process since, if replacement occurs, there is a strong possibility that the same item may be selected more than once. In sum, *simple random sampling* from a finite population is a process of sampling that gives each possible sample of size n an equal probability of being selected.

8.3   The Arithmetic of Random Sampling

An important consideration in simple random sampling is determining the total number of possible samples of a given size that can be extracted from a finite population. Once this number is known, we can then set out to list the elements appearing in each of these hypothetical samples as well as calculate probabilities concerning the same.

Given a finite population of size N, let us determine the number of ways of selecting a sample of size n. If we sample with replacement, then the first element can be selected in N different ways and the probability of choosing any one element on the first draw is $\frac{1}{N}$; the second element can also be selected in N different ways and, here too, the probability of selecting any one element on the second draw is $\frac{1}{N}$, and so on. Thus the total number of samples of size n that can be selected with replacement is

$$\underbrace{N \cdot N \cdots N}_{\text{taken } n \text{ times}} = N^n. \tag{8.9}$$

Here the order of the elements in each sample matters; that is, each sample is a permutation of n elements. Thus n ordered sample units have been drawn from the N units in the population in a way that gives each of the population units the same constant probability $\frac{1}{N}$ of being drawn. Moreover, each ordered sample has probability $\left(\frac{1}{N}\right)^n$ of being drawn.

Example 8.3.1 If the population consists of $N = 4$ distinct objects labeled, A, B, C, and D, then a random sample of size $n = 2$ with replacement can be drawn by requiring that the first of the two places in the sample be any one of the four letters (each has the same probability of $\frac{1}{4}$ of being chosen) and that the second of the two places in the sample be any one of the four letters (again with an equal probability of $\frac{1}{4}$). The total number of ordered samples obtained in this manner is thus $N^n = (4)^2 = 16$, and each of these samples has probability $\left(\frac{1}{N}\right)^n = \left(\frac{1}{4}\right)^2 = \frac{1}{16}$ of being drawn (see the accompanying array of ordered samples presented in part (a) of Figure 8.2). ■

Example 8.3.2 To solidify our understanding of the salient features of simple random sampling with replacement, let us assume that we have a vessel containing three red and seven blue balls. When a ball is selected at random, we will denote the event *a red (blue) ball is chosen* as $R(B)$. If X denotes the population random variable, then the population probability (relative frequency) distribution is given in panel (a) of Table 8.1. Let us now extract a random sample of size $n = 2$ from this population with replacement, where the random variable $X_1(X_2)$ describes the outcome obtained on the first (second) draw. Under sampling with replacement, X_1 and X_2 are independent random variables. The possible samples drawn and their associated probabilities are depicted in panel (b) of Table 8.1. For instance, $P(X_1 = R \cap X_2 = R) = \left(\frac{3}{10}\right) \cdot \left(\frac{3}{10}\right) = \frac{9}{100}$. We may consequently determine the probability distribution of X_1 (panel (c) of Table 8.1) by

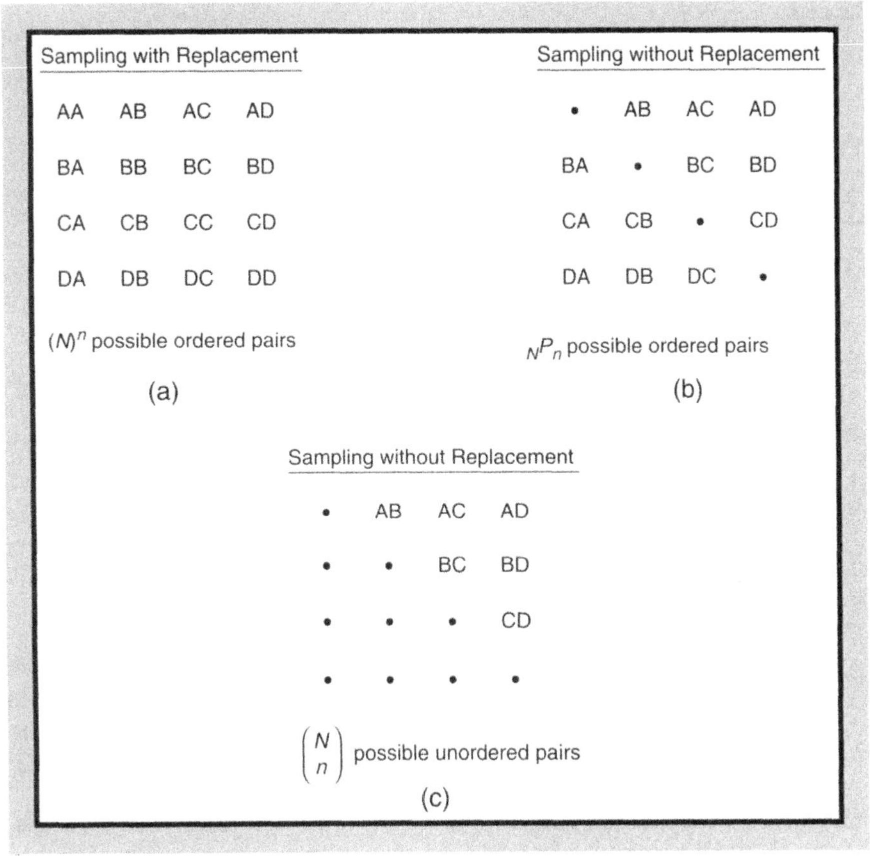

Sampling with Replacement | Sampling without Replacement

AA   AB   AC   AD   •   AB   AC   AD

BA   BB   BC   BD   BA   •   BC   BD

CA   CB   CC   CD   CA   CB   •   CD

DA   DB   DC   DD   DA   DB   DC   •

$(N)^n$ possible ordered pairs $_NP_n$ possible ordered pairs

(a)   (b)

Sampling without Replacement

•   AB   AC   AD

•   •   BC   BD

•   •   •   CD

•   •   •   •

$\binom{N}{n}$ possible unordered pairs

(c)

Figure 8.2   (a) $(N)^n$ possible ordered pairs; (b) $_NP_n$ possible ordered pairs; (c) $\binom{N}{n}$ possible unordered pairs.

noting that

$$P(X_1 = R) = \frac{9}{100} + \frac{21}{100} = \frac{3}{10};$$

$$P(X_1 = B) = \frac{21}{100} + \frac{49}{100} = \frac{7}{10}.$$

Similarly, since

$$P(X_2 = R) = \frac{9}{100} + \frac{21}{100} = \frac{3}{10};$$

$$P(X_2 = B) = \frac{21}{100} + \frac{49}{100} = \frac{7}{10},$$

the probability distribution of X_2 is easily obtained (panel (d) of Table 8.1). Clearly the probability distributions of X_1 and X_2 are identical to each other and to the

Table 8.1

a.		b.	
Population Probability Distribution		**Sampling with Replacement**	
X	Probability	X_1, X_2	Probability
R	3/10	R, R	$(3/10) \cdot (3/10) = 9/100$
R	7/10	R, B	$(3/10) \cdot (7/10) = 21/100$
	1	B, R	$(7/10) \cdot (3/10) = 21/100$
		B, B	$(7/10) \cdot (7/10) = 49/100$
			1

c.		d.	
Probability Distribution of X_1		**Probability Distribution of X_2**	
X_1	Probability	X_2	Probability
R	3/10	R	3/10
B	7/10	B	7/10
	1		1

distribution of the population random variable X. In general, when we extract a random sample of size n from a finite population with replacement, we obtain n independent and identically distributed sample random variables $X_1, \ldots, X_n$, each having the same probability distribution as the parent population. ∎

Next, if sampling is undertaken without replacement, there are N choices for the first sample element (and each of these can be selected with probability $\frac{1}{N}$), $N-1$ choices for the second (each having a probability of $\frac{1}{N-1}$ of being drawn), and so on. So when we have N items in the population and we extract samples of size n without replacement, there will be a total of $N(N-1)(N-2)\cdots(N-(n-1)) = {}_N P_n$ possible ordered samples. In addition, each of these ordered samples is equally likely with probability $\frac{1}{N} \cdot \frac{1}{N-1} \cdot \frac{1}{N-2} \cdots \frac{1}{(N-(n-1))} = \frac{1}{{}_N P_n}$ of being drawn, and the probability is n/N that any member of the population will be included in the sample.

Example 8.3.3 If the finite population again consists of $N = 4$ items marked $A, B, C,$ and D, then drawing random samples of size $n = 2$ without replacement results in $N(N-(n-1)) = 4(4-1) = 12(= {}_4 P_2)$ possible samples, each having probability $\frac{1}{{}_4 P_2} = \frac{1}{12}$ of being drawn (see array (b) of Figure 8.2 wherein the sample pairs $AA, BB, CC,$ and DD have been deleted). ∎

If we disregard the order in which the n elements appear in a sample (e.g., AB and BA may be viewed as the same sample), then each sample obtained without replacement is a combination of n different items and thus each sample has $n!$ different orderings. Hence $\frac{[N(N-1)(N-2)\cdots(N-(n-1))]}{n!} = \binom{N}{n}$ is the number of different samples of size n obtainable from N elements without replacement when order is neglected. Moreover, each of these unordered samples is equally likely with probability $1/\binom{N}{n}$ of being chosen and the probability that any member of the population will be included in the sample is $\frac{n}{N}$.

Example 8.3.4 Again taking $N = 4$ (we have the same population elements A, B, C, and D) and $n = 2$, there are

$$\frac{N(N - (n - 1))}{n!} = \binom{N}{n} = \binom{4}{2} = \frac{4!}{2!2!} = 6$$

possible samples of size two when we disregard order (see array (c) of Figure 8.2). Moreover, each of these samples has the same probability $\left(\frac{1}{6}\right)$ of being drawn. ∎

In sum, under a regime of simple random sampling from a finite population of size N, all possible samples of a given size n are equally likely to be selected. If samples are taken with replacement, each possible sample has a probability of $\left(\frac{1}{N}\right)^n$ of being drawn; if sampling occurs without replacement, each possible sample has a probability of $\frac{1}{_N P_n}$ of occurring when the order in which the items are selected matters. If order is not important, each possible sample has a probability of $1/\binom{N}{n}$ of being selected. And, as the preceding discussion has suggested, when we sample without replacement, the process of simple random sampling implies that, on any given trial, each element that has not been previously drawn has an equal probability of being selected.

Example 8.3.5 Returning to the earlier example problem in which we had a container housing three red and seven blue balls, let us now extract a sample of size $n = 2$ from the same, but this time without replacement. (The population distribution is given in Table 8.1a.) Again the random variable $X_1(X_2)$ describes the outcome obtained on the first (second) draw. Since X_1 and X_2 are not independent under sampling without replacement, the probabilities associated with the various possible samples are conditional on the particular color selected on the first draw (see panel (a) of Table 8.2). That is, these probabilities are determined as:

$$P(X_1 = R \cap X_2 = R) = P(X_1 = R) \cdot P(X_2 = R \,|\, X_1 = R) = \frac{2}{9} \cdot \frac{3}{10} = \frac{6}{90};$$

$$P(X_1 = R \cap X_2 = B) = P(X_1 = R) \cdot P(X_2 = B \,|\, X_1 = R) = \frac{7}{9} \cdot \frac{3}{10} = \frac{21}{90};$$

$$P(X_1 = B \cap X_2 = R) = P(X_1 = B) \cdot P(X_2 = R \,|\, X_1 = B) = \frac{3}{9} \cdot \frac{7}{10} = \frac{21}{90};$$

$$P(X_1 = B \cap X_2 = B) = P(X_1 = B) \cdot P(X_2 = B \,|\, X_1 = B) = \frac{6}{9} \cdot \frac{7}{10} = \frac{42}{90}.$$

Table 8.2

a.			b.			c.	
Sampling Without Replacement			**Probability Distribution of X_1**			**Probability Distribution of X_2**	
X_1, X_2	**Probability**		X_1	**Probability**		X_2	**Probability**
R, R	6/90		R	3/10		R	3/10
R, B	21/90		B	7/10		B	7/10
B, R	21/90			1			1
B, B	42/90						
	1						

The probability distribution of X_1 (panel (b) of Table 8.2) can be determined by noting that

$$P(X_1 = R) = \frac{6}{90} + \frac{21}{90} = \frac{3}{10};$$

$$P(X_1 = B) = \frac{21}{90} + \frac{42}{90} = \frac{7}{10}.$$

And since

$$P(X_2 = R) = \frac{6}{90} + \frac{21}{90} = \frac{3}{10};$$

$$P(X_2 = B) = \frac{21}{90} + \frac{42}{90} = \frac{7}{10},$$

the probability distribution of X_2 can also be readily determined (panel (c) of Table 8.2). So even though the sample random variables X_1 and X_2 are not independent, their probability distributions are identical to each other and to the distribution of the parent population. ∎

8.4 The Sampling Distribution of a Statistic

We noted in the preceding section that random samples would be used to make inferences about unknown population characteristics. Such characteristics are termed *parameters*—fixed numerical constants that describe a population distribution of a random phenomenon. For instance, if the population random variable X is exponentially distributed, then we completely specify X's probability density function once we set the value of the scale parameter (λ); if X is normally distributed, we completely specify X's probability density function once the mean (μ) and the standard deviation (σ) are given; and if X is binomially distributed, X's probability mass function is completely described by the probability of a success (p) and the number of trials (n).

A parameter is a characteristic of a population, whereas a *statistic* is a characteristic of a sample that can be used to estimate or determine a parameter. In this regard, we may view a statistic as a function of sample random variables $X_1, \ldots, X_n$ and known constants (e.g., the sample size n) and which is itself a random variable. Hence a statistic cannot contain any unknown parameters.

If we denote a statistic as $T = g(X_1, \ldots, X_n, n)$ (here the sample size is an argument of the real-valued function g), then its realization $t = g(x_1, \ldots, x_n, n)$ is determined once the realizations x_i of the sample random variables $X_i, i = 1, \ldots, n$, are known. Here g is a rule or law of correspondence that tells us how to get t's from sets of x_i's. If the statistic T is used to determine some unknown population parameter θ, then T is called a *point estimator* of θ and its realization t is termed a *point estimate* of θ. (Here a point estimator of θ reports a single numerical value as the estimate of θ. This is in contrast to an interval estimator of θ, which reports a range of values and which will be introduced in a later chapter.) In Chapter 10 we shall study estimation and, in particular, the desirable properties that a *good* estimator should possess.

We noted earlier that inductive statistics involves inferring from sample data something about the characteristics of a population. Because only a limited amount of data is examined, errors in making inferences are unavoidable due to the presence of *sampling error*—which is the difference between the realization (t) of a sample statistic (T) and the corresponding (unknown) population value θ. Sampling error exists because of chance variability in random sampling; that is, the realization of a statistic computed from a random sample of a given size depends upon which population elements are included in the sample and, if we repeat the sampling process and obtain another sample of the same size, we usually get a different realization of T. Hence realizations vary from sample to sample and are expected to be different (to varying degrees) from a population parameter. In order to have confidence in any inference made about a population parameter, we must develop methods for assessing the magnitude of the sampling error connected with the same. And this can be accomplished only in terms of the probability distribution of a statistic. That is, since T is a random variable, it has a probability distribution whose properties will enable us to evaluate the chance of making an erroneous inference. In general, the probability distribution of a statistic that has been generated under random sampling from a given population will be called the *sampling distribution* of the statistic. Once the sampling distribution of a statistic is determined, it can be used to make inferences about a parameter from random samples taken from a population by finding, say, the probability that any realization of the statistic is due solely to chance factors. Hence sampling distributions aid us in discerning the patterns and magnitudes of sampling errors; they describe how a statistic varies, due to chance, under random sampling.

For instance, let T be an estimator for θ with t the realization of T. We take a random sample of size n and record $t(= t_1)$. Then we take a second random sample of the same size and again find $t(= t_2)$. Typically $t_1 \neq t_2$. As we take more and more random samples of the same size n, we build up a relative frequency distribution of various t's or estimates of θ. As the number of random samples gets larger and larger, this relative frequency distribution gets closer and closer to the true theoretical sampling distribution of T, which reflects the inherent sample-to-sample variability of the estimates of θ, and which is generally not of the same form as the population probability density function (probability mass function) $f(x;\theta)$. However, as we shall soon see, a sampling distribution is always related in some specific way to the parent population distribution.

To formalize the preceding discussion, let us extract all possible random samples of a given size from a specific population distribution and compute the corresponding realizations of a statistic T. The resulting relative frequency distribution yields an approximation to the theoretical probability distribution that would have been obtained had the number of samples been *infinite* rather than just *large*. Hence we may define the *sampling distribution of a statistic T* as a theoretical probability distribution that associates with each possible realization t of T its probability of occurrence (or probability density value). As such, it provides us with a theoretical model for the relative frequency distribution of the possible realizations of T that we would obtain under repeated sampling from the same population.

Clearly the form of a sampling distribution of T depends upon, or can be described in terms of, the distribution of the population from which the sample random variables $X_1,\ldots,X_n$ were obtained. Under random sampling, the X_i's, $i = 1,\ldots,X_n$, are independent and identically distributed, where the marginal distributions $f(x_i;\theta)$ of the X_i's, $i = 1,\ldots,X_n$, are of the same form as the population distribution $f(x;\theta)$. Hence the theoretical or sampling distribution of T can be obtained from the underlying population distribution, with the former distribution intimately related to, but distinct from, the latter. In this regard, (8.1), which describes the joint occurrence of n independent and identically distributed sample random variables, depends on $f(x;\theta)$ and n and thus serves as the basis of forming the sampling distribution or probability density function $f(t;\theta,n)$ of T.

8.5   The Sampling Distribution of the Mean

Although we can form the sampling distribution of any measurable statistic under random sampling, we shall concentrate in this section on the sampling distribution of the mean and, in subsequent sections, on the sampling distribution of a proportion and of the variance.

We begin our discussion with an *additivity theorem* (Theorem 8.1) involving a linear combination (Y) of a set of random variables. Specifically;

THEOREM 8.1.   Let $X_1,\ldots,X_n$ be a set of random variables with $a_1,\ldots,a_n$ a set of constants. In addition, let $E(X_i) = \mu_i$ and $V(X_i) = \sigma_i^2, i = 1,\ldots,n$. If the random variable $Y = \sum_{i=1}^{n} a_i X_i$, then:

(a) The mean and variance of Y are, respectively,

$$E(Y) = \sum_{i=1}^{n} a_i \mu_i \tag{8.10}$$

and

$$V(Y) = \sum_{i=1}^{n} a_i^2 \sigma_i^2 + 2\sum\sum_{i<j} a_i a_j \sigma_{ij}, \tag{8.11}$$

where $\sigma_{ij} = COV(X_i, X_j), i = 1,\ldots,n; j = 1,\ldots,n$.

(b) If the random variables $X_i, i = 1,\ldots,n$, are independent, then $\sigma_{ij} = 0$ and (8.11) reduces to

$$V(Y) = \sum_{i=1}^{n} a_i^2 \sigma_i^2. \tag{8.11.1}$$

(c) If the random variables $X_i, i = 1,\ldots,n$, are independent and identically distributed with $\mu_i = \mu$ and $\sigma_i^2 = \sigma^2$ for all $i = 1,\ldots,n$, then

$$E(Y) = \mu \sum_{i=1}^{n} a_i, \tag{8.10.1}$$

$$V(Y) = \sigma^2 \sum_{i=1}^{n} a_i^2. \tag{8.11.2}$$

(d) If the random variables $X_i, i = 1,\ldots,n$, are independent normally distributed random variables, then Y is normally distributed with $E(Y)$ and $V(Y)$ given by (8.10) and (8.11.1), respectively.

(e) If the random variables $X_i, i = 1,\ldots,n$, are independent normally distributed random variables with $\mu_i = \mu$ and $\sigma_i^2 = \sigma^2$ for all $i = 1,\ldots,n$, then Y is normally distributed with $E(Y)$ and $V(Y)$ provided by (8.10.1) and (8.11.2), respectively.

(Note that if $a_i = 1$ for all $i = 1,\ldots,n$, then the sum of n normally distributed independent random variables is normally distributed with its mean (variance) equal to the sum of the individual means (variances).)

8.5.1 Sampling from an Infinite Population

If we extract a random sample of size n from an infinite population (or if we sample with replacement from a finite population), then the sample random variables $X_1,\ldots,X_n$ are independent and identically distributed with $E(X_i) = \mu$ and $V(X_i) = \sigma^2$ for all $i = 1,\ldots,n$, where μ and σ^2 are the population mean and variance respectively and σ^2 is assumed finite. Let us form the statistic

$$\overline{X} = \sum_{i=1}^{n} \frac{X_i}{n} \tag{8.12}$$

or the *sample mean of the* X_i. (If x_i is the realization of X_i, then $\bar{x} = \sum_{i=1}^{n} \frac{x_i}{n}$ is the realization of $\bar{X}$.) Since $\bar{X}$ is a random variable, its relationship to the population random variable X is provided by Theorem 8.2:

THEOREM 8.2. If $X_1, \ldots, X_n$ are independent random variables having the same distribution with mean μ and variance σ^2, then $E(\bar{X}) = \mu$ and $V(\bar{X}) = \frac{\sigma^2}{n}$.

To rationalize this set of results let us consider part (c) of Theorem 8.1 wherein $Y = \bar{X}$, $E(X_i) = \mu_i = \mu$, $V(X_i) = \sigma_i^2 = \sigma^2$, and $a_i = \frac{1}{n}$. Then from (8.10.1) and (8.11.2):

$$E(\bar{X}) = \mu \sum_{i=1}^{n} \frac{1}{n} = \mu; \tag{8.13}$$

$$V(\bar{X}) = \sigma^2 \sum_{i=1}^{n} \frac{1}{n^2} = \frac{\sigma^2}{n}. \tag{8.14}$$

(It can also be demonstrated that: $E\left(\frac{1}{\bar{X}}\right) \approx \frac{1}{\mu}$; $V\left(\frac{1}{\bar{X}}\right) \approx \left(\frac{1}{\mu}\right)^4 V(\bar{X})$.)

Note that these results are quite general and do not depend on the *form* of the population probability density function; these expressions involve only the mean and variance of the population probability density function.

How may we interpret (8.13) and (8.14)? Given that an expectation is the mean of a random variable, it readily is seen that (8.13) informs us that the average of the sample means taken over all possible samples of size n is equal to the population mean. So although the value of $\bar{X}$ determined from a particular sample generally will not equal μ, the average of all possible values of $\bar{X}$ for fixed n will equal μ. Next, from (8.14), the positive square root of $V(\bar{X})$ will be denoted as

$$\sigma_{\bar{X}} = \frac{\sigma}{\sqrt{n}} \tag{8.15}$$

and termed the *standard error of the mean*. It is simply the standard deviation of the distribution of sample means taken over all possible samples of size n drawn from a given population; it essentially serves as a measure of the average sampling error arising when the mean of a single sample is used to estimate the population mean. For $n > 1$, it is always the case that $\sigma_{\bar{X}} < \sigma$.

A special case of Theorem 8.2 emerges when we sample from a population that is normally distributed. That is, for Theorem 8.3:

THEOREM 8.3. If $X_1, \ldots, X_n$ are independent normal random variables with $E(X_i) = \mu_i = \mu$ and $V(X_i) = \sigma_i^2 = \sigma^2$ for all $i = 1, \ldots, n$, then the distribution of the sample mean $\bar{X}$ is normal with $E(\bar{X}) = \mu$ and $V(\bar{X}) = \frac{\sigma^2}{n}$.

So given that the $X_i, i = 1, \ldots, n$, are independent normal sample random variables, the *sampling distribution of the mean is normal* with probability density function

$$f(\bar{x}; \mu, \sigma_{\bar{X}}, n) = \frac{\sqrt{n}}{\sqrt{2\pi}\sigma} e^{-\frac{n}{2}\left(\frac{\bar{x}-\mu}{\sigma}\right)^2}, \quad -\infty < \bar{x} < +\infty. \tag{8.16}$$

The associated cumulative distribution function can then be written as

$$F(\bar{x}; \mu, \sigma_{\bar{X}}, n) = \frac{\sqrt{n}}{\sqrt{2\pi}\sigma} \int_{-\infty}^{\bar{x}} e^{-\frac{n}{2}\left(\frac{t-\mu}{\sigma}\right)^2} dt; \tag{8.17}$$

and if a random sample of size n is taken from a standard normal or $N(0,1)$ population, then the preceding expression simplifies to

$$F(\bar{x}; 0, \sigma_{\bar{X}}, n) = \frac{\sqrt{n}}{\sqrt{2\pi}} \int_{-\infty}^{\bar{x}} e^{-\frac{n}{2}t^2} dt. \tag{8.17.1}$$

8.5.2   Sampling from a Finite Population

The preceding discussion concerning the properties of the sampling distribution of the mean ((8.13) and (8.14)) was framed in terms of simple random sampling from an infinite population (or, equivalently, using random sampling with replacement from a finite population). Let us now examine the properties of the random variable $\bar{X}$ under sampling from a finite or discrete population without replacement.

To this end our experiment consists of randomly selecting a sample of size n from the finite set of numbers $\{C_1, \ldots, C_N\}$, where $N(> n)$ is the population size. Under random sampling without replacement, let X_1 denote the first number selected from the set of N items, X_2 denotes the second number selected from the $N-1$ remaining items, $\ldots$, and X_n represents the nth number selected from the $N-(n-1)$ remaining terms. Hence these sample random variables $X_i, i = 1, \ldots, n$, constitute a random sample of size n taken from this finite population and their joint probability mass function is

$$f(x_1, \ldots, x_n) = \frac{1}{N(N-1)\cdots(N-(n-1))} = \frac{1}{{}_NP_n}, \tag{8.17}$$

where x_i is the realization of the sample random variable $X_i, i = 1, \ldots, n$, and the order in which the items are selected matters. If order is not important, then each possible sample of size n has a probability of $\frac{n!}{{}_NP_n} = \frac{1}{\binom{N}{n}}$ of being selected.

Since in (8.17) the order of the X_i's, $i = 1, \ldots, n$, is not important, it follows that, for $n = 1$, the marginal probability mass function for any individual X_i (take $i = r$) is

$$f(x_r) = \frac{1}{N}, \quad x_r = C_j, \, j = 1, \ldots, N. \tag{8.18}$$

Then

$$E(X_r) = \sum_{j=1}^{N} C_j \left(\frac{1}{N}\right) = \mu \tag{8.19}$$

and

$$V(X_r) = \sum_{j=1}^{N} (C_j - \mu)^2 \left(\frac{1}{N}\right) = \sigma^2 \tag{8.20}$$

are, respectively, the mean and variance of this finite population.

Next, for any ordered pair of realizations of, say, the sample random variables X_r and X_s, the joint marginal probability mass function is

$$f(x_r, x_s) = \frac{1}{N(N-1)}, \tag{8.21}$$

with $x_r = C_j$, $j = 1, \ldots, N$; $x_s = C_h$, $h = 1, \ldots, N$, and $j \neq h$. Then using (8.21), the covariance of X_r and X_s may be written as

$$COV(X_r, X_s) = \sum_{\substack{j=1 \\ (j \neq h)}}^{N} \sum_{h=1}^{N} (C_j - \mu)(C_h - \mu) \left(\frac{1}{N(N-1)}\right).^{3} \tag{8.22}$$

---

[3] In order to simplify this covariance expression, let us make use of the following relationships:

1. $\sum_{j=1}^{N} C_j = N\mu$;

   from (8.20), $\sum_{j=1}^{N} (C_j^2 - 2\mu C_j + \mu^2) = N\sigma^2$ and thus

2. $\sum_{j=1}^{N} C_j^2 = N(\sigma^2 + \mu^2)$;

3. $\sum_{\substack{j=1 \\ (j \neq h)}}^{N} \sum_{h=1}^{N} \mu C_j = \mu \sum_{h \neq 1} C_1 + \mu \sum_{h \neq 2} C_2 + \cdots + \mu \sum_{h \neq N} C_N = \mu(N-1)C_1 + \mu(N-1)C_2 + \cdots$

   $+ \mu(N-1)C_N = \mu(N-1) \sum_{j=1}^{N} C_j = N(N-1)\mu^2$;

   similarly,

4. $\sum_{\substack{j=1 \\ (j \neq h)}}^{N} \sum_{h=1}^{N} \mu C_h = N(N-1)\mu^2$;

Utilizing the expressions presented in the footnote, it is readily demonstrated that (8.22) can be rewritten as

$$COV(X_r, X_s) = \frac{1}{N(N-1)} \left[\sum_{\substack{j=1 \\ (j \neq h)}}^{N} \sum_{h=1}^{N} C_j C_h - N(N-1)\mu^2 \right]$$

$$= \frac{1}{N(N-1)} \left[N^2 \mu^2 - N(\sigma^2 + \mu^2) \right] - \mu^2 = -\frac{\sigma^2}{N-1}. \quad (8.22.1)$$

Armed with this result, we may now state Theorem 8.4:

THEOREM 8.4. If $\overline{X}$ is the mean of a random sample of size n selected from a finite population of size N with mean μ and variance σ^2, then $E(\overline{X}) = \mu$ and $V(\overline{X}) = \frac{\sigma^2}{n} \frac{N-n}{N-1}$, where $\frac{N-n}{N-1}$ is termed the *finite population correction factor*.

To verify these assertions let us employ expressions (8.10), (8.11) of Theorem 8.1 with $Y = \overline{X}$, $a_i = \frac{1}{n}$, $\mu_i = \mu$, $\sigma_i^2 = \sigma^2$, and, from (8.22.1), $\sigma_{ij} = -\frac{\sigma^2}{N-1}$. Then

$$E(\overline{X}) = \sum_{i=1}^{n} \frac{1}{n} \mu = \mu \quad (8.23)$$

---

5. $\sum_{\substack{j=1 \\ (j \neq h)}}^{N} \sum_{h=1}^{N} \mu^2 = \sum_{h \neq 1} \mu^2 + \sum_{h \neq 2} \mu^2 + \cdots + \sum_{h \neq N} \mu^2 = (N-1)\mu^2 + (N-1)\mu^2 + \cdots + (N-1)\mu^2$

$$= N(N-1)\mu^2;$$

and

6. $\sum_{\substack{j=1 \\ (j \neq h)}}^{N} \sum_{h=1}^{N} C_j C_h = \sum_{h \neq 1} C_1 C_h + \sum_{h \neq 2} C_2 C_h + \cdots + \sum_{h \neq N} C_N C_h.$

Since $C_1 + \sum_{h \neq 1} C_h = N\mu$, $\sum_{h \neq 1} C_h = N\mu - C_1$ we have $C_1 \sum_{h \neq 1} C_h = C_1(C_2 + C_3 + \cdots + C_N) = C_1(N\mu - C_1) = C_1 N\mu - C_1^2$.

Similar results hold for the remaining $N-1$ terms on the right-hand side of (6). Hence (6) becomes

(6.1) $\sum_{\substack{j=1 \\ (j \neq h)}}^{N} \sum_{h=1}^{N} C_j C_h = (C_1 N\mu - C_1^2) + (C_2 N\mu - C_2^2) + \cdots + (C_N N\mu - C_N^2) = N\mu \sum_{j=1}^{N} C_j - \sum_{j=1}^{N} C_j^2$

$$= N^2 \mu^2 - \sum_{j=1}^{N} C_j^2.$$

and

$$V(\bar{X}) = \sum_{i=1}^{n} \frac{1}{n^2}\sigma^2 + 2\sum\sum_{i<j}\frac{1}{n^2}\sigma_{ij}$$

$$= \frac{\sigma^2}{n} + \frac{2}{n^2}\left[\sum_{j=2}^{n}\left(-\frac{\sigma^2}{N-1}\right) + \sum_{j=3}^{n}\left(-\frac{\sigma^2}{N-1}\right) + \cdots + \left(-\frac{\sigma^2}{N-1}\right)\right]$$

$$= \frac{\sigma^2}{n} + \frac{2}{n^2}\left(-\frac{\sigma^2}{N-1}\right)\left[(n-1) + (n-2) + \cdots + (n-(n-1))\right]$$

$$= \frac{\sigma^2}{n} + \frac{2}{n^2}\left(-\frac{\sigma^2}{N-1}\right)\left[\frac{n(n-1)}{2}\right] = \frac{\sigma^2}{n}\frac{N-n}{N-1}.^4 \qquad (8.24)$$

So under random sampling without replacement from a finite population, the standard error of the mean now appears as

$$\sigma_{\bar{X}} = \frac{\sigma}{\sqrt{n}}\sqrt{\frac{N-n}{N-1}}. \qquad (8.25)$$

Note that as $N \to \infty$ with n held constant, $\frac{\sigma}{\sqrt{n}}\sqrt{\frac{N-n}{N-1}} \to \frac{\sigma}{\sqrt{n}}$ as expected; and as $n \to N (= \text{constant})$, $\frac{\sigma}{\sqrt{n}}\sqrt{\frac{N-n}{N-1}} \to 0$. In this regard, when N is large relative to n (in particular, if the sample size is less than 5% of the population size or if $\frac{n}{N} < 0.05$), use $\sigma_{\bar{X}} = \frac{\sigma}{\sqrt{n}}$ as an approximation to (8.25).

Why is the finite population correction factor required if n is a sizable proportion of the finite population and we sample without replacement? It is needed simply because as each successive unit is withdrawn from the population, the variability in the same is diminished before the next item is selected (since there is now one fewer item in the population).

---

[4] To derive the term in square brackets, set

$$(n-1) + (n-2) + \cdots + (n-(n-1)) = \sum_{i=1}^{n} n - 1 - 2 - 3 - \cdots - (n-1)$$

$$= \sum_{i=1}^{n} n - (1 + 2 + 3 + \cdots + (n-1)) = n^2 - S_n,$$

where S_n denotes the sum of the first n terms of an *arithmetic progression* with common difference $d = 1$. (An arithmetic progression may be viewed as a set of numbers in which each one after the first is obtained from the preceding one by adding a fixed number d.) If a_1 is the first term and a_n is the nth term of the progression, then $a_n = a_1 + (n-1)d = 1 + (n-1)(1) = n$ and thus $S_n = \frac{n}{2}(a_1 + a_n) = \frac{n}{2}(1+n)$. Hence $n^2 - S_n = n^2 - \frac{n}{2}(1+n) = \frac{n(n-1)}{2}$.

Example 8.5.2.1   To see exactly how the sampling distribution of the mean is constructed, let us take as our population of size $N = 7$ the set of values $X : 1, 2, 3, 4, 5, 6, 7$. Here the mean and standard deviation are, respectively, $\mu = 4$ and $\sigma = 2.16$. If we extract samples of size $n = 3$ without replacement, then, if the order in which the values are selected is not important, we obtain $\binom{7}{3} = 35$ possible random samples, with the probability of choosing any one sample equal to $\frac{1}{35}$. If the sample random variables X_i have realizations x_i, $i = 1, 2, 3$, then each sample can be written as an unordered triple (x_1, x_2, x_3). The 35 samples and their associated sample means are presented in Table 8.3a. If we form the relative frequency distribution of these sample means, then this construct serves as the sampling distribution of the mean for all possible unordered samples of size $n = 3$ taken without replacement from a population of size $N = 7$. This sampling distribution is illustrated in Table 8.3b and incorporates the assumption that the probability of getting a mean of, say, three equals the probability of choosing a sample that has a mean of three. And since three of the 35 samples exhibit a mean of three, the probability that $\overline{X} = 3$ is $\frac{3}{35}$.

As anticipated from our preceding discussions concerning the properties of the distribution of sample means,

$$E(\overline{X}) = E(X) = \mu = 4;$$

$$\sigma_{\overline{X}} = \frac{\sigma}{\sqrt{n}} \sqrt{\frac{N - n}{N - 1}} = \frac{2.16}{\sqrt{3}} \sqrt{\frac{7 - 3}{7 - 1}} = 1.02.$$

(Note that the finite population correction factor was included in the calculation of $\sigma_{\overline{X}}$ since the sample size is greater than 5% of the population size.) As these values reveal, the sampling distribution of the mean is centered exactly on the population mean (see Figure 8.3) and $\sigma_{\overline{X}} < \sigma$; that is, the individual sample means or values of $\overline{X}$ are clustered much more closely about the population mean than are the original values of X.

What Table 8.3b gives us is an approximation to the true (unknown) sampling distribution of the mean—the theoretical distribution that associates with realizations $\bar{x}_i$ of $\overline{X}$ the probability mass (density) of each taken over all possible samples of a given size. So for a fixed n, as the number of random samples taken gets larger and larger, the resulting relative frequency distribution tends to get closer and closer to the true sampling distribution of the mean.

Finally, the sampling errors associated with the various $\bar{x}_i$ values are indicated in Table 8.4. As mentioned earlier, these sampling errors reflect the inherent or natural variation in sample means due to chance under random sampling. And since $\sigma_{\overline{X}}$ depicts the average sampling error arising when $\overline{X}$ is used to estimate μ, it is readily demonstrated, via Table 8.4, that $\sum_{i=1}^{13} \frac{|\bar{x}_i - \mu|}{13} = 1.07$ (a value close to the $\sigma_{\overline{X}}$ level obtained earlier using (8.25)). So by studying the sampling distribution of the mean, we can learn something about the patterns of these sampling errors as well as their magnitudes. ∎

Table 8.3

	a.			b.	
Sample Number	**Realizations** (x_1, x_2, x_3)	**Sample Means** $\bar{x}_j, j = 1, \ldots, 35$		**Sampling Distribution of the Mean** $\overline{X}$	$P(\overline{X} = \bar{x}_i)$
1	1, 2, 3	2		2	1/35
2	1, 2, 4	7/3		7/3	1/35
3	1, 2, 5	8/3		8/3	2/35
4	1, 2, 6	3		3	3/35
5	1, 2, 7	10/3		10/3	4/35
6	1, 3, 4	8/3		11/3	4/35
7	1, 3, 5	3		4	5/35
8	1, 3, 6	10/3		13/3	4/35
9	1, 3, 7	11/3		14/3	4/35
10	1, 4, 5	10/3		5	3/35
11	1, 4, 6	11/3		16/3	2/35
12	1, 4, 7	4		17/3	1/35
13	1, 5, 6	4		6	1/35
14	1, 5, 7	13/3			1
15	1, 6, 7	14/3			
16	2, 3, 4	3			
17	2, 3, 5	10/3			
18	2, 3, 6	11/3			
19	2, 3, 7	4			
20	2, 4, 5	11/3			
21	2, 4, 6	4			
22	2, 4, 7	13/3			
23	2, 5, 6	13/3			
24	2, 5, 7	14/3			
25	2, 6, 7	5			
26	3, 4, 5	4			
27	3, 4, 6	13/3			
28	3, 4, 7	14/3			
29	3, 5, 6	14/3			
30	3, 5, 7	5			
31	3, 6, 7	16/3			
32	4, 5, 6	5			
33	4, 5, 7	16/3			
34	4, 6, 7	17/3			
35	5, 6, 7	6			

8.6   A Weak Law of Large Numbers

As (8.13) reveals, the average value of $\overline{X}$ taken over all possible samples of size n equals the population mean μ. Hence the sampling distribution of $\overline{X}$ is centered at μ. And as indicated by (8.15), the scatter or spread of the realizations of $\overline{X}$ (the $\bar{x}_i$'s, $i = 1, \ldots, n$) about μ varies inversely with the sample size n. So when

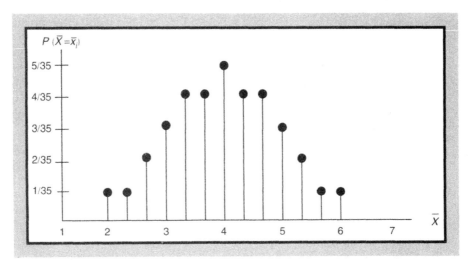

Figure 8.3   Probability mass function for the sampling distribution of $\overline{X}$.

Table 8.4

Sampling Error
$\bar{x}_i - \mu$
$2 - \mu = -2$
$7/3 - \mu = -5/3$
$8/3 - \mu = -4/3$
$3 - \mu = -1$
$10/3 - \mu = -2/3$
$11/3 - \mu = -1/3$
$4 - \mu = 0$
$13/3 - \mu = 1/3$
$14/3 - \mu = 2/3$
$5 - \mu = 1$
$16/3 - \mu = 4/3$
$17/3 - \mu = 5/3$
$6 - \mu = 2$

n is large, the $\bar{x}_i$'s tend to be much more concentrated about μ than when n is small. This latter observation is underscored by what may be called a *weak law of large numbers*. This law addresses the following question: Given that the sample random variables $X_1, \ldots, X_n$ are *finite* in number, can any reliable estimate of μ be made using $\overline{X}$ given that μ is actually the average of an *infinite* number of values taken from the population distribution? As we shall now see, the answer is yes. (In what follows we shall, for purposes of exposition, denote the sample mean computed from a random sample of size n as $\overline{X}_n$ to emphasize its dependence on n.)

Given that the $X_i, i = 1, \ldots, n$, represent a set of independent and identically distributed sample random variables whose individual expectations $E(X_i) = \mu$ exist, the weak law of large numbers essentially informs us that the average of these variables converges to their common mean μ as the number of such variables increases. Thus realizations of $\overline{X}_n$ can be expected to be closer to μ as more sample random variables are introduced.

The weak law of large numbers (as it applies to estimating the population mean μ) essentially states that we can determine a random sample size of (at least) n such that the probability is as close to 1 as desired that the difference between $\overline{X}_n$ and μ will not exceed some arbitrarily small quantity. That is, there exists an n for which the probability is as close to 1 as desired that μ lies within the interval $\overline{X}_n \pm \varepsilon, \varepsilon > 0$. More formally, we have Theorem 8.5, *The Weak Law of Large Numbers.*

THEOREM 8.5. Let $\overline{X}_n$ be the mean of a random sample of size n taken from an infinite population with mean μ and finite variance σ^2. Choose any two small numbers ε and δ such that $\varepsilon > 0$ and $0 < \delta < 1$. If n is any positive integer greater than $\frac{\sigma^2}{\delta \varepsilon^2}$, then

$$P\left(\left|\overline{X}_n - \mu\right| < \varepsilon\right) \geq 1 - \delta. \tag{8.26}$$

(Here $\left|\overline{X}_n - \mu\right| < \varepsilon$ implies that $\overline{X}_n$ deviates from μ by less than ε or that, for ε sufficiently small, $\overline{X}_n$ is arbitrarily close to μ while $P(\cdot) \geq 1 - \delta$ means that, for δ sufficiently small, the indicated probability can be made as close to 1 as desired.)

To rationalize this result let us express (4.19.2) (Chebyshev's inequality) in terms of $\overline{X}_n$ as

$$P\left(\left|\overline{X}_n - \mu\right| < k\sigma_{\overline{X}_n}\right) \geq 1 - \frac{1}{k^2}, \ k > 0. \tag{8.27}$$

Setting $\sigma_{\overline{X}_n} = \frac{\sigma}{\sqrt{n}}$ and $\varepsilon = \frac{k\sigma}{\sqrt{n}}$ we may rewrite this expression as

$$P\left(\left|\overline{X}_n - \mu\right|^2 < \frac{k^2\sigma^2}{n}\right) \geq 1 - \frac{1}{k^2}$$

or, since $\frac{1}{k^2} = \frac{\sigma^2}{n\varepsilon^2}$,

$$P\left(\left|\overline{X}_n - \mu\right|^2 < \varepsilon^2\right) \geq 1 - \frac{\sigma^2}{n\varepsilon^2} \geq 1 - \delta \tag{8.28}$$

for $\delta > \frac{\sigma^2}{n\varepsilon^2}$ or $n > \frac{\sigma^2}{\delta\varepsilon^2}$.

We determined earlier via the frequency limit principle (Section 3.8 of Chapter 3) that the probability of an event can be interpreted as the expected long-run relative frequency of that event. That is, if the trials of a random experiment are independent, then, in the long run, the relative frequency of an event A should approach some number $P(A)$ (the probability of event A). The formalization of this notion is the weak law of large numbers.

Example 8.6.1   In this context, to see how the weak law of large numbers works, suppose that the failure rate for a certain model television set while under extended warranty has been fairly stable over the last few years at about 3%. An appliance store that sells a *large number* of these sets under extended warranty can thus expect the relative frequency of failures in the near future to be very close to 3%. If only a few sets are sold under extended warranty, then variations from 3% (some of them sizeable) will surely occur. However, as more and more sales under extended warranty are considered, the long-run relative frequency of failures should stabilize at about 3%. Hence the weak law of large numbers can thus be interpreted as a statement of *empirical regularity* or *statistical stability*.   ∎

8.7   Convergence Concepts

If a random variable X depends upon a positive integer n, then this dependency can be represented explicitly by writing the variable as X_n. The associated cumulative distribution function for X_n can then be denoted as $F_n(t)$. Let $\{F_n(t)\}$ denote a sequence of cumulative distribution functions indexed by the positive integer n and let $F(t)$ be the cumulative distribution function for the random variable X. Then the random variable X_n *converges in distribution (in law)* to X (or X_n is said to have a *limiting distribution function $F(t)$*) if

$$\lim_{n \to \infty} F_n(t) = F(t)$$

for every t at which $F(t)$ is continuous. This type of convergence will be written as $d\text{-}\lim_{n \to \infty} X_n = X$ or $X_n \xrightarrow{d} X$. In this regard, at points of continuity of $F(t)$, the sequence of cumulative distribution functions $F_n(t)$ converges to $F(t)$ as n increases without bound. What this means is that for each (arbitrarily small) number $\varepsilon > 0$, a positive integer n_ε can be found such that $|F_n(t) - F(t)| < \varepsilon$ for all $n > n_\varepsilon$; that is, the $F_n(t)$'s begin to *pile up* at $F(t)$ beyond some value n_ε of n. We note briefly that the limiting distribution of a random variable X_n may or may not exist; and if it exists, it may be *degenerate*; that is, it has a probability of 1 at a single point $t = a$. Here $t = a$ is a constant that is independent of n.

If the limiting distribution of a random variable X_n is degenerate at $t = a$, then X_n is said to *converge stochastically* to $t = a$ or, alternatively, X_n converges to $t = a$ in probability. In this regard, if X_n converges to a in probability, then a is termed the *probability limit* of X_n and written $p\text{-}\lim_{n \to \infty} X_n = a$ or $X_n \xrightarrow{p} a$.

More formally, the random variable X_n converges stochastically to the constant a if and only if, for every arbitrarily small number $\varepsilon > 0$,

$$\lim_{n \to \infty} P(|X_n - a| < \varepsilon) = 1.^5 \tag{8.29}$$

Here (8.29) reveals that the probability of X_n differing from a by an arbitrarily small amount ε nears 1 as n increases without bound. Looked at in another fashion, if (8.29) holds for every $\varepsilon > 0$, then X_n converges stochastically to the constant a; that is,

$$\lim_{n \to \infty} F_n(t) = \begin{cases} 0, & t < a; \\ 1, & t \geq a \end{cases}$$

and thus X_n has a limiting distribution with cumulative distribution function

$$F(t) = \begin{cases} 0, & t < a; \\ 1 & t \geq a. \end{cases}$$

Note that if X_n is stochastically convergent to a, then $Y_n = X_n - a$ converges stochastically to zero; that is, for every $\varepsilon > 0$, $\lim_{n \to \infty} P(|Y_n| < \varepsilon) = 1$.

We note briefly that if $X_n \xrightarrow{P} a$ and $X'_n \xrightarrow{P} a'$, then:

1. $X_n + X'_n \xrightarrow{P} a + a'$;

2. $X_n \cdot X'_n \xrightarrow{P} a \cdot a'$;

3. $\frac{X_n}{X'_n} \xrightarrow{P} \frac{a}{a'}$;

4. $X_n^{1/2} \xrightarrow{P} a^{1/2}$ if $P(X_n \geq 0) = 1$.

Let us frame the preceding discussion in terms of the random variable $\overline{X}_n$, the mean of a random sample of size n taken from an infinite population with a finite mean μ and variance σ^2. We know from our earlier results that the mean of X_n is μ and its variance is $\frac{\sigma^2}{n}$. For every fixed $\varepsilon = \frac{k\sigma}{\sqrt{n}} > 0$, (8.27) becomes

$$P(|\overline{X}_n - \mu| < \varepsilon) = 1 - \frac{\sigma^2}{n\varepsilon^2}$$

---

[5] Since $P(|\overline{X}_n - a| < \varepsilon) + P(|X_n - a| \geq \varepsilon) = 1$, a necessary and sufficient condition for the stochastic convergence of X_n to a is

$$\lim_{n \to \infty} P(|X_n - a| \geq \varepsilon) = 0, \quad \varepsilon > 0. \tag{8.29.1}$$

and, for fixed $\varepsilon > 0$ and σ^2 finite,

$$\lim_{n \to \infty} P\left(\left|\overline{X}_n - \mu\right| < \varepsilon\right) = \lim_{n \to \infty} \left(1 - \frac{\sigma^2}{n\varepsilon^2}\right) = 1. \tag{8.30}$$

But this result is the gist of the weak law of large numbers (Theorem 8.5); that is, the preceding convergence concepts imply that the limiting cumulative distribution function of $\overline{X}_n$ is degenerate (assigns all of its mass to the single point μ with probability 1) and thus, via (8.30), $\overline{X}_n$ converges stochastically to μ or $\overline{X}_n$ converges to μ in probability. Since (8.30) involves the limit of a sequence of probability statements, the weak law of large numbers does not state that $\overline{X}_n$ is necessarily getting closer and closer to μ as n gets larger and larger. It does state, however, that as n increases without bound, the probability tends to 1 that $\overline{X}_n$ will not differ from μ by more than ε.

A stronger form of convergence is provided by the possibility that X_n *converges to a with probability* 1; that is, for every arbitrarily small number $\varepsilon > 0$,

$$P\left(\lim_{n \to \infty} |X_n - a| < \varepsilon\right) = 1 \tag{8.31}$$

or

$$\lim_{n \to \infty} P\left(|X_m - a| < \varepsilon, \text{all } m > n\right) = 1. \tag{8.31.1}$$

This type of convergence is often referred to as *almost sure (a.s.) convergence.* Thus X_n converges almost surely to a limiting value a if (8.31) holds, where a is now termed the *almost sure limit* of X_n and written $pl\text{-}\lim_{n \to \infty} X_n = a$ or $X_n \overset{a.s.}{\longrightarrow} a$. (Note that convergence with probability 1 implies convergence in probability and, in turn, convergence in probability implies convergence in distribution. However, converse statements do not hold.) Based upon this convergence concept, the statement: "If we extract a random sample of size n from a population with finite mean μ and variance σ^2, the sample mean $\overline{X}_n$ converges to μ with probability 1" is termed the *strong law of large numbers.*

What is the difference between the weak law of large numbers and its strong counterpart? The implication of the weak law of large numbers is that for a sufficiently large (fixed) n, the random variable $\overline{X}_n$ has a high probability of being near μ, but it does not stipulate that $\overline{X}_n$ will remain near μ if n is increased. The strong law of large numbers requires that $\overline{X}_n$ approaches μ with probability 1 so that all but a finite number of the $\overline{X}_n$'s are within a distance ε of μ; that is, only finitely many of the events $\left|\overline{X}_n - \mu\right| > \varepsilon$ will occur. Thus $\overline{X}_n$ approaches μ with probability 1.

We close this section by noting that an *asymptotic distribution* of a random variable X_n is any distribution that closely approximates the actual distribution of X_n for large samples. An asymptotic distribution may thus depend upon the sample size n, whereas a limiting distribution does not.

8.8 A Central Limit Theorem

We noted earlier (Theorem 8.3) that if we extract a random sample from a normal population with mean μ and variance σ^2, then the sampling distribution of the mean (i.e., the random variable $\overline{X}$) is normal with mean μ and variance $\frac{\sigma^2}{n}$. In this instance we can conveniently use the standard normal area table to calculate probabilities involving the random variable $\overline{X}$. To see this, let us assume that we take a random sample of size $n = 100$ from a population that is $N(\mu, \sigma) = N(50, 15)$. What is the probability that the sample mean takes on a value between 48 and 55? Since $\overline{X}$ is $N(\mu, \sigma_{\overline{X}}) = N(\mu, \sigma/\sqrt{n}) = N(50, 1.5)$, it follows that the *standardized sample mean*

$$\overline{Z} = \frac{\overline{X} - \mu}{\sigma_{\overline{X}}} \sim N(0, 1). \tag{8.33}$$

Then, in general, since

$$P(a \leq \overline{X} \leq b) = P\left(\frac{a - \mu}{\sigma_{\overline{X}}} \leq \frac{\overline{X} - \mu}{\sigma_{\overline{X}}} \leq \frac{b - \mu}{\sigma_{\overline{X}}}\right) = P\left(\overline{z}_1 \leq \overline{Z} \leq \overline{z}_2\right),$$

it is easily shown that

$$P(48 \leq \overline{X} \leq 55) = P\left(\frac{48 - 50}{1.5} \leq \frac{\overline{X} - \mu}{\sigma_{\overline{X}}} \leq \frac{55 - 50}{1.5}\right)$$

$$= P\left(-1.33 \leq \overline{Z} \leq 3.33\right) = 0.9082.$$

But what if we do not have any information pertaining to the form of the population probability density function? Can we still use the standard normal area table to calculate probabilities involving $\overline{X}$? Interestingly enough, the answer, under certain conditions, is yes.

The justification for an affirmative answer to the preceding question is provided by what is called a *central limit theorem* (since it specifies the limiting distribution of a measure of central location $\overline{X}$). To see this, suppose we select a simple random sample of size n from some infinite albeit unknown population. Given that the sample random variables X_i, $i = 1, \ldots, n$, are independent and identically distributed, we can determine the approximate sampling distribution of their mean $\overline{X}_n = \sum_{i=1}^{n} \frac{X_i}{n}$ if the sample size is relatively large and the population distribution has mean μ and finite variance σ^2. Under these circumstances the random variable $\overline{X}_n$ will be approximately $N(\mu, \sigma_{\overline{X}})$ for fixed n.

A formalization of this argument is provided by Theorem 8.6, Central Limit Theorem.[6]

---

[6] For a proof of Theorem 8.6 see Appendix 8.A to this chapter.

THEOREM 8.6. Let X_i, $i = 1, \ldots, n$, represent n independent and identically distributed sample random variables with $E(X_i) = \mu$ and $V(X_i) = \sigma^2$ finite and let the random variable $\overline{X}_n = \sum_{i=1}^{n} \frac{X_i}{n}$ have mean $E(\overline{X}_i) = \mu$ and variance $V(\overline{X}_i) = \frac{\sigma^2}{n}$ so that the standardized sample mean is $\overline{Z}_n = \frac{\overline{X}_n - \mu}{\sigma/\sqrt{n}}$.

Then the cumulative distribution function of $\overline{Z}_n$ converges to the $N(0, 1)$ cumulative distribution function as $n \to \infty$ and thus $\overline{X}_n$ is approximately distributed $N(\mu, \sigma_{\overline{X}})$.

If we denote the cumulative distribution function of $\overline{Z}_n$ as $F_{\overline{Z}_n}(t)$, then this theorem indicates that $\overline{Z}_n$ has a limiting standard normal distribution function or

$$\lim_{n \to \infty} F_{\overline{Z}_n}(t) = F(t; 0, 1) = \frac{1}{\sqrt{2\pi}} \int_{-\infty}^{t} e^{-\frac{1}{2}x^2} dx \qquad (8.34)$$

and thus $\overline{X}_n$ has an asymptotic normal distribution with mean μ and variance $\frac{\sigma^2}{n}$. This result is most remarkable and, as a practical matter, allows us to make the following sequence of assertions: If the distribution of $\overline{Z}_n$ converges to $F(t; 0, 1)$, then, for sufficiently large n, the distribution of $\overline{Z}_n$ is approximately $N(0, 1)$ and thus $\overline{X}_n$ is approximately $N(\mu, \sigma_{\overline{X}})$. All this holds regardless of how the sample random variables X_i, $i = 1, \ldots, n$, are distributed. The only requirement is that the unknown population has a mean and a finite variance. Of course, if the population distribution is known, then the distribution of $\overline{X}_n$ is known exactly and not just approximately.

In terms of (8.34), the central limit theorem enables us to infer, for sufficiently large n, that $F_{\overline{Z}_n}(t) \approx F(t; 0, 1)$. For any n, $F_{\overline{X}_n}(t) = F_{\overline{Z}_n}\left(\frac{t - \mu}{\sigma_{\overline{X}}}\right)$. Hence for sufficiently large n,

$$F_{\overline{X}_n}(t) \approx F\left(\frac{t - \mu}{\sigma_{\overline{X}}}; 0, 1\right) = \frac{1}{\sqrt{2\pi}} \int_{-\infty}^{(t-\mu)/\sigma_{\overline{X}}} e^{-\frac{1}{2}x^2} dx; \qquad (8.35)$$

that is, the cumulative distribution function of $\overline{X}_n$ is approximately equal to the standard normal cumulative distribution function regardless of the form of the population distribution from which the sample random variables X_i, $i = 1, \ldots, n$, were obtained.

It should be obvious that the greater the departure of the population distribution from normality, the larger n needs to be in order for $\overline{Z}_n$ to secure an adequate approximation to the standard normal probability model. For highly skewed populations, $n > 30$ usually will render a good approximation to normality so that, given μ and σ^2, $\overline{Z}_n$ can be used to calculate probabilities involving $\overline{X}$. (If the population distribution has high tails and a low center, then $n > 100$ may be required.)

It is the Central Limit Theorem that enables us to conclude that, for each interval from $\bar{z}_1$ to $\bar{z}_2 (\bar{z}_1 < \bar{z}_2)$, the probability that the standardized sample mean $\overline{Z}_n$ assumes a value between $\bar{z}_1$ and $\bar{z}_2$ tends, as $n \to \infty$, to the probability that the standard normal variable Z takes on a value between $\bar{z}_1$ and $\bar{z}_2$. And this in turn allows us to conclude that, as $n \to \infty$,

$$P\left(a \leq \overline{X}_n \leq b\right) = P(\bar{z}_1 \leq \overline{Z}_n \leq \bar{z}_2) \to \frac{1}{\sqrt{2\pi}} \int_{\bar{z}_1}^{\bar{z}_2} e^{-\frac{1}{2}z^2} \, dz$$

$$= F(\bar{z}_2; 0, 1) - F(\bar{z}_1; 0, 1). \tag{8.36}$$

Example 8.8.1   If some population model has a mean of $\mu = 30$ and a standard deviation of $\sigma = 8$ and we select a simple random sample of size $n = 50$ from this distribution, we can determine, say, the probability that $\overline{X}$ will not differ from μ by more than ± 2 units or the probability that $\left|\overline{X} - 30\right| \leq 2$. Then $-2 \leq \overline{X} - 30 \leq 2$ and thus, via the standard normal area table,

$$P(28 \leq \overline{X} \leq 32) = P\left(\frac{28 - 30}{8/\sqrt{50}} \leq \frac{\overline{X} - \mu}{\sigma/\sqrt{n}} \leq \frac{32 - 30}{8/\sqrt{50}}\right)$$

$$= P(-1.76 \leq \overline{Z} \leq 1.76) = 0.9216. \quad \blacksquare$$

Example 8.8.2   Let us extract a simple random sample of size $n = 80$ from a uniform probability density (7.1) defined over the interval $(\alpha, \beta) = (10, 70)$. For this population $\mu = \frac{80}{2} = 40$ and $\sigma = \frac{60}{\sqrt{12}} = 17.32$. Then from the standard normal area table, the probability that $\overline{X}$ takes on a value between, say, 45 and 50 is

$$P\left(45 \leq \overline{X} \leq 50\right) = P\left(\frac{45 - 40}{17.32/\sqrt{80}} \leq \frac{\overline{X} - \mu}{\sigma/\sqrt{n}} \leq \frac{50 - 40}{17.32/\sqrt{80}}\right)$$

$$= P(2.58 \leq \overline{Z} \leq 5.17) \approx 0.005. \quad \blacksquare$$

Under what circumstances is it appropriate to invoke the conclusions of the Central Limit Theorem? In many experiments a measurable characteristic can often be viewed as the *additive result of many independent small factors*, each of which may or may not be readily observable or measurable. Hence the Central Limit Theorem informs us that the normal law is adequate for approximating sums $S_n = \sum_{i=1}^{n} X_i$ of many small independent random variables X_i, $i = 1, 2, \ldots$. It is the *additivity* of these many independent random variables that is of vital importance in that normality obtains from sums of numerous small independent causes or disturbances such as errors of measurement—errors that can be attributed to many independent factors, with each contributing only a small portion to their total impact. So with the sample mean $\overline{X}_n$ expressible as a linear combination of independent and identically distributed sample random variables, the sampling distribution of the mean is taken as approximately normal for large n. Have we already observed this phenomenon in action? You need only check Figure 8.3 to verify that an answer in the affirmative is warranted.

To underscore the importance of the aforementioned additivity notion in assessing the limiting behavior of certain sums of random variables, let $S_n = \sum_{i=1}^{n} X_i$ represent the sum of n independent and identically distributed random variables X_i, $i = 1, 2, \ldots, n$, each with mean μ and variance σ^2. Then $E(S_n) = n\mu$, $V(S_n) = n\sigma^2$, and, as $n \to \infty$, the standardized sum

$$Z'_n = \frac{S_n - n\mu}{\sigma \sqrt{n}}$$

is approximately $N(0, 1)$ distributed; that is, for sufficiently large n and fixed t,

$$P(Z'_n \leq t) \approx F(t; 0, 1) = \frac{1}{\sqrt{2\pi}} \int_{-\infty}^{t} e^{-\frac{1}{2}y^2} \, dy.$$

Note that by virtue of the form of $E(S_n)$ and $V(S_n)$, both the expectation and variance of the distribution of sums S_n become larger as n increases.

Example 8.8.3   This tendency towards normality of S_n for sufficiently large n can readily be demonstrated by examining the probability mass function of S_n for $n = 1, 2, 3, 4$, where, for the sake of concreteness, S_n denotes the number of heads obtained in n tosses of a fair coin. Here each X_i is a Bernoulli random variable with a uniform probability mass function (see Table 8.5) and with mean $E(X_i) = p(= \frac{1}{2})$ and $V(X_i) = p(1 - p)(= \frac{1}{4})$. (In this instance Z'_n above becomes $Z''_n = \frac{S_n - np}{\sqrt{np(1-p)}}$.)

For $n = 1$, the sample space has two simple events $\{E_1 = 0, E_2 = 1\}$ and the probability mass or relative frequency distribution of $S_1 = X_1$ is given in Table 8.6a. When $n = 2$, $S_2 = X_1 + X_2$ and the sample space has four simple events $\{E_1, \ldots, E_4\}$ corresponding to the sums $0 + 0 = 0$, $1 + 0 = 1$, $0 + 1 = 1$, and $1 + 1 = 2$, respectively (see Table 8.6b). For $n = 3, S_3 = X_1 + X_2 + X_3$, the sample space admits eight simple events $\{E_1, \ldots, E_8\}$ (e.g., $0 + 1 + 0 = 1$), and thus S_3's relative frequency distribution is given in Table 8.6c. And when $n = 4, S_4 = X_1 + X_2 + X_3 + X_4$ and the sample space has 16 simple events $\{E_1, \ldots, E_{16}\}$. The probability mass function for this case appears in Table 8.6d.

Table 8.5

X_i	$P(X_i)$
0	1/2
1	1/2
	1

Table 8.6

a.

$S_1 = X_1$	$P(S_1)$
0	1/2
1	1/2
	1

b.

$S_2 = X_1 + X_2$	$P(S_2)$
0	1/4
1	1/2
2	1/4
	1

c.

$S_3 = X_1 + X_2 + X_3$	$P(S_3)$
0	1/8
1	3/8
2	3/8
3	1/8
	1

d.

$S_4 = X_1 + X_2 + X_3 + S_4$	$P(S_4)$
0	1/16
1	4/16
2	6/16
3	4/16
4	1/16
	1

Table 8.7

n	$E(S_i) = np$	$V(S_i) = np(1 - p)$
1	1/2	1/4
2	1	1/2
3	3/2	3/4
4	2	1
⋮	⋮	⋮

Looking to the mean and variance of each of the probability mass functions presented in Table 8.6, it is easily shown in Table 8.7 that $E(S_i)$ and $V(S_i) \to +\infty$ as $n \to +\infty$. ∎

8.9 The Sampling Distribution of a Proportion

Let the discrete random variable X represent the number of successes obtained in n independent and identical trials of a simple alternative experiment. Then X is distributed binomially with probability mass function

$$b(X; n, p) = \binom{n}{X} p^X (1 - p)^{n-X}, \ X = 0, 1, \ldots, n, \tag{8.37}$$

and cumulative distribution function

$$P(X \leq x) = \sum_{i \leq x} \binom{n}{i} p^i (1-p)^{n-i}. \tag{8.38}$$

Here p depicts the probability of a success and $1 - p$ denotes the probability of a failure on any given trial of the simple alternative experiment. Then $E(X) = np$ and $V(X) = np\,(1-p)$.

Additionally, if we view p as the true proportion of successes in the binomial population, then we may use the statistic $\hat{P} = \frac{X}{n}$ to depict the proportion of successes obtained in a simple random sample of size n taken with replacement from this population. Here $\hat{P}$ may be termed the *observed relative frequency of a success* and serves as an estimator of the probability of a success p in the population. The sample realization of $\hat{P}$ will be denoted as $\hat{p} = \frac{x}{n}$, where x is the realized number of successes. Then the probability of obtaining the proportion $\hat{p}$ in a sample of size n is

$$P(X = n\hat{p}) = \binom{n}{n\hat{p}} p^{n\hat{p}} (1-p)^{n-n\hat{p}} \tag{8.39}$$

and the probability of obtaining a proportion less than or equal to $\hat{p}$ in a sample of size n is given by the cumulative distribution function

$$P(X \leq n\hat{p}) = \sum_{i \leq n\hat{p}} \binom{n}{i} p^i (1-p)^{n-i}. \tag{8.40}$$

(Note that if $n\hat{p}$ is not an integer, then we may round down to the nearest integer.)

To calculate the mean and variance of $\hat{P}$, let us view the outcome on the ith trial as corresponding to a separate Bernoulli random variable X_i, where $X_i = 1$ (when a success occurs) or $X_i = 0$ (for a failure). Then the X_i, $i = 1, \ldots, n$, represent a set of independent and identically distributed sample random variables with $E(X_i) = p$, $V(X_i) = p(1-p)$, $i = 1, \ldots, n$, and since $X = \sum_{i=1}^{n} X_i$, $\hat{P} = \frac{X}{n} = \frac{1}{n} \sum_{i=1}^{n} X_i$ is simply the *average number of successes* in the n trials. Since $\hat{P}$ is a random variable, its relationship to the random variable X is provided by Theorem 8.7:

THEOREM 8.7. If $X_1, \ldots, X_n$ are independent Bernoulli random variables having the same mean $E(X_i) = p$ and variance $V(X_i) = p(1-p)$, then $E(\hat{P}) = p$ and $V(\hat{P}) = \frac{p(1-p)}{n}$.

To verify this set of results we need only compute

(a) $E(\hat{P}) = E\left(\dfrac{X}{n}\right) = \dfrac{1}{n} E(X) = \dfrac{1}{n}(np) = p$

(b) $V\left(\hat{P}\right) = V\left(\dfrac{X}{n}\right) = \dfrac{1}{n^2} V(X) = \dfrac{1}{n^2}[np(1-p)] = \dfrac{p\,(1-p)}{n}$

$$\tag{8.41}$$

Here (8.41a) informs us that the average of the sample proportions taken over all possible samples of size n is equal to the population proportion p. Next, from (8.41b), the positive square root of $V(\hat{P})$ will be denoted as

$$\sigma_{\hat{P}} = \sqrt{\frac{p(1-p)}{n}} \tag{8.42}$$

and termed the *standard error of the sample proportion*. This value serves as a measure of how sample proportions vary, due to chance, from sample to sample under random sampling. Note that (8.42) holds if we sample from an infinite population or if we sample with replacement from a finite population. But if we sample without replacement from a finite population of size N and if $n > 0.05N$, then

$$\sigma_{\hat{P}} = \sqrt{\frac{p(1-p)}{n}} \sqrt{\frac{N-n}{N-1}}, \tag{8.42.1}$$

where the term $\sqrt{(N-n)/(N-1)}$ is the *finite population correction factor*.

To construct the sampling distribution of the proportion $\hat{P} = \frac{X}{n}$, let us first determine the probability distribution of the random variable X. Here $X = \sum_{i=1}^{n} X_i$ is a binomial random variable depicting the number of successes obtained in n trials of a Bernoulli experiment, where each X_i, $i = 1, \ldots, n$, is a Bernoulli random variable with realizations 0 or 1. And since the X_i are independent and identically distributed sample random variables, any particular set of realizations obtained in the n trials is the n-tuple $(X_1 = x_1, X_2 = x_2, \ldots, X_n = x_n)$ with probability

$$P(X_1 = x_1, \ldots, X_n = x_n) = \prod_{i=1}^{n} P(X = x_i), \tag{8.43}$$

where $P(X_i = x_i = 1) = p$ and $P(X_i = x_i = 0) = 1 - p$ for all $i = 1, \ldots, n$.

Example 8.9.1   Specifically, suppose we toss a coin (possibly fair) $n = 4$ times in succession and count the number of successes X obtained in the four trials, where a success is defined as the *coin shows heads*. Then $X = X_1 + X_2 + X_3 + X_4$ is a random variable depicting the total number of heads obtained in the four tosses and X_i is the number of heads obtained (either 0 or 1) on the ith toss, $i = 1, 2, 3, 4$. Since there are $N = 2$ possible outcomes (heads (H) or tails (T)) for each of the $n = 4$ tosses of the coin, the total number of possible ordered 4-tuples is $N^n = 2^4 = 16$. These outcome 4-tuples, the realizations of the X_i, the realizations of X, and their probabilities are all depicted in Table 8.8. For instance, at the outcome point (H, H, T, T): $x_1 = x_2 = 1, x_3 = x_4 = 0, x = 2$, and, by virtue of (8.43), $P(x = 2) = p \cdot p \cdot (1-p)(1-p) = p^2(1-p)^2$.

Based on the results presented in Table 8.8, the probability or sampling distribution of X is indicated in Table 8.9a, and the sampling distribution of $\hat{P} = \frac{X}{n} = \frac{X}{4}$

Table 8.8

Outcome Point				Realizations of X_i: x_1	x_2	x_3	x_4	Realization of X: $x = \sum_{i=1}^{4} x_i$	Probability: $P(X = x)$
H	H	H	H	1	1	1	1	4	p^4
H	H	H	T	1	1	1	0	3	$p^3(1-p)$
H	H	T	H	1	1	0	1	3	$p^3(1-p)$
H	H	T	T	1	1	0	0	2	$p^2(1-p)^2$
H	T	H	H	1	0	1	1	3	$p^3(1-p)$
H	T	H	T	1	0	1	0	2	$p^2(1-p)^2$
H	T	T	H	1	0	0	1	2	$p^2(1-p)^2$
H	T	T	T	1	0	0	0	1	$p(1-p)^3$
T	H	H	H	0	1	1	1	3	$p^3(1-p)$
T	H	H	T	0	1	1	0	2	$p^2(1-p)^2$
T	H	T	H	0	1	0	1	2	$p^2(1-p)^2$
T	H	T	T	0	1	0	0	1	$p(1-p)^3$
T	T	H	H	0	0	1	1	2	$p^2(1-p)^2$
T	T	H	T	0	0	1	0	1	$p(1-p)^3$
T	T	T	H	0	0	0	1	1	$p(1-p)^3$
T	T	T	T	0	0	0	0	0	$(1-p)^4$

Table 8.9

a.

Sampling Distribution of $X = \sum_{i=1}^{4} X_i$

X	$P(X = x)$
0	$(1-p)^4$
1	$4p(1-p)^3$
2	$6p^2(1-p)^2$
3	$4p^3(1-p)^2$
4	p^4
	1

b.

Sampling Distribution of the Proportion $\hat{P} = \frac{X}{n} = \frac{X}{4}$

$\hat{P}$	$P(\hat{P} = \frac{x}{4})$
0	$(1-p)^4$
1/4	$4p(1-p)^3$
1/2	$6p^2(1-p)^2$
3/4	$4p^3(1-p)^2$
1	p^4
	1

c.

Sampling Distribution of $\hat{P} = \frac{X}{4}$ when $p = \frac{1}{2}$

$\hat{P}$	$P(\hat{P} = \frac{x}{4})$
0	1/16
1/4	1/4
1/2	3/8
3/4	1/4
1	1/16
	1

d.

Sampling Error $\frac{x}{4} - p = \frac{x}{4} - \frac{1}{2}$

$$0 - \tfrac{1}{2} = -\tfrac{1}{2}$$
$$\tfrac{1}{4} - \tfrac{1}{2} = -\tfrac{1}{4}$$
$$\tfrac{1}{2} - \tfrac{1}{2} = 0$$
$$\tfrac{3}{4} - \tfrac{1}{2} = \tfrac{1}{4}$$
$$1 - \tfrac{1}{2} = \tfrac{1}{2}$$

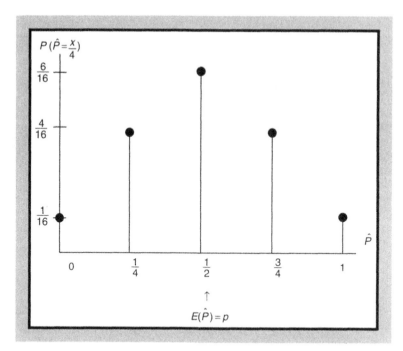

Figure 8.4   Probability mass function for the sampling distribution of $\hat{P} = \frac{X}{4}$ when $p = \frac{1}{2}$.

is given by Table 8.9b. Moreover, if $p = \frac{1}{2}$ (the coin is fair), the distribution of $\hat{P}$ for this special case is depicted in Table 8.9c and illustrated graphically in Figure 8.4. In addition, as required by (8.41a), $E(\hat{P}) = p = \frac{1}{2}$; and from (8.41b), $V(\hat{P}) = \frac{p(1-p)}{n} = \frac{1}{16}$. Note that, by virtue of these results, the sampling distribution of $\hat{P}$: (a) is centered at $p = \frac{1}{2}$; and (b) becomes more concentrated about p as n increases; that is, $V(\hat{P}) \to 0$ as $n \to \infty$.

So for a fixed n, as the number of repetitions of our simple alternative experiment gets larger and larger, the resulting relative frequency distribution gets closer and closer to the true sampling distribution of $\hat{P}$. Finally, the sampling errors associated with the various $\frac{x}{4}$ values are indicated in Table 8.9d. These sampling errors reflect the inherent natural variation in the sample proportions due to chance under random sampling. ∎

We noted earlier that, by virtue of (8.41b), the scatter or spread of the realizations of $\hat{P} = \frac{X}{n}$ (the $\frac{x}{n}$'s) about p varies inversely with the sample size n. Hence the probability of $\hat{P}$ being close to p increases with increasing n. This latter observation sets the stage for a discussion of the *weak law of large numbers for proportions* since $\hat{P} = \frac{X}{n} = \frac{1}{n} \sum_{i=1}^{n} X_i$ is a specialization of the sample mean for $X_i = 0$ or $1, i = 1, \ldots, n$.

In this regard, the average number of successes $\hat{P} = \frac{X}{n}$ in n independent Bernoulli trials should be near p for sufficiently large and fixed n. Hence the weak law of large numbers informs us that as n increases without bound, the probability that the average number of successes $\hat{P}$ deviates from p by less than any preassigned tolerance level $\varepsilon > 0$ tends to 1 or

$$P\left(\left|\hat{P} - p\right| < \varepsilon\right) \to 1 \text{ as } n \to \infty, \tag{8.44}$$

that is, $\hat{P}$ converges stochastically to p. To see this let us express Chebyshev's inequality in terms of $\hat{P}$ as

$$P\left(\left|\hat{P} - p\right| < k\sigma_{\hat{P}}\right) \geq 1 - \frac{1}{k^2}, \; k > 0. \tag{8.45}$$

For $\sigma_{\hat{P}} = \sqrt{\frac{p(1-p)}{n}}$ and $\varepsilon = k\sqrt{\frac{p(1-p)}{n}}$, this expression can be rewritten as

$$P\left(\left|\hat{P} - p\right| < \varepsilon\right) = 1 - \frac{p(1-p)}{n\varepsilon^2}$$

and, for fixed $\varepsilon > 0$,

$$\lim_{n\to\infty} P\left(\left|\hat{P} - p\right| < \varepsilon\right) = 1.$$

Probability statement (8.44) may be written in the alternative form

$$P\left(\left|\hat{P} - p\right| \geq \varepsilon\right) \to 0 \text{ as } n \to \infty, \tag{8.44.1}$$

that is, the probability that the average number of successes $\hat{P}$ differs from p by more than $\varepsilon > 0$ tends to zero as n increases indefinitely. Note that neither (8.44) nor (8.44.1) implies that $\hat{P}$ itself necessarily gets closer and closer to p. But if, as $n \to \infty$, $\hat{P}$ remains near p, then the *strong law of large numbers* applies; that is, the probability is 1 that only finitely many of the terms $|\hat{P} - p| > \varepsilon > 0$ will occur (or all but a finite number of $\hat{P}$'s are within a distance ε of p). Hence the probability is 1 that $\hat{P} - p$ becomes smaller and stays small as $n \to \infty$ or

$$P\left(\lim_{n\to\infty} \left|\hat{P} - p\right| < \varepsilon\right) = 1.$$

Thus $\hat{P}$ converges to p with probability 1 or $\hat{P}$ converges almost surely to p.

We noted earlier that Chebyshev's Theorem provides us with a very general statement about the probability of $\hat{P}$ being near p. A more satisfactory estimate of this probability can be made if we possess knowledge about the form of the sampling distribution of $\hat{P}$. Since $\hat{P}$ represents the average number of successes in a sample of size n, a useful approximation to the form of the probability distribution of $\hat{P}$ is provided, for large n, by the Central Limit Theorem.

Given that the result obtained on each trial of a simple alternative experiment is recorded as the realization x_i of the individual Bernoulli random variable $X_i, i = 1, \ldots, n$, it follows that these X_i's are independent and identically distributed. Hence the Central Limit Theorem applies to their mean $\hat{P} = \frac{X}{n} = \frac{1}{n}\sum_{i=1}^{n} X_i$. Since each individual X_i is a binomial random variable (take $n = 1$) with mean p and standard deviation $\sqrt{p(1-p)}$, it follows that the mean of $\hat{P}$ is p and the standard deviation of $\hat{P}$ is $\sqrt{p(1-p)/n}$. Hence the random variable

$$Z_{\hat{P}} = \frac{\hat{P} - p}{\sqrt{p(1-p)/n}} \tag{8.46}$$

is approximately $N(0, 1)$ for large n.

Looked at in another fashion, we previously noted in Chapter 7 that, by virtue of the De Moivre–Laplace–Gauss Limit Theorem, the standardized binomial random variable $X = \sum_{i=1}^{n} X_i$ has as its limiting distribution the standard normal distribution; that is,

$$\frac{X - np}{\sqrt{np(1-p)}} \xrightarrow{d} Z = N(0, 1).$$

But since

$$\frac{X - np}{\sqrt{np(1-p)}} = \frac{\frac{X}{n} - p}{\sqrt{np(1-p)}/n} = Z_{\hat{P}}$$

(from (8.46)), the De Moivre–Laplace–Gauss limit result carries over to $\hat{P}$ as well so that $\hat{P}$ is approximately $N(p, \sigma_{\hat{P}})$.

Example 8.9.2 Suppose we toss a fair coin $n = 100$ times in succession and count the total number of heads (deemed a success) X obtained. What is the probability that the average number of heads $\hat{P}$ will not differ from $p(= 0.5)$ by more than 3%? In this instance $\sigma_{\hat{P}} = \sqrt{0.5(0.5)/100} = 0.05$. Hence we are interested in finding

$$P(p - 0.03 \leq \hat{P} \leq p + 0.03) = P\left(-\frac{0.03}{0.05} \leq \frac{\hat{P} - p}{\sigma_{\hat{P}}} \leq \frac{0.03}{0.05}\right) = P(-0.6 \leq Z_{\hat{P}} \leq 0.6).$$

Since $Z_{\hat{p}}$ is approximately $N(0,1)$, it is readily demonstrated that the preceding probability simplifies to 0.4515. ∎

8.10   The Sampling Distribution of the Variance

Let us extract a random sample of size n from an infinite population (or, equivalently, sample with replacement from a finite population). Then the sample random variables $X_i, i = 1,\ldots,n$, are independent and identically distributed with $E(X_i) = \mu$ and $V(X_i) = \sigma^2$ for all i, where μ and σ^2 are the population mean and variance, respectively, and it is assumed that σ^2 is finite.

Suppose we are interested in the variability of a given measurable characteristic X associated with some random phenomenon. Then we must examine the properties of the statistic

$$S^2 = \frac{\sum_{i=1}^{n}(X_i - \overline{X})^2}{n-1} \tag{8.47}$$

or the *sample variance* of the X_i. (If x_i is the realization of X_i and $\bar{x}$ is the realization of $\overline{X}$, then $s^2 = \frac{\sum_{i=1}^{n}(x_i-\bar{x})^2}{n-1}$ is the realization of S^2.) In this regard, to determine how sample variances would be expected to vary, due to chance, under random sampling, we can derive empirically the distribution of individual sample variances and then examine its properties. As was the case for the sample mean $\overline{X}$, various general properties of the distribution can be determined.

Since S^2 is a random variable, its relationship to the population random variable X is provided by Theorem 8.8:

THEOREM 8.8.   If $X_1,\ldots,X_n$ are independent sample random variables having the same distribution with mean μ and variance σ^2, then:

$$E(S^2) = \sigma^2; \tag{8.48}$$

$$V(S^2) = \frac{1}{n}\left[\mu_4 - \left(\frac{n-3}{n-1}\right)\sigma^4\right], \quad n > 1, \tag{8.49}$$

where $\mu_4 = E[(X - \mu)^4]$ is the fourth central moment of the population.

We may easily rationalize (8.48) by first noting that

$$\sum_{i=1}^{n}(X_i - \overline{X})^2 = \sum_{i=1}^{n}(X_i + \mu - \mu - \overline{X})^2 = \sum_{i=1}^{n}[(X_i - \mu) - (\overline{X} - \mu)]^2$$

$$= \sum_{i=1}^{n}(X_i - \mu)^2 - n(\overline{X} - \mu)^2.$$

Then

$$E(S^2) = \frac{1}{n-1} E\left[\sum_{i=1}^{n} (X_i - \overline{X})^2\right]$$

$$= \frac{1}{n-1} E\left[\sum_{i=1}^{n} (X_i - \mu)^2 - n(\overline{X} - \mu)^2\right]$$

$$= \frac{1}{n-1}\left[\sum_{i=1}^{n} E(X_i - \mu)^2 - nE(\overline{X} - \mu)^2\right]$$

$$= \frac{1}{n-1}\left[n\sigma^2 - nV(\overline{X})\right]$$

$$= \frac{1}{n-1}\left[n\sigma^2 - n\left(\frac{\sigma^2}{n}\right)\right] = \sigma^2.$$

This result informs us that the average of the sample variances taken over all possible samples of size n is equal to the population variance. Hence the sampling distribution of S^2 is centered at σ^2.

Next, for

$$S^2 = \frac{\sum_{i=1}^{n}(X_i - \overline{X})^2}{n-1} = \frac{\sum_{i=1}^{n}(X_i - \mu)^2 - n(\overline{X} - \mu)^2}{n-1},$$

we seek to determine, using (8.48),

$$V(S^2) = E\left\{[S^2 - E(S^2)]^2\right\} = E\left[(S^2 - \sigma^2)^2\right] = E(S^2)^2 - \sigma^4.$$

Then

$$E(S^2)^2 = E\left[\frac{\sum_{i=1}^{n}(X_i - \overline{X})^2}{n-1}\right]^2 = \frac{1}{(n-1)^2} E\left[\sum_{i=1}^{n}(X_i - \mu)^2 - n(\overline{X} - \mu)^2\right]^2$$

$$= \frac{1}{(n-1)^2} E\left[\left(\sum_{i=1}^{n}(X_i - \mu)^2\right)^2 - 2n(\overline{X} - \mu)^z\right. \tag{8.50}$$

$$\left. \times \sum_{i=1}^{n}(X_i - \mu)^2 + n^2(\overline{X} - \mu)^4\right].$$

Upon taking the expectation of each term within the square brackets of (8.50) separately, it can be shown (with some extensive algebraic manipulations) that:

1. $E\left(\sum_{i=1}^{n}(X_i - \mu)^2\right)^2 = n\left(\mu_4 + (n-1)\sigma^4\right);$

2. $2nE\left((\overline{X} - \mu)^2 \sum_{i=1}^{n}(X_i - \mu)^2\right) = 2\left(\mu_4 + (n-1)\sigma^4\right);$

   and

3. $n^2 E\left[(\overline{X} - \mu)^4\right] = \frac{1}{n}(\mu_4 + 3(n-1)\sigma^4).$

Substituting 1–3 into (8.50) we obtain

$$E(S^2)^2 = \frac{1}{n}\left[\mu_4 + \left(\frac{n^2 - 2n + 3}{n-1}\right)\sigma^4\right]. \tag{8.50.1}$$

Hence from (8.50.1),

$$V(S^2) = E(S^2)^2 - \sigma^4 = \frac{1}{n}\left[\mu_4 - \left(\frac{n-3}{n-1}\right)\sigma^4\right].$$

Note that $V(S^2) \to 0$ as $n \to \infty$.

It is important to note that these results are quite general and independent of the form of the population probability density function. Unfortunately, unless we have prior knowledge concerning the shape of the population density function, we cannot say much about the form of the distribution of sample variances. (However, if we are sampling from a normal population, then much more can be said about the form of the sampling distribution of the variance; as the next chapter will attest, under the assumption of normality of the population density function, the distribution of sample variances is chi-square distributed.)

We saw earlier that the weak law of large numbers could be applied to $\overline{X}$ as an estimator of μ. A similar law applies to S^2 as an estimate of σ^2. (In what follows we shall denote the sample variance computed from a random sample of size n as S_n^2 to emphasize its dependence on n.) To see this we need only note that since the $X_i, i = 1, \ldots, n$, are independent and identically distributed sample random variables whose individual variances exist, the *weak law of large numbers for variances* states that the variance of these variables converges to their common variance σ^2 as the sample size increases. Thus realizations of S_n^2 can be expected to be close to σ^2 as more sample random variables are introduced.

Stated alternatively, the weak law of large numbers for variances implies that there exists a sample size n such that the probability is as close to 1 as desired that the difference between S_n^2 and σ^2 will not exceed some arbitrarily small quantity. More formally, we have Theorem 8.9 (Weak Law Of Large Numbers for Variances).

THEOREM 8.9. Let S_n^2 be the variance of a random sample of size n taken from an infinite population with finite variance σ^2. Then for every fixed $\varepsilon > 0$,

$$\lim_{n\to\infty} P\left(\left|S_n^2 - \sigma^2\right| < \varepsilon\right) = 1, \tag{8.51}$$

that is, S_n^2 converges stochastically to σ^2 (or S_n^2 converges to σ^2 in probability). To see this let us express Chebyshev's inequality (4.19.2) in terms of S_n^2 as

$$P\left(\left|S_n^2 - \sigma^2\right| < k\sqrt{V(S_n^2)}\right) \geq 1 - \frac{1}{k^2}$$

or

$$P\left(\left|S_n^2 - \sigma^2\right|^2 < k^2 V(S_n^2)\right) \geq 1 - \frac{1}{k^2}. \tag{8.52}$$

If we take $\varepsilon = k\sqrt{V(S_n^2)}$, then (8.52) becomes

$$P\left(\left|S_n^2 - \sigma^2\right|^2 < \varepsilon^2\right) \geq 1 - \frac{V(S_n^2)}{\varepsilon^2}. \tag{8.52.1}$$

Then

$$\lim_{n\to\infty} P\left(\left|S_n^2 - \sigma^2\right|^2 < \varepsilon^2\right) = \lim_{n\to\infty}\left(1 - \frac{V(S_n^2)}{\varepsilon^2}\right) = 1$$

since $V(S_n^2) \to 0$ as $n \to \infty$. Hence a sufficient condition that S_n^2 converges in probability to σ^2 is $V(S_n^2) \to 0$ as $n \to \infty$.

Example 8.10.1   To see exactly how the sampling distribution of the variance is determined, let us take as our finite population of size $N = 6$ the set of values $X : 1, 2, 3, 4, 5, 6$. Here the mean and variance of this population are $\mu = 3.5$ and $\sigma^2 = 2.916$, respectively. If we select random samples of size $n = 2$ with replacement, then, if the order in which the values are chosen matters, we obtain $N^n = 6^2 = 36$ possible random samples, with the probability of choosing any one sample equal to $\frac{1}{36}$. Given that the sample random variables X_i have realizations x_i, $i = 1, 2$, each sample can be written as the ordered pair (x_1, x_2). The 36 samples and their associated sample variances are presented in Table 8.10a. If we form the relative frequency distribution of these sample variances, then this construct represents the sampling distribution of the variance for all possible ordered samples of size $n = 2$ taken with replacement from a population of size $N = 6$. This sampling distribution is illustrated in Table 8.10b.

Table 8.10

Sample Number	Realizations (x_1, x_2)	Sample Variances $s_j^2, j = 1, \ldots, 36$
1	1, 1	0
2	1, 2	0.5
3	1, 3	2
4	1, 4	4.5
5	1, 5	8
6	1, 6	12.5
7	2, 1	0.5
8	2, 2	0
9	2, 3	0.5
10	2, 4	2
11	2, 5	4.5
12	2, 6	8
13	3, 1	2
14	3, 2	0.5
15	3, 3	0
16	3, 4	0.5
17	3, 5	2
18	3, 6	4.5
19	4, 1	4.5
20	4, 2	2
21	4, 3	0.5
22	4, 4	0
23	4, 5	0.5
24	4, 6	2
25	5, 1	8
26	5, 2	4.5
27	5, 3	2
28	5, 4	0.5
29	5, 5	0
30	5, 6	0.5
31	6, 1	12.5
32	6, 2	8
33	6, 3	4.5
34	6, 4	2
35	6, 5	0.5
36	6, 6	0

b.

Sampling Distribution of the Variance	
S^2	$P(S^2 = s_i^2))$
0	6/36
0.5	10/36
2	8/36
4.5	6/36
8	4/36
12.5	2/36
	1

As indicated earlier (Theorem 8.8):

$$E(S^2) = \sigma^2 = 2.916;$$

$$V(S^2) = \frac{1}{n}\left[\mu_4 - \left(\frac{n-3}{n-1}\right)\sigma^4\right] = \frac{1}{2}\left[\frac{88.378}{6} + 38.503\right] = 11.616.$$

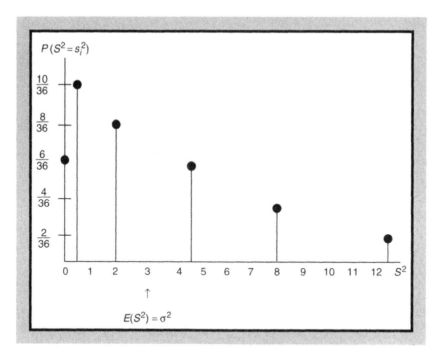

Figure 8.5   Probability mass function for the sampling distribution of S^2.

Clearly the sampling distribution of the variance is centered exactly on the population variance (see Figure 8.5). ■

What Table 8.10b provides us with is an approximation to the true (unknown) *sampling distribution of the sample variance*—a theoretical distribution that associates with realizations s^2 of S^2 the probability mass (density) of each taken over all possible samples of a given size. And for fixed n, as the number of random samples taken gets larger and larger, the resulting relative frequency distribution tends to get closer and closer to the true sampling distribution of the variance.

8.11 A Note on Sample Moments

It was mentioned earlier that for a random sample of size n, the sample mean $\overline{X}_n = \frac{1}{n}\sum_{i=1}^{n} X_i$ and the sample variance $S_n^2 = \frac{\sum_{i=1}^{n}(X_i - \overline{X}_n)^2}{n-1}$ are statistics used to estimate the population mean μ and population variance σ^2, respectively. As we shall now see, $\overline{X}_n$ and S_n^2 can be thought of as special cases of more general concepts called sample moments. In this regard, we may view sample moments as statistics that are used to estimate their population counterparts.

More formally, let $X_1, \ldots, X_n$ constitute a random sample of size n taken from an infinite population (or from some population probability density function).

Then the rth *sample moment about zero* is defined as

$$M'_r = \frac{1}{n} \sum_{i=1}^{n} X_i^r. \tag{8.53}$$

Note that for $r = 1$, $M'_1 = \overline{X}_n$; that is, the first sample moment about zero is simply the sample mean. Note also that the sample variance S_n^2 can be written in terms of sample moments about zero as

$$S_n^2 = \frac{\sum_{i=1}^{n}(X_i - \overline{X}_n)^2}{n - 1} = \frac{1}{n - 1}\left(\sum_{i=1}^{n} X_i^2 - n\overline{X}_n^2\right) = \left(\frac{n}{n - 1}\right)\left[M'_2 - (M'_1)^2\right].$$

We may next specify the rth *sample moment about the mean of X* or the rth *central sample moment* as

$$M_r = \frac{1}{n} \sum_{i=1}^{n} (X_i - \overline{X}_n)^r. \tag{8.54}$$

Remember that the population counterparts of (8.53) and (8.54) are, for X a random variable, $\mu'_r = E(X^r)$ and $\mu_r = E(X - \mu)^r$, respectively (see (4.21) through (4.24)). In addition, M'_r and M_r are currently random variables. However, once a sample is taken, these variables have realizations

$$m'_r = \frac{1}{n} \sum_{i=1}^{n} x_i^r$$

and

$$m_r = \frac{1}{n} \sum_{i=1}^{n} (x_i - \bar{x}_n)^r,$$

respectively, where x_i is the realization of X_i, $i = 1, \ldots, n$, and $\bar{x}_n$ is the realization of $\overline{X}_n$.

We mentioned earlier that the sampling distributions of the mean and variance were centered on the population mean and population variance, respectively, or $E(\overline{X}_n) = \mu$, $E(S_n^2) = \sigma^2$. A similar result holds for the sampling distributions of sample moments taken about zero. That is, the various sample moments M'_r, $r = 1, 2, \ldots$, can be used to estimate their corresponding population moments since the sampling distribution of M'_r is concentrated about μ'_r. This result is incorporated in Theorem 8.10:

THEOREM 8.10. Let $X_1,\dots,X_n$ represent a set of independent and identically distributed sample random variables whose individual moments about zero μ'_r and μ'_{2r} exist. Then the mean and variance of the rth sample moment taken about zero can be expressed in terms of population moments as:

$$E(M'_r) = \mu'_r; \tag{8.55}$$

$$V(M'_r) = \frac{1}{n}\left[\mu'_{2r} - (\mu'_r)^2\right]. \tag{8.56}$$

To verify these results let us first determine

$$E(M'_r) = \frac{1}{n}E\left(\sum_{i=1}^{n}X_i^r\right) = \frac{1}{n}\sum_{i=1}^{n}E(X_i^r) = \frac{1}{n}\sum_{i=1}^{n}\mu'_r = \mu'_r.$$

Next,

$$V(M'_r) = \frac{1}{n^2}V\left(\sum_{i=1}^{n}X_i^r\right)$$

$$= \frac{1}{n^2}\sum_{i=1}^{n}V(X_i^r)$$

$$= \frac{1}{n^2}\sum_{i=1}^{n}\left[E(X_i^{2r}) - E(X_i^r)^2\right]$$

$$= \frac{1}{n^2}\sum_{i=1}^{n}\left[\mu'_{2r} - (\mu'_r)^2\right] = \frac{1}{n}\left[\mu'_{2r} - (\mu'_r)^2\right].$$

For instance, if $X_1,\dots,X_n$ are independent sample random variables having the same distribution with mean μ and variance σ^2, then it is easily shown that $E(M'_1) = \mu'_1 = \mu$ or that $E(\overline{X}_n) = \mu$. In addition, $V(M'_1) = \frac{1}{n}[\mu'_2 - (\mu'_1)^2]$ or $V(\overline{X}_n) = \frac{\sigma^2}{n}$.

Appendix 8.A   Proof of a Central Limit Theorem

In order to offer a proof of a Central Limit Theorem (presented as Theorem 8.6, earlier), first we need to introduce the concept of *convergence of moment-generating functions*. In this regard, suppose that $\{X_i,\ i = 1, 2,\dots, n\}$ is a sequence of random variables whose individual moment-generating functions $m_{X_i}(t)$, $i = 1,\ldots,n$, exist for all $|t| < t_0, t_0 > 0$. Also, suppose that X is a random variable whose moment-generating function $m_X(t)$ exists, also for all $|t| < t_0$, and that

$$\lim_{i \to \infty} m_{X_i}(t) = m_X(t). \qquad (8.A.1)$$

Then $m_X(t)$ is termed the *limiting moment-generating function* of the random variables X_i, $i = 1,\ldots,n$. Moreover, let $F_{X_i}(s)$ depict the cumulative distribution function of X_i and let $F_X(s)$ represent the cumulative distribution function for X. Then if (8.A.1) holds, the cumulative distribution functions of the X_i converge to the cumulative distribution function of X, that is, for all points where F_X is continuous,

$$\lim_{i \to \infty} F_{X_i}(s) = F_X(s). \qquad (8.A.2)$$

So for all t within some suitably restricted neighborhood of zero, convergence of $m_{X_i}(t)$ to $m_X(t)$ implies convergence of $F_{X_i}(s)$ to $F_X(s)$. And since $m_X(t)$ is unique, there is only one probability density function $f_X(x)$ that yields this moment-generating function.

Armed with these considerations, the proof of a Central Limit Theorem that immediately follows is thus based upon demonstrating that, for $|t| < t_0$, the moment-generating function of some random variable $\overline{Z}_n$ tends to that of a standard normal random variable Z as $n \to \infty$. But this, in turn, implies that $\overline{Z}_n$ has a cumulative distribution function that converges to that of a $N(0, 1)$ random variable.

For $\{X_i, i = 1, 2,\ldots,n\}$ a set of independent and identically distributed sample random variables with $E(X_i) = \mu$ and $V(X_i) = \sigma^2$ finite for all i, let $\overline{X}_n = \frac{1}{n}\sum_{i=1}^{n} X_i$ with $E(\overline{X}_n) = \mu$ and $V(\overline{X}_n) = \sigma^2/n$. In addition, let $Z_i = (X_i - \mu)/\sigma$, $i = 1,\ldots,n$, with $\overline{Z}_n = \frac{\overline{X}_n - \mu}{\sigma/\sqrt{n}}$. Since

$$\overline{Z}_n = \frac{\sum_{i=1}^{n} X_i - n\mu}{n\left(\frac{\sigma}{\sqrt{n}}\right)} = \frac{1}{\sqrt{n}} \sum_{i=1}^{n} Z_i$$

and the $Z_i, i = 1,\ldots,n$, constitute a set of independent and identically distributed random variables (since the X_i are independent and identically distributed), it follows from (5.73.4) that

$$m_{\overline{Z}_n}(t) = m_{\sum_i \frac{Z_i}{\sqrt{n}}}(t) = m_{\sum_i Z_i}\left(\frac{t}{\sqrt{n}}\right) = \left[m_Z\left(\frac{t}{\sqrt{n}}\right)\right]^n$$

$$= \left[E\left(e^{tZ/\sqrt{n}}\right)\right]^n, \quad |t| < \sigma\sqrt{n}t_0, \qquad (8.A.3)$$

where $m_Z(t)$ denotes the common moment-generating function of the Z_i's, which is taken to be finite for $|t| < \sigma t_0$, $t_0 > 0$. In addition $E(Z) = 0$ and $E(Z^2) = 1$.

If $e^{tZ/\sqrt{n}}$ is replaced by its Maclaurin's series expansion, we obtain

$$e^{tZ/\sqrt{n}} = 1 + \frac{t}{\sqrt{n}}Z + \frac{t^2}{2!n}Z^2 + \frac{t^3}{3!n^{3/2}}Z^3 + \cdots . \qquad (8.A.4)$$

Given this expression,

$$E\left(e^{tZ/\sqrt{n}}\right) = 1 + \frac{t^2}{2!n} + \frac{t^3}{3!n^{3/2}}E(Z^3) + \cdots \qquad (8.A.5)$$

and thus (8.A.3) becomes

$$m_{\overline{Z}_n}(t) = \left[m_Z\left(\frac{t}{\sqrt{n}}\right)\right]^n = \left\{1 + \frac{1}{n}\left[\frac{t^2}{2!} + \frac{t^3}{3!\sqrt{n}} + \cdots\right]\right\}^n = \left(1 + \frac{a_n}{n}\right)^n, \quad (8.A.6)$$

where the square bracketed term is denoted as a_n. Now, since the terms within a_n save for $\frac{t^2}{2!}$ all contain positive powers of n in their denominators, it follows that as $n \to \infty$, all terms within a_n except $\frac{t^2}{2}$ (remember that t is fixed in value) go to zero as a limit. Hence

$$\lim_{n\to\infty} m_{\overline{Z}_n}(t) = \lim_{n\to\infty}\left(1 + \frac{a_n}{n}\right)^n = e^{t^2/2}. \qquad (8.A.7)$$

Remember that $e^{t^2/2}$ or (7.A.1) is the moment-generating function for a standard normal random variable.

As we alluded earlier, if the sequence of moment-generating functions $m_{\overline{Z}_n}(t) \to m_Z(t) = e^{t^2/2}$, then the limit of the cumulative distributions $F_{\overline{Z}_n}(s)$ must be $F(s; 0, 1)$, the cumulative distribution function corresponding to the moment-generating function $e^{t^2/2}$. So if $\overline{X}_n$ has a limiting $N(0,1)$ cumulative distribution function or

$$\lim_{n\to\infty} F_{\overline{X}_n}(s) = F(s; 0, 1),$$

then $\overline{X}_n$ has an asymptotic normal distribution and thus the limiting distribution of $\overline{X}$ is $N(\mu, \frac{\sigma}{\sqrt{n}})$.

8.12 Exercises

8-1. Suppose a random variable X is $N(\mu, 1.5)$. Given $n = 10$, determine the probability that $\overline{X}$ will not differ from μ by more than 0.5 units. (Hint: $Z = (\overline{X} - \mu)/\sigma_{\overline{X}}$ is $N(0, 1)$.)

8-2. Suppose as in Exercise 8-1 we know that X is $N(\mu, 1.5)$. How large of a sample will be needed if we require that $P(|\overline{X} - \mu| \le 0.5) = 0.95$?

8-3. Suppose that from a population of size $N = 20$ random samples of size $n = 4$ are to be selected. How many possible samples can be obtained? What is the probability that any one of them will be selected?

8-4. List all possible samples of size $n = 2$ given the population $X : A, B, C, D, E, F$. If $A = 3$, $B = 5$, $C = 6 = D$, $E = 7$, and $F = 8$, determine the sampling distribution of the mean.

8-5. Let the random variables $X_1, \ldots, X_4$ be Poisson distributed with mean λ. Determine the joint probability distribution of the $X_i, i = 1, \ldots, 4$. Are these random variables independent?

8-6. If a population variable X is $N(200, 12)$ and the sample size is $n = 16$, find:

(a) $P(\overline{X} > 206)$

(b) $P(\overline{X} \leq 209)$

(c) $P(195 < \overline{X} < 203)$

8-7. Suppose a population variable X is $N(2800, 2500)$. If a random sample of size $n = 36$ is selected, find the probability that the sample mean $\overline{X}$ will not differ from the population mean μ by more than 30. Within what interval would $\overline{X}$ fall 90% of the time under random sampling for samples of size 36? How large of a sample must be taken so that the probability that $\overline{X}$ is within 20 units of μ is 0.95?

8-8. Suppose a population variable X has a mean of 35 and a standard deviation of 10. What is the expression for $\sigma_{\overline{X}}$, the standard error of the mean? If a sample of $n = 2000$ is taken, what is the value of $\sigma_{\overline{X}}$? Use Chebyshev's Theorem to find the probability that the difference between $\overline{X}$ and μ is less than 3 units.

8-9. A population distribution has an unknown mean with $\sigma = 2.5$. How large of a sample must be drawn so that the probability is at least 0.99 that $\overline{X}_n$ will lie within one unit of the population mean?

8-10. Let X be the outcome obtained on the toss of a fair single six-sided die and let Y be the number of heads obtained on three flips of a fair coin. Determine the joint probability mass function. Are X and Y independent random variables? Also, find: $E(X)$; $E(Y)$; $V(X)$; and $V(Y)$. If $W = 2X + 3Y$, find $E(W)$ and $V(W)$.

8-11. Let X_1, X_2, and X_3 depict sample random variables drawn from a population with probability density function $f(x) = 2e^{-2x}, 0 \leq x < +\infty$. What is the joint probability density function of these random variables? What is its generalization to a random sample of size n? What is the moment-generating function of $Y = X_1 + X_2 + X_3$?

8-12. If $X_1, \ldots, X_n$ depicts a random sample of size n taken from the probability density function $f(x; 2) = 2e^{-2x}, 0 \leq x < +\infty$ (see Exercise 8-11), find $P(X_1 > 10, X_2 > 10, \ldots, X_n > 10)$.

8-13. In a sample of $n = 100$ observations the number of successes was found to be $X = 35$. If the population proportion is $p = 0.40$, find:

(a) $P(\hat{P} \leq 0.38)$

(b) $P(\hat{P} \geq 0.50)$

8-14. A nutritionist claims that about 50% of the individuals tasting a new soft drink can detect the presence of a certain artificial sweetener. If the claim is correct, find the probability that out of a random sample of size $n = 70$, 30 or fewer individuals detect the sweetener. What is the probability that more than 60 detect it?

8-15. Suppose a population variable X has an unknown mean μ but a known variance σ^2. A random sample of size n is selected. Are the following quantities statistics?

(a) $\overline{X}$

(b) $\overline{X} + \mu$

(c) $\overline{X} + \sigma^2$

(d) X_3

(e) $X_2 + \mu - \sigma^2$

(f) $\frac{1}{2}\min\{X_i\}$

(g) $\overline{X}/\sqrt{n}$

(h) $\overline{X}/\sigma$

(i) $\log X_i$

8-16. Suppose that X has an unknown mean μ with $\sigma^2 = 3$. How large of a sample must be taken so that the probability will be at least 0.90 that $\overline{X}$ will lie within 0.6 of the population mean? How large of a sample would you need to be 99% certain that $\overline{X}$ is within one standard deviation of μ? (Hint: Use the law of large numbers.)

8-17. Given the letters A, B, C, and D, determine the number of different samples of size $n = 2$ that can be selected:

(a) With replacement

(b) Without replacement

If $n = 3$, recalculate (a) and (b).

8-18. $N = 8$ different items are within a population. If we wish to select a sample of size $n = 3$, how many ways are there of sampling? What is the probability that any one sample will be chosen? What is the probability that any one item will be in the sample?

8-19. Suppose that a population variable X has the values 1, 2, 3, 4, 5, and 6. Determine the sampling distribution of the mean for samples of size $n = 2$.

What is its mean and standard deviation? Repeat this exercise for $n = 3$ and compare your results.

8-20. Which of the following statements are true?

(a) If X is $N(\mu, \sigma)$, then $\overline{X}$ is $N(\mu, \frac{\sigma}{\sqrt{n}})$.

(b) If X is not $N(\mu, \sigma)$, then $\overline{X}$ is approximately $N(\mu, \frac{\sigma}{\sqrt{n}})$ for large n.

(c) If X is not $N(\mu, \sigma)$, then $Y = \sum_{i=1}^{n} X_i$ is approximately $N(n\mu, \sigma\sqrt{n})$ for large n.

8-21. Let $X_1, \ldots, X_n$ be independently distributed Bernoulli $b(X_i; n, p)$ random variables with $0 < p < 1$ for all $i = 1, \ldots, n$, where $E(X_i) = p$ and $V(X_i) = p(1 - p)$. Demonstrate that $\overline{X}_n$ converges in probability to $E(X_i) = p$.

8-22. Let $X_1, \ldots, X_n$ be independently distributed Poisson $p(X; \lambda)$ random variables with $\lambda > 0$ for all $i = 1, \ldots, n$. Demonstrate that $\overline{X}_n$ converges in probability to $E(X_i) = \lambda$.

8-23. Let $X_1, \ldots, X_n$ be independently distributed $N(\mu, 1)$ random variables with $-\infty < \mu < +\infty$. Find a real number A such that $e^{\overline{X}_n}$ converges in probability to A.

8-24. Let X_n be $b(X; n, p)$. Does $1 - (X_n/n)$ converge stochastically to $1 - p$?

8-25. Suppose X_n is $N\left(0, n^{-1/2}\right)$. Verify that $\{X_n\}$ converges to 0 in distribution.

8-26. Based upon past experience, it is known that a certain manufacturing process has a defect rate of 3%. A random sample of size $n = 450$ is taken from a large production run and the CEO wants to determine the probability that the sample proportion of defectives will not exceed 1%. If you are the production manager, what is your response to the CEO?

8-27. Suppose a random variable X is $N(4, 1.5)$. If a random sample of size $n = 25$ is drawn, what is the probability that $\overline{X}$ will exceed 4.5?

8-28. A particular socioeconomic group has a mean annual income of $\mu = \$35,800$ with $\sigma = \$3,500$. A random sample of $n = 250$ individuals from this group is taken. What is the probability that $\overline{X}$ will be within $500 of μ?

8-29. In a mayoral election last year candidate A received 55% of the 8,576 votes cast. If a random sample of $n = 200$ eligible voters had been taken the day before the election, what would have been the probability that candidate A would have received at least 50% of the votes cast?

8-30. Suppose the sampling distribution of the sample variance has been determined from samples of size $n = 4$ taken from a population variable $X : -2, 0, 1, 2, 3, 4, 5, 7$. Find its mean and variance.

8-31. Given the sample values $1, 3, 1, 5, 8$, and 2, find M_1' and M_2'. Use them to determine S^2. Also find M_3 and M_4.

8-32. Demonstrate that the mean $\overline{X}$ of a random sample of size n drawn from a population random variable X, which is $N(\mu, \sigma)$, is distributed as $N(\mu, \sigma/\sqrt{n})$.

8-33. Let a random variable X be binomial $b(X; n, p)$. Demonstrate that the limiting moment-generating function of X is the moment-generating function of a Poisson random variable Y with mean λ. (Hint: Starting with $m_X(t) = (1 - p + pe^t)^n$, take the limit as $n \to \infty$ with $np = \lambda = $ constant.)

8-34. Let $\{X_1, \ldots, X_n\}$ be a set of sample random variables drawn from a distribution that is $N(\sigma, \mu)$. Demonstrate that the limiting moment-generating function of $\overline{X}_n$ is $e^{\mu t}$. How do we interpret this result? If X is not normally distributed, demonstrate that this result still holds.

8-35. Suppose $X_1, \ldots, X_n$ depicts a random sample drawn from the gamma probability density function (7.48). What is the form of the sampling distribution of $\overline{X}$?

8-36. Determine the moment-generating function of the observed relative frequency of a success $\hat{P} = \frac{X}{n}$, where $X = \sum_{i=1}^{n} X_i$ is the sum of n independent Bernoulli random variables X_i ($= 0$ or 1), $i = 1, \ldots, n$.

8-37. Suppose X_1, X_2 are independent random variables with $X_1 \sim N(4, 6.2)$, $X_2 \sim N(5, 7.1)$. If we form a new random variable $Y = 2X_1 - 3X_2$, find $P(Y > 8)$. What is the form of the moment-generating function for Y?

8-38. Let $X_1, \ldots, X_{25}$ constitute a set of sample random variables drawn from a population that is $N(\mu, \sigma) = N(100, 35)$. What is the form of the sampling distribution of $\overline{X}$? Determine the moment-generating function for $\overline{X}$. How is the probability density function for $\overline{X}$ specified? What is $P(\overline{X} < 90)$?

8-39. (Sampling Distribution of the Median) Given a random sample of size n, let us arrange the sample random variables $X_1, i = 1, \ldots, n$, in order of increasing numerical magnitude. Let the resulting set of n values be denoted as $X_{(1)}, X_{(2)}, \ldots, X_{(n)}$, where, for $1 \le i \le n$, $X_{(1)} = \min X_{(i)}, X_{(2)} = 2nd$ smallest $X_i, \ldots, X_{(n)} = \max X_i$. Clearly $X_{(1)} \le X_{(2)} \le \cdots \le X_{(n)}$. Here X_i is termed the *ith order statistic* and is a function of the sample elements X_i. In general, the *order statistics* of a random sample are simply the sample values placed in ascending order. Although the $X_{(i)}, i = 1, \ldots, n$, are indeed random variables, they are not independent since, if $X_i \ge t$, then obviously $X_{(i+1)} \ge t$.

Consider the order statistic, which is (approximately) the $(np)^{th}$ order statistic $X_{(np)}$ for a random sample of size n, $0 < p < 1$. Choose p_n so that np_n is an integer and $n|p_n - p|$ is finite. Then $X_{(np_n)}$ is the $(np_n)^{th}$ order statistic for a random sample of size n. Based upon these considerations, let $X_1, \ldots, X_n$ be a set of independent and identically distributed random variables with common probability density function $f(x)$ and strictly monotone cumulative distribution function $F(t)$, $0 < F(t) < 1$.

Then for γ_p the pth population quantile (the quantile of order p, $0 < p < 1$, is a value γ_p such that $F_{(\gamma P)} = P(X \leq \gamma_p) = p$), the $(np_n)^{th}$ order statistic $X_{(np_n)}$ is asymptotically normal with

$$E(X_{(np_n)}) = \gamma_p,$$

$$V(X_{(np_n)}) = \frac{p(1-p)^2}{nf(\gamma_p)},$$

where $f(\cdot)$ is the population probability density function.

In particular, if $p = 0.5$, then $\gamma_{0.5}$ is the population median. (In terms of order statistics, if $X_1, \ldots, X_n$ denotes a sample of size n, then the *sample median is*

$$md = \begin{cases} X_{((n+1)/2)} & \text{if } n \text{ is odd;} \\ \frac{1}{2}\left(X_{(n/2)} + X_{((n/2)+1)}\right) & \text{if } n \text{ is even.} \end{cases}$$

Then, in the light of the preceding discussion, the sampling distribution of the sample median is asymptotically normal with mean equal to the population median $\gamma_{0.5}$ and variance $1/4nf(\gamma_{0.5})^2$. Moreover, if the population probability density function is normal with mean μ and variance σ^2, then the sampling distribution of the sample median is asymptotically normal with mean μ and variance $1/4nf(\mu)^2 = \pi\sigma^2/2n$.

We may express the standard error of the median as $\sigma_{\gamma_{0.5}} = \sqrt{\pi/2}\left(\sigma/\sqrt{n}\right)$. (Since the standard error of the mean is only $\sigma/\sqrt{n}$, we see that, in terms of large samples taken from a normal population, the sample mean is less variable (and thus more reliable) than the sample median as a measure of central tendency.)

Armed with these considerations, if a random sample of size $n = 100$ is taken from a population that is $N(40, 10)$, determine the probability that the sample median (md) will not differ from the population median ($\gamma_{0.5} = \mu = 40$) by move than ± 5 units.

The Chi-Square, Student's *t*, and Snedecor's *F* Distributions

9.1 Derived Continuous Parametric Distributions

In this chapter we shall deal with three new continuous probability distributions, each of which is *built up* or *derived* from one or more other continuous probability distributions. These new distributions are the chi-square, Student's *t*, and Snedecor's *F*.

As we shall soon see, the chi-square distribution can be constructed as the sum of the squares of *v* independent standard normal random variables, where the key parameter of this distribution, its number of degrees of freedom, is taken to be the number of squared standard normal variates entering into the sum.

The Student's *t* distribution is formed as the ratio between a standard normal random variable and the square root of a chi-square random variable that has been divided by its degrees of freedom, and where these two random variables are taken to be independent. The key parameter of the *t* distribution is its degrees of freedom, which corresponds to the degrees of freedom of its constituent chi-square variable.

Looking to Snedecor's *F* distribution, it is determined by taking the ratio between two independent chi-square random variables that have been divided by their degrees of freedom. Here the key parameters of the *F* distribution are the numbers of degrees of freedom associated with the chi-square random variables in the numerator (stated first) and denominator.

These distributions will be employed in many sampling situations in which, for instance, we are sampling from a normal population and we need to make an inference about the (unknown) population mean or variance. Or if we have two independent random samples from normal populations and we want to compare their (unknown) variances.

9.2 The Chi-Square Distribution

The chi-square distribution arises as a special case of the gamma probability distribution. That is, if in the gamma probability density function (7.48) we set the shape parameter α equal to $\frac{v}{2}$ (v a positive integer) and the scale parameter θ equal to 2, then we obtain the *chi-square probability density function*

$$f(x; v) = \begin{cases} \dfrac{1}{\Gamma(\frac{v}{2})2^{\frac{v}{2}}}x^{\frac{v}{2}-1}e^{-\frac{x}{2}}, & 0 < x < +\infty, v(> 0) \text{ an integer}; \\ 0 & \text{elsewhere} \end{cases} \tag{9.1}$$

associated with the continuous random variable X. Here X is termed a *chi-square random variable with v degrees of freedom* and denoted χ_v^2. Note that (9.1) exhibits but a single parameter, namely v. And as v varies through positive integer values, we obtain a whole family of particular chi-square distributions (see Figure 9.1).

In addition, for X distributed as χ_v^2, the *chi-square cumulative distribution function* appears as

$$F(t; v) = P(X \le t) = \frac{1}{\Gamma(\frac{v}{2})2^{\frac{v}{2}}} \int_0^t x^{\frac{v}{2}-1}e^{-\frac{x}{2}}\,dx, \; x > 0. \tag{9.2}$$

The chi-square probability distribution exhibits the following properties:

1. From (7.50) and (7.51) it follows that the mean and standard deviation of a chi-square variable X are, respectively,

$$E(X) = v; \tag{9.3}$$

$$\sqrt{V(X)} = \sqrt{2v}. \tag{9.4}$$

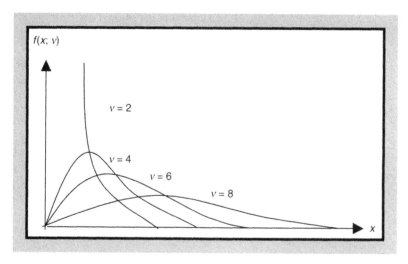

Figure 9.1 A family of chi-square distributions for selected values of v.

2. The chi-square distribution is positively skewed and its coefficient of skewness (from (7.52)) is

$$\alpha_3 = \frac{4}{\sqrt{2v}} > 0 \qquad (9.5)$$

with $\lim_{v \to \infty} \alpha_3 = 0$.

3. If $v > 2$, the chi-square distribution attains its maximum at $x = v - 2$.

4. The chi-square distribution has a peak that is sharper than that of a normal distribution since its coefficient of kurtosis (from (7.53)) is

$$\alpha_4 = 3\left(1 + \frac{4}{v}\right) > 3 \qquad (9.6)$$

with $\lim_{v \to \infty} \alpha_4 = 3$.

5. If the random variable X is chi-square distributed with mean and standard deviation given by (9.3), and (9.4), respectively, then the quantity $Z = (X - v)/\sqrt{2v} \to N(0,1)$ as $v \to \infty$. Moreover, when $v > 30$, chi-square probabilities can be determined via a standard normal approximation and percentiles of the chi-square distribution can be approximated by percentiles of the $N(0,1)$ distribution. In this regard, if X is χ_v^2 with $v > 30$, then it can be shown that the statistic $\sqrt{2X}$ has a probability density function that is approximately $N(\sqrt{2v - 1}, 1)$. Hence the quantity $Z = \sqrt{2X} - \sqrt{2v - 1}$ is approximately $N(0,1)$.

6. The chi-square distribution is said to be *stochastically increasing in its degrees of freedom*; that is, if a random variable X is χ_v^2 and p and q are positive integers such that $p > q$, then for any real $a > 0$, $P(\chi_{v=p}^2 > a) > P(\chi_{v=q}^2 > a)$.

Selected quantiles of the chi-square distribution can be determined from the chi-square table for various values of the degrees of freedom parameter v. From (9.2) and for various cumulative probabilities $1 - \alpha$ (see Figure 9.2), the quantile $\chi_{1-\alpha,v}^2$ satisfies

$$F(\chi_{1-\alpha,v}^2; v) = P(X \le \chi_{1-\alpha,v}^2) = \int_0^{\chi_{1-\alpha,v}^2} f(x; v)dx = 1 - \alpha \qquad (9.7)$$

or, alternatively,

$$P(X > \chi_{1-\alpha,v}^2) = 1 - F(\chi_{1-\alpha,v}^2; v) = \int_{\chi_{1-\alpha,v}^2}^{\infty} f(x; v)dx = \alpha. \qquad (9.7.1)$$

That is, for various degrees of freedom v, $\chi_{1-\alpha,v}^2$ gives the value of χ_v^2 below which the proportion $1 - \alpha$ of the χ_v^2 distribution falls (or $\chi_{1-\alpha,v}^2$ is the value of χ_v^2 above which the proportion α of the distribution is found).

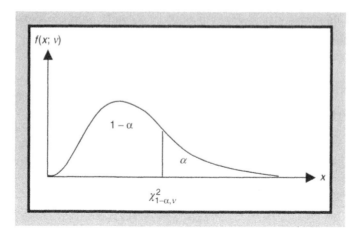

Figure 9.2   The quantile $\chi^2_{1-\alpha,\nu}$ satisfies $P(X \le \chi^2_{1-\alpha,\nu}) = 1 - \alpha$ and $P(X > \chi^2_{1-\alpha,\nu}) = \alpha$.

Example 9.2.1   If in Table A.4 of the Appendix we set $1 - \alpha = 0.90$ and $\nu = 14$, then

$$P(X \le \chi^2_{0.90,14}) = P(X \le 21.064) = 0.90.$$

Similarly:

$$P(X \le \chi^2_{0.10,5}) = P(X \le 1.610) = 0.10;$$
$$P(X \le \chi^2_{0.95,5}) = P(X \le 11.1) = 0.95;$$

and

$$P(X > \chi^2_{0.90,10}) = P(X > 15.987) = 0.10. \quad \blacksquare$$

If $\nu > 30$, property (5), earlier, enables us to set

$$1 - \alpha = P(X \le \chi^2_{1-\alpha,\nu}) = P\left(\sqrt{2X} - \sqrt{2\nu - 1} \le \sqrt{2\chi^2_{1-\alpha,\nu}} - \sqrt{2\nu - 1}\right) \approx P(Z \le z_{1-\alpha}),$$

where Z is $N(0,1)$ and $z_{1-\alpha}$ is the $100(1 - \alpha)^{th}$ percentile of the standard normal distribution (see Figure 9.3). Then $z_{1-\alpha} \approx \sqrt{2\chi^2_{1-\alpha,\nu}} - \sqrt{2\nu - 1}$ or

$$\chi^2_{1-\alpha,\nu} \approx \frac{1}{2}\left(z_{1-\alpha} + \sqrt{2\nu - 1}\right)^2. \tag{9.8}$$

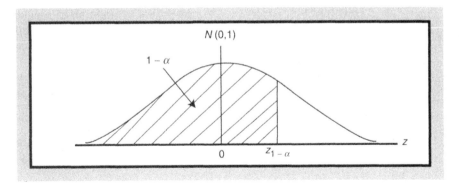

Figure 9.3   For $v > 30$, $P(X \le \chi^2_{1-\alpha,v}) = 1 - \alpha \approx P(Z \le z_{1-\alpha})$.

Example 9.2.2   If $v = 40$ and $1 - \alpha = 0.90$, then from the standard normal area table (Table A.2), we have $0.90 = P(Z \le z_{0.90}) = P(Z \le 1.28)$. Then from (9.8),

$$\chi^2_{0.90,40} \approx \frac{1}{2}\left(1.28 + \sqrt{79}\right)^2 = 51.697.$$

A direct examination of the chi-square table (Table A.4) renders $\chi^2_{1-\alpha,v} = \chi^2_{0.90,40} = 51.805$. Clearly the standard normal approximation to this chi-square percentile using the $N(0,1)$ percentile $z_{0.90} = 1.28$ works quite well.   ∎

The chi-square distribution is of paramount importance when sampling from a normal population or probability density function, as Theorems 9.1 and 9.2 will attest:

THEOREM 9.1.   If the random variable X is $N(\mu, \sigma)$, then $Z = (X - \mu)/\sigma$ is $N(0,1)$ and Z^2 is χ^2_1.

In general,

THEOREM 9.2.   Let $X_1, \ldots, X_n$ constitute a random sample of size n taken from a normal population with mean μ and standard deviation σ. Then $Z_i = (X_i - \mu)/\sigma$, $i = 1, \ldots, n$, are independent standard normal random variables and

$$Y = \sum_{i=1}^{n} Z_i^2 = \sum_{i=1}^{n} \frac{(X_i - \mu)^2}{\sigma^2} \quad \text{is} \quad \chi^2_n.$$

(Note that the number of degrees of freedom (n) corresponds to the number of independent standard normal variables that are squared and summed.)

Theorems 9.3 and 9.4 deal with the *reproductive (additivity) property* of the chi-square distribution.

THEOREM 9.3.   If $X_1, \ldots, X_n$ constitute a set of independent chi-square random variables with $v_1, \ldots, v_n$ degrees of freedom, respectively, then $Y = \sum_{i=1}^{n} X_i$ has a chi-square distribution with $v = \sum_{i=1}^{n} v_i$ degree of freedom.

Here independent chi-square random variables sum to a chi-square random variable and their respective degrees of freedom are also additive; that is, if X_i is $\chi_{v_i}^2$, $i = 1, \ldots, n$, then $\sum_{i=1}^{n} X_i$ is $\chi_{\sum v_i}^2$. Also,

THEOREM 9.4.   If X_i and X_j are independent random variables with X_i distributed as $\chi_{v_i}^2$ and $X_i + X_j$ distributed as χ_v^2, $v > v_i$, then X_j is distributed as $\chi_{v_j=v-v_i}^2$.

Example 9.2.3   Let us assume that a population random variable X is $N(50, 3.2)$. If a random sample of size $n = 25$ is taken from this population, what is the probability that the quantity $S_0^2 = \sum_{i=1}^{n}(X_i - \mu)^2/n < 12$ (squared units)? By virtue of Theorem 9.2, since μ is known (no correction for degrees of freedom is warranted), the quantity

$$Y = \frac{\sum_{i=1}^{n}(X_i - \mu)^2}{\sigma^2} = \frac{n}{\sigma^2} \frac{\sum_{i=1}^{n}(X_i - \mu)^2}{n} = \frac{nS_0^2}{\sigma^2} \tag{9.9}$$

is χ_{25}^2. Then from Table A.4,

$$P(S_0^2 < 12) = P\left(\frac{nS_0^2}{\sigma^2} < \frac{n(12)}{\sigma^2}\right) = P\left(Y < \frac{25(12)}{10.24} = 29.296\right) = 0.75.$$

Here $29.296 \approx \chi_{0.75,25}^2 = \chi_{1-\alpha,v}^2$.   ∎

9.3   The Sampling Distribution of the Variance When Sampling from a Normal Population

In Section 8.10 we discussed the sampling distribution of the variance

$$S^2 = \frac{\sum_{i=1}^{n}(X_i - \overline{X})^2}{n - 1} \tag{9.10}$$

under sampling from a general population. Let us now examine the properties of the sampling distribution of S^2 when sampling is undertaken from a population which is normally distributed. We begin with Theorem 9.5:

THEOREM 9.5.   If S^2 is the variance of a random sample of size n taken from a normal population with mean μ and variance σ^2, then the quantity $Y = \frac{(n-1)S^2}{\sigma^2}$ is distributed as χ_{n-1}^2.

To see this let us rewrite (9.10) as

$$(n-1)S^2 = \sum_{i=1}^{n}(X_i - \mu - \overline{X} + \mu)^2 = \sum_{i=1}^{n}(X_i - \mu)^2 - n(\overline{X} - \mu)^2.$$

Then

$$\frac{(n-1)S^2}{\sigma^2} + \frac{n(\overline{X} - \mu)^2}{\sigma^2} = \frac{\sum_{i=1}^{n}(X_i - \mu)^2}{\sigma^2}$$

or

$$\frac{(n-1)S^2}{\sigma^2} + \left(\frac{\overline{X} - \mu}{\sigma/\sqrt{n}}\right)^2 = \frac{\sum_{i=1}^{n}(X_i - \mu)^2}{\sigma^2}. \tag{9.11}$$

Now, the second term on the left-hand side of (9.11) is the square of a $N(0,1)$ random variable and thus, by Theorem 9.1, is distributed as χ_1^2. Since the right-hand side of (9.11) is χ_n^2 (via Theorem 9.2), it follows from Theorem 9.4 that $\frac{(n-1)S^2}{\sigma^2}$ is distributed as χ_{n-1}^2.

From the preceding theorem and (9.1), the probability density function of $Y = \frac{(n-1)S^2}{\sigma^2}$ can be written as

$$f(y; n-1) = \begin{cases} \dfrac{1}{\Gamma\left(\frac{n-1}{2}\right)2^{\frac{n-1}{2}}} y^{\frac{n-1}{2}-1}e^{-\frac{y}{2}}, & 0 \le y < +\infty, n(>1) \text{ an integer}; \\ 0 & \text{elsewhere}. \end{cases} \tag{9.12}$$

And with Y distributed as χ_{n-1}^2, if follows from (9.3) and (9.4), respectively, that:

$$E(Y) = n - 1; \tag{9.13}$$

$$\sqrt{V(Y)} = \sqrt{2(n-1)}. \tag{9.14}$$

Given $Y = \frac{(n-1)S^2}{\sigma^2}$, it follows that $S^2 = \frac{\sigma^2 Y}{n-1}$ and thus

$$E(S^2) = \frac{\sigma^2}{n-1}E(Y) = \sigma^2; \tag{9.15}$$

$$V(S^2) = \left(\frac{\sigma^2}{n-1}\right)^2 V(Y) = \frac{2\sigma^4}{n-1}. \tag{9.16}$$

So if we are taking random samples of size n from a normal population with mean μ and variance σ^2, it follows that the sampling distribution of S^2 has a mean of σ^2 and a variance equal to $\frac{2\sigma^4}{(n-1)}$.

Example 9.3.1   Let us extract a random sample of size $n = 25$ from a population that is $N(\mu, \sigma)$ with μ unknown but with $\sigma = 3$. What is the probability that S^2 will not differ from σ^2 by more than ± 2 (squared units)? Here we are interested in finding

$$P(7 \leq S^2 \leq 11) = P\left(\frac{(n-1)7}{\sigma^2} \leq \frac{(n-1)S^2}{\sigma^2} \leq \frac{(n-1)11}{\sigma^2}\right)$$

$$= P\left(\frac{24(7)}{9} \leq Y \leq \frac{24(11)}{9}\right) = P(18.66 \leq Y \leq 29.33).$$

Since Y is $\chi^2_{n-1} = \chi^2_{24}$, Table A.4 enables us to find

$$P(18.66 \leq Y \leq 29.33) = P(Y \leq 29.33) - P(Y \leq 18.66) \approx 0.75 - 0.25 = 0.50.$$

Similarly, the probability that S^2 exceeds 6 is found to be

$$P(S^2 > 6) = P\left(\frac{(n-1)S^2}{\sigma^2} > \frac{(n-1)6}{\sigma^2}\right) = P(Y > 16) = 1 - P(Y \leq 16)$$

$$\approx 1 - 0.10 = 0.90. \quad \blacksquare$$

An important theorem that addresses the relationship between the sample variance S^2 and the sample mean $\overline{X}$ when sampling from a normal population is Theorem 9.6:

> **THEOREM 9.6.**   Let $X_1, \ldots, X_n$ constitute a random sample of size n taken from a normal population with mean μ and standard deviation σ. Then the statistics $\overline{X}$, S^2 determined from this sample are independent random variables.

So if we have n independent observations on a population random variable X, then the sample mean and sample variance computed from these observations are independent under the normality of X. This should be intuitively apparent since, with S^2 a measure of dispersion about $\overline{X}$, the dispersion of a random variable is not determined by where the random variable happens to be centered. Hence information about the sample mean in no way dictates the value of the sample variance and conversely.

A specialization of Theorem 9.6 is provided by Theorem 9.6.1:

> **THEOREM 9.6.1.**   If $Z_1, \ldots, Z_n$ is a random sample taken from a standard normal population (the random variable Z is $N(0, 1)$), then:

(a) $\overline{Z}$ is $N(0, \frac{1}{\sqrt{n}})$

(b) $\sum_{i=1}^{n}(Z_i - \overline{Z})^2$ is χ^2_{n-1} (from Theorem 9.5)

(c) $\overline{Z}$ and $\sum_{i=1}^{n}(Z_i - \overline{Z})^2$ are independent random variables

9.4   **Student's *t* Distribution**

We noted in the preceding chapter (e.g., Section 8.7) that under random sampling from a population that is $N(\mu, \sigma)$, with σ known, the standardized sample mean $\overline{Z} = \frac{\overline{X} - \mu}{(\sigma/\sqrt{n})}$ is $N(0, 1)$. But if σ is unknown, then it can be estimated from the sample values $X_1, \ldots, X_n$ as

$$S = \sqrt{\frac{\sum_{i=1}^{n}(X_i - \overline{X})^2}{n - 1}}. \tag{9.17}$$

If we replace σ by S in $\overline{Z}$, then the distribution of the resulting quantity

$$\frac{\overline{X} - \mu}{S/\sqrt{n}} \tag{9.18}$$

is not $N(0, 1)$, even though sampling is undertaken from a normal population.

To determine the *exact* sampling distribution of (9.18) when sampling from a normal distribution $N(\mu, \sigma)$ with *both* μ and σ unknown, let us form the random variable

$$T = \frac{Z}{\sqrt{X/v}}, \quad -\infty < T < +\infty, \tag{9.19}$$

where Z and X are independent random variables, Z is $N(0, 1)$, and X is χ^2_v. In this instance the random variable T is said to follow a *Student's t distribution with v degrees of freedom* and will be denoted as t_v. Its probability density function is given by

$$f(t; v) = \begin{cases} \dfrac{\Gamma\left(\frac{v+1}{2}\right)}{\sqrt{\pi v}\,\Gamma\left(\frac{v}{2}\right)} \left[1 + \left(\frac{t^2}{v}\right)\right]^{-\frac{(v+1)}{2}}, & -\infty < t < +\infty, v(> 0) \text{ an integer;}^{[1]} \\ 0 & \text{elsewhere.} \end{cases} \tag{9.20}$$

Here (9.20) represents a one-parameter family of distributions and, as we vary the parameter v, we obtain an assortment of particular t distributions (see Figure 9.4).

---

[1] Given (9.19), (9.20) is arrived at via a suitable transformation of the joint probability density function of (independent) Z and X

$$f(z, x; v) = \begin{cases} \dfrac{1}{\sqrt{2\pi}} e^{-\frac{1}{2}z^2} \cdot \dfrac{1}{\Gamma(v/2)2^{v/2}} x^{\frac{v}{2}-1} e^{-x/2}, & -\infty < z < +\infty, x > 0, v(> 0) \text{ an integer;} \\ 0 & \text{elsewhere.} \end{cases} \tag{9.21}$$

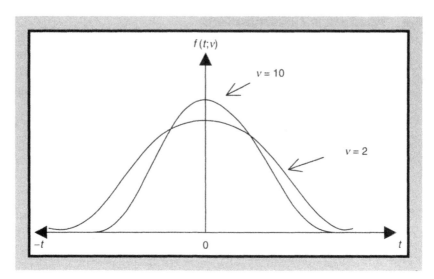

Figure 9.4   A family of t distributions for selected values of v.

In addition, the *Student's t cumulative distribution function* (using (9.20)) appears as

$$F(t;v) = P(T \le t) = \int_{-\infty}^{t} f(t;v)\,dt, \quad -\infty < t < +\infty. \tag{9.22}$$

We have just found that if Z is $N(0,1)$, X is χ_v^2, and Z and X are independent random variables, then $T = \frac{Z}{\sqrt{X/v}}$ is t distributed with v degrees of freedom. To demonstrate that this result is applicable to sampling from a normal population, we need to determine the connection between (9.18) and (9.19). In this regard, we need only note that when sampling from a normal population with σ known, the random variable $\overline{Z} = \frac{(\overline{X} - \mu)}{(\sigma/\sqrt{n})}$ is $N(0,1)$ and, by virtue of Theorem 9.5, the quantity $\frac{(n-1)S^2}{\sigma^2}$ is χ_{n-1}^2. Since $\overline{X}$ and S^2 are independent (Theorem 9.6), it follows that

$$T = \frac{\frac{(\overline{X}-\mu)}{(\sigma/\sqrt{n})}}{\sqrt{\frac{(n-1)\frac{S^2}{\sigma^2}}{n-1}}} = \frac{\overline{X} - \mu}{S/\sqrt{n}} \tag{9.19.1}$$

is t distributed with $n - 1$ degrees of freedom.

The Student's t distribution exhibits the following properties:

1. It is symmetrical about zero, asymptotic to the horizontal axis, and attains its unique maximum at its modal value of $t = 0$.

2. The basic shape of the t probability density function is similar to that of the standard normal distribution, with the tails of the former being a bit wider

than those of the latter. Hence there is a larger area under the tails of the t distribution than under those of the standard normal distribution.

3. For infinite degrees of freedom, the t and the standard normal distribution are one and the same. In fact, for $v \geq 30$, the t distribution closely approximates the standard normal distribution. (This is why the t distribution is termed a *small sample distribution*.) More formally, the quantity

$$T_n = \frac{\sqrt{n}(\bar{X}_n - \mu)}{\sqrt{S_n^2}}$$

has a t distribution with $n - 1$ degrees of freedom, where $\bar{X}_n$ and $S_n^2 = \sum_{i=1}^{n} \frac{(X_i - \bar{X}_n)^2}{(n-1)}$ are, respectively, the mean and variance of a sample of size n taken from a normal distribution with mean μ and (known) standard deviation σ. Then the random variable $T_n \to N(0, 1)$ as $n \to \infty$. Stated alternatively, we may employ (9.20) to demonstrate that the t probability density function converges to the standard normal probability density function for large v or

$$\lim_{v \to \infty} f(t; v) = \frac{1}{\sqrt{2\pi}} e^{-\frac{1}{2}t^2}, \quad -\infty < t < +\infty.$$

In this regard, since the asymptotic distribution of $T = \frac{\bar{X} - \mu}{S/\sqrt{n}}$ is $N(0, 1)$, if $a < b$, then

$$P(a < T < b) \to \frac{1}{\sqrt{2\pi}} \int_a^b e^{-t^2/2} dt \text{ as } n \to \infty.$$

4. The mean and standard deviation of the t distribution are, respectively,

$$E(T) = 0, \quad v > 1; \tag{9.23}$$

$$\sqrt{V(T)} = \sqrt{\frac{v}{v - 2}} > 1, \quad v > 2. \tag{9.24}$$

Given these restrictions on v, it follows that the t distribution does not have moments (about zero) of all orders; that is, if there are v degrees of freedom, then there are only $v - 1$ moments. Thus a t distribution has no mean when $v = 1$; it has no variance when $v \leq 2$. In fact, all odd moments of T are zero.

5. The coefficient of skewness is

$$\alpha_3 = 0; \tag{9.25}$$

and the coefficient of kurtosis is

$$\alpha_4 = 3 + \frac{6}{v-4}, \quad v \geq 4 \tag{9.26}$$

with $\lim_{v \to \infty} \alpha_4 = 3$ (the value under normality).

Selected quantiles of the *t* distribution appear in Table A.3 of the Appendix for various probabilities α (the area in the right-hand tail of the *t* distribution) and degrees of freedom values *v*. That is, the *right-tailed quantile* $t_{\alpha, v}$ is the value of *T* for which

$$P(T \geq t_{\alpha,v}) = \int_{t_{\alpha,v}}^{+\infty} f(t; v)\, dt = \alpha, \quad 0 \leq \alpha \leq 1 \tag{9.26}$$

(see Figure 9.5a). In this regard, we may view $t_{\alpha, v}$ as an *upper percentage point* of the *t* distribution—the point for which the probability of a larger value of *T* is α. By symmetry, the *left-tailed quantile* $-t_{\alpha, v}$ is a *lower percentage point* of

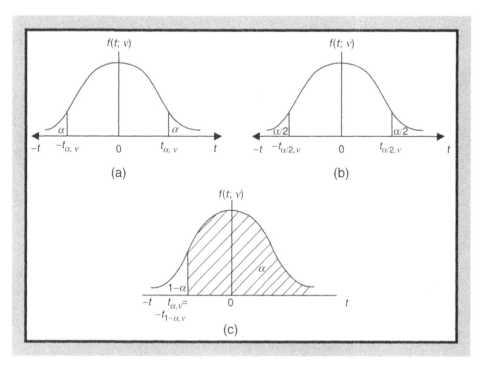

Figure 9.5 (a) $-t_{\alpha,v}$ and $t_{\alpha,v}$ are respectively left- and right-tailed quantiles of the *t* distribution; (b) $\pm t_{\alpha/2,v}$ are two-tailed quantiles of the *t* distribution; (c) For $\alpha > 0.5, t_{\alpha,v} = -t_{1-\alpha,v}$.

the *t* distribution—the point for which the probability of a smaller value is α (see Figure 9.5a) or

$$P(T \le -t_{\alpha,v}) = \int_{-\infty}^{-t_{\alpha,v}} f(t;v)dt = \alpha, \quad 0 \le \alpha \le 1. \tag{9.27}$$

In addition, *two-tailed quantiles* for the *t* distribution appear as $\pm t_{\frac{\alpha}{2},v}$ and satisfy the relationship

$$P\left(T \le -t_{\frac{\alpha}{2},v}\right) + P\left(T \ge t_{\frac{\alpha}{2},v}\right) = \frac{\alpha}{2} + \frac{\alpha}{2} = \alpha. \tag{9.28}$$

Here α is divided between the two tails of the distribution so that $\frac{\alpha}{2}$ is the area in each tail (see Figure 9.5b). For $\alpha > 0.5$, $t_{\alpha,v} = -t_{1-\alpha,v}$ (see Figure 9.5c).

Example 9.4.1   From Table A.3, it is easily seen that:

(a) $t_{\alpha,v} = t_{0.05,20} = 2.086$ or that $P(T \ge 2.086) = 0.05$ (see Figure 9.6a)

(b) $t_{\alpha,v} = t_{0.10,15} = 1.341$ and thus $P(T \le -1.341) = 0.10$ (see Figure 9.6a)

(c) For $\alpha = 0.05$, $t_{\frac{\alpha}{2},v} = t_{0.025,10} = 2.228$ so that

$$P(T \le -2.228) + P(T \ge 2.228) = 0.025 + 0.025 = 0.05 \quad \text{(see Figure 9.6b).} \quad \blacksquare$$

Example 9.4.2   To see how the *t* distribution might be applied, let us extract a random sample of size $n = 25$ from a normal population with unknown mean and standard deviation. Suppose we find that the realization of S is 10. Given this

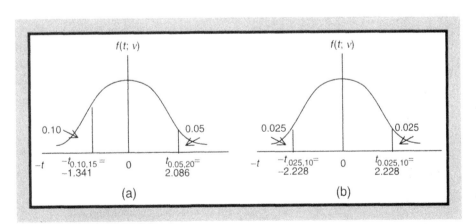

Figure 9.6   (a) $P(T \ge 2.086) = 0.05$, $P(T \le -1.341) = 0.10$); (b) $P(T \le -2.228) + P(T \ge 2.228) = 0.05$.

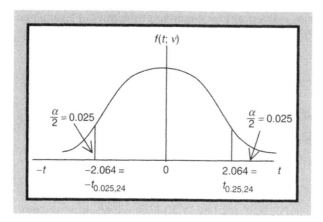

Figure 9.7 $P(-2.064 \leq T \leq 2.064) = 0.95$.

sample result, determine the probability that the sample mean will not differ from the population mean by more than ± 4 units. That is, we seek

$$P(|\overline{X} - \mu| \leq 4) = P\left(-4 \leq \overline{X} - \mu \leq 4\right)$$

$$= P\left(\frac{-4}{s/\sqrt{n}} \leq \frac{\overline{X} - \mu}{S/\sqrt{n}} \leq \frac{4}{s/\sqrt{n}}\right) = P(-2 \leq T \leq 2).$$

Since the random variable T is t distributed with $n - 1 = 24$ degrees of freedom, the preceding probability can be approximated (see Figure 9.7) as $P(-2.064 \leq T \leq 2.064) \approx 1 - 2(0.025) = 0.95$. ∎

9.5   Snedecor's *F* Distribution

Suppose we conduct a sampling experiment whose outcome is a set of observations on two independent chi-square random variables X and Y with degrees of freedom v_1 and v_2, respectively. The joint probability density function of X and Y is

$$g(x, y; v_1, v_2) = \begin{cases} \frac{1}{\Gamma(\frac{v_1}{2})\Gamma(\frac{v_2}{2})2^{(v_1+v_2)/2}} \, x^{\frac{v_1}{2}-1} y^{\frac{v_2}{2}-1} e^{-\frac{1}{2}(x+y)}, \\ \qquad 0 < x, \ y < +\infty, \quad v_1 \text{ and } v_2 \text{ positive integers;} \\ 0 \qquad\qquad\qquad\qquad\qquad \text{elsewhere.} \end{cases} \tag{9.29}$$

Let us define a new random variable

$$F = \frac{X/v_1}{Y/v_2}; \tag{9.30}$$

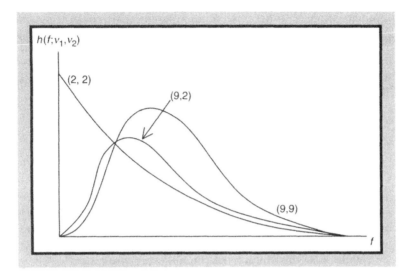

Figure 9.8   A family of F distributions for selected combinations of v_1, v_2 values.

that is, F is the ratio of two independent chi-square random variables, each divided by its degrees of freedom. Then it can be shown (by transforming x to f via (9.30) in the joint probability density function (9.29) and integrating out y) that the *probability density function of F* appears as

$$
h(f; v_1, v_2) = \begin{cases} \dfrac{\Gamma((v_1+v_2)/2)}{\Gamma(v_1/2)\Gamma(v_2/2)} \left(\dfrac{v_1}{v_2}\right)^{v_1/2} \dfrac{f^{\frac{v_1}{2}-1}}{[1+(v_1/v_2)f]^{(v_1+v_2)/2}}, \\ \qquad 0 < f < +\infty, \ v_1 \text{ and } v_2 \text{ positive integers;} \\ 0 \qquad\qquad\qquad\qquad\qquad \text{elsewhere.} \end{cases} \tag{9.31}
$$

Hence the random variable F is said to follow *Snedecor's F distribution with v_1, v_2 degrees of freedom* and will be denoted as F_{v_1,v_2}. Here (9.31) is a two-parameter family of distributions and, as we vary the parameters v_1 and v_2, we generate a whole assortment of particular F distributions (Figure 9.8 illustrates the F distribution for several combinations $((v_1, v_2))$ of degrees of freedom). In addition, the *Snedecor's F cumulative distribution function* is, from (9.31),

$$
H(t; v_1, v_2) = P(F \le t) = \int_0^t h(f; v_1, v_2) \, df, \ f > 0. \tag{9.32}
$$

We have just determined that the ratio of two independent chi-square random variables, each divided by its degrees of freedom, is F_{v_1,v_2}. What is the connection between this result and sampling from a normal population? To answer this question, let $X_1, \ldots, X_{n_1}$ be a random sample consisting of n_1 independent and identically distributed $N(\mu_1, \sigma_1)$ random variables. In addition, let $Y_1, \ldots, Y_{n_2}$ constitute a random sample of n_2 independent and identically distributed $N(\mu_2, \sigma_2)$

random variables. If these two random samples are independent, then, according to Theorem 9.5, the quantities $X = (n_1 - 1)\frac{S_1^2}{\sigma_1^2}$ and $Y = (n_2 - 1)\frac{S_2^2}{\sigma_2^2}$ are independent chi-square random variables with $v_1 = n_1 - 1$ and $v_2 = n_2 - 1$ degrees of freedom, respectively, where $S_1^2(S_2^2)$ is the variance of the first (second) sample. Then from (9.30), the random variable

$$F_{n_1-1, n_2-1} = \frac{\frac{(n_1-1)S_1^2}{\sigma_1^2}/(n_1 - 1)}{\frac{(n_2-1)S_2^2}{\sigma_2^2}/(n_2 - 1)} = \frac{S_1^2/\sigma_1^2}{S_2^2/\sigma_2^2} \tag{9.33}$$

is F distributed with $n_1 - 1$ and $n_2 - 1$ degrees of freedom. As we shall see later on, (9.33) can be used to make inferences about the ratio of variances σ_1^2/σ_2^2 of two independent normal distributions whose means μ_1 and μ_2 are unknown and whose variances σ_1^2 and σ_2^2 are also unknown.

Snedecor's F distribution has the following properties:

(a) The probability density (9.31) of the F random variable (9.30) is not symmetrical in v_1 and v_2. This is why the degrees of freedom in the numerator of F is stated first. In this regard,

(b) If the random variable F ((9.30)) is F_{v_1,v_2}, then the random variable $F' = \frac{1}{F} = \frac{Y/v_2}{X/v_1}$ is F_{v_2,v_1}. Hence $P(F \leq k) = P(F' = \frac{1}{F} \geq \frac{1}{k})$.

(c) The mean and variance of the F distribution are, respectively,

$$E(F) = \frac{v_2}{v_2 - 2}, \quad v_2 > 2; \tag{9.34}$$

$$V(F) = \frac{2v_2^2(v_1 + v_2 - 2)}{v_1(v_2 - 2)^2(v_2 - 4)}, \quad v_2 > 4. \tag{9.35}$$

Thus the F distribution has no mean when $v_2 \leq 2$; it has no variance when $v_2 \leq 4$. (Note that the mean of the random variable F ((9.30)) depends only on the degrees of freedom associated with the chi-square random variable in the denominator.)

(d) The F distribution is skewed positively for any values of v_1 and v_2; and it becomes less skewed as v_1 and v_2 increase without bound. Its coefficient of skewness is

$$\alpha_3 = \sqrt{\frac{8(v_2 - 4)}{(v_1 + v_2 - 2)v_1}} \cdot \frac{(2v_1 + v_2 - 2)}{v_2 - 6}, \quad v_2 > 6. \tag{9.36}$$

(e) The F distribution has a single mode at

$$f = \frac{v_2(v_1 - 2)}{v_1(v_2 + 2)}, \quad v_1 > 2,$$

and as $f \to \infty$, the probability density value $h(f; v_1, v_2) \to 0$.

(f) The F distribution has a peak that is sharper than that of a normal distribution since its coefficient of kurtosis

$$\alpha_4 = 3 + \frac{12\left[(v_2 - 2)^2(v_2 - 4) + v_1(v_1 + v_2 - 2)(5v_2 - 22)\right]}{v_1(v_2 - 6)(v_2 - 8)(v_1 + v_2 - 2)} > 3, \quad v_2 > 8,$$
(9.37)

and $\displaystyle\lim_{v_1, v_2 \to \infty} \alpha_4 = 3.$

(g) The F distribution can be obtained under a suitable transformation of the β distribution ((7.56)). To this end let us set $\alpha = \frac{v_1}{2}$ and $\beta = \frac{v_2}{2}$. Then

$$f(x; v_1, v_2) = \begin{cases} \dfrac{\Gamma\left(\frac{v_1+v_2}{2}\right)}{\Gamma\left(\frac{v_1}{2}\right)\Gamma\left(\frac{v_2}{2}\right)} x^{\frac{v_1}{2}-1}(1-x)^{\frac{v_2}{2}-1}, & 0 < x < 1, \\ & v_1 \text{ and } v_2 \text{ positive integers;} \\ 0 & \text{elsewhere.} \end{cases}$$
(9.38)

Let us define a new random variable F whose values are related to the random variable X by $f = \frac{v_2 x}{v_1(1-x)}$. Then solving for x enables us to transform (9.38) to (9.31). Conversely, if the random variable F is F_{v_1, v_2}, then the random variable $X = \frac{v_1 F / v_2}{[1+(v_1 F / v_2)]}$ has a β distribution with parameters $\alpha = \frac{v_1}{2}$, $\beta = \frac{v_2}{2}$.

(h) If the random variable X is F_{v_1, v_2}, then when $v_2 \to \infty$, the distribution of $v_1 X \to \chi^2_{v_1}$.

(i) If the random variable $T = \frac{Z}{\sqrt{X/v}}$ has a Student's t distribution with v degrees of freedom, then T^2 is $F_{1,v}$. With the t distribution symmetrical about zero, it follows that

$$P(F_{1,v} \le k) = P(T^2 \le k) = P(|T| \le \sqrt{k}),$$

where T has v degrees of freedom. Also, the random variable $\frac{1}{2}\sqrt{v}(F_{v,v}^{1/2} - F_{v,v}^{-1/2})$ has a t distribution with v degrees of freedom.

(j) If a random variable X is $F_{2(n-r+1),2r}$ (r an integer and $0 \le r \le n$) and Y follows a binomial distribution with parameters n and p, then it can be shown that

$$P\left(X > \frac{1-p}{p}\frac{r}{n-r+1}\right) = P(Y \ge r).$$

Various quantiles for the F distribution can be determined for selected cumulative proportions $1 - \alpha$ and for various combinations of the degrees of freedom parameters v_1 and v_2. That is, given (v_1, v_2), the *quantile of order $1 - \alpha$* is the value $f_{1-\alpha,v_1,v_2}$ such that

$$P(F \leq f_{1-\alpha,v_1,v_2}) = \int_0^{f_{1-\alpha,v_1,v_2}} h(f; v_1, v_2) \, df = 1 - \alpha, \ 0 \leq \alpha \leq 1 \qquad (9.39)$$

(see Figure 9.9a). The quantile values $f_{1-\alpha,v_1,v_2}$ usually are tabulated (see Table A.5 of the Appendix) only for selected α's that are less than 0.5; that is, separate tables exist for $1 - \alpha = 0.90, 0.95, 0.975,$ and 0.99, with the columns of each table depicting values of v_1 (the degrees of freedom for the numerator of F) and the rows exhibiting values of v_2 (degrees of freedom for the denominator of F). In this case, since $\alpha < 0.5$, the quantity $f_{1-\alpha,v_1,v_2}$ is termed the *upper α-percentage point* of the F distribution since α is the area in its right-hand tail (see Figure 9.9b).

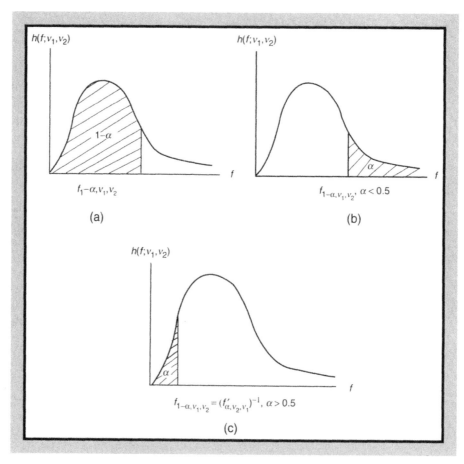

Figure 9.9 (a) $f_{1-\alpha,v_1,v_2}$ is the quantile of order $1 - \alpha$ for the F distribution; (b) For $\alpha < 0.5$, $f_{1-\alpha,v_1,v_2}$ is the upper percentage point of the F distribution; (c) the reciprocal property of the F distribution.

That is, $f_{1-\alpha,v_1,v_2}$ is the value of f such that the probability of a larger value is α. For example, Table A.5 reveals that $f_{0.95,10,6} = 4.06$. Hence 4.06 depicts the upper 5% point of the $F_{10,6}$ distribution.

If $\alpha > 0.5$, then we may find left-sided quantiles or lower percentage points of the F distribution in the following way. We noted earlier (property (b)) that if the random variable F is F_{v_1,v_2}, then the random variable F' is F_{v_2,v_1}. Moreover, for $\alpha > 0.5$,

$$P(F \le f_{1-\alpha,v_1,v_2}) = P\left(F' = \frac{1}{F} > \frac{1}{f_{1-\alpha,v_1,v_2}}\right) = 1 - \alpha \tag{9.40}$$

or

$$P\left(F' = \frac{1}{F} \le \frac{1}{f_{1-\alpha,v_1,v_2}}\right) = \alpha. \tag{9.41}$$

But since F' is F_{v_2,v_1}, it follows that the *quantile of order* α or the *lower α-percentage point* of F' is the value f'_{α,v_2,v_1} such that

$$P(F' \le f'_{\alpha,v_2,v_1}) = \alpha. \tag{9.42}$$

That is, f'_{α,v_2,v_1} is the value of f' such that the probability of a smaller value is α. Since (9.41) and (9.42) are equivalent probability statements, we may conclude that

$$f_{1-\alpha,v_1,v_2} = \frac{1}{f'_{\alpha,v_2,v_1}}, \quad \alpha > 0.5 \tag{9.43}$$

(see Figure 9.9c). This latter equality is known as the *reciprocal property* of the F distribution.

The preceding discussion has indicated that because of the relationship between the random variables F and $F' = \frac{1}{F}$, we need only tabulate upper $(1 - \alpha)$-percentage points of the F distribution. That is, if F is F_{v_1, v_2} and F' is F_{v_2,v_1}, then for $\alpha > 0.5$, the left-hand $100(1 - \alpha)$ percent quantile or lower $(1 - \alpha)$-percentage point of $F, f_{1-\alpha,v_1,v_2}$, is the reciprocal of the upper α-percentage point of F'. In sum, since an F table gives values $f_{1-\alpha,v_1,v_2}$ for $\alpha(< 0.5)$ for the area in the right-hand tail only, an F value with $1 - \alpha(\alpha > 0.5)$ of the area in the left-hand tail is found by the following process:

1. Reverse the order of the degrees of freedom designations v_1, v_2.

2. Find $f_{1-\alpha,v_2,v_1}$, where $\alpha(< 0.5)$ is the area in the right-hand tail.

3. Take the reciprocal of $f_{1-\alpha,v_2,v_1}$ to find $f_{1-\alpha,v_1,v_2}$, $\alpha > 0.5$. So if $f_{1-\alpha,v_1,v_2}$ is the lower α-percentage point of F_{v_1,v_2}, then $f_{1-\alpha,v_1,v_2} = (f'_{\alpha,v_2,v_1})^{-1}$.

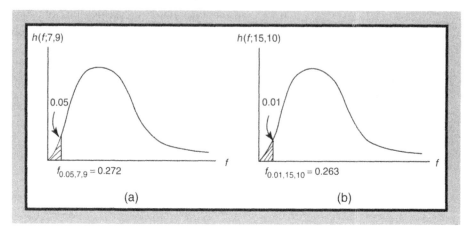

Figure 9.10   (a) $P(F \leq 0.272) = 0.05$; (b) $P(F \leq 0.263) = 0.01$.

Example 9.5.1   For instance, what is the lower 5% point of $F_{7,9}$? From Table A.5, let us find $f'_{0.95,9,7} = 3.68$. Then $f_{0.05,7,9} = \frac{1}{3.68} = 0.272$. Hence

$$P(F \leq f_{0.05,7,9}) = P\left(F \leq \frac{1}{f'_{0.95,9,7}}\right) = P(F \leq 0.272) = 0.05$$

(see Figure 9.10a). Similarly, if F is $F_{15,10}$, what is its lower 1% point? Using Table A.5, we first determine $f'_{0.99,10,15} = 3.80$. Then $f_{0.01,15,10} = \frac{1}{3.80} = 0.263$ or

$$P(F \leq f_{0.01,15,10}) = P\left(F \leq \frac{1}{f'_{0.99,10,15}}\right) = P(F \leq 0.263) = 0.01$$

(see Figure 9.10b).   ∎

9.6 Exercises

Chi-Square Distribution

9-1. Suppose a random variable X is chi-square distributed with $\nu = 8$ degrees of freedom. What is $E(X)$, $V(X)$, α_3, and α_4?

9-2. If X is χ^2_ν, find:

    (a) $\chi^2_{0.90,10}$

    (b) $\chi^2_{0.05,15}$

    (c) $\chi^2_{0.99,5}$

9-3. For $v = 60$ degrees of freedom, use the standard normal approximation to determine the following quantiles of the chi-square distribution:

(a) $\chi^2_{0.90,v}$

(b) $\chi^2_{0.95,v}$

9-4. Suppose X is $N(30, 3)$. For a random sample of size $n = 40$, find $P(S^2 < 10)$, where S^2 is the sample variance.

9-5. Suppose X is $N(\mu, 4)$, where μ is unknown. For $n = 25$, find:

(a) $P(S^2 < 18)$

(b) $P(12 < S^2 < 20)$

9-6. Demonstrate that if the random variable Z is $N(0, 1)$, then $Y = Z^2$ is χ^2_1. (Hint: We need to show that the moment-generating function of Z^2 corresponds to that of a chi-square distribution with 1 degree of freedom.) To this end we may directly find

$$m_{Z^2}(t) = E(e^{tZ^2}) = \frac{1}{\sqrt{2\pi}} \int_{-\infty}^{+\infty} e^{tz^2} e^{-z^2/2} dz = \frac{1}{\sqrt{2\pi}} \int_{-\infty}^{+\infty} e^{-z^2/2(1-2t)^{-1}} dz.$$

Since the integrand can be viewed as the probability density function of a random variable that is $N(0, (1 - 2t)^{-1/2})$, let us rewrite the preceding expression as

$$m_{Z^2}(t) = (1 - 2t)^{-1/2} \int_{-\infty}^{+\infty} \frac{1}{\sqrt{2\pi}(1 - 2t)^{-1/2}} e^{-z^2/2(1-2t)^{-1}} dz$$

$$= (1 - 2)^{-1/2}, \ 0 \le t < \frac{1}{2}.$$

Alternatively, in the moment-generating function for the gamma probability density function derived in Exercise 7.38, set $\alpha = \frac{v}{2} = \frac{1}{2}$ and $\theta = 2$.

9-7. Suppose a random variable X is χ^2_n. Demonstrate that X's moment-generating function is of the form $m_X(t) = (1 - 2t)^{-n/2}, t < \frac{1}{2}$.

9-8. Verify that the moment-generating function derived in Exercise 9-7 is a special case of the gamma moment-generating function.

9-9. Demonstrate that the statement made in Theorem 9.1 is valid.

9-10. Suppose a random variable X_i is $\chi^2_{v_i}$. Determine X_i's moment-generating function. (Hint: Use the moment-generating function for the gamma probability distribution.)

9-11. Demonstrate that Theorem 9.3 is valid by showing that the moment-generating function for $Y = \sum_{i=1}^n X_i$ corresponds to that of a probability distribution that is χ^2_v, where $v = \sum_{i=1}^n v_i$.

9-12. Demonstrate the validity of Theorem 9.2 by using the results offered by Theorems 9.1 and 9.3.

9-13. Suppose $X_1, \ldots, X_n$ depicts a random sample drawn from a population that is $N(\mu, \sigma)$. Demonstrate that $\overline{X}$ and S^2 are independent random variables. (Hint: First show that if $\overline{X}$ is independent of $X_i - \overline{X}$, $i = 1, \ldots, n$, then $\overline{X}$ is independent of $(n - 1)S^2 = \sum_{i=1}^{n} (X_i - \overline{X})^2$.)

9-14. From Theorem 9.5 it was determined that $Y = (n - 1)\frac{S^2}{\sigma^2}$ is χ_{n-1}^2. Determine the moment-generating function for Y.

9-15. Suppose a random variable X is $N(\mu, 1.5)$. If a random sample of size $n = 20$ is drawn from the population distribution, determine the probability that $S^2 > 2.5$ (squared units).

t Distribution

9-16. Suppose a random variable T is t-distributed with $\nu = 10$ degrees of freedom. For $\alpha = 0.05$, find:

(a) $t_{\alpha,\nu}$

(b) $t_{\alpha/2,\nu}$.

What is $V(T)$? Find α_4. What is $P(T \geq 1.729)$ when $\nu = 19$? For $\nu = 14$, find $P(-1.761 \leq T \leq 1.761)$.

9-17. Suppose a random variable X is $N(15, \sigma)$, where σ is unknown. From a sample of size $n = 10$ it has been determined that $\bar{x} = 13$ with $s = 2$. Find $P(\overline{X} \leq 14)$.

9-18. A sample of size $n = 16$ is extracted from a $N(18, \sigma)$ population. If $\bar{x} = 20$ and $s = 2$, find $P(\overline{X} \geq 19)$.

9-19. Find the probability that $\overline{X}$ does not exceed 290 when a sample of size $n = 10$ has been taken from a $N(305, \sigma)$ population and $s = 20$.

9-20. Suppose a random sample with realizations 1,2,3,4,7, and 10 is drawn from a population that is $N(\mu, \sigma)$ with both μ and σ unknown. Determine the (approximate) probability that $\overline{X}$ will be within 3.5 units of the population mean.

F-Distribution

9-21. Suppose a random variable F is F-distributed with $\nu_1 = 6$ and $\nu_2 = 8$ degrees of freedom, respectively. What is $E(F)$ and $V(F)$? Determine α_3, α_4, and the mode of F.

9-22. Suppose the random variable F follows an F-distribution with $\nu_1 = 4$ and $\nu_2 = 8$ degrees of freedom, respectively. For $\alpha = 0.025$, find $f_{1-\alpha,\nu_1,\nu_2}$. If $\alpha = 0.10$, find $f_{1-\alpha,\nu_1,\nu_2}$.

9-23. If F is $F_{4,5}$, what is the lower 1% point? Suppose F is $F_{9,7}$. What is the lower 5% point?

9-24. If F is $F_{10,7}$, find:

   (a) The F-value with 5% of the area to its right

   (b) The area to the right of the value 6.62

10

Point Estimation and Properties of Point Estimators

10.1 Statistics as Point Estimators

We noted in Chapter 8 that a *statistic* is a characteristic of a sample that is used to estimate or determine a parameter θ. In particular, we specified a statistic as some function of the sample random variables $X_1, \ldots, X_n$ (and also of the sample size n), which itself is a random variable that does not depend upon any unknown parameters. Additionally, we denoted a statistic as $T = g(X_1, \ldots, X_n, n)$, where g is a real-valued function that does not depend upon θ or on any other unknown parameter. Then the realization of T, $t = g(x_1, \ldots, x_n, n)$, is determined once the realizations x_i of the sample random variables $X_i, i = 1, \ldots, n$, are known. Thus g is a rule (typically expressed as a formula), which tells us how to get t's from sets of x_i's. If the statistic T is used to determine some unknown population parameter θ (or some function of θ, $\tau(\theta)$), then T is called a *point estimator* of θ, and its realization t is termed a *point estimate* of θ. Hence a point estimator renders a single numerical value as the estimate of θ.

Specifying a statistic T involves a form of *data reduction*; that is, we summarize the information about θ contained in a sample by determining a few essential characteristics of the sample values. Hence, for purposes of making inferences about the parameter θ, we employ only the realization of T rather than the entire set of observed data points. Thus the role of T is that it reduces or condenses the n sample random variables $X_1, \ldots, X_n$ into a single random variable. (For instance, to summarize central tendency, we might reduce the entire set of sample values to simply a single numerical measure—to a statistic or point estimator representing the sample mean.) As will be evidenced, successful point estimation practice involves developing methods of data reduction that: (1) do not ignore any relevant sample information about θ; and, at the same time, (2) tend to discard any irrelevant information concerning the determination of θ.

Since inferences about θ are made on the basis of sample data and thus a limited amount of information is typically processed by T, errors in making inferences are unavoidable because of what is termed *sampling error*—defined as the difference between the sample statistic or estimator T and θ or

$$SE(T,\theta) = T - \theta. \tag{10.1}$$

Since T is a random variable, then so is $SE(T,\theta)$. Sampling error exists because of chance variability in random sampling. Under repeated random sampling (using a fixed n) from a given population, the realizations of T vary from sample to sample and consequently so do the realizations of the sampling errors of T (denoted $SE(t,\theta) = t - \theta$).

In order to make *good* inferences about a population parameter, we must develop methods for assessing the magnitude of the sampling error connected with the same. As we shall now see, this can be done in terms of the probability distribution or so-called sampling distribution of T,[1] which will enable us to discern the pattern of behavior of the sampling errors of T.

To this end, let the random variable X depict a measurable characteristic of some random phenomenon and let $f(x;\theta)$ represent the probability density function associated with the underlying population distribution of X. While θ is an unknown parameter, we may, at times, assume that the form of f is known (e.g., $f(x;\theta)$ may be taken as normally distributed). Let $T = g(X_1,\ldots,X_n,n)$ be a point estimator of θ. (In what follows we shall assume that we have an infinite population or we are sampling with replacement.) If T is to be a *good* point estimator of θ, then we must develop a collection of desirable properties that any such estimator might possess. In order to compile a set of performance criteria that T (or other alternative estimators) might satisfy, our yardstick or standard of *goodness* will be framed in terms of certain key characteristics of the sampling distribution of T (hereafter represented as $h(t;\theta,n)$), such as its mean and variance.

What are the desirable or optimal properties that a *good* point estimator (or sequence of point estimators) may possess? As we will see, these properties typically are classified as small (finite) sample properties or large sample (asymptotic) properties. And all of these properties will be framed in terms of the sampling distribution of a point estimator.

---

[1] Recall that the sampling distribution of T is the relative frequency distribution of the realizations t of T, which emerges under repeated sampling from the same population. Hence, conceptually, a large number of samples of the same size will be taken and t will be determined for each one. Since the resulting t's can be expected to vary randomly about θ, the relative frequency distribution (the sampling distribution of T) will consequently serve as the basis for assessing the *goodness* of a point estimator. In fact, it is the sampling distribution of T that provides information on the long-run performance of T as an estimator of θ.

10.2   Desirable Properties of Estimators as Statistical Properties

Why initiate a discussion of the desirable properties of point estimators in terms of the characteristics of their sampling distributions? First, since estimators are random variables, it is important to study their distributions since it is the statistical properties of the same that determine which estimator is most appropriate in any particular sampling situation. Hence the preference for one estimator over another can be rationalized on the basis of the properties of the sampling distributions of their estimates.

Second, since T is a random variable, we cannot expect that the estimates produced will always equal the parameter θ. If T is a *good* estimator of θ, then obviously its realizations t should be near θ. However, we only know t *ex post*; that is, we do not know the value of t until after the sample is taken. Obviously the *goodness* of an estimator cannot be determined on the basis of examining its realization for a single sample. But if we extract many random samples from the population at hand and construct the relative frequency distribution of the resulting realizations of T (the sampling distribution of T), then we can readily determine if this distribution is *well behaved*. That is, is the sampling distribution of T concentrated about θ? In addition, how closely are the realizations of T concentrated about θ? In this regard, what is important is the *a priori* behavior of an estimator; that is, we are interested in how an estimator is *expected* to behave under repeated sampling from the same population. Looked at from another perspective, an assessment of the *goodness* of a point estimator must be based on the probability distribution of its sampling errors and not on the magnitude of any one realization of that error.

Moreover, if T' is an alternative estimator of θ with realization t', then for one particular sample t' may be closer to θ than t, whereas for another sample it may be t and not t' that is closer to θ. Hence the relative merit of using T versus T' as an estimator of θ hinges upon the statistical properties of the sampling distributions of these estimators. For instance, if we find that the sampling distribution T is more highly concentrated about θ than that of T', then we can typically expect t to be closer to θ than t'. In this circumstance we may consequently be led to conclude that not only is T a *good* estimator of θ, but it is also a *better* estimator of θ than T'.

In sum, the various realizations t of an estimator T are incorporated into the sampling distribution of T, a probability density function $h(t; \theta, n)$ that embodies all relevant statistical information pertaining to T. Hence the process of evaluating the *goodness* of an estimator hinges upon conceptualizing the entire spectrum of realizations that would be obtained if the estimation rule $T = g(X_1, \ldots, X_n, n)$ were applied again and again to many samples taken from a given population. What is relevant is that the process of repeated sampling is considered conceptually and that it is the long-run behavior of an estimator under repeated sampling that matters.

Certain measurable characteristics of h have been deemed *desirable* or *meritorious* as far as the determination of the parameter θ is concerned.

These measurable characteristics thus serve as a guide in selecting the *best* estimator of θ. Although some estimators of θ may attain these desirable properties (or at least come close to attaining them), others may not. It is the satisfaction of certain optimal properties of an estimator that ultimately will provide us with a *best guess* estimate of θ. And it is to the specification of these properties that our attention and efforts now turn. As will be explained in the next few sections, a point estimator of θ may be unbiased, consistent, efficient, sufficient, have minimum variance, and so on.

10.3   Small Sample Properties of Point Estimators

10.3.1   Unbiased, Minimum Variance, and Minimum Mean Squared Error (MSE) Estimators

A *small sample property* of a point estimator is essentially a characteristic of the sampling distribution of the estimator that holds for a *fixed* sample size n. This is in contrast to a *large sample* or *asymptotic property* of a point estimator, which results as $n \to \infty$.

In what follows we shall assume that $X_1, \ldots, X_n$ depicts a set of independent and identically distributed sample random variables taken from the population distribution $f(x; \theta)$ of the random variable X. In addition, let $T = g(X_1, \ldots, X_n, n)$ be a point estimator of θ. Since T is a random variable with probability density function or sampling distribution $h(t; \theta, n)$, it should be intuitively clear that, if T is to serve as a *good* estimator of θ, then the realizations t of T should be *close* to θ; that is, the sampling distribution of T should be concentrated about θ. We shall now consider ways of measuring the *closeness* of an estimator to θ.

Given the nature of the process of random sampling, we cannot expect to find an estimator that is always on target; that is, since T is a random variable, its realizations vary from sample to sample so that T will not always produce estimates that are exactly equal to θ. As we noted in Section 10.1, sampling error $\left(SE(T, \theta) = T - \theta \right)$ is a fact of life and, for any individual sample, the sampling error typically will not be zero. However, it is desirable that an estimator should yield the correct result at least on the average. In this regard, the average value of the sampling error taken over a large number of samples should be zero or $E\left(SE(T, \theta) \right) = E(T) - \theta = 0$.

Let us term the expected value of the sampling error of T its *bias* (denoted $B(T, \theta)$). Then if the bias or expected sampling error $B(T, \theta) = E(T) - \theta$ is (identically) zero for all n and θ or

$$E(T) = \theta, \tag{10.2}$$

then T is said to be an *unbiased estimator* of θ; that is, an unbiased estimator is one whose bias is zero. So if T is an unbiased estimator of θ, its sampling distribution is centered exactly on θ (see Figure 10.1a); that is, as the number of samples increases, the estimator is on target. If, however, the sampling distribution

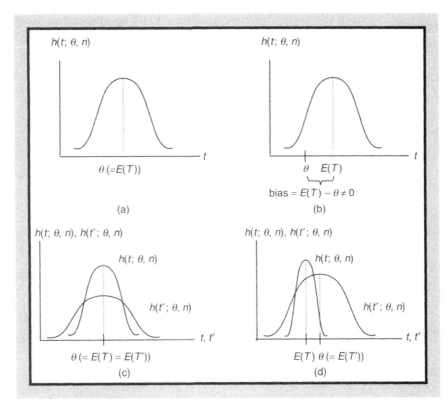

Figure 10.1 (a) T is an unbiased estimator of θ; (b) T is a biased estimator of θ; (c) Both T and T' are an unbiased estimator of θ but $V(T) < V(T')$; (d) A biased estimator T may be preferable to an unbiased estimator T' of θ.

of T is not centered on θ (see Figure 10.1b), then T is termed a *biased estimator* of θ since the bias or $B(T, \theta) = E(T) - \theta$ is different from zero. Here (10.2) implies that obviously $E(T) \neq \theta$. Note that if an estimator T is unbiased, then the average of its realizations taken over many samples is equal to the parameter θ to be estimated. However, this does not mean that the estimator actually has a high probability of being close to θ for any *given* sample. Unbiasedness assures us that, over the long run and for a large number of estimations, errors will not tend to cumulate; that is, on average, T estimates θ without systematic error.

For instance, if the independent and identically distributed sample random variables $X_i, i = 1, \ldots, n$, each have the same mean μ and variance σ^2 as the population random variable X, then:

1. As revealed by Theorem 8.2, if $T = \overline{X}$, then $E(\overline{X}) = \mu$ for any population distribution. Hence the sample mean is an unbiased estimator of the population mean; that is, under repeated sampling from the same population, $\overline{X}$ is, on the average, equal to μ for any μ.

2. Since the sample random variables X_i all have the same distribution as the population random variable X, it follows that for $T = X_i$, $E(X_i) = \mu$ for all $i = 1, \ldots, n$. Hence any X_i serves as an unbiased estimator of μ.

3. If a statistic $T = \sum_{i=1}^{n} a_i X_i$ is a linear combination of the sample random variables X_i, $i = 1, \ldots, n$, then by virtue of Theorem 8.1, $E(T) = \mu \sum_{i=1}^{n} a_i$. If $\sum_{i=1}^{n} a_i = 1$, then T is an unbiased estimator of μ.

4. We noted earlier (Theorem 8.7) that if $T = S^2 = \sum_{i=1}^{n} \frac{(X_i - \bar{X})^2}{n-1}$, then $E(S^2) = \sigma^2$ for any population distribution. Hence the sample variance is an unbiased estimator of the population variance; that is, under repeated sampling from the same population, S^2 is, on the average, equal to σ^2 for any σ^2. (It is important to remember that we are sampling from an infinite population. S^2 is not an unbiased estimator of σ^2 for a finite population.)

5. Since the sample random variables X_i all have the same distribution as the population random variable X, it follows that for $T = X_i$, $V(X_i) = \sigma^2$ for all $i = 1, \ldots, n$. Hence the variance of any X_i serves as an unbiased estimator of σ^2.

6. The quantity $T = S_1^2 = \sum_{i=1}^{n} \frac{(X_i - \bar{X})^2}{n}$ is a biased estimator of the population variance σ^2. To see this let us first write $\sum_{i=1}^{n}(X_i - \bar{X})^2 = \sum_{i=1}^{n} X_i^2 - n\bar{X}^2$. Since

$$V(X_i) = \sigma^2 = E(X_i^2) - \mu^2;$$

$$V(\bar{X}) = \frac{\sigma^2}{n} = E(\bar{X}^2) - \mu^2,$$

it follows that

$$E\left(\sum_{i=1}^{n}(X_i - \bar{X})^2\right) = E\left(\sum_{i=1}^{n} X_i^2 - n\bar{X}^2\right)$$

$$= E\left(\sum_{i=1}^{n} X_i^2\right) - nE(\bar{X}^2)$$

$$= \sum_{i=1}^{n} E(X_i^2) - nE(\bar{X}^2)$$

$$= \sum_{i=1}^{n}(\sigma^2 + \mu^2) - n\left(\frac{\sigma^2}{n} + \mu^2\right) = (n-1)\sigma^2.$$

Hence

$$E(S_1^2) = \frac{1}{n} E\left(\sum_{i=1}^{n}(X_i - \bar{X})^2\right) = \frac{(n-1)\sigma^2}{n} = \sigma^2 - \frac{\sigma^2}{n}.$$

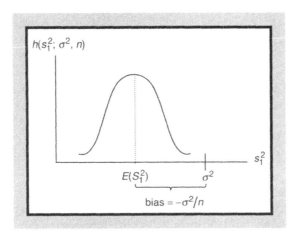

Figure 10.2 S_1^2 is a biased estimator of σ^2.

Clearly S_1^2 is a biased estimator of σ^2, where the extent of the bias is $-\frac{\sigma^2}{n}$ or the negative of the variance of $\overline{X}$ (see Figure 10.2). Thus S_1^2 generally understates σ^2 for finite samples (i.e., S_1^2 is said to be biased downwards).

We may easily correct S_1^2 for the amount of bias by multiplying S_1^2 by $\frac{n}{n-1}$. Then $\left(\frac{n}{n-1}\right) S_1^2 = S^2$ and, as indicated earlier, $E(S^2) = \sigma^2$. This is why we termed the quantity S^2 (and not S_1^2) as the *sample variance*—we divide $\sum_{i=1}^{n}(X_i - \overline{X})^2$ by $n - 1$ rather than by n so that we obtain an unbiased estimator of σ^2. In fact, we can often convert a biased estimator into one that is unbiased by multiplying the former by a suitable function of the sample size n.

It is important to note that we cannot conclude that for every particular sample the realization of S^2 provides us with a closer estimate of σ^2 than the realization of S_1^2. Remember that unbiasedness is an *average property* of an estimator. Hence it is only under repeated sampling from the same population that S^2 is on target on the average. However, for any given sample, the realization of S_1^2 may be closer to σ^2 than the realization of S^2.

If the mean μ of the population random variable X is known, then $S_0^2 = \frac{1}{n}\sum_{i=1}^{n}(X_i - \mu)^2$ is an unbiased estimator of σ^2.

7. Although the sample variance S^2 serves as an unbiased estimator of the population variance σ^2, it does not follow that the sample standard deviation S is generally an unbiased estimator of the population standard deviation σ. To verify this statement we need only remember, by virtue of Theorem 9.5, that the quantity $Y = \frac{(n-1)S^2}{\sigma^2}$ is distributed as χ_{n-1}^2. (Note that the probability density function of Y is provided by (9.1) with $v = n - 1$.) Since

$$S = \left(\frac{\sigma}{(n-1)^{1/2}}\right) Y^{1/2},$$

it follows that

$$E(S) = \frac{\sigma}{(n-1)^{1/2}} E(Y^{1/2})$$

and thus, under a suitable transformation, it can be shown that

$$E(S) = \left[\frac{2^{1/2}\Gamma(n/2)}{\Gamma((n-1)/2)\,(n-1)^{1/2}} \right] \sigma, n > 1.$$

Clearly $E(S) \neq \sigma$ as was stated earlier.

In general, there is no known unbiased estimator of σ except in special cases; that is, the correction factor to make S an unbiased estimator of σ depends upon the form of the underlying probability density function. For example, if X is $N(\mu, \sigma)$, then an approximate unbiased estimator for σ is, for large n,

$$T_\sigma = \left(1 + \frac{1}{4(n-1)} \right) S.$$

8. For X a binomial random variable, Theorem 8.6 informs us that if $T = \hat{P} = \frac{X}{n}$, then $E(\hat{P}) = p$. Hence the sample proportion serves as an unbiased estimator of the population proportion p. That is, $\hat{P}$ is, on the average, equal to p for any p.

9. If T is an unbiased estimator of a parameter θ, it does not follow that every function of $T, j(T)$, is an unbiased estimator of $j(\theta)$.

It may be the case that there are many unbiased estimators for a parameter θ. What is needed then is at least one other criterion for judging the *goodness* of an estimator. To this end, it is reasonable to consider the *variance of an estimator T*, defined as

$$V(T) = E\big[(T - E(T))^2\big] = E(T^2) - E(T)^2. \qquad (10.3)$$

Here $V(T)$ provides us with a measure of the average variation about the mean of T that can be expected to arise under repeated sampling from a given population. Hence it serves as a measure of the dispersion of an estimator about its mean. Moreover, we may define the *standard error of an estimator T* as the positive square root of its variance. It will be denoted as

$$S_T = \sqrt{V(T)}. \qquad (10.4)$$

From (10.3), we can state that T is a *minimum variance estimator* of θ if $V(T) \leq V(T')$, where T' is any alternative estimator of θ.

Given a choice between two unbiased estimators T and T' of θ (here $E(T) = E(T') = \theta$), it seems preferable to choose the one with the smallest variance (see Figure 10.1c) since using T yields a higher probability of obtaining an estimate that is closer to θ. However, if T is a biased estimator of θ and the bias is small, then T may be preferable to an unbiased estimator T' whose variance is large (see Figure 10.1d). The problem is that if the variance of an estimator is large, then we may obtain a sample for which the realization of the estimator is far from the true parameter value. Clearly a tradeoff exists between bias and variance when it comes to determining which estimator is *best* for estimating θ. So if we desire an estimator that typically yields estimates that are *close* to the true θ, then both the bias and variance of an estimator must be taken into account simultaneously when evaluating the merits of an estimator.

A general criterion for assessing the *goodness* of an estimator T of θ in terms of the relative magnitude of its bias and variance is the *mean squared error* (MSE) of T or

$$MSE(T, \theta) = E\left[(T - \theta)^2\right], \tag{10.5}$$

which, in the context of repeated sampling from the same population, is defined as the expected value of the square of its sampling error; that is, it measures the average squared difference between T and the parameter θ or, stated alternatively, it indicates how the realizations of an estimator are, on the average, distributed about θ. Looked at in an alternative fashion, we may think of the quantity $(T - \theta)^2$ as measuring the *loss* from using T as an estimator of θ, and $E[(T - \theta)^2]$ may be viewed as the *average loss* or *risk* of using T as an estimator of θ. If we perform the indicated squaring in (10.5), then we obtain

$$MSE(T, \theta) = E\left(T^2 - 2\theta T + \theta^2\right) = E\left(T^2\right) - 2\theta E(T) + \theta^2$$
$$= V(T) + E(T)^2 - 2\theta E(T) + \theta^2 \tag{10.5.1}$$
$$= V(T) + \left(E(T) - \theta\right)^2 = V(T) + B(T, \theta)^2.$$

Hence the mean squared error of T is expressible as the sum of the variance of T (which measures the *precision* of T) and the square of its bias (indicating the degree of *accuracy* of T). An estimator with *good MSE* properties (in terms of bringing us *close* to θ) is one that has a small combined variance and (squared) bias.

As (10.5.1) reveals, $MSE(T, \theta) \geq V(T)$ since $B(T, \theta) \geq 0$. Here the mean squared error of T measures the dispersion of T about the true θ whereas the variance of T considers its dispersion about $E(T)$. If the mean of T coincides with θ so that T is unbiased, then $E(T) = \theta$ and thus $MSE(T, \theta) = V(T)$.

Example 10.3.1.1   Let $\overline{X}$ and $\tilde{X} = \overline{X} + \frac{1}{n}$ each serve as estimators of the population mean μ. Hence $E(\overline{X}) = \mu$ (as verified earlier) and $E(\tilde{X}) = \mu + \frac{1}{n}$. Clearly $\tilde{X}$ is a biased estimator of μ, where the bias is $\frac{1}{n}$. Moreover, $V(\overline{X}) = V(\tilde{X})$ and $MSE(\overline{X}, \mu) = V(\overline{X})$ and, from (10.5.1), $MSE(\tilde{X}, \mu) = V(\overline{X}) + \left(\frac{1}{n}\right)^2$. Note that the unbiased estimator of μ or $\overline{X}$ has the smallest mean squared error and thus is preferable to $\tilde{X}$. If a variance only criterion were adopted, the two estimators would be deemed equally good. ∎

Next, if we sample from a normal population, then $S^2 = \sum_{i=1}^{n} \frac{(X_i - \bar{X})^2}{n-1}$ is an unbiased estimator for σ^2 and, from (9.16), $V(S^2) = \frac{2\sigma^4}{n-1} = MSE(S^2, \sigma^2)$. As an alternative estimator of σ^2, let us use $S_1^2 = \sum_{i=1}^{n} \frac{(X_i - \bar{X})^2}{n} = \left(\frac{n-1}{n}\right) S^2$. Although S_1^2 is biased downward $\left(E(S_1^2) = \sigma^2 - \frac{\sigma^2}{n}\right)$, it does have a smaller variance than S^2 since

$$V(S_1^2) = \left(\frac{n-1}{n}\right)^2 V(S^2) = \left(\frac{n-1}{n}\right)^2 \frac{2\sigma^4}{n-1} < V(S^2).$$

Hence the biased estimator S_1^2 is a bit more precise than the unbiased estimator S^2. Looking to their mean squared errors, we see that

$$MSE(S^2, \sigma^2) = V(S^2) = \frac{2\sigma^4}{n-1}$$

and, from (10.5.1),

$$MSE(S_1^2, \sigma^2) = V(S_1^2) + B(S_1^2, \sigma^2)^2$$
$$= \left(\frac{n-1}{n}\right)^2 \frac{2\sigma^4}{n-1} + \frac{\sigma^4}{n^2} = (2n-1)\frac{\sigma^4}{n^2}.$$

And since

$$MSE(S^2, \sigma^2) - MSE(S_1^2, \sigma^2) = \left(\frac{2}{n-1} - \frac{(2n-1)}{n^2}\right)\sigma^4 > 0,$$

it follows that the average spread of the realizations of S^2 about σ^2 is a bit larger than the average spread of the realizations of S_1^2 about σ^2. What this discussion indicates is that if T is a biased estimator of θ, it is generally preferable to judge the *goodness* of T on the basis of its mean squared error instead of simply examining its variance alone. By trading off variance for bias, the mean squared error improves.

Let us now summarize some of the preceding considerations concerning criteria for selecting a *best* estimator (in terms of *closeness* to a given parameter) from among several competing estimators.

1. If the estimators are all unbiased, then the *best* estimator is the one with minimum mean squared error and thus with minimum variance. Hence in this case simply comparing the variances of alternative estimators will do.

2. Unless subject to additional qualification, the quality of minimum variance taken by itself is a rather weak property of an estimator. The attribute of minimum variance is most desirable when it is coupled with unbiasedness (and vice versa).

3. If the estimators have mixed attributes in terms of bias and variance, then we must use a selection method that incorporates a tradeoff between variance and (squared) bias, and this involves employing a mean squared error criterion. Specifically, T is a *minimum mean squared error estimator* of θ if $MSE(T, \theta) \leq MSE(T', \theta)$, where T' is any alternative estimator of θ.

4. A biased estimator may be preferable to an unbiased estimator if it has a smaller mean squared error (since the unbiased estimator may display considerable variability relative to the biased one).

Point (3) enables us to conclude that the *best* estimator is one that minimizes the mean squared error over the set of alternative estimators of θ. However, as revealed by (10.5), the mean squared error depends upon the value of θ. This observation is crucial since the criterion of minimal mean squared error generally does not render an estimator that is superior to all other estimators for *all possible* θ's; that is, one estimator may be better than another possible estimator for one value of θ but not for another θ value. Hence there may not be one *best* estimator of θ. What is desirable is an estimator with the uniformly lowest mean squared error for *all* θ's. Unfortunately, such an estimator typically does not exist since the collection of all possible estimators of θ is usually very large. Hence the minimum mean squared error criterion is seldom operational; mean squared error properties are often unknown and difficult to verify. To circumvent this problem, let us restrict our attention to the class of unbiased estimators. By doing so we radically improve the odds of finding an estimator that satisfies the minimum mean squared error (and thus minimum variance) criterion for all θ's within this restricted class of estimators. In what follows then, we shall focus our attention on the class of unbiased estimators and, within it, we shall search for one with minimum variance. (Remember that if T is an unbiased estimator of θ, then $E(T) = \theta$ and $MSE(T, \theta) = V(T)$.)

10.3.2 Efficient Estimators

We may characterize the notion of an unbiased estimator having the smallest variance among the class of all unbiased estimators of θ in the following fashion. Specifically, T is termed an *efficient* (alternatively *minimum variance unbiased* or *best unbiased*) *estimator* of θ if T is unbiased and $V(T) \leq V(T')$ for all possible values of θ, where T' is any other unbiased estimator of θ (see Figure 10.3). Hence T is more efficient than T' since the sampling distribution of T is more closely concentrated about θ than the sampling distribution of T'.

How do we determine whether or not a given estimator is efficient? Although it may be a fairly straightforward task to check an estimator for unbiasedness, such is not the case when it comes to checking for minimum variance. This is

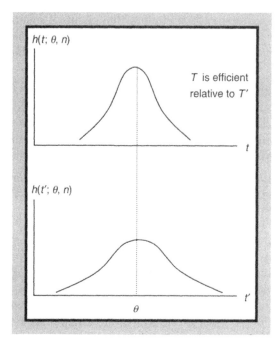

Figure 10.3 T relative to T' is an efficient or best unbiased estimator of θ.

because we need to examine the variance of all possible unbiased estimators of θ. And since there may be many such estimators, it is quite heroic to claim that a given estimator is efficient. Hence, as a practical matter, we shall adopt the following compromise. Instead of looking to state that a given unbiased estimator is *best* in that it has a variance that is less than or equal to the variance of any of any other unbiased estimator, we will simply be satisfied with *pairwise comparisons*; that is, with being able to claim that one unbiased estimator is preferred (on minimum variance grounds) to another. So when we compare two unbiased estimators, we will concern ourselves with their *relative efficiency* (RE). Hence the unbiased estimator with the smaller variance is more efficient relative to the other unbiased estimator.

In view of this strategy, if T and T' are both unbiased estimators of θ and $V(T') \geq V(T)$ for all θ and strict inequality holds for some value of θ, then T is judged *relatively more efficient* than T' in estimating θ. To operationalize this relative efficiency concept, let us work with the ratio of the variances of T and T'. In this regard, let both T and T' be unbiased estimators of θ with variances $V(T)$ and $V(T')$, respectively. Then the *efficiency of T relative to T'* (or the *efficiency of T with respect to T'*) is given by the ratio

$$RE = \frac{V(T)}{V(T')} \times 100 \qquad (10.6)$$

(here RE is expressed as a percentage). We shall adopt the convention that the smaller variance goes into the numerator so that $0 \leq RE \leq 1$. Once RE is determined, this comparison scheme allows us to conclude that $V(T) = V(T') \times RE$ or the variance of T is $RE\%$ of the variance of T'.

Example 10.3.2.1   It can be shown that if the probability density function is normal with mean μ and variance σ^2, then the sampling distribution of the sample median is asymptotically normal with mean μ and variance $\frac{\pi\sigma^2}{2n}$. Moreover, we noted earlier that the variance of the sample mean is only $\frac{\sigma^2}{n}$. Hence, for large samples taken from a normal population, the sample mean is less variable (and thus more reliable) than the sample median as a measure of central location. Let us now frame these observations in terms of the preceding relative efficiency concept.

For a random sample of size n taken from a normal population, both the sample mean $\overline{X}$ and the sample median $\gamma_{0.5}$ are unbiased estimators of the population mean. Thus the efficiency of the sample mean relative to the sample median is

$$RE = \frac{V(\overline{X})}{V(\gamma_{0.5})} \times 100 = \frac{\sigma^2/n}{\pi\sigma^2/2n} \times 100 = \frac{200}{\pi} = 63.66.$$

Thus the amount of variability associated with the sample mean is approximately 64% of the variability associated with the sample median. And since we have just verified that the sampling distribution of the mean is more closely concentrated about the population mean than the sampling distribution of the median, it follows that if only a single sample were to be taken, the more likely it is that the sample mean would be closer to μ than the sample median. So on the basis of this relative efficiency criterion, we may conclude that the sample mean is preferable to the sample median as an estimator of the population mean.   ∎

Let us now consider a generalization of the definition of the relative efficiency of one estimator with respect to another (10.6). For *any* two estimators T and T' (be they biased or unbiased), the efficiency of T relative to T' is

$$RE' = \frac{MSE(T,\theta)}{MSE(T',\theta)} \times 100. \tag{10.7}$$

Here too RE' is expressed as a percentage with $0 \leq RE' \leq 1$ if the smaller mean squared error appears in the numerator. Once RE' is calculated, we see that $MSE(T,\theta) = MSE(T',\theta) \times RE'$ or the mean squared error of T is $RE'\%$ of the mean squared error of T'. Note that if the estimators T and T' are unbiased, then (10.6) serves as a special case of (10.7).

10.3.3   Most Efficient Estimators

We now turn to a more comprehensive approach to finding an efficient or minimum variance unbiased estimator. Our search for any such estimator is facilitated by the notion of finding a lower limit $CR(\theta,n)$ on the variance of any unbiased estimator of a parameter θ; that is, if T is any unbiased estimator of θ, then $V(T) \geq CR(\theta,n)$. This lower bound enables us to determine if a given unbiased estimator has the (theoretically) smallest possible variance in the sense that, if $V(T) = CR(\theta,n)$, then T represents the *most efficient* estimator of θ. In this regard, if we can find an unbiased estimator T for which $V(T) = CR(\theta,n)$, then the *most efficient* estimator is actually a *minimum variance bound estimator*.

To set the stage for the determination of the minimum variance bound for an estimator of θ, let us turn to a concept developed earlier in Section 8.2, namely the joint probability density (mass) function of a random sample. Let us briefly review this notion, treating the continuous case first. We previously noted that if $f(x;\theta)$ depicts the probability density function of the population distribution for a measurable characteristic X, then, under random sampling, the generated outcomes $X_i, i = 1,\ldots,n$, represent a collection of n independent sample random variables. Since the X_i have the same probability density function as the population distribution, their joint probability density function at $x_1,\ldots,x_n$ can be expressed (under independence) as

$$l(x_1,\ldots,x_n;\theta,n) = \prod_{i=1}^{n} f(x_i;\theta), \tag{10.8}$$

where $f(x_i;\theta) = f(x;\theta)$ is the marginal probability density function for the random variable $X_i, i = 1,\ldots,n$.

Next, if the sample random variables $X_1,\ldots,X_n$ are discrete and the value of X_i is determined on the ith trial of a random experiment (under, say, sampling with replacement), then again the X_i's, $i = 1,\ldots,n$, depict a set of n independent sample random variables. If $f(x;\theta)$ represents the probability mass function of the population random variable X, then under independence each X_i has the same probability mass function as the population distribution and thus the probability of the event $\{X_1 = x_1,\ldots,X_n = x_n\}$ is given by the joint probability mass function

$$P(X_1 = x_1,\ldots,X_n = x_n) = l(x_1,\ldots,x_n;\theta,n) = \prod_{i=1}^{n} f(x_i;\theta), \tag{10.9}$$

where $f(x_i;\theta) = f(x;\theta)$ is the marginal probability mass function for $X_i, i = 1,\ldots,n$.

For the continuous case, (10.8) simply yields the value of the joint probability density function at the set of sample realizations $x_1,\ldots,x_n$. In the discrete case, (10.9) gives the probability of the event $\{X_i = x_i, i = 1,\ldots,n\}$. (Actually, in the continuous case, the joint probability density value is proportional to the probability of observing the sample outcome in a suitably restricted neighborhood of the realizations obtained.)

Let us now change the interpretation of equation (10.8) by rewriting it as

$$\mathcal{L}(\theta;x_1,\ldots,x_n,n) = \prod_{i=1}^{n} f(x_i;\theta). \tag{10.8.1}$$

Here (10.8.1) is termed the *likelihood function* of the sample. (In what follows we shall, for the most part, frame our discussion in terms of the likelihood function for a set of continuous sample random variables $X_i, i = 1, \ldots, n$. A parallel treatment can easily be offered for the discrete case.)

Let us briefly discuss the arguments appearing in (10.8) and (10.8.1). We previously formulated the sampling distribution of a statistic as the joint probability density function of the sample random variables ((10.8)) given a *fixed* value of θ in the population probability density function $f(x; \theta)$. However, if we treat the X_i's as *fixed* and the parameter θ as a variable, (10.8.1) provides us with the joint probability density of the given sample realizations as a function of θ. So when we express the joint probability density function of a given sample as a function of θ, we call the resulting equation a likelihood function. In sum, (10.8) depicts a sampling distribution since the X_i's are variable and θ is held constant, and (10.8.1) is termed a *likelihood function* wherein the X_i's have constant realizations $x_i, i = 1, \ldots, n$, and θ is variable; that is (10.8.1) yields the *likelihood* that the random variables $X_1, \ldots, X_n$ assume the realizations $x_1, \ldots, x_n$. So for a set of continuous independent and identically distributed random variables $X_1, \ldots, X_n$, both the theoretical sampling distribution and the likelihood function are represented by the same joint probability density function—only the interpretation differs.

For convenience, we usually work with the logarithm of $\mathcal{L}$ since forming log $\mathcal{L}$ transforms the multiplicative function (10.8.1) into an additive one; that is, $\log \mathcal{L} = \sum_{i=1}^{n} \log f(x_i; \theta)$. This last expression will be termed the *log-likelihood function* of the sample.

We now return to the issue of finding a minimum variance bound (hereafter called the *Cramér-Rao* (1945, 1946) *lower bound* and denoted $CR(\theta, n)$) for an unbiased estimator of a parameter θ. Specifically, if T is an unbiased estimator of θ, then under some very general (regularity) conditions,[2] the variance of T must satisfy the *Cramér-Rao inequality*

$$V(T) \geq \frac{1}{-E\left(\dfrac{\partial^2 \log \mathcal{L}}{\partial \theta^2}\right)} = CR(\theta, n). \tag{10.10}$$

Hence the variance of T is never smaller than $CR(\theta, n)$, which is constant for a fixed n. Note that the form of the population distribution $f(x; \theta)$ from which the random sample is drawn must be known (even though θ itself is unknown).

If T is an unbiased estimator of θ and strict equality holds in (10.10), then T is the *most efficient* or *minimum variance bound estimator* of θ. Hence the realizations of the *most efficient* estimator are those that are most concentrated about θ and thus have the highest probability of being close to θ. Moreover, it is important to note that equality holds in (10.10) if and only if there exists a function

---

[2] Specifically: (a) the sample space S is independent of θ; (b) the derivatives $\partial \mathcal{L}/\partial \theta$, $\partial^2 \mathcal{L}/\partial \theta^2$ exist for all admissible θ; and (c) $E(\partial \log \mathcal{L}/\partial \theta) = 0$.

$\alpha(\theta, n)$ such that

$$\frac{\partial \log \mathcal{L}}{\partial \theta} = \alpha(\theta, n) \left(t\left(x_1, \ldots, x_n, n\right) - \theta \right). \tag{10.11}$$

Let us now consider an assortment of general observations pertaining to the salient features of the Cramér-Rao lower bound:

1. The Cramér-Rao inequality provides us with a lower limit $CR(\theta, n)$ on the variance of an unbiased estimator T of θ. However, (10.10) is only a sufficient and not a necessary condition for an unbiased estimator of θ to be an efficient estimator of that parameter. Hence this inequality does not necessarily imply that the variance of an efficient estimator has to equal the Cramér-Rao lower bound; it may be possible to find an unbiased estimator of θ whose variance is smaller than those of alternative unbiased estimators of θ, but whose variance exceeds $CR(\theta, n)$. Hence there can exist an efficient estimator whose variance does not coincide with the minimum variance bound. In fact, (10.10) may be unattainable. But if the variance of an unbiased estimator of θ attains the Cramér-Rao lower bound, that estimator is termed the *most efficient* or minimum variance bound estimator of θ.

2. The Cramér-Rao lower bound is unique and always exists, but, as noted earlier, an estimator that attains it does not always exist. Moreover, this bound depends only on $f(x; \theta)$ and n and is independent of the estimator under consideration. If for some unbiased estimator T (10.10) holds as a strict equality for *all* values of θ, then T is termed a *uniformly most efficient estimator* of θ.

3. The quantity $-E\left(\frac{\partial^2 \log \mathcal{L}}{\partial \theta^2}\right)$ is termed the *information number* for the Cramér-Rao lower bound. It reflects the amount of information on θ in the realizations $x_1, \ldots, x_n$ and varies inversely with the same; that is, as we obtain more information about θ, the information number increases and thus the lower bound on the variance of any unbiased estimator of θ decreases.

4. The Cramér-Rao lower bound provides us with an operational procedure for identifying specific estimators that can be labeled *most efficient*. That is, if a known unbiased estimator has a variance that equals this bound, then obviously our search for the *most efficient* estimator of θ can be terminated.

5. There are usually many possible unbiased estimators for a given population parameter θ. If a particular unbiased estimator of θ does not attain the Cramér-Rao lower bound, we usually do not know if the estimator is actually the most efficient one at hand. It may be that no unbiased estimator of θ attains the Cramér-Rao lower bound.

6. Let us specify the general *exponential family* of probability density functions as

$$f(x; \theta) = a(\theta)b(x)e^{c(\theta)d(x)}, \quad -\infty < x < +\infty, \tag{10.12}$$

for admissible θ. (For instance, if $\theta = \lambda$ and we set $a(\lambda) = e^{-\lambda}$, $b(x) = \frac{1}{x!}$, $c(\lambda) = \log \lambda$, and $d(x) = x$, then clearly the Poisson probability density function belongs to this exponential family of distributions.) Then the likelihood function for a random sample of size n taken from an exponential population described by (10.12) is

$$\mathcal{L} = \prod_{i=1}^{n} f(x_i;\theta) = a(\theta)^n \left(\prod_{i=1}^{n} b(x_i) \right) e^{c(\theta) \sum_{i=1}^{n} d(x_i)}. \qquad (10.13)$$

Interestingly enough, it can be shown that if T is an unbiased estimator of a parameter θ and the variance of T coincides with the Cramér-Rao lower bound, then the underlying probability density function from which the sample random variables were drawn is a member of the exponential family of distributions. Conversely, if the population probability density function $f(x;\theta)$ is a member of the exponential family of distributions, then there exists an unbiased estimator T of θ such that $V(T) = CR(\theta, n)$.

7. If T is the minimum variance bound estimator of θ and T' is any other unbiased estimator of θ, then the efficiency of T' relative to T is $RE = \left(\frac{V(T)}{V(T')} \right) \times 100, 0 \leq RE \leq 1$. Obviously for the *most efficient* or minimum variance bound estimator of θ we have $RE = 1$. Otherwise $RE < 1$. For instance, given a sample of size n taken from a $N(\mu, \sigma)$ population, we determined earlier that $S^2 = \sum_{i=1}^{n} \frac{(X_i - \bar{X})^2}{n-1}$ is an unbiased estimator of σ^2. Moreover, we also found that S^2 is not a minimum variance bound estimator of σ^2. However, $S_0^2 = \sum_{i=1}^{n} \frac{(X_i - \mu)^2}{n}$ (with μ known) is a minimum variance bound estimator of σ^2. Then the efficiency of S^2 relative to S_0^2 is

$$\frac{V(S_0^2)}{V(S^2)} = \frac{2\sigma^4/n}{2\sigma^4/(n-1)} = \frac{n-1}{n} < 1.$$

If for a given parameter θ a minimum variance bound or *most efficient* estimator does not exist, then it is highly likely that there may be an estimator T' whose variance is smaller than the variance of any other unbiased estimator for all admissible values of θ. We have previously referred to any such estimator as simply an efficient or minimum variance unbiased estimator of θ. So if T' is an efficient estimator of θ and T is any other unbiased estimator of θ, a measure of the efficiency of T relative to T' is $RE = \left(\frac{V(T)}{V(T')} \right) \times 100$.

To see exactly how the Cramér-Rao lower bound is employed, let us consider the following assortment of example problems.

Example 10.3.3.1   Let $X_1, \ldots, X_n$ be a set of independent and identically distributed sample random variables taken from a normally distributed population with probability density function $f(x; \mu, \sigma) = \left(\sqrt{2\pi}\sigma\right)^{-1} e^{-(x-\mu)^2/2\sigma^2}$. Assuming that σ^2 is known, what is the Cramér-Rao lower bound on the variance of any unbiased estimator of μ? Since

$$\mathcal{L}\left(\mu, \sigma^2; x, n\right) = \frac{1}{(2\pi\sigma^2)^{n/2}} e^{-\frac{1}{2} \sum_{i=1}^{n} \frac{(x_i - \mu)^2}{\sigma^2}},$$

$$\log \mathcal{L} = -\frac{n}{2} \log\left(2\pi\sigma^2\right) - \frac{1}{2} \sum_{i=1}^{n} \frac{(x_i - \mu)^2}{\sigma^2}$$

and thus

$$\frac{\partial \log \mathcal{L}}{\partial \mu} = \sum_{i=1}^{n} \frac{x_i - \mu}{\sigma^2}, \quad \frac{\partial^2 \log \mathcal{L}}{\partial \mu^2} = -\frac{n}{\sigma^2}.$$

Then

$$-E\left(\frac{\partial^2 \log \mathcal{L}}{\partial \mu^2}\right) = -E\left(-\frac{n}{\sigma^2}\right) = \frac{n}{\sigma^2}$$

so that, from (10.10),

$$V(T) \geq \frac{1}{-E\left(\dfrac{\partial^2 \log \mathcal{L}}{\partial \mu^2}\right)} = \frac{\sigma^2}{n} = CR(\mu, n).$$

Since $\overline{X}$ is an unbiased estimator of μ and $V(\overline{X}) = \frac{\sigma^2}{n}$, it follows that $T = \overline{X}$ is a minimum variance bound estimator of μ.

Alternatively, let us directly employ (10.11) and write

$$\frac{\partial \log \mathcal{L}}{\partial \mu} = \frac{n}{\sigma^2} (\overline{x} - \mu),$$

where $\alpha(\mu, n) = \frac{n}{\sigma^2}$. Then $\overline{X}$ is a minimum variance bound estimator of μ (given that σ^2 is known) since its variance coincides with $CR(\mu, n)$. ∎

Example 10.3.3.2   Again let $X_1, \ldots, X_n$ be a set of independent and identically distributed sample random variables taken from a normally distributed population. This time, let us assume that μ is known. What is the Cramér-Rao lower bound on the variance of any unbiased estimators of σ^2? From log $\mathcal{L}$ just given,

$$\frac{\partial \log \mathcal{L}}{\partial \sigma^2} = -\frac{n}{2}\frac{1}{\sigma^2} + \frac{1}{2}\sum_{i=1}^{n}(x_i - \mu)^2\left(\frac{1}{\sigma^4}\right),$$

$$\frac{\partial^2 \log \mathcal{L}}{\partial\left(\sigma^2\right)^2} = \frac{n}{2}\frac{1}{\sigma^4} - \sum_{i=1}^{n}(x_i - \mu)^2\left(\frac{1}{\sigma^6}\right).$$

Then

$$-E\left(\frac{\partial^2 \log \mathcal{L}}{\partial\left(\sigma^2\right)^2}\right) = \frac{-n}{2\sigma^4} + \frac{1}{\sigma^6}E\left(\sum_{i=1}^{n}(x_i - \mu)^2\right) = \frac{-n}{2\sigma^4} + \frac{n\sigma^2}{\sigma^6} = \frac{n}{2\sigma^4}$$

and thus, using (10.10),

$$V(T) \geq \frac{1}{-E\left(\dfrac{\partial^2 \log \mathcal{L}}{\partial(\sigma^2)^2}\right)} = \frac{2\sigma^4}{n} = CR(\sigma^2, n).$$

Hence any unbiased estimator T of σ^2 must satisfy this inequality. Since $S_0^2 = \sum_{i=1}^{n}\frac{(X_i - \mu)^2}{n}$ is an unbiased estimator of σ^2 and $V(S_0^2) = \frac{2\sigma^4}{n}$, it follows that $T = S_0^2$ is a minimum variance bound estimator of σ^2. A way of directly finding a minimum variance bound estimator of σ^2 is given by (10.11). To pursue this method, let us express $\frac{\partial \log \mathcal{L}}{\partial \sigma^2}$ as

$$\frac{\partial \log \mathcal{L}}{\partial \sigma^2} = \frac{n}{2\sigma^4}\left(\frac{\sum_{i=1}^{n}(x_i - \mu)^2}{n} - \sigma^2\right).$$

If we set $\alpha(\sigma^2, n) = \frac{n}{2\sigma^4}$, then S_0^2 is the minimum variance bound estimator of σ^2 (assuming that μ is known) since $V(S_0^2)$ coincides with $CR(\sigma^2, n)$. ∎

We noted earlier that if we sample from a normal population wherein both μ and σ^2 are unknown, then an unbiased estimator of σ^2 is $S^2 = \frac{\sum_{i=1}^{n}(X_i - \bar{X})^2}{n-1}$ with $V(S^2) = \frac{2\sigma^4}{n-1}$. Hence $V(S^2)$ exceeds the Cramér-Rao lower bound; that is, if μ is unknown, $CR(\sigma^2, n)$ cannot be attained. So although S^2 is an efficient estimator of σ^2, it is not the *most efficient* estimator of the same since its variance exceeds the Cramér-Rao lower bound.

Example 10.3.3.3 Let $X_1, \ldots, X_n$ be a set of independent and identically distributed sample random variables taken from a population that follows a Poisson distribution with probability mass function $p(x; \lambda) = \frac{e^{-\lambda} \lambda^x}{x!}, x = 0, 1, \ldots; \lambda > 0,$ with $E(X) = V(X) = \lambda$. Forming

$$\mathcal{L}(\lambda; x, n) = \prod_{i=1}^{n} \frac{e^{-\lambda} \lambda^{x_i}}{x_i!} = \frac{e^{-n\lambda} \lambda^{\sum_{i=1}^{n} x_i}}{\prod_{i=1}^{n} x_i!}$$

with

$$\log \mathcal{L} = -n\lambda + \sum_{i=1}^{n} x_i \log \lambda - \sum_{i=1}^{n} \log x_i!,$$

we have

$$\frac{\partial \log \mathcal{L}}{\partial \lambda} = -n + \frac{\sum_{i=1}^{n} x_i}{\lambda}, \quad \frac{\partial^2 \log \mathcal{L}}{\partial \lambda^2} = -\frac{\sum_{i=1}^{n} x_i}{\lambda^2}.$$

Then

$$-E\left(\frac{\partial^2 \log \mathcal{L}}{\partial \lambda^2}\right) = \frac{1}{\lambda^2} \sum_{i=1}^{n} E(x_i) = \frac{1}{\lambda^2} n\lambda = \frac{n}{\lambda}$$

and thus, from (10.10),

$$V(T) \geq \frac{1}{-E\left(\frac{\partial^2 \log \mathcal{L}}{\partial \lambda^2}\right)} = \frac{n}{\lambda} = \frac{\sigma^2}{n} = CR(\lambda, n).$$

Hence any unbiased estimator of λ must satisfy $V(T) \geq \frac{\sigma^2}{n}$. Since the sample mean $\overline{X}$ is an unbiased estimator of $E(X) = \lambda$ and the variance of the sample mean is $\frac{\sigma^2}{n}$, it follows that $V(T) = V(\overline{X}) = \frac{\sigma^2}{n}$ and thus $T = \overline{X}$ is a minimum variance bound estimator of λ.

As an alternative, we may utilize (10.11) to write

$$\frac{\partial \log \mathcal{L}}{\partial \lambda} = -n + \frac{n\overline{x}}{\lambda} = \frac{n}{\lambda}(\overline{x} - \lambda),$$

where $\alpha(\lambda, n) = \frac{n}{\lambda}$. Hence $\overline{X}$ is a minimum variance bound estimator of λ since its variance equals $CR(\lambda, n)$. ∎

Example 10.3.3.4 For X a Bernoulli random variable, its probability mass function can be written as $f(x;p) = P(X = x) = p^x(1-p)^{1-x}, x = 0$ or 1. If we repeat the implied simple alternative experiment n times in succession, then the value of the discrete random variable X_i is determined on the ith trial. Clearly the X_i's, $i = 1,\ldots,n$, constitute a set of independent and identically distributed sample random variables. Given this random sampling scheme, what is the Cramér-Rao lower bound on the variance of any unbiased estimator of p? The joint probability mass function (and thus likelihood function) appears as

$$\mathcal{L}(p;x,n) = p^{\sum\limits_{i=1}^{n} x_i} (1-p)^{n-\sum\limits_{i=1}^{n} x_i}, \quad x_i = 0 \text{ or } 1,$$

and thus

$$\log \mathcal{L} = \left(\sum_{i=1}^{n} x_i\right) \log p + \left(n - \sum_{i=1}^{n} x_i\right) \log(1-p).$$

Then

$$\frac{\partial \log \mathcal{L}}{\partial p} = \left(\frac{1}{p}\right) \sum_{i=1}^{n} x_i - \left(\frac{1}{1-p}\right) \left(n - \sum_{i=1}^{n} x_i\right),$$

$$\frac{\partial^2 \log \mathcal{L}}{\partial p^2} = -\left(\frac{1}{p^2}\right) \sum_{i=1}^{n} x_i - \left(\frac{1}{(1-p)^2}\right) \left(n - \sum_{i=1}^{n} x_i\right).$$

We next form

$$-E\left(\frac{\partial^2 \log \mathcal{L}}{\partial p^2}\right) = E\left[\frac{\sum\limits_{i=1}^{n} x_i}{p^2}\right] + E\left[\frac{n - \sum\limits_{i=1}^{n} x_i}{(1-p)^2}\right]$$

$$= \left(\frac{1}{p^2}\right) \sum_{i=1}^{n} E(x_i) + \left(\frac{1}{(1-p)^2}\right) \left(n - \sum_{i=1}^{n} E(x_i)\right).$$

Now, since $E(x_i) = p$ for all i, it follows that the preceding expression becomes

$$-E\left(\frac{\partial^2 \log \mathcal{L}}{\partial p^2}\right) = \frac{n}{p} + \frac{n}{1-p} = \frac{n}{p(1-p)}.$$

Hence, by virtue of (10.10),

$$V(T) \geq \frac{1}{-E\left(\dfrac{\partial^2 \log \mathcal{L}}{\partial p^2}\right)} = \frac{p(1-p)}{n} = CR(p,n).$$

Since the estimator $T = \frac{X}{n}$ (where X is the number of successes in n trials) is unbiased for p and $V(T) = \frac{p(1-p)}{n}$, it follows that T is a minimum variance bound estimator of p. Again (10.11) provides us with a direct method for finding the *most efficient* estimator for a parameter. In this instance, since $T = \frac{X}{n} = \frac{1}{n}\sum_{i=1}^{n} X_i$ and

$$\frac{\partial \log \mathcal{L}}{\partial p} = \frac{nt}{p} + \frac{n - nt}{(1-p)} = \frac{n}{p(1-p)}(t - p),$$

where $\alpha(p,n) = n/p(1-p)$, we see that T is indeed a minimum variance bound estimator of p with $V(T) = CR(p,n)$. ∎

The preceding set of example problems has indicated that, under certain general regularity conditions (see footnote 2), we may determine if a particular estimator is a minimum variance bound or *most efficient* estimator of a parameter θ. But what if the population probability density function $f(x;\theta)$ does not satisfy the regularity assumptions underlying the existence of the Cramér-Rao lower bound or the lower bound is unattainable for the set of admissible unbiased estimators? If no such estimator attains the Cramér-Rao lower bound, how do we know which estimator from among many possible unbiased estimators of θ is efficient? One way to address these issues is to introduce the concept of sufficiency while restricting our discussion to the class of unbiased estimators.

10.3.4   Sufficient Statistics

Sufficient statistics often can be used to develop estimators that are efficient (have minimum variance among all unbiased estimators). In fact, as we shall now see, if a minimum variance bound or *most efficient* estimator exists, it will be found to be a sufficient statistic. Moreover, if an efficient estimator of θ exists, it is expressible as a function of a sufficient statistic.

To set the stage for a formal definition of a sufficient statistic, let us note first that, quite generally, a sufficient statistic for a parameter θ is one that utilizes all the information about θ that appears in a random sample. So by employing a sufficient statistic for θ, we are adopting a method of data reduction that does not discard any information about θ while condensing the sample data into a single value within which all sample information is contained.

In this regard, suppose that $X_1,\ldots,X_n$ represent a set of sample random variables taken from a population with probability density function $f(x;\theta)$. Then an estimator $T = (X_1,\ldots,X_n,n)$ of a parameter θ is termed *sufficient* if and only if the joint probability density function of $X_1,\ldots,X_n$ given T is independent of θ for any realization t of T; that is, the conditional distribution of $X_1,\ldots,X_n$ given T or $f(X_1,\ldots,X_n|T)$ does not depend on θ. So if T is given, then $X_1,\ldots,X_n$ has nothing further to add about the value of θ and no alternative estimator can contribute any additional information about θ. Thus any inference about θ depends on the sample random variables $X_1,\ldots,X_n$ only through the value of T; computing T enables us to achieve data reduction without losing any information about the parameter θ. In this regard, a sufficient statistic T fully exploits the information content of the sample; and it cannot be improved upon in the sense that there are no additional properties of the sample data not already included in the specification of T.

We may view this definition of a sufficient statistic from another angle by noting that if the value of a sufficient statistic is *known*, then the sample values themselves are superfluous and cannot add any additional information about θ since the joint distribution of the sample random variables given T does not depend on θ. That is, how much additional information about θ can be obtained by sampling from a distribution that does not depend on θ?

For instance, suppose we want to estimate the population mean μ from a set of sample random variables $X_1, \ldots, X_n$ taken from a normal probability density function with known standard deviation. To do so we summarize all the sample information in the statistic $X = \sum_{i=1}^{n} X_i$ and then construct the estimator $\overline{X} = \frac{1}{n} \sum_{i=1}^{n} X_i$. Has this process of reducing the entire data set to the single value X retained all the information about μ in the sample, or has some of the information regarding μ been lost? As we shall soon see, X summarizes *all* the values in the sample into a single number without loss of any information concerning μ and thus is a sufficient statistic for μ and thus the estimator $\overline{X}$ is also sufficient for μ. (Note that the median is not sufficient since it utilizes only the ranking and not all the values of the sample observations.)

In light of the preceding discussion, we may now generally conclude that if T is a sufficient statistic for a parameter θ, then any one-to-one function of T, $T' = \phi(T)$ is also sufficient for θ. For example, we just observed that if $X = \sum_{i=1}^{n} X_i$ is a sufficient statistic for μ, then so is $\overline{X} = \phi(X) = \frac{X}{n}$.

Example 10.3.4.1   To clarify the preceding definition of a sufficient statistic, let the value of the sample random variable $X_i, i = 1, \ldots, n$, be determined on the ith trial of a simple alternative or binomial experiment wherein the probability of a success ($X_i = 1$) is p and the probability of a failure ($X_i = 0$) is $1 - p$. Let $X = \sum_{i=1}^{n} X_i$ be the total number of successes in the n independent trials. If the value of X, and thus $\hat{P} = \frac{X}{n}$, is known, can we gain any additional information about p by examining the value of any alternative estimator that also depends on $X_1, \ldots, X_n$? According to our definition of sufficiency, the conditional distribution of $X_1, \ldots, X_n$ given X must be independent of p. Is it? For $x = n\hat{p}$ the realization of $X = n\hat{P}$, let us form

$$P(X_1 = x_1, \ldots, X_n = x_n | X) = \frac{P(X_1 = x_1, \ldots, X_n = x_n, X = x)}{P(X = x)}$$

$$= \frac{p^x (1-p)^{n-x}}{\binom{n}{x} p^x (1-p)^{n-x}} = \frac{1}{\binom{n}{x}} = \frac{1}{\binom{n}{n\hat{p}}}.$$

Clearly the conditional distribution of the sample random variables given X is free of p. So once X is known, no alternative estimator expressed as a function of the $X_i, i = 1, \ldots, n$, can add any additional information concerning the value of p. Thus X (and consequently $\hat{P}$) summarizes all relevant information about p

by reducing the sample information to a single value (while not discarding any information about p) and thus is deemed sufficient for p. ∎

The definition of a sufficient statistic can tell us how to check to see if a particular estimator is sufficient for a parameter θ, but it does not tell us how to actually go about finding a sufficient statistic (if one exists). To address the issue of operationally determining a sufficient statistic, we turn to Theorem 10.1, the *Fisher-Neyman Factorization Theorem* (1922, 1924, 1935):

THEOREM 10.1.   Let $X_1, \ldots, X_n$ be a random sample taken from a population with probability density function $f(x; \theta)$. The estimator $T = g(X_1, \ldots, X_n, n)$ is a sufficient statistic for the parameter θ if and only if the likelihood function of the sample factors as the product of two nonnegative functions $h(t; \theta, n)$ and $j(x_1, \ldots, x_n, n)$ or

$$\mathcal{L}(\theta; x_1, \ldots, x_n, n) = h(t; \theta, n) \cdot j(x_1, \ldots, x_n, n) \qquad (10.14)$$

for every realization $t = g(x_1, \ldots, x_n, n)$ of T and all admissible values of θ.

Although the function j is independent of θ (it may possibly be a constant), the function h depends on the sample realizations via the estimator T and thus this estimator constitutes a sufficient statistic for θ.

Equation (10.14) is known as the *factorization criterion for a sufficient statistic*, and is useful as a device for discovering sufficient statistics (although it is not useful in determining that a given statistic is not sufficient). As we shall now see, this criterion avoids the potentially difficult and cumbersome procedure of constructing a conditional probability distribution of the sample random variables as required by the direct application of the definition of sufficiency.

To see how the factorization theorem is applied in specific instances, let us consider the following assortment of example problems.

Example 10.3.4.2   Let the sample random variables $X_1, \ldots, X_n$ be drawn from a Poisson population with probability mass function, $p(x; \lambda) = e^{-\lambda} \frac{\lambda^x}{x!}, x = 0, 1, \ldots; \lambda > 0$. To demonstrate that the minimum variance bound estimator $\overline{X}$ is sufficient for λ, let us start with the likelihood function

$$\mathcal{L}(\lambda; x, n) = e^{-n\lambda} \frac{\lambda^{\sum_{i=1}^{n} x_i}}{\prod_{i=1}^{n} x_i!} = \frac{e^{-n\lambda} \lambda^{n\bar{x}}}{\prod_{i=1}^{n} x_i!} = h(\bar{x}; \lambda, n) \cdot j(x_1, \ldots, x_n, n),$$

where $h(\bar{x}; \lambda, n) = e^{-n\lambda} \lambda^{n\bar{x}}$. Note that h depends upon the $X_i, i = 1, \ldots, n$, only through the function or realization $t = \bar{x}$. Moreover, this factorization holds for all x and λ. Hence, by the factorization criterion (10.14), $\overline{X}$ is a sufficient statistic for λ. ∎

Example 10.3.4.3   If the sample random variables $X_1, \ldots, X_n$ are taken from a $N(\mu, \sigma)$ population with probability density function $f(x; \mu, \sigma) = (2\pi\sigma^2)^{-\frac{1}{2}} e^{-\frac{(x-\mu)^2}{2\sigma^2}}, -\infty < x, \mu < +\infty, \sigma > 0$, then, with σ^2 known, the likelihood function factors as

$$
\mathcal{L}(\mu, \sigma^2; x, n) = (2\pi\sigma^2)^{-\frac{n}{2}} e^{-\sum_{i=1}^{n} \frac{(x_i-\mu)^2}{2\sigma^2}} = (2\pi\sigma^2)^{-\frac{n}{2}} e^{-\sum_{i=1}^{n} \frac{(x_i-\bar{x}+\bar{x}-\mu)^2}{2\sigma^2}}
$$

$$
= (2\pi\sigma^2)^{-\frac{n}{2}} e^{-\frac{\sum_{i=1}^{n}(x_i-\bar{x})^2 + n(\bar{x}-\mu)^2}{2\sigma^2}}
$$

$$
= \left[(2\pi\sigma^2)^{-\frac{n}{2}} e^{-\sum_{i=1}^{n} \frac{(x_i-\bar{x})^2}{2\sigma^2}} \right] \left[e^{-\frac{n(\bar{x}-\mu)^2}{2\sigma^2}} \right]
$$

$$
= j(x_1, \ldots, x_n, n) \cdot h(\bar{x}; \mu, n),
$$

where $h(\bar{x}; \mu, n) = e^{-n(\bar{x}-\mu)^2/2\sigma^2}$. Since h is a function of the sample random variables only through $t = \bar{x}$, it follows from the factorization criterion that, for all x and μ, $\overline{X}$ is a sufficient statistic for μ as well as a minimum variance bound estimator for the same. ∎

Example 10.3.4.4   If the sample random variables are extracted from a $N(\mu, \sigma)$ distribution with $\mu = 0$, then the population probability density function appears as $f(x; \mu, \sigma) = (2\pi\sigma^2)^{-1/2} e^{-x^2/2\sigma^2}, -\infty < x, \mu < +\infty, \sigma > 0$, and the likelihood function assumes the form

$$
\mathcal{L}(\mu, \sigma^2; x, n) = (2\pi\sigma^2)^{-\frac{n}{2}} e^{-\sum_{i=1}^{n} \frac{x_i^2}{2\sigma^2}}
$$

$$
= \left[(\sigma^2)^{-\frac{n}{2}} e^{-\frac{ns_0^2}{2\sigma^2}} \right] \left[(2\pi)^{-\frac{n}{2}} \right] = h(s_0^2; \sigma^2, n) \cdot j(x_1, \ldots, x_n, n),
$$

where $h(s_0^2; \sigma^2, n) = (\sigma^2)^{-n/2} e^{-ns_0^2/2\sigma^2}$ and the function j is a constant. (Remember that with μ known, $S_0^2 = \frac{1}{n} \sum_{i=1}^{n}(x_i - \mu)^2$ serves as a minimum variance bound estimator of σ^2.) Since h depends on the $X_i, i = 1, \ldots, n$, only through the function or realization $s_0^2 = \frac{1}{n} \sum_{i=1}^{n} x_i^2$ of $S_0^2 = \frac{1}{n} \sum_{i=1}^{n} X_i^2$, the factorization criterion enables us to conclude that, for all x and σ^2, $\sum_{i=1}^{n} X_i^2$ as well as S_0^2 is sufficient for σ^2. ∎

Example 10.3.4.5 Let $X_1, \ldots, X_n$ represent a random sample taken from a Bernoulli population with probability mass function $f(x; p) = p^x(1-p)^{1-x}, x = 0$ or $1, 0 \leq p \leq 1$. Verify that the minimum variance bound estimator $\hat{P} = \frac{X}{n}$ is also sufficient for p. Given that the likelihood function can be expressed as

$$\mathcal{L}(p; x, n) = p^{\sum\limits_{i=1}^{n} x_i} (1-p)^{n - \sum\limits_{i=1}^{n} x_i}$$

$$= p^{n\hat{p}}(1-p)^{n-n\hat{p}} = h(\hat{p}; p, n) \cdot j(x_1, \ldots, x_n, n),$$

we see that $h(\hat{p}; p, n) = p^{n\hat{p}}(1-p)^{n-n\hat{p}}$ and $j = 1$. Since h depends on the $X_i, i = 1, \ldots, n$, only via the function $n\hat{p} = \sum_{i=1}^{n} x_i$, the factorization criterion deems $\hat{P} = \frac{X}{n}$, where $X = \sum_{i=1}^{n} X_i$, a sufficient statistic for p. ∎

10.3.5 Minimal Sufficient Statistics

It may be the case that there is more than one sufficient statistic associated with the parameter θ of some population probability density (or mass) function. Hence it is only natural to ask whether one sufficient statistic is better than any other such statistic in the sense that it achieves the highest possible degree of data reduction without loss of information about θ. A sufficient statistic that satisfies this requirement will be termed a minimal sufficient statistic. More formally, a sufficient statistic T for a parameter θ is termed a *minimal sufficient statistic* if, for any other sufficient statistic T' for θ, T is a function of T'.

We note briefly that:

1. The entire sample $X_1, \ldots, X_n$ serves as a sufficient statistic for a parameter θ. To see this, let us factor the likelihood function as $\mathcal{L}(\theta; x_1, \ldots, x_n, n) = \mathcal{L}(\theta; t(x_1, \ldots, x_n), n) \cdot j(x_1, \ldots, x_n)$, where $t = (x_1, \ldots, x_n)$ and $j = 1$ for all x. Then, by the factorization theorem, we see that $T = (X_1, \ldots, X_n)$ is a sufficient statistic for θ.

2. If a population probability density (or mass) function $f(x; \theta)$ is a member of the exponential family of distributions given by (10.12), then the factorization criterion (10.14) applied to the likelihood function (10.13) reveals that $\sum_{i=1}^{n} d(x_i)$ is a minimal sufficient statistic for θ.

A general procedure for finding a minimal sufficient statistic is provided by Theorem 10.2, the *Lehmann–Scheffé Theorem* (1950, 1955, 1956):

THEOREM 10.2. Let $\mathcal{L}(\theta; x_1, \ldots, x_n, n)$ denote the likelihood function of a random sample taken from the population probability density function $f(x; \theta)$. Suppose there exists a function $T = g(X_1, \ldots, X_n, n)$ such that, for the two sets of sample realizations $\{x_1,\ldots,x_n\}$ and $\{y_1,\ldots,y_n\}$, the likelihood ratio

$$\frac{\mathcal{L}(\theta;x_1,\ldots,x_n,n)}{\mathcal{L}(\theta;y_1,\ldots,y_n,n)} \tag{10.15}$$

is independent of θ if and only if $g(x_1,\ldots,x_n,n) = g(y_1,\ldots,y_n,n)$.
Then T is a minimal sufficient statistic for θ.

Example 10.3.5.1 Let $X_1,\ldots,X_n$ and $Y_1,\ldots,Y_n$ be two sets of sample random variables taken from a Poisson population with probability mass function $p(x;\lambda) = e^{-\lambda}\lambda^x/x!, x = 0,1,\ldots;\lambda > 0$. From (10.15),

$$\frac{\mathcal{L}(\lambda; x, n)}{\mathcal{L}(\lambda; y, n)} = \frac{e^{-n\lambda}\lambda^{\sum_{i=1}^{n}x_i}/\prod_{i=1}^{n}x_i!}{e^{-n\lambda}\lambda^{\sum_{i=1}^{n}y_i}/\prod_{i=1}^{n}y_i!} = \lambda^{\sum_{i=1}^{n}(x_i-y_i)}\left(\frac{\prod_{i=1}^{n}y_i!}{\prod_{i=1}^{n}x_i!}\right) = \lambda^{n(\bar{x}-\bar{y})}\left(\frac{\prod_{i=1}^{n}y_i!}{\prod_{i=1}^{n}x_i!}\right).$$

Clearly this ratio is free of λ if and only if $\bar{x} = \bar{y}$. Hence $\overline{X}$ is a minimal sufficient statistic for λ. ■

10.3.6 On the Use of Sufficient Statistics

Let us now consider the importance or usefulness of a sufficient statistic (provided, of course, that one exists). As will be seen next, sufficient statistics often can be used to construct estimators that have minimum variance among all unbiased estimators or are efficient. In fact, sufficiency is a necessary condition for efficiency in that an estimator cannot be efficient unless it utilizes all the sample information.

We note first that if an unbiased estimator T of a parameter θ is a function of a sufficient statistic S, then T has a variance that is smaller than that of any other unbiased estimator of θ that is not dependent on S. Second, if T is a minimum variance bound or *most efficient* estimator of θ, then T is also a sufficient statistic (although the converse does not necessarily hold).

This first point may be legitimized by the Rao-Blackwell Theorem (presented next), which establishes a connection between unbiased estimators and sufficient statistics. In fact, this theorem informs us that we may improve upon an unbiased estimator by conditioning it on a sufficient statistic. That is, an unbiased estimator and a sufficient statistic for a parameter θ may be combined to yield a single estimator that is both unbiased and sufficient for θ and has a variance no larger (and usually smaller) than that of the original unbiased estimator.

To see this let T be any unbiased estimator of a parameter θ and let the statistic S be sufficient for θ. Then the Rao-Blackwell Theorem indicates that another estimator T' can be derived from S (i.e., expressed as a function of S) such that T' is unbiased and sufficient for θ with the variance of T' uniformly less than or equal to the variance of T. More formally, we have Theorem 10.3, The *Rao-Blackwell Theorem* (1947):

> **THEOREM 10.3.** Let $X_1, \ldots, X_n$ be a random sample taken from the population probability density function $f(x; \theta)$ and let $T = g(X_1, \ldots, X_n, n)$ be any unbiased estimator of the parameter θ. For $S = u(X_1, \ldots, X_n, n)$ a sufficient statistic for θ, define $T' = E(T/S)$, where T is not a function of S alone. Then $E(T') = \theta$ and $V(T') \leq V(T)$ for all θ (with $V(T') < V(T)$ for some θ unless $T = T'$ with probability 1).

What this theorem tells us is that if we condition any unbiased estimator T on a sufficient statistic S to obtain a new unbiased estimator T', then T' is also sufficient for θ and is a uniformly *better* unbiased estimator of θ. In this regard, if we seek an efficient or minimum variance unbiased estimator of θ, we need only consider estimators that are functions of a sufficient statistic. In particular, an efficient estimator must be a function of a minimal sufficient statistic; that is, although conditioning an unbiased estimator T on a sufficient statistic S helps us obtain a *better* estimator T' of θ in the sense that $V(T') \leq V(T)$, it should be evident that conditioning T on a minimal sufficient statistic S' enables us to obtain an efficient or minimum variance unbiased estimator T'' of θ. In this regard, it is usually the case that any unbiased estimator that is a function of a minimal sufficient statistic is an efficient estimator of θ. Moreover, the resulting efficient estimator is unique.

It was mentioned earlier that minimum variance bound estimators and sufficient statistics are related. In fact, as we shall now see, under certain conditions they are one and the same. To see this let us return to the Factorization Theorem 10.1 and rewrite the factorization criterion or sufficiency condition (10.14) as

$$\log \mathcal{L}(\theta; x_1, \ldots, x_n, n) = \log h(t; \theta, n) + \log j(x_1, \ldots, x_n, n). \qquad (10.14.1)$$

Then the factorization criterion may be respecified as

$$\frac{\partial \log \mathcal{L}}{\partial \theta} = \frac{\partial \log h(t; \theta, n)}{\partial \theta}. \qquad (10.14.2)$$

Here also (10.14.2) provides us with a sufficiency condition for an estimator T.

If the right-hand side of (10.14.2) can be expressed in the same form as the right-hand side of (10.11); that is,

$$\frac{\partial \log h(t; \theta, n)}{\partial \theta} = \alpha(\theta, n) \left(t(x_1, \ldots, x_n, n) - \theta \right), \qquad (10.14.3)$$

then the sufficient statistic T is also a minimum variance bound estimator of the parameter θ. However, if (10.14.2) cannot be written as (10.14.3), then a minimum variance bound estimator of θ does not exist, even though T is a sufficient statistic for θ. But if a minimum variance bound estimator for θ exists, it must, in general, be a function of a sufficient statistic. Clearly the condition for a minimum variance bound estimator to exist (10.11) is much stronger or more restrictive than the sufficiency condition (10.14.2).

Example 10.3.6.1   We noted earlier that if our random sample is taken from a Poisson population, then the likelihood function of the sample can be factored as

$$\mathcal{L}(\lambda; x, n) = \frac{e^{-n\lambda} \lambda^{n\bar{x}}}{\prod\limits_{i=1}^{n} x_i!} = h(\bar{x}; \lambda, n) \cdot j(x_1, \ldots, x_n, n),$$

with $h(\bar{x}; \lambda, n) = e^{-n\lambda} \lambda^{n\bar{x}}$. Then $\log h = -n\lambda + n\bar{x} \log \lambda$ and thus

$$\frac{\partial \log h}{\partial \lambda} = -n + \frac{n\bar{x}}{\lambda} = \frac{n}{\lambda}(\bar{x} - \lambda),$$

where, in terms of (10.14.3), $\alpha(\lambda, n) = \frac{n}{\lambda}$ and $t(x_1, \ldots, x_n, n) - \lambda = \bar{x} - \lambda$. Hence the sufficient statistic $\overline{X}$ is also a minimum variance bound estimator of λ.   ∎

10.3.7   Completeness

Before we continue with our search for efficient or best unbiased estimators of a parameter θ, let us introduce some terminology that will be of considerable importance when it comes to linking the notions of sufficiency and efficiency. This is the concept of completeness. Specifically, let $f(x; \theta)$ represent a family of probability density (mass) functions and let $g(x)$ be a continuous function of x that does not depend on θ. If $E(g(x)) = 0$ for every admissible θ requires that $g(x) = 0$ at each point x where at least one member of the family $f(x; \theta)$ is positive, then the family of probability density (mass) functions $f(x; \theta)$ is termed a *complete family of probability density (mass) functions*. So for a complete family of, say, probability density functions, there are no unbiased estimators of zero other than zero itself.

Example 10.3.7.1   Given the family of Bernoulli probability mass functions $f(x; p) = p^x(1 - p)^{1-x}, x = 0$ or $1; 0 < p < 1$, and $g(x)$ a continuous function of x that does not involve p, suppose that $E(g(x)) = 0$ for every $p, 0 < p < 1$. According to the preceding definition, we must demonstrate that this implies that $g(x) = 0$ at each x at which at least one member of this family of probability mass functions is positive. But each f is positive at $x = 0$ and 1 and thus $E(g(x)) = 0$

implies $g(0) = g(1) = 0, 0 < p < 1$. From

$$E(g(x)) = \sum_{x=0}^{1} g(x)f(x;\theta) = \sum_{x=0}^{1} g(x)p^x(1-p)^{1-x}$$

$$= g(0)(1-p) + g(1)p$$

$$= p(g(1) - g(0)) + g(0) = 0, 0 < p < 1.$$

Since this expression must vanish for all admissible values of p, it follows that $g(1) - g(0) = 0$ and $g(0) = 0$ or $g(1) = g(0) = 0$. ∎

Armed with the completeness concept, let us continue our discussion of suffi-ciency as a vehicle for achieving efficiency. We noted earlier that conditioning an unbiased estimator T on a sufficient statistic S results in a new estimator T' that is a uniformly *better* unbiased estimator of a parameter θ. So if we desire an efficient or best unbiased estimator of θ, then we need only consider unbiased estimators that are functions of a sufficient statistic. But if T' is an improvement over T in the sense that $V(T') \leq V(T)$, how do we know that T' is efficient? If T' is a minimum variance bound estimator of θ, then obviously T' is best unbiased. But if $V(T')$ does not coincide with the Cramér-Rao lower bound, then we obviously need an alternative way of characterizing an efficient estimator.

One possibility is to introduce the concept of an unbiased estimator of zero; that is, an estimator R that satisfies $E(R) = 0$ for all θ. Clearly an unbiased estimator of zero is devoid of any information about θ. Let's see if this type of estimator can be utilized to improve upon the unbiased estimator T. To this end we form $T' = T + \alpha R, \alpha$ a constant, where $E(T) = \theta, E(R) = 0$. Clearly $E(T') = \theta$ so that T' is an unbiased estimator of θ. Is $V(T') \leq V(T)$? To answer this question let us form (using (5.45))

$$V(T') = V(T + \alpha R) = V(T) + \alpha^2 V(R) + 2\alpha \, COV(T, R).$$

If for some particular θ we find that $COV(T, R) < 0$ (or >0), then it can easily be demonstrated that one can determine an α for which $V(T') < V(T)$ so that T cannot be an efficient estimator of θ. Hence whether or not T is an efficient estimator of θ hinges upon the connection between T and any unbiased estimator of zero. In this regard, we can conclude that T is a best unbiased or efficient estimator of θ if and only if T is uncorrelated with *all* unbiased estimators of zero. So if T is efficient, then $COV(T, R) = 0$ for all θ and any estimator R for which $E(R) = 0$. And if the unbiased estimator T is such that $COV(T, R) = 0$ for any R satisfying $E(R) = 0$, then T is efficient.

How do we know that T is uncorrelated with *all* unbiased estimators of zero? One way to ensure that there is no association between T and *all* unbiased estimators of zero is to place a restriction on the family of probability density (mass) functions of the sufficient statistic S. In this regard, since $COV(T,0) = 0$, we can require that the family of probability densities $f(x;\theta,n)$ be complete; that is, for this family there are no unbiased estimators of zero other than zero itself.

In view of the preceding assortment of considerations and perspectives, our analysis concerning the link between sufficiency and efficiency can now be brought to closure. To accomplish this task, we may summarize the previous discussion as follows. Let S be a sufficient statistic for a parameter θ and suppose that T is an unbiased estimator of θ. Under what conditions will the function $T' = l(S) = E(T/S)$ represent a *unique* efficient or minimum variance unbiased estimator of θ? To answer this question we need to utilize the notion of a complete family of probability density (or mass) functions for a sufficient statistic. To this end let $f(s;\theta,n)$ be a family of sampling distributions or probability density functions of S and let $g(S) = w(X_1,\dots,X_n,n)$ be any statistic that is not a function of θ, where $X_1,\dots,X_n$ represent a collection of sample random variables taken from a population with probability density function $f(x;\theta)$. If

$$\left.\begin{array}{l} E\big(g(s)\big) = \int_S g(s) \cdot f(s;\theta,n)ds \equiv 0 \\[4pt] \text{implies } P\big(g(s) \equiv 0\big) = 1 \text{ for all} \\[4pt] \text{admissible } \theta \text{ whenever } f > 0, \end{array}\right\} \qquad (10.16)$$

then the family of probability density functions $f(s;\theta,n)$ is termed complete and the statistic S is called a *complete sufficient statistic*. As (10.16) reveals, S is a complete sufficient statistic if the only unbiased estimator of zero that is a function of S is the statistic $g(S)$ that is identically zero with probability equal to unity.

Hence there can be only one function $T' = l(S) = E(T/S)$ such that $E(T') = \theta$ so that T' is the *unique* minimum variance unbiased or efficient estimator of θ. So if T is an unbiased estimator of θ and T is not a function of S alone, then an efficient estimator of θ can be determined as $T' = l(s) = E(T/S)$, where S is a complete sufficient statistic for θ.

We note briefly that if a population probability density (or mass) function $f(x;\theta)$ belongs to the exponential family of distributions given by (10.12), then it can be demonstrated that $S = \sum_{i=1}^{n} d(x_i)$ is a complete minimal sufficient statistic for the parameter θ. So for this class of distributions, $T' = E(T/S)$ provides the unique best unbiased or efficient estimator for θ.

It should be evident from the preceding discussion that, at times, finding an efficient or best unbiased estimator of a parameter θ is no trivial task. We typically need to specify the form of the population probability density function and, in the case of trying to determine the *most efficient* estimator of θ, demonstrate that a given unbiased estimator has a variance that coincides with the Cramér-Rao lower bound. However, if the form of the population probability density function is unknown or if the Cramér-Rao lower bound is unattainable, then it is very difficult to establish if a given unbiased estimator is efficient.

10.3.8 Best Linear Unbiased Estimators

Instead of looking for an unbiased estimator with minimum variance among all alternative unbiased estimators of θ, an alternative approach is to restrict our search for an efficient estimator to a smaller class of unbiased estimators, namely those unbiased estimators that are linear functions of the sample observations. In this regard, for a set of sample random variables $X_1, \ldots, X_n$, an estimator T is termed a *linear combination* of the $X_i, i = 1, \ldots, n$, if

$$T = \sum_{i=1}^{n} a_i X_i, \tag{10.17}$$

where the a_i are constant for all $i = 1, \ldots, n$.

The notion of the linearity of an estimator now leads us to a specialization of the concept of an efficient estimator of a parameter θ. Specifically, T is termed a *best linear unbiased estimator* (BLUE) of θ if:

(a) T is a linear combination of the sample values $X_i, i = 1, \ldots, n$

(b) $E(T) = \theta$

(c) $V(T) \leq V(T')$, where T' is any other linear unbiased estimator of θ

That is, out of the class of all linear unbiased estimators of θ, T has minimum variance. As usual, the expression for T cannot involve θ or any other population parameter. So if we are interested in finding a BLUE for θ, then we need only find a linear estimator that satisfies properties (b) and (c). Moreover, it is often possible to find any such estimator without knowledge about the form of the population probability density function from which the sample random variables were taken.

Example 10.3.8.1   Let us derive a BLUE for the population mean μ. Suppose that $X_1, \ldots, X_n$ is a collection of sample random variables taken from a population with mean μ and variance σ^2. (Remember that $E(X_i) = \mu$ and $V(X_i) = \sigma^2$ for all $i = 1, \ldots, n$.) What we shall now demonstrate is that the sample mean $\overline{X}$ is a BLUE of μ. To this end let T be a linear unbiased estimator of μ. Then $T = \sum_{i=1}^{n} a_i X_i$ and $E(T) = \sum_{i=1}^{n} a_i E(X_i) = \mu \sum_{i=1}^{n} a_i = \mu$ only if $\sum_{i=1}^{n} a_i = 1$. Next $V(T) = V\left(\sum_{i=1}^{n} a_i X\right) = \sum_{i=1}^{n} a_i^2 V(X_i) = \sigma^2 \sum_{i=1}^{n} a_i^2$.

Hence the problem of finding a BLUE of μ amounts to finding a set of constants $a_1, \ldots, a_n$ such that the variance $V(T) = \sigma^2 \sum_{i=1}^{n} a_i^2$ is as small as possible under the restriction (or constraint) that $\sum_{i=1}^{n} a_i = 1$. But since σ^2 is constant, this is equivalent to solving the constrained optimization problem:

$$\min \left\{ \sum_{i=1}^{n} a_i^2 \right\} \; s.t. \; \sum_{i=1}^{n} a_i = 1, \tag{10.18}$$

where *s.t.* stands for *subject to* and the a_i's, $i = 1, \ldots n$, are the variables or unknowns.[3]

Alternatively, we may easily convert this constrained problem to an unconstrained optimization problem by eliminating one of the variables in the constraint equation (say, a_1) and substituting its value into the objective function $\sum_{i=1}^{n} a_i^2$ and then minimizing the latter quantity. That is, given $a_1 = 1 - \sum_{j=2}^{n} a_j$, we seek to determine the $a_i, i = 1, \ldots, n$, which minimize $M = \left(1 - \sum_{j=2}^{n} a_j\right)^2 + \sum_{j=2}^{n} a_j$. Setting

$$\frac{\partial M}{\partial a_j} = -2\left(1 - \sum_{j=2}^{n} a_j\right) + 2a_j = 0, \; j = 2, \ldots, n$$

(the first-order conditions for a minimum), we have $a_j = 1 - \sum_{j=2}^{n} a_j = a_1$. Hence

$$\sum_{i=1}^{n} a_i = a_1 + \sum_{j=2}^{n} a_j = a_1 + \sum_{j=2}^{n} a_1 = a_1 + (n-1)a_1 = 1$$

or $na_1 = 1$ and thus $a_1 = \frac{1}{n}$.

Then $a_j = \frac{1}{n}, j = 2, \ldots, n$, or $a_i = \frac{1}{n}$ for all i solves (10.18). (It is assumed that the second-order conditions for a minimum of M hold.) Given that the constants that make T unbiased and minimize its variance are $a_i = \frac{1}{n}$ for $i = 1, \ldots, n$, it follows that $T = \sum_{i=1}^{n} \left(\frac{1}{n}\right) X_i = \overline{X}$ and $V(T) = \sigma^2 \sum_{i=1}^{n} \left(\frac{1}{n}\right)^2 = \frac{\sigma^2}{n} = V(\overline{X})$. Thus $\overline{X}$ has the smallest variance among all linear unbiased estimators of μ.   ∎

10.3.9   Jointly Sufficient Statistics

One final set of results for this section concerns the existence of jointly sufficient statistics. Specifically, for $X_1, \ldots, X_n$ a set of sample random variables taken from the population probability density function $f(x; \theta)$, the statistics $S_k = u^k(X_1, \ldots, X_n, n), k = 1, \ldots, r$, are said to be *jointly sufficient* if and only if the joint probability density function of $X_1, \ldots, X_n$ given $S_1, \ldots, S_r$ is independent of θ for any set of realizations $s_k = u^k(x_1, \ldots, x_n, n), k = 1, \ldots, r$; that is, the

---

[3] Forming the Lagrangian function we have $L = \sum_{i=1}^{n} a_i^2 + \lambda\left(1 - \sum_{i=1}^{n} a_i\right)$, where λ is the undetermined (and unrestricted in sign) Lagrange multiplier. Then the first-order conditions for a minimum are

$$\partial L/\partial a_i = 2a_i - \lambda = 0, i = 1, \ldots, n;$$

$$\partial L/\partial \lambda = 1 - \sum_{i=1}^{n} a_i = 0.$$

Solving this system simultaneously yields $\lambda = 2/n$ and $a_i = \lambda/2 = 1/n, i = 1, \ldots, n$. (It is assumed that the second-order conditions for a constrained minimum are satisfied.)

conditional distribution of $X_1, \ldots, X_n$ given $S_1, \ldots, S_r$ or $f(X_1, \ldots, X_n | S_1, \ldots, S_r)$ does not depend on θ. (It is important to note that θ may not simply represent a single unknown population parameter but may, in fact, depict a set of parameters; that is, θ may be the v-tuple $(\theta_1, \ldots, \theta_v)$.)

In general, any collection of one-to-one functions of the set of sufficient statistics $S_1, \ldots, S_r$ is also sufficient. That is, if $S_1, \ldots, S_r$ is a set of jointly sufficient statistics for a parameter θ, then the collection of one-to-one functions of the $S_k, k = 1, \ldots, r$, or

$$S'_j = \phi^j(S_1, \ldots, S_r), j = 1, \ldots, w,$$

is also jointly sufficient for θ. For example, it will be demonstrated shortly that if $S_1 = \sum_{i=1}^{n} X_i$ and $S_2 = \sum_{i=1}^{n} X_i^2$ are jointly sufficient statistics for $\theta_1 = \mu$ and $\theta_2 = \sigma^2$, respectively, then so are

$$S'_1 = \phi^1(S_1, S_2) = \overline{X} = \frac{S_1}{n}$$

and

$$S'_2 = \phi^2(S_1, S_2) = \sum_{i=1}^{n} (X_i - \overline{X})^2 = \sum_{i=1}^{n} X_i^2 - n\overline{X}^2 = S_2 - \frac{1}{n} S_1^2.$$

We next examine a set of statistics for their joint sufficiency by considering Theorem 10.4, the *Generalized Fisher-Neyman Factorization Theorem:*

THEOREM 10.4. Let $X_1, \ldots, X_n$ be a random sample taken from a population with probability density function $f(x; \theta)$. The set of estimators $S_k = u^k(X_1, \ldots, X_n, n), k = 1, \ldots, r$, is jointly sufficient for the parameter θ if and only if the likelihood function of the sample factors as the product of the two nonnegative functions $h(s_1, \ldots, s_r; \theta, n)$ and $j(x_1, \ldots, x_r, n)$ or

$$\mathcal{L}(\theta; x_1, \ldots, x_n, n) = h(s_1, \ldots, s_r; \theta, n) \cdot j(x_1, \ldots, x_n, n) \qquad (10.19)$$

for every set of realizations $s_k = u^k(x_1, \ldots, x_n, n), k = 1, \ldots, r$, of the S_k's, $k = 1, \ldots, r$, and all admissible values of θ.

Here the function j is independent of θ (it may be a constant) and h depends on the sample realizations via the estimators $S_k, k = 1, \ldots, r$, and thus the S_k's, $k = 1, \ldots, r$, constitute a set of jointly sufficient statistics for θ. Equation (10.19) is termed the *factorization criterion for jointly sufficient statistics.*

Example 10.3.9.1 To see how (10.19) is applied, let $X_1, \ldots, X_n$ depict a set of independent and identically distributed sample random variables taken from a normally distributed population with probability density function $f(x; \mu, \sigma) = (\sqrt{2\pi}\sigma)^{-1} e^{-(x-\mu)^2/2\sigma^2}$. Assuming that both μ and σ^2 are unknown, we have

$\theta = (\theta_1, \theta_2) = (\mu, \sigma^2)$. Then as developed earlier, the joint probability density or likelihood function of the sample has the form

$$\mathcal{L}(\mu, \sigma^2; x, n) = \frac{1}{(2\pi\sigma^2)^{n/2}} e^{-\frac{1}{2}\sum\limits_{i=1}^{n}\frac{(x_i-\mu)^2}{\sigma^2}}$$

$$= \frac{1}{(2\pi\sigma^2)^{n/2}} e^{-\frac{1}{2\sigma^2}\left(\sum\limits_{i=1}^{n} x_i^2 - 2\mu\sum\limits_{i=1}^{n} x_i + n\mu^2\right)}$$

$$= \frac{1}{(2\pi\sigma^2)^{n/2}} e^{-\frac{1}{2\sigma^2}(s_2 - 2\mu s_1 + n\mu^2)}$$

$$= h(s_1, s_2; \theta, n) \cdot j(x_1, \ldots, x_n, n),$$

where

$$h(s_1, s_2; \theta, n) = \frac{1}{(2\pi\sigma^2)^{n/2}} e^{-\frac{1}{2\sigma^2}(s_2 - 2\mu s_1 + n\mu^2)}$$

and $j = 1$. Since h depends on the $X_i, i = 1, \ldots, n$, only through the functions $s_1 = \sum_{i=1}^{n} x_i$ and $s_2 = \sum_{i=1}^{n} x_i^2$, it follows that the statistics $S_1 = \sum_{i=1}^{n} X_i, S_2 = \sum_{i=1}^{n} X_i^2$ are jointly sufficient for μ and σ^2, respectively. And since $\overline{X} = \frac{S_1}{n} = S_1'$ and $S^2 = \frac{1}{n-1}\sum_{i=1}^{n}(X_i - \overline{X})^2 = \frac{1}{n-1}\left(\sum_{i=1}^{n} X_i^2 - n\overline{X}^2\right) = \frac{1}{n-1}\left(S_2 - \frac{1}{n}S_1^2\right) = S_2'$ are one-to-one functions of S_1 and S_2, it follows that $\overline{X}$ and S^2 are jointly sufficient for μ and σ^2. ∎

We noted earlier that a minimal sufficient statistic is one that achieves the highest possible degree of data reduction without losing any information about a parameter θ. Moreover, a general procedure for finding a minimal sufficient statistic was provided by the Lehman-Scheffé Theorem (Theorem 10.2). Let us now extend this theorem to the determination of a set of jointly minimal sufficient statistics. We thus have Theorem 10.5, the *generalized Lehman-Scheffé Theorem*:

THEOREM 10.5.   Let $\mathcal{L}(\theta; x_1, \ldots, x_n, n)$ denote the likelihood function of a random sample taken from the probability density function $f(x; \theta)$. Suppose there exists a set of functions $S_k = u^k(X_1, \ldots, X_n, n), k = 1, \ldots, r$, such that, for two sets of sample realizations $\{x_1, \ldots, x_n\}$ and $\{y_1, \ldots, y_n\}$ the likelihood ratio

$$\frac{\mathcal{L}(\theta; x_1, \ldots, x_n, n)}{\mathcal{L}(\theta; y_1, \ldots, y_n, n)} \tag{10.20}$$

is independent of θ if and only if $u^k(x_1, \ldots, x_n, n) = u^k(y_1, \ldots, y_n, n), k = 1, \ldots, r$. Then the $S_k, k = 1, \ldots, r$, represent, jointly, a set of minimal sufficient statistics for θ.

Example 10.3.9.2   Let $X_1, \ldots, X_n$ and $Y_1, \ldots, Y_n$ be two sets of sample random variables taken from a normal probability density function with unknown mean μ and variance σ^2. From (10.20),

$$\frac{\mathcal{L}(\mu, \sigma^2; x, n)}{\mathcal{L}(\mu, \sigma^2; y, n)} = \frac{\left(\frac{1}{\sqrt{2\pi}\sigma}\right)^n e^{-\frac{1}{2\sigma^2}\sum_{i=1}^{n}(x_i - \mu)^2}}{\left(\frac{1}{\sqrt{2\pi}\sigma}\right)^n e^{-\frac{1}{2\sigma^2}\sum_{i=1}^{n}(y_i - \mu)^2}}$$

$$= e^{-\frac{1}{2\sigma^2}\left(\sum_{i=1}^{n}(x_i - \mu)^2 - \sum_{i=1}^{n}(y_i - \mu)^2\right)}$$

$$= e^{-\frac{1}{2\sigma^2}\left[\left(\sum_{i=1}^{n}x_i^2 - \sum_{i=1}^{n}y_i^2\right) - 2\mu\left(\sum_{i=1}^{n}x_i - \sum_{i=1}^{n}y_i\right)\right]}$$

$$= e^{-\frac{1}{2\sigma^2}\left[(s_2^x - s_2^y) - 2\mu(s_1^x - s_1^y)\right]},$$

where $s_2^x = \sum_{i=1}^{n} x_i^2, s_2^y = \sum_{i=1}^{n} y_i^2, s_1^x = \sum_{i=1}^{n} x_i$, and $s_1^y = \sum_{i=1}^{n} y_i$. Clearly this ratio is free of μ and σ^2 if and only if $s_2^x = s_2^y$ and $s_1^x = s_1^y$. Thus $\sum_{i=1}^{n} x_i$ and $\sum_{i=1}^{n} x_i^2$ are jointly minimal sufficient statistics for μ and σ^2, respectively. And by virtue of the argument used in the preceding example problem, it follows that $\overline{X}$ and S^2 are also jointly minimal sufficient statistics for μ and σ^2. ∎

We indicated earlier that the Rao-Blackwell Theorem (Theorem 10.3) serves to establish a link between unbiased estimators and sufficient statistics; that is, we may improve upon an unbiased estimator by conditioning it on a sufficient statistic. We now extend this theorem to a collection of jointly sufficient statistics. Specifically, we state the *generalized Rao-Blackwell Theorem* (Theorem 10.6):

THEOREM 10.6.   Let $X_1, \ldots, X_n$ be a random sample taken from a population probability density function $f(x; \theta)$ and let $T = g(X_1, \ldots, X_n, n)$ be any unbiased estimator of the parameter θ. For $S_k = u^k(X_1, \ldots, X_n, n)$, $k = 1, \ldots, r$, a set of jointly sufficient statistics for θ, define $T' = E(T|S_1, \ldots, S_r)$, where T' is not a function of the S_k's, $k = 1, \ldots, r$, alone. Then $E(T') = \theta$ and $V(T') \leq V(T)$ for all θ (with $V(T') < V(T)$ for some θ unless $T = T'$ with probability 1).

So if we condition any unbiased estimator T on a set of jointly sufficient statistics $S_k, k = 1, \ldots, r$, to obtain a new unbiased estimator T', then T' is also sufficient for θ and is a uniformly *better* unbiased estimator of θ.

10.4 Large Sample Properties of Point Estimators

10.4.1 Asymptotic or Limiting Properties

In this section our aim is to develop *large sample* or *asymptotic properties* of a point estimator of a parameter θ. Strictly speaking, asymptotic properties are actually properties of a sequence of point estimators $\{T_n\}$ indexed by the sample size n; that is, if T_n denotes an estimator based on a sample of size n, then $T_1 = g_1(X_1, 1), T_2 = g_2(X_1, X_2, 2), \ldots, T_n = g_n(X_1, \ldots, X_n, n)$. Hence asymptotic properties of an estimator T_n are *limiting properties* that emerge as $n \to \infty$.

A large sample property of a point estimator is a characteristic of the sampling distribution of the estimator that obtains as $n \to \infty$. That is, as n increases without bound, we will consider a set of *asymptotic distributional properties* of an estimator—properties such as asymptotic unbiasedness, its asymptotic variance, and asymptotic efficiency, which characterize the sampling distribution of an estimator as $n \to \infty$.

We may view the *asymptotic distribution* of an estimator as the shape of the sampling distribution of the estimator as n becomes very large. That is, if the distribution of an estimator tends to become more and more similar in form to some specific or *limiting distribution* for increasing n, then this limiting distribution is taken to be the asymptotic distribution of the estimator. Hence the asymptotic distribution of an estimator T is the shape that the sampling distribution of T assumes as it approaches its limiting form. The asymptotic distribution of T is *not* the sampling distribution of T when the sample size is *infinite* since the distribution of T with an infinite sample size is *degenerate*; that is, the distribution of T reduces or collapses to a single point (possibly θ) with zero variance and with probability mass equal to unity. Hence the asymptotic distribution of T is not the final (degenerate) form that its sampling distribution assumes as $n \to \infty$; it is the form that the sampling distribution of T tends to take immediately before degeneracy occurs.

To focus more sharply on the concept of the limiting form of the sampling distribution of an estimator T, let us briefly review the notion of convergence in distribution introduced earlier in Section 8.6. To this end let $\{T_n\}$ be a sequence of estimators of a parameter θ, where T_n has the cumulative distribution function $F_n(t)$. Then T_n *converges in distribution* to a random variable V with cumulative distribution function $F(t)$ if $\lim_{n \to \infty} F_n(t) = F(t)$ (or $\lim_{n \to \infty} |F_n(t) - F(t)| = 0$) at all continuity points of $F(t)$. (This type of convergence was denoted as $T_n \overset{d}{\longrightarrow} V$.) In this regard, if T_n converges in distribution to V, then $F(t)$ serves as the limiting distribution of T_n.

Although the form of the sampling distribution of an estimator T may be unknown when n is small, this may not be the case when n is large; that is, the sampling distribution of T may have a specific known form (its limiting distribution) when n increases without bound. For example, if $X_1, \ldots, X_n$ constitutes a set of sample random variables taken from an unknown population distribution with finite mean μ and finite variance σ^2, then, via the central limit theorem, $\overline{Z}_n = \frac{\overline{X}_n - \mu}{\sigma/\sqrt{n}} \overset{d}{\longrightarrow} Z$, where Z is $N(0, 1)$. Here $N(0, 1)$ is the limiting distribution of $\overline{Z}_n$ and thus $T_n = \overline{X}_n$ is asymptotically $N\left(\mu, \frac{\sigma}{\sqrt{n}}\right)$. Hence the asymptotic distribution of $\overline{X}_n$, which is determined from the known limiting distribution of a function $\overline{Z}_n$ of $\overline{X}_n$, provides us with an *approximation* to the true sampling distribution of $\overline{X}_n$ (which may or may not be exactly normally distributed).

10.4.2 Asymptotic Mean and Variance

Given that the limiting distribution and its moments exist, the *limiting mean* and *limiting variance* of an estimator T_n are, respectively, the mean and variance of its limiting distribution. Let us see exactly how the limiting mean and limiting variance are characterized. We noted earlier that an estimator T of a parameter θ is unbiased if $E(T) = \theta$. Now, an estimator T may actually be biased for small samples but, as n increases without bound, its bias $B(T_n, \theta) = E(T_n) - \theta$ may decline to zero. If this is the case, then T_n is said to be an *asymptomatically unbiased estimator* of θ. More formally, let us define the *asymptotic expectation* (AE) of an estimator T_n as

$$AE(T_n) = \int T_n h(t; \theta) dT_n, \tag{10.21}$$

where $h(t; \theta)$ is the limiting (sampling) distribution of T_n. Here $AE(T_n)$ is simply the limiting mean of T_n or the mean of the limiting distribution of T_n. In this regard, we may now state that T_n is an asymptotically unbiased estimator of θ if $AE(T_n) = 0$. Hence the *asymptotic bias* of T_n is simply $AE(T_n) - \theta$. Note that if a estimator is unbiased, then it is asymptotically unbiased (since $E(T_n) = \theta$ for all values of n). However, the asymptotic unbiasedness of an estimator does not necessarily imply that the estimator is unbiased.

An alternative condition for asymptotic unbiasedness is that $E(T_n) \to \theta$ as $n \to \infty$ or $\lim_{n \to \infty} E(T_n) = \theta$, where

$$\lim_{n \to \infty} E(T_n) = \lim_{n \to \infty} \int T_n h(t; \theta, n) \, dT_n \tag{10.22}$$

and $h(t; \theta, n)$ is the sampling distribution of T_n. Here (10.22) is simply the limit of the expectation of T_n as $n \to \infty$. Implicit in the condition $\lim_{n \to \infty} E(T_n) = \theta$ is the requirement that, for indefinitely large $n, E(T_n)$ exists. If $E(T_n)$ does not exist, then this alternative notion of asymptotic unbiasedness is undefined. In general, $AE(T_n) = \theta$ does not imply $\lim_{n \to \infty} E(T_n) = \theta$.

Example 10.4.2.1   Let us consider the statistic $Y_n = \overline{X}_n + \frac{1}{n}$ as an estimator of the population mean μ. Since $E(Y_n) = E(\overline{X}_n) + \frac{1}{n} = \mu + \frac{1}{n}$, it follows that Y_n is a biased estimator of μ, where the amount of bias is $B(Y_n, \mu) = \frac{1}{n}$. But since $\lim_{n \to \infty} E(Y_n) = \mu$, we see that Y_n is an asymptotically unbiased estimator for μ. ∎

Since the sampling distribution of an estimator T_n might become degenerate as n increases without bound (i.e., collapses to a single point with probability mass equal to one), the asymptotic variance of T_n is *not* defined as $\lim_{n \to \infty} V(T_n)$ since, under degeneracy, $V(T_n) \to 0$ as $n \to \infty$. Instead, the *asymptotic variance* of T_n

is simply the variance of the asymptotic distribution of T_n or

$$AV(T_n) = \frac{1}{n} \lim_{n \to \infty} E\left\{\sqrt{n}\left(T_n - E(T_n)\right)\right\}^2 \qquad (10.23)$$

given that the indicated individual expectations exist. Thus the asymptotic variance of T_n is determined as the limiting value of the expected squared deviation of T_n about its mean.

10.4.3 Consistency

A desirable asymptotic property of a *good* point estimator T of a parameter θ is that the sampling distribution of T becomes more closely concentrated about θ as the sample size n increases without bound. Here we are considering the behavior of the sampling distribution of T which, conceptually, is constructed from indefinitely large samples. A concept that deals with the limiting behavior of the sampling distribution of T_n as $n \to \infty$ is consistency. Specifically, T is said to be a *consistent estimator* of a parameter θ if the sequence of estimators $\{T_n\}$ converges in probability to θ as n increases without bound $T_n \xrightarrow{p} \theta$; that is, T is a consistent estimator of θ if

    a. $\lim_{n \to \infty} P(|T_n - \theta| < \varepsilon) = 1$ or

                                        (10.24)

    b. $\lim_{n \to \infty} P(|T_n - \theta| \geq \varepsilon) = 0$

for all admissible θ and all real $\varepsilon > 0$.[4]

Equation (10.24a) reveals that the probability that the sampling error $|T_n - \theta|$ is less than any arbitrary small positive constant ε approaches one when n approaches infinity (i.e., as the sample size becomes infinite, T_n will be arbitrarily close to θ with high probability), and (10.24b) indicates that the probability that the sampling error $|T_n - \theta|$ equals or exceeds an arbitrary small positive constant ε approaches zero when n approaches infinity (i.e., as the sample size becomes infinite, the probability that T_n misses θ is small). So if (10.24a) is true, then the sequence of random variables $\{T_n\}$ converges in probability to the constant θ. And if (10.24b) holds, the values in the sequence $\{T_n\}$ that are not close to θ become increasingly unlikely as n grows large. In either case the sampling distribution of T_n collapses upon θ—the limiting distribution of T_n has all of its probability mass concentrated on θ.

To see exactly how the definition of consistency (as incorporated in (10.24)) can be used to actually verify the same for an estimator, let us demonstrate that $\overline{X}$ is a consistent estimator of μ.

---

[4] An alternative version of, say, (10.24a) appears as: a sequence of estimators $\{T_n\}$ is consistent for θ if, for arbitrarily small positive constants ε and δ, a value n_ε of n exists such that if $n > n_\varepsilon$, $P(|T_n - \theta| < \varepsilon) > 1 - \delta$.

Example 10.4.3.1   Let $X_1, \ldots, X_n$ depict a set of sample random variables drawn from a population with a finite mean (μ) and variance (σ^2). From the (4.19.2) version of Chebyshev's inequality we have

$$P(|\overline{X}_n - \mu| < k\sigma_{\overline{X}}) \geq 1 - \frac{1}{k^2}.$$

If we set $k = \dfrac{\varepsilon}{\sigma_{\overline{X}}} = \varepsilon \dfrac{\sqrt{n}}{\sigma}$ (since $\sigma_{\overline{X}} = \dfrac{\sigma}{\sqrt{n}}$), then

$$P(|\overline{X}_n - \mu| < \varepsilon) \geq 1 - \frac{\sigma^2}{n\varepsilon^2}$$

and thus

$$\lim_{n \to \infty} P(|T_n - \theta| < \varepsilon) = 1.$$

Hence, by virtue of (10.24a), the sequence of estimators $\{\overline{X}_n\}$ converges in probability to μ so that $\overline{X}$ is a consistent estimator of μ. This result should be familiar to you since it was actually encountered earlier under the heading of the (weak) law of large numbers, which essentially states that $\overline{X}_n \xrightarrow{p} \mu$. ∎

Example 10.4.3.2   The preceding demonstration that the sample mean is a consistent estimator of the population mean was offered without having to specify the form of the population probability density (mass) function. However, if the form of the population distribution is known, then it may be utilized to also demonstrate that $\{\overline{X}_n\}$ constitutes a sequence of consistent estimators for μ. To see this let us assume that the sample random variables are taken from a population density function, which is $N(\mu, \sigma)$. Then $\{\overline{X}_n\}$ depicts a sequence of $N\left(\mu, \frac{\sigma}{\sqrt{n}}\right)$ random variables. Let us determine

$$P(|\overline{X}_n - \mu| < \varepsilon) = \int_{\mu-\varepsilon}^{\mu+\varepsilon} \frac{1}{\sqrt{2\pi}\,\sigma_{\overline{X}}} e^{-\frac{1}{2}\frac{(\bar{x}_n - \mu)^2}{\sigma_{\overline{X}}^2}}\, d\bar{x}_n = \int_{\mu-\varepsilon}^{\mu+\varepsilon} \frac{\sqrt{n}}{\sqrt{2\pi}\,\sigma} e^{-\frac{n}{2}\frac{(\bar{x}_n - \mu)^2}{\sigma^2}}\, d\bar{x}_n.$$

Set $y = \frac{\bar{x}_n - \mu}{\sigma}$ or $\sigma y = \bar{x}_n - \mu$. Then by virtue of (7.n4.1) (p. 274),

$$P(|\overline{X}_n - \mu| < \varepsilon) = \int_{-\varepsilon}^{\varepsilon} \frac{\sqrt{n}}{\sqrt{2\pi}} e^{-\frac{n}{2}y^2}\, dy.$$

Next, let $y = \dfrac{z}{\sqrt{n}}$ so that, from (7.4n.2) (p. 274),

$$P(|\overline{X}_n - \mu| < \varepsilon) = \int_{-\sqrt{n}\varepsilon}^{\sqrt{n}\varepsilon} \frac{1}{\sqrt{2\pi}} e^{-\frac{1}{2}z^2}\, dz = P(|Z| < \sqrt{n}\varepsilon),$$

where Z is $N(0, 1)$. Then from Chebyshev's inequality (4.19.2), $P(|Z| < \sqrt{n}\varepsilon) \geq 1 - \frac{1}{n\varepsilon^2}$ and thus $\lim_{n\to\infty} P(|Z| < \sqrt{n}\varepsilon) = 1$. Hence $\{\bar{X}_n\}$ is a consistent sequence of estimators for μ.   ■

Example 10.4.3.3   If the random variable X represents the number of successes in n independent trials of a simple alternative experiment, then X is binomially distributed and, if the population probability of a success p is unknown, then, as determined earlier, the best estimator for p is the sample proportion or the observed relative frequency of a success $\hat{P} = \frac{X}{n}$. It should be intuitively clear that, as n increases without bound, the sequence of estimators $\{\hat{P}_n\}$ should approach p; that is, $\{\hat{P}_n\}$ should converge in probability to p (or $\hat{P}_n \xrightarrow{p} p$) and thus $\hat{P}$ is a consistent estimator of p. To see this remember that $\hat{P}$ is a random variable with $E(\hat{P}) = p$ and $\sigma_{\hat{P}} = \sqrt{\frac{p(1-p)}{n}}$. Then using Chebyshev's inequality (4.19.2),

$$P(|\hat{P}_n - p| < k\sigma_{\hat{P}}) \geq 1 - \frac{1}{k^2}.$$

Let $k = \frac{\varepsilon}{\sigma_{\hat{P}}} = \varepsilon \frac{\sqrt{n}}{\sqrt{p(1-p)}}$. Then

$$P(|\hat{P}_n - p| < \varepsilon) \geq 1 - \frac{p(1-p)}{n\varepsilon^2}$$

and thus

$$\lim_{n\to\infty} P(|\hat{P}_n - p| < \varepsilon) = 1$$

so that, as anticipated, $\hat{P}$ is a consistent estimator of p.   ■

A practical way to determine if a particular estimator is consistent is by tracing the behavior of its bias and variance as $n \to \infty$. To this end we offer the following two theorems. We assume that T is an unbiased estimator of θ. Then, by Theorem 10.7:

THEOREM 10.7.   An unbiased estimator T for a parameter θ is also a consistent estimator of θ if $\lim_{n\to\infty} V(T_n) = 0$.

To see this let $E(T_n) = \theta$ and $V(T_n) < \infty$. For real $\varepsilon > 0$, Chebyshev's inequality ((4.19.1)) informs us that $P(|T_n - \theta| \geq \varepsilon) \leq V(T_n)/\varepsilon^2$. If $\lim_{n\to\infty} V(T_n) = 0$, then (10.24b) holds and thus T is a consistent estimator of θ (see Figure 10.4a). (This restriction on the limit of the variance of T_n is only sufficient for T_n to be consistent for θ; that is, T_n can be a consistent estimator of θ even if $V(T_n)$ does not approach zero.)

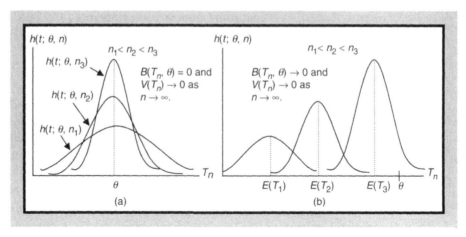

Figure 10.4   (a) T is a consistent estimator for θ if $V(T_n) \to 0$ as $n \to \infty$; (b) T is a consistent estimator for θ if $E(T_n) \to \theta$ and $V(T_n) \to 0$ as $n \to \infty$.

Since $\overline{X}$ is an unbiased estimator of μ, Theorem (10.7) applied to $\sigma_{\overline{X}}^2 = \frac{\sigma^2}{n}$ also reveals that $\overline{X}$ is a consistent estimator of μ. And if this theorem is applied to $\sigma_{\hat{P}}^2 = \frac{p(1-p)}{n}$, then clearly $\hat{P}$ is a consistent estimator of p.

The sample median is not a consistent estimator for μ if the population distribution is not symmetrical. In this regard, the discussion presented in Section 8.9 informs us that if the population probability density function is $N(\mu, \sigma)$, then the sampling distribution of the sample median is asymptotically normally distributed with mean μ and variance $\pi \frac{\sigma^2}{2n}$. Since the sample median in this case is unbiased and its variance approaches zero as $n \to \infty$, Theorem 10.7 reveals that the sample median is a consistent estimator of μ.

If T is not an unbiased estimator of θ, then we have Theorem 10.8:

THEOREM 10.8.   An estimator T for a parameter θ is called a *mean-squared-error-consistent estimator* of θ if $\lim_{n\to\infty} MSE(T_n, \theta) = 0$ or

(a) $\lim\limits_{n\to\infty} B(T_n, \theta) = 0$, where $B(T_n, \theta)$ denotes the bias of T_n

(b) $\lim\limits_{n\to\infty} V(T_n) = 0$

Clearly, mean–squared–error–consistency implies that both the bias and variance of T_n approach zero as $n \to \infty$ so that the sampling distribution of T_n eventually becomes degenerate at θ. Hence T_n comes closer to the unknown θ (T_n is asymptotically unbiased) and becomes more precise as $n \to \infty$. To verify this let $V(T_n) < +\infty$ and, for real $\varepsilon > 0$, let us write Chebyshev's inequality as $P(|T_n - \theta| \geq \varepsilon) \leq MSE(T_n, \theta)/\varepsilon^2$, where $MSE(T_n, \theta) = E(T_n - \theta)^2 = V(T_n) + B(T_n, \theta)^2$. If $\lim\limits_{n\to\infty} V(T_n) = \lim\limits_{n\to\infty} B(T_n, \theta) = 0$, then $\lim\limits_{n\to\infty} MSE(T_n, \theta) = 0$

and thus $\lim_{n\to\infty} P\left(|T_n - \theta| \geq \varepsilon\right) = 0$. Again T is consistent for θ by virtue of (10.24b). In this instance, $\lim_{n\to\infty} MSE(T_n, \theta) = 0$ is a sufficient condition for T to be a consistent estimator of θ in that it is possible to find an estimator whose mean squared error does not approach zero as $n \to \infty$ and yet is consistent (the estimator may not have a finite mean or variance). However, if we restrict our discussion to estimators that possess finite asymptotic means and variances, then the condition $\lim_{n\to\infty} MSE(T_n, \theta) = 0$ is both necessary and sufficient for consistency. Moreover, if an estimator is mean–squared–error–consistent, then it is also consistent in the sense of (10.24). (Hence the conditions given in (10.24) are weaker than those offered in Theorem 10.8. In this regard, an estimator satisfying (10.24) is often termed *weakly consistent*.) In either case the probability limit of T_n is θ.

The preceding discussion informs us that for consistency to hold, the sampling distribution of T_n concentrates perfectly on θ as n approaches infinity; that is, the mean and variance of the sampling distribution of T_n approach θ and zero, respectively, as n increases without bound (see Figure 10.4b).

Example 10.4.3.4 We noted earlier in Section 8.10 that the sample variance $S^2 = \dfrac{\sum\limits_{i=1}^{n}(X_i - \bar{X})^2}{n-1}$ is an unbiased estimator of the population variance σ^2 and that, for a general population probability density function, if μ_4 (the fourth central moment of the population) is finite, then

$$V(S^2) = \frac{1}{n}\left[\mu_4 - \left(\frac{n-3}{n-1}\right)\sigma^4\right], n > 1.$$

Clearly $V(S^2) \to 0$ as $n \to \infty$ and thus, by virtue of Theorem 10.7, S^2 is a consistent estimator of σ^2 or the sequence of estimators $\{S_n^2\}$ converges in probability to σ^2. (See also Theorem 8.8—the (weak) law of large numbers for variances.) It was also mentioned earlier that $S_1^2 = \frac{1}{n}\sum_{i=1}^{n}(X_i - \bar{X})^2$ is a biased (downward) estimator of σ^2. However, S_1^2 is an asymptotically unbiased estimator of σ^2 since its bias approaches zero as n becomes indefinitely large; that is, $\lim_{n\to\infty} B(S_1^2, \sigma^2) = \lim_{n\to\infty}\left(-\frac{\sigma^2}{n}\right) = 0$. And since $MSE(S_1^2, \sigma^2) = (2n - 1)\left(\frac{\sigma^4}{n^2}\right)$, it follows that $\lim_{n\to\infty} MSE(S_1^2, \sigma^2) = 0$ and thus, via Theorem 10.8, S_1^2 is a mean–squared–error–consistent estimator of σ^2. ∎

Additional considerations regarding the notion of consistency of an estimator are:

(a) An estimator whose variance corresponds to the Cramér-Rao lower bound is consistent for a parameter θ; and any such estimator is unbiased, with its variance approaching zero as $n \to \infty$.

(b) Consistency deals with the limiting behavior of T_n as $n \to \infty$; consistency does not imply that the realization of T_n is necessarily close to θ for any given n.

(c) Let T be a consistent estimator for a parameter θ with $\{a_n\}, \{b_n\}$ sequences of constants satisfying $\lim_{n\to\infty} a_n = 1$, $\lim_{n\to\infty} b_n = 0$. Then the sequence of estimators $\{U_n\}$, where $U_n = a_n T_n + b_n$, converges in probability to θ as n increases without bound so that $U = aT + b$ is a consistent estimator of θ.

(d) If T_n is a term of a consistent sequence of estimators $\{T_n\}$, then T_n is said to be a consistent estimator of θ.

(e) Consistency does not imply unbiasedness; but consistency necessarily implies asymptotic unbiasedness. In fact, the distinction between the unbiasedness and consistency of an estimator T_n of a parameter θ hinges upon the treatment of the sample size; that is, for unbiasedness, as the *number of samples* increases, T_n is on target; for consistency, as the *size of a single sample* increases, T_n is on target or T_n converges in probability to θ.

(f) Any continuous function of a consistent estimator is itself a consistent estimator. Specifically, we have Theorem 10.9, the *Slutsky Theorem:*

> **THEOREM 10.9.**   If $T_n \xrightarrow{p} \theta$ and $g(T_n)$ is a continuous function of T_n, then $g(T_n) \xrightarrow{p} g(\theta)$.

Hence consistency is *carried forward* by continuous functions; for example, if T_n is a consistent estimator of θ, then $\log T_n$ is a consistent estimator of $\log \theta$.

(g) Since convergence in probability implies convergence in distribution (see Section 8.6), it follows that if $T_n \xrightarrow{p} \theta$ (T is consistent for θ), then $T_n \xrightarrow{d} \theta$ and thus $AE(T_n) = \theta$ ($=$ constant); that is, T is an asymptotically unbiased estimator of θ. But remember that asymptotic unbiasedness does not necessarily imply consistency. However, consistency and asymptotic unbiasedness are equivalent if T_n is consistent and asymptotically normally distributed.

(h) A special case of convergence in probability is *convergence in mean square:* if an estimator T_n has mean μ_n and variance σ_n^2 and if $\lim_{n\to\infty} \mu_n = \theta$ and $\lim_{n\to\infty} \sigma_n^2 = 0$, then T_n converges in mean square to θ and $T_n \xrightarrow{p} \theta$. Note that convergence in probability does not imply convergence in mean square.

(i) Sample moments are consistent estimators of their corresponding population parameters or moments. That is, given the population random variable X, the statistic $\sum \frac{X_i^k}{n}$ is a consistent estimator of $E(X^k)$ if both the mean and variance of X^k are finite.

10.4.4 Asymptotic Efficiency

We noted earlier that in the finite sample case, other things being equal, an estimator with a small variance is preferable to one that has a large variance. This is also true for indefinitely large samples. In this regard, we now turn to a consideration of the concept of asymptotic efficiency—a property of an estimator that is related to the dispersion of its asymptotic distribution (so that the estimator must possess a finite asymptotic mean and variance). Specifically, if two estimators of a parameter θ are each asymptotically unbiased, the one with the smaller asymptotic variance is said to be *asymptotically more efficient* than the other. And the estimator with the smallest asymptotic variance among *all* asymptotic unbiased estimators of θ is said to be *asymptotically efficient*.

It was previously mentioned (in Theorem 10.8) that if an estimator T for a parameter θ is mean-squared-error consistent (i.e., if $\lim_{n \to \infty} MSE(T_n, \theta) = 0$), then it is also asymptotically unbiased. Hence first and foremost, an estimator must be mean–squared–error–consistent in order to be deemed asymptotically efficient.

According to (10.24), an estimator will be consistent if its sampling distribution concentrates perfectly on θ as $n \to \infty$; that is, $E(T_n) \to \theta$ and $V(T_n) \to 0$ as $n \to \infty$. If T is a mean–squared–error–consistent estimator of θ (hence it is also consistent in the sense of (10.24)), then it will be asymptotically efficient if its variance approaches zero faster than the variance of any other mean–squared–error–consistent estimator of θ.

We may further clarify the preceding comments by formally introducing the property of *asymptotic relative efficiency*. If two estimators T and T' are mean–squared–error–consistent (or at least both are asymptotically unbiased), the ratio of their *large sample variances* is a measure of their relative efficiency; the one with the smaller variance for a given n is relatively more efficient. To operationalize this discussion let us express the *efficiency of T relative to T'* by the ratio

$$RE = \frac{V(T_n)}{V(T_n')} \times 100, \quad 0 \le RE \le 1 \tag{10.25}$$

(where the smaller variance appears in the numerator of this expression). Then the *asymptotic efficiency of T relative to T'* is $RE_0 \times 100$, where

$$RE_0 = \lim_{n \to \infty} \frac{V(T_n)}{V(T_n')}, \quad 0 \le RE_0 \le 1. \tag{10.25.1}$$

And if T is a mean–squared–error–consistent estimator of θ and $RE_0 < 1$ for *any* other such estimator T' of θ, then T is termed asymptotically efficient.

Example 10.4.4.1 It was determined earlier that if we sample from a normal population, $S^2 = \frac{1}{n-1} \sum_{i=1}^{n} (X_i - \overline{X})^2$ and $S_1^2 = \frac{1}{n} \sum_{i=1}^{n} (X_i - \overline{X})^2$ are, respectively, unbiased and asymptotically unbiased estimators of σ^2. Then the efficiency of

S_1^2 relative to S^2 is given by

$$RE = \frac{V(S_1^2)}{V(S^2)} = \frac{\left[\frac{n-1}{n}\right]^2 \left[\frac{2\sigma^4}{n-1}\right]}{\frac{2\sigma^4}{n-1}} = \left(\frac{n-1}{n}\right)^2 \times 100.$$

And since in this instance $RE = 1$, it follows that the asymptotic efficiency of S_1^2 relative to S^2 is 100%; that is, S_1^2 and S^2 are asymptotically equally efficient. ∎

We noted in Section 10.3 that a minimum variance bound or *most efficient* estimator of a parameter θ may not always exist (especially for all sample sizes n). However, if T is unbiased and constitutes a minimum variance estimator of θ, then, as n increases without bound, it may be the case that the asymptotic variance of T equals the Cramér-Rao lower bound $CR(\theta, n)$ given by equation (10.10) (assuming that the form of the population probability density function is known). In this instance T is termed an *asymptotically most efficient estimator* of θ. Thus for very large samples, the variance of an asymptotically efficient estimator of θ can be approximated by the Cramér-Rao lower bound (provided, of course, that this lower bound exists). And if a minimum variance bound estimator of θ does not exist for all n, a large sample approximation to any such estimator can often be provided by an asymptotically efficient estimator of θ. So if an estimator is found to be *most efficient*, then it is also *asymptotically most efficient*. However, the converse of this proposition need not necessarily hold.

10.4.5 Asymptotic Normality

One important characteristic that an estimator of a parameter θ may exhibit is that, in addition to displaying many of the aforementioned desirable properties of a *good* estimator, it also tends to be asymptotically normally distributed. This observation thus leads us to specify what is called a *best asymptotically normal estimator*. To this end let $X_1, \ldots, X_n$ be a set of sample random variables drawn from a population with a finite mean (μ) and variance (σ^2). Then a sequence of estimators $\{T_n\}$ of a parameter θ is termed best asymptotically normal if and only if:

(a) The distribution of $\sqrt{n}(T_n - \theta) \to N(0, V(T_n))$ as $n \to \infty$

(b) $\{T_n\}$ is consistent for θ; that is, for every real $\varepsilon > 0$, $\lim_{n \to \infty} P(|T_n - \theta| > \varepsilon) = 0$ for all admissible θ

(c) For $\{W_n\}$ any other sequence of consistent estimators of θ, the distribution of $\sqrt{n}(W_n - \theta) \to N(0, V(W_n))$

(d) $V(T_n) \leq V(W_n)$ for all admissible θ

Note that since part (c) of this definition requires T_n to be consistent and part (d) specifies that T_n is asymptotically efficient for θ, a best asymptotically normal estimator can alternatively be termed a *consistent asymptotically normal efficient estimator* for θ.

For instance, if random samples are drawn from a population with mean μ and variance σ^2, then the sequence of estimators $\{\overline{X}_n\}$, where $\overline{X}_n = \sum_{i=1}^{n} \frac{X_i}{n}$, is a best asymptotic normal estimator of μ (via the Central Limit Theorem). The issue of $\overline{X}_n$ being *best* for μ was addressed earlier in this chapter. To consider the asymptotic normality of $\overline{X}_n$ we need only remember that Section 8.8 revealed that $Z_n = \sqrt{n}\frac{\overline{X}_n - \mu}{\sigma} \xrightarrow{d} N(0,1)$; that is, under the conditions of the Central Limit Theorem, $\overline{X}_n$ is approximately $N\left(\mu, \frac{\sigma}{\sqrt{n}}\right)$ for essentially *any* n. But as $n \rightarrow \infty, V(\overline{X}_n) = \frac{\sigma}{\sqrt{n}} \rightarrow 0$ and thus the sampling distribution of $\overline{X}_n$ becomes degenerate. To avoid this difficulty, let us work with the random variable $W_n = \sqrt{n}(\overline{X}_n - \mu)$ instead of Z_n. Then paralleling the discussion presented in Theorem 8.6, we see that as $n \rightarrow \infty$, the sequence of random variables $\{W_n\}$ has a sequence of cumulative distribution functions $\{F_n(t)\}$ that converge pointwise to the cumulative distribution function of a $N(0,\sigma)$ random variable or $W_n = \sqrt{n}(\overline{X}_n - \mu) \xrightarrow{d} N(0,\sigma)$. Hence $\overline{X}_n$ converges in distribution to a $N(0,\sigma)$ random variable.

10.5   Techniques for Finding Good Point Estimators

10.5.1   Method of Maximum Likelihood

We now turn to two estimation methods that have been proven successful in yielding point estimators that generally have *good* properties. The first is the *method of maximum likelihood* and the second is the *method of least squares.* Both methods can be classified as data reduction techniques that yield statistics, which are used to summarize sample information. The maximum likelihood method requires knowledge of the form of the population probability density (or probability mass) function, whereas the least squares method does not.

To obtain an estimate of some unknown population parameter θ, let us begin by assuming that the sample random variables $X_1, \ldots, X_n$ have been drawn from a population with probability density function $f(x;\theta)$. Given that the $X_i, i = 1, \ldots, n$, are independent and identically distributed, their joint probability density function appears as

$$l(x_1, \ldots, x_n; \theta, n) = \prod_{i=1}^{n} f(x_i; \theta)$$

wherein θ is fixed and the arguments are the variables $x_i, i = 1, \ldots, n$. But as explained earlier (see Section 8.3), if the $x_i, i = 1, \ldots, n$ are treated as realizations of the sample random variables and θ is variable and no longer held constant, then the preceding expression is termed the likelihood function of the sample and can be rewritten as

$$\mathcal{L}(\theta; x_1, \ldots, x_n, n) = \prod_{i=1}^{n} f(x_i; \theta) \tag{10.26}$$

to emphasize its dependence on θ. For computational convenience $\mathcal{L}$ will be transformed to

$$\log \mathcal{L}(\theta; x_1, \ldots, x_n, n) = \sum_{i=1}^{n} \log f(x_i; \theta). \tag{10.26.1}$$

Here (10.26) represents, in terms of θ, the *a priori* probability of obtaining the observed random sample. And as θ varies over some admissible range for fixed realization x_i, $i = 1, \ldots, n$, the said probability does likewise. In short, the (log) likelihood function expresses the probability of the observed random sample as a function of θ. (Note that if the sample random variables $X_i, i = 1, \ldots, n$, are drawn with replacement from a discrete population with probability mass function $P(X = x) = f(x; \theta)$ and the realization x_i appears n_j times, $j = 1, \ldots, k$, with $\sum_{j=1}^{k} n_j = n$, then the likelihood function assumes the form

$$\mathcal{L}(\theta; x_1, \ldots, x_k, n) = \prod_{j=1}^{k} f(x_j; \theta)^{n_j} \tag{10.27}$$

with

$$\log \mathcal{L}(\theta; x_1, \ldots, x_k, n) = \sum_{j=1}^{k} n_j \log f(x_j; \theta).) \tag{10.27.1}$$

So with θ treated as a variable in $\log \mathcal{L}$, the method of maximum likelihood is grounded in the *principle of maximum likelihood*: select as an estimate of θ that value of the parameter (call it $\hat{\theta}$) that maximizes the probability of observing the given random sample. But this means that we are looking for a value of $\theta, \hat{\theta}$, for which

$$\log \mathcal{L}(\hat{\theta}; x_1, \ldots, x_n, n) \geq \log \mathcal{L}(\theta; x_1, \ldots, x_n, n)$$

for all admissible values of θ. Hence to find $\hat{\theta}$, we need only maximize $\log \mathcal{L}$ with respect to θ. In this regard, if $\mathcal{L}$ is a twice-differentiable function of θ, then a necessary or first-order condition for $\log \mathcal{L}$ to attain a maximum at $\theta = \hat{\theta}$ is

$$\left. \frac{d \log \mathcal{L}}{d\theta} \right|_{\theta = \hat{\theta}} = 0. [5] \tag{10.28}$$

---

[5] In what follows we shall deal primarily with $\log \mathcal{L}$ rather than with $\mathcal{L}$ itself since $\log \mathcal{L}$ is a strictly monotonic function of θ and the maximum of $\mathcal{L}$ occurs at the same θ as the maximum of $\log \mathcal{L}$.

Hence all we need to do is set $\partial \log \mathcal{L}/\partial \theta = 0$ and solve for the value of $\theta, \hat{\theta}$, which makes this derivative vanish. (It is assumed that a solution to (10.28) exists and is unique.) If $\hat{\theta} = g(x_1, \ldots, x_n, n)$ is the value of θ that maximizes $\log \mathcal{L}$, then $\hat{\theta}$ will be termed the *maximum likelihood estimate* of θ; it is the realization of the *maximum likelihood estimator* $\hat{T} = g(X_1, \ldots, X_n, n)$ and represents the parameter value *most likely* to have generated the sample realizations $x_i, i = 1, \ldots, n$. Note that $\hat{\theta}$ is not the *true* value of θ but simply a rule that tells us how to calculate θ in terms of the sample realizations $x_i, i = 1, \ldots, n$.

If θ is held fixed, then the population density $f(x; \theta)$ is fully specified. But if the x_i's are held fixed and θ is variable, then we seek to determine from which density (as indexed by θ) the given set of sample values was most likely to have been drawn; that is, we want to determine from which density the likelihood is largest that the sample was obtained. This determination can be made by finding the value of $\theta, \hat{\theta}$, for which $\hat{\mathcal{L}} = \mathcal{L}(\hat{\theta}; x_1, \ldots, x_n, n) = \max_\theta \mathcal{L}(\theta; x_1, \ldots, x_n, n)$—$\hat{\theta}$ thus makes the probability of getting the observed sample greatest in the sense that it is the value of θ that would generate the observed sample *most often*.

To help reinforce your understanding of the preceding set of theoretical/ definitional considerations surrounding the method of maximum likelihood, we offer the following assortment of example problems.

Example 10.5.1.1 Suppose we have a biased coin for which one side is four times as likely to turn up on any given flip as the other. After $n = 3$ tosses we must determine if the coin is biased in favor of heads (H) or in favor of tails (T). If we define a success as getting heads on any flip of the coin, then the probability that heads occurs will be denoted as p (and thus the probability of tails on any single flip is $1 - p$). For this problem $\theta = p$ so that the Bernoulli probability mass function can be written as $P(X = x; \theta) = f(x; \theta) = p^x(1 - p)^{1-x}$. So if heads occurs, $x = 1$ and $f(1; p) = p$; and if tails occurs, $x = 0$ and $f(0; p) = 1 - p$. Since one side of the coin is four times as likely to occur as the other, the possible values of $\theta = p$ are $\frac{1}{5}$ or $\frac{4}{5}$.[6] For each of these θ's, the associated Bernoulli probability distributions are provided by Table 10.1.

Hence maximizing the logarithm of the likelihood function is equivalent to maximizing the likelihood function itself since $d \log \mathcal{L}/d\theta = (1/\mathcal{L}) \, d\mathcal{L}/d\theta = 0$ implies $d\mathcal{L}/d\theta = 0$. In addition, it is assumed that $\hat{\theta}$ is an element of the interior of the set of admissible θ's and that the sufficient or second-order condition for a maximum of $\log \mathcal{L}$ at $\hat{\theta}$ is satisfied, that is,

$$\left. \frac{d^2 \log \mathcal{L}}{d\theta^2} \right|_{\theta = \hat{\theta}} < 0. \tag{10.29}$$

[6] Since the odds of one side are *4 to 1* relative to the other, we have $y + 4y = 1$ or $y = \frac{1}{5}$. Hence $\frac{1}{5}$ is the probability of one side occurring and $\frac{4}{5}$ is the probability of the other side occurring. In general, if the odds in favor of some event A occurring are a to b, then $P(A) = \frac{a}{a+b}$; and the odds against A must be $P(\overline{A}) = \frac{b}{a+b}$.

Table 10.1

a. $\theta = \frac{1}{5}$			b. $\theta = \frac{4}{5}$	
X	$P\left(X = x; \frac{1}{5}\right) = f\left(x; \frac{1}{5}\right)$		X	$P\left(X = x; \frac{4}{5}\right) = f\left(x; \frac{4}{5}\right)$
0	$f\left(0; \frac{1}{5}\right) = \frac{4}{5}$		0	$f\left(0; \frac{4}{5}\right) = \frac{1}{5}$
1	$f\left(1; \frac{1}{5}\right) = \frac{1/5}{1}$		1	$f\left(1; \frac{4}{5}\right) = \frac{4/5}{1}$

Suppose we toss the combination (H, T, H) so that $x_1 = 1, x_2 = 0$, and $x_3 = 1$. Let us express the likelihood function for $n = 3$ as

$$\mathcal{L}(p; x_1, x_2, x_3, n = 3) = \prod_{i=1}^{3} f(x_i; p) = p^{x_1 + x_2 + x_3}(1 - p)^{3 - (x_1 + x_2 + x_3)}.$$

Then

$$\mathcal{L}(p; 1, 0, 1, n = 3) = p^2(1 - p).$$

Clearly the probability of the observed sample is a function of $\theta = p$. For, $\theta = p = \frac{1}{5}$,

$$\mathcal{L}\left(\frac{1}{5}; 1, 0, 1, n = 3\right) = \left(\frac{1}{5}\right)^2 \left(\frac{4}{5}\right) = \frac{4}{125};$$

and for $\theta = p = \frac{4}{5}$,

$$\mathcal{L}\left(\frac{4}{5}; 1, 0, 1, n = 3\right) = \left(\frac{4}{5}\right)^2 \left(\frac{1}{5}\right) = \frac{16}{125}.$$

So if the coin is biased toward heads, $\theta = p = \frac{4}{5}$ and thus the probability of the event (H, T, H) is $\frac{16}{125}$; and if it is biased toward tails, $\theta = p = \frac{1}{5}$ so that the probability of observing (H, T, H) is $\frac{4}{125}$. Since the maximum of the likelihood functions is $\frac{16}{125}$, the maximum likelihood estimate of p is $\hat{p} = \frac{4}{5}$, and this estimate yields the largest *a priori* probability of the given event (H, T, H); that is, it is the value of p that renders the observed sample combination (H, T, H) *most likely*. ∎

Example 10.5.1.2   Next, suppose that in $n = 5$ drawings (with replacement) from a vessel containing a large number of red and black balls we obtain two red and three black balls. What is the best estimate of the proportion of red balls in the vessel? Let p (respectively, $1 - p$) denote the probability of getting a red (respectively, black) ball from the vessel on any draw. Clearly the desired proportion of red balls must coincide with p. Under sampling with replacement, the various drawings yield a set of independent events and thus the probability of obtaining the given sequence of events is the product of the probabilities of the individual drawings. With two red and three black balls, the probability of the *given* sequence of events is $p^2(1 - p)^3$. But the observed sequence of outcomes is only one way of getting two red and three black balls. The total number of ways of getting two red and three black balls is $\binom{5}{2} = 10$. Hence the implied binomial probability is $b(2; 5, p) = 10p^2(1 - p)^3$. Since the number of red balls is fixed, this expression is a function of p—the likelihood function of the sample. Thus the likelihood function for the observed number of red balls is

$$\mathcal{L}(p; 2, 5) = 10p^2(1 - p)^3, \; 0 \leq p \leq 1. \tag{10.29}$$

Hence the maximum likelihood method has us choose the value for p, which makes the observed outcome of two red and three black balls the most probable outcome. To make our choice of p, let us perform the following experiment: we specify a whole range of possible p's and select the one that maximizes $\mathcal{L}$. That is, we will chose the p, $\hat{p}$, that maximizes the probability of getting the actual sample outcomes. Hence $\hat{p}$ *best explains the realized sample.* All this is carried out in Table 10.2. For instance, if $p = 0.1$, then $\mathcal{L}(0.1; 2.5) = 10(0.1)^2(0.9)^3 = 0.0729$. The remaining entries are determined in a similar fashion. So as p varies from 0 to 1, we find that the value of p which maximizes the likelihood function $\mathcal{L}$ is $\hat{p} = 0.4$ (see Figure 10.5). Hence the maximum likelihood estimator of p is the hypothetical population value that is most likely to have generated the observed (fixed) sample values; it is the value of p that maximizes $\mathcal{L}$. It should be no surprise that $\hat{p}$ equals the realization of $\hat{P} = \frac{X}{n}$, the observed relative frequency of a success or the observed sample proportion, where X is the observed number of successes. A more formal approach to solving a problem such as this is provided by the following direct application of the maximum likelihood method. ∎

Table 10.2

p	$\mathcal{L}(p; 2, 5)$
0.0	0.0000
0.1	0.0729
0.2	0.2048
0.3	0.3087
($\hat{p} =$)0.4	0.3456
0.5	0.3125
0.6	0.2304
0.7	0.1323
0.8	0.0552
0.9	0.0081
1.0	0.0000

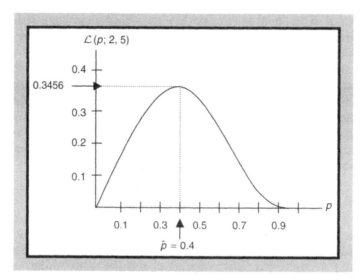

Figure 10.5 $\hat{p} = 0.4$ maximizes the likelihood function $\mathcal{L}(p; 2, 5)$.

Example 10.5.1.3   For a binomial distribution, the random variable X is defined as the number of successes obtained in n independent trials of a simple alternative experiment, where $X = 0, 1, \ldots, n$. Let p be the probability of a success (assumed constant from trial to trial). If we observe exactly $X = x$ successes in the n trials, find the maximum likelihood estimator of the binomial parameter p. Since the likelihood function for the observed number of successes x is given by the probability that $X = x$, we can write

$$\mathcal{L}(p; x, n) = \frac{n!}{x!(n-x)!} p^x (1-p)^{n-x}, 0 \le p \le 1,$$

and thus

$$\log \mathcal{L} = \log \left[\frac{n!}{x!(n-x)!} \right] + x \log p + (n-x) \log(1-p).$$

Then from (10.28), setting

$$\frac{d \log \mathcal{L}}{dp} = \frac{x}{p} - \frac{n-x}{1-p} = 0$$

renders the maximum likelihood estimate $\hat{p} = \frac{x}{n}$. (Note that $d^2 \log \mathcal{L}/dp^2 < 0$ for $\hat{p} < 1$.) Hence the maximum likelihood estimator for p is $\hat{P} = \frac{X}{n}$, the fraction of successes in the n trials. And as concluded earlier, $\hat{P}$ serves as an unbiased estimator of p. Thus the proportion of successes in a random sample of size n from a binomial population is the maximum likelihood estimator of the probability of a success in the population. (Alternatively, if $X_1, \ldots, X_n$ are taken to be independent and identically distributed Bernoulli random variables, then $f(x_i; p) = p^{x_i}(1-p)^{1-x_i}$ and thus

$$\mathcal{L}(p; x_1, \ldots, x_n, n) = \prod_{i=1}^{n} p^{x_i}(1-p)^{1-x_i} = p^y(1-p)^{n-y},$$

where $y = \sum_{i=1}^{n} x_1$. Then, provided $0 < y < n$, $\frac{\partial \mathcal{L}}{\partial p} = 0$ yields $\hat{p} = \frac{y}{n} = \frac{x}{n}$ as previously determined, since $X_i = 0$ or 1, $i = 1, \ldots, n$.) For instance, from (10.29),

$$\log \mathcal{L} = \log 10 + 2 \log p + 3 \log(1-p)$$

and thus

$$\frac{d \log \mathcal{L}}{dp} = \frac{2}{p} - \frac{3}{1-p} = 0$$

or $\hat{p} = 0.4$ as obtained in the preceding example problem. ∎

Example 10.5.1.4 Suppose the sample random variables $X_1, \ldots, X_n$ are now drawn from a population that follows an exponential distribution with probability density function

$$f(x; \lambda) = \begin{cases} \lambda e^{-\lambda x}, & \lambda > 0, \quad x \geq 0; \\ 0 & \text{elsewhere.} \end{cases}$$

What is the maximum likelihood estimator of λ? For $f(x_i; \lambda) = \lambda e^{-\lambda x_i}$, the likelihood function appears as

$$\mathcal{L}(\lambda; x_1, \ldots, x_n, n) = \lambda^n e^{-\lambda \sum_{i=1}^{n} x_i}$$

with

$$\log \mathcal{L} = n \log \lambda - \lambda \sum_{i=1}^{n} x_i.$$

From

$$\frac{d \log \mathcal{L}}{\partial \lambda} = \frac{n}{\lambda} - \sum_{i=1}^{n} x_i = 0$$

we obtain $\hat{\lambda} = \left(\frac{1}{n} \sum_{i=1}^{n} x_i \right)^{-1} = \bar{x}^{-1}$. Hence the maximum likelihood estimator of λ is the random variable $\hat{T} = \overline{X}^{-1}$ (the inverse of the sample mean). ∎

Example 10.5.1.5 It is important to note that a maximum likelihood estimator of a parameter θ can exist even if there is no admissible θ for which $d \log \mathcal{L}/d\theta = 0$, so that the calculus cannot always be used to obtain a maximum likelihood estimator for θ. For instance, if the sample random variables $X_1, \ldots, X_n$ are taken from a population that is uniformly distributed, then the implied probability density function appears as

$$f(x;\theta) = \begin{cases} \theta^{-1}, & 0 < x < \theta; \\ 0 & \text{elsewhere.} \end{cases}$$

(Here $\theta = \beta - \alpha$ in equation (7.1).) Hence

$$f(x_i;\theta) = \begin{cases} \theta^{-1}, & 0 < x_i < \theta; \\ 0 & \text{elsewhere,} \end{cases}$$

$i = 1, \ldots, n$, and thus $\mathcal{L}(\theta; x_1, \ldots, x_n, n) = \prod_{i=1}^{n} f(x_i;\theta) = \theta^{-n}$. What is the maximum likelihood estimator of θ ? Since $\mathcal{L} = \theta^{-n}$ is a monotonically decreasing function of θ, it follows that $\frac{d\mathcal{L}}{d\theta} \neq 0$ for any $0 < \theta < +\infty$. But since $\mathcal{L}$ increases when θ decreases and θ cannot be less than any of the observed sample values, it must be the case that the value of θ that maximizes $\mathcal{L}$ is $\hat{\theta} = x_{(n)} = \max x_i, i = 1, \ldots, n$, the largest of the sample realizations. Hence the maximum likelihood estimator of θ is $\hat{T} = X_{(n)}$, the nth order statistic. Although $X_{(n)}$ is not an unbiased estimator for θ, it can be transformed into an unbiased estimator by multiplying $X_{(n)}$ by $\frac{n+1}{n}$; that is, $\hat{T}' = \left(\frac{n+1}{n}\right) X_{(n)}$ is unbiased for θ. ∎

Example 10.5.1.6 Let $X_1, \ldots, X_n$ represent a set of sample random variables drawn from a normal distribution with probability density function $f(x; \mu, \sigma) = (\sqrt{2\pi}\sigma)^{-1} e^{-\frac{(x-\mu)^2}{2\sigma^2}}, -\infty < x + \infty$. Let us find the maximum likelihood estimators for μ and σ^2. To do so requires that we generalize (10.28); that is, if the likelihood function depends upon h parameters $\theta_1, \ldots, \theta_h$, then the first-order conditions for $\log \mathcal{L}(\theta_1, \ldots, \theta_h; x_1, \ldots, x_n, n)$ to attain a maximum at $\theta_j = \hat{\theta}_j, j = 1, \ldots, h$, are

$$\left. \frac{\partial \log \mathcal{L}}{\partial \theta_j} \right|_{\theta_j = \hat{\theta}_j} = 0, j = 1, \ldots, h. \tag{10.28.1}$$

Hence we need only set $\partial \log \mathcal{L}/\partial \theta_j = 0$, $j = 1, \ldots, h$, and solve the resulting simultaneous equation system for the *maximum likelihood estimates* $\hat{\theta}_j = g^j(x_1, \ldots, x_n, n)$. Here the $\hat{\theta}_j$'s are the realizations of the *maximum likelihood estimators* $\hat{T}_j = g^j(X_1, \ldots, X_n, n), j = 1, \ldots, h$ (provided, of course, that the appropriate set of second-order conditions for a maximum of log $\mathcal{L}$ are satisfied). The maximum likelihood estimates of μ and σ^2 are, respectively, the values $\hat{\mu}$ and $\hat{\sigma}^2$ for which

$$\mathcal{L}(\mu, \sigma^2; x_1, \ldots, x_n, n) = \prod_{i=1}^{n} f(x_i; \mu, \sigma^2) = (2\pi\sigma^2)^{-\frac{n}{2}} e^{-\frac{1}{2\sigma^2} \sum_{i=1}^{n} (x_i - \mu)^2}$$

or

$$\log \mathcal{L} = -\frac{n}{2} \log(2\pi) - \frac{n}{2} \log \sigma^2 - \frac{1}{2\sigma^2} \sum_{i=1}^{n} (x_i - \mu)^2$$

attains a maximum. Then from (10.28.1),

$$(a) \quad \frac{\partial \log \mathcal{L}}{\partial \mu} = \frac{1}{\sigma^2} \sum_{i=1}^{n} (x_i - \mu) = 0;$$

$$(b) \quad \frac{\partial \log \mathcal{L}}{\partial \sigma^2} = -\frac{n}{2\sigma^2} + \frac{1}{2\sigma^4} \sum_{i=1}^{n} (x_i - \mu)^2 = 0. \tag{10.30}$$

From (10.30a), $\sum_{i=1}^{n} x_i = n\mu$ or $\hat{\mu} = \frac{1}{n} \sum_{i=1}^{n} x_i = \bar{x}$ (the maximum likelihood estimate of the mean μ of a normal population is the realization of the mean estimator $\bar{X} = \frac{1}{n} \sum_{i=1}^{n} X_i$). And from (10.30b) and $\hat{\mu} = \bar{x}$ we obtain $\hat{\sigma}^2 = \frac{1}{n} \sum_{i=1}^{n} (x_i - \mu)^2 = \frac{1}{n} \sum_{i=1}^{n} (x_i - \bar{x})^2$ (the maximum likelihood estimate of the population variance σ^2 of a normal population is the realization of the variance estimator $S_1^2 = \frac{1}{n} \sum_{i=1}^{n} (X_i - \bar{X})^2$). As indicated earlier, $\bar{X}$ is an unbiased estimator of μ and S_1^2 is a biased estimator of σ^2. Hence the method of maximum likelihood may not always yield unbiased estimates of population parameters. ∎

Having discussed the particulars of the maximum likelihood method for determining point estimators of parameters, let us now turn to an inventory of their (desirable) properties. Specifically:

(a) Maximum likelihood estimators may or may not be unbiased. For example, under random sampling from a normal population, we found (in Example 10.5.1.6) that $\bar{X}$ was unbiased for μ but that S_1^2 was a biased estimator for σ^2.

(b) If the maximum likelihood estimator $\hat{\theta}$ of a parameter θ is unique, then $\hat{\theta}$ will be a function of a sufficient statistic. So if a sufficient statistic for θ exists, it can always be found by the method of maximum likelihood. To see this let $T = g(X_1, \ldots, X_n, n)$ be a sufficient statistic for a parameter θ. Then the maximum likelihood estimator $\hat{\theta}$ is always a function of T in that $\hat{\theta}$ depends on the sample observations only through the value of T. That is, if T is a sufficient statistic for θ, then by virtue of the Fisher–Neyman Factorization

Theorem, the likelihood function can be factored as (10.14) and thus $\log \mathcal{L} = \log h(t; \theta, n) + \log j(x_1, \ldots, x_n, n)$. Since $\log j$ does not depend on θ, maximizing $\log \mathcal{L}$ with respect to θ is equivalent to maximizing $\log h$ with respect to θ. And since $\log h$ depends on the sample observations only through T, $\hat{\theta}$ depends on the same only through T. So if T is sufficient for θ, then the maximum likelihood estimator $\hat{\theta}$ is always a function of T.

Keep in mind, however, that the dependence of $\hat{\theta}$ on a sufficient statistic does not guarantee that a maximum likelihood estimator will always be unbiased or that it will have a variance that attains the Cramér-Rao lower bound. For example, it can easily be demonstrated that $S_0^2 = \frac{1}{n} \sum_{i=1}^{n}(X_i - \mu)^2$ is a maximum likelihood estimator of σ^2 when the sample random variables are drawn from a normal population with known mean μ. Moreover, S_0^2 is unbiased for σ^2 and the variance of S_0^2 equals the Cramér-Rao lower bound. However, $S_1^2 = \frac{1}{n} \sum_{i=1}^{n}(X_i - \overline{X})^2$ is also a maximum likelihood estimator of σ^2 when sampling from a normal population. But this estimator is biased and does not have a variance that coincides with the Cramér-Rao lower bound.

(c) The maximum likelihood technique generally will yield an efficient or minimum variance unbiased estimator of a parameter θ, if one exists. This follows from the preceding set of comments concerning sufficiency. That is, if a maximum likelihood estimator $\hat{\theta}$ can be found and $\hat{\theta}$ is unbiased for θ (or perhaps $\hat{\theta}$ can be transformed to become unbiased for θ), then $\hat{\theta}$ will typically be an efficient or best unbiased estimator of θ. So if an efficient estimator of θ exists, the method of maximum likelihood can be used to find it. For example, under random sampling from a normal population, $\overline{X}$ is the maximum likelihood estimator of μ. Since $\overline{X}$ is unbiased, it is best unbiased or efficient for μ. Under the same sampling scheme, S_1^2 is a maximum likelihood estimator of σ^2. But S_1^2 is biased for σ^2 and thus cannot be the best unbiased estimator of σ^2. However, $S^2 = \left(\frac{n}{n-1}\right) S_1^2$ is unbiased for σ^2. So when a maximum likelihood estimator $\hat{\theta}$ for θ is biased, if there exists a best unbiased estimator for θ, then it must be some variant of $\hat{\theta}$.

(d) The maximum likelihood method yields estimators that possess the *invariance property*, which essentially states that maximum likelihood estimation is invariant under a transformation of parameters. More formally, for $\hat{\theta}$ a maximum likelihood estimator of θ, if $\phi(\theta)$ is a single-valued function of θ (for each θ there is *unique* $\phi(\theta)$), then the maximum likelihood estimator of $\phi(\theta)$ is $\phi(\hat{\theta})$. For instance, if our maximum likelihood estimator of θ is $\hat{T}$, then:

- Our estimator of θ^2 is $\hat{T}^2$

- Our estimator of $\theta + k$ is $\hat{T} + k, k$ a constant

- Our estimator of $k\theta$ is $k\hat{T}, k$ a constant

- Our estimator of $\alpha\theta + \beta$ is $\alpha\hat{T} + \beta, \alpha$ and β constants, and so on

So if a maximum likelihood estimate has already been obtained for a parameter θ, and an estimate is desired for $\phi(\theta)$, there is no need to reestimate the model—$\phi(\hat{\theta})$ will suffice. Hence under this invariance property of maximum likelihood estimators, we can, for computational convenience, reparameterize the likelihood function in any fashion that is suitable for our purposes. For instance, in Example 10.5.1.4 (the probability density function is exponential in form), we could set $\lambda = \frac{1}{\tau}$. Then $\log \mathcal{L} = n \log\left(\frac{1}{\tau}\right) - \frac{1}{\tau}\sum_{i=1}^{n} x_i$. Setting $\partial \log \mathcal{L}/\partial \tau = 0$ yields $\hat{\tau} = \bar{x}$. Then $\hat{\lambda} = \hat{\tau}^{-1} = \bar{x}^{-1}$. In a similar vein we may note that if $S_1^2 = \frac{1}{n}\sum_{i=1}^{n}(X_i - \bar{X})^2$ is the (biased) maximum likelihood estimator of σ^2 when sampling from a normal population, then, via the invariance property, the maximum likelihood estimator of σ is $S_1 = \left[\frac{1}{n}\sum_{i=1}^{n}(X_i - \bar{X})^2\right]^{1/2}$. Hence under the invariance property, it makes no difference whether we maximize the likelihood function as a function of θ or as a function of ϕ (θ). However, for the class of unbiased estimators, the invariance property may not be satisfied. For example, although $\left(\frac{n}{n-1}\right)S_1^2$ is an unbiased estimator for σ^2, the positive square root of this estimator is not an unbiased estimator for σ. Hence unbiasedness and invariance may not always be compatible properties of a maximum likelihood estimator or of its transform.

We next turn to a collection of asymptotic properties of maximum likelihood estimators. In particular:

(e) Maximum likelihood estimators are consistent estimators of population parameters. In this regard, for $X_1,\ldots,X_n$ a set of sample random variables taken from a population with probability density function $f(x;\theta)$, let $\hat{\theta}_n$ denote the maximum likelihood estimator of θ. Then under certain regularity conditions on f (and thus on $\mathcal{L}$),[7] $\hat{\theta}$ is said to be a consistent estimator of a parameter θ if the sequence of estimator $\{\hat{\theta}_n\}$ converges in probability to θ as n increases without bound or if

$$\lim_{n \to \infty} P(|\hat{\theta}_n - \theta| < \varepsilon) = 1$$

for every real $\varepsilon > 0$ and every admissible θ.

---

[7] The usual set of *regularity conditions* imposed upon the likelihood function when asymptotic properties of maximum likelihood estimators are discussed, and which guarantee that $\partial \log \mathcal{L}/\partial \theta = 0$ has a solution $\hat{\theta}$ that asymptotically approaches θ_0 (the true or exact value of θ) are:

(a) The range of x is independent of θ.

(b) The probability density function $f(x;\theta)$ possesses derivatives with respect to θ of at least third order, and these derivatives are bounded by integrable functions of x. This requirement holds for all x and all θ within an open interval containing θ.

(c) $E(\partial \log \mathcal{L}/\partial \theta) = 0$ and $E(\partial \log \mathcal{L}/\partial \theta)^2 = E(-\partial^2 \log \mathcal{L}/\partial \theta^2)$.

(d) $\partial \log f(x;\theta)/\partial \theta|_{\theta=\theta_0}$ has a positive and finite variance.

(f) Under the preceding set of regularity conditions on the population probability density function (and thus on $\mathcal{L}$), maximum likelihood estimators of population parameters are generally asymptotically efficient; that is, as $n \to \infty$, the sequence of maximum likelihood estimators $\{\hat{\theta}_n\}$ for θ asymptotically attains the Cramér-Rao lower bound. In this regard, the maximum likelihood estimator is the *most efficient* estimator of θ since it has minimum asymptotic variance. Under this property, the Cramér-Rao lower bound provides a convenient variance approximation for maximum likelihood estimators of θ.

(g) Let $X_1, \ldots, X_n$ depict a set of sample random variables drawn from a population probability density function $f(x; \theta)$ and let f satisfy the preceding regularity conditions specified in footnote 7. Then the maximum likelihood estimator $\hat{\theta}_n$ is asymptotically normal with mean θ (hence $\hat{\theta}_n$ is asymptotically unbiased) and variance corresponding to the Cramér-Rao lower bound. Moreover, the sequence of maximum likelihood estimators $\{\hat{\theta}_n\}$ for θ is best asymptotically normal.

10.5.2 Method of Least Squares

A second technique for obtaining a *good* (point) estimator of a parameter θ is the method of least squares. Least squares estimators are determined in a fashion such that desirable properties of an estimator are essentially *built into them* by virtue of the process by which they are constructed. In this regard, least squares estimators are *BLUE*; that is, *best linear unbiased estimators*. (Remember *best* means that out of the class of all unbiased linear estimators of θ, the least squares estimators have minimum variance, and thus minimum mean squared error.) Additionally, least squares estimators have the advantage that knowledge of the form of the population probability density function is not required.

Example 10.5.2.1   Let us determine a least squares estimator for the population mean μ. Given that the sample random variables $X_1, \ldots, X_n$ are independent and identically distributed, it follows that $E(X_i) = \mu$ and $V(X_i) = \sigma^2$ for all $i = 1, \ldots, n$. Now, let $X_i = \mu + \varepsilon_i$, $i = 1, \ldots, n$, where ε_i is an *observational random variable* with $E(\varepsilon_i) = 0$ and $V(\varepsilon_i) = \sigma^2$. Under random sampling, ε_i accounts for the difference between X_i and its mean μ. Then the *principle of least squares* directs us to choose μ (the X_i's are fixed, $i = 1, \ldots, n$) so as to minimize the sum of the squared deviations between the observed X_i values and their mean μ; that is, we should choose μ so as to minimize $\sum_{i=1}^{n} \varepsilon_i^2 = \sum_{i=1}^{n}(X_i - \mu)^2$. To this end we set

$$\frac{\partial \sum_{i=1}^{n} \varepsilon_i^2}{\partial \mu} = 2 \sum_{i=1}^{n}(X_i - \mu)(-1) = 0$$

so as to obtain $\hat{\mu} = \overline{X}$. (Note that $\frac{\partial^2}{\partial \mu^2} \sum_{i=1}^{n} \varepsilon_i^2 = 2n > 0$ as required for a minimum.) Hence the least squares estimator of the population mean μ is the sample mean $\overline{X}$. ∎

In general, the least squares method is highly appropriate for estimating moments about zero of a probability distribution. For a population random variable X, its rth moment about zero is $\mu'_r = E(X^r), r = 0,1,2,\ldots$. Then by the just provided argument, the principle of least squares has us choose μ'_r so as to minimize the quantity $\sum_{i=1}^{n} \varepsilon_i^2 = \sum_{i=1}^{n}(X_i^r - \mu'_r)^2$. Setting $\frac{\partial}{\partial \mu'} \sum_{i=1}^{n} \varepsilon_i^2 = 0$ yields the least squares estimator $\mu'_r = \frac{1}{n}\sum_{i=1}^{n} X_i^r$. Under some very mild restrictions (the mean and variance of X^r are finite), these estimators are generally consistent and asymptotically normal. Whereas least squares estimators are limited to the class of linear estimators, this is not the case for maximum likelihood estimators.

10.6 Exercises

10-1. Let $X_1,\ldots,X_n$ constitute a random sample taken from an exponential distribution with probability density function

$$f(x;\theta) = \begin{cases} \frac{1}{\theta}e^{-x/\theta}, & x > 0, \theta > 0; \\ 0 & \text{elsewhere.} \end{cases}$$

Which of the following point estimators for θ are unbiased:

$$T_1 = X_2, T_2 = \frac{X_1 + X_3}{2}, \quad \text{or} \quad T_3 = \overline{X}?$$

Find the efficiency of T_1 relative to T_3.

10-2. Suppose $\{X_1, X_2, \ldots, X_n\}$ is a set of sample random variables taken from a population with mean μ and variance σ^2 and that $T_1 = \overline{X}$ and $T_2 = \frac{X_1 + X_n}{2}$ are point estimators of μ. Are T_1 and/or T_2 unbiased? Find the efficiency of T_1 relative to T_2.

10-3. Suppose $\{X_1, X_2, \ldots, X_n\}$ is a set of sample random variables taken from a uniform distribution on the interval $(\theta, \theta + 1)$. Is $T_1 = \overline{X} - 0.5$ an unbiased estimator of θ? Is $T_2 = 2\overline{X}$ an unbiased estimator of θ?

10-4. Suppose $\{X_1, X_2, \ldots, X_n\}$ is a set of sample random variables taken from a population that is $N(\mu, \sigma)$. Is $T = \frac{(X_1 - \overline{X})^2}{n}$ an unbiased estimator of σ^2?

10-5. Suppose that T_1 and T_2 are independent unbiased estimators of a parameter θ, with $V(T_1) = \sigma_1^2$ and $V(T_2) = \sigma_2^2$. If $T_3 = \alpha T_1 + (1 - \alpha)T_2$, α constant, is a new unbiased point estimator of θ, how should α be chosen to minimize $V(T_3)$?

10-6. Suppose $\{X_1, X_2, \ldots, X_n\}$ depicts a set of sample random variables drawn from a population for which $E(X_i) = \mu$ and $V(X_i) = \sigma^2$. Let $T_1 = \sum_{i=1}^{n} \frac{X_i}{n+1}$ and $T_2 = \sum_{i=1}^{n} \frac{X_i}{n-1}$ be possible estimators for μ. Compare:

(a) $E(T_1)$ and $E(T_2)$

(b) $MSE(T_1, \mu)$ and $MSE(T_2, \mu)$

10-7. Let $\{X_1, X_2, \ldots, X_n\}$ be a set of independent and identically distributed sample random variables drawn from a binomial population with unknown parameter p. Two possible estimators for p are $T_1 = \frac{X}{n}$ and $T_2 = \frac{X+1}{n+1}$, where X is the observed number of successes. Are both of these estimators unbiased? Compare the mean square error of each.

10-8. The statistics $\overline{X}$ and S^2 are unbiased estimators of μ and σ^2, respectively, (for all μ and σ^2) for $\{X_1, X_2, \ldots, X_n\}$ a set of sample random variables. Find the mean square errors of $\overline{X}$ and S^2.

10-9. Let $\{X_1, X_2, \ldots, X_n\}$ represent a set of Bernoulli random variables. For $\overline{X}$ an estimator of unknown p (the probability of a success), find $MSE(\overline{X}, p)$.

10-10. A set of sample random variables $\{X_1, X_2, X_3\}$ is taken from a population with mean μ and variance σ^2. Find the mean squared error of the following estimators for μ:

(a) $T_1 = (X_1 + X_2 + 2X_3)/4$

(b) $T_2 = (2X_1 + X_2 + X_3)/3$

(Hint: Use the relationships $V(T) = E(T^2) - E(T)^2$ and $E(X_i^2) = V(X_i) + E(X_i)^2 = \sigma^2 + \mu^2$.)

10-11. Let $\{X_1, X_2, \ldots, X_n\}$ constitute a set of sample random variables taken from a general population probability density function with $E(X_i) = \mu' = \mu$, $E(X_i^2) = \mu'_2$, and $E(X_i^4) = \mu'_4$ all finite. Verify that $S^2 = \sum_{i=1}^{n} \frac{(X_i - \overline{X})^2}{n-1}$ is a consistent estimator of σ^2.

10-12. Let $\{X_1, X_2, \ldots, X_n\}$ depict a set of sample random variables drawn from the probability density function

$$f(x; \theta) = \begin{cases} \theta x^{\theta-1}, & 0 < x < 1, \theta > 0; \\ 0 & \text{elsewhere.} \end{cases}$$

Find a consistent estimator for θ.

10-13. Let $\{X_1, X_2, \ldots, X_n\}$ be a collection of sample random variables taken from the probability density function $f(x; \theta) = \frac{1}{2}(1 + \theta x), -1 < x, \theta < 1$. Find a consistent estimator for θ.

10-14. Let $\{X_1, X_2, \ldots, X_n\}$ depict a set of sample random variables drawn from a $N(\mu, 1)$ population. For each random variable Y_n presented, determine if there exists a real number c such that $Y_n \xrightarrow{p} c$ as $n \to \infty$:

(a) $Y_n = e^{\bar{X}_n}$

(b) $Y_n = e^{\bar{X}_n^2 - 2\bar{X}_n}$

10-15. Verify that if $X_1, \ldots, X_n$ depicts a random sample from a population for which both $E(X^k)$ and $V(X^k)$ exist, then $\frac{1}{n}\sum_{i=1}^{n} X_i^k$ is a consistent estimator of $E(X^k)$. Use this result to demonstrate that $S_1^2 = \frac{1}{n}\sum_{i=1}^{n}(X_i - \bar{X})^2$ is a consistent estimator of σ^2.

10-16. Suppose $\{X_1, X_2, \ldots, X_n\}$ is a set of sample random variables drawn from a $N(\mu, \sigma)$ population. Consider two estimators for μ:

$$T_1 = \sum_{i=1}^{n} \frac{X_i}{n+1}, \quad T_2 = \frac{1}{2}X_i + \frac{1}{2n}\sum_{i=2}^{n} X_i.$$

Which of these two estimators has the largest bias? Are these estimators efficient? Are T_1 and T_2 asymptotically unbiased? Are they consistent? Are they asymptotically efficient?

10-17. Determine a lower bound for the probability that $\hat{P} \left(= \frac{X}{n}\right)$ lies within an ε-neighborhood of $p(= E(\hat{P}))$. For what value of n will this probability exceed the value $1 - \delta$?

10-18. Suppose $\{X_1, X_2, \ldots, X_n\}$ is a set of sample random variables drawn from the exponential probability density function $f(x; \theta) = \left(\frac{1}{\theta}\right)e^{-x/\theta}, 0 < x < +\infty$. Verify that $\bar{X}$ is a sufficient statistic for θ.

10-19. Let $X_1, \ldots, X_n$ be a random sample drawn from a $N(\mu, \sigma)$ probability density function. Maximum likelihood estimators for μ and σ^2 are $\bar{X}$ and $S_1^2 = \frac{1}{n}\sum_{i=1}^{n}(X_i - \bar{X})^2$, respectively. Demonstrate that $\bar{X}$ is an unbiased estimator for μ and S_1^2 is a biased estimator for σ^2. Find the mean squared error for both $\bar{X}$ and S_1^2.

10-20. Let $X_1, \ldots, X_n$ constitute a random sample drawn from a uniform distribution on $\left[\theta - \frac{1}{4}, \theta + \frac{1}{4}\right]$. Is X_i an unbiased estimator for θ?

10-21. Suppose $\{X_1, X_2, \ldots, X_n\}$ is a set of independent sample random variables drawn from the probability mass function $g(x; \theta) = \theta(1 - \theta)^{x-1}$, $x = 1, 2, \ldots, 0 < \theta < 1$. Find a sufficient statistic for θ.

10-22. Let $\{X_1, X_2, \ldots, X_n\}$ be a set of sample random variables for each of the following probability density functions. Find a sufficient statistic for $\theta > 0$:

(a) $f(x; \theta) = 2\theta^{-2}x, 0 < x < \theta$

(b) $f(x; \theta) = \theta x^{\theta-1}, 0 < x < 1, \theta > 0$

10-23. Suppose $X_1, \ldots, X_n$ is a random sample taken from a variable X that is $N(\mu, \sigma)$. Express $P\left(|\bar{X}_n - \mu| < \varepsilon\right)$ as an integral that converges to 1 as $n \to \infty$.

10-24. Suppose a random variable X has a probability density function of the form

$$f(x; \lambda) = \begin{cases} \lambda e^{-\lambda x}, & x \geq 0, \lambda > 0; \\ 0 & \text{elsewhere.} \end{cases}$$

For $X_1, \ldots, X_n$ a random sample drawn from this exponential distribution, let $T = \sum_{i=1}^{n} \frac{X_i}{n}$ be an estimator for λ. Find $MSE(T, \lambda)$.

10-25. Let $\{X_1, X_2, \ldots, X_n\}$ depict a set of sample random variables drawn from the probability density function $f(x; \theta) = 2\theta^{-2}x, 0 < x < \theta$. Find a best unbiased estimator for θ^2.

10-26. Suppose $\{X_1, \ldots, X_n\}$ is a set of sample random variables drawn from the probability density function:

$$f(x; \theta) = \theta/(1 + x)^{\theta+1}, 0 < x < \theta < +\infty.$$

Find a sufficient statistic for θ.

10-27. Given that $X_1, \ldots, X_n$ depicts a random sample drawn from a variable X that is $N(0, \sigma)$, find a sufficient statistic for σ^2.

10-28. Suppose $X_1, \ldots, X_n$ is a random sample taken from a population with known mean $\mu = \mu_0$ and unknown variance σ^2. Is $S_0^2 = \frac{1}{n} \sum_{i=1}^{n} (X_i - \mu_0)^2$ an unbiased estimator of σ^2?

10-29. Let $X_1, \ldots, X_n$ be a random sample taken from a Bernoulli probability distribution $\left(P(X_i = 1) = p, P(X_i = 0) = 1 - p, \text{with } p \text{ unknown}\right)$. Find a minimal sufficient statistic for p.

10-30. Given the probability density function presented in Exercise 10–12, find a minimal sufficient statistic for θ.

10-31. Suppose $\{X_1, X_2, \ldots, X_n\}$ is a set of sample random variables taken from a population distribution that is $N(\mu, \sigma)$. Demonstrate that $\sum_{i=1}^{n} (X_i - \bar{X})^2$ is a minimal sufficient statistic for σ^2. Use this statistic to find an efficient estimator for σ^2.

10-32. Let $\{X_1, X_2, \ldots, X_n\}$ be a set of sample random variables drawn from the probability density function $f(x; \theta) = \theta(1 - \theta)^{x-1}, x = 1, 2, \ldots; 0 < \theta < 1$. Find a minimal sufficient statistic for θ.

10-33. Let $X_1, \ldots, X_n$ represent a random sample taken from a population with mean μ and variance σ^2. Let $\bar{X}_n = \frac{1}{n} \sum_{i=1}^{n} X_i$ and $S_n^2 = \frac{1}{(n-1)} \sum_{i=1}^{n} (X_i - \bar{X})^2$ depict sequences of estimators for μ and σ^2, respectively. Verify that $\{\bar{X}_n\}$

and $\{S_n^2\}$ are mean–squared–error–consistent sequences of estimators for μ and σ^2, respectively.

10-34. It is known that for X a $N(\mu, \sigma)$ random variable, the maximum likelihood estimator of σ^2 is $\hat{T} = S_1^2 = \frac{1}{n}\sum_{i=1}^{n}(X_i - \overline{X})^2$ and that T is a biased estimator of σ^2. Demonstrate that the bias goes to zero as $n \to \infty$.

10-35. Suppose $X_1, \ldots, X_n$ constitutes a random sample drawn from a $N(\mu, \sigma)$ population. Does $S^2 = \frac{1}{n-1}\sum_{i=1}^{n}(X_i - \overline{X})^2$ converge to σ^2 in probability?

10-36. Suppose $X_1, \ldots, X_n$ depicts a random sample drawn from a binomial population. In addition, let $X = \sum_{i=1}^{n} X_i$ success be observed. Is $\hat{P} = \frac{X}{n}$ a consistent estimator for the binomial parameter p?

10-37. Let $X_1, \ldots, X_n$ represent a random sample drawn for the exponential probability density function $f(x; \lambda) = \lambda e^{-\lambda x}, x \geq 0, \ \lambda > 0$. Find an asymptotically efficient estimator for λ.

10-38. For $X_1, \ldots, X_n$ a random sample drawn from the probability density function $f(x; \theta) = \frac{1}{\theta}e^{-x/\theta}, x \geq 0, \theta > 0$, find a minimal sufficient statistic for θ.

10-39. Let $X_1, \ldots, X_n$ depict a random sample drawn from the exponential probability density function $f(x; \theta) = \frac{1}{\theta}e^{-x/\theta}, x \geq 0, \theta > 0$. Find an efficient or minimum variance unbiased estimator for $V(X_i)$.

10-40. Let $X_1, \ldots, X_n$ depict a random sample drawn from the Poisson probability density function $f(X; \theta) = \theta^X e^{-\theta}/X!, \ x = 0, 1, 2, \ldots$. It is known that $T = \sum_{i=1}^{n} X_i$ is a sufficient statistic for the mean $\theta > 0$. Demonstrate that $\frac{T}{n} = \overline{X}$ is an efficient statistic for θ.

10-41. Let $S_1^2 = \frac{1}{n}\sum_{i=1}^{n}(X_i - \overline{X})^2$ denote the variance of a random sample of size $n(>1)$ drawn from a $N(\mu, \theta)$ probability density function. The estimator $T = \frac{n}{n-1}S_1^2$ is unbiased since $E(T) = \theta$. Is the statistic T the most efficient estimator for θ?

10-42. Let $\{X_1, X_2, \ldots, X_n\}$ be a collection of sample random variables drawn from the probability mass function $f(x; \theta) = \frac{\theta^x e^{-\theta}}{x!}, x = 0, 1, 2, \ldots, 0 < \theta < +\infty$. Verify that the family of distributions of $X = \sum_{i=1}^{n} X_i$ is complete.

10-43. Let $\{X_1, X_2, \ldots, X_n\}$ constitute a set of sample random variables from a Poisson probability density function with parameter (mean) λ. Find the maximum likelihood estimator of θ.

10-44. Let $\{X_1, X_2, \ldots, X_n\}$ be a set of sample random variables drawn from the probability density function

$$f(x; \theta) = \begin{cases} (1 + \theta)x^\theta, & 0 \leq x \leq 1, \theta > 0; \\ 0 & \text{elsewhere.} \end{cases}$$

Find the maximum likelihood estimator of θ.

10-45. Consider a family of probability density functions of the form $f(x;\theta) = \frac{1}{\theta}, 0 < x < \theta$. Is this family complete?

10-46. Suppose we draw a random sample of size n from a multinomial distribution with probability mass function (6.24). Additionally, suppose there occurs x_j outcomes of type $A_j, j = 1, \ldots, k$. Determine the maximum likelihood estimators for the parameters $p_j, j = 1, \ldots, k$. (Hint: Maximize the likelihood function subject to the restriction $\sum_{j=1}^{k} p_j = 1$.)

10-47. Suppose $X_1, \ldots, X_n$ constitutes a random sample drawn from a population that is $N(\mu, 1)$. Find the maximum likelihood estimator of μ.

10-48. Suppose $\{X_1, X_2, \ldots, X_n\}$ represents a set of sample random variables drawn from each of the following probability density functions. Find a maximum likelihood estimator of θ:

(a) $f(x;\theta) = \frac{1}{\theta} x^{\frac{1}{\theta}-1}, 0 < x < 1, \theta > 0$

(b) $f(x;\theta) = \theta x^{\theta-1}, 0 \le x \le 1, \theta > 0$

(c) $f(x;\theta) = \frac{1}{\theta^2} x e^{-x/\theta}, x > 0, \theta > 0$

10-49. Let $\{X_1, X_2, \ldots, X_n\}$ represent a set of sample random variables drawn from the probability density function

$$f(x;\theta_1,\theta_2) = \begin{cases} \frac{1}{\theta_2} e^{-(x-\theta_1)^2/\theta_2}, & -\infty < \theta_1 \le x < +\infty, 0 < \theta_2 < +\infty, \\ 0 & \text{elsewhere.} \end{cases}$$

Find maximum likelihood estimates for θ_1 and θ_2. Are these estimates jointly sufficient statistics?

10-50. Suppose $\{X_1, X_2, \ldots, X_n\}$ constitutes a set of sample random variables drawn from the gamma probability density function (7.4.8). Determine maximum likelihood estimates of α, θ. (Note: Do not expect to find a closed form solution.)

10-51. Suppose $X_1, \ldots, X_n$ depicts a random sample drawn from the probability density function $f(x,\theta) = \frac{1}{\theta} e^{-x/\theta}, x \ge 0, \theta > 0$. Find a minimum variance bound estimator for θ.

10-52. Let $\{X_1, X_2, \ldots, X_n\}$ depict a set of sample random variables drawn from a $N(\theta, 1)$ population. Find a best unbiased estimator of θ^2. Does the variance of this estimator attain the Cramér-Rao lower bound?

10-53. Let $X_1, \ldots, X_n$ depict a random sample drawn from the probability density function $f(x;\theta) = \theta x^{\theta-1}, 0 < x < 1, \theta > 0$. Find the minimum variance bound for an unbiased estimator of θ.

10-54. Let $\{X_1, X_2, \ldots, X_n\}$ be a set of sample random variables drawn from the *Cauchy probability density function* $f(x; \theta) = \{\pi [1 + (x - \theta)^2]\}^{-1}$, $-\infty < x, \theta < +\infty$. Verify that the Cramér-Rao lower bound is $\frac{2}{n}$.

10-55. Let $\{X_1, X_2, \ldots, X_n\}$ be a set of sample random variables taken from a geometric distribution with probability mass function $f(X; p) = (1 - p)^{X-1}p, X = 1, 2, 3, \ldots, 0 < p < 1$. Determine the maximum likelihood estimator of p.

10-56. Suppose $X_1, \ldots, X_n$ is a random sample drawn from the probability mass function $f(x; \theta) = P(X = x) = \theta(1 - \theta)^{x-1}, x = 1, 2, \ldots, 0 < \theta < 1$. Is this family of distributions complete?

10-57. Suppose $X_1, \ldots, X_n$ is a random sample taken from a binomial probability mass function $b(X; n, p)$. Is the statistic $X = \sum_{i=1}^{n} X_i$ complete for p?

10-58. Method of Moments Technique for Finding Point Estimators:
Let $\{X_1, X_2, \ldots, X_n\}$ be a set of sample random variables taken from a distribution with a probability density function $f(x; \theta_1, \ldots, \theta_k)$. Let $\mu_r' = E(X^r)$ denote the rth moment of the *distribution* with $M_r' = \sum_{i=1}^{n} \frac{X_i^r}{n}$ representing the rth moment of the *sample*, $r = 1, 2, 3, \ldots$. Then the *method of moments* has us set $E(X^r)$ to M_r' for as many values of $r = 1, 2, \ldots$, as needed in order to obtain enough equations to uniquely solve for $\theta_1, \ldots, \theta_k$. For $X_1, \ldots, X_n$ taken from $f(x; \theta) = \theta x^{\theta-1}, 0 < x < 1; 0 < \theta < +\infty$, obtain an estimate of θ by the method of moments. (Here we need only $r = 1$.)

10-59. Suppose $X_1, \ldots, X_n$ represents a random sample drawn from a uniform distribution with probability density function $u(x; 0, \theta) = \frac{1}{\theta}, 0 < x < \theta, 0$ elsewhere. Use the method-of-moments technique to estimate the parameter θ.

10-60. Demonstrate that the method-of-moments estimator for θ determined in Exercise 10-59 is a consistent estimator of θ.

10-61. Let $\{X_1, X_2, \ldots, X_n\}$ correspond to a set of sample random variables taken from a distribution that is $N(\mu, \sigma)$. Find estimates of μ and σ using the method of moments. (Hint: First determine method-of-moments estimates of μ and σ^2.)

10-62. Let $\{X_1, X_2, \ldots, X_n\}$ depict a set of sample random variables from the Poisson probability mass function $f(x; \lambda) = \frac{\lambda^x e^{-\lambda}}{x!}, x = 0, 1, 2, \ldots, 0 < \lambda < +\infty$. Determine the method-of-moments estimator of λ.

10-63. If $\{X_1, X_2, \ldots, X_n\}$ is a set of sample random variables taken from the probability density function $f(x; \theta) = \theta e^{-\theta x}, 0 < x < +\infty$, find the method-of-moments estimator of θ.

10-64. Let $\{X_1, X_2, \ldots, X_n\}$ represent a set of sample random variables drawn from the probability mass function $f(x; \theta) = \theta^x(1 - \theta)^{1-x}, x = 0, 1$ and $0 \leq \theta \leq \frac{1}{2}$. Find the method-of-moments estimator of θ.

10-65. Let $X_1, \ldots, X_n$ be a random sample taken from the probability density function $f(x; \theta) = (\theta + 1)x^\theta, 0 < x < 1$. Find a method-of-moments estimator for θ.

10-66. Suppose $X_1, \ldots, X_n$ is a random sample drawn from a probability density function that is $N(\mu, \sigma)$. Find a constant c such that cS is an unbiased estimator for σ. (Hint: $c = \dfrac{\left[\sqrt{\frac{n-1}{2}}\,\Gamma\left(\frac{n-1}{2}\right)\right]}{\Gamma\left(\frac{n}{2}\right)}$.)

Interval Estimation and Confidence Interval Estimates

11.1 Interval Estimators

In the preceding chapter we discussed issues concerning the determination of a point estimator $T = g(X_1, \ldots, X_n, n)$ for a parameter θ of a population distribution $f(x; \theta)$. As indicated therein, the realization $t = g(x_1, \ldots, x_n, n)$ of T reports a single numerical value as the estimate of θ. Although a point estimator may possess one or more of an assortment of desirable properties (as defined in Chapter 10), its realization alone does not offer any assessment of how precisely the parameter θ has been estimated by the chosen sample. What we need to do is accompany t by some measure of the possible error associated with our estimate of θ. That is, we need to bracket t with an interval that carries with it some *assurance* that θ is a member of the interval.

Our approach to estimation in this chapter is to now report a range of likely values for θ rather than a single point estimate. This range of values is termed an *interval estimate* and typically is referred to as a *confidence interval*—a range of values that enables us to state, with a certain *degree of confidence,* that the reported interval contains θ. The role of a confidence interval is to indicate how precisely θ has been estimated from the sample; that is, the narrower the interval, the more precise the estimate.

We noted in the previous chapter that if T is a point estimator of θ, then, for any particular sample, its realization t will not equal θ precisely; that is, from equation (10.1), $\theta = t \pm SE(t, \theta)$, where $SE(t; \theta)$ is the realized sampling error of T. Hence to determine an interval estimate for θ, we must specify how large this allowance for sampling error must be. Clearly any adjustment for the effects of sampling error must take into account the form or characteristics of the sampling distribution of $T, h(t; \theta, n)$, as well as how confident we choose to be that the interval contains θ.

To this end, let us bracket T by the *error bound*, $\pm \Delta T$, where this error bound will be taken as some multiple k of the standard error of the sampling

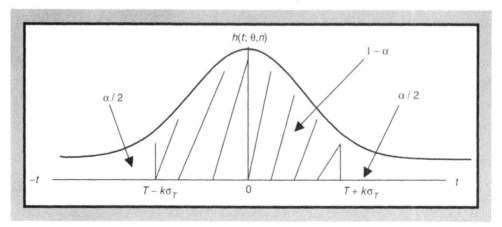

Figure 11.1   The confidence probability is $1 - \alpha = P(T - k\alpha_T < T < T + k\alpha_T)$.

distribution of T; that is, the likely range of differences between T and θ is $\pm \Delta T = \pm k\sigma_T$, so that we now may conclude that θ lies somewhere within the interval determined by $T \pm k\sigma_T$. Although this interval indicates something about the variability of T as an estimator of θ, it does not carry with it any measure of the degree of uncertainty associated with T. However, we may readily quantify this uncertainty, in probability terms, by finding the area under the sampling distribution of T between $T - k\sigma_T$ and $T + k\sigma_T$; that is,

$$P(T - k\sigma_T < T < T + k\sigma_T) = \int_{T-k\sigma_T}^{T+k\sigma_T} h(t;\theta,n)dt = 1 - \alpha, \qquad (11.1)$$

where $1 - \alpha$ is termed the *confidence probability*. (Figure 11.1 illustrates this probability when the sampling distribution of T is symmetrical about zero.)

An examination of (11.1) reveals that $(T - k\sigma_T, T + k\sigma_T)$ is actually a *random interval* that varies from sample to sample. In this regard, we may define an *interval estimator* for a parameter θ as a random interval of values with a given probability $1 - \alpha$ of containing θ. And since $T \pm k\sigma_T$ are random variables, $1 - \alpha$ is a probability associated with the sample random variables $X_1, \ldots, X_n$ before the sample is drawn, and not with the sample realizations $x_1, \ldots, x_n$ after the sample is drawn. Hence the confidence probability $1 - \alpha$ is the probability of obtaining a sample such that the interval, once realized, contains θ. However, once the sample is taken and the sample realizations are obtained, we no longer have a random interval and thus $1 - \alpha$ is no longer a probability; $1 - \alpha$ will simply be called our *degree of confidence* or the *confidence coefficient*, and is given a long-run relative frequency interpretation, that is, if many samples of size n were taken from the same population distribution and the corresponding realizations of the interval $T \pm k\sigma_T$ were obtained, then, in the long run, as the number of samples increases without limit, $100(1 - \alpha)\%$ of the realized intervals would contain θ, and $100\alpha\%$ of them would not. So, although $T \pm k\sigma_T$ is a random interval with confidence probability $1 - \alpha$, its realization is a confidence interval with confidence coefficient $1 - \alpha$. Thus there is no probability associated with the realized interval save for the values 1 or 0 (either the realized interval contains θ or it does not).

11.2   Central Confidence Intervals

It should be evident that (11.1) yields the probability that the random variable T assumes a value in the interval $T \pm k\sigma_T$. However, to use any such interval for making inferences about θ, we need to find a way to determine k as well as to ultimately transform (11.1) into a probability statement that brackets θ. This is readily achieved if we specify, in advance, the confidence coefficient $1 - \alpha$ and then use the sampling distribution $h(t; \theta, n)$ to find lower and upper confidence limits $L_1(T)$ and $L_2(T)$, respectively, for θ, where these confidence limits depend upon the sample random variables $X_1, \ldots, X_n$ via T (T is thus typically taken to be a sufficient statistic for θ) and $L_1(T) \leq L_2(T)$ for all sample points $(X_1, \ldots, X_n)$. To this end let T be a sufficient point estimator for θ obtained from a random sample of size n. Moreover, let $\left(L_1(T), L_2(T)\right)$ represent a *central confidence interval*; that is, the confidence limits $L_1(T)$ and $L_2(T)$ are chosen so that the area under each tail of $h(t; \theta, n)$ is $\frac{\alpha}{2}$ (see Figure 11.1). This centrality restriction conveniently provides us with a sufficient condition for the existence of unique lower and upper confidence limits $L_1(T)$ and $L_2(T)$, respectively.

For concreteness sake, suppose $\theta = \mu$ and $T = \overline{X}$ and that we extract a random sample of size n from a normal probability density function with unknown mean μ and known standard deviation σ. Then from Theorem 8.3 and (8.16), the sampling distribution of $\overline{X}$ has the form

$$h(\bar{x}; \mu, \sigma_{\bar{x}}, n) = \frac{\sqrt{n}}{\sqrt{2\pi}\sigma} e^{-\frac{n}{2}[(\bar{x}-\mu)/\sigma]^2}, \quad -\infty < \bar{x} < +\infty,$$

where $\sigma_{\bar{x}} = \frac{\sigma}{\sqrt{n}}$. Then

$$P(L_1 < \overline{X} < L_2) = \frac{\sqrt{n}}{\sqrt{2\pi}\sigma} \int_{L_1}^{L_2} e^{-\frac{n}{2}[(\bar{x}-\mu)/\sigma]^2} d\bar{x} = 1 - \alpha \qquad (11.2)$$

and, under the change of variable $\bar{z} = (\bar{x} - \mu)/\sigma_{\bar{x}}$, the preceding expression becomes (for $d\bar{x} = \sigma_{\bar{x}} d\bar{z}$)

$$\frac{1}{\sqrt{2\pi}} \int_{(L_1-\mu)/\sigma_{\bar{x}}}^{(L_2-\mu)/\sigma_{\bar{x}}} e^{-\frac{\bar{z}^2}{2}} d\bar{z} = 1 - \alpha. \qquad (11.3)$$

For a given α, the lower and upper limits of integration in (11.3) are, respectively, $-z_{\alpha/2}$ and $z_{\alpha/2}$ (under the centrality restriction) since $\overline{Z}$ is $N(0, 1)$; that is,

$$-z_{\alpha/2} = \frac{L_1 - \mu}{\sigma_{\bar{x}}} \quad \text{or} \quad L_1 = \mu - z_{\alpha/2}\sigma_{\bar{x}};$$

and

$$z_{\alpha/2} = \frac{L_2 - \mu}{\sigma_{\bar{x}}} \quad \text{or} \quad L_2 = \mu + z_{\alpha/2}\sigma_{\bar{x}}.$$

Hence (11.2) becomes $P(\mu - z_{\alpha/2}\sigma_{\bar{x}} < \overline{X} < \mu + z_{\alpha/2}\sigma_{\bar{x}}) = 1 - \alpha$ or, equivalently,

$$P\left(\underbrace{\overline{X} - z_{\alpha/2}\sigma_{\bar{x}}}_{L_1(\overline{X})} < \mu < \underbrace{\overline{X} + z_{\alpha/2}\sigma_{\bar{x}}}_{L_2(\overline{X})}\right) = 1 - \alpha. \tag{11.4}$$

As was the case with (11.1), this equation is a probability statement about a random interval, with $1 - \alpha$ serving as the confidence probability. However, once $\overline{X}$ is replaced by its realization $\bar{x}$, (11.4) becomes a *confidence statement* concerning the constant μ (we no longer have a random interval) with realized confidence limits

$$\ell_1(\bar{x}) = \bar{x} - z_{\alpha/2}\sigma_{\bar{x}}; \; \ell_2(\bar{x}) = \bar{x} + z_{\alpha/2}\sigma_{\bar{x}} \tag{11.5}$$

and *confidence coefficient* $1 - \alpha$. (Note that k in (11.1) above is thus $z_{\alpha/2}$.) Hence we are thus able to conclude that we may be $100(1 - \alpha)\%$ confident that the population mean μ lies between $\ell_1(\bar{x})$ and $\ell_2(\bar{x})$. In this regard, we fix the value of the sample mean at $\bar{x}$ and then we determine the range of values of μ (the confidence interval $(\ell_1(\bar{x}), \ell_2(\bar{x}))$) that make this sample mean most plausible.

11.3   The Pivotal Quantity Method

The procedure that we have just employed to determine a $100(1 - \alpha)\%$ confidence interval for the population mean μ is termed the *pivotal quantity method*. In general, to find a $100(1 - \alpha)\%$ confidence interval for a parameter θ, this technique has us first find a *pivotal quantity* that possesses the following characteristics: (1) it is a random variable that is expressible as a function of the sample random variables $X_1, \ldots, X_n$ and the unknown parameter θ (θ is, in fact, the only unknown parameter in the specification of the pivotal quantity), and (2) it has a sampling distribution that does not depend upon θ (or upon any other parameter). Then a probability statement involving the pivotal quantity (with confidence probability $1 - \alpha$) is transformed into a confidence statement concerning θ (with confidence coefficient $1 - \alpha$).

More specifically, the pivotal quantity method (for obtaining a central confidence interval) proceeds as follows:

(a) Let $X_1, \ldots, X_n$ denote a set of sample random variables taken from a population probability density function $f(x; \theta)$, where θ is an unknown parameter, and let T serve as a point estimator of θ, where T is usually sufficient for θ.

(b) Let the random variable $Y = g(T; \theta)$ have a known sampling distribution that does not depend upon θ (or upon any other parameter) so that Y serves as a pivotal quantity.

(c) Then for any fixed quantity $1 - \alpha, 0 < \alpha < 1$, there will exist values y_1 and y_2 (depending on $1 - \alpha$) such that $P(y_1 < Y < y_2) = 1 - \alpha$.

(d) This probability statement with confidence probability $1 - \alpha$ is then transformed into an equivalent probability statement $P(L_1(T) < \theta < L_2(T)) = 1 - \alpha$.

(e) If T is replaced by its realization t, we then obtain the confidence statement $P\left(\ell_1(t) < \theta < \ell_2(t)\right) = 1 - \alpha$ with confidence coefficient $1 - \alpha$ and thus a $100(1 - \alpha)\%$ confidence interval for θ is $\left(\ell_1(t), \ell_2(t)\right)$.

In short, the pivotal quantity method requires that for any sample realization $(x_1, \ldots, x_n)$, the inequality $y_1 < g(t; \theta) < y_2$ can be *pivoted* (i.e., transformed or *inverted*) as $\ell_1(t) < \theta < \ell_2(t)$; that is, if

$$y_1 < g(t; \theta) < y_2 \quad \text{if and only if} \quad \ell_1(t) < \theta < \ell_2(t)$$

for each sample realization $(x_1, \ldots, x_n)$, then $\left(\ell_1(t), \ell_2(t)\right)$ is a $100(1 - \alpha)\%$ confidence interval for θ. This interval thus gives the fraction of time, under repeated sampling from the same population, that the realized intervals will contain θ. In general, it can be shown that a pivotal quantity exists if we sample from a population that possesses a continuous cumulative distribution function $F(s; \theta)$; and if this cumulative distribution function is continuous in θ for each s, then the pivotal quantity can (at least theoretically) be inverted to find confidence limits $\ell_1(t)$ and $\ell_2(t)$. The next section demonstrates exactly how the pivotal quantity method works.

11.4   A Confidence Interval for μ Under Random Sampling from a Normal Population with Known Variance

Let $X_1, \ldots, X_n$ depict a set of sample random variables drawn from a normal population with unknown mean μ but known variance σ^2. Then, as indicated earlier, the best estimator for μ is $T = \overline{X}$, where $\overline{X}$ is $N(\mu, \sigma_{\bar{x}})$ and $\sigma_{\bar{x}} = \frac{\sigma}{\sqrt{n}}$ (Theorem 8.3). Hence from (8.33), the standardized sample mean $\overline{Z} = \frac{\overline{X} - \mu}{\sigma / \sqrt{n}}$ is $N(0, 1)$ and thus, from an adaptation of (7.31),

$$P(-z_{\alpha/2} < \overline{Z} < z_{\alpha/2}) = P\left(-z_{\alpha/2} < \frac{\overline{X} - \mu}{\sigma / \sqrt{n}} < z_{\alpha/2}\right) = 1 - \alpha. \tag{11.6}$$

Here $\overline{Z}$ serves as a pivotal quantity (it clearly has a known sampling distribution that does not depend on μ) in this expression and thus (11.6) can easily be pivoted or transformed into (11.4) or

$$P\left(\underbrace{\overline{X} - z_{\alpha/2}\frac{\sigma}{\sqrt{n}}}_{L_1} < \mu < \underbrace{\overline{X} + z_{\alpha/2}\frac{\sigma}{\sqrt{n}}}_{L_2}\right) = 1 - \alpha. \tag{11.7}$$

As indicated earlier, (L_1, L_2) is a random interval with confidence probability $1 - \alpha$; that is, if we extracted many random samples of size n from our normal population and formed (L_1, L_2) for each of them, then, in the long run, $100(1 - \alpha)\%$ of these intervals would contain μ and $100\alpha\%$ of them would not.

If $\overline{X}$ is now replaced by its realization $\bar{x}$, (11.7) becomes

$$P\left(\underbrace{\bar{x} - z_{\alpha/2}\frac{\sigma}{\sqrt{n}}}_{\ell_1} < \mu < \underbrace{\bar{x} + z_{\alpha/2}\frac{\sigma}{\sqrt{n}}}_{\ell_2}\right) = 1 - \alpha. \tag{11.7.1}$$

Here (11.7.1) is a confidence statement with confidence coefficient $1 - \alpha$. Hence a *$100(1 - \alpha)\%$ confidence interval for μ* is

$$(\ell_1, \ell_2) = \left(\bar{x} - z_{\alpha/2}\frac{\sigma}{\sqrt{n}}, \bar{x} + z_{\alpha/2}\frac{\sigma}{\sqrt{n}}\right), \tag{11.8}$$

that is, we may be $100(1 - \alpha)\%$ confident that $\ell_1 < \mu < \ell_2$. In this regard, if $1 - \alpha = 0.95$ and many samples of size n were taken from a normal population and (ℓ_1, ℓ_2) was computed for each of them, then, in the long run, 95% of these intervals would bracket μ and 5% of them would not. As the preceding sentence indicates, a confidence statement such as $P(\ell_1 < \mu < \ell_2) = 1 - \alpha$ actually pertains to the *behavior of samples* since μ is a fixed parameter.

Example 11.4.1 Suppose we take a random sample of size $n = 100$ from a normal population with μ unknown but with $\sigma = 15$ and we find that $\bar{x} = 77$. Then from (11.8), 90%, 95%, and 99% confidence intervals for μ are, respectively:

(a) $1 - \alpha = 0.90, \frac{\alpha}{2} = 0.05$, and thus $z_{\alpha/2} = z_{0.05} = 1.645$. Thus $(\ell_1, \ell_2) = (74.532, 79.468)$. Hence we may be 90% confident that μ lies between 74.532 and 79.468.

(b) $1 - \alpha = 0.95, \frac{\alpha}{2} = 0.025$, and thus $z_{0.025} = 1.96$. Then $(\ell_1, \ell_2) = (74.060, 79.940)$. We are now 95% confident that μ falls between 74.060 and 79.940.

(c) $1 - \alpha = 0.99, \frac{\alpha}{2} = 0.005$ so that $z_{0.005} = 2.58$. Then $(\ell_1, \ell_2) = (73.130, 80.870)$. Hence we may be 99% confident that μ lies between 73.130 and 80.870. ∎

What factors influence the *width (w) of a confidence interval?* From (11.7.1) or (11.8) we can easily determine $w = \ell_2 - \ell_1 = 2z_{\alpha/2}\frac{\sigma}{\sqrt{n}}$. Hence, as a general proposition, we see that w varies directly with $1 - \alpha$ (it affects the magnitude of $z_{\alpha/2}$) and σ; w varies inversely with n.

We noted at the outset of this chapter that the role of a confidence interval is to determine the degree of precision associated with our estimation of a population parameter θ; that is, the narrower the interval, the more precise the estimate of θ will be. In this regard, and in view of our current discussion, how precisely have we estimated the population mean μ? To answer this question, let us rewrite the left-hand side of (11.7.1) as

$$P\left(|\bar{x} - \mu| < z_{\alpha/2}\frac{\sigma}{\sqrt{n}}\right) = 1 - \alpha. \tag{11.9}$$

Since $\bar{x} - \mu$ can be either positive or negative, let us attach both a plus and a minus sign to $z_{\alpha/2}\frac{\sigma}{\sqrt{n}}$, where now $z_{\alpha/2}$ can be thought of as a *100(1 − α)% reliability coefficient;*[1] it serves as a measure of the *degree of reliability* that we place on our interval estimate. On the basis of these considerations, the *degree of precision* of $\bar{x}$ as an estimate of μ can now be expressed as: we are within $\pm z_{\alpha/2}\frac{\sigma}{\sqrt{n}}$ units of μ with $100(1 - \alpha)\%$ reliability. Hence this notion of precision is simply the *confidence interval half-width.* That is, since the width is $w = \ell_2 - \ell_1 = 2z_{\alpha/2}\frac{\sigma}{\sqrt{n}}$, the half width is $\frac{w}{2} = z_{\alpha/2}\frac{\sigma}{\sqrt{n}}$.

There is yet a third way to interpret a confidence interval estimate of a parameter θ (in this case μ). Since an inequality appears in the confidence statement given by (11.9), this result can be given an *error bound* interpretation; that is, we may state that we are $100(1 - \alpha)\%$ confident that $\bar{x}$ will not differ from μ by more than $\pm z_{\alpha/2}\frac{\sigma}{\sqrt{n}}$ units.

Example 11.4.2   Given the 95% confidence interval for μ determined in part (b) of the previous example problem, we may summarize our three equivalent interpretations of $(\ell_1, \ell_2) = (74.060, 79.940)$ as follows:

1. We may be 95% confident that the population mean μ lies between $\ell_1 = 74.060$ and $\ell_2 = 79.940$.

2. We are within $\pm z_{\alpha/2}\frac{\sigma}{\sqrt{n}} = \pm 2.940$ units of μ with 95% reliability.

3. We are 95% confident that $\bar{x}$ will not differ from μ by more than ± 2.940 units.

Any of these three statements can be used to depict how precisely we have estimated μ on the basis of our sample results. ■

---

[1] We may view the notion of *reliability* as a *long-run* concept that emerges under repeated sampling from the same population and thus has the same long-run relative frequency interpretation as the confidence coefficient $1 - \alpha$.

Given that we are sampling from a normal population with unknown mean μ but known variance σ^2, we can easily address the problem of estimating μ with a prespecified or target level of precision and reliability. To do so will require that we calculate a particular sample size n that will enable us to attain our chosen level of precision and reliability. Starting from the expression for the confidence interval half-width or $\frac{w}{2} = z_{\alpha/2}\frac{\sigma}{\sqrt{n}}$, we can readily solve for n as

$$n = \left(\frac{z_{\alpha/2}\sigma}{\frac{w}{2}}\right)^2. \tag{11.10}$$

Here this expression (called the *sample size requirements formula*) gives the sample size required to estimate μ with a degree of precision of $\pm\frac{w}{2}$ with $100(1-\alpha)\%$ reliability.

Example 11.4.3   If $\sigma = 15$ and we desire to estimate μ to within $\pm\frac{w}{2} = \pm 2$ units of its true value with 99% reliability, then the sample size needed to achieve this objective is

$$n = \left(\frac{2.58 \times 15}{2}\right)^2 = 374.42 \approx 375$$

(note that we typically round our result for n up to the next highest integer). Hence 375 represents the sample size required for a degree of precision of ± 2 units with 99% reliability.   ∎

A couple of comments pertaining to (11.10) are warranted:

1. A glance at the structure of (11.10) reveals that, for fixed values of $1 - \alpha$ and σ, if we desire to double our degree of precision (i.e., we choose to be *twice as precise*), then we must quadruple the sample size.

2. This formula requires that σ is known. If σ is unknown, then we may approximate the standard deviation of our normal population as $\sigma \approx range/6$ since most (about 99% via the empirical rule) of the observations on the population characteristic or random variable X will fall within an interval involving 3 standard deviations on either side of μ.

11.5   A Confidence Interval for μ Under Random Sampling from a Normal Population with Unknown Variance

We found in Chapter 9 that when sampling from a normal population with both the mean (μ) and variance (σ^2) unknown, the random variable $T = \frac{\bar{X}-\mu}{S/\sqrt{n}}$ ((9.19.1)) has a t distribution with $v = n - 1$ degrees of freedom. Then from (9.28) and Figure 9.5.6, we may form

$$P(-t_{\alpha/2,n-1} < T < t_{\alpha/2,n-1}) = P\left(-t_{\alpha/2,n-1} < \frac{\overline{X} - \mu}{S/\sqrt{n}} < t_{\alpha/2,n-1}\right) = 1 - \alpha,$$

$$(11.11)$$

where $T = \frac{\overline{X}-\mu}{S/\sqrt{n}}$ thus serves as a pivotal quantity. Then proceeding as earlier (see the derivation underlying (11.9)), the preceding probability statement may be transformed or pivoted into

$$P\left(\underbrace{\overline{X} - t_{\alpha/2,n-1}\frac{S}{\sqrt{n}}}_{L_1} < \mu < \underbrace{\overline{X} + t_{\alpha/2,n-1}\frac{S}{\sqrt{n}}}_{L_2}\right) = 1 - \alpha. \qquad (11.12)$$

As defined previously, (L_1, L_2) is a random interval with confidence probability $1 - \alpha$. However, once $\overline{X}$ and S are replaced by their realizations $\bar{x}$ and s, respectively, in (11.12), a *100(1 − α)% confidence interval for* μ is

$$(\ell_1, \ell_2) = \left(\bar{x} - t_{\alpha/2,n-1}\frac{s}{\sqrt{n}}, \bar{x} + t_{\alpha/2,n-1}\frac{s}{\sqrt{n}}\right), \qquad (11.13)$$

that is, we may be 100(1−α)% confident that $\ell_1 < \mu < \ell_2$. Here, too, the quantity $\pm t_{\alpha/2,n-1}\frac{s}{\sqrt{n}}$ may be termed our *error bound* on $\bar{x}$ as an estimate of μ or the *degree of precision* involved in using $\bar{x}$ as an estimate of μ.

It must be emphasized that since T is a *small sample statistic*, (11.13) must be employed if $n \leq 30$ and σ^2 is unknown and is estimated by S^2 (see Figure 11.2).

Example 11.5.1 Suppose that from a random sample of size $n = 25$ it is determined that $\bar{x} = 118$ and $s = 9.5$. Find a 90% confidence interval for μ. Here $1 - \alpha = 0.90, t_{\alpha/2,n-1} = t_{0.05,24} = 1.711$, and thus, from (11.13), we obtain $118 \pm 1.711\left(\frac{9.5}{\sqrt{25}}\right)$ or 118 ± 3.251. Hence $(\ell_1, \ell_2) = (114.749, 121.251)$ so that we may be 90% confident that the population mean μ lies between 114.749 and 121.251. Stated alternatively, we may be 90% confident that the sample mean will not differ from the population mean by more than ± 3.251 units. ∎

11.6 A Confidence Interval for σ^2 Under Random Sampling from a Normal Population with Unknown Mean

We found earlier, by virtue of Theorem 9.5, that under random sampling from a normal population with unknown mean and variance, the quantity $Y = (n-1)\frac{S^2}{\sigma^2}$ is distributed as χ^2_{n-1}, where S^2 is the sample variance determined from a random

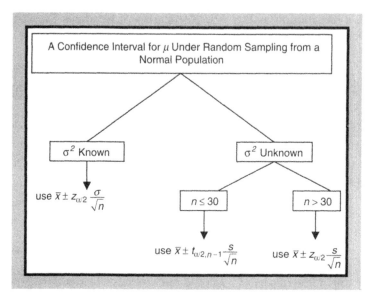

Figure 11.2 A confidence interval for μ under random sampling from a normal population.

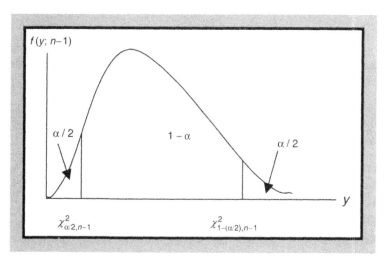

Figure 11.3 $P\left(\chi^2_{\alpha/2,n-1} < Y < \chi^2_{1-(\alpha/2),n-1}\right) = 1 - \alpha.$

sample of size n and $n - 1$ depicts degrees of freedom associated with the chi-square random variable Y. Then from Figure 11.3, we seek to determine quantiles $\chi^2_{\alpha/2,n-1}$ and $\chi^2_{1-(\alpha/2),n-1}$ such that

$$P\left(\chi^2_{\alpha/2,n-1} < Y < \chi^2_{1-(\alpha/2),n-1}\right) = P\left(\chi^2_{\alpha/2,n-1} < \frac{(n-1)S^2}{\sigma^2} < \chi^2_{1-(\alpha/2),n-1}\right) = 1 - \alpha,$$

(11.14)

where $Y = (n-1)\frac{S^2}{\sigma^2}$ serves as our pivotal quantity. We may now pivot or transform (11.14) to the probability statement

$$P\left(\frac{1}{\chi^2_{\alpha/2,n-1}} > \frac{\sigma^2}{(n-1)S^2} > \frac{1}{\chi^2_{1-(\alpha/2),n-1}}\right) = P\left(\underbrace{\frac{(n-1)S^2}{\chi^2_{1-(\alpha/2),n-1}}}_{L_1} < \sigma^2 < \underbrace{\frac{(n-1)S^2}{\chi^2_{\alpha/2,n-1}}}_{L_2}\right)$$

$$= 1 - \alpha. \tag{11.15}$$

Although (L_1, L_2) is a random interval with confidence probability $1 - \alpha$, once S^2 is replaced by its sample realization s^2 in (11.15), we obtain a *100(1 − α)% confidence interval for σ^2* of the form

$$(\ell_1, \ell_2) = \left(\frac{(n-1)s^2}{\chi^2_{1-(\alpha/2),n-1}}, \frac{(n-1)s^2}{\chi^2_{\alpha/2,n-1}}\right). \tag{11.16}$$

Hence we may be $100(1 - \alpha)\%$ confident that $\ell_1 < \sigma^2 < \ell_2$. Moreover, the preceding confidence interval for σ^2 can easily be converted into a confidence interval for σ by simply taking the square root of the lower and upper confidence limits for σ^2 and then forming the interval $(\sqrt{\ell_1}, \sqrt{\ell_2})$; that is, a *100(1 − α)% confidence interval for σ* is

$$\left(\sqrt{\ell_1}, \sqrt{\ell_2}\right) = \left(\sqrt{\frac{(n-1)s^2}{\chi^2_{1-(\alpha/2),n-1}}}, \sqrt{\frac{(n-1)s^2}{\chi^2_{\alpha/2,n-1}}}\right) \tag{11.17}$$

so that we may be $100(1 - \alpha)\%$ confident that $\sqrt{\ell_1} < \sigma < \sqrt{\ell_2}$.

It was mentioned in Chapter 9 (property (5), p. 351) that for degrees of freedom $\upsilon > 30$, chi-square probabilities can be determined via a standard normal approximation and thus percentiles of the chi-square distribution can be approximated by percentiles of the $N(0, 1)$ distribution. In fact, if Y is χ^2_υ with $\upsilon > 30$, then the quantity $Z = \sqrt{2Y} - \sqrt{2\upsilon - 1}$ is approximately $N(0, 1)$. So for large samples we have

$$P(-z_{\alpha/2} < Z < z_{\alpha/2}) = P\left(-z_{\alpha/2} < \sqrt{2Y} - \sqrt{2\upsilon - 1} < z_{\alpha/2}\right)$$

$$= P\left(-z_{\alpha/2} < \sqrt{\frac{2(n-1)S^2}{\sigma^2}} - \sqrt{2\upsilon - 1} < z_{\alpha/2}\right) \approx 1 - \alpha. \tag{11.18}$$

Since $2(n - 1) \approx 2\upsilon - 1$ for large n, it follows that (11.18) can be rewritten as

$$P\left(-z_{\alpha/2} < \sqrt{2(n-1)}\left(\frac{S}{\sigma} - 1\right) < z_{\alpha/2}\right) \approx 1 - \alpha, \tag{11.18.1}$$

where $\sqrt{2(n-1)}\left(\frac{S}{\sigma}-1\right)$ now serves as the pivotal quantity. Then under a suitable pivot operation or transformation of (11.18.1) we ultimately obtain

$$P\left(\underbrace{\frac{S\sqrt{2(n-1)}}{z_{\alpha/2}+\sqrt{2(n-1)}}}_{L_1}<\sigma<\underbrace{\frac{S\sqrt{2(n-1)}}{-z_{\alpha/2}+\sqrt{2(n-1)}}}_{L_2}\right)\approx 1-\alpha,\qquad (11.19)$$

where (L_1,L_2) is a random interval with confidence probability $1-\alpha$.

If S is replaced by its realization in (11.19), then a large sample (approximate) $100(1-\alpha)\%$ confidence interval for σ is

$$(\ell_1,\ell_2)=\left(\frac{s\sqrt{2(n-1)}}{z_{\alpha/2}+\sqrt{2(n-1)}},\frac{s\sqrt{2(n-1)}}{-z_{\alpha/2}+\sqrt{2(n-1)}}\right);\qquad (11.20)$$

and a large sample (approximate) $100(1-\alpha)\%$ confidence interval for σ^2 is

$$(\ell_1^2,\ell_2^2)=\left(\frac{2(n-1)s^2}{\left(z_{\alpha/2}+\sqrt{2(n-1)}\right)^2},\frac{2(n-1)s^2}{\left(-z_{\alpha/2}+\sqrt{2(n-1)}\right)^2}\right).\qquad (11.21)$$

This discussion is summarized in Figures 11.4, and 11.5.

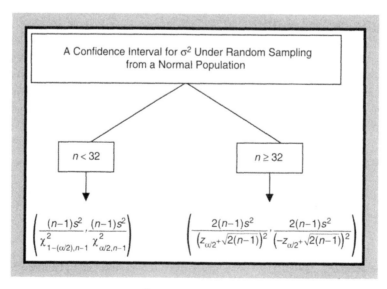

Figure 11.4   A confidence interval for σ^2 under random sampling from a normal population.

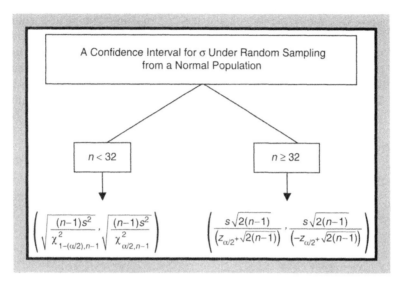

Figure 11.5   A confidence interval for σ under random sampling from a normal population.

Example 11.6.1   Suppose we extract a random sample of size $n = 20$ from a normal population and that we obtain $s^2 = 72.25$. Find a 95% confidence interval for σ^2. With $1 - \alpha = 0.95, \chi^2_{\alpha/2,n-1} = \chi^2_{0.025,19} = 8.9$ and $\chi^2_{1-(\alpha/2),n-1} = \chi^2_{0.975,19} = 32.9$. Then from (11.16), a 95% confidence interval for σ^2 is $(\ell_1, \ell_2) = (41.725, 154.242)$; that is, we may be 95% confident that the population variance lies between 41.725 and 154.242. If we take the square root of these confidence limits for σ^2, then we obtain a 95% confidence interval for σ or $6.459 < \sigma < 12.419$.

If, however, we had taken a sample of size $n = 160$ from the same population and obtained $s^2 = 74.65$, then, according to (11.21), our large sample (approximate) 95% confidence interval for σ^2 would appear as $(\ell_1^2, \ell_2^2) = (60.648, 94.304)$.

Hence we are now 95% confident that the population variance is contained within the interval $(60.648, 94.304)$. Again passing to square roots, a large sample 95% confidence interval for the population standard derivation is $7.787 < \sigma < 9.711$. ∎

11.7   A Confidence Interval for p Under Random Sampling from a Binomial Population

We noted in Section 6.5 that the essential characteristics of a binomial experiment are: (1) X is a discrete random variable; (2) we have a simple alternative experiment (on the ith trial the random variable X_i has two possible values—0 (for a failure) or 1 (for a success)); (3) the trials are identical and independent; and (4) p, the probability of a success, is constant from trial to trial.

In what follows we shall now view p as the *proportion of success in the population*. To estimate p, let us extract a random sample of size n from a binomial population (we perform a binomial experiment), where n is taken to be large. The rationale for this restriction on n follows from the DeMoivre-Laplace-Gauss Limit Theorem, which essentially states that a standardized binomial random variable has as its limiting distribution the standard normal distribution; that is,

$$\frac{X-np}{\sqrt{np(1-p)}} \xrightarrow{d} Z = N(0,1).$$

As revealed in the preceding chapter, the best estimator for p is the *sample proportion of successes*, $\hat{P} = \frac{X}{n}$ where X is the observed number of successes in a sample of size n, $E(\hat{P}) = p$, $V(\hat{P}) = p(1-p)/n$, and the best estimator for $V(\hat{P})$ is $S^2(\hat{P}) = \hat{P}(1-\hat{P})/n$. Hence, by the DeMoivre-Laplace-Gauss Limit Theorem,

$$Z_{\hat{P}} = \frac{\hat{P} - E(\hat{P})}{\sqrt{S^2(\hat{P})}} = \frac{\hat{P} - p}{\sqrt{\hat{P}(1-\hat{P})/n}} \xrightarrow{d} Z = N(0,1).$$

So for large n,

$$P\left(-z_{\frac{\alpha}{2}} < Z_{\hat{P}} < z_{\frac{\alpha}{2}}\right) = P\left(-z_{\frac{\alpha}{2}} < \frac{\hat{P} - p}{\sqrt{\dfrac{\hat{P}(1-\hat{P})}{n}}} < z_{\frac{\alpha}{2}}\right) \approx 1 - \alpha. \qquad (11.22)$$

Given that $Z_{\hat{P}}$ serves as a pivotal quantity (it has a known sampling distribution which is independent of p), (11.22) can be transformed to

$$P\left(\underbrace{\hat{P} - z_{\frac{\alpha}{2}}\sqrt{\frac{\hat{P}(1-\hat{P})}{n}}}_{L_1} < p < \underbrace{\hat{P} + z_{\frac{\alpha}{2}}\sqrt{\frac{\hat{P}(1-\hat{P})}{n}}}_{L_2}\right) \approx 1 - \alpha. \qquad (11.23)$$

Thus the probability that p lies within the random interval (L_1, L_2) is $1 - \alpha$.

Once $\hat{P}$ is replaced by its realization $\hat{p}(= \frac{x}{n}$, where x is the realized number of successes in the sample),we obtain as a large sample (approximate) $100(1 - \alpha)\%$ *confidence interval for p*

$$(\ell_1, \ell_2) = \left(\hat{p} - z_{\frac{\alpha}{2}}\sqrt{\frac{\hat{p}(1-\hat{p})}{n}}, \hat{p} + z_{\frac{\alpha}{2}}\sqrt{\frac{\hat{p}(1-\hat{p})}{n}}\right), \qquad (11.24)$$

that is, we may be $100(1 - \alpha)\%$ confident that $\ell_1 < p < \ell_2$. For instance, if $1 - \alpha = 0.90$ and many samples of size n were extracted from a binomial population and (ℓ_1, ℓ_2) was determined for each of them, then, in the long run, 90% of these intervals would contain p and 10% of them would not.

This procedure for estimating p via a confidence interval is essentially a *large-sample* approximation technique. In this regard, to obtain a good approximation, we require that $n \geq 50$ or, alternatively, $n\hat{p} \geq 5$ *and* $n(1 - \hat{p}) \geq 5$ (i.e., both of these inequalities must be satisfied).

Example 11.7.1   Suppose that a random sample of size $n = 200$ is taken from a binomial population and that $x = 83$ successes are realized. Then from (11.24), an (approximate) 99% confidence interval for the true or population proportion of successes p can be determined as follows. Since $1 - \alpha = 0.99$, $\frac{\alpha}{2} = 0.005$ and thus $z_{\frac{\alpha}{2}} = z_{0.005} = 2.58$. Also, $\hat{p} = \frac{x}{n} = \frac{83}{200} = 0.42$ and thus $(\ell_1, \ell_2) = (0.33, 0.51)$. Hence we may be 99% confident that p lies between 0.33 and 0.51.   ∎

How precisely have we estimated p? To answer this question, let us rewrite the left-hand side of (11.23), with $\hat{p}$ replacing $\hat{P}$, as

$$P\left(|p - \hat{p}| < z_{\frac{\alpha}{2}} \sqrt{\frac{\hat{p}(1 - \hat{p})}{n}} \right) \approx 1 - \alpha. \tag{11.25}$$

Hence the *degree of precision* of $\hat{p}$ as an estimate of p is $\pm z_{\alpha/2}\sqrt{\frac{\hat{p}(1-\hat{p})}{n}}$; that is, we are within $\pm z_{\alpha/2}\sqrt{\frac{\hat{p}(1-\hat{p})}{n}}$ units of p with $100(1 - \alpha)\%$ reliability. As indicated earlier, this degree of precision is simply the *confidence interval half-width* $\frac{w}{2}$. And since an inequality appears in the confidence statement given by (11.25), we may also offer an *error bound* interpretation of this result. That is, we may assert with $100(1 - \alpha)\%$ confidence that $\hat{p}$ will not differ from p by more than $\pm z_{\alpha/2}\sqrt{\frac{\hat{p}(1-\hat{p})}{n}}$ units.

Example 11.7.2   On the basis of the preceding discussion, it should be evident that we now have three equivalent ways of interpreting a $100(1 - \alpha)\%$ confidence interval for p. To see this, let us examine the 99% confidence interval for p determined earlier.

Specifically, given $(\ell_1, \ell_2) = (0.33, 0.51)$:

1. We may be 99% confident that the population proportion p lies between $\ell_1 = 0.33$ and $\ell_2 = 0.51$.

2. We are within $\pm z_{\alpha/2}\sqrt{\frac{\hat{p}(1-\hat{p})}{n}} = \pm 0.09$ units of p with 99% reliability.

3. We are 99% confident that $\hat{p}$ will not differ from p by more than ± 0.09 units.   ∎

We now turn to the problem of estimating p with a prespecified or desired level of precision and reliability. This involves determining the sample size n that will enable us to attain our target level of precision and reliability. To this end let us express Chebyshev's inequality (4.19.2) in terms of $\hat{P}$ as

$$P\left(\left|\hat{P} - p\right| < k\sigma_{\hat{p}}\right) \geq 1 - \frac{1}{k^2}, \quad k > 0.$$

A glance back at (11.25) reveals that $k\sigma_{\hat{p}}$ is simply an indicator of how precisely we have estimated p using $\hat{P}$ or $k\sigma_{\hat{p}} = \frac{w}{2}$, where $\sigma_{\hat{p}} = \sqrt{\frac{p(1-p)}{n}}$. If k is taken to be the $100(1-\alpha)\%$ reliability coefficient $z_{\alpha/2}$, then $P(|\hat{P} - p| < z_{\alpha/2}\sigma_{\hat{p}}) \approx 1 - \alpha$ and thus

$$\frac{w}{2} = z_{\alpha/2}\sqrt{\frac{p(1-p)}{n}}.$$

Upon solving this equality for n we thus obtain the *sample size requirements formula*

$$n = \frac{(z_{\alpha/2})^2 p(1-p)}{\left(\frac{w}{2}\right)^2}. \tag{11.26}$$

Here this expression represents the sample size required for a degree of precision of $\pm\frac{w}{2}$ with $100(1-\alpha)\%$ reliability. Note that this formula involves p as an argument. But since p is unknown and is, in fact, the parameter to be estimated, we need an independent estimate of p in order to determine n. If no such estimate exists, then what is commonly done in practice is to set $p = \frac{1}{2}$ in (11.26) (since $V(\hat{P}) = \frac{p(1-p)}{n}$ attains its maximum at $p = \frac{1}{2}$). So for $p = \frac{1}{2}$, the value of n that emerges turns out to be a bit larger than that actually needed to attain the chosen level of precision and reliability.

Example 11.7.3   If we choose to estimate p to within $\pm 5\%$ of its true value with 99% reliability, then, according to (11.26), the sample size required (assuming $p = \frac{1}{2}$) is $n = (2.58)^2(0.5)(0.5)/(0.05)^2 \approx 666$.   ∎

The confidence limits given by (11.24) are appropriate when n is very large. But if n is relatively small, then (11.24) can be improved upon as follows. First, since the binomial random variable X is discrete, a *continuity correction factor* of the form $\frac{1}{2n}$ can be introduced into (11.24) so that a *100(1 − α)% confidence interval for p* is

$$(\ell_1, \ell_2) = \left[\left(\hat{p} - \frac{1}{2n}\right) - z_{\frac{\alpha}{2}}\sqrt{\frac{\hat{p}(1-\hat{p})}{n}}, \left(\hat{p} + \frac{1}{2n}\right) + z_{\frac{\alpha}{2}}\sqrt{\frac{\hat{p}(1-\hat{p})}{n}}\right]. \tag{11.27}$$

A second procedure for improving upon (11.24) is to offer the observation that, for large n, the possible values of p corresponding to the limits $\hat{P} \pm z_{\frac{\alpha}{2}}\sigma_{\hat{p}}$ must satisfy the relationship

$$\left| p - \hat{P} \right| = z_{\alpha/2}\sqrt{\frac{p(1-p)}{n}} \tag{11.28}$$

and thus we may use this expression to construct a quadratic confidence interval for p; that is, (11.28) can be rewritten as a quadratic equation in terms of p as

$$\left(p - \hat{P} \right)^2 = z_{\alpha/2}^2 \left(\frac{p(1-p)}{n} \right)$$

or, for $\hat{P} = \frac{X}{n}$,

$$(n + z_{\alpha/2}^2)p^2 - (2X + z_{\alpha/2}^2)p + \frac{X^2}{n} = 0. \tag{11.29}$$

Then solving for p via the quadratic formula[2] and simplifying ultimately renders, at the sample realization $\hat{p} = \frac{x}{n}$, the *100(1 − α)% confidence interval for p*

$$\frac{n}{n + z_{\alpha/2}^2} \left[\hat{p} + \frac{z_{\alpha/2}^2}{2n} \pm z_{\alpha/2}\sqrt{\frac{\hat{p}(1-\hat{p})}{n} + \left(\frac{z_{\alpha/2}}{2n} \right)^2} \right]. \tag{11.30}$$

Example 11.7.4   We determined earlier that, for $\hat{p} = \frac{83}{200} = 0.42$, a 99% confidence interval for the population proportion p is $(\ell_1, \ell_2) = (0.33, 0.51)$. Let us refine this calculation somewhat by applying (11.27) and (11.30) in turn. First, from (11.27), we obtain $(\ell_1, \ell_2) = (0.3275, 0.5125)$. And from (11.30), $(\ell_1, \ell_2) = (0.3346, 0.5104)$.   ∎

11.8   Joint Estimation of a Family of Population Parameters

We now turn to the problem of making joint inferences from the same set of sample observations. Although we have constructed *separate* $100(1 - \alpha)\%$ confidence intervals for the population mean μ and population variance σ^2, we cannot state with $100(1 - \alpha)\%$ confidence that the results for *both* μ and σ^2 are jointly correct. If the inferences on the mean and variance were independent (e.g., different samples were used), the probability of both sets of results holding jointly would simply be $(1 - \alpha)^2$. However, these inferences are not independent since they are derived from the same set of sample observations.

---

[2] We may find the zeros of the quadratic function $ax^2 + bx + c$ or, equivalently, solve the quadratic equation $ax^2 + bx + c = 0(a \neq 0)$ by employing the quadratic formula $x = \frac{-b \pm \sqrt{b^2 - 4ac}}{2a}$.

What we need is a procedure that would provide us with $100(1 - \alpha)\%$ confidence concerning the correctness of the entire set or family of estimates. To develop any such assurance, we need to distinguish between the notion of a *statement confidence coefficient* and that of a *family confidence coefficient*. A statement confidence coefficient is associated with a confidence statement for an individual parameter—it indicates the long-run proportion of realized intervals (ℓ_1, ℓ_2) that would contain the true population parameter under repeated random sampling from the same population. (Here each sample provides us with an estimate of a single parameter.) By contrast, if an entire family of parameters is estimated for each random sample selected from a given population, then a family confidence coefficient indicates the long-run proportion of families of realized intervals that would contain the true values of the entire family of parameters under repeated sampling from the same population. Hence a family confidence coefficient pertains to the simultaneous reliability of the entire family of confidence statements concerning the true values of a family of parameters.

More specifically, if we jointly estimate μ and σ^2 (our family of parameters) from the same set of sample realizations and if the family confidence coefficient were, say, 0.95, then we would conclude that if many random samples were selected and interval estimates for *both* μ and σ^2 were made from each sample (our family of interval estimates), then, in the long-run, 95% of the samples would yield a family of realized intervals containing the true values of μ and σ^2 simultaneously, and 5% of the samples would yield a family of realized intervals excluding either one or both of the true population parameters.

If both μ and σ^2 are to be estimated jointly from the same sample, then we need a confidence coefficient (our family confidence coefficient) that will ensure that the complete set of estimates is correct or reliable. In what follows we shall explore the Bonferroni method of determining joint confidence intervals (or joint confidence statements) having a family confidence coefficient. The procedure is straightforward: we simply adjust each statement confidence coefficient to a level higher than $1 - \alpha$ so that the family confidence coefficient is at least $1 - \alpha$.

To examine the rationale underlying the Bonferroni technique, let us extract a random sample of size n from a normal population with unknown mean and variance and consider the individual confidence limits determined for μ and σ^2 earlier ((11.13) and (11.16), respectively):

$$CI_1 : \bar{x} - t_{\alpha/2, n-1} \frac{s}{\sqrt{n}} < \mu < \bar{x} + t_{\alpha/2, n-1} \frac{s}{\sqrt{n}};$$

$$CI_2 : \frac{(n-1)s^2}{\chi^2_{1-(\alpha/2), n-1}} < \sigma^2 < \frac{(n-1)s^2}{\chi^2_{\alpha/2, n-1}}.$$

What is the probability that both sets of confidence limits (the family of interval estimates) contain the parameters μ and σ^2 simultaneously?

To answer this question, let us construct the following events:

$$A_1 = \{\mu \notin CI_1\}, A_2 = \{\sigma^2 \notin CI_2\};$$

where $P(A_1) = P(A_2) = \alpha$. By the general addition rule for probabilities (3.1), $P(A_1 \cup A_2) = P(A_1) + P(A_2) - P(A_1 \cap A_2)$ and thus $1 - P(A_1 \cup A_2) = 1 - P(A_1) - P(A_2) + P(A_1 \cap A_2)$. From corollary 1 and the first of DeMorgan's Laws,

$$1 - P(A_1 \cup A_2) = P(\overline{A_1 \cup A_2}) = P(\overline{A}_1 \cap \overline{A}_2) = P(\mu \in CI_1 \text{ and } \sigma^2 \in CI_2),$$

that is, *both* confidence intervals contain their associated true parameter values. Hence

$$P(\overline{A}_1 \cap \overline{A}_2) = 1 - P(A_1) - P(A_2) + P(A_1 \cap A_2).$$

Since $P(A_1 \cap A_2) \geq 0$, the preceding expression can be rewritten as the *Bonferroni inequality*

$$P(\overline{A}_1 \cap \overline{A}_2) \geq 1 - P(A_1) - P(A_2) = 1 - \alpha - \alpha = 1 - 2\alpha, \qquad (11.31)$$

where the right-hand side of this inequality is termed the *Bonferroni lower bound*. So if $\mu \in CI_1$ and $\sigma^2 \in CI_2$ hold individually with, say, 95% confidence, then (11.31) establishes a family confidence coefficient of at least $1 - 0.05 - 0.05 = 0.90$ that both CI_1 and CI_2 determined from the same sample contain μ and σ^2 simultaneously.

To obtain a family confidence coefficient of at least $1 - \alpha$ for estimating μ and σ^2 jointly, we first estimate μ and σ^2 separately with statement confidence coefficients of $1 - \frac{\alpha}{2}$ each (the area in each tail of the appropriate sampling distribution is thus $\frac{\alpha}{4}$.) Hence the Bonferroni lower bound is $1 - \frac{\alpha}{2} - \frac{\alpha}{2} = 1 - \alpha$ so that the $100(1 - \alpha)\%$ family or *Bonferroni joint confidence limits* for μ and σ^2 are:

$$BCI_1 : \bar{x} - t_{\alpha/4,n-1} \frac{s}{\sqrt{n}} < \mu < \bar{x} + t_{\alpha/4,n-1} \frac{s}{\sqrt{n}};$$

$$BCI_2 : \frac{(n-1)s^2}{\chi^2_{1-(\alpha/4),n-1}} < \sigma^2 < \frac{(n-1)s^2}{\chi^2_{\alpha/4,n-1}}. \qquad (11.32)$$

In sum, use BCI_1 with $t_{\alpha/4,n-1}$ to get a $100(1 - \alpha/2)\%$ confidence interval for μ; and use BCI_2 with $\chi^2_{1-(\alpha/4),n-1}$ and $\chi^2_{\alpha/4,n-1}$ to get a $100(1 - \alpha/2)\%$ confidence interval for σ^2. Then the joint or Bonferroni confidence set is $BC = \{BCI_1 \cap BCI_2\}$ and $P\left((\mu, \sigma^2) \varepsilon BC\right) = P(\mu \varepsilon BCI_1 \text{ and } \sigma^2 \varepsilon BCI_2) = 1 - \alpha$.

It is useful to reiterate that if the family confidence coefficient is to be $1 - \alpha$, then each statement confidence coefficient should be $1 - \frac{\alpha}{2}$. But this implies that $\frac{\alpha}{2}$ must be divided between the two tails of each estimator's sampling distribution. Hence finding $t_{\alpha/4,n-1}$ or $\chi^2_{1-(\alpha/4),n-1}$ and $\chi^2_{\alpha/4,n-1}$ is required. In general, if r joint confidence interval estimates are desired with a family confidence coefficient of $1 - \alpha$, then determining each separate interval estimate with a statement confidence coefficient of $1 - \frac{\alpha}{r}$ will do.

Example 11.8.1 Suppose we extract a random sample of size $n = 25$ from a normal population with unknown mean and variance and we find that $\bar{x} = 17$ and $s^2 = 7.5$. Find a set of Bonferroni joint confidence limits for μ and σ^2 when the family confidence coefficient is $1 - \alpha = 0.90$. Here $\alpha = 0.10$, $\frac{\alpha}{4} = 0.025$ so that $t_{\frac{\alpha}{4}, n-1} = t_{0.025, 24} = 2.064$ and $\chi^2_{1-\frac{\alpha}{4}, n-1} = \chi^2_{0.975, 24} = 39.4$ while $\chi^2_{\frac{\alpha}{4}, n-1} = \chi^2_{0.025, 24} = 12.4$. Then from (11.32):

$$BCI_1 : 17 \pm 2.064 \left(\frac{2.738}{\sqrt{25}} \right) \text{ or } 15.870 < \mu < 18.130;$$

$$BCI_2 : \frac{24(7.5)}{39.4} < \sigma^2 < \frac{24(7.5)}{12.4} \text{ or } 4.568 < \sigma^2 < 14.516.$$

Thus we are at least 90% confident that μ lies within the interval (15.870, 18.130) *and* σ^2 lies within the interval (4.568, 14.516). ∎

11.9 Confidence Intervals for the Difference of Means When Sampling from Two Independent Normal Populations

Let $\{X_1, \ldots, X_{n_X}\}$ and $\{Y_1, \ldots, Y_{n_Y}\}$ be two sets of sample random variables taken from independent normal distributions with means μ_X and μ_Y and variances σ^2_X and σ^2_Y, respectively. Our objective is to determine a $100(1 - \alpha)\%$ confidence interval for the difference of means $\mu_X - \mu_Y$. In order to do so, we must first examine the characteristics of the sampling distribution of the difference between two sample means.

To this end, let X and Y be independent random variables, where X is $N(\mu_X, \sigma_X)$ and Y is $N(\mu_Y, \sigma_Y)$. In order to use $\bar{X} - \bar{Y}$ as an estimator for $\mu_X - \mu_Y$, we need to detail the properties of the distribution of $\bar{X} - \bar{Y}$ given that we are sampling from two independent normal populations.

We first assume that the population variances are equal and known; that is, $\sigma^2_X = \sigma^2_Y = \sigma^2$ is known. By virtue of Theorem 8.3, $\bar{X}$ is $N\left(\mu_X, \sigma/\sqrt{n_X}\right)$ and $\bar{Y}$ is $N\left(\mu_Y, \sigma/\sqrt{n_Y}\right)$. Since $\bar{X}$ and $\bar{Y}$ are independent normally distributed random variables, an application of part (d) of Theorem 8.1 to $\bar{X}$ and $\bar{Y}$ (with $a_1 = 1$ and $a_2 = -1$ in equations (8.10) and (8.11.1)) leads us to conclude that $\bar{X} - \bar{Y}$ is normally distributed with mean $\mu_X - \mu_Y$ and variance $\left(\frac{\sigma^2_X}{n_X}\right) + \left(\frac{\sigma^2_Y}{n_Y}\right) = \sigma^2 \left(\frac{1}{n_X} + \frac{1}{n_Y}\right)$.

Then with σ^2 known, the quantity

$$Z_{\Delta\mu} = \frac{(\bar{X} - \bar{Y}) - (\mu_X - \mu_Y)}{\sigma \sqrt{\frac{1}{n_X} + \frac{1}{n_Y}}} \tag{11.33}$$

[Sampling from two independent normal populations with equal and known variances]

is $N(0, 1)$.

If the population variances are unequal but still assumed to be known, then the quantity

$$Z'_{\Delta\mu} = \frac{(\bar{X} - \bar{Y}) - (\mu_X - \mu_Y)}{\sqrt{\dfrac{\sigma_X^2}{n_X} + \dfrac{\sigma_Y^2}{n_Y}}} \tag{11.34}$$

[Sampling from two independent normal populations with unequal but known variances]

is still $N(0, 1)$.

If the common variance $\sigma^2 (= \sigma_X^2 = \sigma_Y^2)$ is unknown, then we need to examine the distribution of $\bar{X} - \bar{Y}$ under sampling from two independent normal populations with unknown but equal variances. Let us specify the individual sample variances as

$$S_X^2 = \frac{\sum\limits_{i=1}^{n_X}(X_i - \bar{X})^2}{n_X - 1}, \quad S_Y^2 = \frac{\sum\limits_{i=1}^{n_Y}(Y_i - \bar{Y})^2}{n_Y - 1}.$$

Then from Theorem 9.5, the quantities $U_X = \frac{(n_X-1)S_X^2}{\sigma^2}$ and $U_Y = \frac{(n_Y-1)S_Y^2}{\sigma^2}$ are chi-square random variables with $n_X - 1$ and $n_Y - 1$ degrees of freedom, respectively. Moreover, since U_X and U_Y are independent chi-square random variables, it follows, via Theorem 9.3, that the quantity

$$U = U_X + U_Y = \frac{(n_X - 1)S_X^2}{\sigma^2} + \frac{(n_Y - 1)S_Y^2}{\sigma^2} \tag{11.35}$$

is chi-square distributed with $k = n_X + n_Y - 2$ degrees of freedom.

Now, according to (9.19), the ratio of a $N(0, 1)$ random variable to the square root of chi-square random variable divided by its degrees of freedom v is (under the independence of these random variables) t distributed with v degrees of freedom. Hence the ratio of $Z_{\Delta\mu}$ to $\sqrt{U/k}$ or

$$T = \frac{Z_{\Delta\mu}}{\sqrt{U/k}} \tag{11.36}$$

follows a t distribution with $k = n_X + n_Y - 2$ degrees of freedom.

Upon substituting the expressions appearing in (11.33) and (11.35) into (11.36) and simplifying yields

$$T_{\Delta\mu} = \frac{(\bar{X} - \bar{Y}) - (\mu_X - \mu_Y)}{S_p\sqrt{\dfrac{1}{n_X} + \dfrac{1}{n_Y}}} \tag{11.36.1}$$

[Sampling from two independent normal populations with equal but unknown variances] where

$$S_p = \sqrt{\frac{(n_X - 1)S_X^2 + (n_Y - 1)S_Y^2}{k}} \tag{11.37}$$

denotes the *pooled estimator* of the common standard deviation σ and $k = n_X + n_Y - 2$ depicts *pooled degrees of freedom*. Note that the term under the radical in (11.37) (i.e., the *pooled variance* S_p^2) is simply a weighted average of the two sample variances S_X^2 and S_Y^2, where the weights are the degrees of freedom associated with the independent chi-square variables U_X and U_Y.

We next consider the instance where the population variances are *unequal and unknown*. Then it can be shown that when σ_X^2 and σ_Y^2 in $Z'_{\Delta\mu}$ (11.34) are replaced by their unbiased estimators S_X^2 and S_Y^2, respectively, the resulting quantity

$$T'_{\Delta\mu} = \frac{(\overline{X} - \overline{Y}) - (\mu_X - \mu_Y)}{\sqrt{\dfrac{S_X^2}{n_X} + \dfrac{S_Y^2}{n_Y}}} \tag{11.38}$$

[Sampling from two independent normal
populations with unequal but unknown variances]

is approximately t distributed with degrees of freedom given approximately by

$$\phi = \frac{\left(\dfrac{S_X^2}{n_X} + \dfrac{S_Y^2}{n_Y}\right)^2}{\left(\dfrac{S_X^2}{n_X}\right)^2 \left(\dfrac{1}{n_X + 1}\right) + \left(\dfrac{S_Y^2}{n_Y}\right)^2 \left(\dfrac{1}{n_Y + 1}\right)} - 2.^3 \tag{11.39}$$

(If ϕ is not an integer, then it must be rounded to the nearest integer value.)

Having discussed the particulars of the sampling distribution of $\overline{X} - \overline{Y}$ under a variety of sampling assumptions, we now turn to the problem of finding confidence interval estimates for $\mu_X - \mu_Y$. The following special cases are of considerable interest and hinge upon the assumptions made about the population variances σ_X^2 and σ_Y^2.

---

[3] Equations (11.38) and (11.39) are connected with the so-called *Behrens-Fisher problem*, for which there is no exact solution. For details concerning this problem see, for instance: Y.V. Linnik, "Latest Investigation on Behrens-Fisher problem," *Sankhya* (28 A), 1966 (pp. 15–24); J.S. Mehta and R. Srinivasan, "On the Behrens-Fisher Problem," *Biometrica* (57), 1990 (pp. 649–655); and H. Scheffé, "Practical Solutions of the Behrens-Fisher Problem," *J. Amer. Stat. Assn.* (65), 1970 (pp. 1501–1508).

11.9.1   Population Variances Known

σ_X^2 and σ_Y^2 *are known.* As indicated earlier, the random variable $Z'_{\Delta\mu}$ is $N(0,1)$. Then proceeding as in Section 11.4, with $Z'_{\Delta\mu}$ serving as a pivot quantity, the probability statement

$$P(-z_{\alpha/2} < Z'_{\Delta\mu} < z_{\alpha/2}) = 1 - \alpha \qquad (11.40)$$

can be pivoted or transformed to

$$P\left(\underbrace{(\overline{X} - \overline{Y}) - z_{\alpha/2}\sqrt{\frac{\sigma_X^2}{n_X} + \frac{\sigma_Y^2}{n_Y}}}_{L_1} < \mu_X - \mu_Y < \underbrace{(\overline{X} - \overline{Y}) + z_{\alpha/2}\sqrt{\frac{\sigma_X^2}{n_X} + \frac{\sigma_Y^2}{n_Y}}}_{L_2}\right)$$

$$= 1 - \alpha. \qquad (11.41)$$

Here (L_1, L_2) is a random interval with confidence probability $1 - \alpha$. Once $\overline{X}$ and $\overline{Y}$ are replaced by their sample realizations $\bar{x}$ and $\bar{y}$, respectively, a *100(1 − α)% confidence interval for* $\mu_X - \mu_Y$ given known population variances is

$$(\ell_1, \ell_2) = \left((\bar{x} - \bar{y}) - z_{\frac{\alpha}{2}}\sqrt{\frac{\sigma_X^2}{n_X} + \frac{\sigma_Y^2}{n_Y}}, (\bar{x} - \bar{y}) + z_{\frac{\alpha}{2}}\sqrt{\frac{\sigma_X^2}{n_X} + \frac{\sigma_Y^2}{n_Y}}\right), \qquad (11.42)$$

that is, we may be $100(1 - \alpha)\%$ confident that $\ell_1 < \mu_X - \mu_Y < \ell_2$.

How precisely have we estimated $\mu_X - \mu_Y$? We may express the *degree of precision* of $\bar{x} - \bar{y}$ as an estimate of $\mu_X - \mu_Y$ as: we are within $\pm z_{\frac{\alpha}{2}}\sqrt{\frac{\sigma_X^2}{n_X} + \frac{\sigma_Y^2}{n_Y}}$ units of $\mu_X - \mu_Y$ with $100(1 - \alpha)\%$ reliability. Stated alternatively, using an *error bound* interpretation: we are $100(1 - \alpha)\%$ confident that $\bar{x} - \bar{y}$ will not differ from $\mu_X - \mu_Y$ by more than $\pm z_{\frac{\alpha}{2}}\sqrt{\frac{\sigma_X^2}{n_X} + \frac{\sigma_Y^2}{n_Y}}$ units.

Note that if $\sigma_X^2 = \sigma_Y^2 = \sigma^2$, then (11.34) reduces to (11.33) and thus, for known and equal population variances, a *100(1 − α)% confidence interval for* $\mu_X - \mu_Y$ is

$$(\bar{x} - \bar{y}) \pm z_{\alpha/2}\sigma\sqrt{\frac{1}{n_X} + \frac{1}{n_Y}}. \qquad (11.42.1)$$

11.9.2   Population Variances Unknown But Equal

σ_X^2 and σ_Y^2 *are unknown but equal.* In this instance the quantity $T_{\Delta\mu}$ (from (11.36.1)) serves as a pivot in the probability statement

$$P(-t_{\alpha/2,k} < T_{\Delta\mu} < t_{\alpha/2,k}) = 1 - \alpha, \quad k = n_X + n_Y - 2. \qquad (11.43)$$

Upon pivoting, this expression is transformed into

$$P\left(\underbrace{(\bar{X} - \bar{Y}) - t_{\frac{\alpha}{2},k} S_p \sqrt{\frac{1}{n_X} + \frac{1}{n_Y}}}_{L_1} < \mu_X - \mu_Y < \underbrace{(\bar{X} - \bar{Y}) + t_{\frac{\alpha}{2},k} S_p \sqrt{\frac{1}{n_X} + \frac{1}{n_Y}}}_{L_2}\right)$$

$$= 1 - \alpha. \tag{11.44}$$

Here also (L_1, L_2) is a random interval with confidence probability $1 - \alpha$. Then replacing $\bar{X}, \bar{Y}$, and S_p by their sample realizations enables us to obtain, for the case of unknown but equal population variances, the *100(1 − α)% confidence interval for* $\mu_X - \mu_Y$

$$(\ell_1, \ell_2) = \left((\bar{x} - \bar{y}) - t_{\alpha/2,k} S_p \sqrt{\frac{1}{n_X} + \frac{1}{n_Y}}, \ (\bar{x} - \bar{y}) + t_{\alpha/2,k} S_p \sqrt{\frac{1}{n_X} + \frac{1}{n_Y}}\right). \tag{11.45}$$

Thus we may be $100(1 - \alpha)\%$ confident that $\ell_1 < \mu_X - \mu_Y < \ell_2$.

The discussion in Section 8.11.1 pertaining to the precision of $\bar{X} - \bar{Y}$ as an estimator of $\mu_X - \mu_Y$ and the error bound interpretation of the confidence interval half width can also be offered for (11.44).

Note that if both n_X and n_Y are large (n_X and n_Y each >30), then (11.45) can be replaced by

$$(\bar{x} - \bar{y}) \pm z_{\alpha/2} S_p \sqrt{\frac{1}{n_X} + \frac{1}{n_Y}}. \tag{11.45.1}$$

11.9.3   Population Variances Unknown and Unequal

σ_X^2 and σ_Y^2 *are unknown and unequal.* For this situation $T'_{\Delta\mu}$ appearing in (11.38) will be the pivot quantity in the probability statement

$$P(-t_{\alpha/2,\phi} < T'_{\Delta\mu} < t_{\alpha/2,\phi}) = 1 - \alpha, \tag{11.46}$$

where ϕ (representing degrees of freedom for the t statistic) is defined by (11.39). Then proceeding as in the preceding two cases, for unknown and unequal population variances, an (approximate) *100(1 − α)% confidence interval for* $\mu_X - \mu_Y$ is

$$(\ell_1, \ell_2) = \left((\bar{x} - \bar{y}) - t_{\alpha/2,\phi} \sqrt{\frac{s_X^2}{n_X} + \frac{s_Y^2}{n_Y}}, \ (\bar{x} - \bar{y}) + t_{\alpha/2,\phi} \sqrt{\frac{s_X^2}{n_X} + \frac{s_Y^2}{n_Y}}\right), \tag{11.47}$$

where $\bar{x}$ and $\bar{y}$ are the realized means of the two population random variables X and Y, respectively, and s_X^2 and s_Y^2 are their respective realized variances. Hence we may be $100(1 - \alpha)\%$ confident that approximately $\ell_1 < \mu_X - \mu_Y < \ell_2$.

Example 11.9.1 Suppose the random variable X is $N(\mu_X, \sqrt{14})$ and the random variable Y is $N(\mu_Y, \sqrt{11})$, with X and Y also taken to be independent. Let $n_X = 50$ and $n_Y = 60$ and suppose we find that $\bar{x} = 87$ and $\bar{y} = 77$. Then, since the population variances $\sigma_X^2 = 14$ and $\sigma_Y^2 = 11$ are known, we may find a 95% confidence interval for $\mu_X - \mu_Y$ using (11.42). That is, for $z_{0.025} = 1.96, (\ell_1, \ell_2) = (8.666, 11.334)$; that is, we may be 95% confident that the true or population difference in means $\mu_X - \mu_Y$ lies between 8.666 and 11.334.

Next, let $n_X = 15, n_Y = 13, \bar{x} = 24, \bar{y} = 20$, and suppose $s_X^2 = 8.2$ and $s_Y^2 = 9.7$ are the sample realizations of S_X^2 and S_Y^2, respectively, where S_X^2 and S_Y^2 are unbiased estimators for their respective unknown population counterparts σ_X^2 and σ_Y^2. If we assume that $\sigma_X^2 = \sigma_Y^2$, then a 95% confidence interval for $\mu_X - \mu_Y$ can be determined from (11.45). To this end, from (11.37),

$$S_p = \sqrt{\frac{(15-1)(8.2) + (13-1)(9.7)}{15 + 13 - 2}} = 2.982.$$

Then with $t_{\alpha/2,k} = t_{0.025,26} = 2.056$, we have $(\ell_1, \ell_2) = (1.682, 6.318)$.

And if σ_X^2 and σ_Y^2 are unknown and there is no compelling evidence to indicate that they are equal, then we must employ (11.47) to find an (approximate) 95% confidence interval for $\mu_X - \mu_Y$. Using the preceding set of sample information we have, from (11.39),

$$\phi = \frac{\left(\dfrac{8.2}{15} + \dfrac{9.7}{13}\right)^2}{\left(\dfrac{8.2}{15}\right)^2 \left(\dfrac{1}{16}\right) + \left(\dfrac{9.7}{13}\right)^2 \left(\dfrac{1}{14}\right)} - 2 = 28.58 - 2 \approx 27.$$

Then $t_{\alpha/2,\phi} = t_{0.025,27} = 2.052$ and thus $(\ell_1, \ell_2) = (1.669, 6.331)$, approximately. ∎

At this point in our discussion of confidence interval estimation of $\mu_X - \mu_Y$ it is important to review the basic assumptions underlying our two-sample procedures. Specifically, we are assuming that:

1. We have *independent random samples*. This assumption is indispensable—it must hold if we are to make probability (and thus confidence) statements about $\mu_X - \mu_Y$.

2. We have *normal populations*. This is a relatively weak assumption; that is, if the samples are sufficiently large, then, by virtue of the Central Limit Theorem, substantial departures from normality will not greatly affect our confidence probabilities.

3. We have *equal variances*. If the population variances are not the same (whether they are known or not), then our confidence interval for $\mu_X - \mu_Y$ is only approximate.

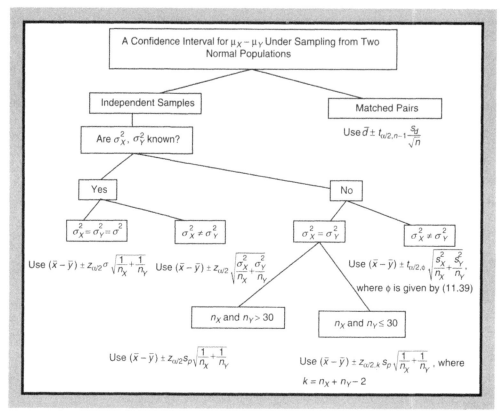

Figure 11.6   A confidence interval for $\mu_X - \mu_Y$ under sampling from two normal populations.

The confidence interval results for this section are summarized in the left-hand branch of Figure 11.6.

11.10 Confidence Intervals for the Difference of Means When Sampling from Two Dependent Populations: Paired Comparisons

The inferential techniques of the preceding section were based upon random sampling from two *independent* (normal) populations. We now turn to the case where the samples are *dependent*. In particular, the samples from the two populations will be *paired*; that is, each observation in the first sample is related in some specific way to exactly one observation in the second sample, so that the two samples are obviously not independent. However, before we develop the statistical machinery for analyzing paired samples, we need to consider some preliminary definitional material concerning the rudiments of experimental design. This discussion will enable us to better understand why pairing is important and if it should be done at all.

We previously defined an *experiment* as any process of observation; that is, it is any proposed or planned process by which observations are generated or data are collected. Hence the *design of an experiment* dictates the specific way and conditions under which the data are collected—it is basically the method of choosing a sample. For instance, we may face an experimental situation in which there exists a well-defined target population and we wish to examine (possibly by a survey) some measurable characteristic of that population. Here we may choose to conduct a survey of the weights of men in a certain age group in a particular locale; or take a survey of children in the 1–6 years of age group in a certain city to determine the extent of their exposure to lead-based products; or perhaps conduct a public opinion poll to assess the popularity of a given candidate for political office.

Alternatively, we may face an experimental situation in which the target population does not actually exist and yet we want to investigate a characteristic of some hypothetical population that possibly could exist. Experiments in the fields of medicine, psychology, agriculture, chemistry, and so on are indicative of this approach to experimental design. For example, we may be interested in monitoring the recovery time (in days) of patients undergoing a new type of surgical process; or we might want to assess the effects of different approaches to teaching basic algebra. In these circumstances, specialized or customized designs, involving the control of extraneous factors and/or the external environment, may be appropriate. Hence the elementary sampling units might be chosen in some particular or restrictive way that ostensibly enhances the efficiency or information content of the experiment over and above that which could be obtained by simple random sampling alone.

An important component of any experimental design is the notion of a *treatment*, which may be defined as any procedure whose effect(s) we want to measure (or simply observe) and compare; for example, determining a patient's response to different types (and/or dosages) of medications; assessing the effects of temperature variations when glazing a painted surface; testing the residual sharpness of different makes of saw blades after exposure to sustained cutting of the same material; determining the effect of a variety of exercise programs (or diets) on the elderly, and so on.

Suppose we want to compare the effects of two treatments that are applied to similar *experimental units*. We may randomly assign a treatment to each experimental unit with the proviso that each treatment is to be applied to the same number of experimental units. An experimental design of this type is commonly known as a *completely randomized design*.

After the treatments are applied, our goal is to measure the effects of each treatment on each experimental unit to which it was applied. We thus have a set of outcomes or responses that may be classified into two groups according to the treatment received. On the basis of this particular design, consequently we may assume that:

1. The outcomes constitute independent random samples from the distribution of *all* such outcomes.

2. The responses of *all* experimental units to which a given treatment could be applied are normally distributed (or, at least, that the number of experimental units receiving each treatment is large).

3. The variance of the outcomes of the experimental units receiving one of the treatments is the same as that of the experimental units receiving the other treatment.

Clearly the key assumptions supporting the use of the two-sample methods presented in Section 11.9 are satisfied. Hence these techniques for making *group comparisons* may consequently be used herein to estimate the difference between the effects of the two treatments; that is, any difference between the means of the two groups of responses mirrors the difference between the effects of the two treatments.

It should be apparent, by virtue of the structure of, say, (11.45), that the precision of our interval estimate of $\mu_X - \mu_Y$, for fixed sample sizes n_X and n_Y, varies inversely with S_p^2, the pooled estimator of the common variance σ^2. Here S_p^2 is a measure of the unexplained variation among experimental units that receive similar treatment. One way to possibly reduce this variation, and thus increase the precision of our estimate of the difference in means for fixed n_X and n_Y values, is to pair the sample observations. That is, if the variation in the treatment outcomes between the members of any pair is less than the variation between corresponding members of different pairs, then the precision of our estimate of $\mu_X - \mu_Y$ can be enhanced. This can be accomplished by randomizing the two treatments over the two members of each pair (e.g., by flipping a coin) in a fashion such that each treatment is applied to one and only one member of each pair.

The result of this *paired experiment* is that we can obtain, for each pair of outcomes, an estimate of the difference between treatments effects, and thus variation between pairs is not included in our estimate of the common variance. So if the intrapair outcome variation is smaller than its interpair counterpart, then we should expect a variance smaller than that which we would obtain if a completely randomized (groups) design were implemented.

It was mentioned at the outset of this section that the random samples drawn from two populations would turn out to be dependent rather than independent. In this regard, samples can be dependent: (a) by virtue of their nature; that is, they are intrinsically dependent; or (b) by deliberate design.

For instance, in case (a), readings are taken on some measurable characteristic from the same experimental units at two different points in time and the means of the measurements at the two times are compared. For example, a researcher might conduct before and after weighings of a herd of cattle to determine the effectiveness of a new feed mixture that purports to accelerate weight gain. Here the weights will be *paired*; that is, for a particular steer, its weight at the start of the experiment (this measurement is an observation in the first sample) will be compared to its weight at the end of the experiment (this latter measurement is thus an observation in the second sample).

In case (b), the experimental units in the two samples may be paired to eliminate extraneous (and possibly confounding) factors or effects that are of no particular interest to the researcher. For instance, the director of a high-school driver education program may be interested in trying a new driver-training method that uses a high-tech driving simulator. Two new classes, each of the same size, are being scheduled for the next term so that the new technology-based technique can be tested. The old teaching method will be used in the first class and the simulator-based method will be employed in the second class. Although over 100 students have signed up for the driver training course, the director deems it prudent to limit the size of each of the two classes to 10 students because each of the various simulator modules takes a considerable amount of time to complete. So out of a pool of over 100 students, the director selects 10 pairs of students, where the individuals comprising each pair are as similar as possible (i.e., same sex, same year in school, similar grade point average, similar extracurricular and athletic interests, similar after-school work and activities, etc.) so that the effect of any extraneous factors can be minimized. After making sure that the two members of any pair are as alike as possible in all respects save for the method of driving instruction, the director then flips a coin to determine which student is assigned to the first class and which to the second class. Again the two samples (in this case classes) will be dependent since the assignments of the students to the two classes were not completely random in that they were deliberately made in pairs.

In general, grouping the experimental units into pairs by design, and thus according to the overall commonality of, say, environmental or extraneous factors, enables us to purge much of the outcome variation due to these factors from our estimate of the common variance σ^2 and thus the precision of our estimate of $\mu_X - \mu_Y$ is increased. And as previously stated, the pairs must be chosen such that the outcome variation among the pairs exceeds that occurring between the experimental units within the pairs.

As we shall now see, the analysis of the results of a paired observation experiment (in which the two samples are not chosen independently and at random) reduces to an application of a single-sample technique. In this regard, suppose our paired experiment yields n pairs of observations denoted by $(X_1, Y_1), (X_2, Y_2), \ldots, (X_n, Y_n)$. Here X_i and $Y_i, i = 1, \ldots, n$, are members of the sample pair (X_i, Y_i), where the X_i sample random variable is drawn from the first population and the sample random variable Y_i is drawn from the second population. (We may view the X_i's as depicting a set of sample outcomes for the first treatment and the Y_i's as representing a set of sample responses for the second or follow-up treatment.)

For the ith pair of sample random variables, let $D_i = X_i - Y_i, i = 1, \ldots, n$, be the difference between the two random variables X_i and Y_i making up that pair, where D_i is taken to be the ith observation on the random variable D. Each D_i thus provides us with a measure of the difference between the effectiveness of the two treatments. Then the best estimators for the mean and variance of D are

$$\bar{D} = \sum_{i=1}^{n} \frac{D_i}{n}, \quad S_D^2 = \sum_{i=1}^{n} \frac{(D_i - \bar{D})^2}{n-1},$$

respectively. Here $\bar{D}$ serves as an estimator of the mean difference between the effects of the first and second treatments and S_D^2, an estimator of the variance of the differences in treatment effects, excludes any variation due to extraneous factors (provided, of course, that pairing has been effective).

If we assume that the D_i's, $i = 1,\dots,n$, constitute a single random sample from a normal population with mean $\mu_D = \mu_X - \mu_Y$ and variance σ_D^2, then the population random variable D, whose values are the paired differences $D_i = X_i - Y_i$ in the population, has a sampling distribution that is $N(\mu_D, \sigma_D)$. Hence the statistic

$$T = \frac{\bar{D} - \mu_D}{S_D/\sqrt{n}}$$

[Sampling from two dependent populations]

follows a t-distribution with $n - 1$ degrees of freedom. Note that this t statistic for paired comparisons, has only $n - 1$ degrees of freedom as compared to the t statistic for group comparisons, which was found to be $k = n_X + n_Y - 2 = 2(n-1)$ (since $n_X = n_Y = n$). However, if pairing is warranted, this loss in degrees of freedom is compensated for by the reduction in variance (due to the elimination of the effects of extraneous factors) and the concomitant increase in the precision of our estimate of $\mu_D = \mu_X - \mu_Y$. (Of course, if there are no extraneous factors to guard against, then we actually *lose* information by pairing the two samples—only $n - 1$ and not $2(n - 1)$ degrees of freedom are used to estimate σ^2.)

Then proceeding as in section 11.5, a $100(1 - \alpha)\%$ *confidence interval for* $\mu_D = \mu_X - \mu_Y$ is

$$(\ell_1, \ell_2) = \left(\bar{d} - t_{\alpha/2,n-1}\frac{s_D}{\sqrt{n}}, \bar{d} + t_{\alpha/2,n-1}\frac{s_D}{\sqrt{n}}\right), \qquad (11.48)$$

where $\bar{d}$ and s_D are the sample realizations of $\bar{D}$ and S_D, respectively. (This result constitutes the right-hand branch appearing in Figure 11.6).

Example 11.10.1   A group of 15 voters was asked to rate the suitability of a particular candidate for political office on a scale from 1 to 5 (with 5 being the highest rating). After the rating was conducted, the group watched a 40-minute

Table 11.1

Ratings of a Political Candidate		
Before (x)	**After (y)**	**Difference ($d = x - y$)**
4	4	0
4	5	−1
3	3	0
5	5	0
5	5	0
5	4	1
2	4	−2
3	5	−2
4	5	−1
3	3	0
3	4	−1
5	4	1
5	4	1
4	4	0
4	4	0

videotape of the candidate's responses to 10 questions pertaining to an assortment of important social/economic issues. At the conclusion of the videotape, all 15 voters were again asked to rate the candidate on the same 1 to 5 scale. The results of both ratings appear in Table 11.1. How precisely have we estimated the true difference between the before- and after-mean ratings of the candidate by the group? Let $\alpha = 0.05$. From the realized differences $d_i, i = 1, \ldots, 15$, appearing in Table 11.1 we have

$$\bar{d} = \sum_{i=1}^{15} \frac{d_i}{n} = -\frac{4}{15} = -0.266,$$

$$s_D^2 = \frac{\sum_{i=1}^{15} d_i^2}{n-1} - \frac{\left(\sum_{i=1}^{15} d_i\right)^2}{n(n-1)} = \frac{14}{14} - \frac{(-4)^2}{15(14)} = 0.924$$

and thus, from (11.48) (here $t_{\alpha/2,n-1} = t_{0.025,14} = 2.145$), $(\ell_1, \ell_2) = (-0.798, 0.266)$. Hence we may be 95% confident that $\mu_D (= \mu_X - \mu_Y)$ lies between −0.798 and 0.266. Alternatively, we may conclude that we are within ±0.532 units of μ_D with 95% reliability. ∎

11.11 Confidence Intervals for the Difference of Proportions When Sampling from Two Independent Binomial Populations

Let $\{X_1, X_2, \ldots, X_{n_X}\}$ and $\{Y_1, Y_2, \ldots, Y_{n_Y}\}$ be two sets of sample random variables extracted from two independent binomial populations, where p_X and p_Y are the *proportions of successes* in the first and second binomial populations, respectively. In addition, let X and Y be independent random variables representing the observed number of successes in the samples of size n_X and n_Y, respectively. Our objective is to determine a $100(1-\alpha)\%$ confidence interval for the difference in proportions $p_X - p_Y$. Moreover, the said confidence interval is only approximate and holds for large samples. To accomplish this task, we must first specify the characteristics of the sampling distribution of the difference between two sample proportions.

On the basis of our discussion in Section 11.7, the best estimators for p_X and p_Y are the *sample proportions of successes* $\hat{P}_X = \frac{X}{n_X}$ and $\hat{P}_Y = \frac{Y}{n_Y}$, respectively, with $E(\hat{P}_X) = p_X$, $E(\hat{P}_Y) = p_Y$, $V(\hat{P}_X) = \frac{p_X(1-p_X)}{n_X}$, and $V(\hat{P}_Y) = \frac{p_Y(1-p_Y)}{n_Y}$. Hence the best estimator for $p_X - p_Y$ is $\hat{P}_X - \hat{P}_Y$ with $E(\hat{P}_X - \hat{P}_Y) = p_X - p_Y$ and the best estimators for $V(\hat{P}_X)$ and $V(\hat{P}_Y)$ are, respectively, $S^2(\hat{P}_X) = \frac{\hat{P}_X(1-\hat{P}_X)}{n_X}$ and $S^2(\hat{P}_Y) = \frac{\hat{P}_Y(1-\hat{P}_Y)}{n_Y}$.

Since $\hat{P}_X$ and $\hat{P}_Y$ are independent random variables, we have, by virtue of (5.47) (with $a = b = c$), $V(\hat{P}_X - \hat{P}_Y) = V(\hat{P}_X) + V(\hat{P}_Y)$, where this variance expression is estimated by $S^2(\hat{P}_X - \hat{P}_Y) = S^2(\hat{P}_X) + S^2(\hat{P}_Y)$. Then according to the De Moivre-Laplace-Gauss Limit Theorem,

$$Z_{\Delta p} = \frac{(\hat{P}_X - \hat{P}_Y) - E(\hat{P}_X - \hat{P}_Y)}{\sqrt{S^2(\hat{P}_X - \hat{P}_Y)}} = \frac{(\hat{P}_X - \hat{P}_Y) - (p_X - p_Y)}{\sqrt{S^2(\hat{P}_X) + S^2(\hat{P}_Y)}} \xrightarrow{d} Z = N(0,1).$$

$$[\text{Sampling from two independent binomial populations}] \qquad (11.49)$$

So for large n,

$$P(-z_{\alpha/2} < Z_{\Delta p} < z_{\alpha/2}) \approx 1 - \alpha \qquad (11.50)$$

Upon substituting (11.49) (the pivotal quantity) into (11.50), the resulting probability statement can be transformed into

$$P\left(\underbrace{\left(\hat{P}_X - \hat{P}_Y\right) - z_{\alpha/2}\sqrt{S^2\left(\hat{P}_X\right) + S^2\left(\hat{P}_Y\right)}}_{L_1} < p_X - p_Y \right.$$

$$< \underbrace{\left(\hat{P}_X - \hat{P}_Y\right) + z_{\alpha/2}\sqrt{S^2\left(\hat{P}_X\right) + S^2\left(\hat{P}_Y\right)}}_{L_2} \Bigg) \approx 1 - \alpha. \qquad (11.51)$$

Thus the probability that $p_X - p_Y$ lies within the random interval (L_1, L_2) is $1 - \alpha$.

Let us next: (1) replace $\hat{P}_x$ and $\hat{P}_y$ by their sample realizations $\hat{p}_x(= x/n_x)$ and $\hat{p}_y(= y/n_y)$, where x (respectively, y) is the realized number of successes in the first (respectively, second) sample; and (2) replace $S^2(\hat{P}_X)$ and $S^2(\hat{P}_Y)$ by their corresponding sample realizations $s^2(\hat{p}_X) = \frac{\hat{p}_X(1-\hat{p}_X)}{n_X}$ and $s^2(\hat{p}_Y) = \frac{\hat{p}_Y(1-\hat{p}_Y)}{n_Y}$, respectively. Then (11.51) renders as a large sample (approximate) *$100(1 - \alpha)\%$ confidence interval for $p_X - p_Y$*

$$(\hat{p}_X - \hat{p}_Y) \pm z_{\alpha/2}\sqrt{\frac{\hat{p}_X(1 - \hat{p}_X)}{n_X} + \frac{\hat{p}_Y(1 - \hat{p}_Y)}{n_Y}}, \qquad (11.52)$$

that is, we may be $100(1 - \alpha)\%$ confident that $\ell_1 < p_X - p_Y < \ell_2$, where (ℓ_1, ℓ_2) is the sample realization of (L_1, L_2).

It was mentioned earlier that we would be undertaking a large sample approach to constructing an approximate confidence interval for the difference in population proportions $p_X - p_Y$. In this regard, the method should yield a good approximation to normality provided that n_X and n_Y are each ≥ 25. This technique should not be used if n_X and n_Y are both < 25.

Example 11.11.1   A random sample was taken from each of two independent binomial populations, yielding the following results: $\hat{p}_X = \frac{x}{n_X} = \frac{18}{30} = 0.600$ and $\hat{p}_Y = \frac{y}{n_Y} = \frac{10}{35} = 0.286$. Hence our best estimate of $p_X - p_Y$, the difference between the two population proportions of successes, is $\hat{p}_X - \hat{p}_Y = 0.314$. How precisely have we estimated $p_X - p_Y$? Using (11.52), with $\alpha = 0.05$, we obtain $(\ell_1, \ell_2) = (0.053, 0.575)$; that is, we may be approximately 95% confident that the true difference in the two population proportions lies between 0.053 and 0.575. Hence we are within ± 0.261 units of the true difference $p_X - p_Y$ with 95% reliability.   ■

11.12 Confidence Interval for the Ratio of Two Variances When Sampling from Two Independent Normal Populations

Suppose we select random samples of sizes n_X and n_Y from two independent normal populations $N(\mu_X, \sigma_X)$ and $N(\mu_Y, \sigma_Y)$, respectively, having unknown means and variances. Our objective is to construct a $100(1 - \alpha)\%$ confidence interval for the ratio of population variances σ_Y^2/σ_X^2.

We noted earlier ((9.33)) that the random variable

$$F_{n_X-1,n_Y-1} = \frac{S_X^2/\sigma_X^2}{S_Y^2/\sigma_Y^2} \tag{11.53}$$

is F distributed with $n_X - 1$ and $n_Y - 1$ degrees of freedom, where the sample variances S_X^2 and S_Y^2 will serve as the best estimators of the population variances σ_X^2 and σ_Y^2, respectively.

Given (11.53), we now seek to find lower and upper quantile values $\frac{1}{f_{1-(\alpha/2),n_Y-1,n_X-1}}$ and $f_{1-(\alpha/2),n_X-1,n_Y-1}$, respectively, of the F distribution such that

$$P\left(\frac{1}{f_{1-(\alpha/2),n_Y-1,n_X-1}} < F_{n_X-1,n_Y-1} < f_{1-(\alpha/2),n_X-1,n_Y-1}\right) = 1 - \alpha. \tag{11.54}$$

Upon substituting (11.53) into (11.54) and pivoting we obtain

$$P\left[\underbrace{\left(\frac{1}{f_{1-(\alpha/2),n_Y-1,n_X-1}}\right)\left(\frac{S_Y^2}{S_X^2}\right)}_{L_1} < \frac{\sigma_Y^2}{\sigma_X^2} < \underbrace{(f_{1-(\alpha/2),n_X-1,n_Y-1})\left(\frac{S_Y^2}{S_X^2}\right)}_{L_2}\right] = 1 - \alpha. \tag{11.55}$$

Thus the probability that σ_Y^2/σ_X^2 lies within the random interval (L_1, L_2) is $1 - \alpha$. And if we replace S_X^2 and S_Y^2 by their sample realizations in (11.55), a $100(1-\alpha)\%$ confidence interval for σ_Y^2/σ_X^2 is

$$(\ell_1, \ell_2) = \left(\left(\frac{1}{f_{1-(\alpha/2),n_Y-1,n_X-1}}\right)\left(\frac{s_Y^2}{s_X^2}\right), (f_{1-(\alpha/2),n_X-1,n_Y-1})\left(\frac{s_Y^2}{s_X^2}\right)\right) \tag{11.56}$$

that is, we may be $100(1 - \alpha)\%$ confident that $\ell_1 < \frac{\sigma_Y^2}{\sigma_X^2} < \ell_2$.

Example 11.12.1   Suppose a random sample has been drawn from each of two independent normal populations, yielding the following results:

Sample 1	Sample 2
$n_X = 5$	$n_Y = 7$
$s_X^2 = 14.6$	$s_Y^2 = 20.3$

Find a 90% confidence interval for $\frac{\sigma_Y^2}{\sigma_X^2}$. Since $\alpha = 0.10$, we first determine that

$$f_{1-\frac{\alpha}{2}, n_X - 1, n_Y - 1} = f_{0.95,4,6} = 4.53, \quad f_{1-\frac{\alpha}{2}, n_Y - 1, n_X - 1} = f_{0.95,6,4} = 6.16,$$

Then from (11.56), $(\ell_1, \ell_2) = (0.226, 6.299)$. Hence we may be 90% confident that $\frac{\sigma_Y^2}{\sigma_X^2}$ lies between 0.226 and 6.299. ∎

11.13   Exercises

11-1. The times spent at the pump were recorded for $n = 65$ randomly selected customers at a small independent gas station. It was found that $\bar{x} = 4.6$ minutes with $s = 2.06$ minutes. Find a 95% confidence interval for the true average time spent at the pump. How precisely have we estimated μ?

11-2. Let X (the length of life in hours of a certain brand of 100-watt frosted light bulb) be $N(\mu, 500)$. A random sample of $n = 20$ bulbs was tested (until they all burned out), yielding a mean life of $\bar{x} = 1357$ hours. Find a 95% confidence interval for μ. Explain your result. Suppose X is not assumed to be normal but we have $\sigma = 500$. Can we still find a 95% confidence interval for μ? What if we increase n to 150 and find $\bar{x} = 1400$ and $s = 5.57$? What are the new 95% confidence limits. What theorem legitimizes them?

11-3. A manufacturer of adhesives has developed a new super latex cement that was tested on $n = 12$ different industrial rubber compounds. The drying times to adhesion (in seconds) were as follows:

81	103	100	67	63	71
90	91	85	88	71	73

Find a 95% confidence interval for the true mean drying time for this new cement.

11-4. Let X be the amount of distillate per day (in pounds) produced by a chemical process at ACE Laboratories and suppose that X is approximately normally distributed. A sample of $n = 10$ consecutive days yielded the following amounts (in pounds): 2.3, 4.1, 5.6, 3.1, 4.0, 5.1, 2.7, 3.6, 4.2, and 5.9. Find the 95% confidence interval for the true average amount of distillate per day.

11-5. A random of sample size $n = 16$ from a distribution of the form $N(\mu, 25)$ yielded a mean of $\bar{x} = 75$. Find a 99% confidence interval for μ. What is the confidence interval if σ is unknown and $s = 25$?

11-6. A certain type of electronic measurement has a standard deviation of 9 units. How large of a sample should be taken so that the 95% confidence interval for the mean will not exceed 4 units in width?

11-7. A random sample is to be drawn from a population that is $N(\mu, \sigma)$, with σ^2 known. One objective is to use $\overline{X}$ to estimate μ to within k units with $100(1 - \alpha)\%$ reliability. Solve for the sample size n required to meet these conditions from the following probability statement: $P(|\overline{X} - \mu| \le k) = 1 - \alpha$.

11-8. Suppose we wish to estimate the average daily yield (in terms of telephone inquiries) of a new magazine ad for a certain product. We want a degree of precision of ± 3 telephone calls with 95% reliability. If it is known that $\sigma = 10$ calls, find the required sample size. If σ is unknown but evidence indicates that the calls are approximately normally distributed with a range of about 60 calls, recalculate n.

11-9. In a sample of $n = 100$ high pressure castings, 15 were found to exhibit one or more defects. Find a 99% confidence interval for the population percent of all defective castings.

11-10. About $\frac{9}{10}$ of $n = 400$ persons of voting age interviewed were in favor of funding a new public safety program. Determine a 95% confidence interval for the fraction of the population who are in favor of the program.

11-11. Under random polling of the eligible voters of a certain city it was determined that out of $n = 350$ voters, 200 of them favor candidate A. Find an approximate 95% confidence interval for p, the true fraction of eligible voters favoring this candidate.

11-12. A psychology experiment is structured so that either response A or response B is forthcoming from a certain stimulus. If p is the true proportion of subjects providing response B, determine the sample size needed to estimate p to within ± 0.87 units with 90% reliability. Suppose a pilot study indicated that p is about 0.7. If no prior estimate of p is available, recalculate n.

11-13. Suppose the sponsor of a popular television show wants to determine the proportion of the viewing public watching the show to within 3% with 95% reliability. How large of a sample should be taken? Assume that past experience dictates that $p = 0.45$. If this known value of p is not available, what is the required sample size?

11-14. If the population mean μ is to be estimated by a $100(1 - \alpha)\%$ confidence interval and we are sampling without replacement from a finite population and $n/N > 0.05$, then a measure of how precisely we have estimated μ is provided by $\pm$ error bound, where the error bound is

$$\frac{w}{2} = z_{\alpha/2} \frac{\sigma}{\sqrt{n}} \sqrt{\frac{N - n}{N - 1}}.$$

Solving this expression for n renders

$$n = \frac{mN}{m+N-1}, \quad \text{where} \quad m = \left(\frac{z_{\alpha/2}\sigma}{w/2}\right)^2.$$

What is the interpretation of n? Suppose X is $N(\mu, 20)$ and the population size is $N = 2300$. How large of a sample is needed so that the 95% confidence interval for the mean does not exceed 4 units in width?

11-15. If the population proportion p is to be estimated by a $100(1-\alpha)\%$ confidence interval and we are sampling from a binomial population of size N without replacement, then a measure of how precisely we have estimated p is determined by $\pm$ error bound, where the error bound is

$$\frac{w}{2} = z_{\alpha/2}\sqrt{\frac{p(1-p)}{n}\left(\frac{N-n}{N-1}\right)}.$$

Solving this expression for n yields

$$n = \frac{mN}{m+N-1}, \quad \text{where} \quad m = \frac{z_{\alpha/2}\,p(1-p)}{(w/2)^2}$$

What is the interpretation of n? (Note: since p is unknown, use $\hat{p}$; if no prior or pilot estimate such as $\hat{p}$ is available, then use $p = \frac{1}{2}$.) Suppose that a union shop involving $N = 2500$ members is interested in assessing union support for a new contract involving enhanced benefits rather than wage increases. To estimate the proportion p in favor of the contract, how large a sample is required so that the 95% confidence interval for p will not exceed 5%?

11-16. Suppose a random sample of size $n = 15$ is taken from a normally distributed population. It is determined that $\sum_{i=1}^{15}(x_i - \bar{x}) = 131.43$. Find a 95% confidence interval for σ^2. What is the 95% confidence interval for σ?

11-17. Suppose the diameters of steel rods X (expressed in hundreds of millimeters) are approximately normally distributed. Given the following set of sample observations: 533, 532, 535, 540, 544, 536, 541, 538, 533, 537, determine a 95% confidence interval for σ^2. What is the 95% confidence interval for σ?

11-18. From a sample of size 20 taken from an approximately normal population it was determined that $s^2 = 5.36$. Find a 90% confidence interval for σ.

11-19. A random sample of 17 high school students was selected from the senior class. It was found that the standard deviation of their summer hourly wage rates was $4.35. Find a 90% confidence interval for the true variance of the hourly wage rate for all seniors who had summer jobs. What assumption is required in order to find the said interval?

11-20. A researcher wants to check the variability of a device used to measure the light intensity (in lumens) emitted from a certain welding machine. Nine independent measurements are recorded: 546, 609, 550, 585, 570, 577, 580, 540, and 561. If the measurements recorded by this instrument are approximately normally distributed, find a 95% confidence interval for σ^2.

11-21. Suppose that $n_X = 10, n_Y = 8, \bar{x} = 70.5, \bar{y} = 77.3, \sigma_X = 22,$ and $\sigma_Y = 19$. If the samples are independent and drawn from two normal populations, find a 95% confidence interval for $\mu_X - \mu_Y$.

If σ_X^2 and σ_Y^2 are unknown and estimated by $s_X^2 = 5$ and $s_Y^2 = 3.7$, respectively, with $n_X = 20$ and $n_Y = 25$, find a 95% confidence interval for $\mu_X - \mu_Y$ given that $\bar{x} = 65.6$ and $\bar{y} = 58.9$.

If $\sigma_X^2 = \sigma_Y^2 = \sigma^2$ and σ^2 is to be estimated using a pooled estimator of the common variance, find a 95% confidence interval for $\mu_X - \mu_Y$ when $\bar{x} = 18.1, \bar{y} = 17.6, n_X = 10, n_Y = 15, s_X^2 = 7,$ and $s_Y^2 = 9$.

If the preceding assumption that $\sigma_X^2 = \sigma_Y^2 = \sigma^2$ is untenable, what is the 95% confidence interval for $\mu_X - \mu_Y$?

11-22. Assume that miles per gallon of gasoline for imported passenger cars (of a given engine displacement) are normally distributed with a known standard deviation of 3.1 miles per gallon, and a comparable figure for domestic vehicles having the same displacement is 4.7 miles per gallon. Suppose that $n_X = 42$ imported and $n_Y = 50$ domestic cars are sampled and their average miles per gallon are 23.5 and 20.1, respectively. Find a 99% confidence interval for the difference $\mu_X - \mu_Y$.

11-23. Resolve the preceding exercise if it is known that $\sigma_X^2 = \sigma_Y^2 = \sigma^2$.

11-24. The ABC corporation claims that its new gasoline additive for passenger cars will enhance the mileage per tankfull of gasoline if a can of their product is added to a full tank of gasoline. Five vehicles were tested, yielding the following results (miles per tankfull):

Vehicle	1	2	3	4	5
Before additive	370	385	375	380	378
After additive	387	397	380	392	389

Find a 95% confidence interval for the difference in the means of these two data sets.

11-25. Two soda vending machines in two separate wings of an office building are tested monthly for average fill. Each machine is set to dispense 8 oz. of soda. On a particular day a service technician obtains the following results (in ounces):

Machine 1	8.5	8.1	8.3	8.7	8.4	8.5
Machine 2	8.2	7.9	8.1	8.0	7.7	

Find a 90% confidence interval for the mean difference in the amount that each machine dispenses. First assume that it is known that $\sigma_1^2 = \sigma_2^2 = \sigma^2$. If this assumption cannot be legitimately made, recalculate the said interval.

11-26. Independent random samples of sizes $n_X = 8$ and $n_Y = 10$ were extracted from normal populations having unknown means. If it is known that $\sigma_X^2 = 2.5$ and $\sigma_Y^2 = 4.8$, find a 95% confidence interval for $\mu_X - \mu_Y$.

Sample 1: 3, 8, 6, 5, 10, 7, 7, 6

Sample 2: 1, 5, 8, 8, 6, 7, 6, 9, 4, 3

What is the 95% confidence interval for $\mu_X - \mu_Y$ if the population variances are unknown?

11-27. Two brands of comparable electric drills (call them A and B) are each guaranteed for 1 year under normal use. For $n_1 = 45$ brand A drills, 10 failed to operate in under 1 year and for $n_2 = 48$ brand B drills, 16 failed before the warranty period ended. Determine a 99% confidence interval for the true difference between the failure proportions during the warranty period. How precisely have we estimated $\mu_1 - \mu_2$?

11-28. A horticulturalist has been using brand A fertilizer on a certain variety of plant. A new brand B fertilizer has just been marketed, which purports to be superior to brand A. Two independent groups of $n_1 = n_2 = 8$ new plantings are to be treated (one with the brand A fertilizer and the other with the brand B fertilizer) and the times (in days) taken to achieve a certain height are recorded:

Brand A: 30 35 30 31 39 39 35 29

Brand B: 30 33 29 33 35 28 30 29

Find a 99% confidence interval for the true mean difference in growth time (given that these times are approximately normally distributed and the variances are approximately equal).

11-29. A time and motion expert is interested in comparing two new methods of completing a certain task. The participants are divided into two groups of equal size and, for each group, their times (in minutes) to completion of the task will be monitored. It is anticipated that the completion times for each group will have a range of 10 minutes. If an estimate of the difference in mean time to completion of the task is recognized to be within ± 2 minutes with 95% reliability, how large of a sample is needed (for each group)? What is the sampling assumption being made?

11-30. Two spray-bottle rug cleaners were tested for their ability to remove a specific type of stain. The first was deemed successful in $x = 37$ out of $n_X = 70$ independent trials and the second was judged successful in $y = 62$ out of $n_Y = 85$ independent trials. If p_X (respectively, p_Y) is the true proportion of times the first (respectively, the second) cleaner is successful at removing the specific stain, determine an approximate 95% confidence interval for $p_X - p_Y$.

11-31. A recent World Health Services poll revealed that $x = 976$ out of $n_X = 1900$ African women had their first child before the age of 15 years, and $y = 732$ out of $n_Y = 1850$ Asian women had their first child before they were 15 years old. If p_X and p_Y are the respective proportions of African and Asian women giving birth before age 15, find a 95% confidence interval for $p_X - p_Y$.

11-32. In a set of $n_X = 150$ trials of a random experiment, $x = 75$ successes were observed; in a second set of $n_Y = 250$ trials (performed independently of the first set of trials), $y = 105$ successes were observed. Determine a 95% confidence interval for $p_X - p_Y$.

11-33. An experiment was conducted to compare the reaction times of a group of 10 volunteers (from a psychology class) to a bright light versus the sound of a loud air horn. When signaled with either the light or horn, the individual was asked to press a button to turn off the light or to turn off the horn. A special timing device monitored the reaction times (expressed in seconds):

Subject	Light(X)	Horn(Y)
1	0.31	0.29
2	0.37	0.30
3	0.41	0.38
4	0.33	0.35
5	0.21	0.25
6	0.38	0.29
7	0.38	0.31
8	0.47	0.37
9	0.95	0.21
10	0.25	0.22

Find a 95% confidence interval for $\mu_D = \mu_X - \mu_Y$.

11-34. A group of 10 enrollees are weighed at the beginning and at the end of a special combined diet and exercise program. Let the random variable (expressed in pounds) depict the weight difference (postprogram less pre-program). Suppose D is approximately normally distributed with values: $-5.7, 1.0, -6.2, -9.8, -3.0, 2.4, -11.1, -1.2, -8.2$, and 4.0. Find a 90% confidence interval for $\mu_D = \mu_X - \mu_Y$.

11-35. Suppose that two independent random samples of sizes $n_X = 10$ and $n_Y = 15$, respectively, are drawn from normal populations and that $s_X^2 = 0.238$ and $s_Y^2 = 0.149$. Find a 95% confidence interval for σ_X^2/σ_Y^2. What is the 95% confidence interval for σ_X/σ_Y?

11-36. A machine designed to dispense a specific amount of soap powder had a sample variance of $s_1^2 = 0.714$ for $n_1 = 17$ fills. A similar machine was tested for $n_2 = 25$ fills and it was determined that $s_2^2 = 0.438$. (Assume the fill amounts for these machines are approximately normally distributed.) Find a 95% confidence interval for σ_1^2/σ_2^2.

11-37. Random samples of sizes $n_X = n_Y = 5$ were taken from normal distributions having unknown variances σ_X^2 and σ_Y^2, respectively. Find:

(a) A 95% confidence interval for σ_X^2

(b) A 99% confidence interval for σ_Y^2

(c) A 95% confidence interval for σ_X^2/σ_Y^2

    Sample 1: 3.7, 8.6, 4.2, 5.5, 6.2
    Sample 2: 2.2, 1.89, 2.6, 2.82, 3.1

11-38. A random sample of size $n = 18$ is taken from a normal population with the result that $\bar{x} = 15$ and $s^2 = 5.3$. Determine a set of Bonferroni joint confidence limits for μ and σ^2 for a family confidence coefficient of $1 - \alpha = 0.95$.

11-39. Suppose $X_1, \ldots, X_n$ depicts random sample drawn from an exponential probability density $f(x; \lambda) = \lambda e^{-\lambda x}, x > 0, \lambda > 0$. It can be demonstrated that a $100(1-\alpha)$% confidence interval for λ has the form $(\ell_1, \ell_2) = \left(\frac{\chi_{\alpha/2}^2}{2n\bar{x}}, \frac{\chi_{1-(\alpha/2)}^2}{2n\bar{x}} \right)$. Given this result, let us assume that a random sample has been drawn from an exponential probability density with unknown λ. If $\sum_{i=1}^{n} x_i = 23,250$, determine a 95% confidence interval for λ.

11-40. Suppose $\overline{X}, \overline{Y}$ are the sample means determined from random samples of sizes n_X, n_Y, respectively, taken from two separate population distributions represented by the random variables X, Y. Verify that if $\overline{X}, \overline{Y}$ are normally and independently distributed, then $\overline{X} - \overline{Y}$ is $N(\mu_{\overline{X}-\overline{Y}}, \sigma_{\overline{X}-\overline{Y}})$, where $\mu_{\overline{X}-\overline{Y}} = \mu_{\overline{X}} - \mu_{\overline{Y}}$ and $\sigma_{\overline{X}-\overline{Y}} = \sqrt{(\sigma_X^2/n_X) + (\sigma_Y^2/n_Y)}$.

11-41. Suppose $\hat{P}_X, \hat{P}_Y$ are sample proportions determined by drawing n_X, n_Y items, respectively, from two separate binomial populations with parameters p_X, p_Y. In addition, let n_X, n_Y be sufficiently large so that $\hat{P}_X, \hat{P}_Y$

are approximately normally distributed. Verify that $\hat{P}_X - \hat{P}_Y$ is approximately $N(\mu_{\hat{P}_X - \hat{P}_Y}, \sigma_{\hat{P}_X - \hat{P}_Y})$, where $\mu_{\hat{P}_X - \hat{P}_Y} = \mu_{\hat{P}_X} - \mu_{\hat{P}_Y}$ and $\sigma_{\hat{P}_X - \hat{P}_Y} = \sqrt{\hat{P}_X(1 - \hat{P}_X)/n_X + \hat{P}_Y(1 - \hat{P}_Y)/n_Y}$.

11-42. Suppose $X_1, \ldots, X_n$ is a random sample drawn from a Poisson probability mass function $f(x; n\lambda) = e^{-n\lambda} \frac{(n\lambda)^X}{X!}, X = 0, 1, 2, \ldots, \lambda > 0$. If $Y = \sum_{i=1}^{n} X_i \sim f(x; n\lambda)$ and y denotes the sample realization of Y, then it can be shown that a $100(1 - \alpha)\%$ confidence interval for λ is

$$(\ell_1, \ell_2) = \left(\frac{1}{2n} \chi^2_{1-(\alpha/2), 2y}, \frac{1}{2n} \chi^2_{\alpha/2, 2(y+1)} \right).$$

For $n = 15$ and $y = 10$, determine a 95% confidence interval for λ.

11-43. Confidence Intervals for Quantiles:

Let $X_1, \ldots, X_n$ represent a set of sample random variables taken from a population of size N. For the quantile Γ_P, let γ_p be its sample realization, where p depicts the proportion of the population below γ_p (e.g., for quartiles, $p = 0.25j, j = 1, 2, 3$; for deciles, $p = 0.10j, j = 1, \ldots, 9$; and for percentiles, $p = 0.01j, j = 1, \ldots, 99$) and $\sigma_p = \sqrt{\left(1 - \frac{n}{N}\right) \frac{p(1-p)}{n}}$.

Then we may form $(L_1, L_2) = (p - k\sigma_p, p + k\sigma_p)$ and thus, from the sample observations (which have been placed in an increasing sequence $x_{(1)}, \ldots, x_{(n)}$), let us determine the point ℓ_1 below which the proportion L_1 of the sample frequencies fall; and the point ℓ_2 below which the proportion L_2 of the sample frequencies lie. Then if $k = 2$, an approximate 95% confidence interval for Γ_P is (ℓ_1, ℓ_2).

For sufficiently large $n (> 50)$, $\pm k$ may be replaced by the $N(0, 1)$ upper and lower percentage points $\pm z_{\alpha/2}$ to give $(\ell_1, \ell_2) = (p - z_{\alpha/2}\sigma_p, p + z_{\alpha/2}\sigma_p)$ so that now a $100(1 - \alpha)\%$ confidence interval for Γ_P is (ℓ_1, ℓ_2). To determine an interval estimate for the median, an alternative approach is to first order the realizations of the sample random variables by increasing magnitude as $x_{(1)}, \ldots, x_{(n)}$. Then for $n > 50$, a $100(1 - \alpha)\%$ confidence interval for the population median is $\ell_1 = x_{(h)} \leq \text{median} \leq x_{(n-h+1)} = \ell_2$, where h is approximated as $h = \frac{n - z_{\alpha/2}\sqrt{n}-1}{2}$. Suppose a set of $n = 60$ observations taken from a population of size $N = 2000$ have been arranged by order of magnitude as:

5, 5, 7, 7, 7, 8, 9, 9, 10, 10, 10, 10, 11, 14, 14,

20, 20, 22, 22, 25, 27, 27, 30, 30, 30, 35, 38, 38, 39, 40,

41, 45, 45, 50, 50, 50, 50, 53, 54, 55, 55, 58, 59, 61, 61, 61,

62, 64, 64, 64, 67, 71, 71, 80, 85, 85, 89, 90, 90, 90.

Find a 95% confidence interval for the median. Also, find a 99% confidence interval for the first quartile Q_1.

11-44. Confidence Interval for the Coefficient of Variation:
We previously expressed the population coefficient of variation as $CV = \frac{\sigma}{\mu}$. Given a collection of sample random variables $X_1, \ldots, X_n$, let us denote the realization of CV as $cv = \frac{s}{\bar{x}}$. Then for $n \geq 25$ and $cv < 0.40$, a $100(1 - \alpha)\%$ confidence interval for CV is

$$\ell_1 = \frac{cv}{1 + z_{\alpha/2}\sqrt{\dfrac{1 + 2cv^2}{2(n+1)}}} \leq CV \leq \ell_2 = \frac{cv}{1 - z_{\alpha/2}\sqrt{\dfrac{1 + 2cv^2}{2(n+1)}}}.$$

Suppose that from a random sample of size $n = 75$ it was found that $\bar{x} = 127, s = 15.5$, and $cv = 0.122$. Determine a 95% confidence interval for CV.

11-45. Confidence Interval for a Population Total:
Let $\{X_1, \ldots, X_n\}$ constitute a set of sample random variables taken from a population of size N. We are interested in determining an interval estimate for the population total $\tau = \sum_{j=1}^{N} X_j$. Let $N\overline{X}$ serve as an estimator for τ. Then for $N\bar{x}$ the sample realization of $N\overline{X}$, a $100(1 - \alpha)\%$ confidence interval for τ is

$$N\bar{x} \pm z_{\alpha/2}\sqrt{\left(1 - \frac{n}{N}\right)\frac{(Ns)^2}{n}},$$

where s is the realized sample standard deviation. Suppose that from a random sample of size $n = 50$ taken from a population of size $N = 750$ it is found that $\bar{x} = 81.2$ and $s = 9.8$. Find a 90% confidence interval for the population total.

12

Tests of Parametric Statistical Hypotheses

12.1 Statistical Inference Revisited

In Chapters 10 and 11 we examined the rudiments of statistical inference by considering the essentials of point and interval estimation; that is, we attempted to make inferences about unknown population parameters based on sample data. Such inferences took the form of confidence intervals. In effect, once a confidence interval was estimated, it essentially specified all distributions that might reasonably explain the population characteristic or variable being studied since each value in the interval effectively specifies a different distribution. This approach is basically one side of the *inferential coin*. We now turn to the other side of this coin by considering an area of statistics that is intimately related to estimation, namely the process of statistical hypothesis testing. In this regard, we now consider a set of procedures that can be used to determine whether one distribution (or set of distributions) is more reasonable than another in describing some population characteristic.

We may view a *statistical hypothesis* as an assertion about some unknown feature (or features) of a population that *might* be true; that is, it is a conjecture about the population distribution of a random variable, where the random variable may be discrete or continuous. The unknown feature of a probability density function (or probability mass function) may be its functional form (e.g., a population characteristic can be adequately modeled by a specific family of distributions) or, if the form is known, the value of one (or more) of its parameters. For instance, a statistical hypothesis may correspond to a statement regarding the form of a probability mass function; for example, the discrete random variable X follows a binomial distribution. Alternatively, if the form of, say, a probability density function is known, then a statistical hypothesis may refer to the numerical value of one of its parameters; for example, the continuous random variable X is $N(\mu, 1)$ and it is hypothesized that "$\mu = 10$." In what follows we shall assume that the form of a population distribution is known and thus concentrate exclusively on *parametric hypotheses*; that is, hypotheses that claim that a certain population parameter assumes a given numerical value. (The next chapter will consider nonparametric or distributional hypotheses and so-called goodness-of-fit tests.)

Let us now consider a brief overview of the essence of statistical hypothesis testing. Specifically, how does the process of statistical hypothesis testing work? Since a statistical hypothesis is a statement about some unknown population parameter, we must decide whether the assertion made is supported by experimental evidence obtained via random sampling. The decision as to whether the sample evidence supports the claim is based on chance factors; that is, the hypothesis will be rejected if it has a low probability of being correct in the light of sample evidence. And if a sample realization leads us to conclude that the probability is high that the claim is correct, then the hypothesis will not be rejected.

12.2 Fundamental Concepts for Testing Statistical Hypotheses

We may define a *simple hypothesis* as one that uniquely specifies the distribution (either a probability density function or probability mass function) from which the sample is drawn; that is, it specifies the functional form of the distribution as well as the values of *all* parameters. This is in contrast to a *composite hypothesis*, which does not uniquely determine a distribution. Thus a simple hypothesis posits a single probability distribution for a sample and a composite hypothesis specifies more than one possible distribution for a sample. Moreover, a statistical hypothesis may be *exact* (it posits a single numerical value for a parameter, e.g., $\mu = 10$) or *inexact* (here a range of values for some parameter is specified, e.g., $\mu \geq 10$).

Conducting a hypothesis test requires two complementary hypotheses called the *null hypothesis* and the *alternative hypothesis*. The null hypothesis is a theory or assumption that we want to test and that is *assumed to be true*. Hence, it is an indication of what we think the population looks like. In contrast, the alternative hypothesis tells us what the population would look like if the null hypothesis is not true—it essentially represents the negation or converse of the null hypothesis. Hence the objective of a hypothesis test is to determine, on the basis of a random sample extracted from the population, which of the two complementary hypothesis is true (even though the actual parameter value may be different from those specified by the null *and* alternative hypotheses). The null and alternative hypotheses will be denoted as H_0 and H_1, respectively. So given the null hypothesis H_0, we must always have a hypothesis H_1 counter to it; otherwise H_0 is not a testable hypothesis. In this regard, we say that the null hypothesis is *tested against* the alternative hypothesis. Incidentally, the word *null* is to be interpreted in the following context—it means that there is *no difference* between the hypothesized value and the true value of some population parameter.

To further examine the relationship between the null and alternative hypotheses, let θ be any population parameter and let $\mathcal{P}$ denote its *parameter space* or set of admissible values for θ. For instance, if a random variable X is $N(\mu, \sigma)$, then the parameter space for μ is $\mathcal{P} = \{\mu / -\infty < \mu < +\infty\}$ and the parameter space for σ is $\mathcal{P} = \{\sigma / 0 < \sigma < +\infty\}$. And if the random variable X is $b(X; n, p)$, then the parameter space for p is $\mathcal{P} = \{p / 0 \leq p \leq 1\}$ and the parameter space for n is $\mathcal{P} = \{n / n = 1, 2, 3, \ldots\}$.

In very general terms then, the form of the null and alternative hypothesis is $H_0 : \theta \in \mathcal{P}_0; H_1 : \theta \in \mathcal{P}_1$, where $\mathcal{P}_0$ and $\mathcal{P}_1$ are subsets of $\mathcal{P}$, $\mathcal{P}_1 = \overline{\mathcal{P}_0} = \mathcal{P} - \mathcal{P}_0$, $\mathcal{P}_0 \cap \mathcal{P}_1 = \emptyset$, and typically $\mathcal{P} = \mathcal{P}_0 \cup \mathcal{P}_1$. Thus $\mathcal{P}_1$ is the complement of $\mathcal{P}_0$ (it contains those elements of $\mathcal{P}$ that are not in $\mathcal{P}_0$), $\mathcal{P}_0$ and $\mathcal{P}_1$ are mutually exclusive or have no elements in common, and $\mathcal{P}_0$ and $\mathcal{P}_1$ taken together usually exhaust all elements within $\mathcal{P}$. (We can often assume that $\mathcal{P}_0$ and $\mathcal{P}_1$ are collectively exhaustive and thus, since they are also mutually exclusive, form a partition of $\mathcal{P}$.) Moreover, if, say, H_0 is a simple hypothesis (here we are assuming that the functional form of the associated probability distribution is known), then it constitutes a single point in $\mathcal{P}$ or $\mathcal{P}_0$ is a singleton; and if, say, H_1 is a composite hypothesis, then $\mathcal{P}_1$ determines an interval in $\mathcal{P}$. For example:

(a) Let the random variable X be distributed as $N(\mu, \sigma)$:

- If μ is unknown but σ is known, then $H_0 : \mu = \mu_0$ is a simple hypothesis and $H_1 : \mu > \mu_0$ is a composite hypothesis. Here H_0 and H_1 are mutually exclusive but not exhaustive. Note also that H_0 is exact and H_1 is inexact. (If we had written $H_0 : \mu \leq \mu_0$, then both the null and alternative hypotheses would be composite, and both would be inexact.)

- If μ is unknown but σ is known, then $H_0 : \mu = \mu_0$ is simple and $H_1 : \mu \neq \mu_0$ is composite. Now H_0 and H_1 are both mutually exclusive and collectively exhaustive. In addition, H_0 is exact while H_1 is inexact.

- If both μ and σ are unknown, the $H_0 : \mu = \mu_0$ is composite. It is also exact.

(b) Let the random variable X be distributed as $b(X; n, p)$:

- If n is given and p is unknown, then $H_0 : p = p_0$ is a simple hypothesis and $H_1 : p < p_0$ is composite. Also, H_0 is exact and H_1 is inexact.

- If neither n nor p is specified, then $H_0 : p = p_0$ is composite and exact.

You might have noted already that the null hypothesis H_0 *always contains at least an equality*. This is so that a reference probability distribution can be specified given that H_0 is assumed to be true. That is, if the random variable X is $N(\mu, \sigma)$ with σ known, then under $H_0 : \mu = \mu_0$, the *null value* μ_0 is used to form the test function $N(\mu_0, \sigma)$. Hence $\mu = \mu_0$ determines a unique probability density function to be used in conducting the hypothesis test.

12.3   **What Is the Research Question?**

In practice, since the null and alternative hypotheses are complementary, support for one proposition or theory is obtained by demonstrating lack of support for its converse; that is, one of these rival hypotheses survives empirical testing because the data contradicts the other. So although the null hypothesis is typically billed as the *hypothesis to be tested* and either rejected or not rejected on the basis of sample evidence, the actual research objective is usually to *obtain support for the alternative hypothesis*. That is, the null hypothesis is the proposition that we wish, in a sense, to *disprove*.

This being the case, although designation between the null and alternative hypotheses is somewhat arbitrary, we generally specify the null hypothesis in an exact manner and postulate *no effect* of some treatment or the *absence of effect* of some research variable. The alternative hypothesis is typically much more fluid and is formulated as a range of values that would result if the treatment or variable being studied did exhibit an effect. Hence the null hypothesis is often structured in a way that is opposite to what we want to establish or believe to be true. Then the alternative hypothesis is simply framed in opposition to the null hypothesis. (In fact, these observations reinforce the notion that, as indicated earlier, equality is always a part of the null hypothesis—it is easier to disprove an exact or specific hypothesis relative to an inexact or nonspecific one.)

By way of accepted methodology, if we want to establish some theory or proposition, we term it the alternative hypothesis and then formulate the contrary (null) hypothesis as a *throw-away*, which is considered as true or valid unless the sample outcome deems it highly improbable. Given that the null hypothesis is essentially expendable (it is a *statistical straw man* or considered to be *data fodder*), then obviously we are compelled to reject it in the face of sample evidence that strongly favors the alternative hypothesis and clearly renders the assertion given by the null hypothesis unreasonable.

For instance, suppose a chemical company is testing a new and improved fertilizer for, say, soybeans. The presumption is that the new fertilizer will do better, in terms of its average yield (bushels per acre) than the old one. Once the experimental results are obtained, the company seeks to determine if its presumption is valid. Hence the null hypothesis is stated as

H_0: the new fertilizer produces the same mean yield as the old one

and the alternative hypothesis is framed as

H_1: the mean yield for the new fertilizer is greater than that for the old one

Here the company wants to disprove the null hypothesis in favor of the alternative hypothesis. Note that the null hypothesis is exact, but the alternative hypothesis is less so.

12.4 **Decision Outcomes**

Once the null and alternative hypotheses have been formulated, there are two possible decisions (based on the sample outcome) that we can ultimately make regarding the null hypothesis H_0—we can reject H_0 or not reject H_0. What are the consequences of these decisions with respect to the characteristics of the prevailing state of nature or true situation within the population? First, after observing the sample, we may

$$Reject\ H_0\ \ given: \begin{cases} H_0\ \text{true} \rightarrow \text{incorrect decision}; \\ H_0\ \text{false} \rightarrow \text{correct decision} \end{cases}$$

or, second, we

$$Do\ Not\ Reject\ H_0\ \ given: \begin{cases} H_0\ \text{true} \rightarrow \text{correct decision}; \\ H_0\ \text{false} \rightarrow \text{incorrect decision.} \end{cases}$$

(Note that the second option has been expressed as *do not reject H_0* rather than *accept H_0*. This is because the null hypothesis is regarded as valid unless the data dictates otherwise and thus, if the sample evidence does not support the rejection of the null hypothesis, it only means that *the data has not made its case.* We shall return to this point later on.)

Whether the decision is to reject or not reject the null hypothesis given the actual state of nature, the possibility exists (because of random fluctuations in sampling) that we may make an incorrect decision or error. In this regard, we shall identify two types of potential errors:

Type I Error (TIE)—Rejecting H_0 when it is true

Type II Error (TIIE)—Not rejecting H_0 when it is false

Stated alternatively in terms of our parameter set notation, for $H_0: \theta \in \mathcal{P}_0$ and $H_1: \theta \in \mathcal{P}_1$:

TIE—Rejecting H_0 given $\theta \in \mathcal{P}_0$

TIIE—Not rejecting H_0 given $\theta \in \mathcal{P}_1$

These outcomes or errors are summarized in Table 12.1a.

Next, let us specify the *risks associated with these incorrect decisions* as the probabilities of committing Type I and Type II Errors. Based upon the preceding definitions of these errors, it should be apparent that they are actually conditional probabilities. That is:

(a) $P(TIE) = P(\text{reject } H_0 | H_0 \text{ true}) = \alpha,\ 0 \le \alpha \le 1$;

(b) $P(TIIE) = P(\text{do not reject } H_0 | H_0 \text{ false}) = \beta,\ 0 \le \beta \le 1.$

$$(12.2)$$

Table 12.1

		a. Decision Outcomes State of Nature					**b.** Risks of Wrong Decisions State of Nature	
		H_0 true	H_0 false				H_0 true	H_0 false
Action taken on the basis of sample information	Reject H_0	Incorrect Decision (TIE)	Correct Decision	Action taken on the basis of sample information	Reject H_0		$P(TIE) = \alpha$	P(correct decision) = $1 - \beta$
	Do not reject H_0	Correct Decision	Incorrect Decision (TIIE)		Do not reject H_0		P(correct decision) = $1 - \alpha$	$P(TIIE) = \beta$

Hence the magnitude of the risk associated with rejecting a true null hypothesis is α (called the α-*risk*) and the risk associated with not rejecting a false null hypothesis is β (termed the β-*risk*) (see Table 12.1b). These α- and β-risks or error probabilities essentially serve as indexes of the *seriousness* of making Type I and Type II Errors. Moreover, Table 12.1b also incorporates the *probabilities of correct decisions* or

$$
\begin{align}
\text{(a)} \quad & P(\text{do not reject } H_0 | H_0 \text{ true}) = 1 - \alpha. \\
\text{(b)} \quad & P(\text{reject } H_0 | H_0 \text{ false}) = 1 - \beta.
\end{align} \tag{12.3}
$$

Note that the sum $\alpha + \beta$ is essentially meaningless. However, as Table 12.1b indicates, since one of the two possible actions or conclusions must be obtained in testing the null hypothesis against the alternative hypothesis (we either reject the null hypothesis or do not reject it), and since the events in each column are complementary, the probabilities within the same must sum to unity; that is,

$$
\begin{align}
\text{(a)} \quad & P(\text{reject } H_0 | H_0 \text{ true}) + P(\text{do not reject } H_0 | H_0 \text{ true}) = \alpha + (1 - \alpha) = 1; \\
\text{(b)} \quad & P(\text{reject } H_0 | H_0 \text{ false}) + P(\text{do not reject } H_0 | H_0 \text{ false}) = (1 - \beta) + \beta = 1.
\end{align} \tag{12.4}
$$

12.5   Devising a Test for a Statistical Hypothesis

Given a null hypothesis H_0 and an alternative hypothesis H_1, we need to specify a *decision rule* that will tell us when to either reject H_0 (the sample outcome favors H_1) or not reject H_0 (the sample outcome favors H_0 itself). This decision rule constitutes a *test of a statistical hypothesis*; that is, a test is simply a criterion that informs us when the null hypothesis should be rejected or not rejected in the face of sample evidence. (It is important to note that a test is specified before the random sample is drawn. This is so that the selection of a decision rule is not influenced by the sample.) Hence the notions of a *test* of a statistical hypothesis and a *decision rule* that tells us whether to reject H_0 or not, are taken to be synonymous.

To conduct a test of a statistical hypothesis we need to specify a *test statistic* T or a random variable whose sampling distribution is known under the assumption that the null hypothesis is true. Here $T = g(X_1, \ldots, X_n, n)$, which is taken to be a *good* estimator of the parameter θ, maps the realizations $x_1, \ldots, x_n$ of the sample random variables $X_1, \ldots, X_n$ into some admissible realization t of T. For some values of the test statistic T the null hypothesis H_0 will be rejected and, for others, it will not be rejected. Hence we may partition the range of g or the entire set $\mathcal{T}$ of realizations or *experimental values* of T (called the *sample space of the test*) into two separate and distinct regions: a *critical region* or region of rejection $\mathcal{R}$, which specifies the sample realizations of T for which the null hypothesis H_0 is rejected (hence $\mathcal{R}$ contains the sample outcomes least favorable to H_0); and a *region of nonrejection* $\overline{\mathcal{R}} = \mathcal{T} - \mathcal{R}$ (it contains the sample outcomes most favorable to H_0 or the sample realizations of T for which H_0 is not rejected). By virtue of the nature of a partition, $\mathcal{T} = \mathcal{R} \cup \overline{\mathcal{R}}$ and $\mathcal{R} \cap \overline{\mathcal{R}} = \emptyset$. In this regard, a test may be viewed as a partition of $\mathcal{T}$ into $\mathcal{R}$ and $\overline{\mathcal{R}}$. Hence prescribing a critical region is equivalent to prescribing a test of a hypothesis.

How do we locate $\mathcal{R}$ and $\overline{\mathcal{R}}$? The boundary between $\mathcal{R}$ and $\overline{\mathcal{R}}$ is called the *critical value* of the test statistic and will be denoted as t_c. Here the critical value of T is determined *a priori* (i.e., before the sample is drawn) and depends upon the form of the sampling distribution of T, the alternative hypothesis made, and the seriousness of the consequences associated with making an incorrect decision. As we shall see later, the problem of constructing a test of a null hypothesis corresponds to the problem of choosing a critical region for the test; that is, a test is specified by its critical region; and this is accomplished by determining the critical value of the test statistic.

We may establish a connection between the regions of rejection and non-rejection and the risks associated with incorrect decisions as follows. Let the area under the sampling distribution of T and above the critical region correspond to the probability of a Type I Error or to α; that is, α will represent the *size* of the critical region $\mathcal{R}$ and will be termed the level of significance of the test. Here the context of the phrase *level of significance* is simply that the sample evidence warrants rejection of the null hypothesis at the α *level*. So based upon the preceding discussion, (12.2a,b) can now be modified as:

$$\text{(a)} \quad \alpha = P(TIE) = P(T \in \mathcal{R} \mid H_0 \text{ true});$$
$$\text{(b)} \quad \beta = P(TIIE) = P(T \in \overline{\mathcal{R}} \mid H_0 \text{ false}).$$

(12.2.1)

(Note that if, in general, we are testing $H_0 : \theta \in \mathcal{P}_0$ versus $H_1 : \theta \in \mathcal{P}_1$, then the preceding probabilities may be rewritten as

$$\text{(a)} \quad \alpha(\theta) = P(TIE) = P(T \in \mathcal{R} / \theta \in \mathcal{P}_0);$$
$$\text{(b)} \quad \beta(\theta) = P(TIIE) = P(T \in \overline{\mathcal{R}} / \theta \in \mathcal{P}_1).$$

(12.2.2)

As (12.2.1a) reveals, α, the size of $\mathcal{R}$, is the probability of obtaining a sample realization of T that falls into $\mathcal{R}$ when the null hypothesis H_0 is true; and from (12.2.1b), β is the probability of obtaining a sample realization of T that does not fall into $\mathcal{R}$ when H_0 is not true.

More formally, for a population parameter θ, if the null hypothesis is simple or $H_0 : \theta = \theta_0$, then for, say, T a continuous random variable, the sampling distribution or probability density function of T given that H_0 is true is completely determined and can be written as $f(t; \theta_0, n)$. Hence the probability that the sample will reject a true null hypothesis is

$$P(TIE) = P(T \in \mathcal{R} \,|f, \theta_0, n) = \int_{t \in \mathcal{R}} f(t; \theta_0, n)dt = \alpha. \qquad (12.5)$$

Here α has a long-run relative frequency interpretation (as does β to follow). That is to say, if the null hypothesis H_0 were true and if we were to repeatedly draw samples of size n from the population, then, in the long run, we would reject H_0 approximately $100\alpha\%$ of the time.

What about the determination of the probability of a Type II Error or β? We noted earlier that the null hypothesis H_0 must be tested against an alternative hypothesis H_1; that is, any hypothesis test should be able to discriminate between H_0 and H_1. Given $H_0 : \theta = \theta_0$, let the alternative hypothesis also be simple or $H_1 : \theta = \theta_1$. (As indicated earlier, if H_0 is true, then H_1 is false; and if H_0 is false, then H_1 is true.) If H_0 is false, then the sampling distribution of T can be written as $f(t; \theta_1, n)$ and thus, since the region of nonrejection of H_0 is $\overline{\mathcal{R}} = \mathcal{T} - \mathcal{R}$, the probability that the sample will not reject a false null hypothesis is

$$P(TIIE) = P(T \in \overline{\mathcal{R}} \,| f, \theta_1, n) = \int_{t \in \overline{\mathcal{R}}} f(t; \theta_1, n)dt = \beta. \qquad (12.6)$$

Clearly β depends upon the alternative hypothesis made.

For instance, if again $H_0 : \theta = \theta_0$ and $H_1 : \theta = \theta_1$ with $\mathcal{R} = \{t / t_1 \le t \le t_2\}$, then, from (12.5,6),

$$\alpha = \int_{t_1}^{t_2} f(t; \theta_0, n)dt = \text{area } A$$

(see Figure 12.1); and since $\overline{\mathcal{R}} = \{t / -\infty < t < t_1\} \cup \{t / t_2 < t < +\infty\}$,

$$\beta = \int_{-\infty}^{t_1} f(t; \theta_1, n)dt + \int_{t_2}^{+\infty} f(t; \theta_1, n)dt = \text{area } B + \text{area } C.$$

So if the null hypothesis H_0 were false and if we were to repeatedly extract random samples of size n from a population, then, in the long run, we would fail to reject H_0 approximately $100\beta\%$ of the time.

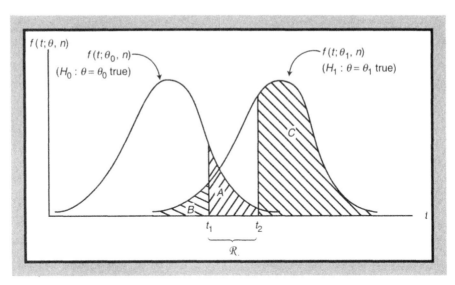

Figure 12.1 $P(TIE) = \alpha =$ area A; $P(TIIE) = \beta =$ area $B +$ area C.

12.6 The Classical Approach to Statistical Hypothesis Testing

How do we go about constructing *good* hypothesis tests? Moreover, what is our yardstick for assessing *goodness*? A convenient gauge of the *goodness* of a test is provided by the sizes of the α- and β-risks associated with making incorrect decisions. As a practical matter, how should these errors be treated?

When testing statistical hypotheses, the classical approach deems committing a Type I Error as a much more serious decision outcome than committing a Type II Error. In this regard, we select, in advance, the maximum tolerable size of a Type I Error α and then, for a fixed random sample size n, construct a test that minimizes the size of the Type II Error β. Note that for a fixed n it is not possible to a priori select values for *both* α and β when constructing a test for a null hypothesis against some alternative hypothesis (we shall return to this point later on). For a given n, α and β are inversely related; that is, the size of β will normally increase as the size of α decreases. For instance, it is readily seen from Figure 12.1 that if the length of interval $\mathcal{R}$ shrinks in size, then $\alpha =$ area A decreases and $\beta =$ area $B +$ area C increases. Moreover, for most hypothesis tests, both α and β will decrease as n increases. Hence we cannot make α extremely small without concomitantly increasing β; one pays for a declining α-risk in terms of an increasing β-risk. In the light of this trade-off between the α- and β-risk, we typically opt to construct a test that has the smallest size Type II Error β among all other tests exhibiting the same size Type I Error or significance level α. In sum, to construct a *good* test, we shall limit our consideration to tests that (tightly) control the probability of a Type I Error (it is set at some maximum tolerable level—usually 10% or 5% or

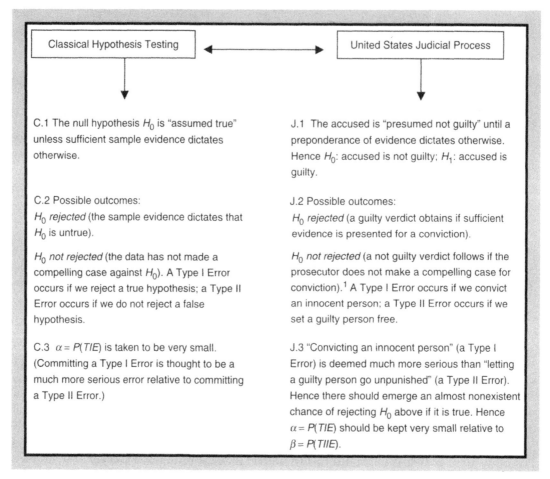

Figure 12.2   The classical approach to hypothesis testing contrasted with the U.S. judicial process.

even 1%). Then within this class of tests we search for the test that exhibits the smallest probability of a Type II Error.

It is interesting to note that the classical approach to statistical hypothesis testing parallels the workings of the judicial process carried out in the United States (see Figure 12.2).

It is important to note that since α and β vary inversely, α cannot be reduced to zero (an innocent person is never convicted) without increasing β to one (every

---

[1] However, this verdict does not necessarily imply that the accused is "innocent." It simply reflects a lack of substantive evidence needed for a conviction.

accused person is set free). The only way that *both* α and β can be reduced simultaneously is by increasing the body (and quality) of evidence (this is analogous to increasing the sample size under classical hypothesis testing).

12.7 Types of Tests or Critical Regions

Most hypothesis testing situations dictate that the null hypothesis H_0 is simple but that the alternative hypothesis H_1 is composite. In this instance H_0 specifies a particular distribution from which a sample might be drawn, and H_1 specifies a collection of alternative distributions that might possibly generate an observed data set. Given this observation, we now proceed to identify three basic cases involving the testing of a simple null hypothesis against composite alternatives. We then examine the implied critical regions (see Figure 12.3).

It is the type of test that *locates* the critical region $\mathcal{R}$; and this is dictated by the sense of direction indicated in the alternative hypothesis made. Moreover, it is the magnitude of the α-risk that determines the *size* of $\mathcal{R}$.

For instance, let us assume that our test statistic T is a continuous random variable with sampling distribution or probability density function expressed as $f(t; \theta, n)$. Then for Case I, given a random sample of size n, we should reject the null hypothesis H_0 when t, the sample realization of T, is *sufficiently greater than* the null value θ_0. Here *sufficiently greater than* implies that the difference between t and θ_0 or $t - \theta_0$ is much greater than that which could be attributed

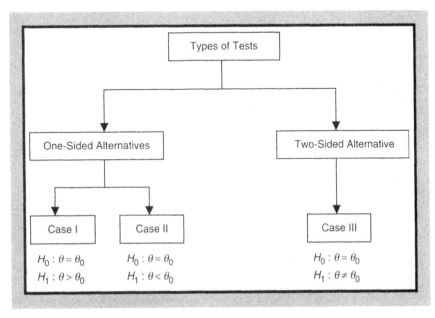

Figure 12.3 Case I: $H_0 : \theta = \theta_0$, $H_1 : \theta > \theta_0$; Case II: $H_0 : \theta = \theta_0$, $H_1 : \theta < \theta_0$; Case III: $H_0 : \theta = \theta_0$, $H_1 : \theta \neq \theta_0$.

to chance factors or to random fluctuations of T. If $\alpha = P(TIE)$ represents the size of the critical region $\mathcal{R}$, then $\mathcal{R} = \{t/t \geq t_c\}$ is located at the upper or right-hand tail of the sampling distribution of T (since H_0 is rejected for *large* values of t), where t_c (the critical value of T) corresponds to the right-tail quantile of the sampling distribution of T; that is, t_c is the value of T for which $P(T \geq t_c) = \alpha =$ area under $f(t; \theta_0, n)$ from t_c to $+\infty$ (see Figure 12.4a). So if H_0 is taken to be true, then it is rejected for realizations t that are *extreme in the H_1 direction*. Here we regard as *extreme* or *very different from θ_0* those t's which, if H_0 were true, would occur by chance only very rarely or with a probability of size α. In this regard, if

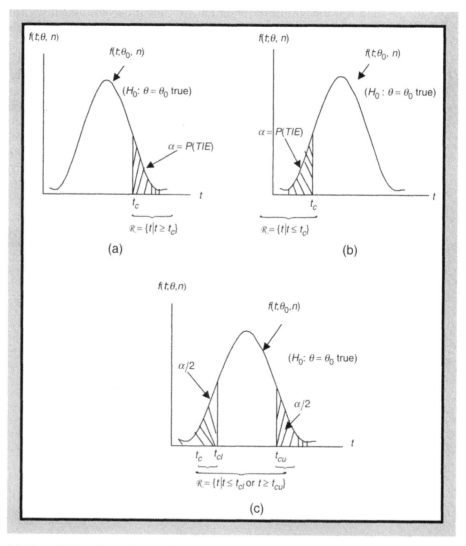

Figure 12.4 (a) The critical region $\mathcal{R}$ for a left-hand tail test; (b) The critical region $\mathcal{R}$ for a right-hand tail test; (c) The critical region $\mathcal{R}$ for a two-tail test.

H_0 were true and we repeated this test a large number of times, then, in the long run, we would incorrectly reject H_0 $100\alpha\%$ of the time.

Looking to Case II, for a given sample of size n, we should reject the null hypothesis H_0 when t is *sufficiently less than* θ_0; that is, the difference $|t - \theta_0|$ is much greater than that which would be expected to occur by chance. For a specified α-risk, $\mathcal{R} = \{t/t \leq t_c\}$ is now located at the lower or left-hand tail of the sampling distribution of T (since now H_0 is rejected for *small* values of t), where t_c corresponds to the left-tail quantile of the sampling distribution of T. Hence t_c is the value of T for which $P(T \leq t_c) = \alpha = $ area under $f(t; \theta_0, n)$ from $-\infty$ to t_c (see Figure 12.4b). Given that H_0 is assumed true, it is now rejected if the sample realization t falls short of θ_0 by *too much*. Hence a realized value of T that is considerably smaller than θ_0 may be deemed, if H_0 were true, as highly unlikely if it occurs with a probability of size α.

Finally, for Case III, we should reject the null hypothesis H_0 if the realized value of the test statistic T is *sufficiently different* from θ_0; that is, if H_0 were true, then it would be rejected for realizations t of T that are *extreme in either direction* or for which $|t - \theta_0|$ is *very large*. Hence realizations t that are much larger than θ_0 or much smaller than θ_0 constitute sample evidence against H_0 and in favor of H_1. For a chosen α, the critical region $\mathcal{R} = \{t/t \leq t_{cl} \text{ or } t \geq t_{cu}\}$ is determined at *both tails* of the sampling distribution of T; that is, α is divided equally between the two tails of the probability density function $f(t; \theta_0, n)$ so that the area under each tail is obviously $\frac{\alpha}{2}$ (see Figure 12.4c). In this regard, t_{cl} is the lower-tail quantile of the sampling distribution of T (t_{cl} is the value of T for which $P(T \leq t_{cl}) = \frac{\alpha}{2}$) and t_{cu} is the upper-tail quantile of the sampling distribution of T(t_{cu} represents the value of T for which $P(T \geq t_{cu}) = \frac{\alpha}{2}$). In sum, this particular test (or $\mathcal{R}$) is structured so that we would reject H_0 if a realization of T turned out to be either extremely low or extremely high in relation to θ_0 so that its occurrence by chance would be very unlikely (the said occurrence would be with a probability of only $100\alpha\%$).

The preceding discussion pertaining to the specification of a critical region for a hypothesis test enables us to now offer a definition of the term *statistical significance*. Specifically, an outcome of a random experiment is deemed statistically significant if it is highly unlikely that the said outcome has occurred solely because of chance factors; that is, it is the result of some legitimate or systematic experimental effect.

12.8   The Essentials of Conducting a Hypothesis Test

Within the classical vein of hypothesis testing, this brief section outlines, in general terms, a stepwise procedure for conducting a statistical hypothesis test on a population parameter θ. In the next few sections more specific hypothesis tests concerning some important parameters will be offered. To this end we have:

1. Formulate the null hypothesis H_0 (assumed true) and the alternative hypothesis H_1 (it determines the location of the critical region $\mathcal{R}$).

The alternative hypothesis is usually a proposition that we want to establish, whereas the null hypothesis, which is stated as the exact opposite of the alternative hypothesis, is expendable and, if deemed highly improbable in the light of sample evidence, is to be rejected.

2. Specify the level of significance α (it determines the size of $\mathcal{R}$).

Here $\alpha = \mathcal{P}$ (Type I Error) typically is taken to be 0.01 or 0.05 or 0.10. It indicates the (small) amount of risk we are willing to undertake in incorrectly rejecting H_0 when it is true.

3. Select the test statistic T whose sampling distribution is known under the assumption that H_0 is true.

4. Find $\mathcal{R}$. This involves determining t_c, the critical value of T.

Given α, t_c corresponds to a particular quantile of the sampling distribution of T. For instance, if T is approximately normal, then t_c will represent a quantile z_c of the Z or $N(0,1)$ distribution. In this regard, since z_c comes from the $N(0,1)$ table, it is sometimes called the *tabular value*.

5. For $H_0 : \theta = \theta_0$, compute the sample realization of the appropriate standardized test statistic

$$\frac{T - \theta_0}{\sigma_T}. \tag{12.7}$$

For example, if T is approximately normal, then $\frac{T-\theta_0}{\sigma_T}$ is approximately $N(0,1)$. Given a fixed sample size n, let z_t, called the *calculated z*, denote the realization of (12.7), which is to be compared to z_c.

6. DECISION RULE: If the sample realization of (12.7) is an element of $\mathcal{R}$, then we reject H_0.

For instance, if $\frac{T-\theta_0}{\sigma_T}$ is approximately $N(0,1)$, then we will reject H_0 if $z_t \in \mathcal{R}$.
   In the sections to follow, it is to be implicitly understood that this general process of executing a statistical hypothesis test is being implemented.

12.9 Hypothesis Test for μ Under Random Sampling from a Normal Population with Known Variance

Let $\{X_1, \ldots, X_n\}$ constitute a set of sample random variables taken from a normal population with unknown mean μ but known variance σ^2. Our objective is to construct hypothesis tests for each of the following sets of hypotheses:

Case I	Case II	Case III
$H_0 : \mu = \mu_0$	$H_0 : \mu = \mu_0$	$H_0 : \mu = \mu_0$
$H_1 : \mu > \mu_0$	$H_1 : \mu < \mu_0$	$H_1 : \mu \neq \mu_0$

As determined earlier, the best estimator for μ is $T = \overline{X}$, where $\overline{X}$ is $N(\mu, \sigma_{\bar{x}})$ and $\sigma_{\bar{x}} = \frac{\sigma}{\sqrt{n}}$ (see Theorem 8.3). Moreover, under the null hypothesis H_0, $\overline{X}$ is $N(\mu_0, \sigma_{\bar{x}})$ and, from (8.33), the standardized sample mean $\overline{Z} = (\overline{X} - \mu_0)/(\sigma/\sqrt{n})$ is $N(0, 1)$ or standard normal.

For Case I, given the one-sided alternative $H_1 : \mu > \mu_0$, the critical region of size α is under the right-hand tail of the $N(0, 1)$ distribution since we require that $P(\overline{Z} \geq z_\alpha) = \alpha$ (see Figure 12.5a). Hence the Case I critical region $\mathcal{R}$ involves that portion of the z-axis such that $\bar{z} \geq z_\alpha$ or

$$\frac{\bar{x} - \mu_0}{\sigma/\sqrt{n}} \geq z_\alpha, \tag{12.8}$$

where $\bar{z}$ is the realized value of $\overline{Z}$ given that $\bar{x}$ is the sample realization of $\overline{X}$. Upon rearranging this inequality we have

$$\bar{x} \geq \mu_0 + z_\alpha \frac{\sigma}{\sqrt{n}}, \tag{12.8.1}$$

that is, H_0 is rejected when $\bar{x}$ exceeds μ_0 by *too much*, where our measure of *too much* is $z_\alpha \frac{\sigma}{\sqrt{n}}$. (*Too much* simply means that the difference $\bar{x} - \mu_0$ exceeds that which could reasonably be attributed to random or sampling fluctuations in the estimator $T = \overline{X}$.) Hence the preceding inequalities serve as the basis for the Case I decision rule for rejecting H_0 relative to H_1 : reject H_0 if (12.8) holds.

Thus, in this instance, H_0 is rejected when the *calculated z* is greater than or equal to the *tabular z* value.

Looking to Case II, for the one-sided alternative $H_1 : \mu < \mu_0$, the critical region of size α is now under the left-hand tail of the $N(0, 1)$ distribution with $P(\overline{Z} \leq -z_\alpha) = \alpha$ (see Figure 12.5b). In this instance the Case II critical region $\mathcal{R}$ involves that portion of the z-axis such that $\bar{z} \leq -z_\alpha$ or

$$\frac{\bar{x} - \mu_0}{\sigma \sqrt{n}} \leq -z_\alpha \tag{12.9}$$

or

$$\bar{x} \leq \mu_0 - z_\alpha \frac{\sigma}{\sqrt{n}}, \tag{12.9.1}$$

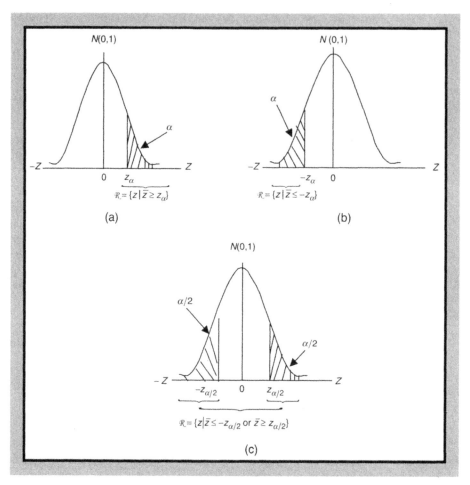

Figure 12.5   (a) The critical region $\mathcal{R}$ for the one-sided alternative $H_1 : \mu > \mu_0$; (b) The critical region $\mathcal{R}$ for the one-sided alternative $H_1 : \mu < \mu_0$; (c) The critical region $\mathcal{R}$ for the two-sided alternative $H_1 : \mu \neq \mu_0$.

that is, H_0 is rejected when $\bar{x}$ falls short of μ_0 by at least $z_\alpha \frac{\sigma}{\sqrt{n}}$ (our indicator of *too much*). Note that (12.9) can be converted to

$$\frac{-(\bar{x} - z_0)}{\sigma/\sqrt{n}} \geq z_\alpha. \tag{12.10}$$

Here (12.10) transforms the critical region from the left-hand tail to the right-hand tail of the $N(0,1)$ distribution. In this regard, we shall, at times, find it convenient to formulate the Case II decision rule as: reject H_0 relative to H_1

when (12.10) is satisfied. (Now H_0 is rejected when the negative of the *calculated* z equals or exceeds the *tabular z*.)

Finally, for Case III, the two-sided alternative hypothesis $H_1 : \mu \neq \mu_0$ implies, for a critical region of size α, that $P(\overline{Z} \leq -z_{\alpha/2}) = \frac{\alpha}{2}$ and $P(\overline{Z} \geq z_{\alpha/2}) = \frac{\alpha}{2}$ or $P(|\overline{Z}| \geq z_{\alpha/2}) = \frac{\alpha}{2}$; that is, α is divided equally between the two tails of the $N(0, 1)$ distribution (see Figure 12.5c). Hence the critical region $\mathcal{R}$ for Case III is that portion of the z-axis such that $\bar{z} \leq -z_{\alpha/2}$ or $\bar{z} \geq z_{\alpha/2}$; that is,

$$\frac{\bar{x} - \mu_0}{\sigma\sqrt{n}} \leq -z_{\alpha/2} \quad \text{or} \quad \frac{\bar{x} - \mu_0}{\sigma\sqrt{n}} \geq z_{\alpha/2}. \tag{12.11}$$

Let us rewrite (12.11) as

$$\bar{x} \leq \mu_0 - z_{\alpha/2}\frac{\sigma}{\sqrt{n}} \quad \text{or} \quad \bar{x} \geq \mu_0 + z_{\alpha/2}\frac{\sigma}{\sqrt{n}}. \tag{12.11.1}$$

Then as these inequalities reveal, H_0 is rejected when the realized sample mean $\bar{x}$ exceeds (respectively, falls short of) μ_0 by *too much*, where now the measure of *too much* is $z_{\alpha/2}\frac{\sigma}{\sqrt{n}}$. Looked at in another fashion, the inequalities in (12.11) together imply

$$\left| \frac{\bar{x} - \mu_0}{\sigma/\sqrt{n}} \right| \geq z_{\alpha/2}. \tag{12.12}$$

Here (12.12) forms the basis for the Case III decision rule for rejecting H_0 relative to H_1 : reject H_0 if (12.12) holds.

Example 12.9.1   Suppose that a manufacturing process produces a large number of a particular type of electronic component having a mean operating life of 2,000 hours and a standard deviation of 300 hours. The research and development department introduces a new process that it claims is more efficient than the old one. The new process is run and a random sample of $n = 100$ components exhibits a mean life of $\bar{x} = 2,100$ hours (with the standard deviation unchanged). Is the new process really more efficient than the old one or did the increase in average operating life arise simply because of chance factors? (Looked at in another fashion, we seek to determine if the random sample of 100 components came from the distribution characterizing the original process whose mean is 2,000 or from an alternative distribution whose mean is 2,100.)

Let the null hypothesis be that the new process is just as efficient as the old one or $H_0 : \mu = \mu_0 = 2,000$. Research and development seeks to demonstrate that their claim is justified. Hence they structure the alternative hypothesis as $H_1 : \mu > 2,000$. (Clearly this problem is structured as Case I, earlier.) Let $\alpha = 0.05$. Since $\bar{x}$ is approximately normal and n is large, it follows that the critical region $\mathcal{R} = \{z/\bar{z} \geq z_{0.05}\} = \{z/\bar{z} \geq 1.645\}$.

For the *calculated z* we have

$$\bar{z} = \frac{\bar{x} - \mu_0}{\frac{\sigma}{\sqrt{n}}} = \frac{2100 - 2000}{\frac{300}{\sqrt{100}}} = 3.33.$$

Then according to our decision rule (12.8), we reject H_0 in favor of H_1 at the 5% level; that is, the new process is indeed better than the old one. In this regard, if we were to perform this experiment a large number of times, then, in the long run, we would make a Type I Error or incorrectly reject the null hypothesis less than 5% of the time. We shall return to this latter point in Section 12.10. ■

Example 12.9.2   In a certain locale the mean income last year for the head of household was \$55,000. Has there been a change in average income for this year? A sample of $n = 200$ households revealed this year's mean income to be \$58,000 with a standard deviation of \$2,700. If we set $H_0 : \mu = \mu_0 = \$55,000$, then, since average income can either increase or decrease over the course of a year, let $H_1 : \mu \neq \$55,000$. (Note that we are now faced with a Case III hypothesis test.) Again with $\overline{X}$ approximately normal and n large, the $N(0,1)$ distribution can be utilized to find the critical region. To this end, for $\alpha = 0.05$, $\mathcal{R} = \{z/\bar{z} \leq -z_{0.025}$ or $\bar{z} \geq z_{0.025}\} = \{z/|\bar{z}| \geq z_{0.025}\} = \{z/|\bar{z}| \geq 1.96\}$.

Looking to the *calculated z* we have

$$\left|\frac{\bar{x} - \mu_0}{\sigma/\sqrt{n}}\right| = \left|\frac{58,000 - 55,000}{2,700/\sqrt{200}}\right| = 15.71.$$

Since we are well within the critical region (H_0 is consequently rejected), it follows that the difference between last year's and this year's average income is highly significant; that is, there has been a statistically significant change (obviously an increase) in the mean head of household income level. ■

The preceding set of decision rules for the Case I, Case II, and Case III hypothesis tests ((12.8), (12.9), and (12.12), respectively) were all stated in terms of comparing a calculated or realized $\overline{Z}$ value with some critical (tabular) z value. However, as an alternative to this approach, we may simply compare the realized sample mean $\bar{x}$ against a *critical sample mean* $\bar{x}_c$. For instance, let us denote the right-hand side of (12.8.1) as $\bar{x}_c$. Then for Case I, we would reject $H_0: \mu = \mu_0$ relative to $H_1: \mu > \mu_0$ if $\bar{x} \geq \bar{x}_c$, where $\bar{x}_c$ depicts the upper-tail critical value of $\bar{x}$. Here $\mathcal{R} = \{\bar{x}|\bar{x} \geq \bar{x}_c\}$. Similarly, for Case II, equation (12.9.1) tells us to reject $H_0: \mu = \mu_0$ relative to $H_1: \mu < \mu_0$ if $\bar{x} \leq \bar{x}_c$, where $\bar{x}_c$ represents the right-hand side of (12.9.1) and is now understood to represent the lower-tail critical value of $\bar{x}$. For this case $\mathcal{R} = \{\bar{x}|\bar{x} \leq \bar{x}_c\}$. Finally, for Case III, (12.11.1) has us reject $H_0 : \mu = \mu_0$ against $H_1 : \mu \neq \mu_0$ if $\bar{x} \leq \bar{x}_{cl}$ or $\bar{x} \geq \bar{x}_{cu}$, where

$$\text{(a)} \quad \bar{x}_{cl} = \mu_0 - z_{\frac{\alpha}{2}} \frac{\sigma}{\sqrt{n}},$$

$$\text{(b)} \quad \bar{x}_{cu} = \mu_0 + z_{\frac{\alpha}{2}} \frac{\sigma}{\sqrt{n}} \tag{12.13}$$

denote lower- and upper-tail critical values of $\bar{x}$, respectively. In this instance $\mathcal{R} = \{\bar{x} \,|\, \bar{x} \leq \bar{x}_{cl} \text{ or } \bar{x} \geq \bar{x}_{cu}\}$. It is interesting to note that the lower- and upper-tail critical values of $\bar{x}$ given by (12.13a,b) are actually the lower- and upper-confidence limits ℓ_1 and ℓ_2, respectively, given by (11.7.1). In this regard, a $100(1 - \alpha)\%$ confidence interval for μ contains all those null values that *would not be rejected* at the $100\alpha\%$ level in a two-tail test involving $H_0 : \mu = \mu_0$ versus $H_1 : \mu \neq \mu_0$. Correspondingly, all null values found outside of a $100(1 - \alpha)\%$ confidence interval for μ *would be rejected* at the $100\alpha\%$ level in a Case III two-tail test.

Example 12.9.3   In the preceding example problem we rejected $H_0 : \mu = \$55,000$ in favor of $H_1 : \mu \neq \$55,000$ at the 5% level. Then from (12.13a,b) we have, respectively,

$$\bar{x}_{cl} = 55,000 - 1.96 \frac{2,700}{\sqrt{200}} = 54,625.80,$$

$$\bar{x}_{cu} = 55,000 + 1.96 \frac{2,700}{\sqrt{200}} = 55,374.20.$$

Hence any μ_0 lying within the 95% confidence interval $(\ell_1, \ell_2) = (\bar{x}_{cl}, \bar{x}_{cu}) = (54,625.80, 55,374.20)$ would not be rejected under a two-tail test at the 5% level. If we choose $\mu_0 = \$58,500$ for our null value, then H_0 would be rejected at the 5% level since it lies outside of the 95% confidence interval $(\bar{x}_{cl}, \bar{x}_{cu})$. In sum, a $100(1 - \alpha)\%$ confidence interval may be regarded as a collection of hypothesized values that would not be rejected at the $100\alpha\%$ level when we have a two-sided alternative hypothesis. ■

12.10   Reporting Hypothesis Test Results

Our previous (classical) approach to reporting the results of parametric hypothesis tests consisted of choosing (*a priori*) a tolerable α-risk, drawing a random sample, and then, on the basis of the sample results, informing you as to whether or not the null hypothesis H_0 was rejected. If α is *small*, then the decision to, say, reject H_0 offers reasonable assurance that we have not acted incorrectly and thus reached the wrong conclusion. But if α is *large*, we would not be easily persuaded that the decision to reject H_0 is the correct one.

   As an alternative to this approach, let us report the *p-value* of a test. Whereas α may be termed a *chosen* level of significance, a *p*-value is the *realized* level of significance; that is, it is a statistic that represents the probability of obtaining a value of the test statistic T that is at least as extreme as that realized by the sample given that the null hypothesis is true. In this regard, we may view a p-value as the most extreme value of α for which a true null hypothesis is rejected. Unlike the α-risk, the p-value is not arbitrary; it is, in fact, *data sensitive* in that it depends upon the sample results.

Let's see exactly how a p-value is obtained for a variety of test situations. Suppose we are faced with a Case I test involving $H_0 : \theta = \theta_0$ versus $H_1 : \theta > \theta_0$. If H_0 is to be rejected in favor of H_1 for large values of our test statistic T, then the p-value associated with t, the sample realization of T, is

$$p\text{-value} = P(T \geq t/H_0 \text{ true})$$

$$= P(T \text{ assumes a value at least as large as } t \text{ when } H_0 \text{ is true}). \quad (12.14)$$

Hence the p-value for this test amounts to the area under the sampling distribution of T to the right of t and in the direction of H_1. This realized level of significance is thus the *smallest* level of significance α for which H_0 would be rejected. So if H_0 were true, there would be a $100p\%$ chance of observing a T at least as large as t.

For Case II (testing $H_0 : \theta = \theta_0$ against $H_1 : \theta < \theta_0$), we reject H_0 in favor of H_1 for small values of T so that, in this instance, the

$$p\text{-value} = P(T \leq t/H_0 \text{ true})$$

$$= P(\text{assumes a value at least as small as } t \text{ when } H_0 \text{ is true}). \quad (12.15)$$

In this instance the p-value is determined as the area under the sampling distribution of T to the left of t and in the direction of H_1. Here also this attained level of significance provides us with the smallest α for which the null hypothesis is rejected. With H_0 true, we have a $100p\%$ chance of observing a T at least as small as t.

Finally, for Case III, if we test $H_0 : \theta = \theta_0$ relative to $H_1 : \theta \neq \theta_0$, then H_0 is rejected in favor of H_1 for very small or very large values of T. Hence the

$$p\text{-value} = P(|T| \geq t/H_0 \text{ true}) = P(T \leq -t \quad \text{or} \quad T \geq t/H_0 \text{ true})$$

$$= P(T \leq -t/H_0 \text{ true}) + P(T \geq t/H_0 \text{ true}). \quad (12.16)$$

By virtue of this equation, it is evident that for a two-sided alternative hypothesis, the p-value can be found by determining, say, the area under the sampling distribution of T to the right of t and then doubling it; that is,

$$p\text{-value} = 2P(T \geq t/H_0 \text{ true}). \quad (12.16.1)$$

Thus this p-value gives us twice the probability of obtaining a value of our test statistic as large as t, or larger, if H_0 were true.

So if the p-value for a particular test is very small, then it is highly unlikely that the sample was extracted from a population for which the null hypothesis

$H_0 : \theta = \theta_0$ is true. Hence we may comfortably reject H_0 in favor of H_1. Thus a p-value may be thought of as a measure of the strength of the sample evidence against H_0; it reflects the extent to which the sample results and H_0 disagree. So the closer t is to θ_0, the larger the p-value for the test. Likewise, the greater the disparity between t and θ_0, the smaller the p-value for the test.

As a convenient rule of thumb we have:

$$\text{if } \alpha \geq p\text{-value, then reject } H_0;$$
$$\text{if } \alpha < p\text{-value, do not reject } H_0, \tag{12.17}$$

where the α-risk of the test results is chosen subjectively by you, and thus reflects your own comfort level associated with the possibility of incorrectly rejecting a true null hypothesis.

Example 12.10.1   Suppose that for a test of $H_0 : \mu = \mu_0 = 100$ versus $H_1 : \mu > 100$ we find, using (12.8), that the realized $\overline{Z}$ is $\bar{z} = 2.47$ (see Figure 12.6a). Then from the standard normal area table, we find that the area under the $N(0,1)$ distribution from 0 to 2.47 is 0.4932. Hence the remaining area under the right-hand tail of this curve is 0.0068. This latter number is thus the p-value for the test. Hence the probability that we would obtain a $\overline{Z}$ value as large as 2.47, or larger, given that H_0 is true is only $0.0068 = P(\overline{Z} \geq 2.47/H_0 : \mu = 100)$. Hence there is less than a 1% chance of making a Type I Error. Thus the null hypothesis can safely be rejected since it is highly unlikely that we would have obtained a $\bar{z}$ of this high a magnitude if H_0 were true. Now, if the threshold of your comfort zone for committing a Type I Error is $\alpha \leq 0.05$, then, by virtue of (12.17), we see that H_0 is again safely rejected.

Next, suppose we are testing $H_0 : \mu = \mu_0 = 100$ against $H_1 : \mu \neq 100$ and the sample yields $|\bar{z}| = 1.25$ (see Figure 12.6b). Since the area under the upper tail of the $N(0,1)$ distribution is 0.1056, it must be doubled to obtain the p-value

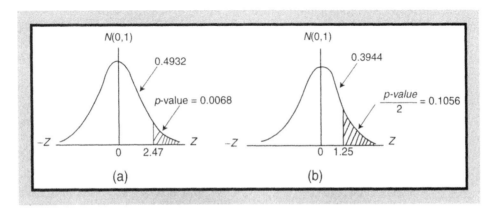

Figure 12.6   Determining p-values for (a) one- and (b) two-tailed tests.

since we are performing a two-tail test; that is, the p-value $= 2\,(0.1056) = 0.2112$ or $P(|\overline{Z}| \geq 1.25/H_0 : \mu = 100) = 0.2112$. Hence the probability that we would obtain a $|\overline{Z}|$ value at least as large as 1.25 if H_0 were true is in excess of 21%.

With a chance of making a Type I Error this large, we would not opt to reject H_0. It is quite likely that the sample came from a population having $\mu = 100$. ∎

It is important to note that although a p-value is not the same thing as an α-risk (it does not have the same long-run relative frequency interpretation), it is, however, defined in terms of (i.e., compared to) $\alpha = P(TIE)$ by virtue of (12.17). Remember that α is chosen in advance of testing; the p-value is realized from the sample and thus emerges as an *ex post* test result. In addition, a p-value provides you with more information about the results of a hypothesis test than that offered by α alone. For instance, what if we were told that, for a given hypothesis test, the null hypothesis was rejected at the $100\alpha\% = 10\%$ level. Although we know that the value of our test statistic lies within the critical region $\mathcal{R}$, we do not know how deeply into $\mathcal{R}$ it is located. In fact, the deeper the realized test statistic lies within $\mathcal{R}$, the more likely it is that we have made a correct decision; that is, the p-value serves as an index of the *strength* of our decision to reject H_0; it does so by indicating how likely it is for the sample to realize a T that is greater than or equal to (or less than or equal to) t when H_0 is true.

12.11 Determining the Probability of a Type II Error β

In this section we shall develop procedures for assessing the magnitude of a Type II Error (defined as not rejecting a false null hypothesis) associated with a hypothesis test. When H_1 is composite, $\beta = P(TIIE) = P(\text{do not reject } H_0/H_0$ false) varies with each μ satisfying H_1, where H_1 is assumed to be true. Hence β is not selected *a priori* as is α—it is, in fact, unknown. If we assume a value for μ that is supported by H_1, then we can calculate β.

For convenience, let us assume that we are sampling from a $N(\mu, \sigma)$ population, where the population mean μ is unknown but the population variance σ^2 is known. Based upon our previous discussion, we know that $\overline{X}$ is the best estimator of μ and that $\overline{X}$ is $N(\mu, \sigma_{\overline{X}})$, where $\sigma_{\overline{X}} = \sigma/\sqrt{n}$. Let us first test $H_0 : \mu = \mu_0$ against $H_1 : \mu > \mu_0$ (our Case I test situation given earlier). Under H_0, $\overline{X}$ is $N(\mu_0, \sigma_{\overline{X}})$.

Given the definition of a Type II Error, it is possible to calculate the probability of any such error only for specific μ's, for which H_0 is false or H_1 is true; that is, to determine β, we need to know the true mean μ, assuming that it is not μ_0. In this regard, from (12.8.1), the region of rejection is $\mathcal{R} = \{\bar{x}/\bar{x} \geq \bar{x}_c\}$, and the region of nonrejection is $\overline{\mathcal{R}} = \{\bar{x}/\bar{x} < \bar{x}_c\}$, where $\bar{x}_c = \mu_0 + z_\alpha \frac{\sigma}{\sqrt{n}}$. Now, since we can determine only β for values of μ satisfying H_1, let us choose $\mu = \mu_1 > \mu_0$. So with H_0 false, the correct distribution of $\overline{X}$ is $N(\mu_1, \sigma_{\overline{X}})$. We thus have the two essential ingredients needed for calculating β: (1) $\overline{\mathcal{R}}$; and (2) a $\mu = \mu_1 > \mu_0$ consistent with H_1. That is to say, β corresponds to the probability that the realization of $\overline{X}$ resides in $\overline{\mathcal{R}}$ given that the true mean is not μ_0 but is, instead, μ_1. Then

$$\beta = P(TIIE) = P(\overline{X} \in \overline{\mathcal{R}}/\mu = \mu_1 \text{ true})$$

$$= P(\overline{X} < \bar{x}_c/\mu = \mu_1 \text{ true})$$

$$= P\left(\overline{Z} < \frac{\bar{x}_c - \mu_1}{\sigma_{\overline{X}}} = \frac{(\mu_0 - \mu_1) + z_\alpha \frac{\sigma}{\sqrt{n}}}{\frac{\sigma}{\sqrt{n}}} \,\middle|\, \mu = \mu_1 \text{ true}\right), \quad (12.18)$$

where $\overline{Z}$ is $N(0,1)$ (see Figure 12.7a). In graphical terms then, β is always the area under the $N(\mu_1, \sigma_{\overline{X}})$ distribution taken over the nonrejection region $\overline{\mathcal{R}}$.

It is easily seen from expression (12.18) that β depends upon (a) the sample size n; (b) the size of the α-risk selected; (c) the population standard deviation σ; and (d) the difference between the null value μ_0 and the true mean μ_1 subsumed

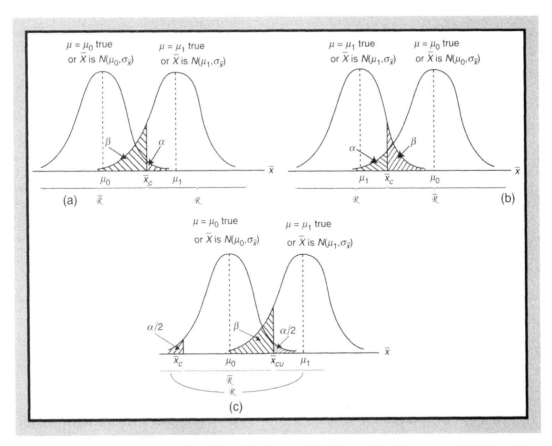

Figure 12.7 (a) Determining $P(TIIE) = \beta$ for the one-sided alternative $H_1 : \mu > \mu_0$; (b) Determining $P(TIIE) = \beta$ for the one-sided alternative $H_1 : \mu < \mu_0$; (c) Determining $P(TIIE) = \beta$ for the two-sided alternative $H_1 : \mu \neq \mu_0$.

under H_1 or $\mu_0 - \mu_1$. More specifically, in terms of its sensitivity to individual changes in these arguments, β (a) varies inversely with n (with σ, α, and $\mu_0 - \mu_1$ held fixed); (b) varies inversely with α (for fixed σ, n, and $\mu_0 - \mu_1$ values); (c) varies directly with σ (with n, α, and $\mu_0 - \mu_1$ fixed); and (d) varies inversely with $\mu_0 - \mu_1$ (holding n, α, and σ constant).

Case (d) merits some additional attention. If the difference $\mu_0 - \mu_1$ is small, then it will be difficult to determine which is the true value of μ and thus β will be relatively large (i.e., the probability of not rejecting H_0 when the true mean is μ_1 will tend to be large). And if $\mu_0 - \mu_1$ is large, then β will be relatively small—it will be easier to discriminate between μ_0 and μ_1 in determining the true population mean μ.

Next, let us test $H_0 : \mu = \mu_0$ versus $H_1 : \mu < \mu_0$ (Case II). Then from (12.9.1), the critical region $\mathcal{R} = \{\bar{x}/\bar{x} \leq \bar{x}_c\}$, $\overline{\mathcal{R}} = \{\bar{x}/\bar{x} > \bar{x}_c\}$, and now $\bar{x}_c = \mu_0 - z_\alpha \frac{\sigma}{\sqrt{n}}$ with $\overline{X}$ distributed as $N(\mu_0, \sigma_{\overline{X}})$. Here also β can be determined only for values of μ satisfying H_1. Hence we can set $\mu = \mu_1 < \mu_0$ (clearly H_0 is false and thus $\overline{X}$ is $N(\mu_1, \sigma_{\overline{X}})$). Then the probability of not rejecting a false null hypothesis is

$$
\begin{aligned}
\beta = P(TIIE) &= P\left(\overline{X} \in \overline{\mathcal{R}} / \mu = \mu_1 \text{ true}\right) \\
&= P(\overline{X} > \bar{x}_c / \mu = \mu_1 \text{ true}) \\
&= P\left(\overline{Z} > \frac{\bar{x}_c - \mu_1}{\sigma_{\overline{X}}} = \frac{(\mu_0 - \mu_1) - z_\alpha \frac{\sigma}{\sqrt{n}}}{\frac{\sigma}{\sqrt{n}}} \middle| \mu = \mu_1 \text{ true}\right) \quad (12.19)
\end{aligned}
$$

(see Figure 12.7b).

Finally, when testing $H_0 : \mu = \mu_0$ against the Case III alternative $H_1 : \mu \neq \mu_0$, we previously determined that $\mathcal{R} = \{\bar{x}/\bar{x} \leq \bar{x}_{cl} \text{ or } \bar{x} \geq \bar{x}_{cu}\}$ and $\overline{\mathcal{R}} = \{\bar{x}/\bar{x}_{cl} < \bar{x} < \bar{x}_{cu}\}$, where $\bar{x}_{cl} = \mu_0 - z_{\alpha/2} \frac{\sigma}{\sqrt{n}}$ and $\bar{x}_{cu} = \mu_0 + z_{\alpha/2} \frac{\sigma}{\sqrt{n}}$ given that $\overline{X}$ is $N(\mu_0, \sigma_{\overline{X}})$. Let $\mu = \mu_1 \neq \mu_0$ be true (H_1 is satisfied and thus H_0 is false so that $\overline{X}$ is now $N(\mu_1, \sigma_{\overline{X}})$).

Hence the probability of not rejecting H_0 given that it is false is

$$
\begin{aligned}
\beta = P(TIIE) &= P\left(\overline{X} \in \overline{\mathcal{R}} | \mu = \mu_1 \text{ true}\right) \\
&= P(\bar{x}_{cl} < \overline{X} < \bar{x}_{cu} | \mu = \mu_1 \text{ true}) \\
&= P\left(\frac{\bar{x}_{cl} - \mu_1}{\sigma_{\overline{X}}} < \overline{Z} < \frac{\bar{x}_{cu} - \mu_1}{\sigma_{\overline{X}}} \middle| \mu = \mu_1 \text{ true}\right) \\
&= P\left(\frac{(\mu_0 - \mu_1) - z_{\alpha/2} \frac{\sigma}{\sqrt{n}}}{\frac{\sigma}{\sqrt{n}}} < \overline{Z} < \frac{(\mu_0 - \mu_1) + z_{\alpha/2} \frac{\sigma}{\sqrt{n}}}{\frac{\sigma}{\sqrt{n}}} \middle| \mu = \mu_1 \text{ true}\right) \quad (12.20)
\end{aligned}
$$

(see Figure 12.7c).

Example 12.11.1   Suppose X is $N(\mu, 15)$ and that $H_0 : \mu = \mu_0 = 20$ and $H_1 : \mu > 20$. Hence under H_0 the distribution of $\overline{X}$ is $N(20, \sigma_{\overline{X}})$. Given a sample of size $n = 25$, $\overline{X}$ is $N(20, 3)$ and $\overline{Z} = (\overline{X} - 20)/3$ is $N(0, 1)$. For $\alpha = 0.05$, the critical region appears as $\mathcal{R} = \{z / \overline{z} \geq z_\alpha\} = \{z / \overline{z} \geq 1.645\}$ or, via (12.8.1), $\mathcal{R} = \{\overline{x} / \overline{x} \geq \overline{x}_c = 20 + 1.645(3) = 24.935\}$. Hence the region of nonrejection is $\overline{\mathcal{R}} = \{\overline{x} / \overline{x} < \overline{x}_c = 24.935\}$ (see Figure 12.8a). Suppose the true population mean is $\mu = \mu_1 = 28$ so that the correct distribution of $\overline{X}$ is $N(28, \sigma_{\overline{X}})$. Then from (12.18), we are interested in finding the probability of not rejecting a false null hypothesis ($\mu = \mu_1 = 28$ satisfies H_1) or

$$\beta = P\left(\overline{Z} < \frac{(\mu_0 - \mu_1) + z_\alpha \frac{\sigma}{\sqrt{n}}}{\frac{\sigma}{\sqrt{n}}} \Bigg| \mu = 28 \text{ true}\right)$$

$$= P\left(\overline{Z} < \frac{-8 + 1.645(3)}{3} = -1.0216 | \mu = 28 \text{ true}\right) = 0.1539.$$

Hence there is a 15.4% chance of not rejecting the null hypothesis $H_0 : \mu = 20$ when the true mean is 28. It is important to remember that we do not know the actual value of the true population mean μ; that is, we do not know what the true distribution of $\overline{X}$ happens to be ($\mu_0 = 20$ and $\mu_1 = 28$ are *guesses*). In this regard, if H_0 is true, then $\overline{X}$ is $N(\mu_0, \sigma_{\overline{X}})$ and the test will lead to an incorrect conclusion

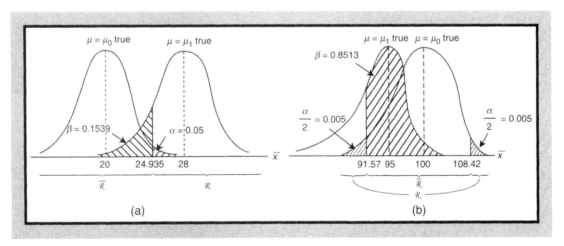

Figure 12.8   (a) The region of rejection $\mathcal{R} = \{\overline{x} / \overline{x} \geq 24.935\}$ and the region of non-rejection $\overline{\mathcal{R}} = \{\overline{x} / \overline{x} < 24.935\}$; (b) The region of rejection $\mathcal{R} = \{\overline{x} / \overline{x} \leq 91.57 \text{ or } \overline{x} \geq 108.42\}$ and the region of non-rejection $\overline{\mathcal{R}} = \{\overline{x} | 91.57 < x < 108.42\}$.

$100\alpha\% = 5\%$ of the time; and if H_1 is true, then $\overline{X}$ is $N(\mu_1, \sigma_{\overline{X}})$ and the test will produce an incorrect result $100\beta\% = 15.4\%$ of the time. ∎

Example 12.11.2   Given that X is $N(\mu, 40)$, the sample size is $n = 150$, and the α-risk is 0.01, let us consider a test of $H_0 : \mu = \mu_0 = 100$ versus $H_1 : \mu \neq 100$. In this instance the critical region has the form $\mathcal{R} = \{z/|\bar{z}| \geq z_{\alpha/2}\} = \{z/|\bar{z}| \geq 2.58\}$. Alternatively, from (12.13), $\mathcal{R} = \{\bar{x}/\bar{x} \leq \bar{x}_{cl} \text{ or } \bar{x} \geq \bar{x}_{cu}\} = \{\bar{x}/\bar{x} \leq 91.57 \text{ or } \bar{x} \geq 108.42\}$, where

$$\bar{x}_{cl} = 100 - 2.58 \left(\frac{40}{\sqrt{150}} \right) = 91.57,$$

$$\bar{x}_{cu} = 100 + 2.58 \left(\frac{40}{\sqrt{150}} \right) = 108.42.$$

Then the region of nonrejection $\overline{\mathcal{R}} = \{\bar{x}/\bar{x}_{cl} < \bar{x} < \bar{x}_{cu}\} = \{\bar{x}/91.57 < \bar{x} < 108.42\}$ (see Figure 12.8b). If the true population mean is $\mu = \mu_1 = 95$, what is the probability of not rejecting H_0? Clearly $\mu_1 = 95$ satisfies H_1 (thus H_0 is false) and is also an element of $\overline{\mathcal{R}}$. Hence, via (12.20),

$$\beta = P\left(\frac{(\mu_0 - \mu_1) - z_{\alpha/2}\frac{\sigma}{\sqrt{n}}}{\frac{\sigma}{\sqrt{n}}} < \overline{Z} < \frac{(\mu_0 - \mu_1) + z_{\alpha/2}\frac{\sigma}{\sqrt{n}}}{\frac{\sigma}{\sqrt{n}}} \Bigg| \mu = 95 \text{ true} \right)$$

$$= P\left(\frac{5 - 2.58(3.27)}{3.27} = -1.05 < \overline{Z} < \frac{5 + 2.58(3.27)}{3.27} = 4.11 \Big| \mu = 95 \text{ true} \right)$$

$$= 0.8513.$$

In this instance we have slightly in excess of an 85% chance of not rejecting $H_0 : \mu = 100$ when the true mean equals 95. So even for a sample of size $n = 150$, the test will not easily detect a difference between μ_0 and μ_1 of 5 units. ∎

Suppose that in the preceding example problem we had realized a sample mean of $\bar{x} = 102$. Then our calculated $\bar{z}$ value would be

$$|\bar{z}| = \left| \frac{\bar{x} - \mu_0}{\sigma/\sqrt{n}} \right| = \left| \frac{102 - 100}{40/\sqrt{150}} \right| = 0.612.$$

Since this realized Z is not a member of $\mathcal{R} = \{z/|z| \geq 2.58\}$, we cannot reject the null hypothesis. But then this means that by not rejecting H_0, we cannot determine the magnitude of the Type II Error that may have been made.

We noted at the outset of this section that once α was chosen, we could determine the unknown β-risk or $P(TIIE)$ by choosing a value of $\mu = \mu_1$ that is consistent with the alternative hypothesis H_1. The β forthcoming from this process was found to depend upon α, n, σ, and $\mu_0 - \mu_1$. This observation provides us with a hint as to how we might determine the sample size n needed to support desired levels of *both* α and β.

Why would any such calculation be important? If in our hypothesis testing process we fail to reject the null hypothesis H_0, then we do so for a given value of n. But if n had been larger, we might have been able to reject H_0 given that the enlarged sample now provides us with sufficient information to do so. To avoid having any doubts about the correctness of a decision to not reject H_0, we may opt to specify, in advance of any testing, the maximum tolerable levels of both α and β and then find the value of n for which these levels are satisfied.

To this end suppose we are faced with testing $H_0 : \mu = \mu_0$ against $H_1 : \mu > \mu_0$. Given that both α and β are to be specified *a priori*, with β evaluated for some $\mu = \mu_1 > \mu_0$ satisfying H_1, the test ultimately depends upon the sample size n and $\bar{x}_c$, the critical value of $\bar{x}$. From the definitions of α and β, we may express each of these error probabilities in terms of n and $\bar{x}_c$ as

$$\alpha = P(TIE) = P(\overline{X} \in \mathcal{R} | \mu = \mu_0 \text{ true})$$

$$= P(\overline{X} \geq \bar{x}_c | \mu = \mu_0 \text{ true}),$$

$$= P\left(\overline{Z} \geq \frac{\bar{x}_c - \mu_0}{\sigma/\sqrt{n}} = z_\alpha | \mu = \mu_0 \text{ true}\right), \quad (12.21)$$

where z_α satisfies $P(Z \geq z_\alpha) = \alpha$; and

$$\beta = P(TIIE) = P(\overline{X} \in \overline{\mathcal{R}} | \mu = \mu_1 \text{ true})$$

$$= P(\overline{X} < \bar{x}_c | \mu = \mu_1 \text{ true})$$

$$= P\left(\overline{Z} < \frac{\bar{x}_c - \mu_1}{\sigma/\sqrt{n}} = -z_\beta | \mu = \mu_1 \text{ true}\right) \quad (12.22)$$

with $-z_\beta$ chosen so that $P(Z < -z_\beta) = \beta$.

From (12.21) and (12.22) we have, respectively,

$$z_\alpha = \frac{\bar{x}_c - \mu_0}{\sigma/\sqrt{n}}, \quad -z_\beta = \frac{\bar{x}_c - \mu_1}{\sigma/\sqrt{n}}.$$

Upon eliminating $\bar{x}_c$ from these latter two expressions we ultimately obtain

$$n = \left[\frac{\sigma(z_\alpha + z_\beta)}{\mu_1 - \mu_0} \right]^2. \quad (12.23)$$

(You can easily verify that for $H_0 : \mu = \mu_0$ versus $H_1 : \mu < \mu_0$, exactly the same formula (12.23) for n obtains.)

If we are testing $H_0 : \mu = \mu_0$ against the two-sided alternative $H_1 : \mu \neq \mu_0$, then (12.23) must be modified to

$$n = \left[\frac{\sigma(z_{\alpha/2} + z_\beta)}{\mu_1 - \mu_0} \right]^2,$$ (12.24)

where $z_{\alpha/2}$ is chosen so that $P(|\overline{Z}| \geq z_{\alpha/2}) = \dfrac{\alpha}{2}$.

Example 12.11.3   Suppose that from a population that is $N(\mu, 13)$, we desire to extract a random sample for a test of $H_0 : \mu = \mu_0 = 100$ versus $H_1 : \mu < 100$. For this test we choose as our maximum tolerable error probabilities $\alpha \leq 0.05$ and $\beta \leq 0.05$. How large a sample will be needed if $\mu_1 - \mu_0 = -6$ (i.e., if $\mu_1 = 94$)? From (12.23) it is easily determined that

$$n = \left[\frac{13(1.645 + 1.645)}{-6} \right]^2 = 50.81.$$

Hence a sample of size 51 will do. Note that if n is not an integer, then we must *round up* the value of n to the next larger integer. That is to say, a sample of at least $n = 51$ is required to ensure that $\alpha \leq 0.05$ and $\beta \leq 0.05$ when $\mu = \mu_1 = 94$ is the true mean.

If in this example we had chosen $H_1 : \mu \neq 100$, then, via (12.24),

$$n = \left[\frac{13(1.96 + 1.645)}{-6} \right]^2 = 61.01.$$

For this revised alternative hypothesis $n = 62$ is required.   ■

12.12 Hypothesis Tests for μ Under Random Sampling from a Normal Population with Unknown Variance

We noted in Section 12.9 that if X is $N(\mu, \sigma)$ and the population variance σ^2 is known, then the appropriate test statistic for conducting hypothesis tests involving the population mean μ is the standard normal variate $\overline{Z} = \frac{\overline{X} - \mu}{\sigma/\sqrt{n}}$. Moreover, if σ^2 is unknown (its best estimator is the sample variance $S^2 = \frac{1}{n-1} \sum_{i=1}^{n} (X_i - \overline{X})^2$), then, via the Central Limit Theorem, the statistic $\frac{\overline{X} - \mu}{S/\sqrt{n}}$ is approximately standard normal for large n. In this instance the test procedures for μ introduced in Section 12.9 still apply. However, if σ^2 is unknown and the sample size n is small ($n \leq 30$), then the standard normal distribution is no longer applicable when testing hypotheses about μ. For this latter case, under random sampling from a normal population, with both the mean (μ) and variance (σ^2) unknown and n small, the statistic $T = \frac{\overline{X} - \mu}{S/\sqrt{n}}$ follows a t distribution with $\nu = n - 1$ degrees of freedom.

For instance, given the Case I test situation ($H_0 : \mu = \mu_0$ and $H_1 : \mu > \mu_0$), the critical region $\mathcal{R}$ involves that portion of the t-axis such that $t \geq t_{\alpha,n-1}$; that is, $\mathcal{R} = \{t/t \geq t_{\alpha,n-1}\}$ or

$$\frac{\bar{x} - \mu_0}{s/\sqrt{n}} \geq t_{\alpha,n-1}, \qquad (12.25)$$

where t is the sample realization of T, $s = \sqrt{\frac{\sum_{i=1}^{n}(x_i - \bar{x})^2}{n-1}}$ is the realized value of the sample standard deviation, and the α-risk satisfies $P(T \geq t_{\alpha,n-1}) = \alpha$. Hence we will reject H_0 in favor of H_1 if (12.25) obtains.

For a Case II test ($H_0: \mu = \mu_0$ and $H_1: \mu < \mu_0$), we now require that $P(T \leq -t_{\alpha,n-1}) = \alpha$ so that $\mathcal{R} = \{t/t \leq -t_{\alpha,n-1}\}$ or, proceeding as before,

$$\frac{-(\bar{x} - \mu_0)}{s/\sqrt{n}} \geq t_{\alpha,n-1}. \qquad (12.26)$$

Hence, alternatively, we reject H_0 relative to H_1 when (12.26) is our test result.

Under Case III ($H_0: \mu = \mu_0$ is tested against $H_1: \mu \neq \mu_0$), we require that $P(T \leq -t_{\alpha/2,n-1}) = P(T \geq t_{\alpha/2,n-1}) = \frac{\alpha}{2}$ so that $\mathcal{R} = \{t/t \leq -t_{\alpha/2,n-1}$ or $t \geq t_{\alpha/2,n-1}\} = \{t/|t| \geq t_{\alpha/2,n-1}\}$ or

$$\left|\frac{\bar{x} - \mu_0}{s/\sqrt{n}}\right| \geq t_{\alpha/2,n-1}. \qquad (12.27)$$

Then H_0 is rejected in favor of H_1, provided (12.27) holds.

Example 12.12.1   Suppose we wish to test $H_0 : \mu = \mu_0 = 20$ against $H_1 : \mu > 20$ at the 5% level. From a sample of $n = 15$ items taken from a normal population we find that $\bar{x} = 25$ and $s = 8$. Can we reject H_0? Here $\mathcal{R} = \{t/t \geq t_{0.05,14} = 1.761\}$. Since the calculated t value

$$\frac{\bar{x} - \mu_0}{s/\sqrt{n}} = \frac{25 - 20}{8/\sqrt{15}} = 2.42$$

is an element of $\mathcal{R}$, we will reject H_0 in favor of H_1; that is, we can safely conclude at the 5% level of significance that the true or population mean exceeds a value of 20.

What is the p-value associated with this test? For this one-sided alternative, the p-value is given by p-value $= P(T \geq 2.42 | H_0 \text{ true})$.

A glance at Table A.3 reveals that, for 14 degrees of freedom, 2.42 falls between $t_{0.025,14} = 2.145$ and $t_{0.01,14} = 2.624$. Thus we reject H_0 for $\alpha = 0.025$ but not for $\alpha = 0.01$. Hence the p-value for this test is such that $0.01 < p < 0.025$. So if H_0 were true, there would be a 1% to 2.5% chance of observing a value of the t statistic at least as large as 2.42. ∎

If, in the preceding example problem, we had tested $H_0 : \mu = \mu_0 = 20$ against $H_1 : \mu \neq 20$ with $\alpha = 0.05$, then $\mathcal{R} = \{t/|t| \geq t_{0.025,14} = 2.145\}$. Since the absolute value of the calculated t lies in $\mathcal{R}$, the null hypothesis is again rejected in favor of the alternative hypothesis at the 5% level of significance.

For this two-sided alternative the p-value appears, from (12.16), as

$$p\text{-value} = P(|T| \geq 2.42|H_0 \text{ true}) = P(T \leq -2.42|H_0 \text{ true}) + P(T \geq 2.42|H_0 \text{ true}).$$

Again looking to Table A.3 we find, for 14 degrees of freedom, that 2.42 lies between $t_{0.05,14} = 2.145$ and $t_{0.02,14} = 2.624$. Hence H_0 is rejected for $\alpha = 0.05$ but not for $\alpha = 0.02$ so the p-value must satisfy $0.02 < p < 0.05$. Assuming that H_0 is true, there would be a 2% to 5% chance of observing a calculated T value as large as 2.42 or larger. Note that, from (12.16.1), the p-value bounds for the two-tail test are $2(0.01 < p < 0.025)$ or $0.02 < p < 0.05$, where the parenthetical expression depicts the p-value limits for the one-sided alternative hypothesis presented first.

We previously calculated the probability of a Type II Error, β, under the assumption that we are sampling from a normal population with known variance σ^2. In this circumstance we were able to determine β using the standard normal ($N(0,1)$) area table. We simply found the probability that the realization of $\overline{X}$ was an element of $\overline{R}$ given that the null hypothesis H_0 was false. However, for small samples and for σ^2 unknown, the task of finding β is not as straightforward as it was when we employed the $N(0,1)$ distribution. The problem is that the distribution of the test statistic when H_0 is false is not exactly the same as that of the *usual* t statistic $T = \frac{\overline{X}-\mu}{S/\sqrt{n}}$, which will be termed the *central t statistic*. If μ_1 satisfies H_1, then the quantity

$$Y = \frac{\overline{X} - x_0}{S/\sqrt{n}}$$

is called the noncentral t statistic. Clearly the difference between T and Y is that x_0 replaces μ in T, where x_0 is an arbitrary constant (ostensibly for which H_1 is true). (If X is $N(\mu,\sigma)$ and $E(X) = \mu = x_0$, then obviously T and Y coincide.)

Although tables for the noncentral t distribution are readily available, suffice it to say that for $n > 30$, a determination of β under a central t test using $x_0 = \mu_1$ true can be made via the standard normal approximation to T.

12.13   Hypothesis Tests for *p* Under Random Sampling from a Binomial Population

To test hypotheses concerning p, the proportion of successes in the population, let us extract a random sample of size n from a binomial population, where the sample size n is taken to be large. Then according to the DeMoivre-Laplace-Gauss

Limit Theorem, the standardized binomial random variable has as its limiting distribution the standard normal distribution or

$$\frac{X - np}{\sqrt{np(1-p)}} = \frac{\frac{X}{n} - p}{\sqrt{\frac{p(1-p)}{n}}} \xrightarrow{d} Z = N(0,1), \tag{12.28}$$

where the random variable X is the observed number of successes in our sample of size n and $0 < p < 1$.

We know from our earlier analysis of binomial experiments that the best estimator for p is $\hat{P} = \frac{X}{n}$. Furthermore, the sampling distribution of $\hat{P}$ has as its mean and variance $E(\hat{P}) = p$ and $V(\hat{P}) = \frac{p(1-p)}{n}$, respectively, and the best estimator for $V(\hat{P})$ is $S^2(\hat{P}) = \frac{\hat{P}(1-\hat{P})}{n}$. If we again invoke the DeMoivre-Laplace-Gauss Limit Theorem, then a possible test statistic is

$$Z_{\hat{P}} = \frac{\hat{P} - E(\hat{P})}{\sqrt{S^2(\hat{P})}} = \frac{\hat{P} - p}{\sqrt{\frac{\hat{P}(1-\hat{P})}{n}}} \xrightarrow{d} Z = N(0,1) \tag{12.28.1}$$

However, as we shall now see, there is no need to estimate $\sqrt{V(\hat{P})}$ from sample data. The null value of p is all that is needed.

Let us denote the null hypothesis as $H_0 : p = p_0$, where p_0 is the null value of p. Since H_0 is assumed to be true, it follows that, under the null hypothesis, $V(\hat{P}) = \frac{p_0(1-p_0)}{n}$. Hence our test statistic is obtained from (12.28) simply by replacing p by p_0; that is, our test statistic is the quantity

$$Z_{\hat{P}}^0 = \frac{\hat{P} - p_0}{\sqrt{\frac{p_0(1-p_0)}{n}}}, \tag{12.29}$$

which, via (12.28), is approximately $N(0,1)$ for large n. (The sampling distribution of $\hat{P}$ is approximately normal if the conditions $np \geq 5$ and $n(1-p) \geq 5$ are met. As a practical matter, we generally require that $n \geq 25$. If we suspect that p is close to either 0 or 1, then a sample size of at least 50 is required.)

Tests for the following three sets of hypothesis will be considered in turn:

Case I	Case II	Case III
$H_0 : p = p_0$	$H_0 : p = p_0$	$H_0 : p = p_0$
$H_1 : p > p_0$	$H_1 : p < p_0$	$H_1 : p \neq p_0$

For Case I, the critical region corresponding to the one-sided alternative $H_1 : p > p_0$ is located under the right-hand tail of the $N(0,1)$ distribution since we require that $P(Z_{\hat{P}}^0 \geq z_\alpha) = \alpha$. Hence $\mathcal{R} = \{z | z_{\hat{P}}^0 \geq z_\alpha\}$ or we are interested in the set of z values for which

$$\frac{\hat{p} - p_0}{\sqrt{\dfrac{p_0(1 - p_0)}{n}}} \geq z_\alpha, \tag{12.30}$$

where $z_{\hat{P}}^0$ is the realized value of $Z_{\hat{P}}^0$ given that $\hat{p} = \frac{x}{n}$ is the sample realization of $\hat{P}$. Hence we reject H_0 provided (12.30) holds.

Looking to Case II, for the one-sided alternative $H_1 : p < p_0$, the critical region of size α is now located under the left-hand tail of the standard normal distribution with $P(Z_{\hat{P}}^0 \leq -z_\alpha) = \alpha$; that is, $\mathcal{R} = \{z / z_{\hat{P}}^0 \leq -z_\alpha\}$ or, under our convention, we reject H_0 when

$$\frac{-(\hat{p} - p_0)}{\sqrt{\dfrac{p_0(1 - p_0)}{n}}} \geq z_\alpha. \tag{12.31}$$

Finally, for Case III, the two-sided alternative hypothesis $H_1 : p \neq p_0$ implies, for a critical region of size α, that $P(|Z_{\hat{P}}^0| \geq z_{\alpha/2}) = \frac{\alpha}{2}$; that is, α is divided equally between the two tails of the $N(0, 1)$ distribution. Here $\mathcal{R} = \{z / z_{\hat{P}}^0 \leq -z_{\alpha/2}$ or $z_{\hat{P}}^0 \geq z_{\alpha/2}\}$. Under this requirement we shall reject H_0 if

$$\left| \frac{\hat{p} - p_0}{\sqrt{\dfrac{p_0(1 - p_0)}{n}}} \right| \geq z_{\alpha/2}. \tag{12.32}$$

Example 12.13.1 Suppose we find $X = 80$ successes in a random sample of size $n = 200$ taken from a binomial population. Let us test $H_0 : p = p_0 = 0.50$ against $H_1 : p \neq 0.50$ using $\alpha = 0.01$. Then $z_{\alpha/2} = z_{0.005} = 2.58$ and thus $\mathcal{R} = \{z / z_{\hat{P}}^0 \leq -2.58$ or $z_{\hat{P}}^0 \geq 2.58\}$. From (12.32) we have

$$\left| \frac{\dfrac{80}{200} - 0.50}{\sqrt{\dfrac{0.50(0.50)}{200}}} \right| = 2.86.$$

Since the *calculated* z exceeds the *tabular* z at the 1% level of significance (we are within the critical region), the null hypothesis is rejected in favor of the alternative hypothesis.

What is the p-value associated with this test? Since our sample yielded a calculated $Z_{\hat{P}}^0$ value of $|z_{\hat{P}}^0| = 2.86$, it follows that the area in the upper tail of the $N(0, 1)$ distribution equals 0.0021. However, we know that this area must be doubled since we are performing a two-tail test. Thus the p-value $= 0.0042$

or $P(|Z_{\hat{P}}^0| \geq 2.86/H_0: p = 0.50) = 0.0042$. Hence the probability that we would obtain a realized $|Z_{\hat{P}}|$ value at least as large as 2.86 if H_0 were true is only about 0.4%. Hence there is less than a 1% chance of committing a Type I Error. We therefore have a highly significant test result—it is very unlikely that we would have obtained a calculated $|Z_{\hat{P}}^0|$ value of this high a magnitude if H_0 were true. ∎

Formulas (12.23) and (12.24) presented in Section 12.11 allowed us to determine the sample size n needed to support chosen α- and β-risks when conducting hypothesis tests involving the population mean μ. Equation (12.23) holds for one-sided alternative hypotheses whereas (12.24) was used when a two-sided alternative hypothesis was chosen. These formulas are easily modified to accommodate the case where a hypothesis test involves a population proportion p.

Suppose we undertake a Case I test of $H_0 : p = p_0$ against $H_1 : p > p_0$ (this analysis will also hold when we have the Case II alternative $H_1 : p < p_0$) and that the maximum tolerable error risks α and β are to be specified *a priori*, where β must be determined for some $p = p_1 > p_0$ satisfying H_1. Let us express both α and β in terms of the sample size n and the critical value of the test statistic $\hat{P}$. To this end we shall rewrite (12.30) as

$$\hat{p} \geq p_0 + z_\alpha \sqrt{\frac{p_0(1 - p_0)}{n}}.$$

If we represent the right-hand side of this inequality as $\hat{p}_c$, then for the Case I scenario we would reject $H_0 : p = p_0$ relative to $H_1 : p > p_0$ if $\hat{p} \geq \hat{p}_c$, where $\hat{p}_c$ denotes the (upper-tail) critical sample proportion. Thus $\mathcal{R} = \{\hat{p}/\hat{p} \geq \hat{p}_c\}$.

We may now express

$$\alpha = P(TIE) = P\left(\hat{P} \in R|p = p_0 \text{ true}\right)$$

$$= P\left(\hat{P} \geq \hat{p}_c|p = p_0 \text{ true}\right)$$

$$= P\left(Z_{\hat{P}} \geq \frac{\hat{p}_c - p_0}{\sqrt{\frac{p_0(1-p_0)}{n}}} = z_\alpha|p = p_0 \text{ true}\right), \qquad (12.33)$$

where z_α satisfies $P(Z \geq z_\alpha) = \alpha$; and

$$\beta = P(TIIE) = P(\hat{P} \in \overline{\mathcal{R}}|p = p_1 \text{ true})$$

$$= P(\hat{P} < \hat{p}_c|p = p_1 \text{ true})$$

$$= P\left(Z_{\hat{P}} < \frac{\hat{p}_c - p_1}{\sqrt{\frac{p_1(1-p_1)}{n}}} = -z_\beta|p = p_1 \text{ true}\right) \qquad (12.34)$$

with $-z_\beta$ chosen so that $P(Z < -z_\beta) = \beta$.

From (12.33) and (12.34) we have, respectively,

$$z_\alpha = \frac{\hat{p}_c - p_0}{\sqrt{\frac{p_0(1-p_0)}{n}}}, \quad -z_\beta = \frac{\hat{p}_c - p_1}{\sqrt{\frac{p_1(1-p_1)}{n}}}.$$

If we eliminate $\hat{p}_c$ from each of these expressions we obtain

$$n = \left[\frac{z_\alpha \sqrt{p_0(1-p_0)} + z_\beta \sqrt{p_1(1-p_1)}}{p_1 - p_0} \right]^2. \tag{12.35}$$

Had we been testing $H_0 : p = p_0$ against $H_1 : p \neq p_0$ (Case III), then (12.35) must be replaced by

$$n = \left[\frac{z_{\alpha/2} \sqrt{p_0(1-p_0)} + z_\beta \sqrt{p_1(1-p_1)}}{p_1 - p_0} \right]^2, \tag{12.36}$$

where $z_{\alpha/2}$ satisfies $P(|Z| \geq z_{\alpha/2}) = \alpha/2$.

Example 12.13.2   Let us assume that we are sampling from a binomial population and that we are interested in obtaining a random sample for a test of $H_0 : p = p_0 = 0.30$ against $H_1 : p > 0.30$. It has been decided that the α- and β-risks should not exceed 5%. How large of a sample will be needed if $p_1 = 0.50$ (or if $p_1 - p_0 = 0.20$)? From (12.35) we have

$$n = \left[\frac{1.645\sqrt{0.30(0.70)} + 1.645\sqrt{0.50(0.50)}}{0.20} \right]^2 = 62.1.$$

We thus conclude that a sample of size 63 will meet the stated test specifications; that is, a sample of at least 63 is needed to ensure that $\alpha \leq 0.05$ and $\beta \leq 0.05$ when $p = p_1 = 0.50$ is the true population proportion.   ∎

12.14   Hypothesis Tests for σ^2 Under Random Sampling from a Normal Population

To test hypotheses concerning the population variance σ^2, let us extract a random sample of size n from a normal population with unknown mean μ and variance σ^2. We know from our earlier analysis of point and interval estimation that the best estimator for σ^2 is S^2, the sample variance. We also determined that the quantity $Y = (n-1)S^2/\sigma^2$ is distributed as χ^2_{n-1}, where $n-1$ depicts degrees of freedom associated with the chi-square random variable Y.

Let us denote the null hypothesis as $H_0 : \sigma^2 = \sigma_0^2$, where σ_0^2 is the null value of σ^2. Given that H_0 is assumed to be true, it follows that under H_0, the quantity

$$\frac{(n-1)S^2}{\sigma_0^2} \tag{12.37}$$

is χ_{n-1}^2.

Tests for the following three sets of hypotheses will be conducted in turn:

Case I	Case II	Case III
$H_0 : \sigma^2 = \sigma_0^2$	$H_0 : \sigma^2 = \sigma_0^2$	$H_0 : \sigma^2 = \sigma_0^2$
$H_1 : \sigma^2 > \sigma_0^2$	$H_1 : \sigma^2 < \sigma_0^2$	$H_1 : \sigma^2 \neq \sigma_0^2$

For Case I, the critical region corresponding to the one-sided alternative hypothesis $H_1 : \sigma^2 > \sigma_0^2$ is located under the right-hand tail of the chi-square distribution since we require that $P(Y > \chi_{\alpha,n-1}^2) = \alpha$ is the $100\alpha\%$ fractile of the chi-square distribution; that is, it is the value beyond which $100\alpha\%$ of the area under the chi-square distribution with $n-1$ degrees of freedom is found. Hence $\mathcal{R} = \{y/y \geq \chi_{\alpha,n-1}^2\}$ (see Figure 12.9a); that is, we are interested in the set of chi-square values

$$\frac{(n-1)s^2}{\sigma_0^2} \geq \chi_{\alpha,n-1}^2, \tag{12.38}$$

where s^2 is the sample realization of S^2. Hence we reject H_0 if (12.38) is satisfied.

Looking to the Case II one-sided alternative $H_1 : \sigma^2 < \sigma_0^2$, the critical region of size α is now located under the left-hand tail of the chi-square distribution with $P(Y < \chi_{1-\alpha,n-1}^2) = \alpha$. Now, $\mathcal{R} = \{y/y \leq \chi_{1-\alpha,n-1}^2\}$ so that we are interested in the set of chi-square values for which

$$\frac{(n-1)s^2}{\sigma_0^2} \leq \chi_{1-\alpha,n-1}^2 \tag{12.39}$$

(see Figure 12.9b). In this instance we reject H_0 if (12.39) holds.

As far as Case III is concerned, the two-sided alternative hypothesis $H_0 : \sigma^2 \neq \sigma_0^2$ implies, for an α-risk for which $P(Y < \chi_{1-(\alpha/2),n-1}^2$ or $Y > \chi_{\alpha/2,n-1}^2) = \alpha$, that the corresponding critical region has the form $\mathcal{R} = \{y/y \leq \chi_{1-(\alpha/2),n-1}^2$ or $y \geq \chi_{\alpha/2,n-1}^2\}$ (see Figure 12.9c). In this instance we are led to focus on the set of chi-square values for which

$$\frac{(n-1)s^2}{\sigma_0^2} \leq \chi_{1-(\alpha/2),n-1}^2 \quad \text{or} \quad \frac{(n-1)s^2}{\sigma_0^2} \geq \chi_{\alpha/2,n-1}^2. \tag{12.40}$$

Given this structure for $\mathcal{R}$, we may now reject H_0 when (12.40) obtains.

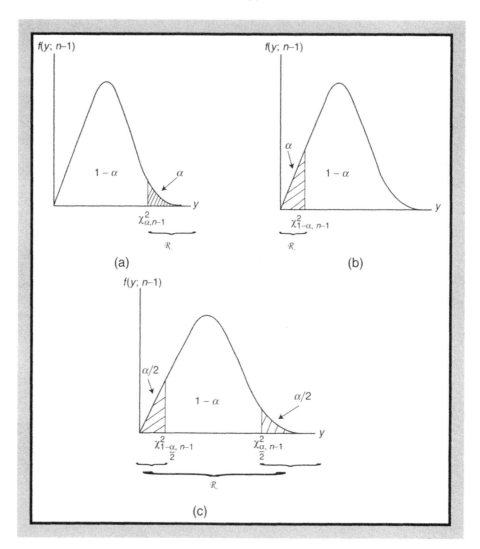

Figure 12.9   (a) The critical region $\mathcal{R}$ for the one-sided alternative $H_1 : \sigma^2 > \sigma_0^2$; (b) The critical region $\mathcal{R}$ for the one-sided alternative $H_1 : \sigma^2 < \sigma_0^2$; (c) The critical region $\mathcal{R}$ for the two-sided alternative $H_1 : \sigma^2 \neq \sigma_0^2$.

Example 12.14.1   Suppose that a particular lot of finished machine parts is claimed to have a diameter variance no larger than $0.008(mm)^2$. A random sample of $n = 15$ such parts yielded $s^2 = 0.010(mm)^2$. The milling department head decides to test $H_0 : \sigma^2 = \sigma_0^2 = 0.008$ against $H_1 : \sigma^2 > 0.008$ at the $\alpha = 0.05$ level. Given that the finished part diameters are normally distributed, $\frac{(n-1)s^2}{\sigma_0^2}$ is χ_{n-1}^2 and $\mathcal{R} = \{y | y \geq \chi_{0.05,14}^2 = 23.6848\}$. Since $(n-1)\frac{s^2}{\sigma_0^2} = \frac{14(0.010)}{0.008} = 17.50$ does not fall within the critical region, we do not reject the null hypothesis at the 5% level; that is, for the chosen α-risk, there is not sufficient sample evidence to indicate that $\sigma^2 > 0.008$.

What is the p-value associated with the observed or realized level of the preceding test statistic? For 14 degrees of freedom, the p-value $= P(Y \geq y|H_0$ true$) = P(Y \geq 17.50|\sigma^2 = 0.008)$. From the chi-square table (Table A.4) we find that for 14 degrees of freedom, $\chi^2_{0.25,14} = 17.1170$. Hence the p-value is a bit smaller than 0.25 so that the probability that we would obtain a chi-square value as large as 17.5, or larger, given that H_0 is true is about 25%. Hence there is a rather sizeable chance of making a Type I Error. Thus the null hypothesis should not be rejected; it is highly likely that we could obtain a realized chi-square value of this low magnitude if H_0 were true. So for any $\alpha < 0.25$, the null hypothesis cannot be rejected.

What if we had tested $H_0 : \sigma^2 = \sigma_0^2 = 0.008$ against $H_1 : \sigma^2 \neq 0.008$ (again with $y = 17.50$). What is the p-value associated with this revised test? For a two-tail test, one half of the p-value is found in the upper tail of the chi-square distribution and amounts to slightly less than 0.25 so that p-value/2 < 0.25 or p-value < 0.50. So if the null hypothesis were true, there would be almost a 50% chance of observing a value of the chi-square statistic at least as large as 17.50. So with the chance of making a Type I Error this large, we should not reject the null hypothesis. It is quite likely that the sample came from a population with $\sigma^2 = 0.008$. ∎

We note briefly that if the assumption of normality of the parent population is violated, then the results of the preceding set of chi-square tests for variances should be interpreted with caution. However, for large samples (say, $n \geq 100$), these tests are fairly reliable if the population is approximately normal.

12.15 The Operating Characteristic and Power Functions of a Test

Let us assume that the alternative hypothesis H_1 is composite or that, in general, $H_1 : \theta \in \mathcal{P}_1$. Then for a fixed sample size n and α-risk, the size of the Type II Error, β, varies for different values of θ subsumed under H_1. In this regard, let us express the magnitude of the Type II Error as a function of θ or as $\beta(\theta), \theta \in \mathcal{P}_1$. Here $\beta(\theta)$, which will be termed the *operating characteristic (OC) function* of the test, yields the probability that a sample realization t of the test statistic T *will not lie within the critical region* given that the parameter value is $\theta \in \mathcal{P}_1$; that is,

$$\beta(\theta) = P\big(T \in \overline{\mathcal{R}}|\theta \in \mathcal{P}_1\big) = P(\text{do not reject } H_0|H_0 \text{ false}) \qquad (12.41)$$

Next, let us define the *power* of a test as the (*a priori*) probability of detecting a false null hypothesis $H_0 : \theta \in \mathcal{P}_0$ or the probability that H_0 will be rejected when the parameter value is $\theta \in \mathcal{P}_1$. By virtue of equation (12.4b), the expression

$$P(\theta) = 1 - \beta(\theta) = P(T \in \mathcal{R}|\theta \in \mathcal{P}_1) = P(\text{reject } H_0|H_0 \text{ false}) \qquad (12.42)$$

is termed the *power function* of the test; it represents the probability that the realization t of T *will fall within the critical region* given that $\theta \in \mathcal{P}_1$. Here $P(\theta)$ is defined for a particular probability density function (or probability mass function), and the value of $P(\theta)$ at a given θ yields the power of the test at that parameter point. The graph of $P(\theta)$ is called the *power curve*; it exhibits the relationship between power and the true value of $\theta \in \mathcal{P}_1$ under H_1.

In general then, since $1 = P(TIIE) + \text{power}$,

$$P(\theta) = \begin{cases} \alpha(\theta) = P(TIE) & \text{if } \theta \in \mathcal{P}_0; \\ 1 - \beta(\theta) = 1 - P(TIIE) & \text{if } \theta \in \mathcal{P}_1. \end{cases} \qquad (12.43)$$

Hence the power function contains all relevant information about a test having critical region $\mathcal{R}$ and thus serves as an index of the capability of a test to discriminate between a true and a false hypothesis. It thus constitutes a means for evaluating the performance of a given test or the relative performance of two competing tests. If we seek to test $H_0 : \theta = \theta_0$, against $H_1 : \theta = \theta_1$, (i.e., $\mathcal{P}_0 = \{\theta_0\}$ and $\mathcal{P}_1 = \{\theta_1\}$), then (12.43) specializes to

$$P(\theta) = \begin{cases} \alpha & \text{if } \theta = \theta_0; \\ 1 - \beta & \text{if } \theta = \theta_1. \end{cases} \qquad (12.43.1)$$

So when both the null and alternative hypotheses are simple, α and β are well-defined and unique probabilities.

What does a typical power curve look like? If, for instance, the null and alternative hypotheses are $H_0: \theta = \theta_0$ and $H_1: \theta \neq \theta_0$, respectively, then $P(\theta) = 1 - \beta(\theta)$ is illustrated in Figure 12.10a. Here we plot the power $P(\theta)$ for each alternative θ covered under H_1 to obtain the power curve. When we test a simple null hypothesis against a composite alternative hypothesis, the probability of a Type II Error, β, is no longer a single number for a given α-risk and sample size; β depends upon the alternative value of the parameter θ chosen. For values of θ near θ_0, $\beta(\theta)$ will be high and thus the power of the test $P(\theta) = 1 - \beta(\theta)$ will be low. When θ is far removed from θ_0, $\beta(\theta)$ will be low and thus the tests power $P(\theta)$ will be high; that is, the more easily the test can discriminate between H_0 and H_1.

In general, the *ideal* (albeit unattainable) *power curve* assumes a value of zero for $\theta \in \mathcal{P}_0$ and unity for $\theta \in \mathcal{P}_1$ (since we do not want to reject H_0 if it is true, and we do want to reject H_0 if it is not true). If again $H_0 : \theta = \theta_0$ and $H_1 : \theta \neq \theta_0$, then the ideal power curve is depicted as Figure 12.10b; that is, ideally, the test would detect a departure from θ_0 with complete certainty or with $P(\theta) = 1$ for all θ included in H_1. And since this ideal is, as indicated before, unattainable,

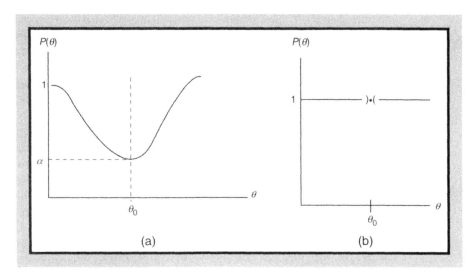

Figure 12.10   (a) The power curve $P(\theta) = 1 - \beta(\theta)$ when $H_0 : \theta \neq \theta_0, H_1 : \theta \neq \theta_0$; (b) The ideal power curve when $H_0 : \theta = \theta_0, H_1 : \theta \neq \theta_0$.

we seek, at best, a *good* test; that is, one whose power curve assumes a value near zero for most $\theta \in \mathcal{P}_0$ and near one for most $\theta \in \mathcal{P}_1$.

Why is the power function important and how is it used? It is typically the case that the null hypothesis H_0 is tested against a composite alternative hypothesis H_1; that is, the parameter set $\mathcal{P}_1$ contains more than just a single element. This being the case, we do not know which particular alternative value of $\theta \in \mathcal{P}_1$ will hold if H_0 is not true. In general, since many different (and thus competing) tests of $H_0 : \theta \in \mathcal{P}_0$ versus $H_1 : \theta \in \mathcal{P}_1$ can be constructed, with a new test emerging for each new value of θ chosen from $\mathcal{P}_1$, the value of the size of the Type II Error, β, will depend upon the particular alternative value of θ being utilized. Hence it is necessary to compare the β's for all possible values of $\theta \in \mathcal{P}_1$ if we are to determine how *good* one test is relative to all other alternative tests. Hence to choose among competing tests of H_0 against H_1, we express β as a function of $\theta, \beta(\theta)$. Here $\beta(\theta)$ represents the probability that $T \in \overline{\mathcal{R}}$ given θ; that is, $\beta(\theta)$ is the probability that the realization t of the test statistic T will not fall within the critical region when $\theta \in \mathcal{P}_1$ is true. However, to work exclusively with $\mathcal{R}$, we calculate $P(\theta) = 1 - \beta(\theta)$—the probability that $T \in \mathcal{R}$ given θ or that the sample realization of T lies within the critical region when θ is true.

We noted in Section 12.6 that the classical approach to hypothesis testing involves choosing a small value for $\alpha = P(TIE)$ and then finding, for a fixed sample size n, the critical region $\mathcal{R}$ for which $\beta = P(TIIE)$ is at a minimum.[2] But since $P(\theta) = 1 - \beta(\theta)$, this means that we must choose $\mathcal{R}$ to maximize $P(\theta)$ for

---

[2]   Given that the α-risk (the probability of committing a Type I Error) is pegged in advance or controlled for, any time we reject the null hypothesis H_0 we immediately know the probability that we have made

all θ implied by H_1. So for all tests having the same α-risk, we seek the test whose power function is closest to the ideal power function or, equivalently, we seek the test with the smallest β value, which is the same as determining a test that maximizes the power $P(\theta)$. In sum, the power function assists us in identifying *good* tests when the alternative hypothesis is composite. So given a set $\mathcal{P}_1$ of alternative competing values of θ, the power function enables us to determine their plausibility relative to the null value θ_0.

What does it mean for a test to have *high power*? It was mentioned earlier that the power of a test represents its ability to detect a true alternative hypothesis or $P(\theta) = P(\text{accept } H_1 | \theta \in \mathcal{P}_1)$. That is, the notion of *power* reflects a test's ability to discriminate between a hypothetical situation ($H_0 : \theta = \theta_0$) and the true one ($H_1 : \theta = \theta_1 (\neq \theta_0)$). A test of *high power* has a much greater chance of detecting a false null hypothesis H_0 than one of low power. So for a given sample size and α-risk the larger the difference between θ_0 and θ_1 (and thus the smaller the probability of not rejecting H_0 when it is false), the *more powerful* is the test of H_0. In sum, since for each $\theta \neq \theta_0$ the value of the power function is the probability of making a correct decision, we desire to have such values as close to unity as possible; that is, we desire a test of *high power*.

We previously specified the power function $P(\theta) = P(\text{reject } H_0 | H_0 \text{ false})$ as (12.43), where $\beta(\theta) = P(TIIE) = P(\text{do not reject } H_0 | H_0 \text{ false})$. If $\theta = \mu$, then the power function may be determined by first calculating β for various values of μ satisfying the alternative hypothesis H_1 and then subtracting each of the resulting β's from unity to obtain the corresponding powers at those μ's. Then a plot of $P(\mu) = 1 - \beta(\mu)$ versus μ traces out the power curve.

Example 12.15.1   Suppose that X is $N(\mu, 15)$. For $n = 25$, $\bar{X}$ is $N(\mu, 3)$ and $\bar{Z} = \frac{\bar{X} - \mu}{3}$ is $N(0, 1)$. Let us consider $H_0 : \mu = \mu_0 = 20$ versus $H_1 : \mu > 20$ with $\alpha = 0.05$. Then from (12.8.1), $\mathcal{R} = \{\bar{x} | \bar{x} \geq \bar{x}_c = 20 + 1.645(3) = 24.935\}$ and thus $\bar{\mathcal{R}} = \{\bar{x} | \bar{x} < \bar{x}_c = 24.935\}$. To calculate β, we need to choose $\mu = \mu_1$ which satisfies H_1 or, from (12.18),

$$\beta(\mu) = P\left(\bar{Z} < \frac{24.935 - \mu_1}{3} \Big| \mu = \mu_1 \text{ true}\right). \tag{12.44}$$

For instance, if $\mu = \mu_1 = 25$, then $\beta(25) = P(\bar{Z} < -0.021 | \mu = 25 \text{ true}) = 0.4920$. And for $\mu = 21$, $\beta(21) = P(\bar{Z} < 1.31 | \mu = 21 \text{ true}) = 0.9049$.

In like fashion various other values of $\beta(\mu)$ for μ satisfying H_1 are computed using (12.44) and are presented in Table 12.2. For instance, if the true mean were $\mu = \mu_1 = 19$, then we would correctly reject the false null hypothesis about 2% of the time; and if the true mean were $\mu = \mu_1 = 24$, then we would correctly reject the false null hypothesis about 38% of the time. The graph of the power function or the power curve is provided by Figure 12.11a. It should be apparent that as the

---

an error. However, this is not the case for a Type II Error; i.e., if we do not reject H_0 and it is false, then we do not know the probability that we have made an error since the β-risk is not controlled for.

Table 12.2

	Determination of $\beta(\mu)$ and $P(\mu) = 1 - \beta(\mu)$ for a One-Sided Alternative Hypothesis	
μ	$\beta(\mu) = P(TIIE)$	Power: $P(\mu) = 1 - \beta(\mu)^*$
19	0.9761	0.0239
20	0.9495	$0.0505(= \alpha)$
21	0.9049	0.0951
22	0.8340	0.1660
23	0.7389	0.2611
24	0.6217	0.3783
25	0.4920	0.5080
26	0.3632	0.6368
27	0.2451	0.7549
28	0.1533	0.8467
29	0.0869	0.9131
30	0.0455	0.9545

*If the true mean were equal to μ, we would correctly reject the false null hypothesis about $100P(\mu)\%$ of the time.

difference between the null value μ_0 under H_0 and the true value μ_1 under H_1 increases, the power of the test concomitantly increases.

A glance at Figure 12.11a reveals that for $\mu > 26$, it is highly likely that the test will reject H_0 and for $\mu < 24$, it is highly unlikely that the test will reject H_0. For $\mu = 25$, there is about a 50-50 chance that H_0 will be rejected. Remember that the *power of a test* reflects its likelihood of detecting the true alternative

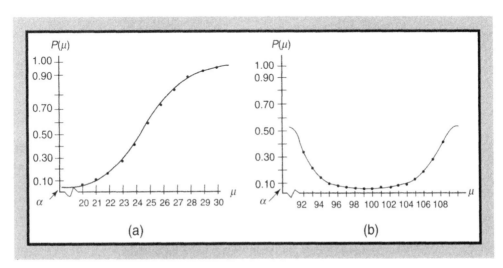

(a)   **(b)**

Figure 12.11   (a) The power curve for selected values of μ appearing in Table 12.2; (b) The power curve for selected values of μ appearing in Table 12.3.

hypothesis. So the further μ_1 under H_1 gets from μ_0 under H_0, the easier it becomes to discriminate between μ_1 and μ_0. Since *power* for a given $\mu = \mu_1$ value yields the probability of making no error, we prefer the test with the highest power associated with it.

Note that since $\mathcal{R} = \{\bar{x}|\bar{x} \geq 24.935\}$, we may use (12.42) directly by forming

$$P(\mu) = P(\overline{X} \geq 24.935|\mu = \mu_1 \text{ true})$$

$$= P\left(\overline{Z} \geq \frac{24.935 - \mu_1}{3}\middle| \mu = \mu_1 \text{ true}\right) = 1 - F\left(\frac{24.935 - \mu_1}{3}\right), \quad (12.45)$$

where again μ_1 must be chosen to satisfy the alternative hypothesis H_1. (Remember that $F(-z) = 1 - F(z)$, where F represents the standard normal cumulative distribution function.) So how is (12.45) to be interpreted? The power function for the test of $H_0 : \mu = 20$ is the function $P(\mu)$ whose value at $\mu = \mu_1$ gives the probability that the test will reject H_0 if in fact $\mu = \mu_1$ is true. ∎

We next turn to the determination of the power function for a test involving a two-sided alternative hypothesis.

Example 12.15.2   For X a $N(\mu, 40)$ random variable, $n = 150$, $\alpha = 0.01$, $H_0 : \mu = 100$ and $H_1 : \mu \neq 100$, let us employ (12.42) to obtain the associated two-tailed power function. For a two-sided alternative hypothesis we know that $\mathcal{R} = \{\bar{x}|\bar{x} \leq \bar{x}_{cl} \text{ or } \bar{x} \geq \bar{x}_{cu}\}$, where, from (12.13),

$$\bar{x}_{cl} = 100 - 2.58\left(\frac{40}{\sqrt{150}}\right) = 91.57, \quad \bar{x}_{cu} = 100 + 2.58\left(\frac{40}{\sqrt{150}}\right) = 108.42.$$

Then from (12.42),

$$P(\mu) = P(\overline{X} \leq 91.57 \text{ or } \overline{X} \geq 108.42|\mu = \mu_1 \text{ true})$$

$$= P(\overline{X} \leq 91.57|\mu = \mu_1 \text{ true}) + P(\overline{X} \geq 108.42|\mu = \mu_1 \text{ true})$$

$$= P\left(\overline{Z} \leq \frac{91.57 - \mu_1}{3.266}\middle| \mu = \mu_1 \text{ true}\right) + P\left(\overline{Z} \geq \frac{108.42 - \mu_1}{3.266}\middle| \mu = \mu_1 \text{ true}\right),$$

$$(12.46)$$

where $\overline{Z} = (\overline{X} - \mu)/(\sigma/\sqrt{n})$ is $N(0, 1)$ and $\mu = \mu_1$ satisfies H_1. If $\mu = \mu_1 = 98$, $P(98) = P(\overline{Z} \leq -1.968|\mu = 98 \text{ true}) + P(\overline{Z} \geq 3.190|\mu = 98 \text{ true}) = 0.0244 + 0.0000 = 0.0244$. And for $\mu = \mu_1 = 105$, $P(105) = P(\overline{Z} \leq -4.112|\mu = 105 \text{ true}) + P(\overline{Z} \geq 1.047|\mu = 105 \text{ true}) = 0.0000 + 0.1469 = 0.1469$. Table 12.3 contains other computed values of the power function for a variety of μ's satisfying H_1. A plot of $P(\mu)$ versus μ then enables us to trace out the power curve for this test (see Figure 12.11b). ∎

Table 12.3

Determination of $P(\mu)$ for a Two-Sided Alternative Hypothesis	
μ	Power: $P(\mu)^*$
92	0.4483
93	0.3300
94	0.2297
95	0.1469
96	0.0869
97	0.0485
98	0.0244
99	0.0136
100	$0.0100\,(=\alpha)$
101	0.0136
102	0.0244
103	0.0485
104	0.0869
105	0.1469
106	0.2297
107	0.3300
108	0.4483

*If the true mean were equal to μ, we would correctly reject the false null hypothesis about $100P(\mu)\%$ of the time.

Example 12.15.3   Suppose we extract a random sample of size $n = 100$ from a binomial population. Given $\alpha = 0.05$, let us determine the power function for the test of $H_0: p = p_0 = 0.80$ against $H_1: p < 0.80$. Under H_0, we know that $Z_{\hat{p}}^0$ (given by (12.29)) is approximately $N(0, 1)$ so that the standard normal distribution will be used to obtain the power of the test for various values of p satisfying H_1. Let $\mathcal{R} = \{\hat{p}/\hat{p} \leq \hat{p}_c\} = \{\hat{p}/\hat{p} \leq 0.7342\}$, where $\hat{p}_c = p_0 - z_\alpha \sqrt{\frac{p_0(1-p_0)}{n}} = 0.80 - 1.645\sqrt{\frac{0.80(0.20)}{100}} = 0.7342$. Then from (12.42) the power of this test is

$$P(p) = P(\hat{P} \leq 0.7342 | p = p_1 \text{ true}) = P\left(Z_{\hat{p}} \leq \left.\frac{0.7342 - p_1}{\sqrt{\frac{p_1(1-p_1)}{n}}}\right| p = p_1 \text{ true}\right).$$

$$(12.47)$$

If $p_1 = 0.70$, then from (12.47), $P(0.70) = P(Z_{\hat{p}} \leq 0.7467 | p = 0.70 \text{ true}) = 0.7703$. Table 12.4 exhibits other computed values of the power function for various p's satisfying H_1. A plot of $P(p)$ against p traces out the power curve for this test (see Figure 12.12). ∎

We close this section by offering a few comments concerning the main factors that affect the power function, namely the level of significance α and the sample

Table 12.4

Determination of $P(p)$ for a One-Sided Alternative Hypothesis	
p	Power: $P(p)$*
0.64	0.9750
0.67	0.9131
0.70	0.7703
0.73	0.5398
0.76	0.2757
0.80	$0.0500\,(=\alpha)$
0.83	0.0054
0.86	0.0000

*If the true proportion were equal to p, we would correctly reject the false null hypothesis about $100P(p)\%$ of the time.

size n. These remarks are framed in the context of conducting hypothesis tests for the population mean μ.

First, for a fixed n and a given alternative hypothesis H_1, as α increases, the power of the test increases *over all possible true values of μ satisfying H_1* (see Figure 12.13a,b) and thus $\beta = P(TIIE)$ decreases. For example, as α increases, $z_{\alpha/2}$ decreases and thus the critical values $\bar{x}_{cl}$ and $\bar{x}_{cu}$ change in a fashion such that $\mathcal{R}$ gets larger and $\overline{\mathcal{R}}$ gets smaller (here $\bar{x}_{cl}$ increases and $\bar{x}_{cu}$ decreases).

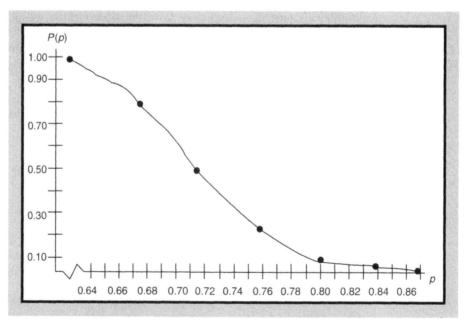

Figure 12.12   The power curve for selected values of p appearing in Table 12.4.

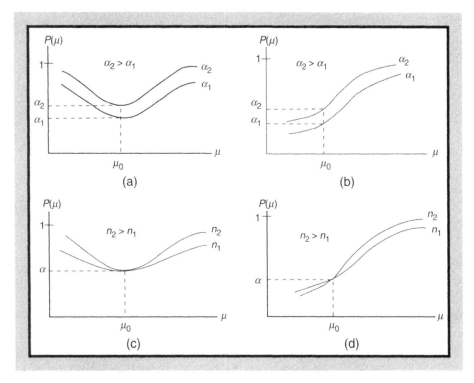

Figure 12.13   The main factors affecting the power function are α and n.

Next, for a fixed α and a given H_1, as n increases, the size of the Type II Error, β, is reduced. But since *power* $= 1 - \beta$, the power function will thus increase as n increases (see Figure 12.13c,d).

In fact, as n increases, the *shape* of the two-tailed power function is affected; its shape tends toward that of the ideal power function depicted in Figure 12.10b. Here also $\bar{x}_{cl}$ increases and $\bar{x}_{cu}$ decreases (or $\mathcal{R}$ gets larger and $\overline{\mathcal{R}}$ gets smaller) since $\sigma_{\bar{x}} = \frac{\sigma}{\sqrt{n}}$ decreases as n increases. This latter observation leads to an important point regarding the way in which the power of a test might be enhanced—it is best to increase the power of the test by increasing n rather than by increasing the α-risk.

The preceding comments enable us to gain additional insight as to why the power function is important—it enables us to determine, for fixed n, whether or not there exists a satisfactory trade-off between the α- and β-risk associated with a test; and, for a given α-risk, whether or not the trade-off between the β-risk level and sample size n is tolerable.

One final consideration merits our attention. It is important to note that the respective powers of one- and two-tailed tests will be different given the same α-risk and the same true alternative μ_1 under H_1. For instance, if a one-tail test is used and the true alternative μ is in the direction of the critical region $\mathcal{R}$, then the

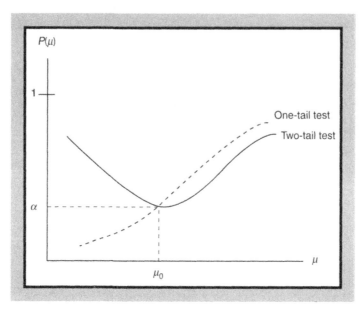

Figure 12.14 The powers of one- and two-tailed tests will be different given the same α and μ under H_1.

one-tail test is more powerful than the two-tail test taken over all relevant values of μ (see Figure 12.14).

12.16 Determining the Best Test for a Statistical Hypothesis

How do we construct what will be called a *best* test of a parametric hypothesis? To answer this question let us first assume that both the null and alternative hypotheses are simple in form. Then for a random sample of size n, we choose to test $H_0: \theta = \theta_0$ against $H_1: \theta = \theta_1$ so that the parameter space $\mathcal{P}$ has only two elements or $\mathcal{P} = \{\theta_1, \theta_2\}$. (For this type of test, both $\alpha = P(TIE)$ and $\beta = P(TIIE)$ are well-defined unique values.) In effect, we seek to determine if our sample random variables $X_1, \ldots, X_n$ came from one of two unique distributions that are completely specified as $f^0(x; \theta_0)$ or $f^1(x; \theta_1)$, respectively. Hence we are equivalently testing $H_0 : X_i$ is distributed as $f^0(x; \theta_0)$, against $H_1 : X_i$ is distributed as $f^1(x; \theta_1)$, $i = 1, \ldots, n$. As we shall now see, the specification of a test as *best* is ultimately based upon the characteristics of its power function $P(\theta)$.

Constructing the *best* test is equivalent to determining the best critical region $\mathcal{R}$ of size α. In this regard, we should choose $\mathcal{R}$ so that $\alpha = P(\theta_0)$ is a (small) fixed number and $\beta = P(TIIE)$ is as small as possible. But since $\beta = 1 - P(\theta_1)$, we seek the test or critical region which maximizes $P(\theta_1)$; that is, we seek a *most powerful* test. In sum, to obtain the best test of $H_0 : \theta = \theta_0$ versus $H_1 : \theta = \theta_1$, select the test with the smallest β-risk among all tests with Type I Error not exceeding some specified α-risk level.

More formally, a best or most powerful test must satisfy the following.

Best Test Criterion: A test with power function $P(\theta)$ of $H_0 : \theta = \theta_0$ versus $H_1 : \theta = \theta_1$ is a *most powerful test of size* α if and only if:

(a) $P(\theta_0) = \alpha$

(b) $P(\theta_1) > \overline{P}(\theta_1)$ for any alternative test

    with power function $\overline{P}$ for which $\overline{P}(\theta_0) \leq \alpha$

$$(12.48)$$

Hence a most powerful test of size α itself has size α under H_0 and, under H_1, has the largest power among all other tests of at most size α. Equivalently, a most powerful test of size α itself has size α and has the smallest β among all other tests of at most size α. To implement this criterion, we simply fix the size of the Type I Error, α, and minimize the size of the Type II Error, β.

It is instructive to note that the preceding Best Test Criterion can be recast in terms of specifying a best or most powerful critical region $\mathcal{R} \subset \mathcal{T}$, where $\mathcal{T}$ is the entire set of realizations of an estimator T of θ. To this end we equivalently have the following.

Best Critical Region Criterion: A critical region $\mathcal{R} \subset \mathcal{T}$ is termed a *most powerful critical region of size* α for testing $H_0 : \theta = \theta_0$ versus $H_1 : \theta = \theta_1$ if, for every subset $\mathcal{A} \subset \mathcal{T}$ for which $P_{\mathcal{A}}(\theta_0) = P((X_1, \ldots, X_n) \in \mathcal{A}/\theta = \theta_0 \text{ true}) = \alpha$:

(a) $P_{\mathcal{R}}(\theta_0) = P((X_1, \ldots, X_n) \in \mathcal{R}/\theta = \theta_0 \text{ true}) = \alpha$

(b) $P_{\mathcal{R}}(\theta_1) = P((X_1, \ldots, X_n) \in \mathcal{R}/\theta = \theta_1 \text{ true}) \geq P_{\mathcal{A}}(\theta_1)$

    $= P((X_1, \ldots, X_n) \in \mathcal{A}/\theta = \theta_1 \text{ true})$

$$(12.49)$$

Hence a most powerful critical region of size α itself has size α under H_0 and, under H_1, has the greatest power among all other subsets $\mathcal{A} \subset \mathcal{T}$ having size α.

The following theorem (known as the Neyman-Pearson Lemma) serves as a guide for finding a best or most powerful test of size α by providing us with a set of sufficient conditions for the existence of a best or most powerful critical region of size α when both the null and alternative hypothesis are simple in form. In particular, the theorem is used to determine the *form* of the critical region $\mathcal{R}$ for best tests (the *actual* $\mathcal{R}$ depends upon the value of α chosen). That is, we seek the $\mathcal{R}$ that is *most powerful* with respect to the alternative hypothesis H_1.

To set the stage for the Neyman-Pearson Lemma, let $X_1, \ldots, X_n$ denote a set of sample random variables taken from a population with a probability density function $f(x; \theta)$. Then the joint probability density function of $X_1, \ldots, X_n$ is the likelihood function $\mathcal{L}(\theta; x_1, \ldots, x_n, n) = \mathcal{L}(\theta; x, n) = \prod_{i=1}^{n} f(x_i; \theta)$. Given $\mathcal{L}$, the theorem employs the *likelihood ratio*

$$\lambda = \Lambda(x_1, \ldots, x_n, n) = \frac{\mathcal{L}_0}{\mathcal{L}_1} = \frac{\mathcal{L}(\theta_0; x, n)}{\mathcal{L}(\theta_1; x, n)}. \qquad (12.50)$$

We may now state Theorem 12.1, the Neyman-Pearson Lemma.

THEOREM 12.1.   Let $X_1, \ldots, X_n$ be a random sample from the probability density function $f(x; \theta)$, where the value of the parameter θ is restricted to only one of two possible elements of the parameter set $\mathcal{P} = \{\theta_0, \theta_1\}$ and $\alpha = P(TIE)$ is fixed. For k a positive constant and $\mathcal{R} \in \mathcal{T}$, let:

(a) $P\big((X_1, \ldots, X_n) \in \mathcal{R}/\theta = \theta_0 \text{ true}\big) = \alpha$

(b) $\lambda = \dfrac{\mathcal{L}_0}{\mathcal{L}_1} \leq k$ if $(X_1, \ldots, X_n) \in \mathcal{R}$ and

$$\lambda = \frac{\mathcal{L}_0}{\mathcal{L}_1} \geq k \ \text{ if } \ (X_1, \ldots, X_n) \in \overline{\mathcal{R}} = \mathcal{T} - \mathcal{R} \qquad (12.51)$$

Then $\mathcal{R}$ is a best critical region of size α (or the test corresponding to the critical region $\mathcal{R}$ is a most powerful test of size α) for testing $H_0 : \theta = \theta_0$ versus $H_1 : \theta = \theta_1$.

Here k must be *chosen* so as to make $\mathcal{R}$ a critical region of size α. As required by (12.51b), the likelihood ratio λ should be small for sample realizations within $\mathcal{R}$ (which yield Type I Errors when $\theta = \theta_0$ and correct decisions when $\theta = \theta_1$); and λ should be large for sample realizations within $\overline{\mathcal{R}}$ (which yield correct decisions when $\theta = \theta_0$ and Type II Errors when $\theta = \theta_1$).

For given α and n, the Neyman-Pearson Lemma provides us with a method for determining the best or most powerful critical region and the best test of the simple null versus simple alternative hypothesis specified earlier. The theorem guarantees, among all possible critical regions of size α, the smallest possible β. So for fixed α and n, if a critical region $\mathcal{R}$ minimizes the value of β, then it is termed the *best critical region of size* α. And since a best critical region maximizes the power of the test, it is alternatively called a *most powerful critical region of size* α; and the corresponding test is termed a *most powerful test of size* α. To see exactly how Theorem 12.1 is utilized, let us consider the following set of example problems.

Example 12.16.1   Suppose that $\{X_1, \ldots, X_n\}$ constitutes a set of sample random variables selected from a probability density function given by

$$f(x; \theta) = \begin{cases} \theta e^{-\theta x}, & \theta > 0, \ x \geq 0; \\ 0 & \text{elsewhere.} \end{cases} \qquad (12.52)$$

Then the likelihood function for the sample random variables appears as

$$\mathcal{L}(\theta; x_1, \ldots, x_n, n) = \prod_{i=1}^{n} f(x_i; \theta) = \theta^n e^{-\theta \sum\limits_{i=1}^{n} x_i}. \qquad (12.53)$$

To find the most powerful critical region of size α for a test of, say, $H_0 : \theta = 1$ versus $H_1 : \theta = 2$, let us employ (12.51b); that is, let us set

$$\lambda = \frac{\mathcal{L}_0}{\mathcal{L}_1} = \frac{e^{-\sum\limits_{i=1}^{n} x_i}}{2^n e^{-2\sum\limits_{i=1}^{n} x_i}} \leq k \quad \text{or} \quad e^{\sum\limits_{i=1}^{n} x_i} \leq k2^n.$$

Then upon transforming this expression to logarithms, the form of $\mathcal{R}$ for the most powerful critical region is

$$\sum_{i=1}^{n} x_i \leq \ln(k2^n) = k'. \tag{12.54}$$

Hence $\mathcal{R} = \{y | y = \sum_{i=1}^{n} x_i \leq k'\}$; that is, the best critical region, as given by (12.54), constitutes that portion of the y-axis under the left-hand tail of the sampling distribution of Y with $\theta = 1$. Here k is chosen so that $P(Y \leq k' | H_0$ true$) = \alpha$. We are thus provided with the best or most powerful test of $H_0 : \theta = 1$ versus $H_1 : \theta = 2$. ∎

Example 12.16.2 Let the set of sample random variables $X_1, \ldots, X_n$ be drawn from a normal distribution with unknown mean μ and known variance σ^2. Our objective is to determine the best critical region of size $\alpha = 0.05$ for testing $H_0 : \mu = \mu_0$, against $H_1 : \mu = \mu_1$, where $\mu_1 > \mu_0$. Given that the likelihood function for the sample random variables appears as

$$\mathcal{L}(\mu; x_1, \ldots, x_n, n) = \frac{1}{(\sqrt{2\pi}\sigma)^n} e^{-\frac{1}{2\sigma^2} \sum\limits_{i=1}^{n} (x_i - \mu)^2},$$

it is easily shown that (12.51b) requires

$$\lambda = \frac{\mathcal{L}_0}{\mathcal{L}_1} = e^{\frac{1}{2\sigma^2} \left[\sum\limits_{i=1}^{n} (x_i - \mu_1)^2 - \sum\limits_{i=1}^{n} (x_i - \mu_0)^2 \right]} \leq k. \tag{12.55}$$

Upon transforming (12.55) to logarithms we have

$$\sum_{i=1}^{n} (x_i - \mu_1)^2 - \sum_{i=1}^{n} (x_i - \mu_0)^2 \leq 2\sigma^2 \ln k$$

or, upon simplifying the left-hand side of this expression,

$$n(\mu_1^2 - \mu_0^2) - 2(\mu_1 - \mu_0) \sum_{i=1}^{n} x_i \leq 2\sigma^2 \ln k. \tag{12.55.1}$$

Since $\mu_1 - \mu_0 > 0$, (12.55.1) may be rewritten as

$$\sum_{i=1}^{n} x_i \geq \frac{n(\mu_1^2 - \mu_0^2) - 2\sigma^2 \ln k}{2(\mu_1 - \mu_0)}$$

or

$$\bar{x} \geq \frac{n(\mu_1^2 - \mu_0^2) - 2\sigma^2 \ln k}{2n(\mu_1 - \mu_0)} = \bar{x}_c. \tag{12.56}$$

As promised by Theorem 12.1, (12.56) specifies the *form* of the best critical region of size α for testing H_0 versus H_1. Since $\mu_1 > \mu_0$, the best critical region of size α is located under the right-hand tail of the sampling distribution of $\bar{X}$ and appears as $\mathcal{R} = \{\bar{x}/\bar{x} \geq \bar{x}_c\}$, where $\bar{x}_c$ denotes the critical sample mean or critical value of $\bar{x}$ (see (12.8.1)). That is, for H_0 true, $P(\bar{X} \geq \bar{x}_c | \mu = \mu_0 \text{ true}) = 0.05$ as required and thus, since $\bar{X}$ is $N(\mu_0, \frac{\sigma}{\sqrt{n}})$,

$$P\left(\bar{Z} \geq \frac{\bar{x}_c - \mu_0}{\sigma/\sqrt{n}} \middle| \mu = \mu_0 \text{ true}\right) = P(Z \geq 1.645 | \mu = \mu_0 \text{ true}) = 0.05$$

or, equivalently, $\mathcal{R} = \{z | z \geq 1.645\}$. So with $\frac{\bar{x}_c - \mu_0}{\sigma/\sqrt{n}} = 1.645$, it follows that $\bar{x}_c = \mu_0 + 1.645\frac{\sigma}{\sqrt{n}}$ (in general, $\bar{x}_c = \mu_0 + z_\alpha \frac{\sigma}{\sqrt{n}}$). So once α is specified, the precise value of $\bar{x}_c$ is easily obtained. (It should be apparent that if we had stipulated $\mu_1 < \mu_0$, then Theorem 12.1 or the Neyman-Pearson Lemma yields a best or most powerful critical region of the form $\mathcal{R} = \{\bar{x} | \bar{x} \leq \bar{x}_c\}$, where now $\bar{x}_c = \mu_0 - z_\alpha \frac{\sigma}{\sqrt{n}}$.)

Whereas the probability of rejecting H_0 when H_0 is true is α, the probability of rejecting H_0 when H_0 is false is the power of the test at $\mu = \mu_1$; for example for $\mu_0 = 0$, $\mu_1 = 1$, and $\sigma = 1$, $\bar{X}$ is $N(0, \frac{1}{\sqrt{n}})$ and thus

$$P(\mu) = P(\bar{X} \geq \bar{x}_c | \mu = \mu_1 \text{ true}) = \int_{\bar{x}_c}^{+\infty} \frac{1}{\sqrt{2\pi}(1/\sqrt{n})} e^{-\frac{1}{2}\left(\frac{\bar{x}-1}{1/\sqrt{n}}\right)^2} d\bar{x}.$$

(Note that $P(\mu) = P(\bar{X} \geq \bar{x}_c | \mu = \mu_0 \text{ true}) = \alpha$.) For $n = 36$ and $\alpha = 0.05$, $\bar{x}_c = \mu_0 + z_\alpha \frac{\sigma}{\sqrt{n}} = \frac{1.645}{6} = 0.2741$. Hence the power of this best or most powerful test of H_0 versus H_1 when H_1 is true is

$$P(\mu) = \int_{0.2741}^{+\infty} \frac{1}{\sqrt{2\pi}(1/\sqrt{n})} e^{-\frac{1}{2}\left(\frac{\bar{x}-1}{1/\sqrt{n}}\right)^2} d\bar{x}$$

$$= \int_{0.2741-1}^{+\infty} \frac{1}{\sqrt{2\pi}(1/\sqrt{n})} e^{-\frac{1}{2}\left(\frac{\bar{x}}{1/\sqrt{n}}\right)^2} d\bar{x} \qquad \text{(see (7.n4.1))}$$

$$= \frac{1}{\sqrt{n}} \int_{(0.2741-1)\sqrt{n}}^{+\infty} \frac{1}{\sqrt{2\pi} \, (1/\sqrt{n})} e^{-\frac{1}{2}z^2} dz \qquad \text{(see (7.n4.2))}$$

$$= \frac{1}{\sqrt{2\pi}} \int_{-4.3554}^{+\infty} e^{-\frac{1}{2}z^2} dz = 0.9999$$

since Z is $N(0,1)$. Then $\beta(\mu_1) = 1 - P(\mu_1)$ would be less than 1%. ∎

Note that the application of the Neyman-Pearson Lemma leads to the same form for the critical region as that obtained for the Case I set of hypotheses in Section 12.9. Hence the Case I critical region derived in Section 12.9 is, via (12.56), a best critical region of size α for $H_0 : \mu = \mu_0$ versus $H_1 : \mu > \mu_0$.

A similar conclusion emerges when we consider the Case II set of hypotheses also introduced in Section 12.9; that is, the Neyman-Pearson Lemma enables us to conclude that a best critical region of size α for $H_0 : \mu = \mu_0$ versus $H_1 : \mu < \mu_0$ is of the form $\mathcal{R} = \{\bar{x}/\bar{x} \le \bar{x}_c\}$, where again $\bar{x}_c = \mu_0 - z_\alpha \frac{\sigma}{\sqrt{n}}$. Hence our intuitive specification of an upper-tail critical region for Case I and a lower-tail critical region for Case II is now legitimized by the Neyman-Pearson Lemma.

Before considering some additional examples pertaining to the application of the Neyman-Pearson Lemma, let us introduce some additional terminology. Specifically, a test of the null hypothesis $H_0 : \theta = \theta_0$ is termed a *uniformly most powerful test of size* α if, for each θ subsumed under the alternative hypothesis H_1, it is at least as powerful as any other test of at most size α. In this regard, the power function of any such test is at least as large at that of any other test of at most size α for every θ covered by H_1. And since $\beta = P(TIIE) = 1 - power$, it follows that, for a fixed α, a uniformly most powerful test is one that has a β smaller than any other test regardless of the value of θ, which happens to be true under the alternative hypothesis. If a uniformly most powerful test exists, then, among all tests of at most size α, the uniformly most powerful test has the greatest chance of rejecting the null hypothesis if rejection is warranted. And it is the Neyman-Pearson Lemma that offers a way of finding uniformly most powerful tests, provided that such tests exist.

The Neyman-Pearson Lemma holds for the case where both the null and alternative hypotheses are of the simple variety. If we opt to test $H_0 : \theta = \theta_0$ against the composite alternative $H_1 : \theta > \theta_0$ (or $H_1 : \theta < \theta_0$), there is no general theorem like the Neyman-Pearson Lemma that can be utilized to locate a best or most powerful critical region. However, the Neyman-Pearson Lemma can be utilized to obtain a best critical region for any *single value* of $\theta_1 > \theta_0$ covered by H_1. Since in most cases the form of $\mathcal{R}$ for the most powerful test does not depend upon the particular θ_1 chosen, it follows that when a test obtained via the Neyman-Pearson Lemma maximizes the power for every $\theta > \theta_0$, it is said to be a uniformly most powerful test of size α for $H_0 : \theta = \theta_0$ against $H_1 : \theta > \theta_0$ (or $H_1 : \theta < \theta_0$).

For example, if we are to determine a uniformly most powerful critical region of size α for the test $H_0 : \mu = \mu_0$ against $H_1 : \mu > \mu_0$ (μ is unknown and σ^2 is assumed to be known), our starting point is the search for the most powerful critical region of size α for the test of $H_0 : \mu = \mu_0$ versus $H_1 : \mu = \mu_1$ for some particular $\mu_1 > \mu_0$. Then according to the Neyman-Pearson Lemma, the most powerful critical region of size α is an upper-tail region of the form $\mathcal{R} = \{\bar{x} | \bar{x} \geq \bar{x}_c\}$, where the precise value of $\bar{x}_c$ is determined by fixing $\alpha = P(\overline{X} \geq \bar{x}_c | \mu = \mu_0 \text{ true})$. And since the form of $\mathcal{R}$ does not depend upon μ_1, any $\mu > \mu_1$ leads to exactly the same type of critical region. Hence we have found the uniformly most powerful critical region of size α, and thus the uniformly most powerful test of size α of $H_0 : \mu = \mu_0$ versus $H_1 : \mu > \mu_0$.

The preceding argument also applies to the test of $H_0 : \mu = \mu_0$ against $H_1 : \mu < \mu_0$. In this instance we obtain the lower-tail critical region $\mathcal{R} = \{\bar{x} | \bar{x} \leq \bar{x}_c\}$ as the uniformly most powerful critical region of size α for all $\mu_1 < \mu_0$ and thus the test of $H_0 : \mu = \mu_0$ versus $H_1 : \mu < \mu_0$ is the uniformly most powerful test of size α.

In sum, a critical region $\mathcal{R}$ is a uniformly most powerful critical region of size α for testing a simple null hypothesis H_0 against a one-sided composite alternative hypothesis H_1 if $\mathcal{R}$ is a best or most powerful critical region of size α for testing H_0 against *each* simple hypothesis covered by H_1. But this means that the power function of the test corresponding to the said critical region should be at least as great as the power function of any other test with the same α-risk taken over *each* simple hypothesis covered by H_1; that is, the test associated with critical region $\mathcal{R}$ is a uniformly most powerful test of size α for testing simple H_0 against composite H_1.

If we wish to test $H_0 : \theta = \theta_0$ against $H_1 : \theta \neq \theta_0$, no uniformly most powerful test exists; that is, there is no single critical region that represents the most powerful test for all values of $\theta = \theta_1 \neq \theta_0$. Hence the Neyman-Pearson Lemma is of no value for two-sided alternative hypotheses.

It is also the case that the Neyman-Pearson Lemma cannot be used when testing hypotheses about a single parameter θ when the population probability density function contains other unspecified parameters. For example, when we tested $H_0 : \mu = \mu_0$ against an alternative hypothesis (simple or composite) and the sample random variables were extracted from a probability density function that was specified as $N(\mu, \sigma)$, we always assumed that the population variance σ^2 was known. But if σ^2 is unknown, then $H_0 : \mu = \mu_0$ does not uniquely specify the form of the population probability density function and thus the null hypothesis is not simple in form as required by the Neyman-Pearson Lemma.

Up to this point in our derivation of hypothesis tests we have been assuming that the null hypothesis is of the simple variety. However, it may be the case that the null hypothesis is composite; for example we may test $H_0 : \theta \leq \theta_0$, against $H_1 : \theta > \theta_0$ (or $H_0 : \theta \geq \theta_0$ against $H_1 : \theta < \theta_0$). Again let the population probability density function contain only a single unspecified parameter, namely θ. For this test, let us define the α-risk to be the probability of a Type I Error when $\theta = \theta_0$; that is, α is the value of the power function for $\theta = \theta_0$ true. Hence this α level is taken to be the maximum value of the power function for $\theta \leq \theta_0$. In this regard, the test of $H_0 : \theta = \theta_0$ against $H_1 : \theta > \theta_0$ is also the uniformly most powerful test for $H_0 : \theta \leq \theta_0$ versus $H_1 : \theta > \theta_0$.

Example 12.16.3   To continue with our set of examples pertaining to the application of the Neyman-Pearson Lemma (Theorem 12.1), let us posit that a collection of sample random variables $X_1, \ldots, X_n$ is drawn from the Bernoulli probability density function

$$f(X;p) = \begin{cases} p^X(1-p)^{1-X}, & X = 0 \quad \text{or} \quad 1; \\ 0 & \text{elsewhere,} \end{cases} \tag{12.57}$$

where the unknown parameter p represents Bernoulli probability and $0 < p < 1$. Let us derive a uniformly most powerful critical region for testing $H_0 : p = p_0$ against $H_1 : p > p_0$ for some specific α. Given that the likelihood function for the sample random variables appears as

$$\mathcal{L}(p; x_1, \ldots, x_n, n) = p^{\sum_{i=1}^{n} x_i} (1-p)^{n - \sum_{i=1}^{n} x_i},$$

it follows from (12.51) that for some value $p_1 > p_0$,

$$\lambda = \frac{\mathcal{L}_0}{\mathcal{L}_1} = \left(\frac{p_0}{p_1}\right)^{\sum_{i=1}^{n} x_i} \left(\frac{1-p_0}{1-p_1}\right)^{n - \sum_{i=1}^{n} x_i} \leq k$$

or

$$\left[\frac{p_0(1-p_1)}{p_1(1-p_0)}\right]^{\sum_{i=1}^{n} x_i} \left(\frac{1-p_0}{1-p_1}\right)^{n} \leq k.$$

Upon transforming this latter expression to logarithms, we have

$$\sum_{i=1}^{n} x_i \ln\left[\frac{p_0(1-p_1)}{p_1(1-p_0)}\right] + n \ln\left(\frac{1-p_0}{1-p_1}\right) \leq \ln k.$$

or

$$\sum_{i=1}^{n} x_i \geq \frac{\ln k - n \ln\left(\frac{1-p_0}{1-p_1}\right)}{\ln\left[\frac{p_0(1-p_1)}{p_1(1-p_0)}\right]} = k'$$

(here $\ln[p_0(1-p_1)/p_1(1-p_0)] < 0$ since $[p_0(1-p_1)/p_1(1-p_1)] < 1$). Hence the Neyman-Pearson Lemma informs us that the best or most powerful critical region for testing $H_0 : p = p_0$ against $H_1 : p = p_1$ appears as $\mathcal{R} = \{\sum_{i=1}^{n} x_i | \sum_{i=1}^{n} x_i \geq k'\}$; that is, we reject H_0 if the number of successes $\sum_{i=1}^{n} x_i$ is larger than k'. Moreover, for each $p_1 > p_0$, the preceding argument holds; for example, if $p_2 > p_0$, then $\mathcal{R}$ is a most powerful critical region of size α for testing $H_0 : p = p_0$ versus $H_1 : p = p_2$.

Thus $\mathcal{R}$ must be a uniformly most powerful critical region of size α for testing $H_0 : p = p_0$, against $H_1 : p > p_0$. (It is important to note that since X is a discrete random variable, it may not be possible to specify a test that has an α equal to some predetermined level. Hence a *best* test will be the one for which the $P(TIE) = P(\sum_{i=1}^n X_i \geq k'/H_0 \text{ true})$ is closest to the chosen α). ∎

Example 12.16.4 Suppose that $\{X_1, \ldots, X_n\}$ represents a set of sample random variables taken from a $N(0, \sigma)$ distribution, where $\sigma (> 0)$ is unknown. Our objective is to determine a uniformly most powerful test or critical region of size α for testing $H_0 : \sigma^2 = \sigma_0^2$, against $H_1 : \sigma^2 > \sigma_0^2$. For this collection of sample random variables the likelihood function has the form

$$\mathcal{L}(\sigma^2; x_1, \ldots, x_n, n) = \left(\frac{1}{2\pi\sigma^2}\right)^{\frac{n}{2}} e^{-\frac{1}{2\sigma^2}\sum_{i=1}^n x_i^2}.$$

Then for $\sigma_1^2 > \sigma_0^2$, (12.51) requires that

$$\lambda = \frac{\mathcal{L}_0}{\mathcal{L}_1} = \left(\frac{\sigma_1^2}{\sigma_0^2}\right)^{\frac{n}{2}} e^{-\left(\frac{\sigma_1^2 - \sigma_0^2}{2\sigma_0^2\sigma_1^2}\right)\sum_{i=1}^n x_i^2} \leq k$$

or, upon transforming to logarithms,

$$\sum_{i=1}^n x_i^2 \geq \left(\frac{2\sigma_0^2\sigma_1^2}{\sigma_1^2 - \sigma_0^2}\right)\left[\frac{n}{2}\ln\left(\frac{\sigma_1^2}{\sigma_0^2}\right) - \ln k\right] = k'.$$

Then via the Neyman-Pearson Lemma, the best or most powerful critical region for testing $H_0 : \sigma^2 = \sigma_0^2$, against $H_1 : \sigma^2 = \sigma_1^2$ is $\mathcal{R} = \{\sum_{i=1}^n x_i^2 | \sum_{i=1}^n x_i^2 \geq k'\}$. We know from Theorem 9.2 that, under H_0, the quantity $Y_0 = \sum_{i=1}^n X_i^2/\sigma_0^2$ is χ_n^2. Hence we require that $P(Y_0 \geq \frac{k'}{\sigma_0^2}|H_0 \text{ true}) = \alpha$. (For instance, if $n = 25$, $\alpha = 0.05$, and $\sigma_0^2 = 4$, then from Table A.4, we can easily obtain $k'/4 = 37.6525$ or $k' = 150.61$. Then $\mathcal{R} = \{\sum_{i=1}^n x_i^2 | \sum_{i=1}^n x_i^2 \geq 150.61\}$ is a best critical region of size $\alpha = 0.05$ for testing $H_0 : \sigma^2 = 4$, against $H_1 : \sigma^2 = \sigma_1^2 > 4$.) Now, we know from our preceding discussion that $\mathcal{R}$ is a best or most powerful critical region of size α for *each* $\sigma_1^2 > \sigma_0^2$; for example, if $\sigma^2 = \sigma_2^2 > \sigma_0^2$, then $\mathcal{R}$ is a most powerful critical region of size α when testing $H_0 : \sigma^2 = \sigma_0^2$ versus $H_1 : \sigma^2 = \sigma_2^2$. In this regard, $\mathcal{R}$ must be a uniformly most powerful critical region of size α for testing $H_0 : \sigma^2 = \sigma_0^2$ against $H_1 : \sigma^2 > \sigma_0^2$. ∎

One final example pertaining to the specification of uniformly most powerful critical regions or tests is warranted.

Example 12.16.5 We noted earlier that for a two-sided alternative hypothesis, no uniformly most powerful test exists; that is, there is no single critical region that represents the most powerful test for all values of a parameter covered by any such alternative hypothesis. To verify this assertion, let our collection of sample random variables be extracted from a $N(\mu, 1)$ population with μ unknown. To test $H_0 : \mu = \mu_0$, against $H_1 : \mu \neq \mu_0$, let us select some $\mu_1 \neq \mu_0$ and set $\sigma = 1$ in (12.55.1) so as to obtain

$$n(\mu_1^2 - \mu_0^2) - 2(\mu_1 - \mu_0) \sum_{i=1}^{n} x_i \leq 2 \ln k. \tag{12.58}$$

If $\mu_1 > \mu_0$, then (12.58) can be rewritten as

$$\bar{x} \geq \frac{\mu_1 + \mu_0}{2} - \frac{\ln k}{n(\mu_1 - \mu_0)} \tag{12.59}$$

(remember that $\mu_1^2 - \mu_0^2 = (\mu_1 - \mu_0)(\mu_1 + \mu_0)$); and if $\mu_1 < \mu_0$, then (12.58) becomes

$$\bar{x} \leq \frac{\mu_1 + \mu_0}{2} - \frac{\ln k}{n(\mu_1 - \mu_0)}. \tag{12.60}$$

So for $\mu_1 > \mu_0$, (12.59) specifies an upper-tail most powerful critical region of size α for testing $H_0 : \mu = \mu_0$, against $H_1 : \mu = \mu_1$; and for $\mu_1 < \mu_0$, (12.60) determines a lower-tail most powerful critical region of size α for testing $H_0 : \mu = \mu_0$, against $H_1 : \mu = \mu_1$. Clearly, we have no uniformly most powerful critical region of size α. ∎

12.17 Generalized Likelihood Ratio Tests

We determined in the preceding section that the Neyman-Pearson Lemma would enable us to identify the best or most powerful critical region (and thus the best or most powerful test) when both the null and alternative hypotheses were simple in form and the population probability density function involved only a single unknown parameter θ; that is, all other relevant parameters had specified values. When the population distribution involves more than a single unknown parameter, then a more general method, also involving a ratio of likelihood functions, is needed to derive *good* tests of hypotheses in the sense that $\beta = P(TIIE)$ is small for a given $\alpha = P(TIE)$. The technique that follows is appropriate for simple or composite hypotheses and can be applied in the instance where multiple unspecified parameters exist in the population probability density function.

The generalized likelihood ratio technique offered next is basically an extension, via the method of maximum likelihood, of the Neyman-Pearson method for specifying critical regions or tests. Generalized likelihood ratio tests typically yield best or most powerful critical regions of a fixed size α.

Suppose that the sample random variables $X_1, \ldots, X_n$ are drawn from a population probability density function involving an unknown parameter θ as well as other unknown parameters. Then the likelihood function of the sample is also a function of θ as well as other unspecified parameters. As denoted earlier ((12.1)), consider $H_0 : \theta \in \mathcal{P}_0$ versus $H_1 : \theta \in \mathcal{P}_1$, where $\mathcal{P}_1 = \overline{\mathcal{P}_0} = \mathcal{P} - \mathcal{P}_0$, $\mathcal{P}_0 \cap \mathcal{P}_1 = \emptyset$, and $\mathcal{P} = \mathcal{P}_0 \cup \mathcal{P}_1$. Here we are assuming that at least one of these two competing hypotheses, H_0 and H_1, is composite.

Let us employ the sample realizations $x_1, \ldots, x_n$ to estimate all unknown parameters by the method of maximum likelihood (see Section 8.5). If $\hat{\mathcal{P}}_0$ depicts the parameter set containing the maximum likelihood estimates of all unknown parameters subject to the restriction $\theta \in \mathcal{P}_0$ and $\hat{\mathcal{P}}$ is the parameter set containing the maximum likelihood estimates of all unknown parameters determined for any $\theta \in \mathcal{P}$, then, provided that maximum likelihood estimates exist, the *generalized likelihood ratio* can be written as

$$\hat{\lambda} = \Gamma(x_1, \ldots, x_n, n) = \frac{\max\limits_{\mathcal{P}_0} \mathcal{L}(\theta)}{\max\limits_{\mathcal{P}} \mathcal{L}(\theta)} = \frac{\mathcal{L}(\hat{\mathcal{P}}_0)}{\mathcal{L}(\hat{\mathcal{P}})}, \quad 0 \le \hat{\lambda} \le 1, \tag{12.61}$$

where $\hat{\lambda}$, the sample realization of the likelihood ratio test statistic $\Gamma(X_1, \ldots, X_n, n)$, serves as the basis for conducting the *Generalized Likelihood Ratio Test*:

A likelihood ratio test of $H_0 : \theta \in \mathcal{P}_0$ against $H_1 : \theta \in \mathcal{P}_1$ has as its critical region the set $\mathcal{R} = \{\hat{\lambda} | \hat{\lambda} \le k\}$, where, for H_0 simple, the constant k, $0 \le k \le 1$, is chosen so that the size of $\mathcal{R}$ is α; and for H_0 composite, k is chosen so that $P(TIE) \le \alpha$ for all $\theta \in \mathcal{P}_0$ and $P(TIE) = \alpha$ for at least one value of $\theta \in \mathcal{P}_0$.

(For $\mathcal{P} = \{\theta_0, \theta_1\}$, the generalized likelihood ratio $\hat{\lambda}$ does not reduce to the simple likelihood ratio λ given by (12.50).) Note that the attained maximum in the numerator of (12.61) is computed over all parameters under the null hypothesis and thus represents a constrained maximum of the likelihood function, where the constraint set is $\mathcal{P}_0$; the attained maximum in the denominator is determined for all parameters over $\mathcal{P}$ and also depicts a constrained maximum of the likelihood function, where the constraint set is $\mathcal{P}$. If, however, the maximum of the likelihood function occurs at an interior point of $\mathcal{P}$, then the denominator is simply the likelihood function evaluated at the ordinary maximum likelihood parameter estimates. Given that $0 \le \hat{\lambda} \le 1$, a value of $\hat{\lambda}$ near zero indicates that the likelihood of the sample is small under H_0 relative to its value under H_1. Hence the sample outcome favors H_1 over H_0 and thus we are inclined to reject H_0. A value of $\hat{\lambda}$ near unity implies that we may place considerable faith in the reasonableness of H_0; that is, the likelihood of the sample cannot be increased much by allowing the parameters to assume values other than those dictated by H_0. Hence we should not be inclined to reject H_0.

Example 12.17.1   To see exactly how the generalized likelihood ratio technique works, let us extract a set of sample random variables $\{X_1, \ldots, X_n\}$ from a $N(\mu, \sigma)$ population, where the mean μ is unknown but the variance σ^2 is known. We seek to determine the critical region for a test of $H_0 : \mu = \mu_0$ versus $H_1 : \mu \neq \mu_0$. Using the preceding parameter-set notation, we have $\mathcal{P}_0 = \{\mu_0\}$, $\mathcal{P}_1 = \{\mu| -\infty < \mu(\neq\mu_0) < +\infty\}$, and $\mathcal{P}$ is the set of all real numbers. Given that $\mathcal{P}_0$ has only one element μ_0, it follows that $\hat{\mathcal{P}}_0 = \{\mu_0\}$ (no maximum likelihood estimate of μ is required). And since $\mathcal{P}$ is the set of all real numbers, $\hat{\mathcal{P}}$ contains the ordinary or unconstrained maximum likelihood estimate $\hat{\mu}$ of μ or $\hat{\mathcal{P}} = \{\bar{x}\}$. For the normal probability density function, the likelihood function for the sample random variables is

$$\mathcal{L}(\mu; x_1, \ldots, x_n, n) = (2\pi\sigma^2)^{-\frac{n}{2}} e^{-\frac{1}{2\sigma^2} \sum\limits_{i=1}^{n}(x_i - \mu)^2}.$$

Then from (12.61),

$$\hat{\lambda} = \frac{\mathcal{L}(\hat{\mathcal{P}}_0)}{\mathcal{L}(\hat{\mathcal{P}})} = \frac{e^{-\frac{1}{2\sigma^2} \sum\limits_{i=1}^{n}(x_i - \mu_0)^2}}{e^{-\frac{1}{2\sigma^2} \sum\limits_{i=1}^{n}(x_i - \bar{x})^2}}. \tag{12.62}$$

Since $\sum_{i=1}^{n}(x_i - \mu_0)^2 = \sum_{i=1}^{n}(x_i - \bar{x})^2 + n(\bar{x} - \mu_0)^2$, (12.62) becomes

$$\hat{\lambda} = e^{-\frac{n}{2\sigma^2}(\bar{x} - \mu_0)^2}. \tag{12.62.1}$$

Then according to the generalized likelihood ratio test procedure, the critical region involves $\hat{\lambda} = e^{-\frac{n}{2\sigma^2}(\bar{x} - \mu_0)^2} \leq k$ or $(\bar{x} - \mu_0)^2 \geq -\frac{2\sigma^2}{n} \ln k = k'$, where k' is chosen so that the size of the critical region is α. To this end, let us replace the preceding inequality by $|\bar{x} - \mu_0| \geq k'' = \sqrt{k'}$. Now, as k varies between 0 and 1, k'' varies between 0 and $+\infty$. Hence the generalized likelihood ratio test rejects the null hypothesis if $\bar{x}$ differs from μ_0 by more than a specified amount. That is, since the random variable $\overline{X}$ is $N(\mu, \frac{\sigma}{\sqrt{n}})$, (12.11.1) has us choose $k'' = z_{\alpha/2}\frac{\sigma}{\sqrt{n}}$ so that H_0 is rejected if $\bar{x} \leq \mu_0 - z_{\alpha/2}\frac{\sigma}{\sqrt{n}}$ or $\bar{x} \geq \mu_0 + z_{\alpha/2}\frac{\sigma}{\sqrt{n}}$ (see (12.12)). ∎

Example 12.17.2   Suppose that the set of sample random variables $\{X_1, \ldots, X_n\}$ is taken from a $N(\mu, \sigma)$ population, where both the mean μ and variance σ^2 are unknown. Suppose we decide to test $H_0 : \mu = \mu_0$ against $H_1 : \mu > \mu_0$. In terms of the preceding notation, $\mathcal{P}_0 = \{\mu_0\}$, $\mathcal{P}_1 = \{\mu|\mu > \mu_0\}$, and $\mathcal{P} = \{\mu|\mu \geq \mu_0\}$. Given these parameter sets, we need to determine both $\mathcal{L}(\hat{\mathcal{P}}_0)$ and $\mathcal{L}(\hat{\mathcal{P}})$. First, let us find the maximum likelihood estimates of μ and σ^2 under the restriction that $\mu \in \mathcal{P}_0$. Since $\mathcal{P}_0$ contains only a single parameter value, it follows that $\mu = \mu_0$ (as in the preceding example, no maximum likelihood estimate is required). Also, when $\mu = \mu_0$, the maximum likelihood estimate of σ^2 is $\hat{\sigma}_0^2 = \frac{1}{n}\sum_{i=1}^{n}(x_i - \mu_0)^2$ (see (10.30b) and the accompanying discussion). Second, to find the maximum likelihood estimates of μ and σ^2 over $\mathcal{P}$, we see from (10.30a) that the ordinary or unrestricted maximum likelihood estimate of μ is $\hat{\mu} = \bar{x}$. (However, the maximum likelihood estimate of μ over $\mathcal{P}$ is $\hat{\mu} = \max\{\bar{x}, \mu_0\}$; that is, if the maximum of $\mathcal{L}$ occurs at an interior point of $\mathcal{P}$, then $\hat{\mu} = \mu_0$.) And from (10.30b), the maximum likelihood estimate of σ^2 in $\mathcal{P}$ is $\hat{\sigma}^2 = \frac{1}{n}\sum_{i=1}^{n}(x_i - \bar{x})^2$. Given that the population random variable is normally distributed, the likelihood function for the sample random variables is

$$\mathcal{L}(\mu, \sigma; x_1, \ldots, x_n, n) = (2\pi\sigma^2)^{-\frac{n}{2}} e^{-\frac{1}{2\sigma^2}\sum_{i=1}^{n}(x_i - \mu)^2}.$$

To determine the likelihood ratio (12.61), $\mathcal{L}(\hat{\mathcal{P}}_0)$ is obtained by evaluating $\mathcal{L}$ at $\mu = \mu_0$ and $\sigma^2 = \hat{\sigma}_0^2$ and $\mathcal{L}(\hat{\mathcal{P}})$ is obtained by evaluating $\mathcal{L}$ at $\mu = \hat{\mu} = \bar{x}$ and $\sigma^2 = \hat{\sigma}^2$; that is,

$$\mathcal{L}(\hat{\mathcal{P}}_0) = (2\pi)^{-n/2}(\hat{\sigma}_0^2)^{-n/2}e^{-n/2}, \quad \mathcal{L}(\hat{\mathcal{P}}) = (2\pi)^{-n/2}(\hat{\sigma}^2)^{-n/2}e^{-n/2},$$

so that

$$\hat{\lambda} = \frac{\mathcal{L}(\hat{\mathcal{P}}_0)}{\mathcal{L}(\hat{\mathcal{P}})} = \left(\frac{\hat{\sigma}^2}{\hat{\sigma}_0^2}\right)^{\frac{n}{2}} = \begin{cases} \left(\dfrac{\sum_{i=1}^{n}(x_i - \bar{x})^2}{\sum_{i=1}^{n}(x_i - \mu_0)^2}\right)^{\frac{n}{2}} & \text{if} \quad \bar{x} > \mu_0; \\ 1 & \text{if} \quad \bar{x} \leq \mu_0. \end{cases} \qquad (12.63)$$

Let us assume that $\bar{x} > \mu_0$ so that $\hat{\lambda} \leq k < 1$. Then the critical region for the generalized likelihood ratio test is $\hat{\lambda} \leq k$ or, from (12.63),

$$\frac{\sum_{i=1}^{n}(x_i - \bar{x})^2}{\sum_{i=1}^{n}(x_i - \mu_0)^2} < k^{2/n} = k'$$

or

$$\frac{n(\bar{x} - \mu_0)^2}{\sum_{i=1}^{n}(x_i - \bar{x})^2} > \frac{1}{k'} - 1 = k''$$

(see the derivation of (12.62.1)) or

$$\frac{n(\bar{x} - \mu_0)^2}{\dfrac{\sum_{i=1}^{n}(x_i - \bar{x})^2}{(n-1)}} = \frac{n(\bar{x} - \mu_0)^2}{s^2} > (n-1)k''$$

so that we ultimately obtain

$$\frac{\sqrt{n}(\bar{x} - \mu_0)}{s} > \sqrt{(n-1)k''} = k'''. \tag{12.64}$$

Under the null hypothesis $H_0: \mu = \mu_0$, we may view (12.64) as the sample realization of the test statistic

$$T = \frac{\sqrt{n}(\overline{X} - \mu_0)}{S},$$

where T follows the Student's t distribution with $n-1$ degrees of freedom (see (9.19.1)). Hence this generalized likelihood ratio test is equivalent to the ordinary t test for the population mean (Section 12.12). ∎

Example 12.17.3   Continuing with our generalized likelihood ratio tests, let us assume that the collection of sample random variables $X_1, \ldots, X_n$ is drawn from a $N(\mu, \sigma)$ population, where the population mean μ is known and the population variance σ^2 is unknown. Suppose we decide to test $H_0: \sigma^2 \leq \sigma_0^2$, against $H_1: \sigma^2 > \sigma_0^2$. We now have $\mathcal{P}_0 = \{\sigma^2 | \sigma^2 \leq \sigma_0^2\}$, $\mathcal{P}_1 = \{\sigma^2 | \sigma^2 > \sigma_0^2\}$, and $\mathcal{P} = \{\sigma^2 | 0 < \sigma^2 < +\infty\}$. Given that the population probability density function is normal, the likelihood function for the set of sample random variables is

$$\mathcal{L}(\sigma^2; x_1, \ldots, x_n, n) = (2\pi\sigma^2)^{-\frac{n}{2}} e^{-\frac{1}{2\sigma^2} \sum_{i=1}^{n}(x_i - \mu)^2}.$$

To determine $\mathcal{L}(\hat{\mathcal{P}}_0)$, we must maximize the likelihood function $\mathcal{L}$ subject to the restriction that $\sigma^2 \leq \sigma_0^2$. We know that the ordinary maximum likelihood estimates of μ and σ^2 are $\hat{\mu} = \bar{x}$ and $\hat{\sigma}^2 = \frac{1}{n}\sum_{i=1}^{n}(x_i - \bar{x})^2$, respectively. So if $\hat{\sigma}^2 \leq \sigma_0^2$, then $\hat{\sigma}^2$ is the maximizing value for σ^2; and if $\hat{\sigma}^2 > \sigma_0^2$, then the maximizing value of σ^2 is σ_0^2. Hence

$$\mathcal{L}(\hat{\mathcal{P}}_0) = \begin{cases} (2\pi\hat{\sigma}^2)^{-\frac{n}{2}} e^{-\frac{n}{2}} & \text{if} \quad \hat{\sigma}^2 \leq \sigma_0^2; \\ (2\pi\sigma_0^2)^{-\frac{n}{2}} e^{-\frac{1}{2\sigma_0^2} \sum_{i=1}^{n}(x_i - \bar{x})^2} & \text{if} \quad \hat{\sigma}^2 > \sigma_0^2. \end{cases}$$

Since $\mathcal{P}$ is the set of all admissible σ^2's, $\hat{\mathcal{P}}$ contains the ordinary maximum likelihood estimates of μ and σ^2 or $\hat{\mu} = \bar{x}$ and $\hat{\sigma}^2 = \frac{1}{n}\sum_{i=1}^{n}(x_i - \bar{x})^2$, respectively. Thus $\mathcal{L}(\hat{\mathcal{P}}) = (2\pi\hat{\sigma}^2)^{-n/2} e^{-n/2}$. Then from (12.61)

$$\hat{\lambda} = \frac{\mathcal{L}(\hat{\mathcal{P}}_0)}{\mathcal{L}(\hat{\mathcal{P}})} = \begin{cases} 1 & \text{if} \quad \hat{\sigma}^2 \leq \sigma_0^2; \\ \left(\frac{\hat{\sigma}^2}{\sigma_0^2}\right)^{\frac{n}{2}} e^{\frac{n}{2} - \frac{1}{2\sigma_0^2} \sum_{i=1}^{n}(x_i - \bar{x})^2} & \text{if} \quad \hat{\sigma}^2 > \sigma_0^2. \end{cases} \tag{12.65}$$

Suppose $\hat{\sigma}^2 > \sigma_0^2$ so that $\hat{\lambda} \leq k < 1$. Then the critical region for the generalized likelihood ratio test is $\hat{\lambda} \leq k$ or, from (12.65),

$$\left(\frac{\sum_{i=1}^{n}(x_i - \bar{x})^2}{n\sigma_0^2} \right)^{\frac{n}{2}} e^{\frac{n}{2} - \frac{1}{2\sigma_0^2}\sum_{i=1}^{n}(x_i-\bar{x})^2} \leq k. \tag{12.65.1}$$

Let us set $y = \frac{\sum_{i=1}^{n}(x_i-\bar{x})^2}{n\sigma_0^2}$. Then from (12.65.1), the expression $\hat{\lambda} = y^{n/2}e^{\frac{n}{2} - \frac{n}{2}y}$ has a maximum at $y = 1$ ($\hat{\lambda}' = 0$ and $\hat{\lambda}'' = -\frac{n}{2} < 0$ at $y = 1$). Then we should reject H_0 if $y = \frac{\hat{\sigma}^2}{\sigma_0^2} > 1$ and $y^{\frac{n}{2}}e^{\frac{n}{2} - \frac{n}{2}y} \leq k$. But $y^{\frac{n}{2}}e^{\frac{n}{2} - \frac{n}{2}y} \leq k$ is equivalent to $y > c$. So if H_0 is true, $\frac{\sum_{i=1}^{n}(X_i-\bar{X})^2}{\sigma_0^2}$ is a chi-square random variable with $n - 1$ degrees of freedom (see Theorem 9.2) and thus $P\left(Y = \frac{\sum_{i=1}^{n}(X_i-\bar{X})^2}{n\sigma_0^2} > c\right) = P(\chi_{n-1}^2 > cn)$. Given that we require $P(TIE) = \alpha$, let us choose $cn = \chi_{1-\alpha,n-1}^2$, where $\chi_{1-\alpha,n-1}^2$ is the $100(1 - \alpha)$ percentile of the chi-square distribution with $n - 1$ degrees of freedom (see (9.7.1)). Then if we take $c' = \sigma_0^2\chi_{1-\alpha,n-1}^2$, we will reject H_0 if $\sum_{i=1}^{n}(x_i-\bar{x})^2 > c'$. For example, if we choose to test $H_0 : \sigma^2 \leq 0.60$ versus $H_1 : \sigma^2 > 0.60$ with $n = 30$ and $\alpha = 0.05$, then $\chi_{1-\alpha,n-1}^2 = \chi_{0.95,29}^2 = 42.5569$. Hence we will reject H_0 if $\sum_{i=1}^{n}(x_i - \bar{x})^2 > 0.60(42.5569) = 25.53$. ∎

We note briefly that if the preceding test involves $H_0 : \sigma^2 \geq \sigma_0^2$ versus $H_1 : \sigma^2 < \sigma_0^2$, then we would reject H_0 if $\sum_{i=1}^{n}(x_i - \bar{x})^2 < \sigma_0^2\chi_{\alpha,n-1}^2$. And if we are testing $H_0 : \sigma^2 = \sigma_0^2$ against $H_1 : \sigma^2 \neq \sigma_0^2$, then we would reject H_0 if $\sum_{i=1}^{n}(x_i - \bar{x})^2 < \sigma_0^2\chi_{\alpha/2,n-1}^2$ or $\sum_{i=1}^{n}(x_i - \bar{x})^2 > \sigma_0^2\chi_{1-(\alpha/2),n-1}^2$.

Example 12.17.4 Suppose that the sample random variables $X_1,\ldots,X_n$ are taken from a Bernoulli probability density function (12.57) with unknown population parameter p, $0 \leq p \leq 1$, and that we are to test $H_0 : p = p_0$ against $H_1 : p \neq p_0$. Here $\mathcal{P}_0 = \{p_0\}$, $\mathcal{P}_1 = \{p | 0 \leq p(\neq p_0) \leq 1\}$, and $\mathcal{P} = \{p | 0 \leq p \leq 1\}$. Then $\hat{\mathcal{P}}_0 = \{p_0\}$, and $\hat{\mathcal{P}}$ contains the ordinary or unconstrained maximum likelihood estimate of p or $\hat{p} = \frac{x}{n} = \bar{x}$, where $x = \sum_{i=1}^{n}x_i$ is the realized number of successes in a sample of size n. Since the likelihood function of the sample random variables appears as

$$\mathcal{L}(p; x_1,\ldots,x_n, n) = p^{\sum_{i=1}^{n}x_i}(1-p)^{n-\sum_{i=1}^{n}x_i},$$

we have

$$\mathcal{L}(\hat{\mathcal{P}}_0) = p_0^{\sum_{i=1}^{n}x_i}(1-p_0)^{n-\sum_{i=1}^{n}x_i},$$

and

$$\mathcal{L}(\hat{\mathcal{P}}) = \bar{x}^{\sum\limits_{i=1}^{n} x_i} (1 - \bar{x})^{n - \sum\limits_{i=1}^{n} x_i}.$$

Then from (12.61),

$$\hat{\lambda} = \frac{\mathcal{L}(\hat{\mathcal{P}}_0)}{\mathcal{L}(\hat{\mathcal{P}})} = \left(\frac{p_0}{\bar{x}}\right)^{\sum\limits_{i=1}^{n} x_i} \left(\frac{1 - p_0}{1 - \bar{x}}\right)^{n - \sum\limits_{i=1}^{n} x_i}. \qquad (12.66)$$

As required by the generalized likelihood ratio test, the critical region has the form $\hat{\lambda} \le k$ or

$$\sum_{i=1}^{n} x_i \left[\ln\left(\frac{p_0}{\bar{x}}\right) - \ln\left(\frac{1 - p_0}{1 - \bar{x}}\right)\right] \le \ln k - n \ln\left(\frac{1 - p_0}{1 - \bar{x}}\right) = k'$$

or $x = \sum_{i=1}^{n} x_i \ge k''$ if $p_0 < \bar{x}$. Similarly, $x = \sum_{i=1}^{n} x_i \le k'''$ if $p_0 > \bar{x}$. So under H_0, the generalized likelihood ratio test has us reject H_0 if the realized number of successes exceeds k'' or falls short of k'''. Since we require that $P(TIE) = \alpha$, we will reject H_0 if $x \ge k_{\alpha/2}$ or $x \le k'_{\alpha/2}$, where $k_{\alpha/2}$ is the *smallest* integer for which $\sum_{y=k_{\alpha/2}}^{n} b(y; n, p_0) \le \frac{\alpha}{2}$ ($b(y; n, p_0)$ via (6.14.1) is the probability of obtaining y successes in n trials of a simple alternative experiment when $p = p_0$) and $k'_{\alpha/2}$ is the *largest* integer for which $\sum_{y=0}^{k'_{\alpha/2}} b(y; n, p_0) \le \frac{\alpha}{2}$.

If our test involves $H_0 : p \le p_0$ versus $H_1 : p > p_0$, then the critical region implied by the generalized likelihood ratio criterion is $\mathcal{R} = \{x | x \ge k_\alpha\}$, where k_α is the *smallest* integer for which $\sum_{y=k_\alpha}^{n} b(y; n, p_0) \le \alpha$. But if we are to test $H_0 : p \ge p_0$ against $H_1 : p < p_0$, then the critical region has the form $\mathcal{R} = \{x | x \le k'\}$, where k'_α is the *largest* integer for which $\sum_{y=0}^{k'_\alpha} b(y; n, p_0) \le \alpha$. ∎

We noted earlier in the statement of the generalized likelihood ratio test that if the null hypothesis H_0 is simple, then the constant k is chosen so that the size of the critical region or $P(TIE) = \alpha$. In this instance, if $g(\lambda)$ depicts the (continuous) probability density function of the statistic $\Gamma(X_1, \ldots, X_n, n)$, then, under H_0, k must satisfy the requirement $\int_0^k g(\lambda/H_0 \ true)d\lambda = \alpha$ provided that g does not depend upon any unknown parameters. However, it may be the case that the generalized likelihood ratio test statistic $\Gamma(X_1, \ldots, X_n, n)$ does not exhibit a probability density function having a well-known form (such as the standard normal or t distribution). As we shall now see, for large samples, we can obtain an approximation to the sampling distribution of $\Gamma(X_1, \ldots, X_n, n)$ by using the chi-square statistic to determine a critical region of size α; that is, we have a way of choosing k so that $P(\Gamma \le k/\theta = \theta_0 \ true) = \alpha$. Specifically, Theorem 12.2

provides the form of the asymptotic distribution of the generalized likelihood ratio statistic Γ:

THEOREM 12.2.   Let the sample random variables $X_1, \ldots, X_n$ be drawn from a population probability density function $f(x; \theta)$, which obeys certain regularity conditions (expressed mainly in terms of restrictions on the derivatives with respect to θ of the sample likelihood function $\mathcal{L}$). For $H_0 : \theta \in \mathcal{P}_0$ true, the statistic $-2 \ln \Gamma(X_1, \ldots, X_n, n)$ converges to a chi-square distribution as $n \to \infty$; that is, the limiting distribution of $-2 \ln \Gamma$ is χ_v^2, where degrees of freedom v = number of parameters assigned specific numerical values under the null hypothesis H_0.

In this regard, for large n, the statistic $-2 \ln \Gamma$ has approximately a chi-square distribution with v degrees of freedom. Then $P(TIE)$ is approximately α if n is large; that is, $\lim_{n \to \infty} P(reject\ H_0 / \theta \in \mathcal{P}_0) = \alpha$ so that the critical region $\mathcal{R} = \{\hat{\lambda} / -2 \ln \Gamma(x_1, \ldots, x_n, n) = -2 \ln \hat{\lambda} \geq \chi_{1-\alpha, v}^2\}$ specifies an *asymptotic test of size α*, where $\chi_{1-\alpha, v}^2$ is the $100(1 - \alpha)$ percentile of the chi-square distribution with v degrees of freedom. As the structure of $\mathcal{R}$ reveals, the rejection of $H_0 : \theta \in \mathcal{P}_0$ for small values of $\hat{\lambda}$ is equivalent to the rejection of H_0 for large values of $-2 \ln \hat{\lambda}$ since $-2 \ln \hat{\lambda}$ increases as $\hat{\lambda}$ decreases.

Looking to the value of degrees of freedom v, we note first that for $H_0 : \mu = \mu_0$, $v = 1$; for example, from (12.61.1), $-2 \ln \hat{\lambda} = \frac{n}{\sigma^2} (\bar{x} - \mu_0)^2 = (\frac{\bar{x} - \mu_0}{\sigma / \sqrt{n}})^2$ is actually a realization of a random variable having a chi-square distribution with 1 degree of freedom. And if $H_0 : \mu = \mu_0$ and $\sigma^2 = \sigma_0^2$, then $v = 2$. In general, if the population probability density function depends upon r parameters $\theta_1, \ldots, \theta_r$ and $H_0 : \theta_1 = \theta_1', \theta_2 = \theta_2', \ldots, \theta_s = \theta_s'$, $s < r$, then $v = s$.

Example 12.17.5   Suppose we choose to test $H_0 : p = p_0 = 0.4$ against $H_1 : p \neq 0.4$ with $n = 130$, $x = 78$, and $\alpha = 0.05$. In this instance the critical region is $\mathcal{R} = \{\hat{\lambda} / -2 \ln \hat{\lambda} > \chi_{0.95, 1}^2 = 3.8416\}$. Since $\bar{x} = \frac{78}{130} = 0.6$, it follows from (12.66) that

$$\hat{\lambda} = \left(\frac{0.4}{0.6}\right)^{78} \left(\frac{0.6}{0.4}\right)^{52} = 0.000026$$

and thus $-2 \ln \hat{\lambda} = 21.08 > 3.8416$. So by virtue of the asymptotic decision rule incorporated in the specification of $\mathcal{R}$, we reject the null hypothesis at the 5% level. ∎

We end this section by briefly discussing a few important characteristics of hypothesis tests. First, it should be intuitively reasonable to expect that a test should be more likely to reject $H_0 : \theta \in \mathcal{P}_0$ if $\theta \in \mathcal{P}_1$ than if $\theta \in \mathcal{P}_0$. That is, a test with power function $P(\theta)$ is said to be *unbiased* if $P(\theta_1) \geq P(\theta_0)$ for every $\theta_1 \in \mathcal{P}_1$ and $\theta_0 \in \mathcal{P}_0$. Hence the probability of rejecting H_0 when it is false is at least as large as the probability of rejecting H_0 when it is true. If we are testing a simple null hypothesis $H_0 : \theta = \theta_0$ against a composite alternative, then a test is unbiased if the power function assumes its minimum value at $\theta = \theta_0$. Hence the probability of rejecting H_0 when it is false is at least as large as the probability of rejecting H_0 when it is true. If we are testing a simple null hypothesis $H_0 : \theta = \theta_0$ against a composite alternative, then a test is unbiased if the power function assumes its minimum value at $\theta = \theta_0$. Hence the probability of rejecting H_0 is least when H_0 is true.

Example 12.17.6 Suppose that we draw a set of sample random variables $\{X_1, \ldots, X_n\}$ from a $N(\mu, \sigma)$ population with μ unknown and σ^2 known and we decide to perform a generalized likelihood ratio test of $H_0 : \mu \leq \mu_0$ versus $H_1 : \mu > \mu_0$. Here $\mathcal{P}_0 = \{\mu | \mu \leq \mu_0\}$, $\mathcal{P}_1 = \{\mu | \mu > \mu_0\}$, and $\mathcal{P} = \varepsilon \mu | -\infty < \mu < +\infty\}$. We note first that the unconstrained maximum likelihood estimate $\hat{\mu}$ of μ over $\mathcal{P}$ is simply $\hat{\mu} = \bar{x}$ and thus $\hat{\mathcal{P}} = \{\bar{x}\}$. Additionally, if $\bar{x} < \mu_0$, then $\hat{\mathcal{P}}_0 = \{\bar{x}\}$; and if $\bar{x} \geq \mu_0$, then $\hat{\mathcal{P}}_0 = \{\mu_0\}$. Hence

$$
\mathcal{L}(\hat{\mathcal{P}}_0) = \begin{cases} (2\pi\sigma^2)^{\frac{n}{2}} e^{-\frac{1}{2\sigma^2} \sum\limits_{i=1}^{n} (x_i - \mu_0)^2} & \text{if} \quad \bar{x} \geq \mu_0; \\[4mm] (2\pi\sigma^2)^{-\frac{n}{2}} e^{-\frac{1}{2\sigma^2} \sum\limits_{i=1}^{n} (x_i - \bar{x})^2} & \text{if} \quad \bar{x} < \mu_0 \end{cases}
$$

and

$$
\mathcal{L}(\hat{\mathcal{P}}) = (2\pi\sigma^2)^{-\frac{n}{2}} e^{-\frac{1}{2\sigma^2} \sum\limits_{i=1}^{n} (x_i - \bar{x})^2}.
$$

Then from (12.61), (12.62), and (12.62.1),

$$
\hat{\lambda} = \frac{\mathcal{L}(\hat{\mathcal{P}}_0)}{\mathcal{L}(\hat{\mathcal{P}})} = \begin{cases} e^{-\frac{n}{2\sigma^2}(\bar{x} - \mu_0)^2} & \text{if} \quad \bar{x} \geq \mu_0; \\[2mm] 1 & \text{if} \quad \bar{x} < \mu_0. \end{cases}
$$

Then according to the generalized likelihood ratio test criterion, we reject the null hypothesis if $\hat{\lambda} \leq k$ or $\frac{\bar{x} - \mu_0}{\sigma/\sqrt{n}} \geq k'$. So for $k < 1$, rejecting H_0 when $\hat{\lambda} \leq k$ is equivalent to rejecting H_0 when $\bar{z}_0 = \frac{\bar{x} - \mu_0}{\sigma/\sqrt{n}} \geq k'$. Then the power function for this test is

$$
P(\mu) = P\left(\bar{Z}_0 = \frac{\overline{X} - \mu_0}{\sigma/\sqrt{n}} \geq k' \right) = P\left(\frac{\overline{X} - \mu}{\sigma/\sqrt{n}} \geq k' + \frac{\mu_0 - \mu}{\sigma/\sqrt{n}} \right)
$$

$$
= \left(\bar{Z} \geq k' + \frac{\mu_0 - \mu}{\sigma/\sqrt{n}} \right),
$$

where $\overline{Z}$ is $N(0, 1)$. For fixed μ_0, $P(\mu)$ is an increasing function of μ so that $P(\mu) > P(\mu_0)$ for all $\mu > \mu_0$. (Note that $P(\mu_0) = P(TIE) = \alpha$ if $P(\bar{Z} \geq k') = \alpha$.) Hence this test is unbiased. ∎

A glance back at the sequence of example problems presented in this section reveals that the generalized likelihood ratio test typically yields sufficient statistics since the parameters appearing in $\hat{\lambda}$ ((12.61)) are replaced by their maximum likelihood estimates. In this regard, let $T = g(X_1, \ldots, X_n, n)$ be a sufficient statistic for a parameter θ with probability density function $h(t; \theta, n)$. Then according to the Fisher-Neyman Factorization Theorem (Theorem 10.1), the likelihood function of the sample random variables $X_1, \ldots, X_n$ factors as $\mathcal{L}(\theta; x_1, \ldots, x_n, n) = h(t; \theta, n) \cdot j(x_1, \ldots, x_n n) = h(g(x_1, \ldots, x_n, n); \theta, n) \cdot j(x_1, \ldots, x_n, n)$ ((10.14)), where j does not depend upon θ. This factorization criterion for a sufficient statistic holds for every sample realization t of T and for all admissible θ. Then, from (12.61),

$$\hat{\lambda}_t = \frac{\mathcal{L}(\hat{\mathcal{P}}_0)}{\mathcal{L}(\hat{\mathcal{P}})} = \frac{h(t; \hat{\mathcal{P}}_0, n)}{h(t; \hat{\mathcal{P}}, n)}. \tag{12.61.1}$$

And since maximum likelihood estimators are functions of sufficient statistics, it follows that $\hat{\lambda}$ depends only upon such statistics. Hence the critical region for the generalized likelihood ratio test will be defined as a function of the sufficient statistic T and thus uses all of the information about θ provided by the sample; that is, the test based on T should be equivalent to the test based on the complete set of sample observations or $\hat{\lambda}$ given by (12.61) equals $\hat{\lambda}_t$ provided by (12.61.1).

As a special case of (12.61.1), we see that if a sufficient statistic T exists, then the λ of the Neyman-Pearson Lemma (Theorem 12.1) given by (12.50) can be rewritten as

$$\lambda = \frac{h(t; \theta_0, n)}{h(t; \theta_1, n)}. \tag{12.50.1}$$

Here, too, the critical region for the Neyman-Pearson test is defined as a function of a sufficient statistic T and thus the best or most powerful test against any alternative hypothesis need only be based on T. The final characteristic of hypothesis tests to be considered herein deals with the notion of consistency. Specifically, a *consistent test* is one that exhibits, for large samples, a high probability of rejecting a false null hypothesis. That is, it is a test having, for large samples, high power against any true alternative hypothesis. In this regard, a consistent test is one for which $P(reject\ H_0/H_0\ false) \to 1$ as $n \to \infty$.

12.18 Hypothesis Tests for the Difference of Means When Sampling from Two Independent Normal Populations

Let $\{X_1, \ldots, X_{n_X}\}$ and $\{Y_1, \ldots, Y_{n_Y}\}$ be two sets of sample random variables taken from independent normal distributions with means μ_X and μ_Y and variances σ_X^2 and σ_Y^2, respectively. Moreover, n_X is the number of sample random variables taken from the X population distribution and n_Y depicts the same for the Y population distribution. Suppose we are interested in testing the difference between the two population means μ_X and μ_Y, where the said difference will be denoted as $\mu_X - \mu_Y = \delta_0$. So if $\mu_X > \mu_Y$, then $\delta_0 > 0$; if $\mu_X < \mu_Y$, then $\delta_0 < 0$.

And if $\mu_X = \mu_Y$, then $\delta_0 = 0$. In this regard, we may test the null hypothesis $H_0 : \mu_X - \mu_Y = \delta_0$ against any one of the following three alternative hypotheses:

Case I	Case II	Case III
$H_0 : \mu_X - \mu_Y = \delta_0$	$H_0 : \mu_X - \mu_Y = \delta_0$	$H_0 : \mu_X - \mu_Y = \delta_0$
$H_1 : \mu_X - \mu_Y > \delta_0$	$H_1 : \mu_X - \mu_Y < \delta_0$	$H_1 : \mu_X - \mu_Y \neq \delta_0$

To construct a *good* test of $H_0 : \mu_X - \mu_Y = \delta_0$ against any of the preceding alternatives, we must exploit the characteristics of the sampling distribution of the difference between two sample means (on this account, review Section 11.9). Given that X and Y are independent random variables, where X is $N(\mu_X, \sigma_X)$ and Y is $N(\mu_Y, \sigma_Y)$, and that $\overline{X} - \overline{Y}$ is our estimator for $\mu_X - \mu_Y$, the form of the distribution of our test statistic $\overline{X} - \overline{Y}$ will be specified under a variety of assumptions concerning the variances of X and Y. To this end, various versions of the test now follow.

12.18.1   Population Variances Equal and Known

We first assume that the population variances are equal and known; that is, $\sigma_X^2 = \sigma_Y^2 = \sigma^2$. Then with σ^2 specified, the quantity

$$Z_{\delta_0} = \frac{(\overline{X} - \overline{Y}) - \delta_0}{\sigma\sqrt{\frac{1}{n_X} + \frac{1}{n_Y}}}$$

[sampling from two independent normal

populations with equal and known variances]

is $N(0, 1)$ (see (11.33)). Then for a test conducted at the $\alpha = P(TIE)$ level, the appropriate decision rules for rejecting H_0 relative to H_1 appear as:

(a)   Case I—reject H_0 if $z_{\delta_0} \geq z_\alpha$;

(b)   Case II—reject H_0 if $z_{\delta_0} \leq z_\alpha$; and

(c)   Case III—reject H_0 if $|z_{\delta_0}| \geq z_{\alpha/2}$,   (12.68)

with z_{δ_0} depicting the sample realization of Z_{δ_0}.

12.18.2   Population Variances Unequal But Known

If the population variances are unequal but still assumed to be known, then the quantity

$$Z'_{\delta_0} = \frac{(\overline{X} - \overline{Y}) - \delta_0}{\sqrt{\frac{\sigma_X^2}{n_X} + \frac{\sigma_Y^2}{n_Y}}} \qquad (12.69)$$

[sampling from two independent normal

populations with unequal but known variances]

is still $N(0,1)$ (see (11.34)). Then the decision rules for rejecting the null hypothesis $H_0 : \mu_X - \mu_Y = \delta_0$ relative to the previously stated alternatives at the $100\alpha\%$ level are:

(a)   Case I—reject H_0 if $z'_{\delta_0} \geq z_\alpha$;

(b)   Case II—reject H_0 if $z'_{\delta_0} \leq -z_\alpha$; and

(c)   Case III—reject H_0 if $|z'_{\delta_0}| \geq z_{\alpha/2}$,   (12.70)

where z'_{δ_0} depicts the sample realization of Z'_{δ_0}.

12.18.3 Population Variances Equal But Unknown

If the common variance $\sigma^2(= \sigma_X^2 = \sigma_Y^2)$ is unknown, then we need to consider the distribution of $\overline{X} - \overline{Y}$ under sampling from two independent normal populations with unknown but equal variances. In this circumstance the quantity

$$T_{\delta_0} = \frac{(\overline{X} - \overline{Y}) - \delta_0}{S_P\sqrt{\frac{1}{n_X} + \frac{1}{n_Y}}}$$ (12.71)

[sampling from two independent normal

populations with equal but unknown variances]

follows a t distribution with pooled degrees of freedom $k = n_X + n_Y - 2$ and

$$S_P = \sqrt{\frac{(n_X - 1)S_X^2 + (n_Y - 1)S_Y^2}{k}}$$ (11.37)

denotes the pooled estimator of the common standard derivation σ, where

$$S_X^2 = \frac{\sum_{i=1}^{n_X}(X_i - \overline{X})^2}{n_X - 1}, \quad S_Y^2 = \frac{\sum_{i=1}^{n_Y}(Y_i - \overline{Y})^2}{n_Y - 1}.$$

Now the Case I–Case III decision rules for rejecting $H_0 : \mu_X - \mu_Y = \delta_0$ relative to H_1 at the $100\alpha\%$ level are:

(a)   Case I—reject H_0 if $t_{\delta_0} \geq t_{\alpha,k}$;

(b)   Case II—reject H_0 if $t_{\delta_0} \leq -t_{\alpha,k}$; and

(c)   Case III—reject H_0 if $|t_{\delta_0}| \geq t_{\alpha/2,k}$,   (12.72)

where t_{δ_0} is the sample realization of T_{δ_0}.

It is important to note that this particular t test assumes that:

(a) Both population distributions are normal

(b) The two random samples as well as the observations within each individual sample are independent

(c) The population variances are equal

Assumption (b) is critical and, for large samples, the violation of assumptions (a) and (c) is not all that serious. However, if the sample sizes are small and (c) seems untenable, then, as will be indicated immediately next, the pooled estimate of σ given by (11.37) can be replaced by separate unbiased estimators of σ_X^2 and σ_Y^2 and an approximation to degrees of freedom must be obtained so that, in this instance, the test itself is only approximate.

12.18.4   Population Variances Unequal and Unknown

If the population variances are now assumed to be unequal and unknown, then the quantity

$$T'_{\delta_0} = \frac{(\overline{X} - \overline{Y}) - \delta_0}{\sqrt{\frac{S_X^2}{n_X} + \frac{S_Y^2}{n_Y}}} \tag{12.73}$$

[sampling from two independent normal

populations with unequal and unknown variances]

is approximately t distributed with degrees of freedom given by

$$\Phi = \frac{\left(\frac{S_X^2}{n_X} + \frac{S_Y^2}{n_Y}\right)^2}{\left(\frac{S_X^2}{n_X}\right)\left(\frac{1}{n_X+1}\right) + \left(\frac{S_Y^2}{n_Y}\right)\left(\frac{1}{n_Y+1}\right)} - 2, \tag{11.39}$$

where S_X^2 and S_Y^2 are given in (12.18.3) and the realization ϕ of Φ must be rounded (if required) to the nearest integer value. For this test of $H_0 : \mu_X - \mu_Y = \delta_0$ versus H_1 at the $100\alpha\%$ level, the decision rules for rejecting H_0 relative to H_1 are:

(a)   Case I—reject H_0 if $t'_{\delta_0} \geq t_{\alpha,\phi}$;

(b)   Case II—reject H_0 if $t'_{\delta_0} \leq -t_{\alpha,\phi}$; and   (12.74)

(c)   Case III—reject H_0 if $|t'_{\delta_0}| \geq t_{\alpha/2,\phi}$,

where t'_{δ_0} is the realization of the test statistic T'_{δ_0}.

Example 12.18.1   Suppose that we decide to test $H_0 : \mu_X - \mu_Y = \delta_0 = 0$ (there is no difference between the population means), against $H_0 : \mu_X - \mu_Y \neq \delta_0 = 0$ (the population means are different) at the $\alpha = 0.05$ level. In addition, let $n_X = 25$, $n_Y = 30, \bar{x} = 50, \bar{y} = 58, \sigma_X^2 = 167$, and $\sigma_Y^2 = 200$. Since the population variances are known, (12.69) yields

$$z'_{\delta_0} = \frac{(50 - 58) - 0}{\sqrt{\frac{167}{25} + \frac{200}{30}}} = \frac{-8}{\sqrt{13.34}} = -2.19.$$

Since the critical region $\mathcal{R} = \{z \mid |z| \geq z_{0.025} = 1.96\}$, (12.70b) implies that we should reject H_0 at the 5% level of significance.

Let us now assume that the population variances σ_X^2 and σ_Y^2 are unknown but equal and, in fact, are estimated from the sample observations as $s_X^2 = 173$ and $s_Y^2 = 193$, respectively. Then from (12.71) and (11.37),

$$t_{\delta_0} = \frac{(50 - 58) - 0}{13.56\sqrt{\frac{1}{25} + \frac{1}{30}}} = \frac{-8}{13.56(0.073)} = -2.18,$$

where degrees of freedom $k = 53$. Looking to (12.72b), since $\mathcal{R} = \{t \mid |t| \geq t_{0.025,53} = 2.01\}$, we again reject H_0 at the 5% level.

If the population variances are taken to be unequal and unknown, then (12.73) and (11.39) simplify to $t'_{\delta_0} = \frac{-8}{\sqrt{13.35}} = -2.38$ and $\phi \approx 56$, respectively. This time $\mathcal{R} = \{t \mid |t| \geq t_{0.025,56} = 2.00\}$ so that we reject H_0 at the $\alpha = 0.05$ level of significance. ∎

The hypothesis test results for this section are summarized in the left-hand branch of Figure 12.15.

Suppose that random samples of *equal size* $n(= n_X = n_Y)$ are to be drawn from two independent normal distributions with known but unequal variances. Furthermore, let us test the null hypothesis $H_0 : \mu_X - \mu_Y = \delta_0$ against the alternative hypothesis $H_0 : \mu_X - \mu_Y = \delta_1 > \delta_0$, and suppose we choose to specify *a priori* the maximum tolerable size of the Type I and Type II Errors α and β, respectively. What common sample size n should be chosen? That is, under the aforementioned sampling conditions, we seek to determine the size of each of two random samples required to assure the attainment of prespecified α- and β-risks.

If H_0 is true, $P(\text{reject } H_0 | H_0 \text{ true}) = \alpha$; and if H_0 is false, $P(\text{do not reject } H_0 | H_0 \text{ false}) = \beta$. From these definitions of α and β we may express each of these error probabilities in terms of n and the critical value of $\bar{x} - \bar{y}$ or δ_C, where the critical region is $\mathcal{R} = \{\bar{x} - \bar{y} | \bar{x} - \bar{y} \geq \delta_C\}$; that is,

$$\alpha = P(TIE) = P(\overline{X} - \overline{Y} \in \mathcal{R} | \mu_X - \mu_Y = \delta_0)$$

$$= P(\overline{X} - \overline{Y} \geq \delta_C | \mu_X - \mu_Y = \delta_0)$$

$$= P\left(Z \geq \frac{\delta_C - \delta_0}{\sqrt{\frac{\sigma_X^2 + \sigma_Y^2}{n}}} = z_\alpha \middle| \mu_X - \mu_Y = \delta_0 \right), \tag{12.75}$$

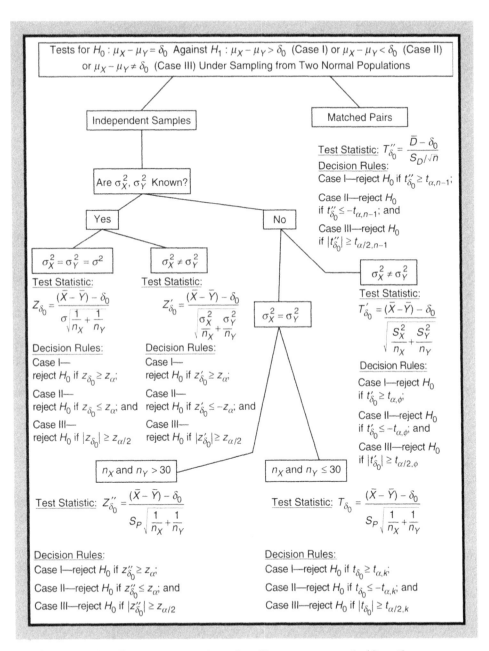

Figure 12.15   Tests for $H_0 : \mu_X - \mu_Y = \delta_0$ against $H_1 : \mu_X - \mu_Y > \delta_0$ (Case I), or $\mu_X - \mu_Y < \delta_0$ (Case II), or $\mu_X - \mu_Y \neq \delta_0$ (Case III) under sampling from two normal populations.

where z_α satisfies $P(Z \geq z_\alpha) = \alpha$; and for $\mu_X - \mu_Y = \delta_1 > \delta_0$ satisfying H_1,

$$\beta = P(TIIE) = P(\overline{X} - \overline{Y} \in \overline{\mathcal{R}}|\mu_X - \mu_Y = \delta_1),$$

$$= P(\overline{X} - \overline{Y} < \delta_C|\mu_X - \mu_Y = \delta_1),$$

$$= P\left(Z < \frac{\delta_C - \delta_1}{\sqrt{\frac{\sigma_X^2 + \sigma_Y^2}{n}}} = -z_\beta|\mu_X - \mu_Y = \delta_1\right), \tag{12.76}$$

with $-z_\beta$ chosen so that $P(Z < -z_\beta) = \beta$.

From (12.75) and (12.76) we have, respectively,

$$z_\alpha = \frac{\delta_C - \delta_0}{\sqrt{\frac{\sigma_X^2 + \sigma_Y^2}{n}}}, \quad -z_\beta = \frac{\delta_C - \delta_1}{\sqrt{\frac{\sigma_X^2 + \sigma_Y^2}{n}}}.$$

Upon eliminating δ_C from these latter two expressions we eventually obtain

$$n = \frac{(\sigma_X^2 + \sigma_Y^2)(z_\alpha + z_\beta)^2}{(\delta_1 - \delta_0)^2}. \tag{12.77}$$

(Note that if n is not an integer, then it should be *rounded up* to the next largest integer.)

If in the preceding pair of hypotheses we replace the right-sided alternative hypothesis by its left-sided counterpart $H_1 : \mu_X - \mu_Y = \delta_1 < \delta_0$, then exactly the same formula ((12.77)) for n obtains. And if we are testing $H_0 : \mu_X - \mu_Y = \delta_0$ against the two-sided alternative $H_1 : \mu_X - \mu_Y = \delta_1 \neq \delta_0$, then an adequate modification to (12.77) can be made by replacing α by $\frac{\alpha}{2}$ or

$$n = \frac{(\sigma_X^2 + \sigma_Y^2)(z_{\frac{\alpha}{2}} + z_\beta)^2}{(\delta_1 - \delta_0)^2}. \tag{12.78}$$

Example 12.18.2   Suppose that X is $N(\mu_X, 10)$ and Y is $N(\mu_Y, 12)$ and that we desire to extract a random sample of the same size n from each of these two independent populations for a test of $H_0 : \mu_X - \mu_Y = \delta_0 = 3$ versus $H_1 : \mu_X - \mu_Y = \delta_1 = 5 > \delta_0$. For this test we decide that the maximum tolerable error probabilities are $\alpha \leq 0.01$ and $\beta \leq 0.05$. How large of a sample from each population will be required? From (12.77),

$$n = \frac{(100 + 144)(2.325 + 1.645)^2}{(5 - 3)^2} \approx 962.$$

Hence a sample of at least $n = 962$ from each population is required to ensure that $\alpha \leq 0.01$ and $\beta \leq 0.05$ when $\delta_1 = 5$ is the true difference in population means. ■

12.19   Hypothesis Tests for the Difference of Means When Sampling from Two Dependent Populations: Paired Comparisons

One of the key assumptions made in the preceding section was that the random samples were drawn from two *independent* (normal) populations. If, however, the samples from the two populations are paired (each observation in the first sample is related in some specific way to exactly one observation in the second sample), then the samples are obviously not independent. For this latter case, as we shall now see, the analysis of the results of a paired observation experiment (in which the two samples are not chosen independently and at random) reduces to the application of a single-sample technique of hypothesis testing.

Following the discussion on paired comparisons offered in Section 11.10, let us assume that our paired experiment yields n pairs of observations denoted by $(X_1, Y_1), (X_2, Y_2), \ldots, (X_n, Y_n)$, where the sample random variables X_i and Y_i, $i = 1, \ldots, n$, are members of the same pair (X_i, Y_i), with X_i drawn from the first population and Y_i drawn from the second population. For the ith pair of sample random variables, let $D_i = X_i - Y_i, i = 1, \ldots, n$, be the difference between the two random variables X_i and Y_i making up that pair, where D_i is taken to be the ith observation on the random variable D. Here each D_i provides us with a measure of the difference between, say, the effectiveness of two separate treatments; for example, the X_i's depict a set of sample outcomes for the first treatment and the Y_i's represent a set of sample outcomes for a second or follow-up treatment.

Let us assume that the D_i's, $i = 1, \ldots, n$, constitute a single random sample from a normal population with mean $\mu_D = \mu_X - \mu_Y$ and variance σ_D^2 (here the D_i's represent a set of independent normally distributed sample random variables with $E(D_i) = \mu_D$ and $V(D_i) = \sigma_D^2$ for all i). Then the population random variable D, whose values are the $D_i = X_i - Y_i, i = 1, \ldots, n$, has a sampling distribution that is $N(\mu_D, \sigma_D)$. Hence the statistic

$$T = \frac{\overline{D} - \mu_D}{S_D / \sqrt{n}} \tag{12.79}$$

follows a t distribution with $n - 1$ degrees of freedom, where

$$\overline{D} = \frac{1}{n} \sum_{i=1}^{n} D_i \quad \text{and} \quad S_D^2 = \frac{1}{n-1} \sum_{i=1}^{n} (D_i - \overline{D})^2.$$

Here $\overline{D}$ serves as an estimator of the mean difference between, say, the effects of the first and second treatments, and S_D^2 is an estimator of the variance of the differences in treatment effects.

For a paired comparison experiment, we may test the null hypothesis $H_0 : \mu_D = \delta_0$ against any one of the three alternative hypotheses:

Case I	Case II	Case III
$H_0 : \mu_D = \delta_0$	$H_0 : \mu_D = \delta_0$	$H_0 : \mu_D = \delta_0$
$H_1 : \mu_D > \delta_0$	$H_1 : \mu_D < \delta_0$	$H_1 : \mu_D \neq \delta_0$

Under H_0, our test statistic is (from (12.79))

$$T''_{\delta_0} = \frac{\overline{D} - \delta_0}{S_D/\sqrt{n}} \tag{12.80}$$

[sampling from two dependent populations]

and the appropriate α-level critical regions are:

(a) Case I—reject H_0 if $t''_{\delta_0} \geq t_{\alpha,n-1}$;

(b) Case II—reject H_0 if $t''_{\delta_0} \leq -t_{\alpha,n-1}$; and   (12.81)

(c) Case III—reject H_0 if $|t''_{\delta_0}| \geq t_{\alpha/2,n-1}$,

where t''_{δ_0} depicts the sample realization of T''_{δ_0}.

Example 12.19.1   An education consulting firm claims that, because of its innovative instructional techniques, high school students taking its review course can, on the average, increase their test scores on a certain national exam by at least 10 points. To support their claim, they offer to allow a random sample of $n = 15$ students from a particular high school to benefit from their course free of charge. Table 12.5 contains the before and after test scores. Is there sufficient evidence to support the firm's claim at the 5% level? Let us test $H_0 : \mu_Y - \mu_X = \delta_0 \geq 10$ against $H_1 : \mu_Y - \mu_X = \delta_0 < 10$.

From the realized differences $d_i = y_i - x_i$, $i = 1,\ldots,15$, appearing in Table 12.5 we have

$$\overline{d} = \frac{1}{n} \sum_{i=1}^{15} d_i = \frac{82}{15} = 5.46,$$

$$s_D^2 = \frac{\sum\limits_{i=1}^{15} d_i^2}{n-1} - \frac{\left(\sum\limits_{i=1}^{15} d_i\right)^2}{n(n-1)} = \frac{938}{14} - \frac{(82)^2}{15(14)} = 34.98.$$

Then from (12.99),

$$t''_{\delta_0} = \frac{\overline{d} - \delta_0}{s_D/\sqrt{n}} = \frac{5.46 - 10}{5.91/\sqrt{15}} = -2.97.$$

Table 12.5

Effectiveness of a Review Course		
Score Before (x)	Score After (y)	Difference $(d = y - x)$
80	82	2
82	82	0
71	81	10
77	79	2
75	77	2
90	95	5
92	95	3
81	90	9
70	80	10
70	82	12
80	85	5
90	85	-5
70	83	13
60	77	17
98	95	-3

Since $t_{\alpha,n-1} = t_{0.05,14} = 1.761$, we see that -2.97 is an element of the critical region $\mathcal{R} = \{t''_{\delta_0} \mid t''_{\delta_0} \leq -1.761\}$ so that, from (12.81a), we reject H_0 at the 5% level—the average increase in test scores for those taking the review course is significantly below 10 points. ∎

12.20 Hypothesis Tests for the Difference of Proportions When Sampling from Two Independent Binomial Populations

Let $\{X_1, X_2, \ldots, X_{n_X}\}$ and $\{Y_1, Y_2, \ldots, Y_{n_Y}\}$ be two sets of sample random variables taken from two independent binomial populations with p_X and p_Y representing the proportions of successes in the first and second binomial populations, respectively. Additionally, let X and Y be independent random variables representing the observed number of successes in the samples of size n_X and n_Y, respectively. To test hypotheses on the differences in proportions $p_X - p_Y$ (remember that these tests are only approximate and hold only for large samples), we must review the characteristics of the sampling distribution of the difference between two sample proportions (see Section 11.11).

We know from Section 11.7 that the best estimators for p_X and p_Y are the sample proportions of successes $\hat{P}_X = \frac{X}{n_X}$ and $\hat{P}_Y = \frac{Y}{n_Y}$, respectively, with $E(\hat{P}_X) = p_X$, $E(\hat{P}_Y) = p_Y$, $V(\hat{P}_X) = \frac{p_X(1-p_X)}{n_X}$, and $V(\hat{P}_Y) = \frac{p_Y(1-p_Y)}{n_Y}$. Then the best estimators for $V(\hat{P}_X)$ and $V(\hat{P}_Y)$ are, respectively, $S^2(\hat{P}_X) = \frac{\hat{P}_X(1-\hat{P}_X)}{n_X}$ and $S^2(\hat{P}_Y) = \frac{\hat{P}_Y(1-\hat{P}_Y)}{n_Y}$.

Since $\hat{P}_X$ and $\hat{P}_Y$ are independent random variables, $V(\hat{P}_X - \hat{P}_Y) = V(\hat{P}_X) + V(\hat{P}_Y)$, with $S^2(\hat{P}_X - \hat{P}_Y) = S^2(\hat{P}_X) + S^2(\hat{P}_Y)$ serving as its estimator. Then as depicted by (11.49), the quantity

$$Z_{\Delta p} = \frac{(\hat{P}_X - \hat{P}_Y) - (p_X - p_Y)}{\sqrt{S^2(\hat{P}_X) + S^2(\hat{P}_Y)}} \tag{12.82}$$

is approximately $N(0,1)$ for large n_X and n_Y.

Suppose we decide to test the null hypothesis $H_0 : p_X - p_Y = \delta_0$ against any of the three alternative hypotheses given in

Case I	Case II	Case III
$H_0 : p_X - p_Y = \delta_0$	$H_0 : p_X - p_Y = \delta_0$	$H_0 : p_X - p_Y = \delta_0$
$H_1 : p_X - p_Y > \delta_0$	$H_1 : p_X - p_Y < \delta_0$	$H_1 : p_X - p_Y \neq \delta_0$

Under $H_0 : p_X - p_Y = \delta_0 \neq 0$, the quantity

$$Z_{\delta_0} = \frac{(\hat{P}_X - \hat{P}_Y) - \delta_0}{\sqrt{S^2(\hat{P}_X) + S^2(\hat{P}_Y)}} = \frac{(\hat{P}_X - \hat{P}_Y) - \delta_0}{\sqrt{\frac{\hat{P}_X(1-\hat{P}_X)}{n_X} + \frac{\hat{P}_Y(1-\hat{P}_Y)}{n_Y}}} \tag{12.83}$$

[Sampling from two independent binomial populations with $p_X \neq p_Y$]

is $N(0,1)$ (by virtue of (12.82)). Then for a test conducted at the $\alpha = P(TIE)$ level, the appropriate decision rules for rejecting H_0 relative to H_1 are:

(a) Case I —reject H_0 if $z_{\delta_0} \geq z_\alpha$;

(b) Case II—reject H_0 if $-z_{\delta_0} \leq -z_\alpha$; and   (12.84)

(c) Case III—reject H_0 if $|z_{\delta_0}| \geq z_{\alpha/2}$,

where z_{δ_0} is the sample realization of Z_{δ_0}.

But if $H_0 : p_X - p_Y = \delta_0 = 0$ is true (the two population proportions p_X and p_Y are assumed to be equal), then we may take p as their common value. In this case the best estimator of the common proportion p is the *pooled estimator*

$$\hat{P} = \frac{X + Y}{n_X + n_Y},$$

where X and Y are, as defined earlier, the observed number of successes in the two independent random samples. Then

$$S^2(\hat{P}_X - \hat{P}_Y) = \frac{\hat{P}(1 - \hat{P})}{n_X} + \frac{\hat{P}(1 - \hat{P})}{n_Y} = \hat{P}(1 - \hat{P})\left(\frac{1}{n_X} + \frac{1}{n_Y}\right)$$

so that (12.83) becomes

$$Z'_{\delta_0} = \frac{\hat{P}_X - \hat{P}_Y}{\sqrt{\hat{P}(1-\hat{P})\left(\frac{1}{n_X} + \frac{1}{n_Y}\right)}}.$$

[Sampling from two independent binomial populations with $p_X = p_Y = p$]

$$(12.83.1)$$

Here we will reject the null hypothesis of equal population proportions at the α level if

(a) Case I—reject H_0 if $z'_{\delta_0} \geq z_\alpha$;

(b) Case II—reject H_0 if $z'_{\delta_0} \leq -z_\alpha$; and (12.85)

(c) Case III—reject H_0 if $\left|z'_{\delta_0}\right| \geq z_{\alpha/2}$,

with z'_{δ_0} depicting the realized value of Z'_{δ_0}.

Example 12.20.1   A random sample of $n_X = 200$ individuals was taken from one socioeconomic group and it was found that $X = 25$ of them were college graduates. A second random sample of $n_Y = 180$ individuals from another socio-economic group exhibited $Y = 30$ college graduates. Is there any compelling sample evidence to indicate that one group's proportion of college graduates is any different than the others? Here we desire to test $H_0 : p_X - p_Y = \delta_0 = 0$ versus $H_1 : p_X - p_Y = \delta_0 \neq 0$ for, say, $\alpha = 0.01$. Then the relevant sample realizations are

$$\hat{p}_X = \frac{x}{n_X} = \frac{25}{200} = 0.125, \ \hat{p}_Y = \frac{y}{n_Y} = \frac{30}{180} = 0.166, \ \hat{p} = \frac{x+y}{n_X + n_Y} = \frac{55}{380} = 0.145,$$

and, from (12.83.1),

$$z'_{\delta_0} = \frac{0.125 - 0.166}{\sqrt{(0.145)(0.855)\left(\frac{1}{200} + \frac{1}{180}\right)}} = -1.138.$$

For $\alpha = 0.01$, $z_{\alpha/2} = z_{0.005} = 2.58$. Since $\left|z'_{\delta_0}\right| < 2.58$, it is evident that we cannot reject H_0 at the 1% level. In fact, we cannot reject H_0 at any reasonable level of significance since the associated p-value for this test is about $0.2542 = P(|Z| \geq 1.138|H_0 \text{ true})$. ■

12.21   Hypothesis Tests for the Difference of Variances When Sampling from Two Independent Normal Populations

Suppose we extract random samples of sizes n_X and n_Y from two independent normal populations $N(\mu_X, \sigma_X)$ and $N(\mu_Y, \sigma_Y)$, respectively, having unknown means and variances. Our objective is to test the null hypothesis $H_0 : \sigma_X^2 = \sigma_Y^2$ against any one of the following alternative hypotheses:

Case I	Case II	Case III
$H_0 : \sigma_X^2 = \sigma_Y^2$	$H_0 : \sigma_X^2 = \sigma_Y^2$	$H_0 : \sigma_X^2 = \sigma_Y^2$
$H_1 : \sigma_X^2 > \sigma_Y^2$	$H_1 : \sigma_X^2 < \sigma_Y^2$	$H_1 : \sigma_X^2 \neq \sigma_Y^2$

We determined in Section 11.12 that the random variable

$$F_{n_X-1,n_Y-1} = \frac{S_X^2/\sigma_X^2}{S_Y^2/\sigma_Y^2} \tag{11.53}$$

follows an F distribution with $n_X - 1$ and $n_Y - 1$ degrees of freedom, where the sample variances S_X^2 and S_Y^2 serve as the best estimators of the population variances σ_X^2 and σ_Y^2, respectively. Under the null hypothesis, (11.53) reduces to

$$F_{n_X-1,n_Y-1} = \frac{S_X^2}{S_Y^2}. \tag{12.86}$$

Let s_X^2 and s_Y^2 be the sample realizations of S_X^2 and S_Y^2, respectively, with f denoting the sample realization of the appropriate F-test statistic. Then we will reject the null hypothesis of equal population variances at the α level if

(a) Case I—reject H_0 if $f = \dfrac{s_X^2}{s_Y^2} \geq f_{1-\alpha,n_X-1,n_Y-1}$;

(b) Case II—reject H_0 if $f = \dfrac{s_Y^2}{s_X^2} \geq f_{1-\alpha,n_Y-1,n_X-1}$;   (12.87)

(c) Case III—reject H_0 if $f = \dfrac{\text{larger sample variance}}{\text{smaller sample variance}} \geq f_{1-(\alpha/2),v_1,v_2}$,

where v_1 corresponds to numerator degrees of freedom and v_2 denotes denominator degrees of freedom.

Example 12.21.1   Random samples have been drawn from two independent normal populations, yielding the following results:

Sample I	Sample II
$n_X = 10$	$n_Y = 8$
$s_X^2 = 50$	$s_Y^2 = 67$

At the 5% level, is there sufficient sample evidence to substantiate the claim that the variance of the second population is significantly larger than the variance of the first population? To answer this question, let us test $H_0 : \sigma_X^2 = \sigma_Y^2$ against $H_1 : \sigma_X^2 < \sigma_Y^2$ with $\alpha = 0.05$. (Remember that the larger sample variance goes into the numerator of our test statistic.) Given that the realized value of F is

$$f = \frac{s_Y^2}{s_X^2} = \frac{67}{50} = 1.34$$

and the critical value of f is $f_{1-\alpha, n_Y-1, n_X-1} = f_{0.95,7,9} = 3.29$, we see, via (12.87b), that since the realized f does not exceed the critical value of f, we cannot reject H_0 at the 5% level; that is, σ_Y^2 is not significantly larger than σ_X^2. To approximate the p-value $= P(F \geq 1.34 | H_0$ true) for this test, we note that 1.34 lies between $f_{0.50,7,9} = 0.978$ and $f_{0.90,7,9} = 2.51$. Hence $0.10 < p\text{-value} < 0.50$.

Alternatively, suppose our sample information had appeared as

Sample I	Sample II
$n_X = 60$	$n_Y = 60$
$s_X^2 = 2.87$	$s_Y^2 = 1.37$

and we wanted to test $H_0 : \sigma_X^2 = \sigma_Y^2$ against $H_1 : \sigma_X^2 \neq \sigma_Y^2$ at the 5% level. Now

$$f = \frac{s_X^2}{s_Y^2} = \frac{2.87}{1.37} = 2.094$$

and $f_{1-(\alpha/2), n_X-1, n_Y-1} = f_{0.975,59,59} \approx 1.67$ so that, according to (12.87c), we reject H_0 at the 5% level. We can again approximate our p-value $= 2P(F \geq 2.094 | H_0$ true) as $0.002 < p\text{-value} < 0.01$ (i.e., 2.09 lies between $f_{0.995,59,59} = 1.96$ and $f_{0.999,59,59} = 2.25$. Hence the p-value lies between $2(0.001) = 0.002$ and $2(0.005) = 0.01$). ∎

12.22   Hypothesis Tests for Spearman's Rank Correlation Coefficient ρ_S

Let us extract a random sample involving n ordered pairs of observations (X_i, Y_i), $i = 1, \ldots, n$, where the sample random variable X_i is the ith observation on the variable X and the sample random variable Y_i is the ith observations on the variable Y. In addition, as stipulated in Section 12.5, X_i is an ordinal value that depicts the rank associated with some variable W, where the ranks are arranged from low to high or from 1 to n. A similar situation exists for the collection of Y_i values on the variable Y.

From (2.33), we may specify as an estimator for ρ_S the sample rank correlation coefficient

$$R_S = 1 - \left[\frac{6 \sum_{i=1}^{n} D_i^2}{n(n^2 + 1)} \right], \qquad (12.88)$$

where $D_i = X_i - Y_i$, $i = 1, \ldots, n$. The sample realization of R_S is then

$$r_S = 1 - \left[\frac{6 \sum_{i=1}^{n} d_i^2}{n(n^2 + 1)} \right] = \delta, \qquad (12.89)$$

where $d_i = x_i - y_i$ is the sample realization of D_i, $i = 1, \ldots, n$.

For the null hypothesis let us choose $H_0 : \rho_S = 0$ (there is no agreement between ranks described by the X and Y variables). We may then test this null hypothesis against the following selection of alternative hypotheses:

Case I	Case II	Case III
$H_0 : \rho_S = 0$	$H_0 : \rho_S = 0$	$H_0 : \rho_S = 0$
$H_1 : \rho_S > 0$	$H_1 : \rho_S < 0$	$H_1 : \rho_S \neq 0$

For moderately large n (≥ 10), it can be demonstrated that, under $H_0 : \rho_S = 0$, the sampling distribution of R_S is approximately normal with $E(R_S) = 0$ and $V(R_S) = \frac{1}{n-1}$. Hence the standardized random variable

$$Z_S = \frac{R_S - E(R_S)}{\sqrt{V(R_S)}} = R_S \sqrt{n-1} \qquad (12.90)$$

is approximately $N(0, 1)$. So for a test conducted at the α level of significance, a set of decision rules for rejecting H_0 relative to H_1 is:

(a)   Case I—reject H_0 if $z_S \geq z_\alpha$;

(b)   Case II—reject H_0 if $z_S \leq -z_\alpha$; and   (12.91)

(c)   Case III—reject H_0 if $|z_S| \geq z_{\alpha/2}$,

where $z_S = r_S \sqrt{n-1}$ is the sample realization of Z_S.

For example, let us assume that the $n = 10$ rankings appearing in Table 2.13 are a random sample taken from some unspecified population distribution. From this data we found that $r_S = 0.93$. Let us test $H_0 : \rho_S = 0$ versus $H_1 : \rho_S > 0$ using $\alpha = 0.05$. Since $z_S = 0.93\sqrt{9} = 2.79 > z_{0.05} = 1.645$, we reject H_0 in favor of H_1; that is, there exists a statistically significant amount of agreement in the ranks at the 5% level.

12.23 **Exercises**

12-1. The times (in minutes) required to complete a test administered by the human resources department of ABC Co. are approximately $N(78, 12)$. The department would like to ascertain whether applicants for employment who have completed at least two years of college complete the test faster than the norm. Suppose a random sample of 30 applicants finish the test in an average time of 72 minutes. Let $\alpha = 0.05$. Suppose the human resources department simply wanted to detect a difference in the sample average from the norm. What is to be concluded? What is the p-value for the test?

12-2. A soft drink company wants to test-market a new product in $n = 40$ randomly selected convenience stores. The product will be introduced if in excess of 10 cases are sold per week in each store. Experience with similar new products reveals that sales are approximately normally distributed and that the standard deviation should be approximately 3.5 cases per week. Should the new product be marketed? Let $\alpha = 0.05$. What is the p-value for this test?

12-3. A distributor of a particular brand of industrial heat lamps states that their average draw of current is 0.9 amps. A sample of $n = 12$ lamps was tested and it was determined that $\bar{x} = 0.13$ with $s = 0.41$. For $\alpha = 0.05$, should we reject the distributors claim?

12-4. Hot-Shot legal services is attempting to increase reading comprehension (RC) of its new hires. Fifteen of its first-year recruits are given a standard RC test. After the test these individuals take a special course designed to enhance their RC and they are given a second RC test. The difference between the second and first sets of scores had a mean of 7 points and the standard deviation of the difference was 9 points. For $\alpha = 0.10$, has the special course increased RC?

12-5. Last year average monthly expenditure per household on a certain product was $12.97. Has there been a statistically significant change in average household expenditure for this year? A sample of size $n = 25$ yielded $\bar{x} = \$11.22$ with $s = \$0.95$. Use $\alpha = 0.01$.

12-6. In a random sample of 500 tulip bulbs taken from a normal population 476 of them bloomed. For $\alpha=0.05$, would you reject the claim that at least 90% of the bulbs will bloom? What is the p-value for this test?

12-7. From a random sample of 200 voters it was found that 110 were in favor of a particular piece of legislation. Is opinion equally divided on this legislative issue? Use $\alpha = 0.05$. What is the p-value? What sampling assumptions are being made?

12-8. A manufacturer of small machine parts claims that at least 98% of all parts shipped to ACE Industries conform to specifications. In a sample of 250 parts, it was found that 22 did not conform to specifications. For $\alpha = 0.01$, should we reject the manufacturers claim?

12-9. To compare the durabilities of two premium exterior house paints (one from House Depot and the other sold by KK-Mart), nine $4' \times 6'$ panels of each paint were exposed to the elements for a six-month trial period. (The arrangement of the panels was randomly determined.) At the end of the trial period the following reflectivity readings were obtained (higher values represent greater reflectivity):

HD: 9.6, 11.1, 11.6, 9.8, 8.9, 9.5, 10.2, 11.5, 10.1

KK-Mart: 9.4, 10.1, 9.2, 8.0, 10.1, 10.6, 7.2, 8.5, 9.5

For $\alpha = 0.05$, determine if the paints exhibit equal reflectivity. What sampling assumptions are being made?

12-10. Given the following (random) sample results:

Sample 1: $n_X = 16$, $\bar{x} = 25$, $s_X = 7.14$

Sample 2: $n_Y = 10$, $\bar{y} = 29$, $s_Y = 5.19$

can we conclude at the 10% level of significance that $\mu_X < \mu_Y$?

12-11. The following are yields (in bushels/acre) for two different varieties of winter wheat (call them A and B):

A: 62.7, 71.4, 76.7, 59.3, 59.7, 64.7, 69.1, 70.5

B: 69.8, 61.5, 49.9, 53.8, 65.1, 66.7, 47.8, 51.1

Can we conclude at the 5% level that the average yields are equal?

12-12. Ten pairs of twins of a different sex made the following scores on a specialized dexterity test:

Female: 92, 83, 95, 96, 85, 61, 76, 80, 92, 87

Male: 90, 93, 80, 86, 71, 91, 80, 70, 80, 81

Is there any reason to believe that the female will generally make the higher average score? Use $\alpha = 0.05$.

12-13. Two different methods were used to determine the fat content (expressed as a percent) in different samples of premium vanilla ice cream. Both methods were used on scoops taken from the same container. Does one of these methods yield higher readings on the average? Use $\alpha = 0.01$.

Sample:	1	2	3	4	5	6	7	8	9
Method 1:	20	21	18	17	22	21	23	19	16
Method 2:	17	20	18	16	17	20	19	21	21

12-14. A time-and-motion expert feels that she can shorten the workers handling time for 5lb. packages in the shipping department of a candy manufacturer. Suppose $n = 10$ workers are chosen at random. Time-and-motion readings (in minutes) on them before and after the completion of a training program appear as

Sample:	1	2	3	4	5	6	7	8	9	10
Before:	7	7	8	7	9	7	6	8	7	7
After:	6	7	6	5	7	5	4	6	4	4

For $\alpha = 0.05$, has the average handling time per worker for a 5lb. package decreased?

12-15. A random sample of $n = 1000$ persons consisted of 485 females and 515 males. Of the females, 250 were college graduates and 327 of the males had graduated college. Does this sample evidence lead us to conclude that the proportion of college graduates is smaller for females? Use $\alpha = 0.05$. Determine the p-value for this test.

12-16. In a sample of size $n_X = 100$ from one binomial population it was found that $x = 31$, and in a sample of size $n_Y = 150$ from a second binomial population it was found that $y = 50$. Test the hypothesis of no difference in the proportions of successes at the $\alpha = 0.01$ level. What is the p-value for this test?

12-17. A random sample of $n_X = 140$ households from a low-income neighborhood exhibited a head of household registered as a Democrat in $x = 70$ cases. A similar sample of size $n_Y = 130$ taken from a high-income neighborhood had a head of household registered as a Democrat in $y = 48$ cases. Is the proportion of Democratic registrations higher in the low-income area? Use $\alpha = 0.10$.

12-18. Fenway Park needs to purchase spotlights that exhibit long life as well as uniformity of operating life. Past experience dictates that the variance of bulb life should not exceed 230 (hours)^2. A sample of $n = 16$ bulbs is obtained from a new vendor wishing to get the lighting contract. It was found that $\bar{x} = 1020$ hours with $s^2 = 275$ (hours)^2. For $\alpha = 0.05$, does the true or population variance exceed 230?

12-19. A sample of $n = 25$ ball bearings were tested for resistance to heat due to friction. It was found that $s^2 = 150$ (degrees)^2. Is this result consistent with the claim that the true variance will not exceed 118 (degrees)^2? Use $\alpha = 0.01$.

12-20. From a sample of size $n = 30$ it was determined that $s^2 = 14.7$ (inches)^2. Test $H_0 : \sigma^2 \geq 18$ (inches)^2 against $H_1 : \sigma^2 < 18$ (inches)^2 at the 1% level. What is your conclusion?

12-21. Farmers consider uniformity of yield to be an important attribute of an agricultural commodity. Two types of seed (brands X and Y) for growing alfalfa are to be compared. They have about the same mean yield/acre but it was determined for $n_X = n_Y = 15$ that $s_X^2 = 15$ (bushels)^2 and $s_Y^2 = 22$ (bushels)^2. Is the variance of brand X significantly lower than the variance of brand Y? Use $\alpha = 0.10$.

12-22. Two brands of heavy duty fan belts (call them brand X and brand Y) have about the same average durability. However, their uniformity of wear is questionable. A sample of 10 brand X belts and 15 brand Y belts yielded $s_X^2 = 5.576$ (miles)^2 and $s_Y^2 = 8.025$ (miles)^2, respectively. Should a trucking company purchase brand X belts for its fleet of trucks? Use $\alpha = 0.05$.

12-23. A study of two independent samples of sizes $n_X = 15$ and $n_Y = 19$ yielded $s_X^2 = 10.8$ and $s_Y^2 = 15.9$, respectively. Test $H_0 : \sigma_X^2 \geq \sigma_Y^2$ against $H_1 : \sigma_X^2 < \sigma_Y^2$ using $\alpha = 0.01$. State your conclusion.

12-24. Let us assume that the $n = 10$ rankings appearing in Table 2.13 are a random sample taken from some unspecified population distribution. From this data we found that $r_s = 0.93$. Test $H_0 : \rho_s = 0$ versus $H_1 : \rho_s > 0$ using $\alpha = 0.05$.

12-25. From a sample of size $n = 14$ rankings it was determined that the realization of Spearman's rank correlation coefficient was $r_s = 0.75$. Test $H_0 : \rho_s = 0$, versus $H_1 : \rho_s > 0$ using $\alpha = 0.01$.

12-26. In a test of depth perception two judges each ranked $n = 10$ objects in order of their approximate distance (rounded to the nearest yard) from a fixed baseline marker. The results are presented in the accompanying table.

Object	Judge 1	Judge 2
1	5	3
2	4	1
3	3	7
4	2	6
5	1	5
6	7	8
7	9	10
8	8	9
9	10	4
10	6	2

Find r_s. Using $\alpha = 0.05$, test $H_0 : \rho_s = 0$, against $H_1 : \rho_s \neq 0$.

12-27. Given that X is $N(\mu, 10)$, the null hypothesis is $H_0 : \mu = \mu_0 = 27$, the alternative hypothesis is $H_1 : \mu < 27$, and $n = 50$, what is the distribution of $\overline{X}$? For $\alpha = 0.05$, find $\mathcal{R}$ and $\overline{\mathcal{R}}$. Suppose the true mean is $\mu = \mu_1 = 18$. What is the probability (β) of not rejecting H_0 when the true mean is 18?

12-28. Recalculate $\beta = P(TIIE)$ for the preceding problem when:

   (a) n increases to 100

   (b) α decreases to 0.01

   (c) σ increases to 15

   (d) μ_1 decreases to 15

12-29. Suppose that in Exercise 12-27 the alternative hypothesis is replaced by $H_1 : \mu \neq 27$. Find $\mathcal{R}$ and $\overline{\mathcal{R}}$. If the true population mean is 22, what is the probability of not rejecting H_0? What is the probability of not rejecting H_0 if the true mean is 29?

12-30. Suppose X is $N(\mu, 20)$ and that, for $\alpha \leq 0.05$ and $\beta \leq 0.05$, we want to test $H_0 : \mu = \mu_0 = 76$, against $H_1 : \mu > 76$. How large of a sample will be needed if $\mu_1 = 80$? How large of a sample would be required if the alternative hypothesis is replaced by $H_1 : \mu \neq 76$.

12-31. Assume that we are sampling from a binomial population and that we are testing $H_0 : p = p_0 = 45$, against $H_1 : p > 0.45$ using $\alpha = 0.05$ given $n = 200$. What is the distribution of $\hat{P}$? Find $\mathcal{R}$ and $\overline{\mathcal{R}}$. Suppose the true proportion is $p_1 = 0.56$. What is $\beta = P(TIIE)$?

12-32. Recalculate β for the preceding problem when:

   (a) n decreases to 100

   (b) α increases to 0.10

   (c) p_1 increases to 0.60

12-33. Suppose that in Exercise 12-31 the alternative hypothesis is replaced by $H_1 : p \neq 0.45$. Find $\mathcal{R}$ and $\overline{\mathcal{R}}$. If the true population proportion is 0.40, what is the probability of not rejecting H_0? What is the probability of not rejecting H_0 if the true proportion is 0.58?

12-34. Assume that we are sampling from a binomial population and that we are to test $H_0 : p = p_0 = 76$, against $H_1 : p < 0.76$. The α- and β-risks are each not to exceed 10%. How large of a sample will be required if $p_1 = 0.60$?

12-35. Suppose that X is $N(\mu, 20)$ and $n = 40$. Let $H_0 : \mu = \mu_0 = 100$, with $H_1 : \mu < 100$. For $\alpha = 0.01$, find $\mathcal{R}$ and $\overline{\mathcal{R}}$. Find $\beta(\mu)$ for $\mu = \mu_1 = 80, 82, 84, 86, 88, 90, 92, 94, 96, 98$. Determine the power $P(\mu)$ for each of the given μ's. Graph the power curve. What does its shape inform us about the test?

12-36. For X distributed as $N(\mu, 10)$, $n = 130$, $\alpha = 0.05$, $H_0: \mu = 200$, and $H_1: \mu \neq 200$, find $\mathcal{R}$ and $\overline{\mathcal{R}}$. Use (12.42) to determine $P(\mu)$ for a variety of μ's satisfying H_1. Graph and then interpret the power curve.

12-37. Suppose a random sample of size $n = 200$ is extracted from a binomial population. For $\alpha = 0.01$, let us test $H_0: p = p_0 = 0.35$, against $H_1: p > 0.35$. Find $\mathcal{R}$ and $\overline{\mathcal{R}}$. Use (12.42) to specify the power $P(p)$ of this test for a variety of p's. Graph the power function and interpret its shape.

12-38. Suppose the random variable X has a probability density function of the form $f(x; \lambda) = \lambda e^{-\lambda x}$, $x > 0$, $\lambda > 0$. For $H_0: \lambda = \lambda_0$, and $H_1: \lambda = \lambda_1$, $\lambda_0 > \lambda_1$, determine the best test of size α of H_0 versus H_1 for a sample of size n.

12-39. Let $X_1, \ldots, X_n$ denote a set of sample random variables drawn from the probability mass function $p(x; \theta) = \frac{e^{-\theta x}}{X!}$, $X = 0, 1, 2, \ldots, \theta \geq 0$; zero elsewhere. Find the uniformly most powerful test of size α of $H_0: \theta = \theta_0$, against $H_1: \theta > \theta_0$.

12-40. Determine the form of the Neyman-Pearson critical region for testing $H_0: \lambda = \lambda_0$ versus $H_1: \lambda > \lambda_0$ for a set of n sample random variables drawn from the probability mass function $f(x; \lambda) = \frac{e^{-\lambda} \lambda^X}{X!}$, $X = 0, 1, \ldots; \lambda > 0$.

12-41. Suppose $\{X_1, \ldots, X_n\}$ is a set of sample random variables drawn from the probability density function $f(x; \theta) = \theta x^{\theta-1}$, $0 < x < 1$; zero elsewhere. Find the most powerful test of size α of $H_0: \theta = \theta_0 = 1$, against $H_1: \theta = \theta_1 = 2$.

12-42. Let $\{X_1, \ldots, X_n\}$ depict a set of sample random variables taken from the (gamma) probability density function $f(x; \alpha_0, \theta) = \frac{1}{\Gamma(\alpha_0)\theta^{\alpha_0}} x^{\alpha_0-1} e^{-x/\theta}$, $x \geq 0$; $\alpha_0, \theta > 0$. Determine the most powerful test of size α of $H_0: \theta = \theta_0$ versus $H_1: \theta = \theta_1$, with $\theta_1 > \theta_0$.

12-43. Use the Neyman-Pearson Lemma to determine the best critical region for testing $H_0: \theta = \theta_0$ versus $H_1: \theta \neq \theta_1(<\theta_0)$ for a random sample of size n given the probability density function $f(x; \theta) = (1 + \theta)x^\theta$, $0 < x < 1$; zero elsewhere.

12-44. Suppose $\{X_1, \ldots, X_n\}$ is a set of sample random variables drawn from a $N(\theta, 1)$ population. Use the Neyman-Pearson Lemma to determine the best critical region of size α for testing $H_0: \theta = 3$ versus $H_1: \theta = 2$.

12-45. For (12.61), explain why $0 \leq \hat{\lambda} \leq 1$.

12-46. For a set of sample random variables $\{X_1, \ldots, X_n\}$ taken from the probability density function

$$f(x; \lambda) = \begin{cases} 2\lambda x e^{-\lambda x^2}, & x > 0, \lambda > 0; \\ 0 & \text{elsewhere} \end{cases}$$

find a uniformly most powerful test of size α of $H_0 : \lambda = \lambda_0$ versus $H_1 : \lambda > \lambda_0$.

12-47. Suppose $X_1, \ldots, X_n$ depicts a random sample drawn from a $N(\mu, \sigma)$ population, where both μ and σ^2 are unknown. Perform the generalized likelihood ratio test of $H_0 : \mu = \mu_0 = 100$ versus $H_1 : \mu \neq 100$.

12-48. Suppose $X_1, \ldots, X_n$ is a random sample taken from a $N(\mu, \sigma)$ random variable with both μ, σ^2 unknown. Conduct the generalized likelihood ratio test of $H_0 : \sigma^2 \geq \sigma_0^2 = 0.37$, against $H_1 : \sigma^2 < 0.37$.

12-49. Hypothesis Tests for the Coefficient of Variation Under Random Sampling. We may want to determine if a set of sample random variables $\{X_1, \ldots, X_n\}$ could have been drawn from a population having a given coefficient of variation $V_0 = (\sigma / \mu)_0$. For instance, suppose that in 1990 a large cross-section of countries was known to have a coefficient of variation of per capita income levels equal to V_{1990}. Has the coefficient of variation calculated from a random sample of these countries for the year 2000 changed significantly from V_{1990}? If, for instance, there has been a statistically significant decline in the value of this statistic over time, then we have some sample evidence that convergence of per capita incomes has occurred; that is, the countries have become more similar or uniform with respect to per capita income levels.

Suppose we want to test $H_0 : V = V_0$, against the two-tail alternative $H_1 : V \neq V_0$ at the $\alpha = P(TIE)$ level of significance. If the sampled population is $N(\mu, \sigma)$ with $V = \sigma / \mu \leq 0.66$, then, for a sample size $n \geq 10$, the test statistic

$$Z_V = \frac{\sqrt{n-1}\,(S/\bar{X} - V_o)}{V_o \sqrt{V_o^2 + 0.5}}$$

is approximately $N(0, 1)$. Then the critical region $\mathcal{R} = \{z_V / |z_V| \geq z_{\alpha/2}\}$, where z_V is the sample realization of Z_V. For one-tailed alternatives (involving either $H_0 : V \leq V_0$ vs. $H_1 : V \leq V_0$ or $H_0 : V \geq V_0$ versus $H_1 : V < V_0$), this test works well for $n \geq 11$ when sampling from a normal population with $V \leq 0.33$. In this instance $\mathcal{R} = \{z_V | z_V | \geq z_\alpha\}$ for $H_1 : V > V_0$ or $\mathcal{R} = \{z_V \,|z_V| \leq -z_\alpha\}$ when $H_1 : V < V_0$.

A sample of size $n = 25$ is taken from a $N(\mu, \sigma)$ population with the result that $\bar{x} = 1.52, s = 4.67$. For $\alpha = 0.05$, test $H_0 : V = V_0 = 0.35$ against $H_1 : V < 0.35$.

Nonparametric Statistical Techniques

13.1 Parametric vs. Nonparametric Methods

The previous chapter dealt with the classical or traditional theory and execution of tests of statistical hypotheses. As indicated therein, a hypothesis is a statement about the probability distribution of a random variable. Procedurally, we assumed that the population distribution had a specific functional form and then we tested hypotheses concerning its parameter(s). However, instead of dealing with a *parametric hypothesis*, we may actually make a statement about the *form* of the population distribution itself; that is, we can test a *distributional hypothesis*. For example, we previously tested hypotheses about the unknown mean μ of a population random variable X, which was assumed to be $N(\mu, \sigma)$, with the variance σ^2 either known or unknown. Clearly the distributional assumption of normality was made. But if we actually choose to conduct a test of the null hypothesis that X is normally distributed, then obviously a distributional hypothesis has been made. The alternative hypothesis is then that the random variable X is not of the stated form.

If this alternative hypothesis is assumed true, then, unfortunately, the likelihood function of the sample is not well defined and thus the likelihood-ratio methodology developed in Sections 12.16 and 12.17 of Chapter 12 does not apply. In such instances $\beta = P(TIIE)$ cannot be evaluated so that comparisons among alternative tests cannot be made.

Additionally, we also assumed under the classical approach to hypothesis testing that we had a *random sample* of observations on some population characteristic X; that is, the sample random variables $X_1, \ldots, X_n$ were not obtained in any systematic fashion and, in fact, were drawn independently of each other. In this regard, we can also test what we shall call a *randomness hypothesis*. That is, we can determine whether any *given* set of sample realizations $\{x_i, \ldots, x_n\}$ can be taken to be a random sample. Here the null hypothesis is that the sample is random and the alternative hypothesis is that it is not random.

The aforementioned distributional and randomness hypotheses are part of a broader class of testable hypotheses that come under the rubric of *nonparametric statistical techniques*. We now offer a detailed comparison of the parametric versus nonparametric approaches to statistical inference.

As noted in the preceding chapter, the common *modus operandi* for implementing any parametric statistical technique quite generally involves (1) identifying a population parameter θ as the object of our inferential inquiry; (2) extracting a random sample of size n; (3) selecting a *good* point estimator T of θ; and (4) specifying the form and parameters of the sampling distribution of T. Once these steps are taken, confidence limits and hypothesis tests can be formulated. For instance, if it is assumed that the population random variable X (measured on a quantitative scale) is $N(\mu, \sigma)$, with σ unknown and n small, then the best estimator of μ is $\overline{X}$, where $\overline{X}$ follows a t-distribution with $n-1$ degrees of freedom. Clearly in this circumstance the problem of statistical inference is reduced, say, to testing hypotheses about the parameter μ.

However, if we are unsure about the legitimacy of any of the assumptions underlying a specific parametric technique (e.g., it may be the case that the assumption of the normality of the parent population is dubious or the value of one or more of the parameters of the population distribution cannot be fixed), then possibly a nonparametric technique is appropriate for inferences about θ. In fact, if X is measured on a nonquantitative scale, then μ and σ are meaningless. Hence a parametric inference concerning, say, μ is impossible. More specifically, let us examine Table 13.1.

Table 13.1

Parametric Methods	Nonparametric Methods
p.1 Not applicable when dealing with nominal or ordinal data. Most parametric methods are relevant for data measured on a quantitative (interval or ratio) scale.	*n.1* Primarily concerned with nominal or ordinal data taken from a (typically) continuous population distribution. If a nonparametric technique is appropriate for interval or ratio scale data, then *n.2* applies.
p.2 Assumes some knowledge of the form of the population probability distribution (e.g., normality). Often incorporates additional assumptions such as independence of the observations (when conducting one-sample tests) or the homogeneity of variances (when performing two-sample tests).	*n.2* Does not require that the probability distribution of some population characteristic assume any functional form. In this regard, since the population distribution is often unspecified, nonparametric methods are aptly termed distribution-free statistical methods. Clearly distribution-free methods are thus a proper subset of the much broader class of nonparametric methods.

Table 13.1 Cont'd.

Parametric Methods	Nonparametric Methods
p.3 For small random samples, the normality assumption is questionable.	*n.3* Suitable for making inferences about populations with small- to medium-sized random samples ($n \leq 50$) given that the form of the population distribution is unknown.
p.4 Null and alternative hypotheses are precisely stated. Hence conclusions obtained from parametric hypothesis tests have a high degree of information content. (For example, if in a two-sample t test of $H_0: \mu_1 - \mu_2 = 0$ versus $H_1: \mu_1 - \mu_2 \neq 0$ we are led to reject the null hypothesis in favor of the alternative hypothesis at a particular level of significance, then we may conclude that "the two populations differ with respect to central tendency.")	*n.4* The null hypothesis formulated in a nonparametric test is not as precise as its parametric counterpart and thus rarely equivalent to the same. (For example, the nonparametric analog to the parametric null hypothesis of equal population means is H_0: the population distributions are identical. If this null hypothesis is rejected in favor of the alternative hypothesis H_1: the population distributions are not identical, then the question of the nature of the difference between the two populations is left unanswered).
p.5 In instances where both parametric and nonparametric techniques apply (i.e., the classical assumptions hold), the power of a parametric test (its ability to detect differences from hypothesized values) is greater than that of its nonparametric counterpart given n, $\alpha = P(TIE)$, and the true situation. That is, for a given level of significance α and sample size n, the probability of committing a Type II Error is lower with parametric tests relative to nonparametric ones. This being the case, nonparametric tests have less than 100% power efficiency when compared to parametric tests.	*n.5* When both parametric and nonparametric techniques apply, a nonparametric technique requires a larger n (or α) to achieve the same power as a parametric technique. In fact, for sufficiently large samples, the power of a nonparametric test approaches that of its parametric analogue. Thus nonparametric methods require more evidence relative to parametric ones to support comparable inferences. When classical assumptions do not hold, nonparametric tests may be at least as powerful as parametric ones in detecting differences from hypothesized values.

In what follows we shall first consider nonparametric tests on a single sample. Specifically, we shall examine tests for the randomness of an ordered sample data set; the sign test for a proportion as well as for the median of a random sample; and the Wilcoxon signed rank test for the median under random sampling. Two-sample tests on both independent and paired samples will subsequently be offered.

13.2 Tests for the Randomness of a Single Sample

In this section we shall examine the two tests that can be used to assess the randomness or, equivalently, independence of a set of observations when the order in which the observations were obtained is important. To this end, we shall attempt to answer the question: Can a sample that is known to be nonrandomly drawn or generated be treated *as if* it represented a random sample? The tests alluded to are the *runs test* and the *von Neumann ratio test*.

A runs test, as the name implies, is based upon the notion of a *run*—an unbroken sequence of identical outcomes/elements (denoted by letters or special symbols) that are preceded and followed by different outcomes or no outcomes at all. For instance, let us assume that a stamping machine in a metal shop has produced the following ordered sequence of defective (d) and nondefective (n) television chassis components: $n\,n\,n\,n\,n\,/\,d\,/\,n\,n\,/\,d\,/\,n\,n\,n\,/\,d\,d\,/\,n\,n\,n$. According to our definition of a run, this sequence has seven runs, where a slash is used to separate each individual run. That is, each sequence of n's and d's, uninterrupted by the other letter, constitutes a run. As this example indicates, we may view a run as a maximal subsequence of like elements, where a *subsequence* is simply a sequence within a sequence. Here we delimit the maximum number of like items (say, n's) before encountering a different item (d's). Clearly the greatest number of runs that can occur in an arrangement of n items is n; and the smallest number of runs in any such arrangement is 1. It must be emphasized that the order in which the observations occur must be preserved so that the various subsequences of runs can be identified. In addition, the length of each run is irrelevant.

As we shall now see, a runs test is used to determine if the elements within an ordered sequence of events occur in a random fashion, where each element within the sequence assumes one of two possible values (e.g., the n's and d's given earlier). In general, any letters (or symbols) may be used to designate the dichotomy within the sequence generated by a binary process, and any principle that orders these letters may be employed.

The total number of runs exhibited by a given sequence of outcomes is typically a good indicator of its possible *lack of randomness*; that is, the presence of *too few runs* might indicate a specific grouping of items or perhaps a trend; and *too many runs* might reveal some sort of periodic pattern within the sequence. In either instance, either too few or too many runs provides us with evidence supporting the lack of randomness in one set of observations.

The single-sample runs test can be performed using a variety of data types. It is applicable when the observations are measured on a nominal or ordinal scale (e.g., consider the preceding subsequences of n's and d's) as well as on an interval or ratio scale. In this latter instance we typically consider runs above and below some reference value such as the median. This is particularly useful if a time-related trend is suspected to exist in the data set; for example too few runs is indicative of a time trend in the data.

Suppose that we observe an ordered sequence of attributes; for example defective versus nondefective items produced; yes or no answers to a specific question asked in a public opinion poll, and so on. In general, let the two types of outcomes observed be denoted by the letters a and b. Suppose further that within the entire sequence of n items we have n_1 a's and n_2 b's, where $n_1 + n_2 = n$. In addition, let the random variable R denote the number of runs occurring in the ordered sequence of n items and let r be its sample realization.

In the tests for sample randomness that follow, the null hypothesis will always be specified as H_0: the order of the sample data is random. Then the following sets of hypotheses can be formulated:

Case I	Case II	Case III
H_0: the order of the sample data is random	H_0: the order of the sample data is random	H_0: the order of the sample data is random
H_1: too many runs (indicating periodicity)	H_1: too few runs (a specific grouping or trend)	H_1: the order of the sample data is not random

(Here Case III is the typical set of null and alternative hypotheses used when testing randomness.)

At this point in our discussion of the runs test we need to explicitly take account of the sample size. To this end we have:

A. *Small Sample Runs Test.* As a general rule, we shall perform a small-sample runs test when $n_1 \leq 15$ and $n_2 \leq 15$. For Case III (which specifies a two-sided alternative hypothesis) with $\alpha = 0.05$, Table A.10 of the Appendix enables us to identify both lower- and upper-tail critical values r_ℓ and r_u, respectively, of the sampling distribution $f(y)$ of R, which restrict the probability in each tail of $f(y)$ to $\frac{\alpha}{2}$. That is, r_ℓ and r_u are chosen so that

$$P(R \leq r_\ell \text{ or } R \geq r_u) = P(R \leq r_\ell) + P(R \geq r_u) = \sum_{y=2}^{r_\ell} f(y) + \sum_{y=ru}^{n} f(y)$$

$$\approx \frac{\alpha}{2} + \frac{\alpha}{2} = \alpha(= 0.05). \tag{13.1}$$

(Since R is a discrete random variable, the true level of significance α may not be precisely equal to 0.05.) Hence the critical region for this two-tailed test is $\mathcal{R} = \{r | r \leq_l \text{ or } r \geq r_u\}$. So if $r \in \mathcal{R}$, we reject the null hypothesis of random order at the 5% level; that is, we cannot treat the given sample *as if* it were random.

Example 13.2.1 Suppose that an industrial process has generated the following sequence of bolts whose diameters either satisfy the tolerance limits (this outcome is denoted as a) or do not satisfy the stated limits (this outcome is depicted as b).

A sample of $n = 25$ bolts yielded the following sequence of outcomes:

$$aaaa/bb/aaa/b/aaaa/bbb/aa/bb/aa/bb$$

Here $n_1 = 15$, $n_2 = 10$, and the realized number of runs is $r = 10$. Let us test H_0: the order of the process outcomes is random, against H_1: the order of the process outcomes is not random (Case III). For $a = 0.05$, Table A.10 renders $r_\ell = 7$ and $r_u = 18$. Hence we may specify the critical region as $\mathcal{R} = \{r | r \leq 7 \text{ or } r \geq 18\}$. In this regard, since our sample result is such that $r_\ell < 10 < r_u$, we cannot reject the null hypothesis of randomness at the 5% level. Hence the process outcomes can be treated as if they represent a random sample. ∎

We note briefly that for the Case I pair of hypotheses, the critical region is $\mathcal{R} = \{r | r \geq r_u\}$, where $P(R \geq r_u) = \frac{\alpha}{2} = 0.025$. And for Case II, $\mathcal{R} = \{r | r \leq r_l\}$, where $P(R \leq r_l) = \frac{\alpha}{2} = 0.025$.

B. *Large Sample Runs Test.*   If either n_1 or n_2 exceeds 15, the sampling distribution of R can be shown to be approximately normal with

$$E(R) = \frac{2n_1 n_2}{n} + 1, \quad V(R) = \frac{2n_1 n_2 (2n_1 n_2 - n)}{n^2(n-1)}, \tag{13.2}$$

where $n_1 + n_2 = n$. Hence

$$Z_R = \frac{R - E(R)}{\sqrt{V(R)}} \tag{13.3}$$

is $N(0, 1)$ and thus, for a test conducted at the $100\alpha\%$ level, the decision rules for rejecting H_0 relative to H_1 are:

(a) Case I—reject H_0 if $z_R \geq z_\alpha$;

(b) Case II—reject H_0 if $z_R \leq -z_\alpha$; and   (13.4)

(c) Case III—reject H_0 if $|z_R| \geq z_{\alpha/2}$,

where z_R is the sample realization of Z_R.

Example 13.2.2   Suppose an exit poll is undertaken on the day of an important mayoral election in a large eastern city. A sample of $n = 36$ voters were asked if they voted in the last election. Their responses were either "yes" (an a is recorded) or "no" (in which case a b is recorded). Can we act as if the following ordered sequence of responses resulted from a random sampling of voters? For $\alpha = 0.10$, let us test H_0: the order of the sample data is random, against H_1: the order of the sample data is not random (Case III again). The responses are

$$aa/b/aaa/bb/aaa/b/aa/bbbb/a/b/aaa/b/aaaa/bb/a/bb/a/bb$$

Here $n_1 = 20, n_2 = 16$ and $r = 18$. Then from (13.2) we have

$$E(R) = \frac{2(20)(16)}{36} + 1 = 18.77,$$

$$V(R) = \frac{2(20)(16)[2(20)(16) - 36]}{(36)^2(35)} = 8.52,$$

and thus, from (13.3),

$$|z_R| = \left| \frac{18 - 18.77}{\sqrt{8.52}} \right| = 0.26.$$

Since this quantity is not an element of the critical region $\mathcal{R} = \{ z_R | \, |z_R| \geq 1.645\}$, we cannot reject the null hypothesis at the 10% level. Hence the set of sample observations can be treated as if they constitute a random sample. ■

The preceding versions of the runs test have been developed for nonquantitative (nominal or ordinal) or attribute data. However, if a sample consists of a set of numerical values, such as those measured on an interval or ratio scale, then a slightly different version of the runs test can be offered. In particular, we now turn to the specification of a test for randomness that utilizes runs above and below the sample median.

C. *Runs Above and Below the Sample Median.*   Our first step is to determine the median of the data set. We then seek to determine if each observation in the ordered sequence is above or below the median. Looking to the same type of letter code as developed earlier, let a represent the label assigned to a value falling above the median, and let b denote the label assigned to a value which falls below the median. (Numerical values equal to the median are discarded.)

As before, $n_1 =$ total number of a's, $n_2 =$ total number of b's, and R is the total number of runs *above and below the median*. Once these quantities have been realized from the sample data, (13.2) through (13.4) can be used to test the hypotheses concerning, say, trends or cyclical behavior.

Example 13.2.3   Suppose the following ordered sequence of $n = 40$ observations emerges from a particular process:

98	99	62	65	89	87	87	88	91	62	65	71	91	97	98	90	62
66	89	90	71	72	80	90	65	60	59	60	52	57	52	55	67	80
91	32	61	61	71	60

Once these observations are arranged in an increasing sequence, we can easily verify that the sample median equals 71. Then those items above the median will be marked with an a and those items below the median will be tagged with a b. The ordered sequence of a's and b's appears as

aa/bb/aaaaa/bb/aaaa/bb/aaaaa/bbbbbbbbb/aa/bbbb

(Note that the three 71's have been omitted since they coincide with the median.) Thus $n = 37, n_1 = 18, n_2 = 19$ and $r = 10$. Since it appears that most of the a's occur at the beginning of the sequence and a bunching of b's emerges at the end of the same, we shall consider the Case II option for our test of randomness of this series; that is, there appears to be a downward trend to the data set. Let $\alpha = 0.05$. Then from (13.2) through (13.4):

$$E(R) = \frac{2(18)(19)}{37} + 1 = 19.48,$$

$$V(R) = \frac{2(18)(19)[2(18)(19) - 37]}{(37)^2(36)} = 8.98,$$

and

$$z_R = \frac{10 - 19.48}{\sqrt{8.98}} = -3.17.$$

Here $\mathcal{R} = \{z_R | z_R \leq -1.96\}$. Clearly we may reject the null hypothesis of randomness at the 5% level; that is, we have a statistically significant downward trend to this data series since it exhibits too few runs to be viewed as if the order of the observations resulted from a purely random process. ∎

D. *von Neumann Ratio Test.* We now turn to the utilization of the von Neumann ratio to test for the randomness (or independence) of an ordered sample data set. Quite often a collection of observations is indexed by the passage of time; that is, it is time that imposes order upon the data set. In such instances the resulting data set, generated sequentially in time, will be referred to as a *discrete time series*. (It is discrete in the sense that observations are made on a variable X at some fixed interval h.)

A series of observations ordered in time is typically *not* a random sample; for example, annual values of U.S. gross domestic product (GDP) from 1960 to 2000 is not a random series. That is to say, the value of an observation recorded on a variable X in period i may be related to its value in period $i - j, i > j$. In such instances we say that the observations are *serially correlated*. If $j = 1$, then we have *first-order serial correlation*; that is, successive X values are related. More formally, we are interested in determining whether

$$P(X_i = a' \mid X_{i-1} = a) = P(X_i = a'), \tag{13.5}$$

that is does knowledge of the value of X in period $i - 1$ enable us to predict the value of X in period i? If the answer is *no*, then the X series is independent or random and thus lacks first-order serial correlation (however, serial correlation of a higher order may still exist). But if the answer is *yes*, then the X values are first-order serially correlated and thus their order is not independent or random. In general, if inequality holds in (13.5) for some value of i, then serial correlation is present—the data stream is not generated by a random process.

Serial correlation, when it occurs, can be classified as either positive or negative. When *positive serial correlation* is present, we find that high (respectively, low) values of X tend to remain high (respectively, low) for a succession of time periods whereas when *negative serial correlation* manifests itself, successive X values tend to alternate between high and low levels. In this regard, although we shall always work with a null hypothesis that asserts that the time series is random or independent (i.e., H_0: no serial correlation) a variety of alternative hypotheses may be specified:

Case I	Case II	Case III
H_0: no serial correlation;	H_0: no serial correlation;	H_0: no serial correlation;
H_1: negative serial correlation	H_1: positive serial correlation	H_1: serial correlation exists

Given a set of sample random variables $\{X_1, \ldots, X_n\}$, the von Neumann ratio test for randomness or independence of a time series employs the difference between the successive values of a variable X or $X_i - X_{i-1}, i = 2, 3, \ldots, n$. If we square each of the differences $X_i - X_{i-1}$ and pass to its expectation, we have

$$E(X_i - X_{i-1})^2 = E[(X_i - \mu) - (X_{i-1} - \mu)]^2$$

$$= E(X_i - \mu)^2 - 2[E(X_i - \mu)(X_{i-1} - \mu)] + E(X_{i-1} - \mu)^2$$

$$= \sigma^2 - 2COV(X_i, X_{i-1}) + \sigma^2. \tag{13.6}$$

Under the null hypothesis of independence, $COV(X_i, X_{i-1}) = 0$ so that (13.6) reduces to

$$E(X_i - X_{i-1})^2 = 2\sigma^2. \tag{13.6.1}$$

Next, if we average the $n - 1$ squared differences, we obtain what is commonly called the *mean–square–successive difference* (MSSD)

$$MSSD = \frac{1}{n - 1} \sum_{i=2}^{n} (X_i - X_{i-1})^2. \tag{13.7}$$

Then, by virtue of (13.6.1),

$$E(MSSD) = \frac{1}{n-1} E\left[\sum_{i=2}^{n}(X_i - X_{i-1})^2\right] = \frac{1}{n-1}\sum_{i=2}^{n}E(X_i - X_{i-1})^2$$

$$= \frac{1}{n-1}(n-1)(2\sigma^2) = 2\sigma^2$$

and thus $E\left[\frac{MSSD}{\sigma^2}\right] = 2$. If we replace σ^2 by its unbiased estimator $S^2 = \frac{\sum_{i=1}^{n}(X_i - \overline{X})^2}{n-1}$, then the statistic

$$D = \frac{MSSD}{S^2} = \frac{\sum_{i=1}^{2}(X_i - X_{i-1})^2}{\sum_{i=1}^{n}(X_i - \overline{X})^2} \tag{13.8}$$

denotes the *von Neumann ratio of the mean–square–successive difference to the variance.* The sample realization of (13.8) is

$$d = \frac{\sum_{i=2}^{n}(x_i - x_{i-1})^2}{\sum_{i=1}^{n}(x_i - \bar{x})^2}. \tag{13.8.1}$$

Turning to Case III (with its two-sided alternative hypothesis), Table A.20 of the Appendix provides us with both lower- and upper-tail critical values d_ℓ and d_u, respectively, of the sampling distribution $f(y)$ of D, which restrict the probability in each tail of $f(y)$ to either $\frac{\alpha}{2} = 0.01$ or $\frac{\alpha}{2} = 0.05$, where $\alpha (= 0.02$ or $0.10)$ is the level of significance of the test. Thus d_ℓ and d_u are chosen so that

$$P(D \leq d_\ell \text{ or } D \geq d_u) = P(D \leq d_\ell) + P(D \geq d_u) = 0.02 \text{ (or } 0.10)$$

Hence the critical region corresponding to the Case III two-tail test is $\mathcal{R} = \{d | d \leq d_\ell \text{ or } d \geq d_u\}$, where both d_ℓ and d_u are specified for either $\alpha = 0.02$ or $\alpha = 0.10$. So if our sample result is such that $d_\ell < d < d_u$, then we cannot reject the null hypothesis at the stated α level. If you suspect the presence of negative serial correlation (Case I), then the critical region is $\mathcal{R} = \{d | d \geq d_u\}$ with $P(D \geq d_u) = \alpha (= 0.01$ or $0.05)$. And if you suspect that position serial correlation is likely to occur (Case II), then $\mathcal{R} = \{d | d \leq d_\ell\}$, where $P(D \leq d_\ell) = \alpha (= 0.01$ or $0.05)$.

Example 13.2.4   Given the following sample of $n = 15$ yearly observations on a variable X (see Table 13.2), is there any compelling evidence to indicate the presence of positive serial correlation (Case II) in the population? Let $\alpha = 0.01$.

Table 13.2

	Yearly Observations on X			
x_i	$x_i - x_{i-1}$	$(x_i - x_{i-1})^2$	$x_i - \bar{x}$	$(x_i - \bar{x})^2$
15	–	–	3	9
14	−1	1	2	4
15	1	1	3	9
13	−2	4	1	1
12	−1	1	0	0
9	−3	9	−3	9
8	−1	1	−4	16
10	2	4	−2	4
9	−1	1	−3	9
8	−1	1	−4	16
11	3	9	−1	1
12	1	1	0	0
13	1	1	1	1
15	2	4	3	9
16	1	1	4	16
180		39		104

Given that $\bar{x} = \frac{180}{15} = 12$, we find, from (13.8.1), that $d = \frac{39}{104} = 0.375$. From Table A.20 we have $\mathcal{R} = \{d \mid d \leq 0.9880\}$. Since our realized D falls within this critical region, we reject the null hypothesis of no serial correlation at the 1% level in favor of the alternative hypothesis that positive serial correlation exists in the population.

Since Table A.20 holds for only $\alpha = 0.01$ or $\alpha = 0.05$, it might be useful to have the p-value associated with d. To this end, it can be shown that for larger samples, the sampling distribution of D tends to normality with

$$E(D) = \frac{2n}{n-1}, \quad V(D) = \frac{4n^2(n-2)}{(n-1)^3(n+1)}. \tag{13.9}$$

Then from the preceding example problem,

$$P(D \leq d) = P(D \leq 0.375) = P\left(\frac{D - E(D)}{\sqrt{V(D)}} \leq \frac{0.375 - 2.142}{\sqrt{0.2665}}\right)$$

$$= P(Z_D \leq -3.423) = 0.0003,$$

where Z_D is $N(0, 1)$. Hence the p-value is 0.0003; that is, the probability of getting a value of Z_D *this small or smaller* if the null hypothesis is true is virtually zero.

(Remember, however, that our $n = 15$ value is not exactly what we would call *large*.) ∎

13.3   Single-Sample Sign Test Under Random Sampling

Suppose we extract a random sample of size n from a given continuous population, with $\{x_1, \ldots, x_n\}$ serving as the set of realizations of the sample random variables $X_1, \ldots, X_n$. Furthermore, suppose we have a criterion that serves to dichotomize the observations x_i; that is, we have a rule that places each sample realization in either of two mutually exclusive categories. If ties occur (an observation is common to each category), then the observation is eliminated and n is appropriately adjusted downward. Once the dichotomization of the sample is complete, the original (remaining) data points x_i are converted to plus (+) or minus (−) signs. The resulting plus or minus signs can be associated with observations on any of the four measurement scales.

For instance, if we have a data set containing observations measured on a nominal scale that pertain to the sex of an individual, then we may arbitrarily assign males a "+" sign and females a "−" sign. Or if we have a set of observations measured on an interval or ratio scale (the population variable is continuous) and we are interested in testing a hypothesis about the population median (denoted $\gamma_{0.5}$), then any observation that exceeds the null value $\gamma_{0.5}^0$ can be replaced by a + sign; and an observation that falls below the null value $\gamma_{0.5}^0$ can be replaced by a − sign. Values tied with the null value are discarded. Note that if the population is symmetric, then the mean and median coincide and thus testing a hypothesis about the median is equivalent to a hypothesis test concerning the mean.

Once the assignment of plus and minus signs is complete, the sign test is performed under the assumptions that (1) plus and minus signs are equally likely within the population itself; and (2) the sign attached to one observation is independent of the sign attached to any other observation. Additionally, since only the number of plus (minus) signs will be needed in what follows, the order in which the plus and minus signs occur is irrelevant (remember that order was important for the runs test). Note that by virtue of assumption 1, the sign test is a special case of the binomial test for the parameter p; that is, under dichotomization the sign test is equivalent to a test of the Bernoulli hypothesis $H_0: p = 0.50$.

A. *Sign Test for a Proportion.*   In the first version of the sign test that follows, let p^+ and p^- denote the population proportions of plus and minus signs, respectively. Moreover, let $S^+ =$ the number of plus signs and $S^- =$ the number of minus signs. The sample realizations of S^+ and S^-, will be denoted as s^+ and s^-, respectively. Given the following sets of hypotheses:

Case I	Case II	Case III
$H_0: p^+ \leq p^-$	$H_0: p^+ \geq p^-$	$H_0: p^+ = p^-$
$H_1: p^+ > p^-$	$H_1: p^+ < p^-$	$H_1: p^+ \neq p^-$

the specification of our test statistic S (with realization s) will be based upon the alternative hypothesis made. That is,

TEST STATISTIC S UNDER H_1:

(a) Case I—$S = S^+$;

(b) Case II—$S = S^-$; and   (13.10)

(c) Case III—$S = \max\{S^+, S^-\}$.

Hence we may frame, for a given level of significance $\alpha = P(\text{TIE})$, our

DECISION RULE FOR REJECTING H_0 IN FAVOR OF H_1:

(a) Case I—reject H_0 if p-value $= P(X \geq s^+) \leq \alpha$;

(b) Case II—reject H_0 if p-value $= P(X \geq s^-) \leq \alpha$;   (13.11)

(c) Case III—reject H_0 if p-value $= 2P(X \geq s) \leq \alpha$,

where the random variable X is binomially distributed $b(x; n, 0.5)$.

Example 13.3.1   Suppose that a market research firm is interested in determining whether or not more than 75% of the shoppers at a given department store have seen at least one of the store's TV commercials broadcast by a local TV station over the past week. A random sample of 15 shoppers was selected and each was asked if they viewed at least one of the commercials. Twelve responded positively (they each were tagged with a plus sign) and three said no (they received minus signs). At the 5% level, is there any compelling evidence that would lead us to conclude that the proportion of shoppers watching the commercials is greater than 0.75? Let us test $H_0: p^+ \leq 0.75$ against $H_1: p^+ > 0.75$. Since $s^+ = 12$, let us find, via Table A.7 of the Appendix,

$$p\text{-value} = P(X \geq 12 | n = 15, p = 0.5)$$

$$= \sum_{j=12}^{15} b(j; 15, 0.5)$$

$$= \sum_{j=12}^{15} \frac{15!}{j!(15-j)!} (0.5)^j (0.5)^{15-j}$$

$$= 1 - P(X \leq 11 | n = 15, \ p = 0.5) = 1 - 0.9824 = 0.0176.$$

Since this p-value or observed level of significance falls short of $a = 0.05$, we will reject the null hypothesis in favor of the alternative. There is sample evidence to warrant the conclusion that more than 75% of the shoppers have seen at least one of the TV commercials run by the department store. ∎

We next turn to the specification of the single-sample:

B. *Sign Test for the Median.*   (As indicated earlier, if we can assume that the population distribution is symmetrical, then the sign test for the median is equivalent to conducting the sign test for the mean.) Given a set of n independent sample random variables $\{X_1, \ldots, X_n\}$ taken from a continuous population whose values are measured on an interval or ratio scale, we may employ the sample realizations $x_1, \ldots, x_n$ to test any of the following hypotheses concerning the median $\gamma_{0.5}$ of the population:

Case I	Case II	Case III
$H_0: \gamma_{0.5} \leq \gamma_{0.5}^0$	$H_0: \gamma_{0.5} \geq \gamma_{0.5}^0$	$H_0: \gamma_{0.5} = \gamma_{0.5}^0$
$H_1: \gamma_{0.5} > \gamma_{0.5}^0$	$H_1: \gamma_{0.5} < \gamma_{0.5}^0$	$H_1: \gamma_{0.5} \neq \gamma_{0.5}^0$

where $\gamma_{0.5}^0$ is the null value of $\gamma_{0.5}$.

If an observation $x_i > \gamma_{0.5}^0$, then x_i is assigned a plus sign; and if $x_i < \gamma_{0.5}^0$, then x_i is assigned a minus sign. Any observation that is equal to the median is discarded and n is reduced appropriately. As earlier, let our test statistic be determined by (13.10). Then for a given level of significance α, the decision rule for rejecting the null hypothesis in favor of the alternative is again (13.11).

Example 13.3.2   A bank manager claims that the median family income of its customers does not exceed $\$25,000$. To substantiate the claim, the manager has a staff member select a random sample of 20 customer records from the loan department where family income information is collected:

10,000	30,000	17,800	25,000	19,700
15,000	22,500	25,000	32,000	22,600
22,000	23,500	26,400	24,300	24,000
27,000	31,000	27,000	21,000	29,000

Here we wish to test $H_0: \gamma_{0.5} \geq \$25,000$, against $H_1: \gamma_{0.5} < \$25,000$ at, say, the $\alpha = 0.05$ significance level. Since $s = s^- = 11$ and there are two ties, we reduce n to 18. We know from our earlier discussion that under H_0, s^- is binomially distributed with $p \geq 0.5$; and under H_1, s^- follows a binomial distribution with $p < 0.5$. Then from Table A.7, let us determine

$$p\text{-value} = P(X \geq 11 | n = 18, \ p = 0.5)$$

$$= \sum_{j=11}^{18} b(j; 18, 0.5)$$

$$= \sum_{j=11}^{18} \frac{18!}{j!(18-j)!} (0.5)^j (0.5)^{18-j}$$

$$= 1 - P(X \le 10 | n = 18, \ p = 0.5)$$

$$= 1 - 0.7597 = 0.2403.$$

Since this p-value exceeds our chosen α, we cannot reject the null hypothesis; the manger's claim is substantiated by the data. ∎

C. *Large-Sample Sign Test for the Median.* For $n \ge 20$, the normal approximation to the calculation of binomial probabilities (see Section 7.3) can be used to conduct a large-sample sign test for the median. Under H_0, S (determined from (13.10)) follows a binomial distribution with $p = 0.5$ so that $E(S) = np = 0.5n$ and $V(S) = np(1 - p) = 0.25n$. Then our test statistic is the quantity

$$Z_s = \frac{(S - 0.5) - E(S)}{\sqrt{V(S)}} = \frac{(S - 0.5) - 0.5n}{0.5\sqrt{n}}, \tag{13.12}$$

which is $N(0, 1)$, where 0.5 is subtracted from S as a correction for continuity (see (7.32.1)). In view of (13.12), our decision rule for rejecting H_0 in favor of H_1 at the α level of significance is

(a) Case I—reject H_0 if $z_S \ge z_\alpha$;

(b) Case II—reject H_0 if $z_s \le -z_\alpha$; and $\qquad$ (13.13)

(c) Case III—reject H_0 if $|z_s| \ge z_{\alpha/2}$,

where z_s is the sample realization of Z_s.

Example 13.3.3 Suppose that for a sample of size $n = 40$ we wish to test $H_0: \gamma_{0.5} = 150$, against $H_1: \gamma_{0.5} \ne 150$ at the 5% level of significance. Furthermore, suppose we find that $s = s^+ = \max\{s^+, s^-\} = 25$. Then from (13.12),

$$|z_s| = \left| \frac{25 - 0.5 - 20}{3.16} \right| = 1.42.$$

Since $|z_s|$ does not exceed $z_{a/2} = z_{0.025} = 1.96$, we cannot reject the null hypothesis at the 5% level. In fact, it is easily demonstrated that the p-value for this test is 0.1556. ∎

13.4 Wilcoxon Signed Rank Test of a Median

The sign test (for the median of a random sample) that we just examined did not assume that the underlying population was symmetrical. However, if we have sufficient evidence that points to a symmetric population distribution, then we can perform the Wilcoxon signed rank test for the median—a test that is more powerful than the sign test since, in addition to having a criterion that dichotomizes the sample observations, we can also exploit the ranks of the data values rather than simply replacing the same by a sign. To this end, let us extract a set of independent sample random variables $\{X_1, \ldots, X_n\}$ from a continuous symmetric population (although the population is taken to be symmetric, it is not assumed to be normally distributed). Here too the population values are measured on an interval or ratio scale.

Given that x_i depicts the realization of the sample random variable $X_i, i = 1, \ldots, n$, we may test the following hypotheses pertaining to the population median $\gamma_{0.5}$:

Case I	Case II	Case III
$H_0: \gamma_{0.5} \leq \gamma_{0.5}^0$	$H_0: \gamma_{0.5} \geq \gamma_{0.5}^0$	$H_0: \gamma_{0.5} = \gamma_{0.5}^0$
$H_1: \gamma_{0.5} > \gamma_{0.5}^0$	$H_1: \gamma_{0.5} < \gamma_{0.5}^0$	$H_1: \gamma_{0.5} \neq \gamma_{0.5}^0$

where $\gamma_{0.5}^0$ is the null value of $\gamma_{0.5}$. To perform the Wilcoxon signed rank test, let us execute the following steps:

1. Subtract the null value $\gamma_{0.5}^0$ from each x_i; that is, from the differences $y_i = x_i - \gamma_{0.5}^0, i = 1, \ldots, n$. (If any $y_i = 0$, eliminate x_i as in the sign test and reduce n accordingly.)

2. Rank the y_i's in order of increasing absolute value. (If any of the nonzero y_i's are tied in value, then these tied y_i's are given the average rank.)

3. Restore to the rank values $1, \ldots, n$ the algebraic sign of the associated difference y_i. Then the ranks with the appropriate signs attached are called the *signed ranks* $R_i, i = 1, \ldots, n$, where R_i^+ denotes a *rank carrying a positive sign*. The sample realization of R_i^+ will be depicted as r_i^+.

Let us define our test statistic W^+ as the sum of the positive ranks or $W^+ = \sum_{i=1}^{n} R_i^+$.

A. *Small-Sample Wilcoxon Signed Rank Test.* A small-sample Wilcoxon signed rank test will be performed when $n \leq 25$. For Case I and a level of significance α, Table A.11 (Appendix) specifies the right-hand tail critical value w_u of the sampling distribution of W^+, which cuts off $100\alpha\%$ of the upper-tail area or $P(W^+ \geq w_u) = \alpha$. Hence the Case I critical region appears as $\mathcal{R} = \{w^+ | w^+ \geq w_u\}$, where w^+ is the sample realization of W^+. We thus reject H_0 for large values of W^+. Turning to Case II, let w_l denote the left-hand tail critical value, which satisfies $P(W^+ \leq w_l) = \alpha$. Hence the lower-tail critical region has the form $\mathcal{R} = \{w^+ | w^+ \leq w_l\}$; that is, we now reject H_0 for small values of W^+. Finally, for Case III, w_l and w_u are chosen to satisfy $P(W^+ \leq w_l \text{ or } W^+ \geq w_u) = \alpha$. Hence the two-tail critical region is structured as $\mathcal{R} = \{w^+ | w^+ \leq w_l \text{ or } w^+ \geq w_u\}$. We consequently reject the null hypothesis for either very small or very large values of our test statistic.

Table 13.3

	Ranks and Signed Ranks		
$y_i = x_i - 22$	$|y_i|$	**Ranks of $|y_i|$**	**Signed Ranks R_i**
-2	2	2.5	-2.5
3	3	4.5	4.5
10	10	11	11
11	11	12	12
5	5	7.5	7.5
-5	5	7.5	-7.5
-12	12	13.5	-13.5
-3	3	4.5	-4.5
4	4	6	6
9	9	10	10
12	12	13.5	13.5
-2	2	2.5	-12.5
-1	1	1	-1
-7	7	9	-9

Example 13.4.1   To see exactly how Table A.11 is employed, suppose the marketing department for a retail food chain is monitoring the number of *hits* its Web site gets over a two week period. The recorded hits per day are:

20 25 32 33 27 17 10 19 26 31 35 20 21 15

At the 5% level, is there sufficient evidence to indicate that the median number of hits per day differs from 22? That is, we seek to test $H_0: \gamma_{0.5} = \gamma_{0.5}^0 = 22$ against $H_1: \gamma_{0.5} \neq 22$. From Table 13.3 we have $w^+ = \sum_{j=1}^{7} r_j^+ = 4.5 + 11 + 12 + 7.5 + 6 + 10 + 13.5 = 64.5$ (since 2 is given an average rank of 2.5, 3 gets an average rank of 4.5, and 12 gets an average rank of 13.5). Looking to Table A.11 for $n = 14$ (since no ties with the null value $\gamma_{0.5}^0 = 22$ have been discarded), the critical region is structured as $\mathcal{R} = \{w^+ | w^+ \leq 21 \text{ or } w^+ \geq 84\}$. Given that $w^+ = 64.5$ is not a member of $\mathcal{R}$, we cannot reject the null hypothesis—there is insufficient evidence for us to conclude that the median number of site hits differs significantly from 22 at the 5% level.   ∎

B. *Large-Sample Wilcoxon Signed Rank Test.*   For $n \geq 25$, an excellent approximate test can be performed using the standard normal distribution. (Actually, the approximation to the standard normal form is quite good for $n \geq 12$.) In this regard, our standardized test statistic under H_0 is

$$Z_{w^+} = \frac{W^+ - m(m+1)/4}{\sqrt{m(m+1)(2m+1)/24}},$$ (13.14)

where Z_{w+} is $N(0, 1)$, W^+ is the sum of the positive signed ranks, and m is the *final* number of observations which are ranked.

Given (13.14), our decision rule for rejecting H_0 in favor of H_1 at the $100\alpha\%$ level of significance is

(a) Case I—reject H_0 if $z_{w+} \geq z_\alpha$;

(b) Case II—reject H_0 if $z_{w+} \leq -z_\alpha$; and $\qquad$ (13.15)

(c) Case III—reject H_0 if $|z_{w+}| \geq z_{\alpha/2}$,

with z_{w+} denoting the sample realization of Z_{w+}.

Example 13.4.2 The number of daily requests for a brochure made to sales personnel by shoppers concerning the features of a new model SUV at an automobile dealership have been recorded for a four week period:

$$
\begin{array}{cccccc}
20 & 25 & 21 & 20 & 30 & 35 \\
10 & 10 & 9 & 11 & 15 & 20
\end{array}
\qquad
\begin{array}{cccccc}
15 & 16 & 10 & 25 & 30 & 40 \\
10 & 9 & 9 & 15 & 17 & 25
\end{array}
$$

For $\alpha = 0.05$, does the median number of brochures per day differ significantly from 20? Here $H_0: \gamma_{0.5} = \gamma_{0.5}^0 = 20$ and $H_1: \gamma_{0.5} \neq 20$. From Table 13.4 we have $m = 21, w^+ = \sum_{j=1}^8 r_j^+ = 6.5 + 1 + 13.5 + 20 + 6.5 + 13.5 + 21 + 6.5 = 88.5$. (Note that 5 is given an average rank of 6.5, 10 is given an average rank of 13.5, and 11 receives an average rank of 18.) Then from (13.14),

$$
|z_{w+}| = \left| \frac{88.5 - 21(22)/4}{\sqrt{21(22)(43)/24}} \right| = 0.938.
$$

A glance back at (13.15) indicates that, at the 5% level of significance, we cannot reject the null hypothesis since $z_{0.025} = 1.96$. In addition, the *p*-value for this test is about 0.34. ∎

It is important to note that the Wilcoxon signed rank test may be viewed as a test of symmetry about the median. That is, we may test H_0: the population distribution is symmetrical about the median, against, say, H_1: the population is not symmetrical about the median.

We end our discussion of single-sample nonparametric hypothesis tests by a comparison of such tests with their parametric counterparts (provided, of course, that a parametric equivalent test exists). Table 13.5 provides the details.

Our efforts now turn to the development of two-sample nonparametric tests, which, very generally, seek to determine if two population distributions are identical. In particular, for independent samples, we shall consider the runs test and the Mann-Whitney (rank sum) test. For paired comparisons, we have the sign test and the Wilcoxon signed rank test.

Table 13.4

	Ranks and Signed Ranks		
$y_i = x_i - 20$	$|y_i|$	**Ranks of $|y_i|$**	**Signed Ranks R_i**
0	–	–	–
5	5	6.50	6.50
1	1	1.00	1.00
0	–	–	–
10	10	13.50	13.50
15	15	20.00	20.00
−5	5	6.50	−6.50
−4	4	3.00	−3.00
−10	10	13.50	−13.50
5	5	6.50	6.50
10	10	13.50	13.50
20	20	21.00	21.00
−10	10	13.50	−13.50
−10	10	13.50	−13.50
−11	11	18.00	−18.00
−9	9	10.00	−10.00
−5	5	6.50	−6.50
0	–	–	–
−10	10	13.50	−13.50
−11	11	18.00	−18.00
−11	11	18.00	−18.00
−5	5	6.50	−6.50
−3	3	2.00	−2.00
5	5	6.50	6.50

Table 13.5

Nonparametric Tests	**vs.**	**Parametric Equivalents**
(a) Single-Sample Runs Test		(a′) No Parametric Equivalent
(b) Single-Sample Sign Test		(b′) Binomial Test on p or $\overline{Z}$ or t test of $H_0: \mu = \mu_0$ (provided the population is normal)
(c) Single-Sample Wilcoxon Signed Rank Test		(c′) t Test of $H_0: \mu = \mu_0$ (provided the population is normal)

13.5   Runs Test for Two Independent Samples

The runs test presented earlier for a single sample can also be used to compare the identity of two population frequency distributions, given that we have two independent and unpaired random samples of sizes n_1 and n_2, respectively. It is assumed that the underlying population characteristic that these samples represent can be described by a variable that follows a continuous distribution (so that the test applies to data measured on an interval or ratio scale).

Each of the $n = n_1 + n_2$ observations in these samples is assigned a letter code; for example any observation from sample 1 is marked with the letter a, and any observation appearing in sample 2 is tagged with the letter b. Next, let us rank all n observations jointly according to the order of magnitude of their scores, with an a placed below each observation belonging to the first sample and a b placed beneath each value belonging to the second sample. Hence we have an ordered sequence of a's and b's so that we can now conduct a test for the randomness of this arrangement.

We now need to focus on the runs or clusterings of the a's and b's. If the two samples have been drawn from identical populations, then we should expect to see many runs since the n observations from the two samples should be completely intermingled when placed in numerical order. (With identical populations, there will be, at most, $n_1 + n_2$ runs in the ordered sequence.) However, if the two populations are not identical (e.g., they differ with respect to location or central tendency), then we should expect fewer runs in the ordered arrangement. In fact, in the extreme case of a large disparity in central tendency, we should expect only two runs—all the observations in one sample lie below all the data points in the other sample.

In view of this discussion, let us frame the null and alternative hypotheses as:

H_0: the population distributions are identical

H_1: too few runs (the two samples come from populations having, say, unequal means)

Here the alternative hypothesis obviously implies a one-tailed critical region.

If we let R denote the total number of runs appearing in the joint sequence of a's and b's, then it can be shown that for large values of n_1 and n_2, the sampling distribution of R is approximately normal with

$$E(R) = \frac{2n_1 n_2}{n_1 + n_2} + 1 \qquad (13.16)$$

and

$$V(R) = \frac{2n_1 n_2 (2n_1 n_2 - n_1 - n_2)}{(n_1 + n_2)^2 (n_1 + n_2 - 1)}.$$

Hence, under H_0, the distribution of the standardized R or

$$Z_R = \frac{R - E(R)}{\sqrt{V(R)}} \qquad (13.17)$$

is $N(0, 1)$ and thus, for a test conducted at the $100\alpha\%$ level of significance, the critical region is $\mathcal{R} = \{z_R | z_R \leq -z_\alpha\}$ with $P(R \leq -z_\alpha) = \alpha$, where z_R is the sample realization of Z_R. For convenience, we may express our decision rule as

$$\text{Reject } H_0 \text{ in favor of } H_1 \text{ if } z_R \geq -z_\alpha. \tag{13.18}$$

The approximation provided by (13.17) is quite good when both n_1 and n_2 exceed 10.

Example 13.5.1   Suppose the following two samples have been extracted independently and at random from two distinct continuous populations. Can we conclude on the basis of a runs test with $\alpha = 0.05$ that these samples come from identical populations?

Sample 1($n_1 = 12$)

15.2	9.0	17.4	18.0	12.0	13.1	20.6	9.1
10.8	15.3	19.0	7.5				

Sample 2($n_2 = 15$)

19.5	22.0	17.1	19.1	15.6	20.2	21.7	21.9
30.0	15.1	17.3	17.5	31.0	35.2	33.4	

Joint Ranking of Samples 1 and 2 ($n = n_1 + n_2 = 27$)

7.5	9.0	9.1	10.8	12.0	13.1	15.1	15.2	15.3	15.6	17.1	17.3
a	a	a	a	a	a	/b/	a	a/	b	b	b/
17.4	17.5	18.0	19.0	19.1	19.5	20.2	20.6	21.7	21.9	22.0	30.0
a/	b/	a	a/	b	b	b/	a/	b	b	b	b
31.0	33.4	35.2									
b	b	b									

Here the realized value of R is $r = 10$. Then from (13.16) we have

$$E(R) = \frac{2(12)(15)}{27} + 1 = 14.333,$$

$$V(R) = \frac{2(12)(15)\,[2(12)(15) - 27]}{(27)^2(26)} = 6.325$$

so that, from (13.16),

$$z_R = \frac{10 - 14.333}{\sqrt{6.325}} = -1.722.$$

Since $z_\alpha = z_{0.05} = 1.645$, our decision rule (13.18) has us reject the null hypothesis of identical population distributions at the 5% level; that is, it appears that the mean of the first population lies significantly below the mean of the second population. ∎

The version of the runs test just described is very general and thus is actually consistent with a whole host of differences among two continuous population distributions (and not just with differences in central tendency). If we reject the null hypothesis that the populations are identical (in every conceivable way), there may be many reasons why the populations differ—reasons that are not explicitly incorporated in the alternative hypothesis. So, although rejection of the null hypothesis tells us that the populations are not identical, we are left cold when it comes to determining exactly how they differ. The price to be paid for such generality is the test's relatively low power (remember, however, that the exact power of a test can be determined only against specific alternative hypotheses). This runs test should be used as the basis for the preliminary study of two independent random samples or applied when all other types of two-sample tests have either been exhausted or deemed inappropriate.

13.6 Mann-Whitney (Rank-Sum) Test for Two Independent Samples

The Mann-Whitney test, like the runs test of the preceding section, is designed to compare the identity of the two population distributions by examining the characteristics of two independent and unmatched random samples of sizes n_1 and n_2, respectively. Here, too, the population distributions are assumed continuous with the observations measured on an interval or ratio scale. However, unlike the runs test, the Mann-Whitney procedure exploits the numerical ranks of the observations once they have been jointly arranged in an increasing sequence.

In this regard, suppose we arrange the $n = n_1 + n_2$ sample values in an increasing order of magnitude and assign them the ranks $1, \ldots, n$ while keeping track of the source sample from which each observation was selected for ranking; that is, an observation taken from sample 1 is tagged with, say, letter a and an observation extracted from sample 2 is marked with the letter b. (If ties in the rankings occur, we simply resolve them by assigning each of the tied values the average of the ranks that would have been assigned to these observations in the absence of a tie.)

What can the rankings tell us about the population distributions? Let R_1 and R_2 denote the *rank sums* for the first and second samples, respectively. If the observations were selected from identical populations, then R_1 and R_2 should be nearly equal in value. However, if the data points in, say, population 1 tended to be larger than those in population 2 (given equal sample sizes) then obviously R_1 should be appreciably larger than R_2, thus providing evidence that the populations differ in some fashion (typically in central tendency or location). Hence very large or very small values of R_1 imply a separation of the rankings of the sample 1 versus sample 2 observations, thus providing evidence of a shift in the location of one population relative to the other.

The Mann-Whitney test frames the null hypothesis as H_0: the population distributions are identical. This hypothesis may be tested against alternatives specified as:

Case I	Case II	Case III
H_0: the population distributions are identical	H_0: the population distributions are identical	H_0: the population distributions are identical
H_1: population 1 is located to the right of population 2	H_1: population 1 is located to the left of population 2	H_1: the populations differ in location

For our test statistic let us use $R_1 =$ sum of the ranks associated with the observations from sample 1, with r_1 denoting its sample realization. (It is immaterial which sample is designated as sample 1.) To appropriately execute the rank-sum test, we need to take account of the sizes of our two samples. To this end we first consider the following.

A. *Small-Sample Mann-Whitney Rank-Sum Test.*   In general, a small-sample rank-sum test will be performed when $n_1 \leq 20$ and $n_2 \leq 20$. For Case I and a given level of significance α, Table A.12 (Appendix) enables us to locate the upper-tail critical value r_u of the sampling distribution of R_1, which restricts the area in the right-hand tail to α or $P(R_1 \geq r_u) = \alpha$. Hence the Case I critical region is of the form $\mathcal{R} = \{r_1 | r_1 \geq r_u\}$; that is, we reject H_0 for large values of R_1. Looking to Case II, if r_l denotes the lower-tail critical value, which is chosen so that $P(R_1 \leq r_l) = \alpha$, then the lower-tail critical region has the form $\mathcal{R} = \{r_1 | r_1 \leq r_l\}$ (in this instance we reject H_0 for small values of R_1). Finally, for the Case III set of hypotheses, r_l and r_u are chosen so that $P(R_1 \leq r_l$ or $R_1 \geq r_u) = \alpha$ (thus the area in each tail of the sampling distribution of R_1 is $\frac{\alpha}{2}$) and thus the two-tailed critical region amounts to $\mathcal{R} = \{r_1 | r_1 \leq r_l$ or $r_1 \geq r_u\}$; that is, based upon the sample values, we seek to detect a shift in the population distributions in either direction.

Example 13.6.1   A law firm in a large city is trying to choose between two messenger services (call them 1 and 2) for pick-up and delivery of legal forms to/from nearby clients. It uses each service for a two-week period and records the times (in minutes) elapsed after a call requesting a pick-up. These times are displayed in Table 13.6. The office manager feels that service 1 might be more responsive than service 2 (pick-ups are quicker for service 1). Conduct a Mann-Whitney test of this hypothesis at the $\alpha = 0.05$ level. Mirroring Case II we form

H_0: the response times are identically distributed

H_1: service 1's response time distribution lies to the left of service 2's response time distribution

Table 13.6

Response Times in Minutes	
Service 1 ($n_1 = 9$)	**Service 2($n_2 = 8$)**
8	10
15	12
13	15
9	13
10	18
17	19
12	20
13	15
16	

If we tag service 1 observations with an a and service 2 observations with a b, then the following rankings allow us to determine the realized value of R_1 as

$$r_1 = 1 + 2 + 3.5 + 5.5 + 8 + 8 + 11 + 13 + 14 = 66.$$

Service	a	a	a	b	a	b	a	a	b	a	b	b	a	a	b	b	b
Joint Ordering	8	9	10	10	12	12	13	13	13	15	15	15	16	17	18	19	20
Rank	1	2	3.5	3.5	5.5	5.5	8	8	8	11	11	11	13	14	15	16	17

Then from Table A.12, $\mathcal{R} = \{r_1 | r_1 \le r_l = 63\}$. Since 66 lies outside of this critical region, we cannot reject H_0 at the 5% level. In effect, the two messenger services are equally responsive. ∎

      B. *Large-Sample Mann-Whitney Rank-Sum Test.*   To conduct a large-sample rank-sum test, let us employ the test statistic $U =$ number of observations in sample 1 that precede each observation in sample 2, given that all $n = n_1 + n_2$ observations have been jointly arranged in an increasing sequence. For example, given the following two samples containing $n_1 = n_2 = 4$ observations each:

Sample 1: 22   27   15   32

Sample 2: 14   19   25   30

our combined ordered arrangement is

$$14 \quad 15 \quad 19 \quad 22 \quad 25 \quad 27 \quad 30 \quad 32$$
$$b \quad\ a \quad\ b \quad\ a \quad\ b \quad\ a \quad\ b \quad\ a$$

Since the smallest sample 2 observation is 14, $U_1 = 0$ observations from sample 1 precede it. The sample 2 value of 19 is preceded by only $U_2 = 1$ sample 1 value; the sample 2 value of 25 is preceded by $U_3 = 2$ sample 1 values; and the sample 2 value of 30 is preceded by $U_4 = 3$ sample 1 values. Hence the total number of observations in sample 1 that precede each observation in sample 2 is $U = U_1 + U_2 + U_3 + U_4 = 0 + 1 + 2 + 3 = 6$. Clearly the value of U depends upon how the a's and b's are distributed under the ranking process.

Given that the populations are taken to be continuous (theoretically, at least, there are no ties), it can be shown that

$$U = n_1 n_2 + \frac{n_1(n_1 + 1)}{2} - R_1, \tag{13.19}$$

where again R_1 is the sum of the ranks assigned to observations in the first sample and the sample realization of U is denoted as $u = n_1 n_2 + \frac{n_1(n_1+1)}{2} - r_1$. Since under the null hypothesis of identical population distributions we have

$$E(R_1) = \frac{n_1(n_1 + n_2 + 1)}{2} \quad \text{and} \quad V(R_1) = \frac{n_1 n_2(n_1 + n_2 + 1)}{12},$$

it follows that

$$E(U) = \frac{n_1 n_2}{2}$$

and

$$V(U) = \frac{n_1 n_2(n_1 + n_2 + 1)}{12}. \tag{13.20}$$

Then as n_1 and n_2 increase without bound, the asymptotic distribution of

$$Z_u = \frac{U - E(U)}{\sqrt{V(U)}} \tag{13.21}$$

is standard normal or $N(0, 1)$. (Actually, the approximation to normality is quite good for $n_1 \geq 10$ and $n_2 \geq 10$.)

Given this result, for a test conducted at the $100\alpha\%$ level, the decision rules for rejecting H_0 relative to H_1 are:

(a) Case I—reject H_0 if $z_u \leq z_\alpha$;

(b) Case II—reject H_0 if $z_u \geq -z_\alpha$; and $\qquad$ (13.22)

(c) Case III—reject H_0 if $|z_u| \geq z_{\alpha/2}$,

where z_u is the sample realization of Z_u. Note that for the Case I alternative, U will be small when R_1 is large. Hence the critical region in this instance is actually under the left-hand tail of the standard normal distribution or $\mathcal{R} = \{z_u | z_u \leq -z_\alpha\}$. And if R_1 is small, then U will be large so that, for the Case II alternative, the region of rejection is under the right-hand tail or $\mathcal{R} = \{z_u | z_u \geq z_\alpha\}$. Both of these cases are addressed under (13.22a,b).

Example 13.6.2 To see exactly how a large-sample test using (13.21) is employed, suppose that a production manager wants to determine if the quality levels of the outputs produced by two different processes are the same. Process 1 is run for $n_1 = 15$ weeks and process 2 is run for only $n_2 = 10$ weeks and the percentages of defective items per week are recorded as:

Process 1: 5.7 6.5 6.4 7.0 6.0 6.1 7.2 7.3 6.3 6.2 6.0 6.8 6.4 7.1 5.9;

Process 2: 6.6 7.1 7.0 6.2 6.7 7.3 7.1 6.9 7.0 6.8.

Given that the two processes are independent, let us test H_0: the process population distributions are identical, against H_1: the process populations differ in location at the $\alpha = 0.05$ level. As Table 13.7 reveals, column 1 specifies the observation numbers assigned to the pooled sample values listed in increasing order of magnitude in columns 2 and 3. The process ranks are then reported in columns 4 and 5; that is, although ranks from 1 to 25 are assigned to the pooled sample values, they are separately identified for each sample. (Note that observations 3 and 4 are tied for process 1 so that the average rank assigned to this *within-process* tie is 3.5. A *between-process* tie occurs for observations 6 and 7. Hence each of these observations is given an average rank of 6.5. Observations 17, 18, and 19 involve both a within-process and a between-process tie. The average rank assigned to each tied value is 18. Other ties are handled accordingly.)

The realized sum of the ranks for process 1 is $r_1 = 158.5$. Then from (13.19) through (13.21),

$$U = 150 + \frac{15(16)}{2} - 158.5 = 111.5,$$

$$E(U) = \frac{15(10)}{2} = 75,$$

$$V(U) = \frac{15(10)26}{12} = 325,$$

and thus

$$|z_u| = \left| \frac{111.5 - 75}{18.03} \right| = 2.02.$$

Since 2.02 is a member of the critical region $\mathcal{R} = \{z_u | |z_u| \geq 1.96\}$ (see (13.22)), we reject the null hypothesis at the 5% level. It appears, therefore, that the two processes do not exhibit identical population distributions—at this level of significance, the process 1 distribution seems to be shifted to the left of the process 2 distribution. ∎

Table 13.7

		Percentage Defective				
Pooled Observation Number	Process 1	Process 2	Process 1 Ranks	Process 2 Ranks	$g = 7$ Groups of Tied Ranks	t_i
1	5.7		1.0			
2	5.9		2.0			
3	6.0		3.5			
					{3.5, 3.5}	2
4	6.0		3.5			
5	6.1		5.0			
6	6.2		6.5			
					{6.5, 6.5}	2
7		6.2		6.5		
8	6.3		8.0			
9	6.4		9.5			
					{9.5, 9.5}	2
10	6.4		9.5			
11	6.5		11.0			
12		6.6		12.0		
13		6.7		13.0		
14		6.8		14.5		
					{14.5, 14.5}	2
15	6.8		14.5			
16		6.9		16.0		
17		7.0		18.0		
18		7.0		18.0	{18, 18, 18}	3
19	7.0		18.0			
20		7.1		21.0		
21		7.1		21.0	{21, 21, 21}	3
22	7.1		21.0			
23	7.2		22.0			
24		7.3		23.5		
					{23.5, 23.5}	2
25	7.3		23.5			
			158.5			

The preceding example problem has offered a practical and convenient way of handling tied observations (either on a within-sample or on a between-sample basis). If ties happen to be relatively infrequent, the normal approximation (13.21) is very efficient in detecting any shifts in location characterized by Cases I through III. However, if ties are numerous, then a modification of $V(U)$ is in order. Specifically, if there are g distinct sets of tied ranks or observations, with j indexing any one such set, and t_j is the number of tied ranks in set j, then the variance of U

adjusted for ties is

$$V(U) = \frac{n_1 n_2}{2} \left[(n_1 + n_2 + 1) - \frac{\sum_{j=1}^{g} t_j(t_j^2 - 1)}{(n_1 + n_2)(n_1 + n_2 - 1)} \right]. \tag{13.23}$$

Example 13.6.3 Since there exists a considerable number of ties occurring in the process defect rates appearing in Table 13.7, let us use (13.23) to recalculate $V(U)$ and consequently Z_u (13.21). The $g = 7$ groups of tied ranks and the associated t_j's, $j = 1, \ldots, 7 = g$, are presented in columns 6 and 7. Upon finding

$$\sum_{j=1}^{7} t_j(t_j^2 - 1) = 5[2(3)] + 2[3(8)] = 78,$$

(13.23) renders

$$V(U) = \frac{15(10)}{12} \left[\frac{26 - 78}{(25)(24)} \right] = 323.38.$$

Hence from (13.21)

$$|z_u| = \left| \frac{111.5 - 75}{17.98} \right| = 2.03.$$

Obviously the nature of the ties exhibited by this pooled arrangement does not lead us to modify our basic conclusion. ∎

The Mann-Whitney test is considered one of the best nonparametric tests for differences in location (and thus for differences in means or medians) when its asymptotic relative efficiency is compared to that of the parametric t test for differences in means, given that the assumptions underlying the t test (e.g., normality) hold. That is, the asymptotic relative efficiency of the Mann-Whitney statistic relative to the t statistic is at least 95% if the populations are normal; it is at least 86% in the absence of normality if the population distributions are identical. In fact, for continuous and independent distributions, the Mann-Whitney test might perform better than the t test in the absence of the normality.

Stated alternatively, for given α and β risks, the sample size required for a t test is approximately 95% of that required for the Mann-Whitney test if normality holds. For nonnormal populations, the Mann-Whitney test requires fewer observations than the t test to achieve comparable α and β risks.

Moreover, if we reject the null hypothesis (of identical population distributions) when performing a two-sample nonparametric hypothesis test, we are not told exactly how the populations differ. If we reject the null hypothesis (of equal population means) for the two-sample t test, then we know that the populations differ specifically with respect to their means.

13.7   The Sign Test When Sampling from Two Dependent Populations: Paired Comparisons

Following the discussions on paired comparisons presented earlier in Sections 11.10 and 12.19, let us assume that a paired experiment yields n pairs of observations denoted as $(X_1, Y_1), \ldots, (X_n, Y_n)$, where the sample random variables X_i and $Y_i, i = 1, \ldots, n$, are members of the same pair (X_i, Y_i), with X_i drawn from population 1 and Y_i drawn from population 2. For the ith pair of sample random variables, let $D_i = X_i - Y_i$, $i = 1, \ldots, n$, be the difference between the two random variables X_i and Y_i making up that pair, where D_i is taken to be the ith observation on the random variable D. No distributional assumptions concerning D will be made. Let $d_i = x_i - y_i$ denote the sample realization of $D_i, i = 1, \ldots, n$, where x_i (respectively, y_i) is the realized value of X_i (respectively, Y_i). It is assumed that the population probability distributions are continuous, with data measured on an ordinal (ranks are assigned within pairs), interval, or ratio scale.

A comparison of the members of each pair (X_i, Y_i) will produce a "+" sign, "–" sign, or a zero value. In this regard, when the value of an element from population 1 exceeds the value of its paired element from population 2, we assign a + sign to the pair and $d_i > 0$. (A + sign is also warranted if, say, of two experimental treatments, the outcome of treatment 1 is deemed *more effective* than that of treatment 2; or if an outcome from population 1 is *preferred* to an outcome obtained from population 2.) If the value of an element from population 1 falls short of the value of its paired counterpart from population 2, then obviously the pair is given a – sign and $d_i < 0$. Ties will occur if $d_i = 0$. In this instance any pair (X_i, Y_i) producing a tie is eliminated from the sample. Hence only the sign and not the magnitude of the difference between paired items is important in the sign test.

The sign test involving matched pairs is designed to test the null hypothesis H_0: the population distributions are identical. This hypothesis is tested against the following assortment of alternative hypotheses:

Case I	Case II	Case III
H_0: the population distributions are identical	H_0: the population distributions are identical	H_0: the population distributions are identical
H_1: population 1 is located to the right of population 2	H_1: population 1 is located to the left of population 2	H_1: the populations differ in location

(Based upon the preceding discussion, it should be reasonably clear that the sign test actually tests H_0: the median of the D_i's is zero. Then the Case I, Case II, and Case III alternatives are, respectively, the median of the D_i's is positive; the median of the D_i's is negative; and the median of the D_i's is different from zero.)

The motivation underlying the matched pairs sign test is similar to that mentioned earlier for the single-sample sign test. That is to say, under the null hypothesis that X_i and Y_i come from identical populations, if only chance factors dictate which member of a pair gets which of two treatments, then we should expect an equal number of positive and negative differences or $P(D_i > 0) = P(D_i < 0) = 0.5$. Let S^+ represent the number of positive differences or + signs (S^- depicts the number of negative differences or − signs). If the X_i and Y_i values have identical population distributions, then S^+ will follow a binomial distribution with $p = P(D_i > 0) = 0.5$. Then the critical region (or p-value) can be obtained by using either the binomial distribution proper or its standard normal approximation.

For a small sample ($n < 20$) sign test applied to the signs of the d_i's, the test statistic under H_1 is provided by (13.10) and the decision rule for rejecting H_0 in favor of H_1 is given in (13.11). And for a large sample ($n \geq 20$) sign test (again utilizing only the signs of the d_i's), we may employ (13.12) (wherein n is to be interpreted as the number of untied sample pairs (X_i, Y_i) for which $d_i \neq 0$) and (13.13). (Actually, the large sample sign test has been shown to be valid if the number of nonzero differences is at least 10.)

Example 13.7.1 Suppose $n = 17$ students in a photography class were asked to rate, on a scale from 1 to 5, samples of two types of paper used to make prints. Use the sample data presented in Table 13.8 to test H_0: the distributions of ratings are identical, versus H_1: the distributions of ratings are different (and one type of paper is preferred to the other). Let us choose $\alpha = 0.05$. For this test, $n = 15 =$ number of untied pairs of observations and, considering the signs of the differences d_i, we see that $S = S^+ = 10$. Then

$$|z_s| = \left| \frac{(S - 0.5) - E(S)}{\sqrt{V(S)}} \right| = \left| \frac{(10 - 0.5) - 0.5(15)}{0.5\sqrt{15}} \right| = 1.03.$$

Since 1.03 is not an element of $\mathcal{R} = \{ z_s | |z_s| \geq 1.96 \}$, we cannot reject H_0—the distributions of ratings do not differ significantly at the 5% level. ■

As stated earlier, the sign test for matched pairs does not make any assumption about sampling from normal populations or that the D_i's are normally distributed; and it is not necessary to assume that the D_i's are identically distributed as in the paired t test. Continuity of the population distributions is all that is called for. Moreover, the sign test checks whether the proportion of pairs exhibiting a positive difference equals the proportion of pairs with a negative difference. Equivalently, the test addresses the issue of location by testing for a median difference equal to zero.

Table 13.8

Student	Rankings of Photography Paper (1 = lowest; 5 = highest)		
	Paper 1	**Paper 2**	**Sign of Difference d_i**
1	4	4	Tie
2	4	3	+
3	5	5	Tie
4	4	5	−
5	3	4	−
6	4	5	−
7	3	2	+
8	4	2	+
9	4	5	−
10	3	2	+
11	5	4	+
12	5	3	+
13	3	4	−
14	4	2	+
15	4	1	+
16	5	3	+
17	4	2	+

If the D_i's are normally distributed with a common variance, then the ordinary t test is superior to the sign test in detecting a shift in location; that is, the t test for paired comparisons is a more powerful test since it utilizes the numerical magnitude of the D_i's. The sign test for matched pairs should be employed when the assumptions underlying the t test for the same do not hold or are not easily verified as holding.

13.8   Wilcoxon Signed Rank Test When Sampling from Two Dependent Populations: Paired Comparisons

Although the sign test developed in the preceding section considers only the sign of the differences $D_i = X_i - Y_i$ for each matched pair of observations (X_i, Y_i), $i = 1, \ldots, n$, it ignores the notion that not only does each difference have a sign, it also has a rank among all such differences. The Wilcoxon signed rank test exploits the signs as well as the ranks of the differences. In this regard, both population distributions are assumed continuous and, since the ranks of the differences are utilized, each observation must be a point on an interval or ratio scale. Moreover, the D_i's are taken to be independent random variables that follow a distribution that is symmetrical about a common median.

For the Wilcoxon rank sum test for matched pairs, the null hypothesis typically is taken to be H_0: the population distributions are identical. (The Wilcoxon signed-rank procedure can also be considered as a test for symmetry if the only assumption made is that the D_i's are randomly drawn from a continuous distribution.) This hypothesis is then tested against the following alternatives:

Case I	Case II	Case III
H_0: the population distributions are identical	H_0: the population distributions are identical	H_0: the population distributions are identical
H_1: population 1 is shifted to the right of population 2	H_1: population 1 is shifted to the left of population 2	H_1: the populations differ in location

As was the case for the sign test involving paired comparisons, let us determine for each of the pairs of values (X_i, Y_i) the difference $D_i = X_i - Y_i$, $i = 1, \ldots, n$. Let $d_i = x_i - y_i$ denote the sample realization of D_i, where x_i and y_i are the realized values of X_i and Y_i, respectively. If $d_i = 0$, its source sample point (X_i, Y_i) is eliminated from further consideration and n is appropriately adjusted downward.

To perform the Wilcoxon signed-rank test, let us consider the following sequence of steps:

1. Determine d_i for all sample points (X_i, Y_i), $i = 1, \ldots, n$. (Discard any zero valued d_i's.)

2. Rank the d_i's in order of increasing absolute value. (If any of the d_i's are tied in value, then the tied d_i's are given the average rank.)

3. Restore to the rank values $1, \ldots, n$ the algebraic sign of the associated difference d_i. The ranks with the appropriate signs are the *signed ranks* R_i, $i = 1, \ldots, n$. They are of two types: R_i^+ is a rank carrying a + sign and R_i^- depicts a rank carrying a − sign. Furthermore, $W^+ = \sum_{i=1}^{n} R_i^+$ is the *sum of the positive signed ranks* and $W^- = \sum_{i=1}^{n} R_i^-$ is the *sum of the negative signed ranks*. (Their sample realizations are denoted as $w^+ = \sum_{i=1}^{n} r_i^+$ and $w^- = \sum_{i=1}^{n} r_i^-$, respectively.)

Under H_0, positive and negative differences of equal absolute value should occur with equal probability. If we were to order these differences according to their absolute values and rank them from smallest to largest, the expected sums of positive and negative signed ranks should be equal. Sizeable differences in these sums of signed ranks would provide evidence indicating a shift in the location of the two population distributions.

To detect a shift in location, let W^+ serve as our test statistic. For $n \leq 25$, we first look to the following.

A. *Small-Sample Wilcoxon Signed Rank Test.*   For Case I and a level of significance α, Table A.11 provides us with a right-hand tail critical value w_u of the sampling distribution of W^+, which satisfies $P(W^+ \geq w_u) = \alpha$. Thus the

Case I critical region $\mathcal{R} = \{w^+|w^+ \geq w_u\}$ so that H_0 is rejected for very large values of W^+. For Case II, w_l denotes the left-hand tail critical value for which $P(W^+ \leq w_l) = \alpha$. Hence the lower-tail critical region is structured as $\mathcal{R} = \{w^+|w^+ \leq w_l\}$; that is, H_0 is rejected when the realized W^+ is very small. And for Case III, w_l and w_u are chosen so that $P(W^+ \leq w_l$ and $W^+ \geq w_u) = \alpha$. Hence the two-tail critical region dictates that H_0 should be rejected for very small or very large values of W^+ or $\mathcal{R} = \{w^+|w^+ \leq w_l$ or $w^+ \geq w_u\}$.

B. *Large-Sample Wilcoxon Signed Rank Test.*   For $n > 25$, an approximate test using the standard normal distribution is again in order. Hence our standardized test statistic under H_0 is provided by (13.14). (In fact, the approximation to the standard normal form works well for a value of n in excess of 15.) And for a given level of significance α, our decision rule for rejecting H_0 in favor of H_1 is given by (13.15).

Example 13.8.1   Suppose two different processes of enameling are to be compared to determine which one is more efficient in terms of time to finishing. Twenty pairs of artisans are chosen, having been matched by years of experience and skill level. The times to finishing (in minutes) of plates of similar sizes are presented in Table 13.9. For $\alpha = 0.05$, determine if the underlying process distributions differ significantly in terms of time to finishing. Clearly Case III applies.

Given three ties, n is reduced to $17 = m$ in (13.14), with $w^+ = 95$ and thus

$$z_{w^+} = \frac{95 - 17(18)/4}{\sqrt{17(18)(35)/24}} = 0.88.$$

Since 0.88 does not fall within the critical region $\mathcal{R} = \{z_{w^+}|z_{w^+} \geq 1.96\}$, we cannot reject the null hypothesis; that is, at the 5% level of significance, the two process population distributions are essentially identical.   ■

A few additional points pertaining to the Wilcoxon signed rank test are in order. First, if we want to explicitly make inferences about the median $\gamma_{0.5}$ of the D_i's, let us specify the null hypothesis as $H_0: \gamma_{0.5} = \gamma_{0.5}^0$. Then the absolute differences $|d_i| = |x_i - y_i - \gamma_{0.5}^0|$ may be ranked and signed as before. This process assumes, of course, that each D_i is drawn independently from a population of differences that is continuous and symmetric about its median.

Next, in the presence of a large number of ties among the d_i's, the standard deviation of W^+ appearing in the denominator of (13.14) can be adjusted for ties by being replaced by

$$\sqrt{V(W^+)} = \left[\frac{n(n+1)(2n+1)}{24} - \frac{\sum_{j=1}^{g} t_j(t_j^2 - 1)}{48} \right]^{\frac{1}{2}},$$

Table 13.9

		Process Times to Finishing (in minutes)			
Pair	Process 1	Process 2	d_i	$\|d_i\|$	Signed rank of d_i
1	55	50	5	5	16.0
2	53	50	3	3	14.5
3	47	47	0	0	Tie
4	49	48	1	1	5.0
5	50	51	−1	1	−5.0
6	49	50	−1	1	−5.0
7	48	49	−1	1	−5.0
8	47	49	−2	2	−11.5
9	52	50	2	2	11.5
10	51	52	−1	1	−5.0
11	50	51	−1	1	−5.0
12	55	52	3	3	14.5
13	58	50	8	8	17.0
14	50	49	1	1	5.0
15	47	49	−2	2	−11.5
16	49	47	2	2	11.5
17	48	47	1	1	5.0
18	50	50	0	0	Tie
19	50	51	−1	1	−5.0
20	50	50	0	0	Tie

where there are g distinct sets of tied ranks, with j indexing any one such set, and t_j is the number of tied ranks in set j.

Finally, the Wilcoxon signed rank test is perhaps the most powerful of all nonparametric tests that hypothesize identical population distributions when making paired comparisons. Given a sample of matched pairs, this test is useful in determining whether the medians of the population distributions are identical, or whether one such distribution is located to the right or left of the other. And in the instance where the assumption of normality cannot be justified, the Wilcoxon signed rank test should be utilized instead of the parametric t test for matched pairs since the latter explicitly assumes normality. The asymptotic relative efficiency of the Wilcoxon signed rank test for matched pairs, when considered against the standard t test for the same, is at least 86%; and if the D_j's follow a normal distribution, the asymptotic relative efficiency of the Wilcoxon versus the t test is about 95%.

We bring our discussion of nonparametric hypothesis tests to a close by comparing these tests with their parametric counterparts (see Table 13.10).

Table 13.10

Nonparametric Tests	vs.	Parametric Equivalents
A. Independent Samples		*A'. Independent Samples*
(*a*) Two-sample runs test		(*a'*) No parametric equivalent
(*b*) Two-sample Mann-Whitney rank-sum test		(*b'*) *t* test of $H_0: \mu_1 - \mu_2 = 0$ (provided the populations are normal)
B. Matched Pairs		*B'. Matched Pairs*
(*a*) Two-sample sign test		(*a'*) *t* Test of $H_0: \mu_x - \mu_y = 0$ (provided the differences $D_i = X_i - Y_i$ are normal)
(*b*) Two-sample Wilcoxon signed rank test		(*b'*) *t* Test of $H_0: \mu_x - \mu_y = 0$ (provided the differences $D_i = X_i - Y_i$ are normal)

13.9   Exercises

13-1. A group of workers was asked if they favored a new proposed contract that emphasizes fringe benefits rather than hourly wage rate increases. "Yes" answers are coded as "*a*" and "no" answers are coded as "*b*." Can the following sequence of responses be treated as if they depict a random sample? Set $\alpha = 0.05$.

aa/bbb/a/b/aaa/b/aaaa/bb/aa/b/a/bbb/aa/b

13-2. The ACE Dairy Co. has developed what it calls an improved and more flavorful variety of vanilla ice cream. It offers a free sample to its customers and asks them if they would switch to this new variety of vanilla ice cream. A "yes" response is denoted as "*a*" and a "no" answer is recorded as "*b*." Can the following sequence of responses appear to be a random sample? Use $\alpha = 0.05$.

a/bb/aaa/b/a/bbbb/a/b/a/bb/a/bbb/aaa/bb/a/bb/aa/bbb/a/b/aa

13-3. Has the following sequence of observations been generated by a random process? Use $\alpha = 0.05$.

147	130	75	100	92	102	120	122	176	131
89	77	141	91	73	97	138	140	80	100
99	89	107	81	102	91	98	79	106	120
91	87	101	92	107	131	140	99	98	105

13-4. Does the following sequence of daily enrollment figures at Happy Trails Day Camp exhibit negative serial correlation? Use $\alpha = 0.01$.

$$35 \quad 23 \quad 25 \quad 19 \quad 24 \quad 23 \quad 27 \quad 29 \quad 25 \quad 26$$
$$20 \quad 18 \quad 25 \quad 22 \quad 27 \quad 22 \quad 31 \quad 29$$

13-5. Test the following sequence of sales of ice (expressed in bags/day) at a local convenience store over a three-week period for positive serial correlation. Use $\alpha = 0.05$.

$$15 \quad 17 \quad 18 \quad 22 \quad 19 \quad 16 \quad 14 \quad 10 \quad 12 \quad 17 \quad 27$$
$$29 \quad 35 \quad 30 \quad 25 \quad 24 \quad 17 \quad 15 \quad 19 \quad 25 \quad 31$$

13-6. DEF Transportation Inc. is interested in determining whether or not one-half of its bus commuters favor the newly instituted schedule change. A sample of 20 riders was selected at random: 14 indicated that they favored the change and 6 did not approve of the change. Use $\alpha = 0.05$.

13-7. A beginning aerobics class for working women has 27 enrollees and meets three times a week. Over a four-week period the number of absences per class was

$$3 \quad 2 \quad 3 \quad 4 \quad 3 \quad 5 \quad 4 \quad 3 \quad 6 \quad 5 \quad 7 \quad 6$$

Is there any evidence, at the $\alpha = 0.05$ level, that the median number of absences differs from 4?

13-8. A wine and sprits shop is interested in knowing whether more than half of its traffic is exclusively for lottery sales. A random sample of 60 customers revealed that 43 purchased only lottery tickets. Construct a test using $\alpha = 0.05$. Is your result consistent with a large-sample sign test?

13-9. A messenger service van completes its route every 90 minutes. For $n = 15$ runs of this route the number of items picked up for delivery was:

$$23 \quad 28 \quad 15 \quad 40 \quad 27 \quad 14 \quad 16 \quad 12 \quad 23 \quad 36 \quad 26 \quad 33 \quad 24 \quad 31 \quad 44$$

Is there reason to believe that the median number of items accepted for delivery/run exceeds 25? Use $\alpha = 0.05$ to perform Wilcoxon's signed rank test.

13-10. Wheels Unlimited operates a limousine service between the local airport and one of the major hotels in a large city. The round trip typically takes one hour, but one of the new drivers seems to be taking more than the usual or expected time of one hour. Fifteen of her runs (expressed in minutes) are timed:

$$56 \quad 62 \quad 55 \quad 70 \quad 69 \quad 65 \quad 68 \quad 71 \quad 75 \quad 80 \quad 66 \quad 62 \quad 63 \quad 58 \quad 68$$

Use the Wilcoxon signed rank test to determine if, at the 5% level, the median time is above one hour.

13-11. A human resources director feels that workers with at least one year of college will score above 75 points on a certain 100 point test. A group of $n = 30$ workers take the test and generate the following scores:

$$
\begin{array}{cccccccccc}
85 & 88 & 89 & 76 & 91 & 90 & 77 & 68 & 74 & 88 \\
87 & 73 & 72 & 77 & 89 & 95 & 85 & 79 & 77 & 92 \\
93 & 97 & 96 & 98 & 70 & 73 & 75 & 77 & 90 & 99
\end{array}
$$

Is the director correct in his speculation? Use $\alpha = 0.05$ and perform the Wilcoxon large-sample signed rank test.

13-12. A random sample of incomes (expressed in thousands of dollars) is taken from each of two communities (call them A and B):

Community A			Community B			
20	20.5	35	19	52	55	31
46	51.5	40	47	67	71	54
39	45	30	62	55	57	60

Use the runs test to determine if the population distributions are identical. Let $\alpha = 0.01$.

13-13. Suppose 15 additional observations on incomes in each of the communities specified in the preceding exercise have been obtained. Retest using the expanded samples.

Additional Community A Observations					Additional Community B Observations				
25	30	41	22	38	27	35	52	41	61
45	50	39	35	41	41	44	61	32	67.5
44	46.5	27	31	46	38	49	51	56	62.5

13-14. A statistics class is randomly divided into two groups. One uses only the textbook and the other uses the textbook and study guide. The final averages for the two groups are:

Textbook Only				Textbook and Study Guide			
77	63	75	84	79	82	74	91
94	96	87	72	93	98	82	85

Is there any evidence to indicate that those students who used the textbook and study guide earned higher course averages? Use $\alpha = 0.05$ and perform a rank-sum test.

13-15. At a garden center two types of fertilizers (call them A and B) are being tested. Sixteen identical plots are available for the test, with four seedlings planted in each plot. Fertilizer A is randomly applied to eight of the plots and Fertilizer B is applied to the remaining plots. After a month of consistent care the average height (in inches) of the four seedlings in each plot is calculated:

Fertilizer A				Fertilizer B			
12	11.2	13	10.8	12.8	14.2	13.8	12.7
11.5	13.1	12.4	10.1	10.6	15.1	14.9	12.9

For $\alpha = 0.05$, is there any evidence that fertilizer B produces plants of greater height? Use the rank-sum test.

13-16. The number of sun worshipers (expressed in thousands per week) who frequented two separated but nearby beaches (call them A and B) last season, were:

Beach A					Beach B				
(June)	12.7	15	11	16	(June)	12.6	13	12	15
(July)	13.8	12	12.7	13.6	(July)	13.7	12	11	11.2
(Aug)	16.4	15.1	14	13	(Aug)	14	14.6	11	13.1

Given that the beaches are in close proximity to each other and have essentially the same features (parking, food, restrooms, lifeguards, etc.), is there any evidence to indicate that beach A is preferred over beach B? Use the large sample rank-sum test for $\alpha = 0.05$.

13-17. A paint manufacturer wants to determine if its new formula for exterior house paint is superior to the old formula. Sixteen large panels of wood were selected. The panels were paired into sections, and the two paints were randomly applied, one to each section of each panel. After exposure to the elements for nine months, reflectivity readings were taken of each side of each panel. If the old formula is more reflective, an "O" is recorded; if the new formula turns out to be brighter, an "N" is recorded:

$N N N N O O N N N O N N N N O N$

Is there any evidence to indicate that the new formula is superior to the old one? Use $\alpha = 0.05$.

13-18. Recently married couples were asked (independently of each other) to specify (under ideal circumstances) the number of years they would like to wait before having children:

Couple	Wife	Husband
1	0	2
2	0	1
3	1	1
4	0	2
5	1	2
6	2	2
7	1	3
8	3	1
9	1	2
10	0	1

Use the sign test for paired comparisons to determine if these couples have different propensities to wait for children. Let $\alpha = 0.05$.

13-19. A production manager wishes to compare two methods for detecting defects on the surface of a finished metal product. Each method is applied to each of 25 finished items:

Item	Method 1	Method 2	Item	Method 1	Method 2
1	4	3	13	3	4
2	3	3	14	5	4
3	0	1	15	7	7
4	0	1	16	1	3
5	1	2	17	4	5
6	2	2	18	2	3
7	3	2	19	0	1
8	1	3	20	0	1
9	4	5	21	1	1
10	0	2	22	0	3
11	0	2	23	3	3
12	2	3	24	1	2
			25	4	5

Is Method 2 superior to Method 1? Use the sign test with $\alpha = 0.05$.

13-20. Suppose two different inspection procedures are being compared using ten pairs of inspectors, matched by years of experience, and so on. The inspectors within each pair are randomly assigned to one of the two procedures and the numbers of items inspected/hour are:

Pair:	1	2	3	4	5	6	7	8	9	10
Procedure 1:	11	10	9	10	9	12	13	9	8	11
Procedure 2:	15	12	10	12	8	12	15	14	10	13

Is Procedure 2 more effective than Procedure 1? Use the Wilcoxon signed rank test with $\alpha = 0.01$.

13-21. The marketing department of ACE Industries is investigating the sales appeal of two different display formats for a given product—one with and one without animation. The two displays are installed at each of 12 stores for one week and the numbers of items purchased from each display are:

Store	Animated Display	Non animated Display	Store	Animated Display	Non animated Display
1	27	25	7	33	27
2	35	30	8	37	29
3	17	10	9	21	19
4	20	21	10	24	17
5	22	20	11	19	22
6	39	30	12	44	30

Is the animated display more effective than the nonanimated one? Employ the Wilcoxon signed-rank test with $\alpha = 0.05$.

13-22. Resolve Exercise 19 using the large-sample Wilcoxon signed rank test.

Testing Goodness of Fit

14.1 Distributional Hypotheses

In Chapter 12 we assumed that a population probability distribution had a specific functional form. This allowed us to test hypotheses concerning its parameter(s). However, instead of dealing with *parametric hypotheses*, we may choose to make a statement about the *form* of the population probability distribution itself; that is, we can test a *distributional hypothesis*. For instance, we previously tested parametric hypotheses about the unknown mean μ of a population random variable X, which was assumed to be $N(\mu, \sigma)$, with the variance σ^2 either known or unknown. Clearly the *distributional assumption* of normality was made. But if we actually choose to test the null hypothesis that X is normally distributed, then obviously a *distributional hypothesis* has been made. The alternative hypothesis is then that the random variable X is not of the stated form.

An important method of testing a distributional hypothesis is *Pearson's goodness-of-fit test*. As the name implies, we are interested in determining if a set of sample observations can be viewed as values of a population random variable having a given probability distribution. So given some distributional hypothesis, this test enables us to verify if it is supported by the data; that is, we can determine if the population follows a specific probability model.

As we shall now see, the Pearson goodness-of-fit test is based upon the multinomial distribution presented in Section 6.6. Moreover, this test is applicable in testing goodness of fit for either discrete or continuous probability distributions.

14.2 The Multinomial Chi-Square Statistic: Complete Specification of H_0

Let us briefly review the essential features of a multinomial random experiment presented in Section 6.6. We noted earlier that for this type of experiment, each of the n trials results in a k-fold alternative; that is, each trial results in any one of k mutually exclusive and collectively exhaustive outcomes $E_1, \ldots, E_k$ with

609

respective probabilities $P(E_1) = p_1, \dots, P(E_k) = p_k$, where $\sum_{i=1}^{k} p_i = 1$ and the p_i are constant from trial to trial. If the (discrete) random variable $X_i, i = 1, \dots, k$, depicts the number of times the ith outcome type E_i occurs in the n independent and identical trials of the k-fold alternative, then the multinomial random variable $(X_1, \dots, X_k)$ follows a multinomial probability distribution with probability density function given by (6.24). (Remember that the binomial experiment is a special case of the multinomial experiment when $k = 2$.)

If we view E_i as a particular outcome class or *cell*, then we may denote the number of trials in which E_i occurs or the number of observations falling into cell i as $X_i, i = 1, \dots, k$. And if x_i is the sample realization of X_i, then clearly $n = \sum_{i=1}^{k} x_i$.

For instance, suppose a population probability distribution consists of a set of k probabilities or relative frequencies p_i assigned to each of a collection of k mutually exclusive and collectively exhaustive categories or cells. Columns 1 and 2 of Table 14.1 provide us with such a probability distribution. Next, suppose that we are presented with a random sample of $n = 100$ observations collected from some unknown source and each observation is placed within one and only one of the seven cells given in Table 14.1. These *observed (or sample) frequencies* o_i are given in column 3. Does the sample distribution (consisting of columns 1 and 3 in Table 14.1) of 100 items differ from the population or theoretical distribution? That is, can we legitimately presume that the sample of 100 data points has been drawn from the population distribution given in the first two columns of Table 14.1?

To determine if the observed or sample distribution is the same as the given population distribution, let us formulate

H_0: the population distribution and the sample distribution are identical

H_1: the sample distribution differs from the population distribution

Here the null hypothesis specifies the exact or theoretical distribution (our benchmark) and the alternative hypothesis simply indicates some unspecified disparity between the two distributions.

Table 14.1

Category	Population Relative Frequency or Theoretical Probability p_i	Observed Frequency o_i	Expected Frequency Under H_0 $e_i (= np_i^0)$
1	0.14	10	14
2	0.17	18	17
3	0.33	24	33
4	0.21	18	21
5	0.10	15	10
6	0.03	10	3
7	0.02	5	2
	1.00	100	100

Given the relative frequencies specified for the theoretical distribution, the exact probability of observing this particular sample distribution can be obtained from the multinomial probability rule (6.23.1); that is, under H_0 and independent random sampling with replacement, the probability of obtaining a sample distribution exactly like the one appearing in columns 1 and 3 of Table 14.1 is

$$m(x_1 = 10, x_2 = 18, x_3 = 24, x_4 = 18, x_5 = 15, x_6 = 10, x_7 = 5; n = 100,$$

$$p_1 = 0.14, p_2 = 0.17, p_3 = 0.33, p_4 = 0.21, p_5 = 0.10, p_6 = 0.03, p_7 = 0.02)$$

$$= \frac{100!}{10!\,18!\,24!\,18!\,15!\,10!\,5!} \left[(0.14)^{10}(0.17)^{18}(0.33)^{24}(0.21)^{18}\ (0.10)^{15}(0.03)^{10}(0.02)^{5} \right].$$

Obviously this sort of calculation requires a considerable amount of computational effort. Needless to say, an alternative approach (which avoids the direct calculation of exact multinomial probabilities) must be sought, which enables us to easily address problems pertaining to goodness of fit. This alternative computation scheme is based upon the so-called *Pearson multinomial chi-square distribution*. Specifically, we have Theorem 14.1:

THEOREM 14.1.   Let the random variable $(X_1, \ldots, X_k)$ follow a multinomial probability distribution with parameters $n, p_1, \ldots, p_k$. Then for k fixed, as the number of trials n increases without limit, the distribution function $F_U(t)$ of the random variable

$$U = \sum_{i=1}^{k} \frac{(X_i - np_i)^2}{np_i} \tag{14.2}$$

approaches the chi-square distribution function $F_{\chi^2}(t)$ with $k - 1$ degrees of freedom; that is, $\lim_{n \to \infty} F_U(t) = F_{\chi^2}(t)$ for every t.

As this theorem reveals, if the possible outcomes of a random experiment can be decomposed into k mutually exclusive and collectively exhaustive categories with p_i denoting the probability that an outcome falls in category i, then, in n independent and identical trials of this k-fold alternative experiment, with X_i representing the number of outcomes favoring category i (so that $\sum_{i=1}^{k} x_i = n$), the limiting distribution of U is χ^2_{k-1} as $n \to \infty$. (In terms of the notation developed in Chapter 8, $U \xrightarrow{d} \chi^2_{k-1}$ as $n \to \infty$.)

It is important to note that the random variable U ((14.2)) is termed the *Pearson or multinomial chi-square statistic*; it serves as a reasonably good approximation to the ordinary chi-square random variable X *only if n is large*, where, as we have seen from Chapter 9, $E(X) = v$ and $V(X) = 2v$. However, it can be demonstrated that for the Pearson random variable U, $E(U) = v (= k - 1)$ and $V(U) = 2v + \frac{1}{n}\left(\sum_{i=1}^{k} p_i^{-1} - v^2 - 4v - 1\right)$. Hence $V(U) \to V(X)$ only in the limit as $n \to \infty$.

How may we rationalize this result? To keep our analysis as transparent as possible, let us consider the case where $k = 2$; that is, the multinomial random variable (X_1, X_2) admits only two possible outcomes or classes of events. Clearly this is the binomial case considered earlier in Chapter 6. Consistent with our binomial terminology, let us call the first class *success* and the second class *failure*. Then X_1 is the number of successes and $X_2 = n - X_1$ is the number of failures in n trials, where p_1 is the probability of a success and $p_2 = 1 - p_1$ is the probability of a failure. Then from (14.2),

$$U = \frac{(X_1 - np_1)^2}{np_1} + \frac{(X_2 - np_2)^2}{np_2} = \frac{(X_1 - np_1)^2}{np_1} + \frac{[n - X_1 - n(1 - p_1)]^2}{n(1 - p_1)}$$

$$= \frac{(X_1 - np_1)^2}{np_1(1 - p_1)} = V^2. \tag{14.3}$$

By the De Moivre-Laplace-Gauss Limit Theorem of Chapter 7, the quantity $V = \frac{X_1 - np_1}{\sqrt{np_1(1-p_1)}}$ is approximately a $N(0,1)$ random variable for large n. So for large n, V^2 is approximately the square of a $N(0,1)$ random variable and thus, via Theorem 9.1, V^2 is χ_1^2. In general, by the preceding argument, (14.2) is the sum of the squares of $k - 1$ independent random variables, each of which approaches the square of a $N(0,1)$ random variable as n increases without bound. Then by Theorem 9.2, this sum is χ_{k-1}^2; that is, under repeated sampling from the same multinomial population, the statistic (14.2) is approximately χ_{k-1}^2.

The significance of Theorem 14.1 is that it enables us to use (14.2) as a test statistic for conducting a *multinomial goodness-of-fit test* without having to resort to the direct calculation of multinomial probabilities. In this regard, for $(X_1, \ldots, X_k)$ a multinomial random variable, we may test a null hypothesis that specifies the parameters of a multinomial distribution as

$$H_0: p_1 = p_1^0, p_2 = p_2^0, \ldots, p_k = p_k^0,$$

against   (14.4)

$$H_1: p_i \neq p_i^0 \quad \text{for at least one } i = 1, \ldots, k.$$

Then for a random sample of size n taken from a multinomial population, the realization of (14.2) under H_0 is

$$u = \sum_{i=1}^{k} \frac{(x_i - np_i^0)^2}{np_i^0}.$$

Since n is nondecreasing as i goes from 1 to k, we reject H_0 if u is an element of $\mathcal{R} = \{u | u \geq \chi_{1-\alpha, k-1}^2\}$, where $\chi_{1-\alpha, k-1}^2$ is the $100(1 - \alpha)$ percentile of the chi-square distribution with $k - 1$ degrees of freedom and $\alpha = P(TIE)$.

As (14.4) reveals, a goodness-of-fit test is essentially a test of a hypothesis concerning specified cell probabilities $p_i, i = 1, \ldots, k$. Keep in mind, however, that in order to use (14.5) to approximate multinomial probabilities (a) each outcome falls into one and only one cell or category; (b) the outcomes are independent; and (c) the sample size n is large.

To see exactly why this test has an upper tail critical region (we reject H_0 for very large u's) and why (14.5) is structured to actually reveal goodness of fit, let us return to the data presented in Table 14.1. It should be intuitively clear that the notion of *goodness of fit* should be assessed on the basis of the degree of disparity between the sample or empirical distribution and the expected or theoretical distribution given that the latter is specified by the null hypothesis. Under H_0, the *expected or theoretical frequencies* for the k categories or cells are simply $E(X_i) = np_i^0 = e_i, i = 1, \ldots, k$; that is, e_i is the product between the sample size and the hypothesized relative frequency or theoretical probability p_i^0. Then if H_0 is true, $e_i = np_i^0$ should be the expected number of occurrences for cell i under n repeated trials of our k-fold alternative experiment. The expected cell frequencies for the problem at hand are given in column 4 of Table 14.1. In view of these considerations, we may rewrite (14.5) as

$$u = \sum_{i=1}^{k} \frac{(o_i - e_i)^2}{e_i}. \tag{14.5.1}$$

Here u serves as an index of goodness of fit given that H_0 specifies the theoretical distribution that is *fitted* to empirical or sample data. As this expression reveals, each squared difference in observed and expected frequencies is *weighted inversely* by the expected frequency for cell i; that is, any departure from expectation receives relatively more weight if we expect only a few outcomes in that cell than if we expect many such outcomes.

Equation (14.5.1) enables us to determine how well the sample distribution and the expected or theoretical distribution agree. Clearly the difference $o_i - e_i$ serves as a measure of the agreement between the observed and expected frequencies belonging to cell i. Under H_0 or perfect agreement, $\sum(o_i - e_i) = 0$. To avoid this problem, we use the squares $(o_i - e_i)^2$. Then $(o_i - e_i)^2$ is weighted inversely by e_i to gauge its relative importance so that $\frac{(o_i - e_i)^2}{e_i}$ is the individual contribution of cell i to u, with the latter now serving as an overall or global measure of goodness of fit.

When there is a high degree of agreement between the sample and theoretical distributions, u tends to be relatively small and thus it is highly unlikely that it would lie within the critical region and that H_0 would be rejected. However, if there is considerable disagreement between the observed and expected frequencies, then u tends to be relatively large, thus favoring the terms upon which it is likely to fall within the critical region and H_0 is rejected.

As this discussion reveals, the critical region must be located in the upper tail of the chi-square distribution. Given that we are determining the goodness of fit between an empirical and theoretical probability distribution, small realized values of U indicate agreement between the observed and expected frequencies

(H_0 is not likely to be rejected), and large realizations of U indicate disagreement between these frequencies (H_0 is likely to be rejected).

Note that degrees of freedom for the chi-square approximation provided by (14.5.1) is $k - 1$; it amounts to the number of cells (k) less the number of linear restrictions (1) imposed upon the sum of squares $\sum(o_i - e_i)^2$; that is, the constraint $\sum(o_i - e_i) = 0$ must hold. So if the values of any $k - 1$ of these differences are known, then the value of the kth difference is uniquely determined.

The goodness-of-fit test just described can be executed for *any* theoretical population distribution we choose to specify under H_0; all that is required is that the population distribution is discrete or its domain can be partitioned into a collection of class intervals. In fact, for $F_0(t)$ a theoretical cumulative distribution function having all of its parameters completely specified, the Pearson goodness-of-fit test can be applied to determining whether the cumulative distribution function $F(t)$ of a random variable X (discrete or continuous) has the form $F_0(t)$. This is because, for large n, the limiting chi-square distribution of (14.2) is independent of the form of the null distribution $F_0(t)$. That is, under $H_0: F(t) = F_0(t)$, we can determine the probability p_i that an outcome of our k-fold alternative experiment falls into category i. More formally, we state Theorem 14.2:

THEOREM 14.2.   Let the sample random variables $Y_1, \dots, Y_n$ be drawn from a population variable Y having a cumulative distribution function $F_Y(t)$. For the k cells or categories

$$I_1 = \{y | y \le a_1\}, I_2 = \{y | a_1 < y \le a_2\}, \dots,$$

$$I_{k-1} = \{y | a_{k-2} < y \le a_{k-1}\}, \text{ and } I_k = \{y | a_{k-1} < y\},$$

suppose X_i is the number of outcomes occurring in cell I_i, $i = 1, \dots, k$. Then $(X_1, \dots, X_k)$ is a multinomial random variable with parameters n and $p_i = P(Y \in I_i)$, $i = 1, \dots, k$.

Then for any cumulative distribution function $F_Y(t)$, we can test the hypothesis $H_0': F_Y(t) = F_0(t)$ by testing the equivalent hypothesis that $H_0'': p_i = P(Y \in I_i)$, $i = 1, \dots, k$. (Here we are saying that the hypothesis that each Y_i has cumulative distribution function $F_Y(t)$ is equivalent to the hypothesis that the multinomial random variable $(X_1, \dots, X_k)$ generated by $Y_1, \dots, Y_n$ exhibits the parameters $p_i = P(Y \in I_i), i = 1, \dots, k$.) If we reject H_0'', then, based upon the sample evidence, $F_Y(t)$ does not adequately describe the population variable Y; if H_0'' is not rejected, then $F_Y(t)$ is consistent with the sample observations.

To apply Theorem 14.2 we need only note that if X is a discrete random variable, then the intervals I_i, $i = 1, \dots, k$, are all singletons (sets containing a single point) in that they amount to the discrete values that the variable assumes; in the continuous case, we must actually construct a set of classes or intervals. In this latter instance the construction is arbitrary save for the fact that the intervals should be of uniform length and the number of intervals should be such that $e_i = np_i \geq 5$ for $i = 1,\ldots,k$. (Note that adjacent intervals may be combined to meet the requirement that $e_i \geq 5$.) Hence a *cell* is taken to be either a specific numerical value (for a discrete distribution) or an interval (in the case of a continuous distribution).

The foregoing discussion has indicated that the chi-square goodness-of-fit test is based upon the differences between the observed cell frequencies obtained from the sample and the cell frequencies we would expect to obtain if the random variable $(X_1,\ldots,X_k)$ conformed to the assumed theoretical distribution (given by H_0). To see this, let us return to the distributions appearing in Table 14.1.

For $\alpha = 0.05$, let the null and alternative hypotheses be given by (14.1). Then the critical region has the form $\mathcal{R} = \{u|u \geq \chi^2_{0.95,6} = 12.5916\}$ (see Table A.4 of the Appendix) and, from (14.5.1),

$$u = \sum_{i=1}^{7} \frac{(o_i - e_i)^2}{e_i} = \frac{(10 - 14)^2}{14} + \frac{(18 - 17)^2}{17} + \cdots + \frac{(5 - 2)^2}{2} = 27.4179.$$

Since this realization of U lies well within the critical region, we reject H_0 at the 5% level—the sample and theoretical distributions are not identical. In fact, the p-value or observed level of significance for this test is less than 0.1%. (In general, p-value $= P(\chi^2 > u|k - 1$ degrees of freedom).)

Example 14.2.1   Given the absolute frequency distribution appearing in columns 1 and 2 of Table 14.2a, determine, for $\alpha = 0.05$, if the random variable X follows a Poisson distribution with $\lambda = 0.7$. Since X is discrete, we may determine the five Poisson probabilities $p(X;\lambda = 0.7), X = 0,1,2,3,4$, using (6.59) (the Poisson probability mass function) or Table A.8 of the Appendix. These cell probabilities appear in column 3 of Table 14.2a. Since $np_4 = 0.9 < 5$ let us combine the cells for $X = 3$ and $X = 4$.

This yields a new absolute frequency distribution for X and thus a revised set of Poisson probabilities (see Table 14.2b). (Note that since the set of Poisson probabilities must sum to unity, $p_4 = 1 - p_1 - p_2 - p_3 = 0.0341$.) As this revised table reveals, each expected frequency $e_i = np_i > 5, i = 0,1,2,3$. So given Table 14.2b, we may now test

H_0: X is Poisson distributed with $\lambda = 0.7$, against

H_1: X is not Poisson distributed with $\lambda = 0.7$.

In terms of our preceding discussion, this null hypothesis is equivalent to

H'_0: (X_1, X_2, X_3, X_4) is a multinomial random variable with parameters $n = 180$, $p_1 = 0.4966, p_2 = 0.3476, p_3 = 0.1217$, and $p_4 = 0.0341$.

Table 14.2

a.

X	$f(x)(= o_i)$	$p(X; \lambda = 0.7)$	$np_i(= e_i)$
0	90	$p_1 = 0.4966$	89.388
1	53	$p_2 = 0.3476$	62.568
2	25	$p_3 = 0.1217$	21.906
3	9	$p_4 = 0.0284$	5.112
4	3	$p_5 = 0.0050$	0.900
	180		

b.

X	$f(x)(= o_i)$	$p(X; \lambda = 0.7)$	$np_i(= e_i)$
0	90	$p_1 = 0.4966$	89.388
1	53	$p_2 = 0.3476$	62.568
2	25	$p_3 = 0.1217$	21.906
3	12	$p_4 = 0.0341$	6.138
	180		

Since $\alpha = 0.05$, the critical region appears as $\mathcal{R} = \left\{ u | u \geq \chi^2_{0.95,3} = 7.8147 \right\}$. And from (14.5.1), we find

$$u = \sum_{i=1}^{4} \frac{(o_i - e_i)^2}{e_i} = \frac{(90 - 89.388)^2}{89.388} + \frac{(53 - 62.568)^2}{62.568} + \frac{(25 - 21.906)^2}{21.906} +$$

$$\frac{(12 - 6.138)^2}{6.138} = 7.5066.$$

Since this value of u is not a member of $\mathcal{R}$, we cannot reject H_0 at the 5% level—it is highly likely that the observed data set was generated by a Poisson random variable having $\lambda = 0.7$. ∎

14.3 The Multinomial Chi-Square Statistic: Incomplete Specification of H_0

In the preceding section the null hypothesis was taken to be simple in form; that is, it was stated in a fashion that completely specified the parameter(s) of some theoretical population distribution. Such complete specification then enabled us to determine the exact multinomial probabilities $p_1, \ldots, p_k$. For instance, in the example problem provided at the end of Section 14.2, the Poisson parameter under H_0 was given as $\lambda = 0.7$. But what if such complete information is not available? Then the null hypothesis will obviously be incompletely specified and a less restrictive composite hypothesis (such as $H_0: X$ is Poisson distributed) will be tested once certain key parameters have been estimated from the sample data.

To formalize this approach, let us consider Theorem 14.3:

THEOREM 14.3.   Let the sample random variables $Y_1, \ldots, Y_n$ be drawn from a population variable Y having a distribution function $F_Y(t; \theta_1, \ldots, \theta_r)$, where $\theta_j, j = 1, \ldots, r$, depicts an unknown population parameter. For I_i and X_i, $i = 1, \ldots, k$, specified as in Theorem 14.2, define

$$p_i = P(Y \in I_i) = \phi^i(\theta_1, \ldots, \theta_r), \; i = 1, \ldots, k.$$

Let $\hat{\theta}_j$ be the maximum likelihood estimators of $\theta_j, j = 1, \ldots, r$, determined from $X_1, \ldots, X_k$. Then for fixed k and certain regularity conditions on the $\phi^i, i = 1, \ldots, k$, as the number of trials n increases without limit, the distribution function $F_Y(t; \hat{\theta}_1, \ldots, \hat{\theta}_r)$ of the random variable

$$V = \sum_{i=1}^{k} \frac{(X_i - n\hat{p}_i)^2}{n\hat{p}_i} \tag{14.7}$$

approaches the chi-square distribution function with $k - r - 1$ degrees of freedom, where

$$\hat{p}_i = \phi^i(\hat{\theta}_1, \ldots, \hat{\theta}_r), \; i = 1, \ldots, k.$$

Note that we lose one degree of freedom for each parameter estimated. So if for every unknown parameter θ_j the statistic $\hat{\theta}_j, j = 1, \ldots, r$, is a maximum likelihood estimator for θ_j, and if the expected frequencies are determined as functions of these maximum likelihood estimators, then V is approximately chi-square distributed with $k - r - 1$ degrees of freedom.

Given that we want to perform a goodness-of-fit test under a less restrictive hypothesis that lacks the complete specification of relevant population parameters, we need only use the observed frequencies $X_i, i = 1, \ldots, k$, to determine maximum likelihood estimates of the relevant parameters. Once the maximum likelihood estimation is complete, the test proceeds in a fashion similar to that given earlier, under complete parameter specification, with the sample realization of (14.7) appearing as

$$v = \sum_{i=1}^{k} \frac{(o_i - \hat{e}_i)^2}{\hat{e}_i}, \tag{14.7.1}$$

where $\hat{e}_i = n\hat{p}_i, i = 1, \ldots, k$. Then for given H_0 and $\alpha = P(TIE), \mathcal{R} = \{v | v \geq \chi^2_{1-\alpha, k-r-1}\}$.

Example 14.3.1   To see how Theorem 14.3 is applied, let us consider a situation in which the Department of Transportation of a large eastern state is monitoring

Table 14.3

a.

X	$f(x)(= o_i)$	$p(X = x_i; \hat{\lambda} = 1.1)(= \hat{p}_i)$	$n\hat{p}_i(= \hat{e}_i)$
0	56	0.3329	57.59
1	68	0.3662	63.35
2	33	0.2014	34.84
3	11	0.0738	12.76
4	2	0.0203	3.51
5	2	0.0045	0.77
6	1	0.0008	0.13
	$\overline{173}$		

b.

X	o_i	$\hat{p}_i$	$\hat{e}_i$
0	56	0.3329	57.59
1	68	0.3662	63.35
2	33	0.2014	34.84
≥ 3	16	0.0995	17.21
	$\overline{173}$		

motor vehicle fatalities on a particular stretch of road for a 100-day period. It collects information on deaths per accident (X) involving at least one passenger vehicle (see Table 14.3a). Let us test $H_0: X$ is Poisson distributed, versus $H_1: X$ is not Poisson distributed, at the 5% level of significance. As noted earlier (see Chapter 10), the maximum likelihood estimator of the Poisson parameter λ is the sample mean $\overline{X}$ with realization $\bar{x} = \frac{1}{n} \sum_{i=1}^{7} x_i f(x_i) = \frac{191}{173} = 1.1 = \hat{\lambda}$.

Then from Table A.8, the Poisson probabilities $p(X; \hat{\lambda} = 1.1), X = 0, 1, \ldots, 6$, are presented in column 3 of Table 14.3a. (Note that in the absence of a table that displays this specific $\hat{\lambda}$ value, these probabilities may be calculated directly from the Poisson probability mass function (6.59) with $\lambda = \hat{\lambda}$.) Once the Poisson probabilities are obtained, the expected frequencies $\hat{e}_i = n\hat{p}_i = np(X = x_i; \hat{\lambda} = 1.1), i = 1, \ldots, 7$, are given in column 4 of Table 14.3a. Since the expected frequencies for the last three cells are each less than 5, let us combine the last four cells so that each expected frequency turns out to be at least 5 (all this is done in Table 14.3b). Note that $\hat{p}_4 = 1 - \hat{p}_1 - \hat{p}_2 - \hat{p}_3 = 0.0995$ so that these final four probabilities sum to unity. (Remember that our justification for this procedure is: if the sample observations are independent, then the observed cell frequencies $X_1 = o_1, \ldots, X_4 = o_4$ follow a multinomial distribution with parameters $n, p_1, \ldots, p_4$. Hence $E(X_i) = np_i$ and thus the estimates of these expected cell frequencies are $\hat{e}_i = n\hat{p}_i, i = 1, \ldots, 4$.)

Given this revised number of cells and $r = 1$ parameters estimated, degrees of freedom for this test amounts to $k - r - 1 = 4 - 1 - 1 = 2$. Hence the critical

Table 14.4

X	o_i	$\hat{p}_i$	$\hat{e}_i$
3.01–3.02	13	0.1685	42.125
3.03–3.04	27	0.2328	58.200
3.05–3.06	52	0.2759	68.975
3.07–3.08	90	0.2018	50.450
3.09–3.10	38	0.0916	22.900
3.11–3.12	$20\}\,30$	$0.0247\}\,0.0294$	$6.175\}\,7.350$
3.13–3.14	$10\}$	$0.0047\}$	$1.175\}$
	$\overline{250}$		

region is $\mathcal{R} = \left\{ v | v \geq \chi^2_{1-\alpha,k-r-1} = \chi^2_{0.95,2} = 5.9915 \right\}$. Then from (14.7.1),

$$v = \sum_{i=1}^{4} \frac{(o_i - \hat{e}_i)^2}{\hat{e}_i} = \frac{(56 - 57.59)^2}{57.59} + \frac{(68 - 63.38)^2}{63.38} + \frac{(33 - 34.84)^2}{34.84} +$$

$$\frac{(16 - 17.21)^2}{17.21} = 0.5672.$$

Since this value of v is not an element of $\mathcal{R}$, we cannot reject H_0 at the 5% level; it appears that deaths per accident can be viewed as following a Poisson random variable with $\hat{\lambda} = 1.1$.   ■

The final example problem offered in this section deals with a very common application of Pearson's chi-square goodness-of-fit test. Specifically, the test is often used to determine if a sample distribution can be thought of as having been drawn from a normal population. A chi-square test of normality now follows.

Example 14.3.2   A manufacturer of brass fittings wants to determine if the inside diameter (X) of a fitting can be viewed as a normal random variable. To test $H_0: X$ is normally distributed, against $H_1: X$ is not normally distributed, at the $\alpha = 0.05$ level, a random sample of $n = 250$ fittings is collected and the appropriate measurement on each fitting is taken. The resulting measurements (expressed in millimeters) are then used to construct the absolute frequency distribution presented in Table 14.4. As a first step, let us estimate μ and σ (since the null hypothesis is not completely specified) from the sample distribution. As indicated in Chapter 10, maximum likelihood estimators for these parameters are $\overline{X}$ and $S_1 = \sqrt{\frac{n-1}{n}S^2}$, respectively. Then their sample realizations are, from (2.E.1) and (2.E.2), respectively, $\bar{x} = 3.052$ and $s_1 = 0.028$ (see Exercise 2.13).

Our next step is to use $\bar{x}$ and s_1 to determine, via the standard normal area table (Table A.1 of the Appendix), the probability that a $N(3.052, 0.028)$ random variable would assume a value in each of the seven cells appearing in Table 14.4. To this end, the cell probabilities are estimated by using the class boundaries $(3.005, 3.025, \ldots, 3.125, 3.145)$ and, under the requirement that these probabilities must sum to 1, the first and last cells are replaced by $-\infty < X \leq 3.025$ and $3.13 \leq X < +\infty$, respectively. Then for $Z = \frac{X - \bar{X}}{s_1}$:

$$\hat{p}_1 = P(-\infty < X \leq 3.025) = P(-\infty < Z \leq -0.964) = 0.1685;$$

$$\hat{p}_2 = P(3.025 \leq X \leq 3.045) = P(-0.964 \leq Z \leq -0.250) = 0.2328;$$

$$\vdots$$

$$\hat{p}_6 = P(3.105 \leq X \leq 3.125) = P(1.892 \leq Z \leq 2.607) = 0.0247;$$

$$\hat{p}_7 = P(3.125 \leq X < +\infty) = P(2.607 \leq Z < +\infty) = 0.0047.$$

Now, if H_0 is true, then step three has us determine the expected cell frequencies as $\hat{e}_i = n\hat{p}_i = 250\hat{p}_i, i = 1, \ldots, 7$ (see column 4 of Table 14.4). Note that since the expected frequency of the last cell is less than 5, the sixth and seventh cells will be combined so that the last cell (which is now the sixth cell) is configured as $3.11 \leq X < +\infty$. Its observed frequency is thus $o_6 = 30$, its cell probability is $\hat{p}_6 = 0.0294$, and its expected frequency is $\hat{e}_6 = n\hat{p}_6 = 7.350$. In view of these adjustments, our final step, which involves an application of (14.7.1), yields

$$v = \sum_{i=1}^{6} \frac{(o_i - \hat{e}_i)^2}{\hat{e}_i} = \frac{(13 - 42.125)^2}{42.125} + \frac{(27 - 58.200)^2}{58.200} + \cdots + \frac{(30 - 7.350)^2}{7.350}$$

$$= 151.797.$$

For degrees of freedom corresponding to $k - r - 1 = 6 - 2 - 1 = 3$, the critical region $\mathcal{R} = \{v | v \geq \chi^2_{0.95,3} = 7.8147\}$. Since v lies well within $\mathcal{R}$, we can safely reject the null hypothesis of normality at the 5% level of significance. ∎

We end our discussion of the chi-square goodness-of-fit test with a few summary comments/cautions:

1. If X is a discrete random variable and the number of cells or categories is $k = 2$ (so that degrees of freedom amounts to $k - 1 = 1$), the normal approximation to the binomial distribution should be used to test $H_0: p_1 = \frac{X_1}{n}$ of the data is in cell 1 and $p_2 = \frac{X_2}{n} = \frac{n - X_1}{n}$ of the data lies in cell 2.

2. If X is a continuous random variable, then to use this test, the data must be grouped into k class intervals or cells of uniform length, where the number of cells typically ranges between 5 and 10 and each cell has an expected

frequency of at least 5. (This requirement implies that the sample size n is moderately large.)

3. The test tends to be less precise for continuous distributions relative to discrete ones since, in the continuous case, the grouping into artificial intervals is arbitrary, whereas, for the discrete case, the data has an inherent or natural grouping that is dictated by the specific variable values.

4. The test is performed only on frequencies and not on, say, proportions.

5. If the chi-square goodness-of-fit test is being applied to determine if the population is normally distributed, then the grouping of adjacent cells in the right- or left-hand tail of the sample distribution may be warranted to ensure that these extreme expected cell frequencies are at least 5. However, under this procedure, the chi-square test undergoes a loss of power in its ability to detect departures from normality in the tails of the sample distribution. In view of this potential problem, it must be emphasized that extreme care must be taken to construct the cells or class intervals when dealing with continuous distributions.

6. For this test the alternative hypothesis typically lacks specificity (e.g., H_1 in (14.1) does not provide us with any hint as to what the population might look like if the null hypothesis H_0 is untrue). If we do not posit an alternative hypothesis that specifies a particular statistical model, then the power of the multinomial goodness-of-fit test is difficult to ascertain. However, it has been shown that the power of the test tends to unity as $n \to \infty$. Hence for very large samples, the test is almost certain to reject H_0. This is obviously a weakness of this type of goodness-of-fit test.

14.4   The Kolmogorov-Smirnov Test for Goodness of Fit

It was determined in the preceding two sections of this chapter that if our sample data is categorical, or if it can be grouped into categories, then Pearson's chi-square statistic is applicable for testing goodness of fit. Hence the chi-square test is appropriate when the null distribution function $F_0(t)$ under $H_0: F(t) = F_0(t)$ is discrete.

To apply Pearson's chi-square goodness-of-fit test when $F_0(t)$ is continuous, we must approximate $F_0(t)$ by grouping the sample data into an arbitrary number of fictitious class intervals or cells. Clearly any such grouping implicitly requires the sample size n to be fairly large since the expected frequency of each cell must be at least five.

An alternative goodness-of-fit test that is more appropriate for continuous distributions (since it does not require grouping the sample data into contrived cells or intervals), and that is suitable for small n, is the *Kolmogorov-Smirnov (K-S) test*. Unlike the chi-square goodness-of-fit test, the K-S test requires that the null distribution function $F_0(t)$ be completely specified; that is, the functional form as well as all parameters are assumed known so that $H_0: F(t) = F_0(t)$

is simple. Hence the K-S goodness-of-fit test is not applicable when the null hypothesis is composite. (Remember that the Pearson chi-square goodness-of-fit test can be performed under incomplete specification of the population parameters, as indicated in Section 14.3.) In addition, this test cannot be applied in instances when the observations are not inherently numerical; that is, the K-S procedure must be applied to data measured on at least an ordinal scale since, as we shall soon see, the observations must be ordered so that the sample distribution function can be formed. The K-S test is therefore not applicable to categorical data.

Before we examine the particulars of the K-S test, let us consider some additional salient points of comparison between the Pearson chi-square and K-S goodness-of-fit tests. First, the chi-square test is used to determine goodness of fit between observations classified into cells or categories and some theoretical or expected distribution taken over the same categories. Under the restriction that expected cell frequencies must be at least five, adjacent problematic or low-count cells can be pooled to fulfill this requirement. However, when the population distribution is continuous, any such consolidation of cells might result in a loss of information, especially in the tails of the distribution or if there are many narrow intervals. Since the K-S procedure is designed to handle continuous data, any placement of the observations into cells or intervals is purely for convenience sake.

Second, if the n sample values are realizations of a continuous random variable X, as opposed to sampling strictly categorical data, then an assessment of goodness of fit via the K-S method can be made at *each* of the realizations x of X. For the Pearson chi-square goodness-of-fit test involving $k(\leq n)$ distinct cells, only k comparisons are made. So since the K-S test allows us to examine goodness of fit for each of the n sample realizations, instead of only for k classes, the K-S procedure utilizes the sample information in a much more comprehensive fashion relative to the chi-square test. In fact, the test results obtained via the chi-square procedure are affected by the number of cells or intervals as well as by their width, each of which is arbitrarily chosen.

Let us now turn to the particulars of the K-S test for goodness of fit between observed sample data and some given theoretical or expected population distribution. The test addresses the question: Did the sample observations come from some hypothesized population distribution? (Remember that the population random variable must be continuous and the population distribution completely specified.)

As we shall now see, if the n sample observations are realizations of a continuous random variable X, then comparisons can be made between observed and expected cumulative relative frequencies determined at each of these values of X. That is to say, we need to compare the values of a *hypothesized distribution function* $F_0(x)$ with the values of the *sample or empirical distribution function* $S_n(x)$ for all realizations x of X. Hence $S_n(x)$ gives the proportion of the sample realizations that do not exceed some number x for all real x; it provides a consistent estimator for the true or population (unknown) distribution function $F(x)$ for the sample realizations of X.

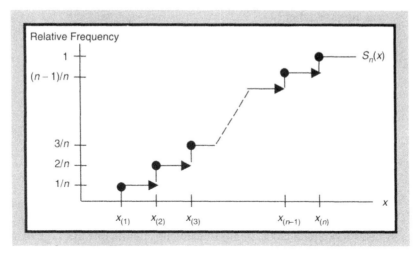

Figure 14.1   An empirical distribution function with step size $\frac{1}{n}$.

Let us consider for the moment the characteristics of the empirical distribution function $S_n(x)$. It amounts to the relative frequency distribution function for a set of sample random variables and, as stated earlier, it exhibits the proportion of sample realizations that are less than or equal to x. In this regard, $S_n(x)$ is a discrete random variable whose graph corresponds to a step function that increases in relative frequency by the amount $\frac{1}{n}$ at each of its points of discontinuity or jump points, where the latter occur at the realizations $x_{(1)}, x_{(2)}, \ldots, x_{(n)}$ of the order statistics $X_{(1)}, X_{(2)}, \ldots, X_{(n)}$ of a random sample.[1] In this regard, the sample or empirical distribution function is defined as

$$S_n(x) = \begin{cases} 0, & x < x_{(1)}; \\ \frac{i}{n}, & x_{(i)} \leq x < x_{(i+1)}, \ i = 1, 2, \ldots, n-1; \\ 1 & x \geq x_{(n)}. \end{cases} \qquad (14.8)$$

The graph of a typical empirical distribution function is given in Figure 14.1, with the size of each step corresponding to $P(X = x_{(i)}) = \frac{1}{n}$ for all i.

Given that $S_n(x)$ is the empirical distribution function of the sample random variables $X_{(1)}, \ldots, X_{(n)}$ taken from the population distribution function $F(x)$, it

---

[1] Given a random sample of size n, let us arrange the sample random variables $X_i, i = 1, \ldots, n$, in order of increasing numerical magnitude, with the resulting set of n values denoted as $X_{(1)}, X_{(2)}, \ldots, X_{(n)}$, where, for $1 \leq i \leq n, X_{(1)} = \min X_i, \ X_{(2)} = 2^{nd}$ smallest $X_i, \ldots, X_{(n)} = \max X_i$. Clearly $X_{(1)} < X_{(2)} < \ldots < X_{(n)}$ with $X_{(i)}$ termed the ith *order statistic*. In general, the order statistics of a random sample are simply the sample values placed in ascending order.

can be shown that

$$P\left(S_n(x) = \frac{j}{n}\right) = \binom{n}{j} F(x)^j \left(1 - F(x)\right)^{n-j}, \; j = 0, 1, \ldots, n,$$

with the mean and variance of $S_n(x)$ corresponding to

$$E\left(S_n(x)\right) = F(x),$$

$$V\left(S_n(x)\right) = \frac{F(x)\left(1 - F(x)\right)}{n},$$

respectively. Moreover, the sample distribution function $S_n(x)$ converges in probability to the true distribution function $F(x)$ for all real x, and the sampling distribution of $S_n(x)$ is asymptotically normal; that is, for large n, a standardized $S_n(x)$ converges in distribution to a standard normal variate.

Since under the null hypothesis the observed sample is from a completely specified theoretical distribution function $F(x)$, it must be the case that $S_n(x) \xrightarrow{P} F(x)$ for all x. Hence it follows that, for large n, the difference between the empirical distribution function and its theoretical counterpart or $|S_n(x) - F(x)|$ should be small save for sampling fluctuations. In fact, the random variable

$$D_n = \sup_x |S_n(x) - F(x)|, \tag{14.9}$$

known as the *Kolmogorov-Smirnov statistic*, will serve as a measure of goodness of fit between $S_n(x)$ and $F(x)$. Here $|S_n(x) - F(x)|$ depicts the vertical distance between the empirical distribution function and the theoretical distribution function at a particular x. Then D_n is the *supremum* or least upper bound of all such distances determined at each x. To ensure that the supremum is actually attained, (14.9) can be replaced by

$$D_n = \max_x \left\{ |S_n(x) - F(x)|, |S_n(x - \varepsilon) - F(x)| \right\}, \varepsilon > 0, \tag{14.9.1}$$

that is, the K-S statistic is the *maximum* vertical distance between the two cumulative relative frequency distributions $S_n(x)$ and $F(x)$. Here the additional term $|S_n(x - \varepsilon) - F(x)|$, the difference between F at x and S_n at the next smaller x value, is included in (14.9.1) since S_n is continuous only from the right. For instance, as Figure 14.2 reveals, distance $AB = |S_n(x') - F(x')|$ and distance $AC = |S_n(x' - \varepsilon) - F(x')| = |S_n(x'') - F(x')|$. Then according to (14.9.1), $D_n = AC$. In addition, (14.9.1) is useful when ties within the sequence of $|S_n(x) - F(x)|$ values are present.

A key property of D_n is that it is a *distribution-free* statistic; that is, the sampling distribution of D_n does not depend upon $F(x)$ provided that $F(x)$ is continuous. Moreover, unlike the Pearson chi-square test, the K-S procedure

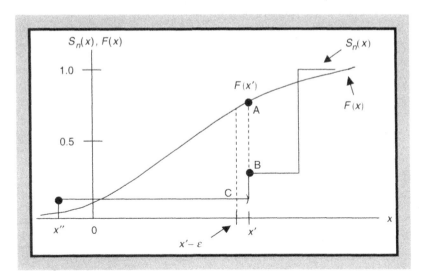

Figure 14.2   The K-S statistic $D_n = AC > AB$.

is *exact* for any sample size in that it does not employ an approximate distribution to test goodness of fit. That is, the exact sampling distribution of D_n is known and has been tabulated (see Table A.14 of the Appendix), whereas the sampling distribution of Pearson's chi-square statistic is only approximately chi-square distributed for any finite n. Hence the K-S procedure is applicable for any sample size n, whereas Pearson's chi-square statistic renders a good chi-square approximation only for large n.

To perform the K-S goodness-of-fit test, suppose we extract a random sample of size n from an unknown distribution function $F(x)$. Let us test $H_0: F(x) = F_0(x)$, against $H_1: F(x) \neq F_0(x)$ for at least one x. Then:

(a) Order the sample realizations from smallest to largest; that is, form the ordered sequence $x_{(1)}, x_{(2)}, \ldots, x_{(n)}$.

(b) Compute the value of the sample or empirical distribution function $S_n(x) = \frac{i}{n}$ for all $i = 1, \ldots, n$ and $x_{(i)} \leq x < x_{(i+1)}$. $S_n(x)$ will abruptly jump by $\frac{1}{n}$ at each ordered x value.

(c) Since the theoretical distribution is completely specified under H_0 (its parameter values are given), find the expected probability $F_0(x)$ at each ordered x value.

So under H_0, our test statistic is

$$D_n^0 = \max_x \left\{ \left| S_n(x) - F_0(x) \right|, \left| S_n(x - \varepsilon) - F_0(x) \right| \right\}, \; \varepsilon > 0, \qquad (14.9.2)$$

with its sample realization denoted as d_n^0, $0 \leq d_n \leq 1$ (see Figure 14.3). (Remember that D_n^0 is a function of the deviations between the observed cumulative

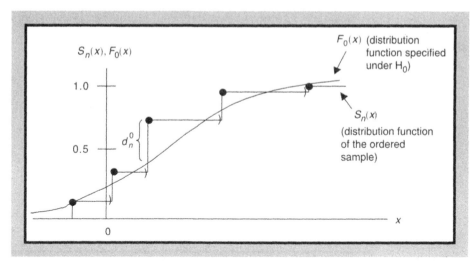

Figure 14.3   Under H_0, d_n^0 is the sample realization of D_n^0.

Table 14.5

One-Sided α	0.10	0.075	0.05	0.025	0.005
Two-Sided α	0.20	0.15	0.10	0.05	0.01
$d_{n,1-\alpha}$	1.07/c	1.14/c	1.22/c	1.36/c	1.63/c

relative frequency distribution and the corresponding cumulative probabilities determined under H_0.)

If H_0 is true, then the absolute differences $|S_n(x) - F_0(x)|$ should be relatively small and attributable to only sampling variation. And if these differences tend to be relatively large, then we have sample evidence against H_0; that is, a large value of d_n^0 casts serious doubt upon H_0. Given that *absolute distances* are taken in (14.9.2) (we ignore their signs), we are essentially performing a two-tail test of goodness of fit; that is, D_n^0 picks up significant deviations in *either direction* between the observed and theoretical distribution functions over all x's. And since we reject H_0 if d_n^0 is large, it follows that, for a given tail probability $\alpha = P(TIE)$, the K-S goodness-of-fit test has a critical region of the form $\mathcal{R} = \{d_n | d_n \geq d_{n,1-\alpha}\}$, where the critical value of $d_n, d_{n,1-\alpha}$, is the upper $100(1 - \alpha)\%$ quantile of the sampling distribution of D_n and is obtained from Table A.14. It is important to note that this table is exact only if the null distribution $F_0(x)$ is continuous; otherwise, these critical values render a conservative test (the chances of rejecting H_0 for a given α are diminished somewhat). For large samples ($n > 40$), some of the critical values $d_{n,1-\alpha}$ determined from the asymptotic approximation to the sampling distribution of D_n are offered in Table 14.5, where $c = \left(n + \sqrt{n/10}\right)^{1/2}$.

We note briefly that in virtually all instances the K-S test is more powerful than Pearson's chi-square test when both are applicable in testing goodness of fit between a sample and a theoretical distribution function. In fact, for $F_0(x)$ a completely specified normal distribution function, the K-S test is asymptotically more powerful than the chi-square test. And when n is small, the K-S procedure, unlike the chi-square test, provides an exact test of $H_0: F(x) = F_0(x)$.

When is Pearson's chi-square test superior to the K-S test for assessing goodness of fit? If $F_0(x)$ is discrete and completely specified, then the chi-square routine provided in Section 14.2 should be used (the sampling distribution of D_n is not exact in this circumstance). Additionally, if $F_0(x)$ is discrete but not completely specified, then the discussion pertaining to the chi-square test given in Section 14.3 applies.

We now turn to some example problems involving the K-S routine for testing goodness of fit.

Example 14.4.1   The run time of a commuter bus in a certain city is thought to be uniformly distributed with time to destination limits of 35 minutes to 50 minutes depending upon morning traffic flow. A random sample of $n = 10$ run times was taken, yielding the following realizations: 37, 42, 48, 30, 38, 39, 49, 47, 40, 41. For $\alpha = 0.05$, determine if the sample is indeed from a uniform distribution. From (7.2), the null distribution function is completely determined as

$$F_0(x; 35, 50) = \begin{cases} 0, & x < 35; \\ \frac{x-35}{15}, & 35 \le x < 50; \\ 1, & x > 50. \end{cases}$$

Hence we shall test $H_0: F(x) = F_0(x; 35, 50)$, against $H_1: F(x) \neq F_0(x; 35, 50)$ for at least one x. Table 14.6 houses all relevant calculations. From (14.9.2), $d_{10}^0 = 0.234$ and from Table A.14, $\mathcal{R} = \{d_{10} | d_{10} \ge d_{10, 0.95} = 0.41\}$. Since d_{10}^0 is not an element of the critical region $\mathcal{R}$, it follows that we cannot reject H_0 at the 5% level—there is no compelling sample evidence that suggests that X is not uniformly distributed between 35 and 50.   ∎

Example 14.4.2   Suppose that five years ago it was determined that a particular population characteristic X is normally distributed with distribution function $F(x; \mu, \sigma) = F(x; 127, 10)$.

Recently, this characteristic was revisited and a random sample of $n = 20$ items was selected, rendering the following realizations: 130, 126, 129, 130, 135, 120, 128, 128, 123, 126, 125, 130, 132, 138, 132, 129, 128, 132, 130, 129. Has there been a significant change in the distribution of X's at the $\alpha = 0.05$ level? Let us test $H_0: F(x) = F_0(x; 127, 10)$, versus $H_1: F(x) \neq F_0(x; 127, 10)$ for at least one x. Our calculations are summarized in Table 14.7. Note that column 3

Table 14.6

Ordered Realizations $x_i, i = 1, ..., 10$	$S_{10}(x) = \frac{i}{10}$ $i = 1, ..., 10$	$F_0(x; 35, 50)$	$\begin{vmatrix} S_{10}(x)- \\ F_0(x; 35, 50) \end{vmatrix}$	$\begin{vmatrix} S_{10}(x - \varepsilon)- \\ F_0(x; 35, 50) \end{vmatrix}$
30	0.10	0.000*	0.100	0.000
37	0.20	0.133	0.067	0.033
38	0.30	0.200	0.100	0.000
39	0.40	0.266	0.134	0.034
40	0.50	0.333	0.167	0.067
41	0.60	0.400	0.200	0.100
42	0.70	0.460	0.234	0.134
47	0.80	0.800	0.000	0.100
48	0.90	0.866	0.034	0.066
49	1.00	0.983	0.067	0.033

*30 is less than the lower limit of 35.

Table 14.7

Ordered Realizations $x_{(i)}, i = 1, ..., 20$	$S_{20}(x) = \frac{i}{20}$, $i = 1, ..., 20$	$z_{(i)} = \frac{x_{(i)} - 127}{10}$, $i = 1, ..., 20$	$F_0(z_{(i)}; 0, 1)$	$\begin{vmatrix} S_{20}(x)- \\ F_0(z_{(i)}; 0, 1) \end{vmatrix}$	$\begin{vmatrix} S_{20}(x - \varepsilon)- \\ F_0(z_{(i)}; 0, 1) \end{vmatrix}$
120	0.05	−0.70	0.2420	0.1920	0.2420
123	0.10	−0.40	0.3446	0.2446	0.2946
125	0.15	−0.20	0.4208	0.2708	0.3208
126	0.20	−0.10	0.4602	0.2602	0.3102
126	0.25	−0.10	0.4602	0.2102	0.2602
128	0.30	0.10	0.5398	0.2398	0.2898
128	0.35	0.10	0.5398	0.1898	0.2398
128	0.40	0.10	0.5398	0.1398	0.1898
129	0.45	0.20	0.5792	0.1292	0.1792
129	0.50	0.20	0.5792	0.0792	0.1292
129	0.55	0.20	0.5792	0.0792	0.0792
130	0.60	0.30	0.6179	0.0179	0.0679
130	0.65	0.30	0.6179	0.0321	0.0179
130	0.70	0.30	0.6179	0.0821	0.0321
130	0.75	0.30	0.6179	0.1321	0.0821
132	0.80	0.50	0.6914	0.1086	0.0586
132	0.85	0.50	0.6914	0.1586	0.1086
132	0.90	0.50	0.6914	0.2086	0.1586
135	0.95	0.80	0.7881	0.1619	0.1119
138	1.00	1.10	0.8643	0.1357	0.0857

exhibits the set of standardized $x_{(i)}$'s and column 4 indicates the corresponding $N(0,1)$ distribution function value. From (14.9.2), $d_{20}^0 = 0.3208$ and, from Table A.14, $\mathcal{R} = \{d_{20}|d_{20} \geq d_{20,0.95} = 0.294\}$. Since d_{20}^0 exceeds the critical value of 0.294, we reject the null hypothesis at the 5% level—there has been a significant change in the distribution of the X characteristic from its previous normal form. ■

Example 14.4.3 Suppose that we are faced with conducting a goodness-of-fit test for normality given that the data set under scrutiny has been grouped into $k(= 8)$ classes (see Table 14.8). It is believed that this sample data set has been drawn from a normal population with a distribution function of the form $F(x; \mu, \sigma) = F(x; 30, 6)$. In this regard, for $\alpha = 0.05$, let us test $H_0: F(x) = F_0(x; 30, 6)$, against $H_1: F(x) \neq F_0(x; 30, 6)$ for at least one x. Here the values of the sample or empirical distribution function $S_{33}(x)$ appear in column 4 of Table 14.8 and the values of the expected or theoretical distribution function

Table 14.8

1. Classes of X	2. Observed Absolute Frequency o_j, $j = 1, ..., 8$	3. Observed Relative Frequency $\frac{o_j}{33}$, $j = 1, ..., 8$	4. Cumulative Observed Relative Frequency $S_{33}(x)$
0–9	1	0.0303	0.0303
10–19	3	0.0909	0.1212
20–29	5	0.1515	0.2727
30–39	10	0.3030	0.5757
40–49	7	0.2121	0.7878
50–59	4	0.1212	0.9090
60–69	2	0.0606	0.9696
70–79	1	0.0303	0.9999
	33		

5. Upper Real Class Limits $x_{(j)}, j = 1, ..., 8$	6. $z_{(j)} = \frac{x_{(j)} - 30}{6}$ $j = 1, ..., 8$	7. $F_0\left(z_{(j)}; 0, 1\right)$	8. $\left|S_{33}(x) - F_0(z_{(i)}; 0, 1)\right|$	9. $\left|S_{33}(x - \varepsilon) - F_0(z_{(i)}; 0, 1)\right|$
9.5	− 3.41	0.0003	0.0300	0.0003
19.5	− 1.75	0.0400	0.0812	0.0097
29.5	− 0.08	0.4681	0.1954	0.3469
39.5	1.58	0.9430	0.3673	0.6703
49.5	3.25	0.9994	0.2116	0.2116
59.5	4.91	1.0000	0.0910	0.0910
69.5	6.58	1.0000	0.0304	0.0304
79.5	8.25	1.0000	0.0001	0.0001

(determined at the upper boundaries of the class intervals) are provided by column 7. Again looking to (14.9.2) we have $d_{33}^0 = 0.6703$. With $\mathcal{R} = \{d_{33}|d_{33} \geq d_{33,0.95} = 0.231\}$, it is readily seen that we may reject H_0 at the 5% level. The observed data set should not be viewed as having been extracted from the null distribution function given earlier. ∎

14.5 The Lilliefors Goodness-of-Fit Test for Normality

We noted in the preceding section that if we wanted to determine if a random sample could be thought of as having been selected from a normal population, then the Kolmogorov-Smirnov goodness-of-fit test was applicable if the null distribution function $F_0(x)$ was completely specified; that is, all of its parameters, specifically μ and σ, were given. However, if the parameters μ and σ within $F_0(x)$ are unknown and must be estimated from sample data (their estimators are $\overline{X}$ and $S = \left[\sum_{i=1}^n (X_i - \overline{X})^2/(n-1)\right]^{1/2}$, respectively), then a modification of the table of critical values of the sampling distribution of D_n must be made. The modified table was developed by Lilliefors so that the value of the test statistic D_n depends upon $\bar{x}$ and s (the sample realizations of $\overline{X}$ and S, respectively) instead of μ and σ.

Lilliefors' test for a normal distribution with unspecified mean and standard deviation is a distribution-free technique that tests the composite null hypothesis

H_0: the random sample is from a normal distribution function with unknown mean and variance, against

H_1: the random sample is not from a normal distribution function

From the realizations $x_{(i)}$ of the order statistics $X_{(i)}, i = 1, \ldots, n$, let us form the *standardized ordered realizations,* $z_{(i)} = \frac{x_{(i)} - \bar{x}}{s}, i = 1, \ldots, n$. Then the two-sided test statistic

$$\hat{D}_n = \max\left\{\left|S_n(z) - \hat{F}(z_{(i)};0,1)\right|, \left|S_n(z-\varepsilon) - \hat{F}(z_{(i)};0,1)\right|\right\}, \varepsilon > 0, \quad (14.9.3)$$

with realization $\hat{d}_n$, is computed from the $z_{(i)}$'s; that is, $S_n(z)$ is the empirical distribution function of the $z_{(i)}$'s and $\hat{F}(z_{(i)};0,1)$ is the standard normal distribution function with mean 0 and standard deviation 1. Clearly $\hat{D}_n$ represents the maximum vertical distance between these two expressions.

As before, we will reject the null hypothesis if $\hat{d}_n \geq d_{n,1-\alpha}$, where $d_{n,1-\alpha}$ depicts the $100(1 - \alpha)\%$ quantile of the approximate sampling distribution of $\hat{D}_n$ and is obtained from Table A.13 of the Appendix. For large samples ($n > 30$), some of the critical values $d_{n,1-\alpha}$ determined from an asymptotic approximation to the sampling distribution of $\hat{D}_n$ are presented in Table 14.9.

Table 14.9

α	0.20	0.15	0.10	0.05	0.01
$d_{n,1-\alpha}$	$\dfrac{0.74}{\sqrt{n}}$	$\dfrac{0.77}{\sqrt{n}}$	$\dfrac{0.82}{\sqrt{n}}$	$\dfrac{0.89}{\sqrt{n}}$	$\dfrac{1.04}{\sqrt{n}}$

Table 14.10

$x_i,$ $i = 1,...,10$	$z_{(i)} = \dfrac{x_{(i)}-\bar{x}}{s},$ $i = 1,...,10$	$S_{10}(z)$	$\hat{F}(z;0,1)$	$\left\| \begin{array}{c} S_{10}(z)- \\ \hat{F}(z;0,1) \end{array} \right\|$	$\left\| \begin{array}{c} S_{10}(z-\varepsilon)- \\ \hat{F}(z;0,1) \end{array} \right\|$
100	-2.55	0.10	0.0054	0.0946	0.0054
145	-0.38	0.20	0.3520	0.1520	0.2520
148	-0.23	0.30	0.4090	0.1090	0.2090
150	-0.14	0.40	0.4443	0.0443	0.1443
156	0.15	0.50	0.5596	0.0596	0.1596
160	0.34	0.60	0.6331	0.0331	0.1331
162	0.46	0.70	0.6772	0.0228	0.0772
165	0.59	0.80	0.7224	0.0776	0.0224
170	0.83	0.90	0.7967	0.1033	0.0033
172	0.93	1.00	0.8238	0.1762	0.0762

Example 14.5.1 Let us determine, for $\alpha = 0.05$, if the following random sample has been extracted from a normal distribution: 100, 150, 172, 145, 160, 170, 156, 165, 148, 162. To this end, we want to test H_0 versus H_1 as given at the outset of this section. Since $\bar{x} = 152.80$ and $s = 20.65$, the $z_{(i)}$'s are determined in column 2 of Table 14.10. Then from (14.9.3), $\hat{d}_{10} = 0.2520$. And since $\mathcal{R} = \{\hat{d}_{10}|\hat{d}_{10} \geq d_{10,0.95} = 0.258\}$, we cannot reject H_0 at the 5% level of significance; that is, the normal distribution function seems to be a reasonable representation of the true unknown distribution function. ■

14.6 The Shapiro-Wilk Goodness-of-Fit Test for Normality

The *Shapiro-Wilk (S-W) test for normality* of an unknown population distribution function is a useful alternative to the chi-square and Kolmogorov-Smirnov tests for goodness of fit. The test is relatively powerful when testing for departures from normality. However, unlike the chi-square and K-S procedures, the S-W test requires a considerable amount of computational effort.

Suppose we extract a set of sample random variables $\{X_1,\ldots,X_n\}$ from an unknown distribution function $F(x)$. As was the case with the preceding Lilliefors goodness-of-fit test, we seek to test the composite null hypothesis

H_0: the random sample is from a normal distribution function with unknown mean and variance, against

H_1: the random sample is not from a normal distribution function

To obtain the S-W test statistic, let us first determine the order statistics of the preceding sample random variables or $X_{(1)}, X_{(2)}, \ldots, X_{(n)}$. Next, for our sample of size n, let us consult Table A.18 of the Appendix so as to obtain the constant coefficients $a_1, a_2, \ldots, a_k$, where k is (approximately) $\frac{n}{2}$. As a last step we form the *S-W test statistic*

$$W = \frac{\left[\sum_{j=1}^{k} a_j (X_{(n-j+1)} - X_{(j)}) \right]^2}{\sum_{i=1}^{n} (X_i - \overline{X})^2}, \tag{14.9.4}$$

where the random variable $\overline{X}$ is the sample mean and the sample realization of W is expressed as

$$w = \frac{\left[\sum_{j=1}^{k} a_j (x_{(n-j+1)} - x_{(j)}) \right]^2}{\sum_{i=1}^{n} (x_i - \bar{x})^2}. \tag{14.9.4.1}$$

For a level of significance $\alpha = P(TIE)$, the S-W test has a critical region of the form $\mathcal{R} = \{w | w \leq w_\alpha\}$, where w_α is the $100\alpha\%$ quantile of the sampling distribution of W (Table A.17 of the Appendix).

Example 14.6.1   Is there any reason to conclude, at the $\alpha = 0.05$ level of significance, that the following sample of $n = 20$ observations was not taken from a normal distribution function: 16, 19, 17, 20, 1, 5, 7, 27, 15, 18, 19, 18, 17, 25, 18, 30, 35, 29, 10, 6? Let H_0 and H_1 be formulated as before. Looking to Table 14.11, column 1 contains the sample realizations of the order statistics $X_{(1)}, \ldots, X_{(20)}$ and column 2 exhibits the a_j's, $j = 1, \ldots, k = 10$, obtained from Table A.18. In addition, $\bar{x} = 17.6$ while $\sum_{i=1}^{20} (x_i - \bar{x})^2 = 1488.80$. Then from (14.9.4.1), $w = \frac{(37.877)^2}{1488.80} = 0.9636$. From Table A.17 we may determine that $\mathcal{R} = \{w | w \leq w_{0.05} = 0.905\}$. Since our realized W value does not lie within the critical region, we have insufficient sample evidence to reject H_0 at the 5% level.   ∎

14.7   The Kolmogorov-Smirnov Test for Goodness of Fit: Two Independent Samples

Let $\{X_1, \ldots, X_n\}$ and $\{Y_1, \ldots, Y_m\}$ be two independent sets of sample random variables of sizes n and m that have been drawn from the unknown distribution functions $F_X(x)$ and $F_Y(x)$, respectively. As was the case for the one-sample K-S goodness-of-fit test, $F_X(x)$ and $F_Y(x)$ are assumed continuous and the random

Table 14.11

$x_{(i)}$, $i = 1, ..., 20$		a_j, $j = 1, ..., 10$	$x_{(n-j+1)} - x_{(j)}$, $j = 1, ..., 10$	$a_j(x_{(n-j+1)} - x_{(j)})$
$x_{(1)} = 1$	$x_{(11)} = 18$	0.4734	$x_{(20)} - x_{(1)} = 34$	16.095
$x_{(2)} = 5$	$x_{(12)} = 18$	0.3211	$x_{(19)} - x_{(2)} = 25$	8.028
$x_{(3)} = 6$	$x_{(13)} = 19$	0.2565	$x_{(18)} - x_{(3)} = 23$	5.900
$x_{(4)} = 7$	$x_{(14)} = 19$	0.2085	$x_{(17)} - x_{(4)} = 20$	4.170
$x_{(5)} = 10$	$x_{(15)} = 20$	0.1686	$x_{(16)} - x_{(5)} = 15$	2.529
$x_{(6)} = 15$	$x_{(16)} = 25$	0.1334	$x_{(15)} - x_{(6)} = 5$	0.667
$x_{(7)} = 16$	$x_{(17)} = 27$	0.1013	$x_{(14)} - x_{(7)} = 3$	0.304
$x_{(8)} = 17$	$x_{(18)} = 29$	0.0711	$x_{(13)} - x_{(8)} = 2$	0.142
$x_{(9)} = 17$	$x_{(19)} = 30$	0.0422	$x_{(12)} - x_{(9)} = 1$	0.042
$x_{(10)} = 18$	$x_{(20)} = 35$	0.0140	$x_{(11)} - x_{(10)} = 0$	0.000
				37.877

variables X and Y are measured on at least an ordinal scale. Our objective then is to determine if these two sets of sample random variables have been drawn from *identical* population distribution functions. This means that the following test does not simply reveal differences in, say, location; it detects any and all differences that may exist between the two distribution functions.

For the *two-sample K-S goodness-of-fit test*, a comparison is made between the empirical distribution functions $S_n(x)$ and $S_m(x)$ of the two samples. To this end, let $x_{(1)}, \ldots, x_{(n)}$ and $y_{(1)}, \ldots, y_{(m)}$ be the realizations of the sets of order statistics $\{X_{(1)}, \ldots, X_{(n)}\}$ and $\{Y_{(1)}, \ldots, Y_{(m)}\}$, respectively. Then the corresponding empirical distribution functions appear as:

$$S_n(x) = \begin{cases} 0 & \text{if } x < x_{(1)}; \\ \frac{i}{n} & \text{if } x_{(i)} \le x < x_{(i+1)}, \, 1 = 1, 2, \ldots, n-1; \\ 1 & \text{if } x \ge x_{(n)} \end{cases}$$

and

$$S_m(x) = \begin{cases} 0 & \text{if } x < y_{(1)}; \\ \frac{j}{m} & \text{if } y_{(j)} \le x < y_{(j+1)}, \, j = 1, 2, \ldots, m-1; \\ 1 & \text{if } x \ge y_{(m)}. \end{cases}$$

Let us pool both sets of realizations of the sample random variables. Then for the combined ordered arrangement of the $n + m$ sample realizations, $S_n(x)$ and $S_m(x)$ exhibit the respective proportions of these realizations that are less than or equal to x. As indicated in Section 14.4, these empirical distribution functions provide consistent estimators for the true or population distribution functions $F_X(x)$ and $F_Y(x)$ for the sample realizations of the X and Y random variables, respectively.

Let us frame the null hypothesis as $H_0: F_X(x) = F_Y(x)$ for all x. Under this composite hypothesis, the population distribution functions are identical and we have two samples drawn from exactly the same population. Then except for sampling variation, there should be fairly close agreement between the two sample or empirical distribution functions. If we test H_0 against the two-sided alternative $H_1: F_X(x) \neq F_Y(x)$ for at least one value of x, then the *K-S two-sample test statistic* is

$$D_{n,m} = \max_x \left| S_n(x) - S_m(x) \right|. \tag{14.10}$$

Its sample realization is denoted as $d_{n,m}$. Given $H_1, D_{n,m}$ depicts the maximum absolute difference between the two empirical distribution functions so that only magnitudes and not the directions of the deviations in this expression are relevant.

For $\alpha = P(TIE)$, we may construct the critical region as $\mathcal{R} = \{d_{n,m} | d_{n,m} \geq d_{n,m,1-\alpha}\}$, where $d_{n,m,1-\alpha}$ is the $100(1 - \alpha)\%$ quantile of the *appropriate* sampling distribution of $D_{n,m}$. That is, if $n = m$, then the relevant critical values appear in Table A.15 of the Appendix; and if $n \neq m$, then the critical values are given in Table A.16. (Note that these sampling distributions are exact if the population distribution functions are continuous.)

Example 14.7.1   Given the following sets of sample realizations:

X: 10, 19, 15, 20, 12, 8, 15, 21 $(n = 8)$

Y: 15, 22, 17, 9, 12, 10, 29, 11, 25, 31 $(m = 10)$

determine, for $\alpha = 0.05$, if these two data sets have been extracted from identical population distribution functions. As Table 14.12 reveals, these two samples are *pooled* and subsequently ordered into a single or combined sample involving $n + m = 18$ data points.

Here $d_{n,m} = d_{8,10} = 0.400$. Looking to Table A.16, the $100(1 - \alpha)\% = 95\%$ quantile of the sampling distribution of $D_{n,m}$ is $\frac{23}{40} = 0.575$ or $\mathcal{R} = \{d_{n,m} | d_{n,m} \geq d_{8,10,0.95} = 0.575\}$. Since our realized $D_{n,m}$ does not fall within $\mathcal{R}$, we cannot reject the null hypothesis at the 5% level; that is, we have no reason to doubt that the two distribution functions $F_X(x)$ and $F_Y(x)$ are identical.   ∎

14.8   Assessing Normality via Sample Moments

Another test for determining if a random sample can be viewed as having been extracted from a normal population is one that utilizes the moments of a distribution. In this regard, the present test for normality exploits the known values of certain key moments of a normal distribution. That is, our test for normality will employ the third and fourth moments about the mean of the sample, where the third moment addresses symmetry and the fourth moment pertains to kurtosis.

Table 14.12

$x_{(i)}, i = 1, ..., 8,$	$y_{(j)}, j = 1, ..., 10$	$S_n(x)$	$S_m(x)$	$\left|S_n(x) - S_m(x)\right|$
5		0.125		$0.125 - 0.000 = 0.125$
8		0.250		$0.250 - 0.000 = 0.250$
	9		0.100	$0.250 - 0.100 = 0.150$
10	10	0.375	0.200	$0.375 - 0.200 = 0.175$
	11		0.300	$0.375 - 0.300 = 0.075$
12	12	0.500	0.400	$0.500 - 0.400 = 0.100$
15	15	0.625	0.500	$0.625 - 0.500 = 0.125$
	17		0.600	$0.625 - 0.600 = 0.025$
19		0.750		$0.750 - 0.600 = 0.150$
20		0.875		$0.875 - 0.600 = 0.275$
21		1.000		$1.000 - 0.600 = 0.400$
	22		0.700	$1.000 - 0.700 = 0.300$
	25		0.800	$1.000 - 0.800 = 0.200$
	29		0.900	$1.000 - 0.900 = 0.100$
	31		1.000	$1.000 - 1.000 = 0.000$

Although tests of the null hypothesis of normality have been performed using either sample skewness or sample kurtosis measures taken individually, we shall now consider a so-called *omnibus or composite sample statistic*, which incorporates both of these measures simultaneously and which can be used to test for the normality of a population distribution. For instance, in Chapter 4 we defined coefficients of skewness and kurtosis as

$$\alpha_3 = \frac{\mu_3}{(\mu_2)^{3/2}} \tag{4.27}$$

and

$$\alpha_4 = \frac{\mu_4}{(\mu_2)^2}, \tag{4.28}$$

respectively, where

$$\mu_r = \frac{\sum_{i=1}^{N}(X_i - \mu)^r}{N}, r = 2, 3, 4,$$

is the rth moment about the mean. (Remember that for a normal distribution, $\alpha_3 = 0$ and $\alpha_4 = 3$.)

Let us now specify a composite test statistic as

$$C = n\left[\frac{\alpha_3^2}{6} + \frac{(\alpha_4 - 3)^2}{24}\right], \tag{14.11}$$

where $c = n\left[(\hat{\alpha}_3^2)/6 + (\hat{\alpha}_4 - 3)/24\right]$ is the sample realization of C, $\hat{\alpha}_3$ is the realized value of α_3, $\hat{\alpha}_4$ is the realization of α_4, and $\frac{6}{n}$ and $\frac{24}{n}$ are the asymptotic variances of $\hat{\alpha}_3$ and $\hat{\alpha}_4$, respectively (Bowman and Shenton (1975)).[2] For large n, C is approximately chi-square distributed with $v = 2$ degrees of freedom. Here C will be employed to test for normality under the null hypothesis

$$H_0: \alpha_3 = 0 \; and \; \alpha_4 = 3. \tag{14.12}$$

The alternative hypothesis is then H_1: the null hypothesis is untrue and the true distribution is χ_2^2. Given these hypotheses, our decision rule is that we reject H_0 if $c \in \mathcal{R} = \{c | c > \chi_{1-\alpha,2}^2\}$, where $\chi_{1-\alpha,2}^2$ is the $100(1-\alpha)$ percentile of the chi-square distribution with 2 degrees of freedom and $\alpha = P(TIE)$.

To calculate C from sample information let us employ, via (4.25.1), (4.26.1), and (4.28.1), the sample realizations of μ_2, μ_3, and μ_4 or

$$\hat{\mu}_2 = \hat{\mu}_2' - \hat{\mu}^2;$$

$$\hat{\mu}_3 = \hat{\mu}_3' - 3\hat{\mu}\hat{\mu}_2' + 2\hat{\mu}^3; and \tag{14.13}$$

$$\hat{\mu}_4 = \hat{\mu}_4' - 4\hat{\mu}\hat{\mu}_3' + 6\hat{\mu}^2\hat{\mu}_2' - 3\hat{\mu}^4,$$

where $\hat{\mu}_r' = \frac{1}{n}\sum_{i=1}^n x_i^r, r = 2,3,4$, is the realized r^{th} moment about zero, $\hat{\mu} = \frac{1}{n}\sum_{i=1}^n x_i$, and x_i is the realization of the sample random variable $X_i, i = 1,\ldots,n$.

Example 14.8.1   Has the following set of $n = 30$ sample realizations (see Table 14.13) been taken from a normal population? Test (14.12) using $\alpha = 0.05$. Let us first determine

$$\hat{\mu} = \frac{175}{30} = 5.833,$$

$$\hat{\mu}_2' = \frac{1,235}{30} = 41.166,$$

$$\hat{\mu}_3' = \frac{9,925}{30} = 330.833,$$

$$\hat{\mu}_4' = \frac{87,599}{30} = 2,919.966.$$

Then from (14.13),

$$\hat{\mu}_2 = 41.166 - (5.833)^2 = 7.143,$$

$$\hat{\mu}_3 = 330.833 - 3(5.833)(41.166) + 2(5.833)^2 = 7.392,$$

---

[2] K.O. Bowman, L.R. Shenton, "Omnibus Test Contours for Departures from Normality Based on $\sqrt{b_1}$ and b_2," *Biometrica*, Vol. 62, Issue 2 (Aug. 1975), pp. 243–250.

Table 14.13

	x_i	x_i^2	x_i^3	x_i^4
	5	25	125	625
	1	1	1	1
	3	9	27	81
	8	64	512	4,096
	4	16	64	256
	2	4	8	16
	2	4	8	16
	4	16	64	256
	5	25	125	625
	6	36	216	1,296
	5	25	125	625
	8	64	512	4,096
	7	49	343	2,401
	10	100	1,000	10,000
	9	81	729	6,561
	11	121	1,331	14,641
	9	81	729	6,561
	12	144	1,728	20,736
	5	25	125	625
	6	36	216	1,296
	6	36	216	1,296
	3	9	27	81
	7	49	343	2,401
	3	9	27	81
	8	64	512	4,096
	7	49	343	2,401
	6	36	216	1,296
	4	16	64	256
	5	25	125	625
	4	16	64	256
Totals	175	1,235	9,925	87,599

$$\hat{\mu}_4 = 2,919.966 - 4(5.833)(330.833) + 6(5.833)^2(41.166)$$

$$- 3(5.833)^4 = 131.859$$

and thus

$$\hat{\alpha}_3 = \frac{\hat{\mu}_3}{(\hat{\mu}_2)^{3/2}} = \frac{7.392}{(7.143)^{3/2}} = 0.3872,$$

$$\hat{\alpha}_4 = \frac{\hat{\mu}_4}{(\hat{\mu}_2)^2} = \frac{131.859}{(7.143)^2} = 2.584.$$

Hence the estimated value of the test statistic C is

$$c = 30 \left[\frac{0.1499}{6} + \frac{(2.584 - 3)^2}{24} \right] = 0.963.$$

Since this realized C value is not an element of $\mathcal{R} = \{c | c > \chi^2_{0.95,2} = 5.991\}$, we cannot reject the null hypothesis—it appears that, at the 5% level of significance, the sample data set was drawn from a normal population. ∎

14.9 Exercises

14-1. ACE Motors surveyed 250 of its customers, asking them to indicate their preferred time for servicing their automobiles and/or trucks: morning (O_1), afternoon (O_2), or Saturday mornings (O_3). The results are $o_1 = 101, o_2 = 60$, and $o_3 = 89$. The service manager speculates that the proportions for the three groups or cells should be $\frac{1}{2}, \frac{1}{4}$, and $\frac{1}{4}$, respectively. Is there any evidence indicating that these proportions differ from 2:1:1? Choose $\alpha = 0.05$.

14-2. John's project for his statistics course was to toss a single six-sided die 500 times and count the number of times each face obtains in order to determine if the die is biased. (Note: A die is considered biased if its faces do not each occur with a relative frequency of 1/6.) The results of John's experiment are

Face Value	1	2	3	4	5	6
Frequency	80	79	88	80	84	89

Does John's die exhibit biased results? Let $\alpha = 0.10$.

14-3. ACE Nice Products has been in operation for one month. It employs four salesman (denoted A, B, C, and D) who have similar territories. Sales calls made for the month are:

Salesman	A	B	C	D
Number of Calls	57	65	60	66

For $\alpha = 0.01$, is the number of calls made uniformly distributed over the sales force?

14-4. Given the following absolute frequency distribution, can we conclude for $\alpha = 0.05$ that it was drawn from a normal population?

X	O_i
3.3–3.4	15
3.5–3.6	35
3.7–3.8	70
3.9–4.0	92
4.1–4.2	60
4.3–4.4	21
4.5–4.6	7

14-5. A recently printed novel was scanned to determine the number of misprints of a certain type per page. The production editor feels that this type of misprint follows a Poisson distribution with $\lambda = 0.50$. For $\alpha = 0.05$, determine if the editor's assertion is reasonable.

Number of Misprints	0	1	2	3	4
Number of Pages	82	37	20	7	2

14-6. A sawmill has its cutting equipment set to produce boards of 8 feet (or 96 inches) in length. The operations manager feels that the "true" length of a board should be normally distributed. A random sample of 295 boards yielded the following absolute frequency distribution:

Actual Length Inches	O_i
95.95–95.96	5
95.97–95.98	32
95.99–96.00	111
96.01–96.02	97
96.03–96.04	42
96.05–96.06	8

For $\alpha = 0.05$, test the manager's "hunch."

14-7. The number of accidents/month on a certain stretch of road is thought to follow a Poisson distribution. Given $\alpha = 0.05$, does the following observed data set (taken over a six year period) support this contention?

Number of Accidents per Month	0	1	2	3	4
O_i	35	20	10	4	3

14-8. A total of 200 students took the same written portion of a driver's education exam:

Score	O_i
70–74	15
75–79	18
80–84	60
85–89	70
90–94	20
95–99	17

If these student drivers can be thought of as a constituting a random sample taken from a population of individuals who would take the exam, can we legitimately hypothesize that the scores earned by the said population would be normally distributed? Take $\alpha = 0.01$.

14-9. A state motor vehicle emission inspection process has been designed so that inspection time is uniformly distributed with limits of 10 and 17 minutes. A sample of $n = 10$ duration times (taken under a variety of traffic conditions) yielded the following times (in minutes):

10.3 11.4 16.7 13.6 9.9 13.2 11.2 16.3 11.8 12.5

Use the Kolmogorov-Smirnov (K-S) test to determine if motor vehicles are inspected at a time that is uniformly distributed between 10 and 17 minutes. Use $\alpha = 0.05$.

14-10. The marketing director for ACE Superstores speculates that daily sales in the housewares department are normally distributed with $\mu = 130$ and $\sigma = 20$. Individual sales at $n = 18$ such stores are recorded:

152   169   102   178   174   160   105   150   148   111
150   149   105   118   162   121   142   129

For $\alpha = 0.05$, determine via the K-S routine if the daily sales data is from the specified normal distribution.

14-11. The scores obtained on a certain financial analysts certification exam are thought to be normally distributed with $\mu = 550$ and $\sigma = 100$. A random sample of $n = 10$ such scores resulted in the values:

500 470 550 580 490 525 495 570 510 530

Use the K-S test for $\alpha = 0.05$ to determine if we can view these observations as having been drawn from the specified normal distribution.

14-12. Given the following absolute frequency distribution, determine if the data set could have been drawn from a normal distribution with $\mu = 75$ and

$\sigma = 15$. Use $\alpha = 0.05$.

Classes of X	O_i
0–19	3
20–39	10
40–59	22
60–79	74
80–99	68
100–119	35
120–139	14
140–159	2

14-13. Using $\alpha = 0.05$, perform a Lilliefors test for normality using the data presented in Exercise 14-10.

14-14. Using $\alpha = 0.05$, perform a Lilliefors test for normality using the data presented in Exercise 14-11.

14-15. Can we conclude at the $\alpha = 0.05$ level that the following set of $n = 22$ data points was extracted from a normal distribution?

80	60	90	100	120	160	150	105	160	155	145
120	140	110	175	87	163	98	110	115	109	177

Employ the Shapiro-Wilk goodness-of-fit-test.

14-16. Two brands of fertilizer (call them A and B) are being tested on two identical plots of land planted with wheat. Each plot is divided into 10 equal sections. Yields per section have been determined for each brand of fertilizer. For $\alpha = 0.05$, use the K-S test to determine if the two samples were drawn from the same population distribution.

Fertilizer A	89.8	99.5	101.2	100.8	95.3	84.7	88.8	98.5	86.1	91.9
Fertilizer B	90.1	85.6	81.5	79.2	82.1	80.2	76.7	93.2	78.3	83.0

14-17. Two different training methods (call them A and B) are being tested at ACE Pharmaceuticals. Two groups of trainees have been selected randomly; one uses Method A and the other uses Method B. At the end of the training period each group is given a proficiency test to determine how well they perform a specific task. The scores are:

Method A	55	70	65	62	81	72	58	50	67	71		
Method B	50	88	84	91	90	78	62	82	75	80	79	83

For $\alpha = 0.05$, use the K-S method to determine if the two groups were drawn from the same population.

14-18. Use the moments of the following data set to determine if observations can be viewed as having been randomly drawn from a population that follows a normal distribution. Use $\alpha = 0.05$.

$$
\begin{array}{cccccccccc}
10 & 6 & 35 & 40 & 14 & 16 & 14 & 16 & 25 & 35 \\
29 & 25 & 29 & 25 & 25 & 50 & 45 & 25 & 16 & 29 \\
\end{array}
$$

Testing Goodness of Fit: Contingency Tables

15.1 An Extension of the Multinomial Chi-Square Statistic

In the Pearson or multinomial chi-square tests offered in Chapter 14 (see Sections 14.2 and 14.3), our (random) sample data was classified in accordance with a single factor or criterion into a set of mutually exclusive and collectively exhaustive categories or cells. As we shall now see, this approach can be applied to situations other than those arising from multinomial experiments. In this regard, suppose our observations are classified according to two or more attributes. Then the question naturally arises: Are these attributes or criteria of classification statistically independent? (Do they lack statistical association?) To answer this query, we shall employ *Pearson's chi-square test for independence* of attributes (or, as it is alternatively called, *Pearson's chi-square test of association*). As will be seen next, this technique can be applied to either quantitative or qualitative data.

Given a particular criterion of classification, suppose we cross-classify our data set according to a second criterion or attribute, where the true or population proportion within each resulting cell is unknown. For instance, given, say, a set of categories on the factor *education* (high school, some college, college degree, graduate degree), we may cross-classify according to a second factor such as sex or income category (low, middle, high). Here the resulting two-way classification scheme is termed a *two-way contingency table*.

15.2 Testing Independence

As just indicated, we are interested in determining whether there is any dependency relationship or association between two population characteristics.

Table 15.1

a.

	B_1	B_2
A_1	(A_1, B_1)	(A_1, B_2)
A_2	(A_2, B_1)	(A_2, B_2)
A_3	(A_3, B_1)	(A_3, B_2)

b.

	B_1	B_2	
A_1	n_{11}	n_{12}	$n_{11} + n_{12} = n_{1\bullet}$
A_2	n_{21}	n_{22}	$n_{21} + n_{22} = n_{2\bullet}$
A_3	n_{31}	n_{32}	$n_{31} + n_{32} = n_{3\bullet}$
	$n_{11} + n_{21} + n_{31} = n_{\bullet 1}$	$n_{12} + n_{22} + n_{32} = n_{\bullet 2}$	$n_{1\bullet} + n_{2\bullet} + n_{3\bullet}$ $= n_{\bullet 1} + n_{\bullet 2} = n$

We thus test

H_0: the two attributes are independent (or the rows and columns represent independent classifications), against   (15.1)

H_1: the two attributes are not independent (or the rows and columns are not independent classifications)

(The alternative hypothesis states that the attributes are dependent or related in some way. However, the extent or nature of the dependency relationship is unspecified.)

Suppose that n independent sample realizations have been obtained from a population of size N and each such realization is classified according to two attributes or criteria, which we shall simply label A and B. (Here we are implicitly assuming that N is *large enough* so that if the sample is taken without replacement, $\frac{n}{N}$ is *small enough* to warrant treating the sample *as if* it were drawn with replacement.) Suppose further that there are r mutually exclusive and collectively exhaustive categories for attribute A $(A_1, A_2, \ldots, A_r)$ and c such categories for attribute B $(B_1, B_2, \ldots, B_c)$.

All n sample realizations can then be classified into a two-way table, with A_i, $i = 1, \ldots, r$, categories for factor A making up the rows and the B_j, $j = 1, \ldots, c$, categories for factor B constituting the columns. The resulting table is termed an $r \times c$ *contingency table*, where each possible combination of factors A and B, (A_i, B_j), $i = 1, \ldots, r$; $j = 1, \ldots, c$, represents a distinct *cell* within the table, for example, a 3×2 contingency table is given by Table 15.1a. In addition, all n sample realizations are to be housed within the $r \times c$ contingency table on the basis of their specific A and B attribute categories. To this and, let n_{ij} denote the number of realizations falling into row i and column j; that is, n_{ij} is the *observed number of items* possessing attribute A_i and B_j in combination.

Let us denote the total number of sample realizations having attribute A_i as

$$n_{i\bullet} = \sum_{j=1}^{c} n_{ij} = i^{th} \text{ row total}$$ (15.2)

(*i* is held fixed and we sum over columns)

(the dot "•" indicates the subscript over which the summation takes place). Similarly, the total number of realizations possessing attribute B_j is

$$n_{\bullet j} = \sum_{i=1}^{r} n_{ij} = j^{th} \text{ column total}. \tag{15.3}$$

$$(j \text{ is held fixed and we sum over rows})$$

Here both (15.2) and (15.3) are termed *marginal totals*. Clearly $n_{1\bullet} + n_{2\bullet} + \cdots + n_{r\bullet} = n = n_{\bullet 1} + n_{\bullet 2} + \cdots + n_{\bullet c}$ or

$$\sum_{i=1}^{r} n_{i\bullet} = n = \sum_{j=1}^{c} n_{\bullet j} \tag{15.4}$$

(see Table 15.1b for the case where $r = 3$ and $c = 2$).

As stated in (15.1), the null hypothesis H_0 is that the two factors A and B are independent. (If H_0 is rejected in favor of H_1, then the A and B attributes are related or statistically dependent.) Under H_0, attributes A and B for the population at large are independent if and only if

$$P(A_i \cap B_j) = P(A_i) \cdot P(B_j) \tag{15.5}$$

for all possible joint events $A_i \cap B_j$, $i = 1, \ldots, r; j = 1, \ldots, c$, or

$$p_{ij} = (p_{i\bullet})(p_{\bullet j}), i = 1, \ldots, r; j = 1, \ldots, c. \tag{15.5.1}$$

Here p_{ij} is the unknown joint probability that an item selected at random from the population will be from the $(i, j)^{th}$ cell, $p_{i\bullet}$ is the unknown marginal probability that an item drawn at random from the population is from category i of characteristic A, and $p_{\bullet j}$ is the unknown marginal probability that a randomly selected item from the population is from category j of characteristic B. Then under the independence assumption,

$$H_0: p_{ij} = (p_{i\bullet})(p_{\bullet j}) \text{ for all } i \text{ and } j$$
$$H_1: p_{ij} \neq (p_{i\bullet})(p_{\bullet j}) \text{ for at least one cell } (i, j) \tag{15.1.1}$$

$i = 1, \ldots, r; j = 1, \ldots, c$.

Since the population marginal probabilities on the right-hand side of (15.5.1) are unknown, let us estimate them from our sample data. To this end, the best (maximum likelihood) estimates of the population marginal probabilities $p_{i\bullet}$ and $p_{\bullet j}$ are the respective sample marginal probabilities

$$\hat{p}_{i\bullet} = \frac{n_{i\bullet}}{n}, \quad i = 1, \ldots, r;$$
$$\hat{p}_{\bullet j} = \frac{n_{\bullet j}}{n}, \quad j = 1, \ldots, c. \tag{15.6}$$

Hence our estimate of the population probability of the joint event $A_i \cap B_j$ is

$$\hat{p}_{ij} = (\hat{p}_{i\bullet})(\hat{p}_{\bullet j}) = \frac{(n_{i\bullet})(n_{\bullet j})}{n^2} \text{ for all } i, j. \tag{15.7}$$

In this regard, the *expected frequency of the joint event* $A_i \cap B_j$ is thus

$$e_{ij} = n\hat{p}_{ij} = \frac{(n_{i\bullet})(n_{\bullet j})}{n} \text{ for all } i, j. \tag{15.8}$$

So in a test of the independence of attributes, the expected frequency in any cell (i, j) is simply the product between the frequency of row i and the frequency of column j (or the product between the marginal total for row i and the marginal total for column j) divided by the sample size n. In this regard, if the proportion of category B_j in the population is $\frac{n_{\bullet j}}{n}$, then the expected number of items having characteristics A_i and B_j is $(\frac{n_{\bullet j}}{n})n_{i\bullet}$; that is, independence means that the proportion of each row total that belongs to the jth column is the same for all rows. So if (15.8) is true, then the observed values are said to occur independently of the row and column categories.

On the basis of (15.8), we may extend Pearson's multinomial chi-square statistic (14.2) to the case where we test for independence among the two factors A and B in an $r \times c$ contingency table. We note first that for n independent trials of a random experiment classified by the categories of attributes A and B specified earlier, if the random variable N_{ij} is the frequency of event $A_i \cap B_j$, then the statistic

$$U = \sum_{i=1}^{r} \sum_{j=1}^{c} \frac{(N_{ij} - np_{ij})^2}{np_{ij}} \tag{15.9}$$

is, for large n, approximately chi-square distributed with $rc - 1$ degrees of freedom. Under (15.1.1) with equations (15.6–15.8) holding, (15.9) becomes *Pearson's chi-square statistic for a test of association*

$$U = \sum_{i=1}^{r} \sum_{j=1}^{c} \frac{(N_{ij} - n\hat{p}_{ij})^2}{n\hat{p}_{ij}}. \tag{15.9.1}$$

So for large n, this quantity is approximately chi-square distributed with $(r - 1)(c - 1)$ degrees of freedom. Let the sample realization of (15.9.1) appear as

$$u = \sum_{i=1}^{r} \sum_{j=1}^{c} \frac{(o_{ij} - e_{ij})^2}{e_{ij}}, \tag{15.9.2}$$

where $o_{ij}(= n_{ij})$ is the realized frequency for the $(i,j)^{th}$ cell. Note that the finite double sum in this expression has us take the sum over all $r \times c$ cells, with the contribution of the $(i,j)^{th}$ cell to the Pearson chi-square statistic being $\frac{(o_{ij}-e_{ij})^2}{e_{ij}}$.

How have the degrees of freedom for (15.9.1 and 15.9.2) been determined? Suppose that the true or underlying joint probability distribution for A and B is completely specified; that is, we have some exact hypothesis that specifies p_{ij} for all i,j. Hence the exact frequency in (15.9) is, as indicated, np_{ij}. Since no parameters have been estimated, degrees of freedom amounts to the number of joint event cells less 1 or $rc - 1$. However, since we are estimating marginal probabilities from sample realizations, we subtract 1 degree of freedom for each parameter estimated. And since we require that $\sum_{i=1}^{r} p_{i\bullet} = 1$ and $\sum_{j=1}^{c} p_{\bullet j} = 1$, there are, respectively, $r - 1$ row parameters and $c - 1$ column parameters to be estimated (e.g., knowing, say, the first $r - 1$ of the probabilities in $\sum_{i=1}^{r} p_{i\bullet} = 1$ uniquely determines the rth probability). Hence it follows that degrees of freedom $= rc - 1 - (r - 1) - (c - 1) = rc - r - c - 1 = (r - 1)(c - 1)$.

As was the case with (14.2), we compare the observed cell frequencies with the expected cell frequencies under H_0. If the deviations between the observed and expected cell frequencies are significantly large (i.e., larger than that attributed to simply sampling variation alone), then we are disposed to reject H_0 for large values of u. Hence the critical region for this upper-tail test is $\mathcal{R} = \{u | u \geq \chi^2_{1-\alpha,(r-1)(c-1)}\}$, where $\chi^2_{1-\alpha,(r-1)(c-1)}$ is the $100(1 - \alpha)$ percentile of the chi-square distribution with $(r - 1)(c - 1)$ degrees of freedom and $\alpha = P(TIE)$.

As a special case of Pearson's chi-square test for independence involving an $r \times c$ contingency table, we are often confronted with performing the test for a 2×2 table (the A and B attributes each have only two categories). To streamline the calculations in this instance, let our 2×2 contingency table be provided by Table 15.2.

Then a particularization of (15.9.2) to the 2×2 case, which includes *Yates' continuity correction*, is

$$u' = \sum_{i=1}^{2} \sum_{j=1}^{2} \frac{\left(|o_{ij} - e_{ij}| - \frac{1}{2}\right)^2}{e_{ij}} \tag{15.10}$$

or, in terms of the actual entries exhibited within Table 15.2,

$$u'' = \frac{n\left(|ad - bc| - \frac{n}{2}\right)^2}{efgh}, \tag{15.10.1}$$

Table 15.2

	B_1	B_2	
A_1	a	b	e
A_2	c	d	f
	g	h	n

where $\mathcal{R} = \{u' \text{ or } u''|u' \text{ or } u'' \geq \chi^2_{1-\alpha,1}\}$. Remember that for either the general $r \times c$ contingency table or the 2×2 special case of the same, in order for the Pearson realizations (15.9.2), (15.10), and (15.10.1) to provide a good approximation to the chi-square distribution with $(r - 1)(c - 1)$ degrees of freedom, each expected cell frequency should be at least 5. If this requirement is not met, then adjacent cells may be combined as is appropriate. It is important to note that it is sometimes argued that if r and c are large, then even if some of the e_{ij}'s are much smaller than 5, the approximation to the chi-square distribution is good provided the e_{ij}'s are all of about the same magnitude.

Example 15.2.1 To see exactly how (15.9.2) is applied, let us assume that the Board of Regents of a medium-sized university is meeting to discuss a proposal from the Athletic Department for upgrading its athletic programs to division 1-A status. The Board has chosen to sample the opinion of the college community to determine whether opinion on this issue is independent of university affiliation. A random sample of 300 so-called *stakeholders* yielded the information (observed frequencies) presented in Table 15.3. For instance: 80 students are in favor of division 1-A athletics; 60 faculty are opposed to the same; 10 staff members are neutral, and so on. Given the information appearing in this table, is opinion independent of university affiliation at the $\alpha = 0.05$ level?

Let $H_0: p_{ij} = p_{i\bullet}p_{\bullet j}$, with $H_1 : p_{ij} \neq p_{i\bullet}p_{\bullet j}$ for at least one cell (i,j), $i = 1,2,3,4$; $j = 1,2,3$, where p_{ij} is the (joint) probability that a stakeholder selected at random will fall into the $(i,j)^{th}$ category; $p_{i\bullet}$ is the (marginal) probability that a randomly selected stakeholder belongs to affiliation category i; and $p_{\bullet j}$ is the (marginal) probability that a stakeholder selected at random has opinion j. By (15.8), the expected frequency of the $(i,j)^{th}$ cell is $\frac{n_{i\bullet}n_{\bullet j}}{n}$ (appearing parenthetically to the right of the observed cell frequency for the $(i,j)^{th}$ cell). For instance, the expected frequency of a staff member opposed to division 1-A athletics is $\frac{n_{4\bullet}n_{\bullet 3}}{n} = \frac{(50)(110)}{300} = 18.33$. Then the realized value of the Pearson chi-square statistic is, from (15.9.2),

$$u = \frac{(80 - 51.66)^2}{51.66} + \frac{(10 - 11.66)^2}{11.66} + \frac{(10 - 70)^2}{70} + \frac{(45 - 36.16)^2}{36.16}$$

$$+ \cdots + \frac{(10 - 5.83)^2}{5.83} + \frac{(30 - 18.33)^2}{18.33} = 146.85.$$

Table 15.3

Affiliation	Opinion			
	In Favor	**Neutral**	**Opposed**	**Totals**
Students	80(51.66)	10(11.66)	10(70)	$100(= n_{1\bullet})$
Alumni	45(36.16)	15(8.16)	10(49)	$70(= n_{2\bullet})$
Faculty	20(41.33)	0(9.33)	60(56)	$80(= n_{3\bullet})$
Staff	10(25.83)	10(5.83)	30(18.33)	$50(= n_{4\bullet})$
Totals	$155(= n_{\bullet 1})$	$35(= n_{\bullet 2})$	$110(= n_{\bullet 3})$	$300(= n)$

Table 15.4

Sex	Opinion		Totals
	Modify	**Do Not Modify**	**Totals**
Male	$157(=a)$	$30(=b)$	$187(=e)$
Female	$98(=c)$	$15(=d)$	$113(=f)$
Totals	$255(=g)$	$45(=h)$	$30(=n)$

Since $r = 4$ and $c = 3$, degrees of freedom is $(r-1)(c-1) = 3 \cdot 2 = 6$. From Table A.4 of the Appendix, the critical value of the chi-square statistic is $\chi^2_{0.95,6} = 12.59$. Since the calculated chi-square value is well within the critical region $\mathcal{R} = \{u | u \geq \chi^2_{0.95,6} = 12.59\}$, we reject the null hypothesis of independence between opinion and university affiliation at the 5% level. In fact, the *p*-value is less than 0.005.   ■

Example 15.2.2   Suppose the results of a company-wide random survey concerning employee satisfaction with the proposed modification of the health insurance plan appear in Table 15.4. Are opinion and sex statistically independent at the $\alpha = 0.10$ level? We again have $H_0: p_{ij} = p_{i\bullet}p_{\bullet j}$, versus $H_1: p_{ij} \neq p_{i\bullet}p_{\bullet j}$ for at least one cell (i,j), $i = 1, 2$; $j = 1, 2$. From (15.10.1),

$$u'' = \frac{300 \left(\left| (157)(15) - (30)(98) \right| - 150 \right)^2}{(187)(113)(255)(45)} = 0.234.$$

Since the critical region $\mathcal{R} = \{u'' | u'' \geq \chi^2_{0.90,1} = 2.71\}$, we cannot reject the null hypothesis of independence between opinion and sex at the 10% level of significance.   ■

15.3   Testing *k* Proportions

Let us consider a special case of (15.9.2). Specifically, we examine the instance in which we want to test for the significance of the differences among the k population proportions $p_i, i = 1, \ldots, k$. That is, suppose we have k independent random samples and that $X_1, \ldots, X_k$ comprise a set of independent binomial random variables with the parameters p_1 and n_1; p_2 and n_2; $\ldots$; p_k and n_k, respectively, where $p_i, i = 1, \ldots, k$, is the (true) proportion of successes for the ith population. Here the random variable $X_i, i = 1, \ldots, k$, depicts the number of successes obtained in the ith sample. Its sample realization is denoted as $x_i, i = 1, \ldots, k$.

Let us arrange the realized number of successes and failures observed in the k independent random samples in the following $k \times 2$ table (see Table 15.5). Here the $2k$ entries within this table are the *observed cell frequencies* o_{ij}, $i = 1, \ldots, k$; $j = 1, 2$.

Table 15.5

Realized Number of Successes and Failures in k Independent Random Samples		
	Successes	**Failures**
Sample 1	$x_1 = o_{11}$	$n_1 - x_1 = o_{12}$
Sample 2	$x_2 = o_{21}$	$n_2 - x_2 = o_{22}$
$\vdots$	$\vdots$	$\vdots$
Sample k	$x_k = o_{k1}$	$n_k - x_k = o_{k2}$

Our objective is to test $H_0: p_1 = p_2 = \cdots = p_k = p_0$, against $H_1: p_i \neq p_0$ for at least one value of $i = 1, \ldots, k$, where p_0 is the null value of p_i.

Under H_0, the expected number of successes for sample i is $n_i p_0, i = 1, \ldots, k$ (since $p_0 = \frac{x_i}{n_i}, i = 1, \ldots, k$); and the expected number of failures for sample i is $n_i(1 - p_0)$ (given that, $1 - p_0 = \frac{n_i - x_i}{n_i}$, $i = 1, \ldots, k$). In this regard, the *expected cell frequencies for column 1* are $e_{i1}^0 = n_i p_0$, $i = 1, \ldots, k$; and the *expected cell frequencies for column 2* are $e_{i2}^0 = n_i(1 - p_0), i = 1, \ldots, k$. So given p_0, (15.9.2) becomes

$$U_{p_0} = \sum_{i=1}^{k} \sum_{j=1}^{2} \frac{(o_{ij} - e_{ij}^0)^2}{e_{ij}^0}, \tag{15.11}$$

where o_{i1} is the collection of observed cell frequencies for the success column; o_{i2} is the set of observed cell frequencies for the failure column; the expected cell frequencies for the success category are $e_{i1}^0 = n_i p_0$; and the expected cell frequencies for the failure category are $e_{i2}^0 = n_i(1 - p_0)$, $i = 1, \ldots, k$. For $\alpha = P(TIE)$ the critical region is $\mathcal{R} = \{u_{p_0} | u_{p_0} \geq \chi_{1-\alpha, k-1}^2\}$.

If p_0 is not specified (we simply test $H_0 = p_1 = p_2 = \cdots = p_k$, against H_1: the p_i's are not all equal), then we may obtain a *pooled estimate of the common proportion of successes* $\hat{p}$ as

$$\hat{p} = \frac{x_1 + x_2 + \cdots + x_k}{n_1 + n_2 + \cdots + n_k}. \tag{15.12}$$

Then from this expression, the expected cell frequency estimates are

$$\hat{e}_{i1}^0 = n_i \hat{p} \quad \text{and} \quad \hat{e}_{i2}^0 = n_i(1 - \hat{p}), i = 1, \ldots, k. \tag{15.13}$$

So when (15.12) is required, (15.11) is replaced by

$$U_{\hat{p}} = \sum_{i=1}^{k} \sum_{j=1}^{2} \frac{(o_{ij} - \hat{e}_{ij}^0)^2}{\hat{e}_{ij}^0}. \tag{15.14}$$

Table 15.6

	Number Favoring Candidate A	Number Favoring Candidate B	
Precinct 1	60	40	$n_1 = 100$
Precinct 2	70	80	$n_2 = 150$
Precinct 3	40	40	$n_3 = 80$
Precinct 4	50	80	$n_4 = 130$

Example 15.3.1   On the basis of the sample realizations presented in Table 15.6, determine if the true or population proportion of voters favoring candidate A (defined as a success) over candidate B is the same across the four precincts polled just before a certain mayoral election. Let $\alpha = 0.05$. Here $H_0: p_1 = p_2 = p_3 = p_4$ is to be tested against H_1: the p_i's are not all equal, $i = 1, 2, 3, 4$.

Given the observed cell frequencies, the pooled estimate of the common proportion of successes (15.12) is

$$\hat{p} = \frac{60 + 70 + 40 + 50}{100 + 150 + 80 + 130} = \frac{220}{460} = 0.478.$$

Then from (15.13), the expected cell frequencies are

$$\hat{e}_{11} = n_1\hat{p} = 100(0.478) = 47.80 \quad \hat{e}_{12} = n_1(1 - \hat{p}) = 100(0.522) = 52.20$$

$$\hat{e}_{21} = n_2\hat{p} = 150(0.478) = 71.70 \quad \hat{e}_{22} = n_2(1 - \hat{p}) = 150(0.522) = 78.30$$

$$\hat{e}_{31} = n_3\hat{p} = 80(0.478) = 38.24 \quad \hat{e}_{32} = n_3(1 - \hat{p}) = 80(0.522) = 41.76$$

$$\hat{e}_{41} = n_4\hat{p} = 130(0.478) = 62.14 \quad \hat{e}_{42} = n_4(1 - \hat{p}) = 130(0.522) = 67.86$$

A substitution of the observed and expected cell frequencies into (15.14) yields

$$u_{\hat{p}} = \frac{(60 - 47.80)^2}{47.80} + \frac{(70 - 71.70)^2}{71.70} + \cdots + \frac{(40 - 41.76)^2}{41.76} + \frac{(80 - 67.86)^2}{67.86} = 10.75.$$

Since $u_{\hat{p}}$ is an element of $\mathcal{R} = \{u_{\hat{p}} | u_{\hat{p}} \geq \chi^2_{1-\alpha, k-1} = \chi^2_{0.95, 3} = 7.815\}$, we reject the null hypothesis of equal proportions at the 5% level; that is, the proportion of voters favoring candidate A over candidate B is not the same across the four precincts polled.   ∎

15.4   Testing for Homogeneity

We next turn to an additional application of Pearson's equation (15.9.2). Section 15.2 employed this expression to test for the independence of two attributes or factors, each characterized by a variety of categories that, when cross-classified, formed an $r \times c$ contingency table. Our objective was to determine if any dependency relationship or association existed between the two population characteristics.

The *test of homogeneity* that follows is constructed so as to determine if *two or more independent random samples are drawn from the same population*. This test is a simple extension of the chi-square test of independence in that it also uses cross-classified data as well as the same realized test statistic (15.9.2). For instance, a sampling plan may be implemented multiple times, where the observations are taken from a given population at, say, different points in time. Here we are interested in determining if the population remains stable over time. Or we may take separate subsamples, one for each of a variety of constituencies at a given point in time, and ask whether or not these subgroups have the same distribution of characteristics or traits and thus can be thought of as coming from the same population.

In testing for independence, *a single random sample of size n was extracted from one specific population.* However, in testing for homogeneity among samples, *two or more independent random samples are obtained—one from each of the individual populations represented.* In this regard, in tests of independence, *all* marginal frequencies or totals are randomly determined for a fixed n. In tests of homogeneity, the row totals $n_{i\bullet}, i = 1, \ldots, r$, are actually fixed sample sizes that are chosen quantities.

As before, for attributes A and B, let the categories $A_1, \ldots, A_r$ and $B_1, \ldots, B_c$ constitute the rows and columns, respectively, of the $r \times c$ two-way table depicted in Table 15.7.

Additionally, let the row totals, $n_{i\bullet} = n_i, i = 1, \ldots, r$, represent the sizes of r separate independent random samples. Suppose each realization within each sample is classified into one of the B attribute categories $B_1, \ldots, B_c$; for example, let the A attribute be fixed at A_i and let the category B_j be observed. Then the cell probability p_{ij} is the conditional probability $p_{ij} = P(B = B_j | A = A_i)$, $i = 1, \ldots, r; j = 1, \ldots, c$.

Table 15.7

Attribute A	Attribute B				Totals
	B_1	B_2	$\ldots$	B_c	
A_1	n_{11}	n_{12}	$\ldots$	n_{1c}	$n_{1\bullet} = n_1$ (fixed)
A_2	n_{21}	n_{22}	$\ldots$	n_{2c}	$n_{2\bullet} = n_2$ (fixed)
$\vdots$	$\vdots$	$\vdots$	$\ldots$	$\vdots$	$\vdots$
A_r	n_{r1}	n_{r2}	$\ldots$	n_{rc}	$n_{r\bullet} = n_r$ (fixed)
Totals	$n_{\bullet 1}$	$n_{\bullet 2}$	$\ldots$	$n_{\bullet c}$	n

Now, our objective is to test the *homogeneity hypothesis*

H_0: the r samples come from the same population, against

H_1: the r samples do not come from the same population

$$(15.15)$$

More specifically, the null hypothesis states that the probability of observing category B_j is the same regardless of the row population sampled; that is, we have

H_0: $p_{ij} = P(B = B_j | A = A_i) = p_j$ for all i, $j = 1, \ldots, c$;

H_1: $p_{ij} = P(B = B_j | A = A_i) \neq p_j$ for at least one value of i, $j = 1, \ldots, c$.

$$(15.16)$$

It is instructive to note that the null and alternative hypotheses expressed in (15.16) may be stated alternatively as

H_0: $(p_{i1}, p_{i2}, \ldots, p_{ic}) = (p_1, p_2, \ldots, p_c)$ for all i;

H_1: $(p_{i1}, p_{i2}, \ldots, p_{ic})$ are not the same for all i.

$$(15.16.1)$$

Then it is readily seen from the null hypothesis stated in (15.16.1) that:

$$p_{11} = p_{21} = \cdots = p_{r1} = p_1$$
$$p_{12} = p_{22} = \cdots = p_{r2} = p_2$$
$$\vdots$$
$$p_{1c} = p_{2c} = \cdots = p_{rc} = p_c$$

$$(15.16.2)$$

Under H_0, the best estimate of the proportion of category B_j (taken over all r samples) is $\frac{n_{\bullet j}}{n}, j = 1, \ldots, c$, where $n_{\bullet j}$ is the marginal total for column B_j. So for sample i, the *expected frequency of the joint event* $A_i \cap B_j$ is

$$e_{ij} = \left(\frac{n_{\bullet j}}{n} \right) n_i, \, j = 1, \ldots, c. \qquad (15.17)$$

Given (15.17), the realization of *Pearson's chi-square statistic for a test of homogeneity* is, from (15.9.2),

$$u = \sum_{i=1}^{r} \sum_{j=1}^{c} \frac{(o_{ij} - e_{ij})^2}{e_{ij}}, \qquad (15.18)$$

where $o_{ij} (= n_{ij})$ is the realized frequency for the $(i,j)^{th}$ cell. As determined earlier, the critical region corresponding to an upper tail test is $\mathcal{R} = \{u | u \geq \chi^2_{1-\alpha, (r-1)(c-1)}\}$, where $\chi^2_{1-\alpha, (r-1)(c-1)}$ is the $100(1-\alpha)$ percentile of the chi-square distribution with $(r-1)(c-1)$ degrees of freedom and $\alpha = P(TIE)$.

Example 15.4.1 Assume that four independent random samples are taken from the student body of a small single-gender college: $n_1 = 100$ Freshman; $n_2 = 95$ Sophomores; $n_3 = 95$ Juniors; and $n_4 = 90$ Seniors. Each student chosen is asked to offer an opinion of either Yes, No, or Indifferent to the question of whether the college should "go co-ed." The realizations of the poll are presented as Table 15.8.

Let us test the null hypothesis of homogeneity (or no difference) of opinion among the four groups against the alternative hypothesis of differences of opinion among these groups or

H_0: the four samples come from the same population, versus

H_1: the four samples are drawn from different populations

In terms of (15.16.2), we seek to test

H_0: $P(Yes/Fr.) = P(Yes/Soph.) = P(Yes/Jr.) = P(Yes/Sr.) = p_1$

$\quad P(No/Fr.) = P(No/Soph.) = P(No/Jr.) = P(No/Sr.) = p_2$

$\quad P(Indiff./Fr.) = P(Indiff./Soph.) = P(Indiff./Jr.) = P(Indiff./Sr.) = p_3$

against

H_1: not all equalities stated in H_0 are true.

Under H_0, the best estimates of the proportions responding Yes, No, and Indifferent for Freshman are, from (15.17), respectively,

$$e_{11} = \frac{270}{380}(100) = 71.05 \text{ (Yes)}$$

$$e_{12} = \frac{65}{380}(100) = 17.11 \text{ (No)}$$

$$e_{13} = \frac{45}{380}(100) = 11.84 \text{ (Indifferent)}.$$

Table 15.8

Year	Opinion			Total
	Yes	No	Indifferent	
Freshman	60 (71.05)	20 (17.11)	20 (11.84)	100 ($= n_1$)
Sophomore	60 (67.5)	30 (16.25)	5 (11.25)	95 ($= n_2$)
Junior	70 (67.5)	10 (16.25)	15 (11.25)	95 ($= n_3$)
Senior	80 (63.95)	5 (15.39)	5 (10.66)	90 ($= n_4$)
Total	270	65	45	380

These expected frequencies appear parenthetically to the right of the observed frequencies given in row 1 of Table 15.8. The remaining expected frequencies are determined and exhibited in a similar fashion. Let us choose $\alpha = 0.05$. Then for $(r - 1)(c - 1) = 6$, $\mathcal{R} = \{u | u \geq \chi^2_{0.95,6} = 12.59\}$. Looking to (15.18),

$$u = \frac{(60 - 71.05)^2}{71.05} + \frac{(20 - 17.11)^2}{17.11} + \frac{(20 - 11.84)^2}{11.84} + \cdots + \frac{(5 - 10.66)^2}{10.66} = 43.86.$$

Clearly we may reject the null hypothesis of homogeneity at the 5% level—the four samples are drawn from different populations. ■

15.5   Measuring Strength of Association in Contingency Tables

We noted in Section 15.2 that Pearson's chi-square statistic is used in testing for association between two attributes A and B. To execute the test we hypothesize independence among the A and B factors and then, for a given level of significance $\alpha = P(TIE)$, we see if we can reject the null hypothesis of independence in favor of the alternative hypothesis that the A and B characteristics are not independent but, at the chosen level of significance, exhibit statistical association.

If the null hypothesis is actually rejected, then what does the notion of lack of independence really tell us? As mentioned earlier, if n is large enough, virtually any degree of statistical association will lead to the rejection of the null hypothesis; that is, even the slightest degree of departure from strict independence of the A and B attributes typically is uncovered by the test.

Although detecting statistical association is important, we need to take the next logical step in our analysis of the A and B attributes and ask the question: How strong is the association between the A and B categories? In this regard, we need to go beyond a simple test of significance and devise a measure of the strength of association between the A and B factors.

Let us first consider a measure of association for a two-way contingency table. One of the most common measures of the strength of statistical association between attributes A and B in a test of independence is the *odds ratio* (or *cross-product ratio*)

$$\psi = \frac{p_{11}p_{22}}{p_{12}p_{21}}, \tag{15.19}$$

where p_{ij} is the true or population probability of an observation falling into the $(i, j)^{th}$ cell (see Table 15.9). In this expression, for row 1, the odds that the response or outcome is in column 2 and not in column 1 is $\psi_1 = \frac{p_{12}}{p_{11}}$; and for row 2, the odds that the response or outcome is in column 2 and not in column 1 is $\psi_2 = \frac{p_{22}}{p_{21}}$. Here $\psi_i \geq 0$, with $\psi_i > 1$ if response 2 is more likely than response 1, $i = 1, 2$. Then the ratio of these odds is $\psi = \frac{\psi_2}{\psi_1}$ or (15.19).

Table 15.9

	Attribute A	
	A_1	A_2
B_1	p_{11}	p_{12}
B_2	p_{21}	p_{22}

Attribute B

When the null hypothesis of independence of the A and B attributes holds, then $\psi_1 = \psi_2$ so that $\psi = 1$. If $1 < \psi < +\infty$, we have positive association between the A and B attributes in that the odds of the column 2 outcome are higher for row 2 than for row 1. And if $0 \le \psi < 1$, there exists negative association between the A and B characteristics in that the odds of the column 2 outcome are higher for row 1 than for row 2.

It should be intuitively clear that an appropriate estimate of ψ in (15.19) is

$$\hat{\psi} = \frac{n_{11}n_{22}}{n_{12}n_{21}}, \tag{15.20}$$

where p_{ij} is estimated by the observed relative frequency $\frac{n_{ij}}{n}$, $n_{ij} \ge 0$, $i = 1,2$; $j = 1,2$.

Under random sampling, $\ln \hat{\psi}$ is asymptotically normally distributed with asymptotic mean $\ln \psi$ and asymptotic standard deviation estimated by

$$\hat{\sigma}(\ln \hat{\psi}) = \left(\frac{1}{n_{11}} + \frac{1}{n_{12}} + \frac{1}{n_{21}} + \frac{1}{n_{22}} \right)^{1/2}. \tag{15.21}$$

Then an approximate or asymptotic $100(1 - \alpha)\%$ confidence interval for $\ln \psi$ is

$$\ln \hat{\psi} \pm z_{\alpha/2}\hat{\sigma}(\ln \hat{\psi}). \tag{15.22}$$

If we represent the lower and upper confidence limits determined in (15.22) as l and u, respectively, then the corresponding $100(1 - \alpha)\%$ confidence interval for ψ is

$$(e^l, e^u). \tag{15.22.1}$$

If the $100(1 - \alpha)\%$ confidence interval for ψ excludes the value 1, then $\hat{\psi}$ departs significantly from 1 at the $\alpha\%$ level so that $H_0: \psi = 1$ can safely be rejected in favor of the two-sided alternative $H_1: \psi \ne 1$. The two-sided p-value for this test (under H_0: $\ln \psi = 0$) can be determined from the standard normal variate value $z = \ln \hat{\psi}/\hat{\sigma}(\ln \hat{\psi})$.

Example 15.5.1   Let us return to the sample data set presented earlier in the
2×2 contingency table involving opinion and sex (see Table 15.4). From (15.20),

$$\hat{\psi} = \frac{157(15)}{98(30)} = 0.80.$$

Since $\hat{\psi} < 1$, we have negative association between the row and column variables
or attributes; that is, for this random sample, females are less likely to report
"do not modify" than are males. In fact, the odds of a person choosing "do not
modify" are 0.8 times lower for females than for males. (If it had turned out that,
say, $\hat{\psi} = 1.66$, then positive association would be exhibited by the row and column
variables. In this instance we would conclude that the odds of choosing "modify"
would be 1.66 times higher for males than for females.) Since from (15.21) we
have

$$\hat{\sigma}(\ln \hat{\psi}) = \left(\frac{1}{157} + \frac{1}{30} + \frac{1}{98} + \frac{1}{15} \right)^{1/2} = 0.341,$$

it follows from (15.22) that for $1 - \alpha = 0.95$, an approximate 95% confidence
interval for $\ln \psi$ is $-0.223 \pm 1.96(0.341)$ or $(l = -0.891, u = 0.445)$. Then a 95%
confidence interval for ψ is, from (15.22.1), $(0.410, 1.560)$. Since 1 is a member of
the confidence interval for ψ (and 0 is within the confidence interval for $\ln \psi$), it
is highly plausible that opinion and sex are not statistically independent.   ∎

If any of the observed cell frequencies in a 2×2 contingency table is very
small, then an alternative estimator of ψ is

$$\tilde{\psi} = \frac{(n_{11} + 0.5)(n_{22} + 0.5)}{(n_{12} + 0.5)(n_{21} + 0.5)}. \tag{15.20.1}$$

It has been shown that $\tilde{\psi}$ and $\hat{\sigma}(\ln \tilde{\psi})$ have smaller asymptotic bias and mean
square error than $\hat{\psi}$ and $\hat{\sigma}(\ln \hat{\psi})$, respectively. In most instances the confidence
interval determined via (15.22) will be very similar to that obtained by adding 0.5
to each observed cell frequency.

We next turn to the specification of indexes of contingency and predictive
association. One way to depict statistical association in an $r \times c$ contingency table
is by *Cramer's phi-squared statistic*

$$\varphi^2 = \frac{U}{n(q-1)}, \quad 0 \leq \varphi^2 \leq 1, \tag{15.23}$$

obtained by dividing Pearsons's chi-square statistic U by its theoretical maximum value $n(q-1)$, where n is the sample size, and $q = \min\{r, c\}$. Here (15.23)
measures the *overall* strength of association in a sample; that is, for $\varphi^2 = 0$, the

sample exhibits complete independence between the A and B attributes; when $\varphi^2 = 1$, there exists complete dependence between these attributes or factors.

Example 15.5.2   From an examination of the chi-square test results determined earlier for the 4×3 contingency table involving university affiliation and opinion (see Table 15.3), it is easily shown that the realized value of Cramer's phi-squared statistic is $\varphi^2 = \frac{140.13}{300(3)} = 0.16$. ∎

Next, let us consider a measure of association that exploits the notion that information about one attribute may enhance our ability to predict the other attribute. That is, association between attributes A and B is measured by the reduction in the probability of error in predicting, say, A, given knowledge concerning which B category has obtained. Given that our goal is to predict one attribute from information pertaining to the other attribute, two situations present themselves. First, if the A and B attributes are independent, then we cannot expect to do any better in predicting, say, A, from knowledge of B than if no information concerning the B attribute was at hand. But if the A and B attributes are in some sense functionally related, then obviously knowledge of B reduces the probability of error in predicting A.

Given the joint probability distribution for attributes A and B, suppose an item is drawn at random from this distribution and we find that category B_j has occurred. If knowing that B_j has occurred enables us to reduce the probability of error in predicting A, then we say that there is *predictive association* between categories A and B; that is, if P(error in predicting A/B known) $<$ P(error in predicting A/B unknown), then we may specify an *index of predictive association for A given B* as

$$\lambda_{A/B} = 1 - \frac{P(\text{error in predicting } A/B \text{ known})}{P(\text{error in predicting } A/B \text{ unknown})}. \tag{15.24}$$

Here this expression represents the proportionate reduction in the probability of error in predicting the A category when the B category is known. (In this regard, if $\lambda_{A/B}$ is, say, 0.15, then knowing the B category reduces the probability of error in predicting the A category by 15% on the average.) If knowing the B category does not reduce the probability of error in predicting A, then $\lambda_{A/B} = 0$; that is, there is no predictive association between A and B. But if $\lambda_{A/B} = 1$, there is complete predictive association between these attributes in that no error is made at all when the B category is given. It is important to note, however, that even if $\lambda_{A/B} = 0$, some statistical association may exist between the A and B attributes; that is, we cannot conclude that A and B are completely independent when $\lambda_{A/B} = 0$. And when there is perfect association between the A and B attributes, then complete predictive association emerges so that $\lambda_{A/B} = 1$.

Given an $r \times c$ contingency table, an estimate of $\lambda_{A/B}$ can be obtained from the sample frequency counts $n_{ij}, i = 1, \ldots, r; j = 1, \ldots, c$; that is, for items drawn at random from a sample of size n, the *estimated proportionate reduction in the probability of error in predicting A given B is*

$$\hat{\lambda}_{A/B} = \frac{\sum\limits_{j=1}^{c} \max\limits_{i} n_{ij} - \max\limits_{i} n_{i\bullet}}{n - \max\limits_{i} n_{i\bullet}}, \quad 0 \leq \hat{\lambda}_{A/B} \leq 1, \tag{15.25}$$

where n_{ij} is the frequency observed in the $(i,j)^{th}$ cell, $\max\limits_{i} n_{ij}$ is the largest observed frequency in column j, and $\max\limits_{i} n_{i\bullet}$ is the largest marginal frequency among the rows A_i. So when A and B are functionally related, we expect to experience an improvement in our ability to predict the A attribute from knowledge of the B attribute. And we can quantify the degree of improvement according to the proportionate reduction in the probability of error index (15.25).

It should be obvious that (15.25) is an asymmetric measure of predictive association in that it is only relevant for predicting the A category when the B category is independently determined or known; that is, the B attribute is the independent variable in the functional relationship between A and B. Let us now reverse the conditional roles of the A and B attributes. Specifically, if knowing that A_i has occurred enables us to reduce the probability of error in predicting B, then again predictive association exists between the A and B attributes. An *index of predictive association for B given A* is provided by

$$\lambda_{B/A} = 1 - \frac{P(\text{error in predicting } B/A \text{ known})}{P(\text{error in predicting } B/A \text{ unknown})}; \tag{15.26}$$

it renders the proportionate reduction in the probability of error in predicting the B category when the A category is known.

In terms of the sample frequency counts $n_{ij}, i = 1, \ldots, r; j = 1, \ldots, c$, within an $r \times c$ contingency table determined from a sample of size n, the *estimated proportionate reduction in the probability of error in predicting B given A* is

$$\hat{\lambda}_{B/A} = \frac{\sum\limits_{i=1}^{r} \max\limits_{j} n_{ij} - \max\limits_{j} n_{\bullet j}}{n - \max\limits_{j} n_{\bullet j}}, \quad 0 \leq \hat{\lambda}_{B/A} \leq 1, \tag{15.27}$$

where $\max\limits_{j} n_{ij}$ is the largest observed frequency in row A_i and $\max\limits_{j} n_{\bullet j}$ is the largest marginal frequency among the columns B_j.

Now, it may be the case that A is predictable from B but B is not predictable from A so that, in general, $\hat{\lambda}_{A/B} \neq \hat{\lambda}_{B/A}$. Hence determining the values of both (15.25) and (15.27) is important once we have rejected the null hypothesis of independence in a chi-square test of association in an $r \times c$ contingency table. This is because a significant chi-square test result in and of itself conveys no information concerning the strength of the (asymmetric) predictive association that may exist between the A and B attributes. It is essential to determine if knowing something about the A attribute can enhance our ability to make an informed assessment of the B attribute and vice versa.

Under random sampling, $\hat{\lambda}_{B/A}$ is normally distributed with asymptotic mean $\lambda_{B/A}$ and asymptotic standard deviation estimated by

$$
\hat{\sigma}(\hat{\lambda}_{B/A}) = \left[\frac{\left(n - \sum\limits_{i=1}^{r} \max\limits_{j} n_{ij}\right)\left(\sum\limits_{i=1}^{r} \max\limits_{j} n_{ij} + \max\limits_{j} n_{\bullet j} - 2\max\limits_{i} n_{i*}\right)}{\left(n - \max\limits_{j} n_{\bullet j}\right)^3} \right]^{1/2}, \quad (15.28)
$$

where $\max\limits_{i} n_{i*}$ is the largest of the r row frequencies observed within the column containing $\max\limits_{j} n_{\bullet j} = n_{\bullet *}$. Then for a given $1 - \alpha$ value ($\alpha = P(TIE)$), an approximate or asymptotic $100(1 - \alpha)\%$ confidence interval for $\lambda_{B/A}$ is

$$
\hat{\lambda}_{B/A} \pm z_{\alpha/2}\hat{\sigma}(\hat{\lambda}_{B/A}). \quad (15.29)
$$

If the resulting confidence interval contains zero, then the sample estimate $\hat{\lambda}_{B/A}$ is not significantly different from zero so that there is no compelling statistical evidence of the existence of predictive association (involving the A to B direction) between the A and B attributes.

Similarly, an approximate or asymptotic $100(1 - \alpha)\%$ confidence interval for $\lambda_{A/B}$ is

$$
\hat{\lambda}_{A/B} \pm z_{\alpha/2}\hat{\sigma}(\hat{\lambda}_{A/B}), \quad (15.30)
$$

with

$$
\hat{\sigma}(\hat{\lambda}_{A/B}) = \left[\frac{\left(n - \sum\limits_{j=1}^{c} \max\limits_{i} n_{ij}\right)\left(\sum\limits_{j=1}^{c} \max\limits_{i} n_{ij} + \max\limits_{i} n_{i\bullet} - 2\max\limits_{j} n_{*j}\right)}{(n - \max\limits_{i} n_{i\bullet})^3} \right]^{1/2}, \quad (15.31)
$$

where $\max\limits_{j} n_{*j}$ is the largest of the c column frequencies observed within the row containing $\max\limits_{i} n_{i\bullet} = n_{*\bullet}$.

Example 15.5.3   Given the random sample data set on opinion (attribute B) and university affiliation (attribute A) presented in Table 15.3, and replicated as Table 15.3.1, determine the degree of predictive association between these two attributes. In addition, find 95% confidence intervals for both $\lambda_{A/B}$ and $\lambda_{B/A}$. First, from (15.25),

$$
\hat{\lambda}_{A/B} = \frac{(80 + 15 + 60) - 100}{300 - 100} = 0.275,
$$

Table 15.3.1

	Opinion (Attribute B)			
	80	10	10	$100 = n_{1\bullet}$
Affiliation (Attribute A)	45	15	10	$70 = n_{2\bullet}$
	20	0	60	$80 = n_{3\bullet}$
	10	10	30	$50 = n_{4\bullet}$
	$155 = n_{\bullet 1}$	$35 = n_{\bullet 2}$	$110 = n_{\bullet 3}$	$n = 300$

that is, knowing the B category reduces the probability of error in predicting the A category by 27.5% on the average. And from (15.31),

$$\hat{\sigma}(\hat{\lambda}_{A/B}) = \left[\frac{(300 - 155)(155 + 100 - 2(80))}{(300 - 100)^3} \right]^{1/2} = 0.041.$$

Then a 95% confidence interval for $\lambda_{A/B}$ is, from (15.30), $0.275 \pm 1.96(0.041)$ or $(0.195, 0.355)$. Since 0 is not a member of this interval, it follows that $\hat{\lambda}_{A/B}$ is significantly different from zero at the $\alpha = 0.05$ level.

Next, if we reverse the conditional roles of the A and B attributes, then, from (15.27),

$$\hat{\lambda}_{B/A} = \frac{(80 + 45 + 60 + 30) - 155}{300 - 155} = 0.414.$$

Thus knowing the A category reduces the probability of error in predicting the B category by 41.4% on the average and, from (15.28),

$$\hat{\sigma}(\hat{\lambda}_{B/A}) = \left[\frac{(300 - 215)(215 + 155 - 2(80))}{(300 - 155)^3} \right]^{1/2} = 0.077$$

so that a 95% confidence interval for $\lambda_{B/A}$ is, from (15.29), $0.414 \pm 1.96(0.077)$ or $(0.263, 0.565)$. Since 0 is not a member of this interval, we may conclude that $\hat{\lambda}_{B/A}$ is significantly different from zero at the $\alpha = 0.05$ level. We have thus determined that a statistically significant degree of predictive association exists in the B to A direction as well as in the A to B direction. ∎

15.6   Testing Goodness of Fit with Nominal-Scale Data: Paired Samples

Suppose we have n mutually independent bivariate sample random variables (X_i, Y_i) with realizations $(x_i, y_i), i = 1, \ldots, n$. Suppose further that the X_i and

Table 15.10

		Y_i **Realizations**	
		$y_i = 1$	$y_i = 0$
X_i Realizations	$x_i = 1$	n_{11} (number of pairs with $x_i = y_i = 1$)	n_{12} (number of pairs with $x_i = 1,\ y_i = 0$)
	$x_i = 0$	n_{21} (number of pairs with $x_i = 0,\ y_i = 1$)	n_{22} (number of pairs with $x_i = y_i = 0$)

Y_i variables are measured on a nominal scale, with two categories or possible values for all X_i and $Y_i, i = 1, \ldots, n$. Given that the data are dichotomous, we can use a set of 0's and 1's to code the realizations of X_i and Y_i for all i. That is, each $x_i = 0$ or 1; each $y_i = 0$ or 1. Hence the possible values of (x_i, y_i) are $(0,0)$, $(0,1)$, $(1,0)$, and $(1,1)$. These possible pairs of values can be summarized in a 2×2 table (see Table 15.10).

The chi-square test that follows, known as the *McNemar test*, attempts to detect a difference between the probability of the $(x_i = 0,\ y_i = 1)$ combination and the probability of the $(x_i = 1,\ y_i = 0)$ combination. Here the combinations $(x_i = 1,\ y_i = 1)$ and $(x_i = 0,\ y_i = 0)$ are ignored since, for the purpose of the test, these combinations are considered to be *ties*. In this regard, we seek to test

$$H_0 : P(X_i = 0,\ Y_i = 1) = P(X_i = 1,\ Y_i = 0),$$

against

$$H_1 : P(X_i = 0,\ Y_i = 1) \neq P(X_i = 1,\ Y_i = 0).$$

In terms of the observed sample frequencies presented in Table 15.10, we may test

H_0: each of the sample proportions $\frac{n_{11}+n_{12}}{n}$ and $\frac{n_{11}+n_{21}}{n}$ serve as estimators of the same population proportion, against

H_1: each of the indicated sample proportions do not serve as estimators of the same population proportion

or, in terms of the observed cell frequencies n_{12} and n_{21}, let us test

$H_0 : r = 1$, against $H_1 : r \neq 1$,

where the population ratio $r = \frac{p_{12}}{p_{21}}$ is estimated by $\hat{r} = \frac{n_{12}}{n_{21}}$.

For $n_{12} + n_{21} > 20$ and 1 degree of freedom, the sample realization of the chi-square test statistic incorporating Yates' continuity correction can be written as

$$\chi^2 = \frac{(|n_{12} - n_{21}| - 1)^2}{n_{12} + n_{21}} \tag{15.32}$$

Table 15.11

		Brand B		
		Chipping	**No chipping**	
Brand A	Chipping	$n_{11} = 40$	$n_{12} = 15$	55
	No chipping	$n_{21} = 10$	$n_{22} = 5$	15
		50	20	70

and, for a given $\alpha = P(TIE)$, the critical region or region of rejection is $\mathcal{R} = \{\chi^2 | \chi^2 \geq \chi^2_{1-\alpha,1}\}$, where $\chi^2_{1-\alpha,1}$ is the $100(1 - \alpha)\%$ quantile of the chi-square distribution with 1 degree of freedom. (Note that (15.32) is equivalent to the realization

$$\chi = \frac{|n_{12} - n_{21}| - 1}{\sqrt{n_{12} + n_{21}}} \qquad (15.32.1)$$

of a standard normal variate with $\mathcal{R} = \{z/z \geq z_{\alpha/2}\}$.)

If $n_{12} + n_{21} \leq 20$, then the binomial test should be performed. In this instance $n = n_{12} + n_{21}, x = n_{12}$ (or $= n_{21}$) serves as the realized number of successes, and the null and alternative hypotheses are $H_0: p = \frac{1}{2}$ and $H_1: p \neq \frac{1}{2}$, respectively.

It is important to mention that the McNemar test *is not* a 2×2 contingency test even though it uses the chi-square variate. This is because the chi-square contingency test hypothesizes independence between the rows and columns of the contingency table whereas the McNemar test is specifically designed to exhibit association between the rows and columns of the said table.

Example 15.6.1 We wish to test at the $\alpha = 0.05$ level whether two different nail polishes (call them brands A and B) are equally effective at resisting chipping over a 24-hour period. In a random sample of $n = 70$ women, brand A was applied to one hand and brand B was applied to the other hand, with the choice of the hand getting, say, brand A determined in a random fashion by tossing a coin. The results of the test are presented in Table 15.11. Here H_0: the proportion of women experiencing *chipping* is the same for both brands; H_1: the proportion of women experiencing *chipping* is not the same for both brands. Given that $n_{12} + n_{21} = 25 > 20$, (15.32) renders $\chi^2 = \frac{(|15-10|-1)^2}{25} = 0.64$. Since 0.64 is not an element of $\mathcal{R} = \{\chi^2 | \chi^2 \geq \chi^2_{0.95,1} = 3.841\}$, we cannot reject the null hypothesis; brands A and B are equally resistant to chipping. ■

The McNemar test is very useful when *testing for a significant change* in outcomes or responses by recording the same before and after some event or treatment. That is, our test situation can be viewed as one in which for each pair (X_i, Y_i), $i = 1, \ldots, n, X_i$ depicts the condition or state of some experimental unit before the treatment and Y_i depicts the condition of the experimental unit after the treatment.

Table 15.12

		After		
		Coach of Team A $(y_i = 1)$	**Coach of Team B** $(y_i = 0)$	
Before	Coach of Team A $(x_i = 1)$	$40 (= n_{11})$	$20 (= n_{12})$	60
	Coach of Team B $(x_i = 0)$	$10 (= n_{21})$	$10 (= n_{22})$	20
		50	30	80

Example 15.6.2 Prior to the playing of a nationally televised college football game involving teams A and B, a random sample of $n = 80$ college football coaches were asked to state their preferences for the coach of team A or the coach of team B to be elected *coach of the year*. Sixty favored the coach of team A and 20 favored the coach of team B. After the game, the same 80 coaches were asked to cast their vote again, with 50 now favoring the coach of team A and 30 favoring the coach of team B. The voting results are summarized in Table 15.12. Let us test H_0: the population voting preference was not changed by the game, against H_1: there has been a change in the proportion of voters who favor the coach of team A relative to the coach of team B. Given the sample results, our realized test statistic is, from (15.32), $\chi^2 = \frac{(|20-10|-1)^2}{30} = 2.70$. Then for, say, $\alpha = 0.05$, the critical region $\mathcal{R} = \{\chi^2 | \chi^2 \geq \chi^2_{0.95,1} = 3.841\}$. Since this realized chi-square statistic is not an element of $\mathcal{R}$, we cannot reject H_0 at the 5% level; voter preference has not been significantly changed by the outcome of the game. ∎

15.7 Exercises

15-1. A group of males and females is asked to state their preference for driving either a mini-van or SUV on a daily basis. The responses are indicated here. Are preference and sex independent? What is the p-value?

		Preference	
		Mini-van	**SUV**
Sex	Male	10	30
	Female	20	20

15-2. Random samples of high school students from urban, suburban, and rural areas are asked if they are planning to attend an institution of higher learning. Are the categories of planning for college and location independent?

		Location		
		Urban	**Suburban**	**Rural**
Planning	Yes	15	40	25
for College	No	35	10	25

Determine the *p*-value.

15-3. A department manager at one of the ACE Department Stores is interested in the relationship between sales force training and customer satisfaction with the service as indicated by customer responses to a brief survey taken randomly at time of purchase.

		Customer Satisfaction		
		Poor	**Good**	**Excellent**
	1 day	10	15	2
Days of	2 days	7	10	9
Training	3 days	5	15	8
	4 days	3	20	10

Are days of training and customer satisfaction independent? What is the *p*-value?

15-4. A sociologist is interested in determining if there is a relationship between family income and type of college attended. A random sample of families having at least one child in some sort of college yielded the following results:

		Type of College		
		Private	**Public**	**Community**
Family	Low	5	10	18
Income	Middle	8	15	10
	High	20	8	6

Are type of college and family income independent? Determine the *p*-value.

15-5. In a large city four independent random samples of city employees (composed of workers categorized as clerical, sanitation, transportation, and buildings and grounds) were taken in order to determine their attitude concerning a new sick-leave policy.

		Attitude		
		In Favor	**Neutral**	**Opposed**
Type of Employee	Clerical	27	33	15
	Sanitation	13	41	21
	Transportation	25	10	40
	Bldg. & Grnds.	11	50	14

Can we conclude that the four samples have been drawn from the same population? Take $\alpha = 0.01$.

15-6. Suppose that in Exercise 15-5 the "Neutral" category is omitted. Is the proportion of city employees in favor of the new policy the same across the four employee types? Use $\alpha = 0.05$.

15-7. Five groups were independently and randomly polled regarding the issue of stricter handgun control. Are the population proportions of individuals favoring stricter handgun control the same across the groups polled? Choose $\alpha = 0.10$.

		Stricter Handgun Control	
		In Favor	**Opposed**
Group	Law Enforcement Officers	37	13
	Hunters	10	40
	High School Teachers	25	25
	Clergy	40	10
	Shopkeepers	30	20

15-8. Determine an estimate of the population odds ratio for Exercise 15-1. Find a 95% confidence interval for the said odds ratio. What is the value of Cramer's phi-squared statistic?

15-9. Find the value of Cramer's phi-squared statistic for the contingency tables appearing in Exercises 15-2, 15-3, and 15-4.

15-10. For Exercise 15-2, find estimates of $\lambda_{A|B}$ and $\lambda_{B|A}$. Also determine 95% confidence intervals for these parameters.

15-11. For Exercise 15-3, determine estimates of $\lambda_{A|B}$ and $\lambda_{B|A}$. Also find 99% confidence intervals for $\lambda_{A|B}$ and $\lambda_{B|A}$.

15-12. For Exercise 15-4, produce estimates of $\lambda_{A|B}$ and $\lambda_{B|A}$. In addition, find 95% confidence intervals for $\lambda_{A|B}$ and $\lambda_{B|A}$.

15-13. Cook-Rite Inc. has developed a spray product called No-Stick that it says prevents food from sticking to the cooking surface of frying pans. Sixty pans of the same size, material, and make are to be tested (at the same heat setting) with No-Stick, with one half of the pan sprayed with No-Stick and the other half uniformly coated with margarine, where the side sprayed with No-Stick is determined in a random fashion.

		No-Stick	
		Food Sticking	**Food not Sticking**
Margarine	Food Sticking	12	20
	Food not Sticking	15	13

For $\alpha = 0.05$, does No-Stick perform better than margarine?

15-14. A random sample of $n = 50$ newspaper business editors were asked if they agreed with the newly elected President's current domestic economic policies. The same group of editors was polled a year later and asked the same question. For $\alpha = 0.05$, has there been a change in the editor's view of the President's current domestic economic policies?

		Second Poll	
		Agree	**Disagree**
First Poll	Agree	20	5
	Disagree	15	10

16

Bivariate Linear Regression and Correlation

16.1 The Regression Model

The purpose of regression analysis is two-fold: (1) we study the functional relationship between two variables X and Y and, if the said relationship is deemed meaningful (typically in a statistical sense), then (2) we seek to forecast or predict a value of one variable (Y) from a given observation on another variable (X). More specifically, as will be seen next, we can predict either the average value of Y for a given value of X or a particular value of Y corresponding to a given value of X. In order to obtain the said predictions, we need to first determine the association between the X and Y variables.

To this end, let us first assume that X and Y are random variables with joint probability density function $f(x, y)$. Then for each realization x of X, there is a conditional distribution of Y given $X = x$ or $h(y|x)$. Then the *regression of Y on X = x* is the mean of Y in the conditional density of Y given $X = x$ and appears as

$$j(x) = E(Y|X = x) = \int_{-\infty}^{+\infty} yh(y|x)dy = \int_{-\infty}^{+\infty} y\frac{f(x, y)}{g(x)}dx, \qquad (16.1)$$

where $g(x) = \int_{-\infty}^{+\infty} f(x, y)dy$ is the marginal distribution of X. Hence $y = j(x)$, the locus of means of the conditional distribution $E(Y|X = x)$ when plotted in the x, y-plane, represents the *regression curve of y on x* for Y given $X = x$. (For an example on the specification of (16.1) see Appendix 7.B.)

Next, if the joint probability density function $f(x, y)$ is unknown, then we can simply assume a particular functional form (typically linear) for the regression curve and then fit the function to a set of observations on the variables X and Y. As will be explained shortly, Y is taken to be a random variable whose values are conditional on a set of *fixed* admissible X values. This second approach to regression analysis (involving random Y, nonrandom X, and a linear regression curve) is the one that will be followed throughout this chapter.

The *basic problem* that we face is to determine the line of *best* fit through a scatter of n sample points $(X_1, Y_1), \ldots, (X_n, Y_n)$. Moreover, once the equation of the implied line has been empirically determined, we must make an assessment of its *goodness of fit* to the data. If the line is indeed found to fit the data *well*, in the sense that it closely approximates the true (but unknown) relationship between X and Y, then it can be used for predictive purposes, with the appropriate prediction error reported.

To solve the regression problem, let us begin by obtaining a sample of n observations on a variable Y: $Y_1, \ldots, Y_n$, where

$$Y_i = \text{systematic component} + \text{random component}, \ i = 1, \ldots, n. \qquad (16.2)$$

Here the *systematic or deterministic part of Y_i* reflects a particular behavioral hypothesis, whereas the random part is unobservable and arises because of some combination of sampling, measurement, and specification error; that is, it covers a *multitude of sins* in that it depicts the influence on Y_i of many omitted variables, each presumably exerting an individually small effect. Moreover, as will be seen later, the X_i's, $i = 1, \ldots, n$, are taken to be nonrandom or predetermined.

If we assume *a priori* a functional relationship between X and Y, and, in particular, that Y is linearly related to X, $Y = \beta_0 + \beta_1 X$, where the unknown population parameters β_0 and β_1 are, respectively, the vertical intercept and slope of the regression line, then (16.2) may be rewritten as

$$Y_i = \beta_0 + \beta_1 X_i + \varepsilon_i, \ i = 1, \ldots, n, \qquad (16.3)$$

where $\beta_0 + \beta_1 X_i$ is the systematic component and ε_i is a stochastic disturbance or random error term. It is apparent that because of the random nature of the ε_i values, we must construct, for purposes of estimation and testing, a probabilistic model that incorporates their requisite properties so that inferences from the set of sample data can be made about the unobserved population regression line.

16.2 The Strong Classical Linear Regression Model

Given that $Y_i, i = 1, \ldots, n$, is assumed to be statistically (and linearly) related to X and ε according to the specification $Y = \beta_0 + \beta_1 X + \varepsilon$, we will refer to X as the *explanatory variable* (remember that X is taken to be nonstochastic, e.g., it is controllable or fully predictable), to ε as a random variable, and to Y as the *explained variable* (Y is also a random variable since ε is stochastic). In this regard, for each fixed X_i there is a whole probability distribution associated with Y_i, the characteristics of which are determined by X_i and the probability distribution of $\varepsilon_i, i = 1, \ldots, n$. For instance, as Figure 16.1 reveals, at each X_i value we cannot expect Y_i to equal $\beta_0 + \beta_1 X_i$ exactly because of the random behavior of ε_i. Hence we only require that, on the average, Y_i equals $\beta_0 + \beta_1 X_i$ since realized values of Y_i may depart radically from the latter. To illustrate this average relationship, we shall indicate that, on the average, the systematic part of Y_i given X_i is $E(Y_i|X_i) = \beta_0 + \beta_1 X_i$; that is, the conditional expectation of Y_i given X_i. (In this regard, β_1 is the average change in Y per unit change in X.)

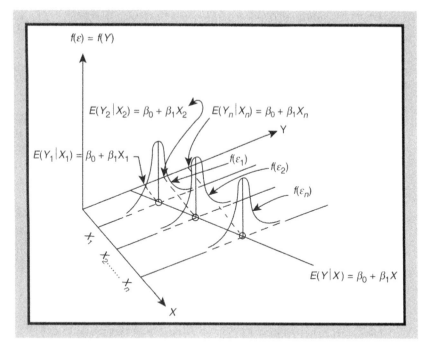

Figure 16.1   The probability distribution of ε_i is concentrated about $E(Y_i|X_i) = \beta_0 + \beta_1 X_i$ under the linearity hypothesis.

Hence the probability distribution of $\varepsilon_i, f(\varepsilon_i)$, is concentrated about $E(Y_i|X_i)$. Thus the regression equation resulting from knowledge of this average relationship is $Y = E(Y|X) + \varepsilon = \beta_0 + \beta_1 X + \varepsilon$; that is, on the average, Y's systematic component is $E(Y|X) = \beta_0 + \beta_1 X$. And as indicated in Figure 16.1, under the linearity hypothesis, each of the n means $E(Y_i|X_i)$ lies on this population regression line. Let us now turn to the specific assumptions concerning the probability distribution of ε.

Given that 16.3 holds:

(A.1) *ε_i is normally distributed for all i.*   Given that ε_i is composed of a diversity of factors that work in opposite directions, we may expect small values of ε_i to occur more frequently than large ones.

(A.2) *ε_i has zero mean or $E(\varepsilon_i) = 0$ for all i.*   Positive deviations from $\beta_0 + \beta_1 X$ are just as likely to occur as negative ones so that the random errors ε_i are distributed about a mean of zero.

(A.3) *The errors ε_i are homoscedastic or $V(\varepsilon_i) = E(\varepsilon_i^2) - E(\varepsilon_i)^2 = E(\varepsilon_i^2) = \sigma_\varepsilon^2 = constant$ for all i.*   Each $f(\varepsilon_i)$ distribution has the *same* constant variance σ_ε^2 whose value is unknown.

(A.4) *Nonautocorrelation of the ε_i's or $COV(\varepsilon_i, \varepsilon_j) = E([\varepsilon_i - E(\varepsilon_i)][\varepsilon_j - E(\varepsilon_j)]) = E(\varepsilon_i \varepsilon_j) = 0, i \neq j$.* (Successive ε_i values are uncorrelated.) In addition, X and ε are uncorrelated or $COV(X_i, \varepsilon_j) = 0$ for all i and j.

(A.5) *X is a nonrandom variable.* The X_i values are fixed; that is, controllable, in repeated sampling from the same population.[1]

Here (A.2) and (A.4) imply that the ε_i's are uncorrelated, and (A.1), (A.2), and (A.4) imply that the ε_i's are independent. Moreover, (A.1) allows us to conduct tests of hypotheses and to construct confidence intervals for β_0, β_1 and $E(Y|X) = \beta_0 + \beta_1 X$ as well as determine prediction intervals for either the average value of Y given some X value (call it X_0) or for a particular value of Y corresponding to X_0.

We noted in (A.5) that the $X_i, i = 1, \ldots, n$, values are held fixed. Hence the sole source of variation in the random variable Y in repeated sampling from the same population is variation in ε so that the probability distribution of Y is identical to that of ε. So as far as the properties of the probability distribution of Y are concerned:

(B.1) $E(Y_i) = E(Y_i|X_i) + E(\varepsilon_i) = \beta_0 + \beta_1 X_i, i = 1, \ldots, n$.

(B.2) $V(Y_i) = E[Y_i - E(Y_i)]^2 = E(\beta_0 + \beta_1 X_i + \varepsilon_i - \beta_0 - \beta_1 X_i)^2 = E(\varepsilon_i^2) = \sigma_\varepsilon^2 =$ constant for all i while $COV(Y_i, Y_j) = 0, i \neq j$.

(B.3) The random variable Y_i is normally distributed since ε_i is normally distributed, $i = 1, \ldots, n$.

(B.4) The random variables Y_i are independent because the ε_i are independent, $i = 1, \ldots, n$.

Having discussed the nature of the variables X, ε, and Y, our next set of objectives is to:

(a) Obtain estimates of β_0, β_1 (denoted as $\hat{\beta}_0, \hat{\beta}_1$, respectively).

(b) Obtain estimates of the variances of $\hat{\beta}_0, \hat{\beta}_1$, and ε.

On the basis of this information we can next:

(c) Test hypotheses concerning the population parameters β_0, β_1.

(d) Construct confidence intervals for the population parameters β_0, β_1.

(e) Test hypotheses concerning the *entire* population regression line; that is, perform a significance test of $\hat{\beta}_0$ and $\hat{\beta}_1$ jointly.

(f) Determine a prediction interval for the average value of Y given X_0.

---

[1] If only A.2–A.5 are considered, then the resulting model is simply termed the *weak classical linear regression model*. If A.1 is also included, then the *strong* case emerges.

(g) Construct a confidence band for the entire population regression line proper.

(h) Determine a prediction interval for a particular value of Y given X_0.

(i) Construct a partitioned sums-of-squares table.

16.3 Estimating the Slope and Intercept of the Population Regression Line

When the values of β_0 and β_1 are estimated on the basis of sample information, we obtain the *sample regression line* $\hat{Y} = \hat{\beta}_0 + \hat{\beta}_1 X$ that serves as our proxy for the *population regression line* $E(X|Y) = \beta_0 + \beta_1 X$, where $\hat{Y}$ is the fitted or estimated value of Y and $\hat{\beta}_0$ and $\hat{\beta}_1$ represent the estimated population parameters. Since most (if not all) of the observed Y_i values will not lie exactly on the sample regression line, the values of Y_i and $\hat{Y}_i$ differ. This difference will be denoted as $e_i = Y_i - \hat{Y}_i, 1, \ldots, n$, and will be termed the ith *residual or deviation* from the sample regression line. Here e_i serves as an estimate of the stochastic or unobserved disturbance ε_i. To see the difference between ε_i and e_i directly, let us examine the hypothetical population and sample regression lines illustrated in Figure 16.2, where Y_i may be viewed as being determined by either the population or sample regression lines

$$\text{(population regression line) } Y_i = E(Y_i|X_i) + \varepsilon_i = \beta_0 + \beta_1 X_i + \varepsilon_i,$$

$$\text{(sample regression line) } Y_i = \hat{Y}_i + e_i = \hat{\beta}_0 + \hat{\beta}_1 X_i + e_i,$$

respectively.

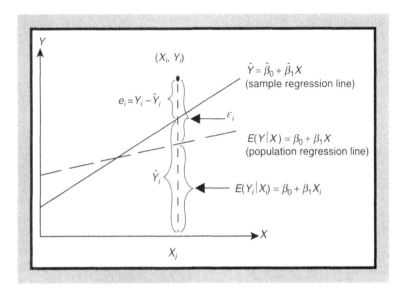

Figure 16.2 The population and sample regression lines.

The decision rule or criterion of goodness of fit to be employed in estimating β_0 and β_1 is depicted by the *principle of least squares*: to obtain the line of best fit, choose $\hat{\beta}_0$ and $\hat{\beta}_1$ so as to minimize the sum of the squared deviations from the sample regression line; that is,

$$\min\left\{\sum_{i=1}^{n} e_i^2 = \sum (Y_i - \hat{Y}_i)^2 = \sum (Y_i - \hat{\beta}_0 - \hat{\beta}_1 X_i)^2 = F(\hat{\beta}_0, \hat{\beta}_1)\right\}.$$

(For convenience the operator $\sum_{i=1}^{n}$ will at times be simplified to $\sum$, where it is to be understood that we always sum over all values of the i index; i.e., $i = 1,\ldots,n$.) Upon setting $\frac{\partial F}{\partial \hat{\beta}_0} = \frac{\partial F}{\partial \hat{\beta}_1} = 0$ (it is assumed that the second-order conditions for a minimum are satisfied), the resulting simultaneous linear equation system

$$\left.\begin{array}{c} \sum e_i = 0 \\[2mm] \sum X_i e_i = 0 \end{array}\right\} \quad \text{or} \quad \left\{\begin{array}{l} \text{(a) } n\hat{\beta}_0 + \hat{\beta}_1 \sum X_i = \sum Y_i \\[2mm] \text{(b) } \hat{\beta}_0 \sum X_i + \hat{\beta}_1 \sum X_i^2 = \sum X_i Y_i \end{array}\right. \tag{16.4}$$

[system of least squares normal equations]

yields the solution

$$\hat{\beta}_0 = \frac{(\sum Y_i)(\sum X_i^2) - (\sum X_i Y_i)(\sum X_i)}{n \sum X_i^2 - (\sum X_i)^2}, \quad \hat{\beta}_1 = \frac{n \sum X_i Y_i - (\sum X_i)(\sum Y_i)}{n \sum X_i^2 - (\sum X_i)^2}.$$

An alternative specification of $\hat{\beta}_0$ and $\hat{\beta}_1$ is in terms of the deviations of X_i and Y_i from their respective means $\overline{X}$ and $\overline{Y}$. That is, if $x_i = X_i - \overline{X}$ (respectively, $y_i = Y_i - \overline{Y}$), $i = 1,\ldots,n$, represents the ith deviation of X_i (respectively, Y_i) from its mean, then

$$\text{(a)} \quad \hat{\beta}_1 = \frac{\sum (X_i - \overline{X})(Y_i - \overline{Y})}{\sum (X_i - \overline{X})^2} = \frac{\sum x_i y_i}{\sum x_i^2},$$

$$\text{(b)} \quad \hat{\beta}_0 = \overline{Y} - \hat{\beta}_1 \overline{X} \quad \text{(from (16.4a)).} \tag{16.5}$$

Note that (16.5b) reveals an important property of the least squares line of best fit $\hat{Y} = \hat{\beta}_0 + \hat{\beta}_1 X$, namely that it passes through the point of means $(\overline{X}, \overline{Y})$ (see Figure 16.3). (For notational convenience, $\hat{\beta}_0$ and $\hat{\beta}_1$ will also be used to denote the sample realizations of these least squares estimators.)

What about the statistical properties of the least squares estimators $\hat{\beta}_0$ and $\hat{\beta}_1$? To answer this question we shall rely upon the *Gauss-Markov Theorem*: If assumptions (A.2) through (A.5) underlying the weak classical linear regression model hold, then, within the class of linear unbiased estimators of β_0 and β_1, the least squares estimators have minimum variance; that is, the least squares estimators are *BLUE* (Best Linear Unbiased Estimators). Hence *BLUE* requires that the estimator $\hat{\beta}_0$ (respectively, $\hat{\beta}_1$) be expressible as a linear combination of

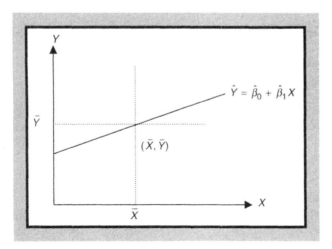

Figure 16.3 The sample regression line passes through the point of means $(\overline{X}, \overline{Y})$.

the $Y_i, i = 1, \ldots, n$, that it be unbiased, and its variance be smaller than that of any alternative linear unbiased estimator of β_0 (respectively, β_1).

It is important to note that if we admit assumption (A.1) to our discussion; that is, if we work with the strong classical linear regression model, then the least squares estimators $\hat{\beta}_0$ and $\hat{\beta}_1$ are maximum likelihood estimators of β_0 and β_1, respectively. (You will verify this assertion in Exercise 16.15.)

Example 16.3.1 Let us assume that a certain manufacturing company has compiled a set of data (see columns 1 and 2 of Table 16.1 and Figure 16.4) pertaining to its last 10 years of operation in a certain market. It has available observations on gross sales (in millions of current dollars) and advertising outlay (also expressed in millions of current dollars). Is there a statistically significant linear relationship between gross sales and advertising expenditure? In addition, if the company is considering a new advertising campaign that would involve an outlay of 12 million dollars in 2005, what is the anticipated level of gross sales for that period?

A glance at columns 1 and 2 of Table 16.1 reveals that with $n = 10$, $\sum X_i = 72.10$, and $\sum Y_i = 122.95$, it follows that $\overline{X} = 7.210$ and $\overline{Y} = 12.295$. On the basis of these values we may determine the entries in columns 3 through 7. And from the indicated column totals, we obtain

$$\hat{\beta}_1 = \frac{\sum x_i y_i}{\sum x_i^2} = \frac{88.7010}{61.7490} = 1.4365,$$

$$\hat{\beta}_0 = \overline{Y} - \hat{\beta}_1 \overline{X} = 12.295 - 1.4365(7.210) = 1.9379.$$

So when X increases by one unit, Y's average increase is 1.4365 units. Moreover, the average value of Y when $X = 0$ is 1.9379. Hence the estimated regression

Table 16.1

	(1)	**(2)**	**(3)**	**(4)**	**(5)**	**(6)**	**(7)**
Year	**Gross Sales** (Y)	**Advertising Outlay** (X)	$x_i = X_i - \bar{X}$ $= X_i - 7.210$	$y_i = Y_i - \bar{Y}$ $= Y_i - 12.295$	x_i^2	y_i^2	$x_i y_i$
1995	7.60	4.00	−3.210	−4.695	10.3041	22.0430	15.0710
1996	8.36	4.50	−2.710	−3.935	7.3441	15.4842	10.6639
1997	8.00	4.60	−2.610	−4.295	6.8121	18.4470	11.2100
1998	9.58	5.00	−2.210	−2.715	4.8841	7.3712	6.0002
1999	11.51	7.00	−0.210	−0.785	0.0441	0.6162	0.1649
2000	13.00	8.00	0.790	0.705	0.6241	0.4970	0.5570
2001	15.10	8.20	0.990	2.805	0.9801	7.8680	2.7770
2002	15.60	9.00	1.790	3.305	3.2041	10.9230	5.9160
2003	16.00	10.50	3.29	3.705	10.8241	13.7270	12.1895
2004	18.20	11.30	4.090	5.905	16.7281	34.8690	24.1515
Column Totals	$\sum Y_i =$ 122.95	$\sum X_i = 72.10$	$\sum x_i = 0$	$\sum y_i = 0$	$\sum x_i^2 =$ 61.7490	$\sum y_i^2 =$ 131.8456	$\sum x_i y_i =$ 88.7010

line is $\hat{Y} = \hat{\beta}_0 + \hat{\beta}_1 X = 1.9379 + 1.4365X$. And if advertising outlay increases to $X_0 = 12$, the predicted value of gross sales, $\hat{Y}_0$, is determined from the estimated regression equation to be $\hat{Y}_0 = \hat{\beta}_0 + \hat{\beta}_1 X_0 = 1.9379 + 1.4365(12) = 19.1759$. ∎

How reliable is the predicted value of Y? The answer to this question obviously hinges upon how well we have estimated the population regression equation itself; that is, does the set of sample data offer sufficient evidence to indicate that advertising outlay serves as a reasonably good explanatory variable as far as the behavior of gross sales is concerned, or can the observed relationship between these variables be attributed solely to chance? In the next three sections we shall explicitly address the issue of *goodness of fit* of the sample regression line.

16.4   Mean, Variance, and Sampling Distribution of the Least Squares Estimators $\hat{\beta}_0$ and $\hat{\beta}_1$

From the Gauss-Markov Theorem we know that $E(\hat{\beta}_0) = \beta_0$ and $E(\hat{\beta}_1) = \beta_1$. Moreover, it can be shown that

$$V(\hat{\beta}_0) = \sigma_\varepsilon^2 \left(\frac{1}{n} + \frac{\bar{X}^2}{\sum x_i^2} \right), V(\hat{\beta}_1) = \frac{\sigma_\varepsilon^2}{\sum x_i^2}, \tag{16.6}$$

where σ_ε^2 is the unknown variance of the random (error) variable ε. (It is evident from these expressions that $V(\hat{\beta}_0)$ and $V(\hat{\beta}_1)$ both vary directly with the spread

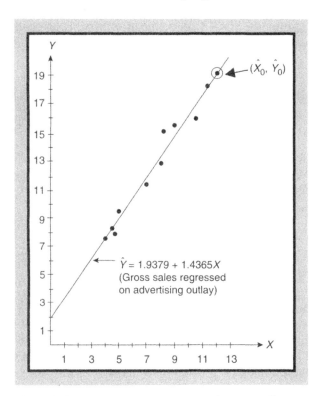

Figure 16.4   Using the least squares regression equation to predict gross sales at an advertising outlay of $X_0 = 12$.

or dispersion of the unobserved ε_i values about the population regression line; that is, with σ_ε^2, and inversely with the variation or dispersion of the X_i values about $\overline{X}$, or $\sum x_i^2$.) In addition,

$$COV(\hat{\beta}_0, \hat{\beta}_1) = E\left(\left[\hat{\beta}_0 - E(\hat{\beta}_0)\right]\left[\hat{\beta}_1 - E(\hat{\beta}_1)\right]\right) = E\left((\hat{\beta}_0 - \beta_0)(\hat{\beta}_1 - \beta_1)\right) = -\overline{X}\left(\frac{\sigma_\varepsilon^2}{\sum x_i^2}\right),$$

where $\hat{\beta}_0 - \beta_0$ and $\hat{\beta}_1 - \beta_1$ are the sampling errors of $\hat{\beta}_0$ and $\hat{\beta}_1$, respectively. Clearly $COV(\hat{\beta}_0, \hat{\beta}_1) < 0$ when $\overline{X} > 0$; that is, the said sampling errors are of opposite sign, to wit an understatement of β_0 is accompanied with an overstatement of β_1, and conversely. It should be apparent that neither of the quantities in (16.6) can be directly employed as an aid in determining just how precisely the population regression line has been estimated since σ_ε^2 is unknown and must be determined from the sample values. To this end we shall use

$$S_\varepsilon^2 = \frac{\sum e_i^2}{n-2} = \frac{\sum (Y_i - \hat{Y}_i)^2}{n-2} = \frac{\sum y_i^2 - \hat{\beta}_1 \sum x_i y_i}{n-2} \tag{16.7}$$

as an unbiased estimator of σ_ε^2. If this estimator for σ_ε^2 is inserted into (16.6), it follows that estimators for the variances of $\hat{\beta}_0$ and $\hat{\beta}_1$ are, respectively,

$$S_{\hat{\beta}_0}^2 = S_\varepsilon^2 \left(\frac{1}{n} + \frac{\overline{X}^2}{\sum x_i^2} \right), \quad S_{\hat{\beta}_1}^2 = \frac{S_\varepsilon^2}{\sum x_i^2}.$$

If we take the positive square root of each of these expressions, then we obtain the estimated standard deviations of $\hat{\beta}_0$ and $\hat{\beta}_1$ or, as they are more commonly called, the estimated *standard errors of the regression coefficients*

$$S_{\hat{\beta}_0} = S_\varepsilon \sqrt{\frac{1}{n} + \frac{\overline{X}^2}{\sum x_i^2}}, \quad S_{\hat{\beta}_1} = \frac{S_\varepsilon}{\sqrt{\sum x_i^2}}. \tag{16.8}$$

In these expressions S_ε is referred to as the *standard error of estimate* and serves as a measure of the average dispersion of the individual sample points about the estimated regression line. Clearly the estimated standard errors of $\hat{\beta}_0$ and $\hat{\beta}_1$ vary directly with S_ε; that is, the greater the dispersion of the sample points about the estimated regression line, the less precise our estimates of β_0 and β_1.

Example 16.4.1   We may continue with the preceding example by using the sample data presented in Table 16.1 to find sample realizations of $S_{\hat{\beta}_0}$ and $S_{\hat{\beta}_1}$. Since from (16.7) we have

$$s_\varepsilon = \sqrt{\frac{\sum y_i^2 - \hat{\beta}_1 \sum x_i y_i}{n - 2}} = \sqrt{\frac{131.8456 - 1.4365(88.7010)}{8}} = 0.7439,$$

it follows from (16.8) that

$$s_{\hat{\beta}_0} = s_\varepsilon \sqrt{\frac{1}{n} + \frac{\overline{X}^2}{\sum x_i^2}} = 0.7439 \sqrt{\frac{1}{10} + \frac{(7.210)^2}{61.7490}} = 0.7219,$$

$$s_{\hat{\beta}_1} = \frac{s_\varepsilon}{\sqrt{\sum x_i^2}} = \frac{0.7439}{\sqrt{61.7490}} = 0.0946. \quad \blacksquare$$

If we now couple assumption (A.1) with the Gauss-Markov results, it can be shown that both $\hat{\beta}_0$ and $\hat{\beta}_1$ are expressible as linear combinations of the independent normal random variables $Y_i, i = 1, \ldots, n$, so that

$$\hat{\beta}_0 \text{ is } N\left(E(\hat{\beta}_0), \sqrt{V(\hat{\beta}_0)}\right) = N\left(\beta_0, \sigma_\varepsilon \sqrt{\frac{1}{n} + \frac{\bar{X}^2}{\sum x_i^2}}\right),$$

$$\hat{\beta}_1 \text{ is } N\left(E(\hat{\beta}_1), \sqrt{V(\hat{\beta}_1)}\right) = N\left(\beta_1, \frac{\sigma_\varepsilon}{\sqrt{\sum x_i^2}}\right)$$

and thus

$$\frac{\hat{\beta}_0 - \beta_0}{\sigma_\varepsilon \sqrt{\frac{1}{n} + \frac{\bar{X}^2}{\sum x_i^2}}} \text{ is } N(0,1) \text{ and } \frac{\hat{\beta}_1 - \beta_1}{\frac{\sigma_\varepsilon}{\sqrt{\sum x_i^2}}} \text{ is } N(0,1). \tag{16.9}$$

If in (16.9) we replace σ_ε by S_ε, then the resulting quantities follow a t distribution with $n - 2$ degrees of freedom (since from a sample of size n we have estimated two population parameters so that the number of independent observations remaining in the sample is $n - 2$); that is,

$$\frac{\hat{\beta}_0 - \beta_0}{S_\varepsilon \sqrt{\frac{1}{n} + \frac{\bar{X}^2}{\sum x_i^2}}} = \frac{\hat{\beta}_0 - \beta_0}{S_{\hat{\beta}_0}} \text{ is } t_{n-2}, \quad \frac{\hat{\beta}_1 - \beta_1}{\frac{S_\varepsilon}{\sqrt{\sum x_i^2}}} = \frac{\hat{\beta}_1 - \beta_1}{S_{\hat{\beta}_1}} \text{ is } t_{n-2}. \tag{16.10}$$

16.5   Precision of the Least Squares Estimators $\hat{\beta}_0, \hat{\beta}_1$: Confidence Intervals

To determine just how precisely the population parameters have been estimated from the sample data, let us construct the probability statements

$$\text{(a)} \quad P\left(-t_{\alpha/2,n-2} \leq \frac{\hat{\beta}_0 - \beta_0}{S_{\hat{\beta}_0}} \leq t_{\alpha/2,n-2}\right) = 1 - \alpha,$$

$$\text{(b)} \quad P\left(t_{\alpha/2,n-2} \leq \frac{\hat{\beta}_1 - \beta_1}{S_{\hat{\beta}_1}} \leq t_{\alpha/2,n-2}\right) = 1 - \alpha, \tag{16.11}$$

where $-t_{\alpha/2,n-2}, t_{\alpha/2,n-2}$ are lower and upper percentage points, respectively, of the t distribution (i.e., values that cut off $\frac{\alpha}{2}$ of the total area under the t distribution at each tail end) and $1 - \alpha$ is the confidence probability. Upon pivoting in (16.11a,b) and passing to sample realizations we obtain the confidence statements

$$\text{(a)} \quad P(\hat{\beta}_0 - t_{\alpha/2,n-2} s_{\hat{\beta}_0} \leq \beta_0 \leq \hat{\beta}_0 + t_{\alpha/2,n-2} s_{\hat{\beta}_0}) = 1 - \alpha,$$

$$\text{(b)} \quad P(\hat{\beta}_1 - t_{\alpha/2,n-2} s_{\hat{\beta}_1} \leq \beta_1 \leq \hat{\beta}_1 + t_{\alpha/2,n-2} s_{\hat{\beta}_1}) = 1 - \alpha, \tag{16.12}$$

with confidence coefficient $1 - \alpha$ so that, from (16.12a), a *$100(1 - \alpha)\%$ confidence interval for β_0* is

$$(\hat{\beta}_0 - t_{\alpha/2,n-2}\, s_{\hat{\beta}_0},\, \hat{\beta}_0 + t_{\alpha/2,n-2}\, s_{\hat{\beta}_0}); \tag{16.13}$$

and from (16.12b), a *$100(1 - \alpha)\%$ confidence interval for β_1* amounts to

$$(\hat{\beta}_1 - t_{\alpha/2,n-2}\, s_{\hat{\beta}_1},\, \hat{\beta}_1 + t_{\alpha/2,n-2}\, s_{\hat{\beta}_1}). \tag{16.14}$$

In this regard, (16.13) and (16.14) inform us that we may be $100(1 - \alpha)\%$ confident that the true regression intercept β_0 (respectively, slope β_1) lies between $\hat{\beta}_0 \pm t_{\alpha/2,n-2}\, s_{\hat{\beta}_0}$ (respectively, $\hat{\beta}_1 \pm t_{\alpha/2,n-2}\, s_{\hat{\beta}_1}$). It should be evident that the narrower the intervals depicted in (16.13) and (16.14), the more precise the estimates of β_0 and β_1, respectively.

Example 16.5.1 How precisely have we estimated β_0 and β_1 from the preceding sample of gross sales-advertising outlay data (see Table 16.1)? If we want a 95% confidence interval, then, since $1 - \alpha = 0.95, \alpha = 0.05$ and thus $\frac{\alpha}{2} = 0.025$. Hence $t_{\alpha/2,n-2} = t_{0.025,8} = 2.306$. From Section 16.4 we found that $s_{\hat{\beta}_0} = 0.7219, s_{\hat{\beta}_1} = 0.0946$. Then from (16.13) we have

$$\hat{\beta}_0 - t_{\alpha/2,n-2}\, s_{\hat{\beta}_0} = 1.9379 - 2.306(0.7219) = 0.2732,$$

$$\hat{\beta}_0 + t_{\alpha/2,n-2}\, s_{\hat{\beta}_0} = 1.9379 + 2.306(0.7219) = 3.6026.$$

Hence we may be 95% confident that the true regression intercept β_0 lies between 0.2732 and 3.6026. And from (16.14) we get

$$\hat{\beta}_1 - t_{\alpha/2,n-2}\, s_{\hat{\beta}_1} = 1.4365 - 2.306(0.0946) = 1.4147,$$

$$\hat{\beta}_1 + t_{\alpha/2,n-2}\, s_{\hat{\beta}_1} = 1.4365 + 2.306(0.0946) = 1.4583.$$

Thus we may be 95% confident the true regression slope β_1 lies between 1.4147 and 1.4583. ∎

16.6 Testing Hypotheses Concerning β_0, β_1

One question that naturally arises once β_0 and β_1 have been estimated is whether or not the variables X and Y are truly linearly related. Equivalently, we may inquire as to whether or not the sample data set exhibits sufficient evidence to indicate that X actually contributes significantly to the prediction of the average value of Y for a given value of X or to the prediction of a particular value of Y corresponding to a given value of X. Conceivably the observed or fitted linear relationship may simply be the result of chance phenomena. To answer these questions we shall test a particular null hypothesis concerning β_1 against an appropriate

alternative hypothesis, where the latter is typically dictated on *a priori* grounds
by some supporting scientific or behavioral (e.g., economic) theory relating X
and Y. The most common type of hypothesis tested is that there is no linear rela-
tionship between X and Y or $E(Y|X) = \beta_0 + \beta_1 X = \beta_0$; that is, the conditional
expectation of Y given X does not depend linearly upon X so that the population regression line is horizontal. Hence the implied null hypothesis of *no linear
relationship between X and Y* is $H_0: \beta_1 = 0$.

As far as possible alternative hypotheses are concerned:

(a) If, say, on an *a priori* basis some specific theory indicates that X and Y
     are positively related, we then choose $H_1: \beta_1 > 0$. In this instance we
     have a one-tail alternative involving the right-hand tail of the t distribu-
     tion. Here the region of rejection is $\mathcal{R} = \{t | t > t_{\alpha, n-2}\}$, where $t_{\alpha, n-2}$ is the
     critical (tabular) value of t. From (16.10), given that H_0 is true, the test
     statistic is $T = \frac{\hat{\beta}_1}{S_{\hat{\beta}_1}}$, where the sample realization of T will be denoted as
     $t = \frac{\hat{\beta}_1}{S_{\hat{\beta}_1}}$. Then if $t > t_{\alpha, n-2}$, we reject H_0 in favor of H_1; that is, at the $100\alpha\%$
     level, we conclude that there exists a statistically significant positive linear
     relationship between X and Y.

(b) But if *a priori* theory indicates that X and Y are negatively related, we then
     have $H_1: \beta_1 < 0$. Relative to this alternative we have a one-tail test involv-
     ing the left-hand tail of the t distribution with $\mathcal{R} = \{t | t < -t_{\alpha, n-2}\}$. Again
     the test statistic is $T = \frac{\hat{\beta}_1}{S_{\hat{\beta}_1}}$. In this regard, if $t < -t_{\alpha, n-2}$ or $-t > t_{\alpha, n-2}$, we
     reject H_0 in favor of H_1 and conclude that, at the $100\alpha\%$ level, there exists
     a statistically significant negative linear relationship between X and Y.

(c) If we are uncertain about the relationship between X and Y, then the appro-
     priate alternative hypothesis is $H_1: \beta_1 \neq 0$. Clearly this alternative implies a
     two-tail test with $\mathcal{R} = \{t | |t| > t_{\alpha/2, n-2}\}$. For this test, if $|t| = |\frac{\hat{\beta}_1}{S_{\hat{\beta}_1}}| > t_{\alpha/2, n-2}$,
     we reject H_0 in favor of H_1 and conclude that, at the $100\alpha\%$ level, there
     exists a statistically significant linear relationship between X and Y.

(Note: If we do not reject H_0, it does not mean that X and Y are *unrelated*
but only that there is no significant *linear* relationship exhibited by the data;
the true underlying relationship between X and Y may be highly nonlinear.)

The occasion may arise when we want to test the null hypothesis $H_0: \beta_1 =
\beta_1^0$, where β_1^0 is some specific or anticipated level of β_1 (not necessarily zero).
If from (16.10) the sample realization of the test statistic T under H_0 is denoted
as $t = \frac{\hat{\beta}_1 - \beta_1^0}{S_{\hat{\beta}_1}}$, then the appropriate set of alternative hypotheses is:

(a) $H_1: \beta_1 > \beta_1^0$. Here $\mathcal{R} = \{t | t > t_{\alpha, n-2}\}$; or

(b) $H_1: \beta_1 < \beta_1^0$. In this case $\mathcal{R} = \{t | t < -t_{\alpha, n-2}\}$; or

(c) $H_1: \beta_1 \neq \beta_1^0$. Then $\mathcal{R} = \{t | |t| > t_{\alpha/2, n-2}\}$.

As far as the true regression intercept is concerned, the relevant null hypotheses to be tested are either H_0: $\beta_0 = 0$ (i.e., the population regression line passes through the origin) or H_0: $\beta_0 = \beta_0^0$, where β_0^0 is some specific or anticipated level of β_0 (not necessarily zero). The appropriate critical region $\mathcal{R}$ and test statistic T are formed in a fashion similar to those specified earlier when hypotheses about β_1 were tested.

Example 16.6.1   On the basis of the least-squares estimates $\hat{\beta}_0$ and $\hat{\beta}_1$ obtained from the sample data on advertising outlay (X) and gross sales (Y), can we conclude, at the 5% level, that the true population parameters β_0 and β_1 are significantly different from zero? As far as the population regression intercept is concerned, H_0: $\beta_0 = 0$ and H_1: $\beta_0 > 0$ (we anticipate *a priori* that even if the company did not advertise, the level of gross sales should be positive). With $t_{\alpha,n-2} = t_{0.05,8} = 1.86$, $\mathcal{R} = \{t|t > 1.86\}$. Since $t = \frac{\hat{\beta}_0}{s_{\hat{\beta}_0}} = \frac{1.9379}{0.7219} = 2.6844 \in \mathcal{R}$, we thus reject H_0 in favor of H_1. Here $0.01 < p\text{-value} < 0.025$.

Is there a significant positive linear relationship between sales and advertising? To answer this question we form H_0: $\beta_1 = 0, H_1$: $\beta_1 > 0$ (we anticipate *a priori* that the effect of the company's promotional activities is to increase sales). Since $\mathcal{R} = \{t|t > 1.86\}$ and $t = \frac{\hat{\beta}_1}{s_{\hat{\beta}_1}} = \frac{1.4365}{0.0946} = 15.1849 \in \mathcal{R}$, we reject H_0 in favor of H_1 ($p\text{-value} < 0.0005$) and conclude that there is a strong (statistically significant) positive linear relationship between X and Y. ∎

Example 16.6.2   At a promotional strategy session of the advertising department one of the account coordinators stated that for every $1 expended on advertising the company would obtain, on the average, about a $2 increase in gross sales. Is this assertion supported by the data at the 5% level? Let us select $H_0 = \beta_1 = \beta_1^0 = 2$ and test it against $H_1 = \beta_1 \neq \beta_1^0 = 2$ (gross sales increases by some figure other than $2). Since $\mathcal{R} = \{t||t| > t_{\alpha/2,n-2} = t_{0.025,8} = 2.306\}$ and $|t| = \left|\frac{\hat{\beta}_1 - \beta_1^0}{s_{\hat{\beta}_1}}\right| = \left|\frac{1.4365-2}{0.0946}\right| = 5.9567 \in \mathcal{R}$, we reject H_0 in favor of H_1, ($p\text{-value} < 0.0005$); that is, on the average, a $1 increase in advertising outlay does not precipitate a $2 increase in gross sales. In fact, the actual increase lies significantly below the $2 level. (Interestingly enough, we could have reached this same conclusion by simply examining the 95% confidence interval for β_1 computed in Section 16.5. Since 2 is not a member of the said interval, β_1 must depart significantly from 2. Hence an explicit or formal test at the $100\alpha\%$ level of H_0: $\beta_1 = \beta_1^0$ against H_1: $\beta_1 \neq \beta_1^0$ is not necessary if the $100(1-\alpha)\%$ confidence interval for β_1 is given.) ∎

One final set of test procedures concerning β_0 and β_1 will be mentioned briefly. We have been performing hypothesis tests on β_0 and β_1 separately. However, we can easily test β_0 and β_1 jointly in order to determine whether or not the entire population regression equation itself is significant; that is, we can determine whether or not β_0 and β_1 are jointly significantly different from zero. To this end we form the joint null hypothesis H_0: $\beta_0 = \beta_1 = 0$ and test it against the joint alternative H_1: $\beta_0 \neq 0, \beta_1 \neq 0$ at the $100\alpha\%$ level. At a particular X_i value, let us compute the difference between the sample and population regression lines as $\hat{Y}_i - E(Y_i|X_i) = \hat{\beta}_0 + \hat{\beta}_1 X_i - (\beta_0 + \beta_1 X_i) = (\hat{\beta}_0 - \beta_0) + (\hat{\beta}_1 - \beta_1)X_i$. Then $\sum[\hat{Y}_i - E(Y_i|X_i)]^2 = \sum[(\hat{\beta}_0 - \beta_0) + (\hat{\beta}_1 - \beta_1)X_i]^2$ serves as a measure of the overall discrepancy between the estimated values of β_0 and β_1 and their true population values for given X_i's. Clearly this sum varies directly with $|\hat{\beta}_0 - \beta_0|$ and $|\hat{\beta}_1 - \beta_1|$. Furthermore, it can be shown that the quantity

$$F = \frac{\sum\left[(\hat{\beta}_0 - \beta_0) + (\hat{\beta}_1 - \beta_1)X_i\right]^2 \Big/ 2}{S_\varepsilon^2} \text{ is } F_{2,n-2}. \tag{16.15}$$

Under H_0, this test statistic becomes

$$F = \frac{\sum(\hat{\beta}_0 + \hat{\beta}_1 X_i)^2 / 2}{S_\varepsilon^2} = \frac{(n\overline{Y}^2 + \hat{\beta}_1 \sum x_i y_i)/2}{S_\varepsilon^2} \tag{16.16}$$

while $\mathcal{R} = \{F | F > f_{1-\alpha,2,n-2}\}$. Here we have a one-tail alternative on the upper tail of the F distribution so that we reject H_0 when the realized F value exceeds $f_{1-\alpha,2,n-2}$.

It is evident that a more general joint test can be performed by stating the joint null hypothesis as H_0: $\beta_0 = \beta_0^0, \beta_1 = \beta_1^0$, where β_0^0 and β_1^0 are specific anticipated values (not necessarily zero) of the true regression intercept and slope, respectively. This hypothesis is to be tested against the joint alternative H_1: $\beta_0 \neq \beta_0^0, \beta_1 \neq \beta_1^0$. Given that H_0 is true, the test statistic may be obtained from (16.15) and written as

$$F = \frac{\left\{n\left[(\hat{\beta}_0 - \beta_0^0) + (\hat{\beta}_1 - \beta_1^0)\overline{X}\right]^2 + (\hat{\beta}_1 - \beta_1^0)^2 \sum x_i^2\right\} \Big/ 2}{S_\varepsilon^2}.$$

Example 16.6.3 Do the estimates obtained for β_0 and β_1 from the sample data on gross sales and advertising expenditure warrant the conclusion that the entire population regression line is significant at the 5% level? To answer this let us set H_0: $\beta_0 = \beta_1 = 0$ and test it against H_1: $\beta_0 \neq 0, \beta_1 \neq 0$ (the entire population regression equation is significant). Since $\mathcal{R} = \{F | F > f_{1-\alpha,2,n-2} = f_{0.95,2,8} = 4.46\}$ and, from (16.16), the realized

$$F = \frac{\left[10\,(12.295)^2 + 1.4365\,(88.7010)\right] \Big/ 2}{0.5533} = 1{,}250.9048 \in \mathcal{R},$$

we conclude that the entire population regression line is statistically significant; that is, $|\hat{\beta}_0|$ and $|\hat{\beta}_1|$ are sufficiently large to warrant rejecting H_0 at the 5% level. ∎

16.7 The Precision of the Entire Least Squares Regression Equation: A Confidence Band

We noted earlier that the purpose of regression analysis is to predict either the average value of Y for a given value of X or a particular value of Y corresponding to a given value of X. If in this section we restrict our discussion to predicting the average value of Y given $X = X_i$, $E(Y|X_i)$, then we can determine just how precisely we have estimated the population regression line $E(Y|X) = \beta_0 + \beta_1 X$ by constructing a confidence band around the sample regression line $\hat{Y} = \hat{\beta}_0 + \hat{\beta}_1 X$. Clearly the narrower the confidence band the more precisely the sample regression line estimates the population regression line. Since $\hat{\beta}_0$ and $\hat{\beta}_1$ are unbiased estimators of β_0 and β_1, respectively, it follows that $E(\hat{Y}) = \beta_0 + \beta_1 X$; that is, an unbiased estimator for the population regression line is the sample regression line. In this regard, since $\hat{Y}_i = \hat{\beta}_0 + \hat{\beta}_1 X_i$ is an estimator for $E(Y_i|X_i) = \beta_0 + \beta_1 X_i$, it can be shown that the variance of $\hat{Y}_i$ may be computed as

$$\sigma_{\hat{Y}_i}^2 = \sigma_\varepsilon^2 \left[\frac{1}{n} + \frac{(X_i - \overline{X})^2}{\sum x_i^2} \right]. \tag{16.17}$$

It is evident that (16.17) cannot be employed directly to ascertain the precision of our estimate of the population regression line since σ_ε^2 is unknown. As before, let us estimate σ_ε^2 by S_ε^2 (see (16.7)). Then from (16.17), the estimated variance of $\hat{Y}_i$ is

$$S_{\hat{Y}_i}^2 = S_\varepsilon^2 \left[\frac{1}{n} + \frac{(X_i - \overline{X})^2}{\sum x_i^2} \right].$$

If we take the positive square root of this expression we obtain the *estimated standard deviation of $\hat{Y}_i$*,

$$S_{\hat{Y}_i} = S_\varepsilon \sqrt{\frac{1}{n} + \frac{(X_i - \overline{X})^2}{\sum x_i^2}}. \tag{16.18}$$

By virtue of assumption (A.1) and the Gauss-Markov results, it can be demonstrated that $\hat{Y}_i = \hat{\beta}_0 + \hat{\beta}_1 X_i$ is a linear combination of normal random variables so that

$$\hat{Y}_i \text{ is } N\left(E(Y_i|X_i), \sqrt{V(\hat{Y}_i)}\right) = N\left(\beta_0 + \beta_1 X_i, \sigma_\varepsilon \sqrt{\frac{1}{n} + \frac{(X_i - \overline{X})^2}{\sum x_i^2}}\right)$$

and thus

$$\frac{\hat{Y}_i - (\beta_0 + \beta_1 X_i)}{\sigma_\varepsilon \sqrt{\dfrac{1}{n} + \dfrac{(X_i - \overline{X})^2}{\sum x_i^2}}} \text{ is } N(0,1). \tag{16.19}$$

If in (16.19) we replace σ_ε by S_ε, the resulting quantity follows a t distribution with $n - 2$ degrees of freedom; that is,

$$\frac{\hat{Y}_i - (\beta_0 + \beta_1 X_i)}{S_\varepsilon \sqrt{\dfrac{1}{n} + \dfrac{(X_i - \overline{X})^2}{\sum x_i^2}}} = \frac{\hat{Y}_i - (\beta_0 + \beta_1 X_i)}{S_{\hat{Y}_i}} \text{ is } t_{n-2}.$$

To determine just how precisely the population regression line has been estimated from the sample data, let us form the probability statement

$$P\left(-t_{\alpha/2, n-2} \leq \frac{\hat{Y}_i - (\beta_0 + \beta_1 X_i)}{S_{\hat{Y}_i}} \leq t_{\alpha/2, n-2}\right) = 1 - \alpha. \tag{16.20}$$

Upon pivoting in (16.20) and employing (16.18) we obtain the confidence statement

$$P\left(\hat{Y}_i - t_{\alpha/2, n-2}\, s_{\hat{Y}_i} \leq \beta_0 + \beta_1 X_i \leq \hat{Y}_i + t_{\alpha/2, n-2}\, s_{\hat{Y}_i}\right) = 1 - \alpha \tag{16.21}$$

so that, from (16.21), a $100(1 - \alpha)\%$ *confidence interval for* $E(Y_i|X_i) = \beta_0 + \beta_1 X_i$ is

$$(\hat{Y}_i - t_{\alpha/2, n-2}\, s_{\hat{Y}_i}, \hat{Y}_i + t_{\alpha/2, n-2}\, s_{\hat{Y}_i}). \tag{16.22}$$

In this regard, (16.22) informs us that we may be $100(1 - \alpha)\%$ confident that the true or population average value of Y_i given X_i lies between $\hat{Y}_i - t_{\alpha/2, n-2} s_{\hat{Y}_i}$ and $\hat{Y}_i + t_{\alpha/2, n-2} s_{\hat{Y}_i}$.

A glance at (16.22) reveals that this confidence interval can be calculated for any X_i within the domain of X. And as we vary X, we can compute a whole set of confidence intervals, with each one corresponding to (i.e., centered around) a point on the sample regression line. The collection of these confidence intervals forms a *confidence band* about the estimated sample regression line. Note that

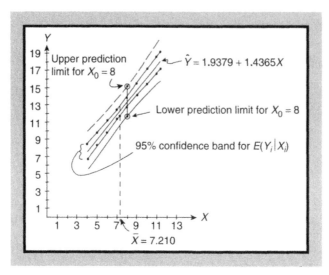

Figure 16.5   A 95% confidence band for $E(Y_i|X_i)$ (see solid lines) and a 95% prediction band for Y_0 (dashed lines).

since $S_{\hat{Y}_i}$ varies directly with $(X_i - \overline{X})^2$, the further X_i lies from $\overline{X}$, the greater the estimated standard deviation of Y_i and thus the wider the confidence band at this point. If we use (16.22) to calculate a confidence interval for each X_i, the locus of end points of these intervals would correspond to two branches of a rectangular hyperbola, indicating that we have more confidence in predicting $E(Y_i|X_i)$ near the center of the X_i's than at the extremes of the range of X values (see Figure 16.5).

We may utilize the results of the preceding discussion to conduct a test of the hypothesis that $E(Y_i|X_i)$ equals some specific or anticipated value E_i^0; that is, the null hypothesis is H_0: $E(Y_i|X_i) = E_i^0$. Given that H_0 is true, the test statistic

$$T = \frac{\hat{Y}_i - E_i^0}{S_{\hat{Y}_i}} \text{ is } t_{n-2}. \tag{16.23}$$

As always, the critical region is determined by the alternative hypothesis offered (see the presentation given in Section 16.6 concerning the various possible alternative hypotheses usually made).

Example 16.7.1   Let us construct a 95% confidence band for the population regression equation given that the estimated regression equation determined from the sales/advertising data set is $\hat{Y} = 1.9379 + 1.4365X$. Using $t_{\alpha/2, n-2} = t_{0.025,8} = 2.306$ and

$$s_{\hat{Y}_i} = 0.7439\sqrt{\frac{1}{10} + \frac{(X_i - 7.210)^2}{61.7490}},$$

Table 16.2

X_i	$\hat{Y}_i$	$s_{\hat{Y}_i}$	95% Confidence Band		Width=Upper Limit Minus Lower Limit
			Lower Limit $(\hat{Y}_i - 2.306 s_{\hat{Y}_i})$	Upper Limit $(\hat{Y}_i + 2.306 s_{\hat{Y}_i})$	
4.00	7.6839	0.3843	6.7977	8.5901	1.7924
5.00	9.1204	0.3148	8.3945	9.8463	1.4518
6.00	10.5569	0.2616	9.9537	11.1602	1.2065
7.00	11.9934	0.2360	11.4492	12.5376	1.0884
$7.21(=\overline{X})$	12.2950	0.2352	11.7526	12.8374	1.0848
$8.00(=X_0)$	13.4299	0.2468	12.8608	13.9990	1.1382
9.00	14.8664	0.2899	14.1979	15.5349	1.3370
10.00	16.3029	0.3537	15.4873	17.1185	1.6312
11.00	17.7394	0.4290	16.7501	18.7289	1.9788
11.30	18.1704	0.4530	17.1258	19.2150	2.0892

we may easily complete Table 16.2 for selected X_i values and thus determine the 95% confidence band. As Figure 16.5 reveals, the confidence band is narrowest at $\overline{X}$, with the width of the individual confidence intervals increasing as we move away from $\overline{X}$ in either direction.

For instance, at a level of, say, advertising outlay of \$10,500,000 ($X_i = 10.5$), is there sufficient evidence for concluding, at the 5% level, that, on the average, gross sales will exceed \$17,000,000 ($E_i^0 = 17.0$)? Here H_0: $E_i^0 \leq 17.0$, H_1: $E_i^0 > 17.0$, and $t_{\alpha,n-2} = t_{0.05,8} = 1.86$. Since $\mathcal{R} = \{t | t > 1.86\}$ and, from (16.23), the realized $T = \frac{17.0212-17.0}{0.3903} = 0.0543 \notin \mathcal{R}$, (p-value > 0.10) it follows that we cannot reject H_0; that is, the predicted average value of gross sales does not lie far enough above $E_i^0 = 17$ to enable us to conclude that gross sales at $X_i = 10.5$ will exceed $E_i^0 = 17$. ■

16.8   The Prediction of a Particular Value of Y Given X

In the preceding section we considered the problem of predicting the average value of Y_i given X_i. Let us now look to the task of predicting a particular Y_i value from a given X_i. Specifically, for $X = X_0$, let us predict or forecast the value of the random variable $Y_0 = \beta_0 + \beta_1 X_0 + \varepsilon_0$. If the true population regression line were known, the predictor of Y_0 would be $E(Y_0|X_0) = \beta_0 + \beta_1 X_0$, a point on the population regression line corresponding to X_0. Since $E(Y_0|X_0)$ is unknown and must be estimated from the sample data, let us use as an estimator $\hat{Y}_0 = \hat{\beta}_0 + \hat{\beta}_1 X_0$, a point on the sample regression line. In this regard, if $\hat{Y}_0$ is used to estimate Y_0, then the forecast error is the random variable $Y_0 - \hat{Y}_0 = \beta_0 + \beta_1 X_0 + \varepsilon_0 - (\hat{\beta}_0 + \hat{\beta}_1 X_0)$. Since $\hat{\beta}_0$ and $\hat{\beta}_1$ are unbiased estimators of β_0 and β_1, respectively, and $E(\varepsilon_0) = 0$, the mean of the forecast error is $E(Y_0 - \hat{Y}_0) = 0$; that is, $\hat{Y}_0$ is an unbiased estimator of Y_0. (Note that $\hat{Y}_0$ is an unbiased estimator for Y_0 as well as for

$E(Y_0|X_0)$.) Moreover, it can be shown that the variance of $Y_0 - \hat{Y}_0$ is $\sigma^2_{(Y_0-\hat{Y}_0)} = \sigma^2_\varepsilon + \sigma^2_{\hat{Y}_0}$, where $\sigma^2_{\hat{Y}_0}$ is determined from (16.17) when $X_i = X_0$. Then

$$\sigma^2_{(Y_0-\hat{Y}_0)} = \sigma^2_\varepsilon \left[1 + \frac{1}{n} + \frac{(X_0 - \overline{X})^2}{\sum x_i^2} \right]. \tag{16.24}$$

Since σ^2_ε is unknown, it will be estimated by S^2_ε (see (16.7)) so that (16.24) becomes

$$S^2_{(Y_0-\hat{Y}_0)} = S^2_\varepsilon \left[1 + \frac{1}{n} + \frac{(X_0 - \overline{X})^2}{\sum x_i^2} \right]. \tag{16.25}$$

Upon taking the positive square root of this expression, we obtain the *estimated standard deviation of the forecast error* $Y_0 - \hat{Y}_0$ or

$$S_{(Y_0-\hat{Y}_0)} = S_\varepsilon \sqrt{1 + \frac{1}{n} + \frac{(X_0 - \overline{X})^2}{\sum x_i^2}}. \tag{16.26}$$

If we invoke assumption (A.1) and employ the Gauss-Markov Theorem, we see that $Y_0 - \hat{Y}_0$ is a linear combination of the normal random variables Y_0 and $\hat{Y}_0$ so that

$$Y_0 - \hat{Y}_0 \text{ is } N\left(E(Y - \hat{Y}_0), \sqrt{V(Y - \hat{Y}_0)} \right) = N\left(0, \sigma_\varepsilon \sqrt{1 + \frac{1}{n} + \frac{(X_0 - \overline{X})^2}{\sum x_i^2}} \right)$$

and thus

$$\frac{Y_0 - \hat{Y}_0}{\sigma_\varepsilon \sqrt{1 + \frac{1}{n} + \frac{(X_0 - \overline{X})^2}{\sum x_i^2}}} \text{ is } N(0,1). \tag{16.27}$$

If in (16.27) σ_ε is replaced by S_ε, the resulting quantity follows a t distribution with $n - 2$ degrees of freedom; that is,

$$\frac{Y_0 - \hat{Y}_0}{S_\varepsilon \sqrt{1 + \frac{1}{n} + \frac{(X_0 - \overline{X})^2}{\sum x_i^2}}} = \frac{Y_0 - \hat{Y}_0}{S_{(Y_0-\hat{Y}_0)}} \text{ is } t_{n-2}.$$

To determine just how precisely Y_0 has been estimated from the sample data, let us form the probability statement

$$P\left(-t_{\alpha/2,n-2} \le \frac{Y_0 - \hat{Y}_0}{S_{(Y_0-\hat{Y}_0)}} \le t_{\alpha/2,n-2} \right) = 1 - \alpha. \tag{16.28}$$

Upon pivoting and passing to sample realizations, (16.28) becomes

$$P\left(\hat{Y}_0 - t_{\alpha/2,n-2}\,S_{(Y_0-\hat{Y}_0)} \le Y_0 \le \hat{Y}_0 + t_{\alpha/2,n-2}\,S_{(Y_0-\hat{Y}_0)}\right) = 1-\alpha \qquad (16.29)$$

so that, from (16.29), a $100(1-\alpha)\%$ *confidence interval for Y_0* is

$$\left(\hat{Y}_0 - t_{\alpha/2,n-2}\,S_{(Y_0-\hat{Y}_0)},\ \hat{Y}_0 + t_{\alpha/2,n-2}\,S_{(Y_0-\hat{Y}_0)}\right). \qquad (16.30)$$

Hence (16.30) informs us that we may be $100(1-\alpha)\%$ confident that the true value of Y_0 lies within the interval determined by $\hat{Y}_0 \pm t_{\alpha/2,n-2}\,S_{(Y_0-\hat{Y}_0)}$.

Upon closely examining (16.30) we can easily see that this confidence interval can be calculated for any $X = X_0$ value. In fact, as X_0 is varied, we can generate a whole set of confidence intervals, the collection of which defines a *prediction band* about the estimated sample regression line. And since $S_{(Y_0-\hat{Y}_0)}$ varies directly with $(X_0 - \overline{X})^2$, the further X_0 lies from $\overline{X}$, the greater is the value of $s_{(Y_0-\hat{Y}_0)}$ and thus the wider the prediction band at X_0. Hence we have more confidence in predicting Y_0 near the center of the X_i's than at the extremes of the range of X values. Note also that since $s_{(Y_0-\hat{Y}_0)} > s_{\hat{Y}_i}$, the prediction band obtained by employing (16.30) is wider than the confidence band obtained from (16.22); that is, for $X = X_0$, it is intuitively clear that the prediction of an individual value of Y, namely Y_0, should have a greater error associated with it than the error arising from an estimate of the average value of Y given $X_0, E(Y_0|X_0)$.

To test the hypothesis that Y_0 equals some specific or anticipated value Y_0^0, let us form $H_0: Y_0 = Y_0^0$. If H_0 is true, the statistic

$$T = \frac{\hat{Y}_0 - Y_0^0}{S_{(Y_0-\hat{Y}_0)}} \text{ is } t_{n-2}. \qquad (16.31)$$

Here, too, the critical region is determined by the alternative hypothesis made (see Section 16.6).

Example 16.8.1   Let us again look to the gross sales/advertising outlay data presented in Table 16.1. If $X_0 = 8$, determine a 95% prediction interval for Y_0. Since $t_{\alpha/2,n-2} = t_{0.025,8} = 2.306$, $\hat{Y}_0 = \hat{\beta}_0 + \hat{\beta}_1 X_0 = 1.9379 + 1.4365(8) = 13.4299$, and

$$S_{(Y_0-\hat{Y}_0)} = 0.7439\sqrt{1 + \frac{1}{10} + \frac{(8-7.210)^2}{61.7490}} = 0.7847,$$

it follows that

$$\hat{Y}_0 - t_{\alpha/2,n-2}\,S_{(Y_0-\hat{Y}_0)} = 13.4299 - 2.306(0.7847) = 11.6203,$$

$$\hat{Y}_0 + t_{\alpha/2,n-2}\,S_{(Y_0-\hat{Y}_0)} = 13.4299 + 2.306(0.7847) = 15.2394.$$

Thus we may be 95% confident that the true value of Y_0 lies between 11.6203 and 15.2394 (see Figure 16.5). If we vary X_0, then we may construct the entire prediction band (see the dashed lines in Figure 16.5) in a fashion similar to that used to derive the confidence band depicted in Table 16.2. ∎

Example 16.8.2   At the 5% level, can we conclude that, for $X_0 = 10$, the predicted value of Y does not exceed 18. Here $H_0: Y_0 \geq 18$ is tested against $H_1: Y_0 < 18$. With $\hat{Y}_0 = 1.9379 + 1.4365(10) = 16.3029$ and $\mathcal{R} = \{t | t < -1.86\}$, we see that, from (16.31), the realized $T = \frac{16.3029 - 18}{0.8237} = -2.0603 \in \mathcal{R}$ ($0.025 <$ p-value < 0.05) so that we will reject H_0 in favor of H_1; that is, Y_0 lies significantly below $Y_0^0 = 18$ at the 5% level. ∎

Example 16.8.3   If we now return to the question posed at the beginning of the final paragraph of Section 16.3, it can be seen that to answer the same, we must determine the precision of the predicted value of Y. We previously found that, on the basis of the estimated linear relationship between X and Y for 1995 to 2004, when $X_0 = 12$, it follows that $\hat{Y}_0 = 19.1759$. Then the 95% lower and upper prediction limits for Y_0 (the true Y value at X_0) are, respectively,

$$\hat{Y}_0 - t_{\alpha/2, n-2} \, s_{(Y_0 - \hat{Y}_0)} = 19.1759 - 2.306(0.9024) = 17.0949,$$

$$\hat{Y}_0 + t_{\alpha/2, n-2} \, s_{(Y_0 - \hat{Y}_0)} = 19.1759 + 2.306(0.9024) = 21.2568.$$

Hence we may be 95% confident that the population value of Y given $X_0 = \$12,000,000$ lies between $17,094,900$ and $21,256,800$. ∎

One important application of the statistic $\frac{Y_0 - \hat{Y}_0}{s_{(Y_0 - \hat{Y}_0)}}$ is the determination of whether or not a new observed data point comes from the same linear relationship as that obtained from the original set of sample points. For example, let us assume that the aforementioned new advertising campaign for 2005 is actually undertaken and that for this particular year the company finds that with an advertising outlay of 12 million dollars, the realized or actual level of gross sales is 22 million dollars. Does this increase in gross sales indicate that there has been a statistically significant change in the underlying linear relationship, or did this additional sample point come from the same linear relationship as before? To determine whether or not there has been a change in the basic structure of the gross sales/advertising outlay equation, let $X_0 = 12$, $Y_0 = 22$, and $\hat{Y}_0 = 19.1759$. At the 5% level, is there sufficient evidence to indicate that Y_0 is significantly greater than $\hat{Y}_0$? Let us test $H_0: Y_0 - \hat{Y}_0 = 0$ against $H_1: Y_0 - \hat{Y}_0 > 0$. Since $\mathcal{R} = \{t | t > 1.86\}$ and, under H_0, the realized $T = \frac{Y_0 - \hat{Y}_0}{s_{(Y_0 - \hat{Y}_0)}} = \frac{22 - 19.1759}{0.9024} = 3.1295 \in$ $\mathcal{R}(0.005 < p$-value $< 0.01)$, we reject H_0 in favor of H_1 and conclude that the underlying linear relationship that held from 1995 to 2004 did not generate the observed pair (12,22).

16.9   Decomposition of the Sample Variation of Y

One useful device for summarizing the regression results is provided by the analysis-of-variance table depicting what we shall call the *partitioned sum of squares*. The purpose of this construct is to ultimately determine why the values of Y change between successive X values. If we invoke assumptions B.1, B.2, and B.4 (see Section 16.2), then the variation in Y can be attributed partly to changes in the X values and partly to the effect of the random disturbance term ε. In this regard, we may ask, how much of the variation in Y can be attributed to the systematic influence of X and how much is attributed to the random effect of the disturbance term?

To answer this question let us begin by determining the sample variation of Y about its mean $\overline{Y}$ (if there is no variation in Y as X increases in value, then all of the Y values lie on the horizontal line $Y = \overline{Y}$). As a concise summary measure of the total variation of the Y_i values about $\overline{Y}$, let us use

$$\sum (Y_i - \overline{Y})^2 = \sum y_i^2. \tag{16.32}$$

For $X = X_i$, Figure 16.6 reveals that $Y_i - \overline{Y} = (\hat{Y}_i - \overline{Y}) + e_i$. Upon substituting this expression into (16.32) we obtain

$$\underbrace{\sum (Y_i - \overline{Y})^2}_{\substack{\text{[Total sum of}\\\text{squares } (SST)]}} = \underbrace{\sum (\hat{Y}_i - \overline{Y})^2}_{\substack{\text{[Regression sum}\\\text{of squares } (SSR)]}} + \underbrace{\sum e_i^2}_{\substack{\text{[Error sum}\\\text{of squares}\\(SSE)]}} + \underbrace{2 \sum (\hat{Y}_i - \overline{Y}) e_i}_{(=0)} \tag{16.33}$$

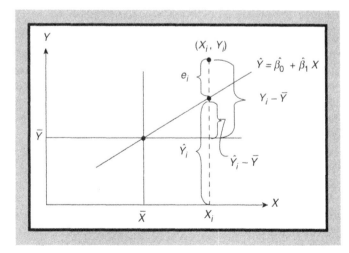

Figure 16.6   Decomposition of $Y_i - \overline{Y}$ into e_i and $\hat{Y}_i - \overline{Y}$.

This expression reveals that the total sum of squares may be partitioned into two parts: (1) the regression sum of squares SSR (which reflects the variation in Y attributed to the linear influence of X); and (2) the error sum of squares SSE (which depicts the variation in Y ascribed to random factors). In this regard, SSR is called the *explained sum of squares (explained SS)* and SSE is referred to as the *unexplained sum of squares (unexplained SS)*.

As far as the computation of the various sums of squares are concerned,

$$SST = \sum y_i^2,$$

$$SSR = \hat{\beta}_1^2 \sum x_i^2 \left(\text{or} = \hat{\beta}_1 \sum x_i y_i\right), \qquad (16.34)$$

$$SSE = \sum y_i^2 - \hat{\beta}_1 \sum x_i y_i.$$

One important application of these sums of squares is the specification of what is called the *sample coefficient of determination*

$$R^2 = \frac{SSR}{SST} = \frac{\text{explained } SS}{\text{total } SS} = \frac{\hat{\beta}_1 \sum x_i y_i}{\sum y_i^2} \qquad (16.35a)$$

or, since $SSR = SST - SSE$,

$$R^2 = 1 - \frac{SSE}{SST} = 1 - \frac{\text{unexplained } SS}{\text{total } SS} = 1 - \frac{\sum y_i^2 - \hat{\beta}_1 \sum x_i y_i}{\sum y_i^2}, \qquad (16.35b)$$

where r^2 denotes its realized value. Here R^2 serves as a measure of *goodness of fit*—it represents the proportion of the total variation in Y that can be explained by the linear influence of X. (When the number of degrees of freedom is small, R^2 is biased upward. To correct for degrees of freedom and thus for the bias, let us compute the *adjusted coefficient of determination* $\bar{R}^2 = 1 - \left(\frac{SSE}{SST}\right)\left[\frac{n-1}{n-2}\right]$. Clearly $\frac{n-1}{n-2} > 1$ so that $\bar{R}^2 < R.^2$ And as $n \to \infty, \bar{R}^2 - R^2 \to 0$.) From (16.35.b) it is evident that $0 \le R^2 \le 1$; that is, when $\sum e_i^2 = \sum y_i^2$ (the sample regression line is $\hat{Y} = \bar{Y}$ and the explained variation in Y or SSR is zero), it follows that $R^2 = 0$ (see Figure 16.7a); and for $\sum e_i^2 = 0$ (the observed points all lie on the sample regression line and the unexplained variation in Y or SSE is zero), we see that $R^2 = 1$ (see Figure 16.7b). The only time that R^2 is undefined is when $\sum y_i^2 = 0$; that is, there is no variation in Y (see Figure 16.7c).

We noted earlier that R^2 is a measure of goodness of fit. Hence $R^2 = 0$ may reflect the fact that a linear function provides a poor fit (here $\sum e_i^2 = \sum y_i^2$) to an essentially nonlinear scatter of points (see Figure 16.7d).

Based upon the sums of squares presented in (16.34), we may construct the *partitioned sums-of-squares (or analysis-of-variance) table* for our regression results as seen in Table 16.3.

Note that the error mean square presented in Table 16.3 is just S_ε^2. So if the underlying population regression equation is truly linear, the error mean square

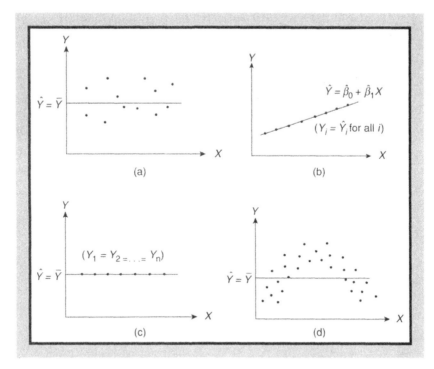

Figure 16.7   Selected special cases for the value of R^2. (See text for details of a–d).

Table 16.3

	Summary Table for the Partitioned Sums of Squares		
Source of Variation	**Sum of Squares (SS)**	**Degrees of Freedom (d.f.)**	**Mean Square (MS) = SS/d.f.**
Regression (explained variability)	(SSR) $\hat{\beta}_1 \sum x_i y_i$	1	(Regression MS) $SSR/1 = \hat{\beta}_1 \sum x_i y_i / 1$ $= \hat{\beta}_1 \sum x_i y_i$
Error (unexplained variability)	(SSE) $\sum y_i^2 - \hat{\beta}_1 \sum x_i y_i$	$n - 2$	(Error MS) $SSE/(n - 2) =$ $\left(\sum y_i^2 - \hat{\beta}_1 \sum x_i y_i \right) / (n - 2)$
Total variability in Y	(SST) $\sum y_i^2$	$n - 1$	$R^2 = SSR/SST;$ $F = \dfrac{\text{Regression MS}}{\text{Error MS}}$ $= [SSR/1] / [SSE/(n - 2)]$

is an estimate of σ_ε^2. Alternatively, the regression mean square term provides us with an estimate of σ_ε^2 only if $H_0: \beta_1 = 0$ is true; that is, only if X is of no use in explaining the variation in Y.

The second important application of the partitioned sums of squares notion is to determine whether or not X contributes significantly to the variation in Y. If $H_0: \beta_1 = 0$ is true (there is no linear relationship between X and Y), then the sole source of variation in Y is the random disturbance term ε since the population regression sum of squares is zero. Now, it can be shown that the statistic

$$\frac{(\hat{\beta}_1 - \beta_1)^2 \sum x_i^2}{\sum e_i^2/(n-2)} \text{ is } F_{1,n-2}.$$

Under $H_0: \beta_1 = 0$, the preceding expression becomes

$$
\begin{aligned}
\frac{\hat{\beta}_1 \sum x_i y_i}{S_\varepsilon^2} &= \frac{\text{regression mean square}}{\text{error mean square}} \\
&= \frac{SSR/1}{SSE/(n-2)} = \frac{R^2/1}{(1-R^2)/(n-2)} \text{ is } F_{1,n-2}.
\end{aligned}
\tag{16.36}
$$

Here the appropriate alternative hypothesis is $H_1: \beta_1 \neq 0$ so that $\mathcal{R} = \left\{ F / \frac{SSR/1}{SSE/(n-2)} > f_{1-\alpha,1,n-2} \right\}$ provides us with a one-tail alternative on the upper tail of the F-distribution. So if the realized value of F exceeds the tabular value, we reject H_0 in favor of H_1 and conclude that there exists a statistically significant linear relationship between X and Y; that is, the observed linear relationship between X and Y did not arise solely because of chance or random factors. It is interesting to note that if we square the T-statistic $T = \frac{\hat{\beta}_1}{S_{\hat{\beta}_1}}$, we obtain the F-statistic presented in (16.36). Thus, under $H_0: \beta_1 = 0$,

$$T^2 = \left(\frac{\hat{\beta}_1}{S_{\hat{\beta}_1}} \right)^2 = \frac{\hat{\beta}_1^2}{S_\varepsilon^2/\sum x_i^2} = \frac{\hat{\beta}_1 \sum x_i y_i}{S_\varepsilon^2} = F.$$

For instance, if $\alpha = 0.05$ and $n - 2 = 10$, then the one-tail F-value $f_{0.95,1,10} = 4.96$ is equivalent to the two-tail t value squared or $(t_{0.025,10})^2 = (2.228)^2 = 4.9639$. Hence the preceding one-sided F-test of $H_0: \beta_1 = 0$ versus $H_1: \beta_1 \neq 0$ is equivalent to the two-sided t test of $H_0: \beta_1 = 0$ versus $H_1: \beta_1 \neq 0$.

Example 16.9.1   Using the preceding set of gross sales/advertising outlay data, construct the analysis-of-variance table for the partitioned sum of squares. What inferences can be made from the information contained within this table? Using Table 16.3 as our guide, we obtain Table 16.3.1.

Table 16.3.1

Summary Table for the Partitioned Sums of Squares			
Source of Variation	**Sum of Squares (SS)**	**Degrees of Freedom (d.f.)**	**Mean Square (MS) = SS/d.f.**
Regression (explained variability)	(SSR) $\hat{\beta}_1 \sum x_i y_i = 1.4365(88.7010)$ $= 4.4266$	1	(Regression MS) $\dfrac{SSR}{1} = 127.4190$
Error (unexplained variability)	(SSE) $\sum y_i^2 - \hat{\beta}_1 \sum x_i y_i$ $= 131.8456 - 127.4190$ $= 4.4266$	$n - 2 = 8$	(Error MS) $\dfrac{SSE}{n-2} = \dfrac{4.4266}{8} = 0.5533$
Total variability in Y			$r^2 = SSR/SST = \dfrac{127.4190}{131.8456}$ $= 0.9664;$
	(SST) $\sum y_i^2 = 131.8456$	$n - 1 = 9$	$F = \dfrac{\text{Regression MS}}{\text{Error MS}} = \dfrac{127.4190}{0.5533}$ $= 230.2892$

Since $r^2 = 0.9664$, we see that approximately 97% of the variation in Y may be explained by the linear influence of X while about 3% is left unexplained. In addition, we may look to the question of whether or not there exists a statistically significant linear relationship between X and Y. That is, for $\alpha = 0.05$, let $H_0: \beta_1 = 0$ and $H_1: \beta_1 \neq 0$ with $\mathcal{R} = \{F | F > f_{1-\alpha,1,n-2} = f_{0.95,1,8} = 5.32\}$. Since the realized $F = $ regression MS/error $MS = 230.2892 \in \mathcal{R}$, we may conclude that the linear relationship between X and Y is highly significant at the 5% level. ■

16.10   The Correlation Model

The purpose of correlation analysis is: (Case A) to determine the degree of covariability between two random variables X and Y (here both X and Y are subject to random errors); or (Case B) if only Y is assumed random and is regressed on X, with the values of the latter variable fixed (see Section 16.2), then correlation serves to measure the goodness of fit of the sample linear regression equation to the scatter of observations on X and Y.

If Case A is of interest, then we need to determine the direction as well as the strength (i.e., the degree of closeness) of the relationship between the random variables X and Y, where X and Y follow a joint bivariate distribution. This will be accomplished by extracting a sample consisting of the n pairs of values $(X_i, Y_i), i = 1, \ldots, n$, from the said distribution. Moreover, once we compute a sample correlation coefficient, we must determine whether or not it serves as a *good* estimate of the underlying degree of covariation within the population.

For Case B, we need to regress Y on X given the scatter of the n sample points $(X_i, Y_i), i = 1, \ldots, n$, under the assumptions of the strong classical linear regression model. Once we obtain the line of best fit and construct the analysis-of-variance table, we will be able to directly determine the sample correlation coefficient as well as test how *good* this measure is as an indicator of goodness of fit of the least squares regression line to the set of observations on X and Y.

To address the Case A problem, let X and Y be random variables, which follow a joint bivariate distribution. If: $E(X)$ and $E(Y)$ depict the means of X and Y, respectively; $S(X)$ and $S(Y)$ represent the standard deviations of X and Y, respectively; and $COV(X, Y)$ denotes the covariance between X and Y, then the *population correlation coefficient*, which serves as a measure of the linear association between X and Y, may be depicted as

$$\rho = \frac{COV(X, Y)}{S(X)S(Y)} = \frac{E\left([X - E(X)][Y - E(Y)]\right)}{S(X)S(Y)}. \tag{16.36}$$

(If we form the standardized variables $\tilde{X} = \frac{X - E(X)}{S(X)}$ and $\tilde{Y} = \frac{Y - E(Y)}{S(Y)}$, then (16.36) can be rewritten as $\rho = COV(\tilde{X}, \tilde{Y})$; that is, we may think of the population correlation coefficient as simply the covariance of the two standardized variables $\tilde{X}$ and $\tilde{Y}$.) As far as the properties of ρ are concerned: it is symmetrical with respect to X and Y (the correlation between X and Y is the same as that between Y and X); it is dimensionless (i.e., a pure number); it is independent of the units or the scale of measurement used; and, by incorporating in its calculation deviations from the means of X and Y, it shifts the origin of the population values to the means of X and Y.

The value of the population correlation coefficient exhibits both the direction and the strength of the linear relationship between the random variables X and Y. That is, if $\rho > 0$, both variables tend to increase or decrease together (we have a *direct relationship* between X and Y), and if $\rho < 0$, an increase in one variable is accompanied by a decrease in the other (there exists an *inverse relationship* between X and Y). Clearly the sign of ρ is determined by the sign of $COV(X, Y)$.

As far as the range of values assumed by ρ is concerned, it can be shown that $|\rho| \leq 1$ or $-1 \leq \rho \leq 1$. In this regard, for $\rho = 1$ or -1, we have perfect positive association or perfect negative association, respectively. If $\rho = 0$, the variables are uncorrelated, thus indicating the absence of any linear relationship between X and Y. (It is important to remember that ρ depicts the strength of the linear relationship between X and Y. If X and Y are independent random variables, then $\rho = 0$ since $COV(X, Y) = 0$. However, the converse is not true; that is, we cannot legitimately infer that X and Y are independent if $\rho = 0$

since the true underlying relationship between X and Y may be highly nonlinear.) And as ρ increases (respectively, decreases) in value from 0 to 1 (respectively, from 0 to -1) the closeness or strength of the linear relationship between X and Y concomitantly increases. It is important to remember that because two random variables may be highly correlated, the association between them does not allow us to infer anything about cause and effect (both X and Y may be related to a third *unobserved* variable, say, W, which *causes* movements in both X and Y) or to predict values of one variable from the other. In addition, no functional relationship between X and Y (as was the case for the preceding regression model involving X and Y) is assumed.

16.11   Estimating the Population Correlation Coefficient ρ

If a sample of size n consisting of the pairs of observations (X_i, Y_i), $i = 1,\ldots,n$, is extracted from the underlying joint bivariate population of true X and Y values, then we may estimate ρ by using the *sample correlation coefficient* (called the *Pearson product-moment coefficient of correlation*)

$$\hat{\rho} = \frac{\frac{1}{n-1}\sum(X_i - \bar{X})(Y_i - \bar{Y})}{\sqrt{\frac{1}{n-1}\sum(X_i - \bar{X})^2}\sqrt{\frac{1}{n-1}\sum(Y_i - \bar{Y})^2}} = \frac{\sum x_i y_i}{\sqrt{\sum x_i^2 \sum y_i^2}}. \qquad (16.37)$$

Although $\hat{\rho}$ is a slightly biased (downward) estimator of ρ for small samples, it is a consistent as well as sufficient estimator. (An unbiased estimator of ρ is $\hat{\rho}^* = \hat{\rho}[1 + (1 - \hat{\rho}^2)/2(n - 4)]$.) Our interpretation of $\hat{\rho}$ is the same as the one advanced earlier for ρ. In this regard, as Figure 16.8a,b reveals, if $\hat{\rho} = 1$ (respectively, -1), we have perfect positive (respectively, negative) linear association between X and Y (here all sample points lie on some imaginary positively or negatively sloped line); if $|\hat{\rho}| < 1$, the random variables are linearly related but to a lesser degree (see Figure 16.8c,d); and if $\hat{\rho} = 0$ (see Figure 16.8e), X and Y are not linearly related at all. And as noted earlier, if $\hat{\rho} = 0$ we may not legitimately conclude that X and Y are not related—the true relationship may be nonlinear (see Figure 16.8f).

Example 16.11.1   Appearing in Table 16.4 is a random sample of $n = 10$ observations taken from a joint bivariate distribution for the random variables X and Y. Are these variables linearly related? If so, in what direction?

If the totals of columns 5, 6, and 7 are inserted into (16.37) we have the realized

$$\hat{\rho} = \frac{\sum x_i y_i}{\sqrt{\sum x_i^2}\sqrt{\sum y_i^2}} = \frac{-22.9870}{\sqrt{7.3690}\sqrt{158.1601}} = -0.6741.$$

With the value of $\hat{\rho} < 0$, it follows that X and Y vary inversely.   ∎

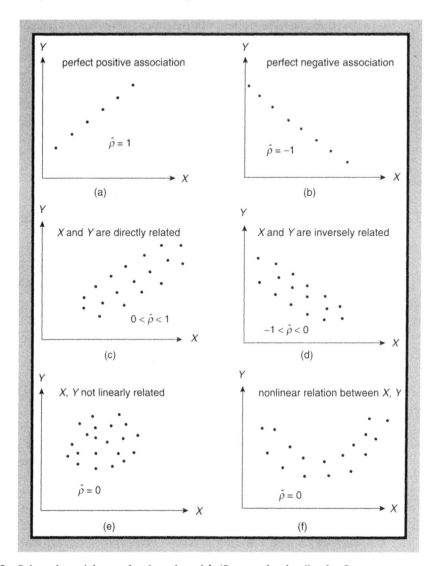

Figure 16.8 Selected special cases for the value of $\hat{\rho}$. (See text for details of a–f).

16.12 Inferences about the Population Correlation Coefficient ρ

The population correlation coefficient ρ may be estimated by $\hat{\rho}$ from a set of observations taken from any joint bivariate distribution relating the random variables X and Y. However, if we want to test hypotheses about ρ or determine confidence intervals for this parameter, then we must strengthen our assumptions concerning the bivariate population at hand. In particular, we shall assume

Table 16.4

(1)	(2)	(3)	(4)	(5)	(6)	(7)
		$x_i = X_i - \overline{X}$ $= X_i - 4.61$	$y_i = Y_i - \overline{Y}$ $= Y_i - 85.27$			
X	Y			x_i^2	y_i^2	$x_i y_i$
4.5	89.5	−0.11	4.23	0.0121	17.8929	−0.4653
3.8	91.1	−0.81	5.83	0.6561	33.9889	−4.7223
3.8	86.9	−0.81	1.63	0.6561	2.6569	−1.3203
3.6	87.0	−1.01	1.73	1.0201	2.9929	−1.7473
3.5	86.2	−1.11	0.93	1.2321	0.8649	−1.0323
4.9	79.2	0.29	−6.07	0.0841	36.0840	−1.7603
5.9	78.0	1.29	−7.27	1.6641	52.8529	−9.3783
5.6	83.1	0.99	−2.17	0.9801	4.7089	−2.1483
4.9	87.5	0.29	2.23	0.0841	4.9729	0.6467
5.6	84.2	0.99	−1.07	0.9801	1.1449	−1.0593
$\sum X_i = 46.1$	$\sum Y_i = 852.7$	$\sum x_i = 0$	$\sum y_i = 0$	$\sum x_i^2 =$ 7.3690	$\sum y_i^2 =$ 158.1601	$\sum x_i y_i =$ −22.9870

that X and Y follow a joint bivariate normal distribution (see Appendix 7.B). In brief, its specific properties are:

(C.1) Both X and Y are random variables with means μ_X and μ_Y, respectively, and possess constant variances σ_X^2 and σ_Y^2, respectively.

(C.2) The individual marginal distributions of both X and Y are $N(\mu_X, \sigma_X)$ and $N(\mu_Y, \sigma_Y)$, respectively. (It must be mentioned that if the individual marginal distributions of two random variables X and Y are normal, it does not necessarily follow that the joint bivariate distribution of X and Y will be normal).

(C.3) The conditional distribution of Y given X is

$$N\left(\mu_X + \rho\sigma_X \tilde{Y}, \sigma_X\sqrt{1 - \rho^2}\right)$$

and the conditional distribution of X given Y is

$$N\left(\mu_Y + \rho\sigma_Y \tilde{X}, \sigma_Y\sqrt{1 - \rho^2}\right),$$

where $\tilde{X} = \frac{X - \mu_X}{\sigma_X}$ and $\tilde{Y} = \frac{Y - \mu_Y}{\sigma_Y}$.

(C.4) The relationship between X and Y is strictly linear and is summarized by the correlation coefficient ρ (we will return to a formal interpretation of C.3 and C.4 when Case B is considered later).

We noted earlier that if X and Y are independent, then $\rho = 0$. It was also mentioned that if X and Y follow a joint bivariate distribution and $\rho = 0$, then we cannot generally conclude that X and Y are independent. Interestingly enough, for a joint bivariate *normal* distribution, independence implies $\rho = 0$ and conversely; that is, zero covariance is equivalent to independence.

Given that X and Y follow a joint bivariate normal distribution, it follows that under $H_0: \rho = 0$,

$$\hat{\rho} \text{ is } N\left(E(\hat{\rho}), \sqrt{V(\hat{\rho})}\right) = N\left(0, \sqrt{\frac{1 - \rho^2}{n - 2}}\right).$$

Then

$$\frac{\hat{\rho} - E(\hat{\rho})}{S(\hat{\rho})} = \frac{\hat{\rho}\sqrt{n - 2}}{\sqrt{1 - \rho^2}} \text{ is } N(0, 1). \tag{16.38}$$

If $\hat{\rho}$ is used to estimate ρ in (16.38), then the resulting quantity

$$T_{\hat{\rho}} = \frac{\hat{\rho}\sqrt{n - 2}}{\sqrt{1 - \hat{\rho}^2}} \text{ is } t_{n-2}. \tag{16.39}$$

We noted earlier that *lack of linear association* is equivalent to independence if X and Y follow a joint bivariate normal distribution. Hence testing lack of linear association is equivalent to testing the independence of X and Y. In this regard, we may test the null hypothesis $H_0: \rho = \rho_0 = 0$ against any of the following alternative hypotheses

Case I	Case II	Case III
$H_0: \rho = 0$	$H_0: \rho = 0$	$H_0: \rho = 0$
$H_1: \rho > 0$	$H_1: \rho < 0$	$H_1: \rho \neq 0$

The corresponding critical regions are determined in the usual fashion; that is, at the $100\alpha\%$ level, our set of decision rules for rejecting H_0 relative to H_1 is:

(a) Case I—reject H_0 if $t_{\hat{\rho}} > t_{\alpha,n-2}$;
(b) Case II—reject H_0 if $t_{\hat{\rho}} < -t_{\alpha,n-2}$; and
(c) Case III—reject H_0 if $|t_{\hat{\rho}}| > t_{\alpha/2,n-2}$,

where $t_{\hat{\rho}}$ is the sample realization of $T_{\hat{\rho}}$.

The test statistic presented in (16.39) is valid only under $H_0: \rho = \rho_0 = 0$. It cannot be used for testing any other hypothesis concerning ρ. This is because when $\rho_0 \neq 0$, the sampling distribution of $\hat{\rho}$ is highly skewed. To circumvent this problem, let us construct an expression involving $\hat{\rho}$ that may be used to transform $\hat{\rho}$ into a test statistic similar to (16.38). To this end we may note that for moderately large samples ($n \geq 20$) taken from a joint bivariate normal population, the function (known as *Fisher's ξ transformation*)

$$\xi = \frac{1}{2} \log_e \left(\frac{1 + \hat{\rho}}{1 - \hat{\rho}} \right) \tag{16.40}$$

is approximately

$$N \left(E(\xi), \sqrt{V(\xi)} \right) = N \left(\frac{1}{2} \log_e \left(\frac{1 + \rho}{1 - \rho} \right), \sqrt{\frac{1}{n - 3}} \right)$$

for any admissible value of ρ. Hence the quantity

$$Z = \frac{\xi - E(\xi)}{S(\xi)} \quad \text{is approximately } N(0, 1) \tag{16.41}$$

and thus, for $H_0: \rho = \rho_0 \neq 0$ (where ρ_0 is some specific or anticipated nonzero value of ρ),

$$\xi_0 = \frac{1}{2} \log_e \left(\frac{1 + \rho_0}{1 - \rho_0} \right)$$

so that we can use the test statistic

$$Z_\xi = \frac{\xi - E(\xi)}{S(\xi)} = \frac{\frac{1}{2} \log_e \left(\frac{1+\hat{\rho}}{1-\hat{\rho}} \right) - \frac{1}{2} \log_e \left(\frac{1+\rho_0}{1-\rho_0} \right)}{1/\sqrt{n - 3}}. \tag{16.42}$$

Again we may test the null hypothesis $H_0: \rho = \rho_0 \neq 0$ against any of the following alternatives

Case I	Case II	Case III
$H_0: \rho = \rho_0$	$H_0: \rho = \rho_0$	$H_0: \rho = \rho_0$
$H_1: \rho > \rho_0$	$H_1: \rho < \rho_0$	$H_1: \rho \neq \rho_0$

Then for a test conducted at the $100\alpha\%$ level, our decision rules for rejecting H_0 in favor of H_1 are:

(a) Case I—reject H_0 if $z_\xi > z_\alpha$;
(b) Case II—reject H_0 if $z_\xi < -z_\alpha$; and
(c) Case III—reject H_0 if $|z_\xi| \geq z_{\alpha/2}$,

where z_ξ is the sample realization of Z_ξ. Note that as far as this general test procedure is concerned (it will also be valid when we construct confidence intervals for ρ later on), since ξ is a one-to-one and onto mapping (for each $\hat{\rho}$ value there exists a unique ξ value and conversely), we can convert a $\hat{\rho}$ value into a ξ value, make an inference about ξ, and then transform the inference concerning ξ back into an inference about ρ.

Example 16.12.1 Given the data appearing in Table 16.4, does our estimate for $\rho, \hat{\rho} = -0.6741$, lie significantly below zero at the 5% level? In addition, are we safe in concluding that $\rho = -0.7$ at this same level? Under H_0: $\rho = 0$, let H_1: $\rho < 0$. Then from (16.39), since

$$|t_{\hat{\rho}}| = \left| \frac{-0.6741\sqrt{8}}{\sqrt{1 - 0.4544}} \right| = 2.5814 \in \mathcal{R} = \{t_{\hat{\rho}} \mid t_{\hat{\rho}} < -1.86\},$$

we may reject H_0 in favor of H_1; that is, at the 5% level the population correlation coefficient lies significantly below zero. (Here $0.01 < p$-value < 0.025.) Next, let us test $H_0 : \rho = \rho_0 = -0.7$ against $H_1 : \rho \neq \rho_0 = -0.7$. Since

$$\xi = \frac{1}{2} \log_e \left(\frac{1 - 0.6741}{1 + 0.6741} \right) = -0.8180$$

and

$$\xi_0 = \frac{1}{2} \log_e \left(\frac{1 - 0.7}{1 + 0.7} \right) = -0.8673$$

(see Table A.9 of the Appendix), we have, from (16.42),

$$|z_{\xi}| = \left| \frac{-0.8180 + 0.8673}{1/\sqrt{7}} \right| = 0.1305 \notin \mathcal{R} = \{z_{\xi} \mid |z_{\xi}| > 1.96\}.$$

Thus we cannot reject H_0 in favor of H_1—at the 5% level, ρ is not significantly different from -0.7. Moreover, p-value $= 0.8966$. ∎

To determine just how precisely the population correlation coefficient has been estimated from the sample data, let us construct the probability statement

$$P\left(-z_{\alpha/2} \leq \frac{\xi - E(\xi)}{S(\xi)} \leq z_{\alpha/2} \right) \approx 1 - \alpha.$$

Upon rearranging this expression and passing to sample realizations we have the confidence statement

$$P\left(\frac{1}{2} \log_e \left(\frac{1 + \hat{\rho}}{1 - \hat{\rho}} \right) - \frac{z_{\alpha/2}}{\sqrt{n-3}} \leq \frac{1}{2} \log_e \left(\frac{1 + \rho}{1 - \rho} \right) \right.$$
$$\left. \leq \frac{1}{2} \log_e \left(\frac{1 + \hat{\rho}}{1 - \hat{\rho}} \right) + \frac{z_{\alpha/2}}{\sqrt{n-3}} \right) \approx 1 - \alpha.$$

Hence an (approximate) $100(1 - \alpha)\%$ confidence interval for the parameter $\frac{1}{2} \log_e \left(\frac{1+\rho}{1-\rho} \right)$ is

$$\left(\frac{1}{2} \log_e \left(\frac{1 + \hat{\rho}}{1 - \hat{\rho}} \right) - \frac{z_{\alpha/2}}{\sqrt{n-3}}, \quad \frac{1}{2} \log_e \left(\frac{1 + \hat{\rho}}{1 - \hat{\rho}} \right) + \frac{z_{\alpha/2}}{\sqrt{n-3}} \right). \qquad (16.43)$$

If the lower and upper confidence limits indicated in (16.43) are denoted as l_ξ and u_ξ, respectively, then these quantities can be used to determine the lower and upper confidence limits l_ρ and u_ρ, respectively, for the parameter ρ by reversing the ξ-transform; that is, upon setting

$$l_\xi = \frac{1}{2} \log_e \left(\frac{1 + l_\rho}{1 - l_\rho} \right), \quad u_\xi = \frac{1}{2} \log_e \left(\frac{1 + u_\rho}{1 - u_\rho} \right) \qquad (16.44)$$

we may read Table A.9 in reverse to find a $100(1 - \alpha)\%$ confidence interval for ρ. That is, if we find l_ξ (respectively, u_ξ) in the body of this table, we can easily determine the corresponding l_ρ (respectively, u_ρ) value so that the (approximate) *100(1 − α)% confidence interval for ρ is (l_ρ, u_ρ).*

Example 16.12.2 Using the data presented in Table 16.4, find a 95% confidence interval for the true or population correlation coefficient ρ given that $\hat{\rho} = -0.6741$. Using (16.43) it is easily seen that an (approximate) 95% confidence interval for $\frac{1}{2} \log_e \left(\frac{1+\rho}{1-\rho} \right)$ is, for $z_{\alpha/2} = z_{0.025} = 1.96$,

$$\left[\frac{1}{2} \log_e \left(\frac{1 + 0.6741}{1 - 0.6741} \right) - \frac{1.96}{\sqrt{7}}, \quad \frac{1}{2} \log_e \left(\frac{1 + 0.6741}{1 - 0.6741} \right) + \frac{1.96}{\sqrt{7}} \right]$$

or, using Table A.9, $l_\xi = -0.8180 - 0.7408 = -1.5588$, $u_\xi = -0.8180 + 0.7408 = -0.0772$. Then

$$-1.5588 = \frac{1}{2} \log_e \left(\frac{1 + l_\rho}{1 - l_\rho} \right),$$

$$-0.0772 = \frac{1}{2} \log_e \left(\frac{1 + u_\rho}{1 - u_\rho} \right)$$

and thus a 95% confidence interval for ρ has as its lower and upper limits $l_\rho = -0.9150$ and $u_\rho = -0.0770$, respectively (here we find 1.5588 and 0.0772 in the body of Table A.9 and read it in reverse so as to obtain the appropriate correlation coefficient values). Thus we may be 95% confident that the population correlation coefficient lies between -0.9150 and -0.0770. ∎

Let us now consider the Case B problem. Our preceding discussion concerning ρ specified that the random variables X and Y followed a joint bivariate distribution (Case A). We then strengthened this assumption by introducing the concept of normality. In this regard, let us, for the moment, retain the requirement that X and Y follow a joint bivariate normal distribution. Since both X and Y are variable, we can regress Y on X and X on Y. In the first instance the X_i's are held fixed (Y is the dependent variable), and in the second the Y_i's are taken to be fixed (X is now the dependent variable). That is, we want the mean of Y given X as well as the mean of X given Y. Hence the implied population regression equations are, respectively,

$$\text{(a)} \quad E(Y|X) = \beta_0 + \beta_1 X,$$
$$\text{(b)} \quad E(X|Y) = \bar{\beta}_0 + \bar{\beta}_1 Y.$$

$$(16.45)$$

To see this let us refer to (C.3) and (C.4). As indicated therein, we may form, for instance,

$$E(Y|X) = \mu_Y + \rho\sigma_Y\tilde{X} = \mu_Y + \rho\sigma_Y\left(\frac{X-\mu_X}{\sigma_X}\right) = \left[\mu_Y - \rho\sigma_Y\frac{\mu_X}{\sigma_X}\right] + \rho\frac{\sigma_Y}{\sigma_X}X$$

$$(16.46)$$

(see Section 16.8). In this regard, if X and Y follow a joint bivariate normal distribution, then ρ^2 can be used as a measure of the goodness of fit of the population regression line to the data points $(X_i, Y_i), i = 1, \ldots, n$, in that it indicates the proportion of the variation in Y explained by the linear influence of X; it also serves as a measure of the covariability between the random variables X and Y. This same conclusion is valid if the assumption of normality is dropped and the random variables X and Y simply follow a joint bivariate distribution. Moreover, if the assumptions of the strong classical linear regression model hold so that only Y is a random variable and X is held fixed, then, since $COV(X, Y)$ does not exist, ρ^2 (or ρ) cannot serve as a measure of covariability—it is only indicative of the goodness of fit of the population regression line to the scatter of data points. In this regard, if the strong classical linear regression model is applied and we have obtained R^2 according to $\frac{SSR}{SST} = \frac{\hat{\beta}_1\sum x_i y_i}{\sum x_i^2}$ (see Table 16.3 of Section 16.8), then we also have obtained R^2 as an estimate of ρ^2. However, R does not serve as an estimate of ρ in (16.36) since the latter is, strictly speaking, undefined.

Are regression and correlation analysis in any way related? The answer is, yes. If we have a sample of n data points $(X_i, Y_i), i = 1, \ldots, n$, then, from (16.46), $\hat{\beta}_1 = \hat{\rho}\left(\frac{S_Y}{S_X}\right)$. Under $H_0: \beta_1 = 0$, we previously used as a test statistic $T = \frac{\hat{\beta}_1}{S_{\hat{\beta}_1}}$. If the preceding expression for $\hat{\beta}_1$ is substituted into this T-statistic, we obtain

$$T = \frac{\hat{\beta}_1}{S_{\hat{\beta}_1}} = \frac{\hat{\rho}\left(\frac{S_Y}{S_X}\right)}{\frac{S_\varepsilon}{\sqrt{\sum x_i^2}}} = \frac{\hat{\rho}\left(\frac{S_Y}{S_X}\right)S_X\sqrt{n-1}\sqrt{n-2}}{\sqrt{\sum y_i^2 - \hat{\beta}_1\sum x_i y_i}}$$

$$= \frac{\hat{\rho}\left(\frac{S_Y}{S_X}\right)S_X\sqrt{n-1}\sqrt{n-2}}{\sqrt{(n-1)S_Y^2 - \hat{\rho}\left(\frac{S_Y}{S_X}\right)S_X S_Y\hat{\rho}(n-1)}} = \frac{\hat{\rho}\sqrt{n-2}}{\sqrt{1-\hat{\rho}^2}}.$$

But this is the T obtained earlier under H_0: $\rho = 0$. Moreover, we previously determined (in Section 16.8) that $F = \frac{\hat{\beta}_1 \sum x_i y_i}{S_\varepsilon^2}$. If $\hat{\beta}_1 = \hat{\rho}\left(\frac{S_Y}{S_X}\right)$ is substituted into this expression we have

$$F = \frac{\hat{\beta}_1 \sum x_i y_i}{S_\varepsilon^2} = \frac{\hat{\rho}\left(\frac{S_Y}{S_X}\right) S_X S_Y \hat{\rho}(n-1)(n-2)}{(n-1)S_Y^2 - \hat{\rho}\left(\frac{S_Y}{S_X}\right) S_X S_Y \hat{\rho}(n-1)} = \frac{\hat{\rho}^2(n-2)}{1 - \hat{\rho}^2},$$

which is the square of the preceding T-statistic. Thus the regression and correlation t tests and the analysis of variance F-test are all equivalent ways of testing for a significant linear relationship between X and Y; that is, testing H_0: $\beta_1 = 0$ is equivalent to testing H_0: $\rho = 0$. In this regard, for the bivariate linear model, if either the t or F-test leads us to reject one of the aforementioned null hypotheses, then the other must be rejected also. Hence we need not perform all three significance tests; only one is necessary and any one will suffice.

16.13   **Exercises**

16-1. Given the following sample data set (with the X values held fixed), find:

  (a) Least squares estimates of β_0 and β_1

  (b) An estimate of σ_ε^2

  (c) Standard error of estimate

  (d) Estimates of the standard errors of the regression coefficients

Y	4	4	6	6	10	8	10	14	10	16
X	2	4	6	8	9	10	11	12	14	16

16-2. Using the data presented in Exercise 16-1:

  (a) Find 95% confidence intervals for β_0 and β_1.

  (b) For $\alpha = 0.01$, test H_0: $\beta_1 = 0$, against H_1: $\beta_1 > 0$.

  (c) Construct a hypothesis test to determine if the population regression line passes through the origin.

  (d) For $\alpha = 0.05$, test H_0: $\beta_1 = \beta_1^0 = 0.65$, against H_1: $\beta_1 < 0.65$.

  (e) Determine if β_0 and β_1 are jointly significantly different from zero.

16-3. The *point elasticity of Y with respect to X* is defined as $\eta = (dY/dX)(X/Y)$. For the population regression equation estimated in Exercise 16-1, find an estimate of the point elasticity coefficient η at the point of means of the variables; that is, find $\hat{\eta} = (d\hat{Y}/dX)(\overline{X}/\overline{Y})$. Interpret your result. (Hint: η serves as an index of the responsiveness of Y to a 1% change in X. So if, say, $\hat{\eta} = 1.35$, then a 1% increase in X leads to a 1.35% increase in Y (the *elastic case*, since $\hat{\eta} > 1$). And if $\hat{\eta} = 0.80$, a 1% increase in X precipitates only a 0.80% increase in Y (the *inelastic case*, since $\hat{\eta} < 1$).

16-4. Using the data presented in Exercise 16-1:

(a) Find a 95% confidence band for the population regression equation using the following X values ($X = 2, 4, 6, 8, 10, 12, 14, 16$).

(b) For $X = 9$, can we conclude at the $\alpha = 0.05$ level that, on the average, Y will exceed 10.5?

(c) If $X_0 = 17$, determine a 95% prediction interval for Y_0.

(d) For $\alpha = 0.05$, can we conclude that, for $X_0 = 17$, the predicted value of Y does not exceed 18?

(e) For $\alpha = 0.05$, determine if the underlying regression relationship generated the point $(X_0, Y_0) = (18, 18)$.

16-5. Using the data presented in Exercise 16-1:

(a) Construct the analysis-of-variance table for the partitioned sum of squares.

(b) What is the value of the coefficient of determination?

(c) Use the F-test with $\alpha = 0.05$ to determine if there exists a significant linear relationship between X and Y.

16-6. For the Exercise 16-1 data set:

(a) Find $\hat{\rho}$ using (16.37).

(b) For $\alpha = 0.01$, test H_0: $\rho = 0$, against H_1: $\rho > 0$.

(c) Using $\alpha = 0.05$, test H_0: $\rho = \rho_0 = 0.85$, against H_1: $\rho < 0.85$.

(d) Find a 95% confidence interval for ρ.

16-7. Given the following sample data set:

Y	4	2	4	6	10	6	8	12	16	14
X	2	4	6	8	10	12	14	16	18	20

(a) Perform a regression and correlation study. Use $\alpha = 0.05$.

(b) For $X_0 = 23$, find a 95% prediction interval for Y_0.

16-8. Our operating assumption in this chapter is that Y is a linear function of X (plus an additive error term). What is important is not that the regression model is linear in the variables but that it must be *linear in the parameters* in order to be estimated. For example:

(a) $Y = \beta_0 + \beta_1 X^{-1}$ is nonlinear in X but linear in the parameters; and

(b) $Y = \beta_0 X^{\beta_1}$ is nonlinear in both X and in the parameters yet $\ln Y = \ln \beta_0 + \beta_1 \ln X$ is linear in the parameters.

For case (a):

1. Show that the slope is everywhere negative and decreases in absolute value as X increases.
2. Graph the function for $X > 0$. Verify that as $X \to 0, Y \to \infty$; and as $X \to \infty, Y \to \beta_0$.
3. What is the interpretation of β_0?
4. Convert Y into a linear estimating equation by introducing a new variable $Z = X^{-1}$. Estimate β_0 and β_1 via least squares using the data in Table A.

For case (b):

1. Show that if $\beta_1 > 0$, the slope is always positive and $Y \to \infty$ as $X \to \infty$. If $\beta_1 > 1$, the slope is monotonically increasing as X increases; if $0 < \beta_1 < 1$, the slope is monotonically decreasing as X increases. If $\beta_1 < 0$, the slope is always negative as X increases.
2. Graph the function for $\beta_1 = -1$ and $X > 0$. What is this type of expression called?
3. Verify that this function has a constant elasticity equal to β_1.
4. Using the data in Table B, estimate this expression via least squares by defining new variables $W = \ln Y$ and $Z = \ln X$. How is β_0 obtained?

A		B	
X	Y	X	Y
1	14	1	2
2	10	2	5
3	8	3	8
4	6	4	12
5	5.8	5	17
6	5.5	6	25
7	5.3	7	37
8	4.9	8	52

16-9. Use the methodology of the preceding exercise and the accompanying data set given here to estimate the parameters β_0 and β_1 of the expression $Y = e^{\beta_0 + \beta_1 X}, Y > 0$, where the vertical intercept is e^{β_0}. The slope of this function is positive (respectively, negative) if $\beta_1 > 0$(respectively, <0). And since $\beta_1 = (1/Y)(dY/dX)$, this function is alternatively referred to as the *constant growth curve*; that is, the proportionate rate of change in Y per unit change in X is the constant β_1.

Y	8	11	14	18	23	31	43	58
X	1	2	3	4	5	6	7	8

16-10. What if instead of regressing a dependent variable Y on an explanatory variable X, we regress the Z-scores of Y on the Z-scores of X? From $Y_i = \beta_0 + \beta_1 X_i + \varepsilon_i$ and $\overline{Y} = \beta_0 + \beta_1 \overline{X} + \bar{\varepsilon}$,

$$\frac{Y_i - \overline{Y}}{\sigma_Y} = \beta_1 \frac{X_i - \overline{X}}{\sigma_X} \frac{\sigma_X}{\sigma_Y} + \frac{\varepsilon_i - \bar{\varepsilon}}{\sigma_Y}$$

or

$$Z_{iY} = \beta_1 \left(\frac{\sigma_X}{\sigma_Y} \right) Z_{iX} + u_i = \beta_1^* Z_{iX} + u_i. \qquad (16.E.1)$$

Applying ordinary least squares to (16.E.1) renders

$$\hat{Z}_{iY} = \hat{\beta}_1^* Z_{iX}.$$

(What assumptions must u_i satisfy?)

The quantity $\hat{\beta}_1^*$ is termed a *standardized regression coefficient* or simply a *beta coefficient*. So although the usual regression slope $\hat{\beta}_1$ represents the average rate of change in Y per unit change in X, we see that the beta coefficient $\hat{\beta}_1^*$ measures the said changes in *standard deviation units*; that is, when X increases by one standard deviation, Y changes by $\hat{\beta}_1^*$ standard deviations on the average. Hence the beta coefficient is independent of units. Note that we can easily recover the usual regression slope as $\hat{\beta}_1 = \hat{\beta}_1^* (\sigma_Y / \sigma_X)$. A glance at the structure of $\hat{\beta}_1^*$ reveals that it is simply the estimated coefficient of correlation between the variables X and Y. For the following data set:

(a) Transform the X and Y variables to Z-scores.

(b) Use least squares to regress the Y Z-score on the X Z-score with the intercept suppressed.

(c) From $\hat{\beta}_1^*$ determine $\hat{\beta}_1$.

(d) Calculate the coefficient of correlation between X and Y directly and verify that it equals $\hat{\beta}_1^*$.

X	1	2	3	4	5	6	7	8
Y	3	4	3	5	6	5	4	7

16-11. Testing for a *structural break*, or testing for the *equality of two regression equations*, can be accomplished by performing the so-called Chow test.[2]

---

[2] Gregory C. Chow, "Tests of the Equality Between Sets of Coefficients in Two Linear Regressions," *Econometrica*, Vol. 28, No. 3, 1960, pp. 591–605.

For instance, suppose we have a sample of n_1 observations on the variables X and Y for one time period and a second sample of n_2 data points on the same variables for another time period (e.g., we can model some measurable characteristic pre 9/11/01 and post 9/11/01). The relevant question is then, "Is there a change (either in intercepts, or slopes, or both) in the response of Y to X between the two periods?" That is, "Can each regression equation be viewed as coming from the same population?" Looked at a third way, "Is the regression relationship between X and Y structurally stable over the two time periods?"

To perform the *Chow test* for equality of two regression equations

$$Y_{i1} = \beta_{01} + \beta_{11}X_{i1} + \varepsilon_{i1}, i = 1, \ldots, n_1; \text{ [Period 1]} \qquad (16.E.2)$$

$$Y_{j2} = \beta_{02} + \beta_{12}X_{j2} + \varepsilon_{j2}, j = 1, \ldots, n_2; \text{ [Period 2]} \qquad (16.E.3)$$

(here the second subscript refers to the subsample or period):

STEP 1.   Pool all $n_1 + n_2 = n$ observations and estimate a single regression equation $Y = \beta_0 + \beta_1 X + \varepsilon$ over the combined sample period. From this regression obtain the residual sum of squares SSE_C with $n_1 + n_2 - 2$ degrees of freedom.

STEP 2.   Estimate the two subperiod regressions (16.E.2) and (16.E.3) separately and obtain the residual sum of squares for each, denoted SSE_1 with degrees of freedom $n_1 - 2$ and SSE_2 with degrees of freedom $n_2 - 2$ respectively. Form $SSE_+ = SSE_1 + SSE_2$ with $n_1 + n_2 - 4$ degrees of freedom.

STEP 3.   To test H_0: $\beta_{01} = \beta_{02}$ and $\beta_{11} = \beta_{12}$, against H_1: H_0 is not true, we use the test statistic

$$F = \frac{(SSE_C - SSE_+)/2}{SSE_+/(n_1 + n_2 - 4)} \sim F_{2, n_1 + n_2 - 4}, \qquad (16.E.4)$$

where f is the sample realization of (16.E.4) and

$$\mathcal{R} = \left\{ f | f > f_{1-\alpha, 2, n_1 + n_2 - 4} \right\}.$$

For the data set given here, employ the Chow test to determine if, for $\alpha = 0.05$, the two regression structures are the same. That is, at this level of significance, can we conclude that the two subsamples have been drawn from the same population regression structure?

Period 1		Period 2	
X	Y	X	Y
88	36	166	90
94	21	176	95
100	18	186	82
105	20	196	104
109	10	211	153
118	12	227	194
127	41	239	175
134	50	250	199
141	43		
155	59		

16-12. *Binary* or *dummy* or *count* variables are used to depict the situation in which some attribute is either present or absent. For instance, in taking the medical history of a patient in a hospital we may ask if he or she is a smoker. If the individual answers yes, we can record a "1"; if the answer is no, we record a "0." In general, a 1 indicates the presence of some attribute and 0 indicates its absence. More formally, if X_i denotes a binary variable, then

$$X_i = \begin{cases} 1 & \text{if an attribute is present;} \\ 0 & \text{if an attribute is not present.} \end{cases}$$

Other instances include the attributes of sex ($X_i = 1$ if the person is male; $X_i = 0$ if the person is female), automobile ownership ($X_i = 1$ if an individual owns an automobile; $X_i = 0$ if not), and so on.

Suppose we are interested in comparing the average salaries of males versus females at the XYZ Corporation, where μ_M (respectively, μ_F) denotes the true or population mean salary level for males (respectively, females). If Y represents employee salaries (assume Y is $N(\mu, \sigma)$, σ = constant, with overall mean μ), then a salary comparison between males and females can be undertaken by formulating the following linear regression model:

$$Y_i = \beta_0 + \beta_1 X_i + \varepsilon_i, i = 1, \ldots, n, \tag{16.E.5}$$

where Y_i is the salary of the ith employee in a sample of size n, X_i is a dummy variable such that

$$X_i = \begin{cases} 1 & \text{if the employee is male (M);} \\ 0 & \text{if the employee is female (F),} \end{cases}$$

and ε_i is a random error term satisfying all of the assumptions of the (strong) classical linear regression model. In this regard, since

$$E(Y_i|X_i = 0) = \mu_F = \beta_0,$$
$$E(Y_i|X_i = 1) = \mu_M = \beta_0 + \beta_1,$$

it follows that

$$\beta_1 = \mu_M - \mu_F;$$

that is, the intercept of the population regression equation measures the mean salary of females and the slope of the same measures the difference between the mean salary of males and the mean salary of females.

So to compare the average salaries of males and females, let us test $H_0: \beta_1 = 0$ (there is no difference between the average salaries of males and females), against any of the following alternative hypotheses:

Case I	Case II	Case III
$H_0: \beta_1 = 0$	$H_0: \beta_1 = 0$	$H_0: \beta_1 = 0$
$H_1: \beta_1 > 0$	$H_1: \beta_1 < 0$	$H_1: \beta_1 \neq 0$
(or $\mu_M > \mu_F$)	(or $\mu_M < \mu_F$)	(or $\mu_M \neq \mu_F$)

The coefficients of (16.E.5) can be estimated via least squares and all of the standard tests and interval estimates apply. (Note that since it can be readily demonstrated that $\hat{\beta}_1 = \bar{Y}_M - \bar{Y}_F$ and $\hat{\beta}_0 = \bar{Y}_F$, where $\bar{Y}_M$ is the sample mean salary of males and $\bar{Y}_F$ is the sample mean salary of females, it follows that testing $H_0: \beta_1 = 0$ is equivalent to the two-sample t-test of $H_0: \mu_M - \mu_F = 0$ developed in Chapter 12.)

For the following data set pertaining to employee absenteeism (Y) for a one-month period at ABC Inc. and smoking (X), determine if smokers have a significantly higher rate of absenteeism relative to non-smokers over this time period using $\alpha = 0.05$. Let

$$X_i = \begin{cases} 1 & \text{if the employee is a smoker;} \\ 0 & \text{if the employee is a nonsmoker.} \end{cases}$$

X	Y (=days absent)
0	3
0	2
1	5
0	2
1	7
1	9
1	11
0	4
0	3
0	2
1	6
0	0
0	1
0	3
1	5

16-13. It is not unusual for a regression study to be performed using a data set containing *outlier observations*; that is, data points that are remote or distinctly separated from the main scatter of observations. Clearly outlier points can have a significant effect on the regression results—they can exhibit inordinately large residuals that, in turn, can impact the location of the estimated regression line.

How can outliers be identified? Moreover, if a sample point is deemed an outlier, should it be retained or eliminated? A data point can be an outlier with respect to its X coordinate, its Y coordinate, or both (see Figure 16.E.1).

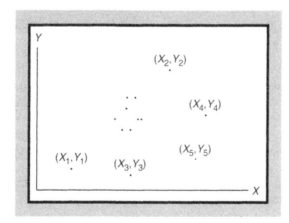

Figure 16.E.1

Points (X_1, Y_1) and (X_2, Y_2) are outliers with respect to both the X and Y coordinates since the said coordinates are well outside of the main scatter of points; (X_3, Y_3) is an outlier with respect to its Y coordinate (its X coordinate is near the middle of the range of X values); and (X_4, Y_4) is outlying with respect to its X coordinate (the Y coordinate is near the middle of the range of Y values).

Outlying points may not be equally influential on the regression results. That is, although points (X_1, Y_1) and (X_2, Y_2) are consistent with the regression equation passing through the nonoutlying points, this will not be the case for (X_5, Y_5).

(a) Identifying outlying X values.

It can be shown that the *variance of the residual* e_i can be expressed as

$$\sigma_{e_i}^2 = \sigma_\varepsilon^2(1 - h_{ii}), \qquad (16.E.6)$$

where

$$h_{ii} = \frac{\sum X_i^2 - 2n\overline{X}X_i + nX_i^2}{n(\sum X_i^2 - n\overline{X}^2)} = \frac{1}{n} + \frac{(X_i - \overline{X})^2}{\sum(X_i - \overline{X})^2} = \frac{1}{n} + \frac{x_i^2}{\sum x_i^2},$$

$$i = 1,\ldots,n, \qquad (16.E.7)$$

with $0 \le h_{ii} \le 1$ and $\sum_{i=1}^{n} h_{ii} =$ number of parameters estimated $= 2$. The quantity h_{ii}, called the *leverage of* X_i, is useful in indicating whether or not X_i is outlying since it essentially measures the distance of X_i from the center of all X values. If X_i exhibits a large leverage value h_{ii}, then it makes a substantial contribution to the determination of the estimated Y value $\hat{Y}_i$; that is, if h_{ii} is large, the more important Y_i is in determining $\hat{Y}_i$. In addition, as (16.E.6) reveals, the larger is h_{ii}, the smaller is the variance of e_i. In this regard, the larger is h_{ii}, the smaller will be the difference $e_i = Y_i - \hat{Y}_i$. (Since h_{ii} and e_i vary inversely, the detection of outliers cannot be left to simply an examination of the least squares residuals e_i.)

A leverage value h_{ii} is deemed *large* if it is more than twice as large as the mean leverage value $\bar{h} = \sum_{i=1}^{n} h_{ii}/n = \frac{2}{n}$. That is, X_i is considered to be an outlier if $h_{ii} > \frac{4}{n}$. Alternatively, it is sometimes suggested that X_i exhibits *high leverage* if $h_{ii} > 0.5$ and X_i exhibits *moderate leverage* if $0.2 \le h_{ii} \le 0.5$.

(b) Identifying outlying Y values.

To detect outlying Y observations we may employ what is called the *studentized deleted residual*

$$d_i^* = e_i \left[\frac{n-3}{SSE(1 - h_{ii}) - e_i^2} \right]^{1/2}. \qquad (16.E.8)$$

(A *deleted residual* is $d_i = Y_i - \hat{Y}_{i(-i)}$, where $\hat{Y}_{i(-i)}$ is the estimated Y value obtained by deleting (X_i, Y_i) and fitting the regression equation to the remaining $n - 1$ observations. Once d_i has been divided by its estimated standard deviation, we obtain (16.E.8).) Here $|d_i^*|$ is compared to, say, the upper 5% quantile of the t distribution; that is, a Y_i value is considered to be an outlier if $|d_i^*| > t_{0.05, n-3}$.

(c) Identifying influential data values

Once it has been discovered that an X and/or Y value is an outlier, it remains to determine whether or not this outlier can be considered as *influential*—in the sense that its exclusion precipitates major changes in the estimated regression equation. A measure of the influence that the ith data point has on the estimated Y value, $\hat{Y}_i$, is provided by

$$(DFFITS)_i = d_i^* \left(\frac{h_{ii}}{1 - h_{ii}} \right)^{1/2}, \qquad (16.E.9)$$

with this quantity essentially representing the ith studentized deleted residual increased or decreased by a scale factor dependent upon the ith leverage value; it amounts to the number of estimated standard deviations that $\hat{Y}_i$ changes when the ith observation (X_i, Y_i) is removed from the data set. (Note that DF stands for the difference between $\hat{Y}_i$ and $\hat{Y}_{i(-i)}$, where, as defined earlier, $\hat{Y}_{i(-i)}$ is the estimated Y value obtained when the ith data point (X_i, Y_i) is eliminated.) In general, X_i is an *influential outlier* if:

1. $|(DFFITS)_i| > 1$ for small to medium-sized samples
2. $|(DFFITS)_i| > 2\sqrt{2/n}$ for large samples

A global or overall measure of the combined influence of the ith data point (X_i, Y_i) on all regression coefficients simultaneously is *Cook's distance measure*

$$D_i = \frac{e_i^2}{2\,MSE} \left[\frac{h_{ii}}{(1 - h_{ii})^2} \right], \qquad (16.E.10)$$

which essentially gauges the combination of the differences in the estimated regression coefficients before and after the ith data point is deleted. The magnitude of D_i is compared to the quantile value of the F distribution with 2 and $n - 2$ degrees of freedom. If the quantile value of $F_{2, n-2}$ is near 50% or more, we may conclude that the ith sample data point has considerable influence on the estimated regression coefficients.

Given the following sample data set, determine if any of the observations is an outlier. (Hint: Graph the $n = 15$ data points and then apply (16.E.7)–(16.E.10).)

X	4	6	7	7	8	9	10	11	12	14	14	16	16	18	18
Y	18	13	1	4	2	8	10	9	12	13	16	15	20	4	18

16-14. Prove the Gauss-Markov Theorem. Hint:

(1) First express $\hat{\beta}_0, \hat{\beta}_1$ as linear functions of the $Y_i, i = 1, \ldots, n$; that is, write

$$\hat{\beta}_1 = \sum_{i=1}^{n} w_i Y_i, \text{ where } w_i = \frac{x_i}{\sum_{i=1}^{n} x_i^2};$$

$$\hat{\beta}_0 = \sum_{i=1}^{n} v_i Y_i, \text{ where } v_i = \frac{1}{n} \bar{X} w_i.$$

(2) Write $\hat{\beta}_1 = \sum_{i=1}^{n} x_i y_i / \sum_{i=1}^{n} x_i^2 = \beta_1 + \sum w_i \varepsilon_i$ and $\hat{\beta}_0 = \beta_0 + (\beta_1 - \hat{\beta}_1)\bar{X} + \bar{\varepsilon}$. Then find $E(\hat{\beta}_1)$, $E(\hat{\beta}_0)$.

(3) Determine the variances of $\hat{\beta}, \hat{\beta}_1$ as

$$V(\hat{\beta}_1) = E\left[(\hat{\beta}_1 - \beta_1)^2\right] = \sigma_\varepsilon^2 / \sum_{i=1}^{n} x_i^2;$$

$$V(\hat{\beta}_0) = E\left[(\hat{\beta}_0 - \beta_0)^2\right] = \sigma_\varepsilon^2 \left(\frac{1}{n} + \frac{\bar{X}^2}{\sum_{i=1}^{n} x_i^2}\right)$$

(given that $E[(\hat{\beta}_1 - \beta_1 \bar{\varepsilon}] = 0$ and $E(\bar{\varepsilon}^2) = \sigma_\varepsilon^2 / n$).

(4) Let $\beta_1' = \sum_{i=1}^{n} a_i Y_i$ be an alternative linear estimator of β_1 and verify that $E(\beta_1') = \beta_1$ under the restrictions $\sum_{i=1}^{n} a_i = 0$ and $\sum_{i=1}^{n} a_i X_i = 1$.

(5) Given these restrictions, $\beta_1' = \beta_1 + \sum_{i=1}^{n} a_i \varepsilon_i$ and $V(\beta_1') = \sigma_\varepsilon^2 \sum_{i=1}^{n} a_i^2$.

(6) To compare β_1' with $\hat{\beta}_1$, set $a_i = w_i + d_i$, d_i constant for all i. Then verify that $V(\beta_1') = V(\hat{\beta}_1) + \sigma_\varepsilon^2 \sum_{i=1}^{n} d_i^2 \geq V(\hat{\beta}_1)$. (A similar line of argumentation holds for $\hat{\beta}_0$.)

16-15. Verify that, for ε normal, the least squares estimators are maximum likelihood estimators.

16-16. Verify that $COV(\hat{\beta}_0, \hat{\beta}_1) = -\bar{X}\sigma_\varepsilon^2 / \sum_{i=1}^{n} x_i^2$.

16-17. Demonstrate that $S_\varepsilon^2 = \sum_{i=1}^{n} e_i^2 / (n-2)$ is an unbiased estimator for σ_ε^2. (Hint: In $d^2 = \frac{1}{n} \sum_{i=1}^{n} e_i^2$ set $e_i = -(\hat{\beta}_1 - \beta_1)x_i + \varepsilon_i - \bar{\varepsilon}$. Then correct for the bias in $E(d^2) = \left(\frac{n-2}{n}\right)\sigma_\varepsilon^2$.)

16-18. Verify that $SST = SSR + SSE$. (Hint: Square $y_i = \hat{\beta}_1 x_i + e_i$ and sum over all i values. Then determine that $\sum_{i=1}^{n} x_i e_i = 0$. The desired result immediately follows.)

16-19. Demonstrate that $0 \leq R^2 \leq 1$. (Hint: Use $\sum_{i=1}^{n} x_i y_i = \hat{\beta}_1 \sum x_i^2$ to verify that $\sum_{i=1}^{n} y_i^2 \geq \sum_{i=1}^{n} e_i^2 \geq 0$.)

Appendix A

Table A.1   Standard Normal Areas (Z is $N(0,1)$)

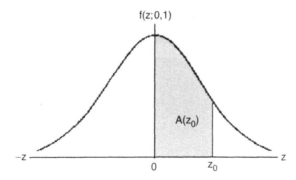

$$A(z_0) = \frac{1}{\sqrt{2\pi}} \int_0^{z_0} e^{-z^2/2} dz.$$

$A(z_0)$ gives the total area under the standard normal distribution between 0 and any point z_0 on the positive z-axis
(e.g., for $z_0 = 1.96$, $A(z_0) = 0.475$).

Figure A.1

z	0.00	0.01	0.02	0.03	0.04	0.05	0.06	0.07	0.08	0.09
0.0	0.0000	0.0040	0.0080	0.0120	0.0150	0.0199	0.0239	0.0279	0.0319	0.0359
0.1	0.0398	0.0438	0.0478	0.0517	0.0557	0.0596	0.0636	0.0675	0.0714	0.0754
0.2	0.0793	0.0832	0.0871	0.0910	0.0948	0.0987	0.1026	0.1064	0.1103	0.1141
0.3	0.1179	0.1217	0.1253	0.1293	0.1331	0.1368	0.1406	0.1443	0.1480	0.1517
0.4	0.1554	0.1591	0.1628	0.1664	0.1700	0.1736	0.1772	0.1808	0.1844	0.1879
0.5	0.1915	0.1950	0.1985	0.2019	0.2054	0.2088	0.2123	0.2157	0.2190	0.2224
0.6	0.2258	0.2291	0.2324	0.2357	0.2389	0.2422	0.2454	0.2486	0.2518	0.2549
0.7	0.2580	0.2612	0.2642	0.2673	0.2704	0.2734	0.2764	0.2794	0.2823	0.2852
0.8	0.2881	0.2910	0.2939	0.2967	0.2996	0.3023	0.3051	0.3078	0.3106	0.3133
0.9	0.3159	0.3186	0.3212	0.3288	0.3264	0.3289	0.3315	0.3340	0.3365	0.3389
1.0	0.3413	0.3438	0.3461	0.3485	0.3508	0.3531	0.3554	0.3557	0.3559	0.3621
1.1	0.3642	0.3665	0.3686	0.3708	0.3729	0.3749	0.3770	0.3790	0.3810	0.3830
1.2	0.3849	0.3869	0.3888	0.3907	0.3925	0.3944	0.3962	0.3980	0.3997	0.4015
1.3	0.4032	0.4049	0.4066	0.4082	0.4099	0.4115	0.4131	0.4147	0.4162	0.4177
1.4	0.4192	0.4207	0.4222	0.4236	0.4251	0.4265	0.4279	0.4292	0.4306	0.4319
1.5	0.4332	0.4345	0.4357	0.4370	0.4382	0.4394	0.4406	0.4418	0.4429	0.4441
1.6	0.4452	0.4463	0.4474	0.4484	0.4495	0.4505	0.4515	0.4525	0.4535	0.4545
1.7	0.4554	0.4564	0.4573	0.4582	0.4591	0.4599	0.4608	0.4616	0.4625	0.4633
1.8	0.4641	0.4649	0.4656	0.4664	0.4671	0.4678	0.4686	0.4693	0.4699	0.4706
1.9	0.4713	0.4719	0.4726	0.4732	0.4738	0.4744	0.4750	0.4756	0.4761	0.4767
2.0	0.4772	0.4778	0.4783	0.4788	0.4793	0.4798	0.4803	0.4808	0.4812	0.4817
2.1	0.4821	0.4826	0.4830	0.4834	0.4838	0.4842	0.4846	0.4850	0.4854	0.4857
2.2	0.4861	0.4864	0.4868	0.4871	0.4875	0.4878	0.4881	0.4884	0.4887	0.4890
2.3	0.4893	0.4896	0.4898	0.4901	0.4904	0.4906	0.4909	0.4911	0.4913	0.4916
2.4	0.4918	0.4920	0.4922	0.4925	0.4927	0.4929	0.4931	0.4932	0.4934	0.4936
2.5	0.4938	0.4940	0.4941	0.4943	0.4945	0.4946	0.4948	0.4949	0.4951	0.4952
2.6	0.4953	0.4955	0.4956	0.4957	0.4959	0.4960	0.4961	0.4962	0.4963	0.4964
2.7	0.4965	0.4966	0.4967	0.4968	0.4969	0.4970	0.4971	0.4972	0.4973	0.4974
2.8	0.4974	0.4975	0.4976	0.4977	0.4977	0.4978	0.4979	0.4979	0.4980	0.4981
2.9	0.4981	0.4982	0.4982	0.4983	0.4984	0.4984	0.4985	0.4985	0.4986	0.4986
3.0	0.4987	0.4987	0.4987	0.4988	0.4988	0.4989	0.4989	0.4989	0.4990	0.4990

Table A.2   Cumulative Distribution Function Values for the Standard Normal Distribution (Z is $N(0,1)$)

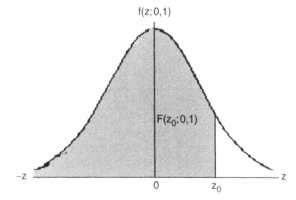

$$F(z_0;0,1) = P(Z \leq z_0) = \frac{1}{\sqrt{2\pi}} \int_{-\infty}^{z_0} e^{-z^2/2} \, dz.$$

$F(z_0;0,1)$ gives the total area under the standard normal distribution from $-\infty$ to any point z_0 on the z-axis (e.g., for $z_0 = 1.96$, $F(1.96;0,1) = 0.975$; for $z_0 = -1.96$, $F(-1.96;0,1) = 0.025$).

Figure A.2

z	0.00	0.01	0.02	0.03	0.04	0.05	0.06	0.07	0.08	0.09
-3.0	0.0014	0.0013	0.0013	0.0012	0.0012	0.0011	0.0011	0.0011	0.0010	0.0010
-2.9	0.0019	0.0018	0.0018	0.0017	0.0016	0.0016	0.0015	0.0015	0.0014	0.0014
-2.8	0.0026	0.0025	0.0024	0.0023	0.0023	0.0022	0.0021	0.0021	0.0020	0.0019
-2.7	0.0035	0.0034	0.0033	0.0032	0.0031	0.0030	0.0029	0.0028	0.0027	0.0026
-2.6	0.0047	0.0045	0.0044	0.0043	0.0042	0.0040	0.0039	0.0038	0.0037	0.0036
-2.5	0.0062	0.0060	0.0059	0.0057	0.0055	0.0054	0.0052	0.0051	0.0049	0.0048
-2.4	0.0082	0.0080	0.0078	0.0076	0.0073	0.0071	0.0070	0.0068	0.0066	0.0064
-2.3	0.0107	0.0104	0.0102	0.0099	0.0096	0.0094	0.0091	0.0089	0.0087	0.0084
-2.2	0.0139	0.0136	0.0132	0.0129	0.0126	0.0122	0.0119	0.0116	0.0113	0.0110
-2.1	0.0179	0.0174	0.0170	0.0166	0.0162	0.0158	0.0154	0.0150	0.0146	0.0143
-2.0	0.0228	0.0222	0.0217	0.0212	0.0207	0.0202	0.0197	0.0192	0.0188	0.0183
-1.9	0.0287	0.0281	0.0274	0.0268	0.0262	0.0256	0.0250	0.0244	0.0239	0.0233
-1.8	0.0359	0.0352	0.0344	0.0336	0.0329	0.0322	0.0314	0.0307	0.0301	0.0294
-1.7	0.0446	0.0436	0.0427	0.0418	0.0409	0.0401	0.0392	0.0384	0.0375	0.0367
-1.6	0.0548	0.0537	0.0526	0.0516	0.0505	0.0495	0.0485	0.0475	0.0465	0.0455
-1.5	0.0668	0.0655	0.0643	0.0630	0.0618	0.0606	0.0594	0.0582	0.0571	0.0559
-1.4	0.0808	0.0793	0.0778	0.0764	0.0749	0.0735	0.0721	0.0708	0.0694	0.0681
-1.3	0.0968	0.0951	0.0934	0.0918	0.0901	0.0885	0.0869	0.0853	0.0838	0.0823
-1.2	0.1151	0.1131	0.1112	0.1094	0.1075	0.1057	0.1038	0.1020	0.1003	0.0985
-1.1	0.1357	0.1335	0.1314	0.1292	0.1271	0.1251	0.1230	0.1210	0.1190	0.1170
-1.0	0.1587	0.1563	0.1539	0.1515	0.1492	0.1469	0.1446	0.1423	0.1401	0.1379
-0.9	0.1841	0.1814	0.1788	0.1762	0.1736	0.1711	0.1685	0.1660	0.1635	0.1611
-0.8	0.2119	0.2090	0.2061	0.2033	0.2005	0.1977	0.1949	0.1922	0.1894	0.1867
-0.7	0.2420	0.2389	0.2358	0.2327	0.2297	0.2266	0.2236	0.2207	0.2177	0.2148
-0.6	0.2743	0.2709	0.2676	0.2643	0.2611	0.2578	0.2546	0.2514	0.2483	0.2451
-0.5	0.3085	0.3050	0.3015	0.2981	0.2946	0.2912	0.2877	0.2843	0.2810	0.2776
-0.4	0.3446	0.3409	0.3372	0.3336	0.3300	0.3264	0.3228	0.3192	0.3156	0.3121
-0.3	0.3821	0.3783	0.3745	0.3707	0.3669	0.3632	0.3594	0.3557	0.3520	0.3483
-0.2	0.4207	0.4168	0.4129	0.4090	0.4052	0.4013	0.3974	0.3936	0.3897	0.3859
-0.1	0.4602	0.4562	0.4522	0.4483	0.4443	0.4404	0.4364	0.4325	0.4286	0.4247
0.0	0.5000	0.4960	0.4920	0.4880	0.4840	0.4801	0.4761	0.4721	0.4681	0.4641

Table A.2   (Contd.)

z	0.00	0.01	0.02	0.03	0.04	0.05	0.06	0.07	0.08	0.09
0.0	0.5000	0.5040	0.5080	0.5120	0.5160	0.5190	0.5239	0.5279	0.5319	0.5359
0.1	0.5398	0.5438	0.5478	0.5517	0.5557	0.5596	0.5636	0.5675	0.5714	0.5753
0.2	0.5793	0.5832	0.5871	0.5910	0.5948	0.5987	0.6026	0.6064	0.6103	0.6141
0.3	0.6179	0.6217	0.6255	0.6293	0.6331	0.6368	0.6406	0.6443	0.6480	0.6517
0.4	0.6554	0.6591	0.6628	0.6664	0.6700	0.6736	0.6772	0.6808	0.6844	0.6879
0.5	0.6915	0.6950	0.6985	0.7019	0.7054	0.7088	0.7123	0.7157	0.7190	0.7224
0.6	0.7257	0.7291	0.7324	0.7357	0.7389	0.7422	0.7454	0.7486	0.7517	0.7549
0.7	0.7580	0.7611	0.7642	0.7673	0.7704	0.7734	0.7764	0.7794	0.7823	0.7852
0.8	0.7881	0.7910	0.7939	0.7969	0.7995	0.8023	0.8051	0.8078	0.8106	0.8133
0.9	0.8159	0.8186	0.8212	0.8238	0.8264	0.8289	0.8315	0.8340	0.8365	0.8389
1.0	0.8413	0.8438	0.8461	0.8485	0.8508	0.8513	0.8554	0.8577	0.8529	0.8621
1.1	0.8643	0.8665	0.8686	0.8708	0.8729	0.8749	0.8770	0.8790	0.8810	0.8830
1.2	0.8849	0.8869	0.8888	0.8907	0.8925	0.8944	0.8962	0.8980	0.8997	0.9015
1.3	0.9032	0.9049	0.9066	0.9082	0.9099	0.9115	0.9131	0.9147	0.9162	0.9177
1.4	0.9192	0.9207	0.9222	0.9236	0.9251	0.9265	0.9279	0.9292	0.9306	0.9319
1.5	0.9332	0.9345	0.9357	0.9370	0.9382	0.9394	0.9406	0.9418	0.9429	0.9441
1.6	0.9452	0.9463	0.9474	0.9484	0.9495	0.9505	0.9515	0.9525	0.9535	0.9545
1.7	0.9554	0.9564	0.9573	0.9582	0.9591	0.9599	0.9608	0.9616	0.9625	0.9633
1.8	0.9641	0.9649	0.9656	0.9664	0.9671	0.9678	0.9686	0.9693	0.9699	0.9706
1.9	0.9713	0.9719	0.9726	0.9732	0.9738	0.9744	0.9750	0.9756	0.9761	0.9767
2.0	0.9772	0.9778	0.9783	0.9788	0.9793	0.9798	0.9803	0.9808	0.9812	0.9817
2.1	0.9821	0.9826	0.9830	0.9834	0.9838	0.9842	0.9846	0.9850	0.9854	0.9857
2.2	0.9861	0.9864	0.9868	0.9871	0.9875	0.9878	0.9881	0.9884	0.9887	0.9890
2.3	0.9893	0.9896	0.9898	0.9901	0.9904	0.9906	0.9909	0.9911	0.9913	0.9916
2.4	0.9918	0.9920	0.9922	0.9925	0.9927	0.9929	0.9931	0.9932	0.9934	0.9936
2.5	0.9938	0.9940	0.9941	0.9943	0.9945	0.9946	0.9948	0.9949	0.9951	0.9952
2.6	0.9953	0.9955	0.9956	0.9957	0.9959	0.9960	0.9961	0.9962	0.9963	0.9964
2.7	0.9965	0.9966	0.9967	0.9968	0.9969	0.9970	0.9971	0.9972	0.9973	0.9974
2.8	0.9974	0.9975	0.9976	0.9977	0.9977	0.9978	0.9979	0.9979	0.9980	0.9981
2.9	0.9981	0.9982	0.9982	0.9983	0.9984	0.9984	0.9985	0.9985	0.9986	0.9986
3.0	0.9987	0.9987	0.9987	0.9988	0.9988	0.9989	0.9989	0.9989	0.9990	0.9990

Table A.3 Quantiles of Student's t Distribution (T is t_v)

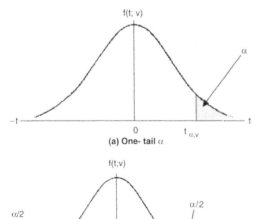

(a) One-tail α

(b) Two-tail α

Given degrees of freedom v, the table gives either: (a) the one-tail $t_{\alpha,v}$ value such that $P(T \geq t_{\alpha,v}) = \alpha$; or (b) the two-tail $\pm t_{\alpha/2,v}$ values for which $P(T \leq -t_{\alpha/2,v}) + P(T \geq t_{\alpha/2,v}) = \alpha/2 + \alpha/2 = \alpha$ (e.g., for $v = 15$ and $\alpha = 0.05$, $t_{0.05,15} = 1.753$ and $t_{0.025,15} = 2.131$).

Figure A.3

	One-tail α					
	0.10	0.05	0.025	0.01	0.005	0.001
	Two-tail α					
v	0.20	0.10	0.05	0.02	0.01	0.002
1	3.078	6.314	12.706	31.821	63.657	318.309
2	1.886	2.920	4.303	6.965	9.925	22.327
3	1.638	2.353	3.182	4.541	5.841	10.215
4	1.533	2.132	2.776	3.747	4.604	7.173
5	1.476	2.015	2.571	3.365	4.032	5.893
6	1.440	1.943	2.447	3.143	3.707	5.208
7	1.415	1.895	2.365	2.998	3.499	4.785
8	1.397	1.860	2.306	2.896	3.355	4.501
9	1.383	1.833	2.262	2.821	3.250	4.297
10	1.372	1.812	2.228	2.764	3.169	4.144
11	1.363	1.796	2.201	2.718	3.106	4.025
12	1.356	1.782	2.179	2.681	3.055	3.930
13	1.350	1.771	2.160	2.650	3.012	3.852
14	1.345	1.761	2.145	2.624	2.977	3.787
15	1.341	1.753	2.131	2.602	2.947	3.733
16	1.337	1.746	2.120	2.583	2.921	3.686
17	1.333	1.740	2.110	2.567	2.898	3.646
18	1.330	1.734	2.101	2.552	2.878	3.610
19	1.328	1.729	2.093	2.539	2.861	3.579
20	1.325	1.725	2.086	2.528	2.845	3.552

Table A.3   (Contd.)

	One-tail α					
	0.10	**0.05**	**0.025**	**0.01**	**0.005**	**0.001**
	Two-tail α					
v	**0.20**	**0.10**	**0.05**	**0.02**	**0.01**	**0.002**
21	1.323	1.721	2.080	2.518	2.831	3.527
22	1.321	1.717	2.074	2.508	2.819	3.505
23	1.319	1.714	2.069	2.500	2.807	3.485
24	1.318	1.711	2.064	2.492	2.797	3.467
25	1.316	1.708	2.060	2.485	2.787	3.450
29	1.311	1.699	2.045	2.462	2.756	3.396
30	1.310	1.697	2.042	2.457	2.750	3.385
40	1.303	1.684	2.021	2.423	2.704	3.307
60	1.296	1.671	2.000	2.390	2.660	3.232
80	1.292	1.664	1.990	2.374	2.639	3.195
100	1.290	1.660	1.984	2.364	2.626	3.174
∞	1.282	1.645	1.960	2.326	2.576	3.090

Table A.4   Quantiles of the Chi-Square Distribution (X is χ_v^2)

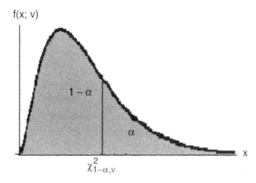

For the cumulative probability $1 - \alpha$ and degrees of freedom v, the quantile $\chi_{1-\alpha,v}^2$ satisfies $F(\chi_{1-\alpha}^2; v) = P(X \leq \chi_{1-\alpha,v}^2) = 1 - \alpha$ or, alternatively, $P(X > \chi_{1-\alpha,v}^2) = 1 - P(X \leq \chi_{1-\alpha,v}^2) = \alpha$ (e.g., for $v = 10$ and $\alpha = 0.05$, $1 - \alpha = 0.95$ and thus $\chi_{0.95,10}^2 = 18.31$).

Figure A.4

v	$1 - \alpha$						
	0.75	**0.90**	**0.95**	**0.975**	**0.99**	**0.995**	**0.999**
1	1.3233	2.7100	3.8400	5.0200	6.6300	7.8800	10.8280
2	2.7726	4.6100	5.9900	7.3800	9.2100	10.6000	13.8160
3	4.1084	6.2500	7.8100	9.3500	11.3400	12.8400	16.2660
4	5.3853	7.7800	9.4900	11.1400	13.2800	14.8600	18.4670
5	6.6257	9.2400	11.0700	12.8300	15.0900	16.7500	20.5150
6	7.8408	10.6400	12.5900	14.4500	16.8100	18.5500	22.4580
7	9.0372	12.0200	14.0700	16.0100	18.4800	20.2800	24.3220
8	10.2188	13.3600	15.5100	17.5300	20.0900	21.9600	26.1250
9	11.3887	14.6800	16.9200	19.0200	21.6700	23.5900	27.8770
10	12.5489	15.9900	18.3100	20.4800	23.2100	25.1900	29.5880

Table A.4   (Contd.)

ν	0.75	0.90	0.95	0.975	0.99	0.995	0.999
	\multicolumn{7}{c}{$1 - \alpha$}						
11	13.7007	17.2800	19.6800	21.9200	24.7300	26.7600	31.2640
12	14.8454	18.5500	21.0300	23.3400	26.2200	28.3000	32.9090
13	15.9839	19.8100	22.3600	24.7400	27.6900	29.8200	34.5280
14	17.1170	21.0600	23.6800	26.1200	29.1400	31.3200	36.1230
15	18.2451	22.3100	25.0000	27.4900	30.5800	32.8000	37.6970
16	19.3688	23.5400	26.3000	28.8500	32.0000	34.2700	39.2520
17	20.4887	24.7690	27.5871	30.1910	33.4087	35.7185	40.7900
18	21.6049	25.9900	28.8700	31.5300	34.8100	37.1600	42.3120
19	22.7178	27.2036	30.1435	32.8523	36.1908	38.5822	43.8200
20	23.8277	28.4100	31.4100	34.1700	37.5700	40.0000	45.3150
21	24.9348	29.6151	32.6705	35.4789	38.9321	41.4010	46.7970
22	26.0393	30.8133	33.9244	36.7807	40.2894	42.7956	48.2680
23	27.1413	32.0069	35.1725	38.0757	41.6384	44.1813	49.7280
24	28.2412	33.1963	36.4151	39.3641	42.9798	45.5585	51.1790
25	29.3389	34.3816	37.6525	40.6465	44.3141	46.9278	52.6200
26	30.4345	35.5631	38.8852	41.9232	45.6417	48.2899	54.0520
27	31.5284	36.7412	40.1133	43.1944	46.9630	49.6449	55.4760
28	32.6205	37.9159	41.3372	44.4607	48.2782	50.9933	56.8920
29	33.7109	39.0875	42.5569	45.7222	49.5879	52.3356	58.3020
30	34.7998	40.2560	43.7729	46.9792	50.8922	53.6720	59.7030
40	45.6160	51.8050	55.7585	59.3417	63.6907	66.7659	73.4020
50	56.3336	63.1671	67.5048	71.4202	76.1539	79.4900	86.6610
60	66.9814	74.4000	79.0800	83.3000	88.3800	91.9500	99.6070
70	77.5766	85.5271	90.5312	95.0231	100.4250	104.2150	112.3170
80	88.1303	96.5782	101.8790	106.6290	112.3290	116.3210	124.8390
90	98.6499	107.5650	113.1450	118.1360	124.1160	128.2990	137.2080
100	109.1410	118.4980	124.3420	129.5610	135.8070	140.1690	149.4490

From E.S. Pearson and H.O. Hartley (1976), *Biometrika Tables for Statisticians*, Vol. I. Reproduced with the kind permission of Oxford University Press.

Table A.5 Quantiles of Snedecor's F Distribution (F is F_{v_1,v_2})

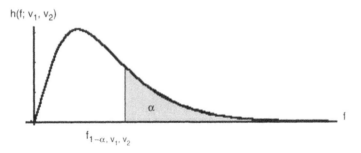

$h(f; v_1, v_2)$

Given the cumulative proportion $1 - \alpha$ and numerator and denominator degrees of freedom v_1 and v_2, respectively, the table gives the upper α-quantile $f_{1-\alpha,v_1,v_2}$ such that $P(F \geq f_{1-\alpha,v_1,v_2}) = \alpha$ (e.g., for $1 - \alpha = 0.95$, $v_1 = 6$, and $v_2 = 10$, $f_{0.95,6,10} = 5.39$).

Figure A.5

$$\alpha = 0.10 \text{ (upper 10\% fractile)}$$

v_2 \ v_1	1	2	3	4	5	6	7	8	9	10	12	15	20	24	30	40	60	120	∞
1	39.86	49.50	53.59	55.83	57.24	58.20	58.91	59.44	59.86	60.19	60.71	61.22	61.74	62.00	62.26	62.53	62.79	63.06	63.33
2	8.53	9.00	9.16	9.24	9.29	9.33	9.35	9.37	9.38	9.39	9.41	9.42	9.44	9.45	9.46	9.47	9.47	9.48	9.49
3	5.54	5.46	5.39	5.34	5.31	5.28	5.27	5.25	5.24	5.23	5.22	5.20	5.18	5.18	5.17	5.16	5.15	5.14	5.13
4	4.54	4.32	4.19	4.11	4.05	4.01	3.98	3.95	3.94	3.92	3.90	3.87	3.84	3.83	3.82	3.80	3.79	3.78	3.76
5	4.06	3.78	3.62	3.52	3.45	3.40	3.37	3.34	3.32	3.30	3.27	3.24	3.21	3.19	3.17	3.16	3.14	3.12	3.10
6	3.78	3.46	3.29	3.18	3.11	3.05	3.01	2.98	2.96	2.94	2.90	2.87	2.84	2.82	2.80	2.78	2.76	2.74	2.72
7	3.59	3.26	3.07	2.96	2.88	2.83	2.78	2.75	2.72	2.70	2.67	2.63	2.59	2.58	2.56	2.54	2.51	2.49	2.47
8	3.46	3.11	2.92	2.81	2.73	2.67	2.62	2.59	2.56	2.54	2.50	2.46	2.42	2.40	2.38	2.36	2.34	2.32	2.29
9	3.36	3.01	2.81	2.69	2.61	2.55	2.51	2.47	2.44	2.42	2.38	2.34	2.30	2.28	2.25	2.23	2.21	2.18	2.16
10	3.29	2.92	2.73	2.61	2.52	2.46	2.41	2.38	2.35	2.32	2.28	2.24	2.20	2.18	2.16	2.13	2.11	2.08	2.06
11	3.23	2.86	2.66	2.54	2.45	2.39	2.34	2.30	2.27	2.25	2.21	2.17	2.12	2.10	2.08	2.05	2.03	2.00	1.97
12	3.18	2.81	2.61	2.48	2.39	2.33	2.28	2.24	2.21	2.19	2.15	2.10	2.06	2.04	2.01	1.99	1.96	1.93	1.90
13	3.14	2.76	2.56	2.43	2.35	2.28	2.23	2.20	2.16	2.14	2.10	2.05	2.01	1.98	1.96	1.93	1.90	1.88	1.85
14	3.10	2.73	2.52	2.39	2.31	2.24	2.19	2.15	2.12	2.10	2.05	2.01	1.96	1.94	1.91	1.89	1.86	1.83	1.80
15	3.07	2.70	2.49	2.36	2.27	2.21	2.16	2.12	2.09	2.06	2.02	1.97	1.92	1.90	1.87	1.85	1.82	1.79	1.76
16	3.05	2.67	2.46	2.33	2.24	2.18	2.13	2.09	2.06	2.03	1.99	1.94	1.89	1.87	1.84	1.81	1.78	1.75	1.72
17	3.03	2.64	2.44	2.31	2.22	2.15	2.10	2.06	2.03	2.00	1.96	1.91	1.86	1.84	1.81	1.78	1.75	1.72	1.69
18	3.01	2.62	2.42	2.29	2.20	2.13	2.08	2.04	2.00	1.98	1.93	1.89	1.84	1.81	1.78	1.75	1.72	1.69	1.66
19	2.99	2.61	2.40	2.27	2.18	2.11	2.06	2.02	1.98	1.96	1.91	1.86	1.81	1.79	1.76	1.73	1.70	1.67	1.63
20	2.97	2.59	2.38	2.25	2.16	2.09	2.04	2.00	1.96	1.94	1.89	1.84	1.79	1.77	1.74	1.71	1.68	1.64	1.61
21	2.96	2.57	2.36	2.23	2.14	2.08	2.02	1.98	1.95	1.92	1.87	1.83	1.78	1.75	1.72	1.69	1.66	1.62	1.59
22	2.95	2.56	2.35	2.22	2.13	2.06	2.01	1.97	1.93	1.90	1.86	1.81	1.76	1.73	1.70	1.67	1.64	1.60	1.57
23	2.94	2.55	2.34	2.21	2.11	2.05	1.99	1.95	1.92	1.89	1.84	1.80	1.74	1.72	1.69	1.66	1.62	1.59	1.55
24	2.93	2.54	2.33	2.19	2.10	2.04	1.98	1.94	1.91	1.88	1.83	1.78	1.73	1.70	1.67	1.64	1.61	1.57	1.53
25	2.92	2.53	2.32	2.18	2.09	2.02	1.97	1.93	1.89	1.87	1.82	1.77	1.72	1.69	1.66	1.63	1.59	1.56	1.52
26	2.91	2.52	2.31	2.17	2.08	2.01	1.96	1.92	1.88	1.86	1.81	1.76	1.71	1.68	1.65	1.61	1.58	1.54	1.50
27	2.90	2.51	2.30	2.17	2.07	2.00	1.95	1.91	1.87	1.85	1.80	1.75	1.70	1.67	1.64	1.60	1.57	1.53	1.49
28	2.89	2.50	2.29	2.16	2.06	2.00	1.94	1.90	1.87	1.84	1.79	1.74	1.69	1.66	1.63	1.59	1.56	1.52	1.48
29	2.89	2.50	2.28	2.15	2.06	1.99	1.93	1.89	1.86	1.83	1.78	1.73	1.68	1.65	1.62	1.58	1.55	1.51	1.47
30	2.88	2.49	2.28	2.14	2.05	1.98	1.93	1.88	1.85	1.82	1.77	1.72	1.67	1.64	1.61	1.57	1.54	1.50	1.46
40	2.84	2.44	2.23	2.09	2.00	1.93	1.87	1.83	1.79	1.76	1.71	1.66	1.61	1.57	1.54	1.51	1.47	1.42	1.38
60	2.79	2.39	2.18	2.04	1.95	1.87	1.82	1.77	1.74	1.71	1.66	1.60	1.54	1.51	1.48	1.44	1.40	1.35	1.29
120	2.75	2.35	2.13	1.99	1.90	1.82	1.77	1.72	1.68	1.65	1.60	1.55	1.48	1.45	1.41	1.37	1.32	1.26	1.19
∞	2.71	2.30	2.08	1.94	1.85	1.77	1.72	1.67	1.63	1.60	1.55	1.49	1.42	1.38	1.34	1.30	1.24	1.17	1.00

Table A.5 (Contd.)

$\alpha = 0.05$ (upper 5% fractile)

v_2 \ v_1	1	2	3	4	5	6	7	8	9	10	12	15	20	24	30	40	60	120	∞
1	161.40	199.50	215.70	224.60	230.20	234.00	236.80	238.90	240.50	241.90	243.90	245.90	248.00	249.10	250.10	251.10	252.20	253.30	254.30
2	18.51	19.00	19.16	19.25	19.30	19.33	19.35	19.37	19.38	19.40	19.41	19.43	19.45	19.45	19.46	19.47	19.48	19.49	19.50
3	10.13	9.55	9.28	9.12	9.01	8.94	8.89	8.85	8.81	8.79	8.74	8.70	8.66	8.64	8.62	8.59	8.57	8.55	8.53
4	7.71	6.94	6.59	6.39	6.26	6.16	6.09	6.04	6.00	5.96	5.91	5.86	5.80	5.77	5.75	5.72	5.69	5.66	5.63
5	6.61	5.79	5.41	5.19	5.05	4.95	4.88	4.82	4.77	4.74	4.68	4.62	4.56	4.53	4.50	4.46	4.43	4.40	4.36
6	5.99	5.14	4.76	4.53	4.39	4.28	4.21	4.15	4.10	4.06	4.00	3.94	3.87	3.84	3.81	3.77	3.74	3.70	3.67
7	5.59	4.74	4.35	4.12	3.97	3.87	3.79	3.73	3.68	3.64	3.57	3.51	3.44	3.41	3.38	3.34	3.30	3.27	3.23
8	5.32	4.46	4.07	3.84	3.69	3.58	3.50	3.44	3.39	3.35	3.28	3.22	3.15	3.12	3.08	3.04	3.01	2.97	2.93
9	5.12	4.26	3.86	3.63	3.48	3.37	3.29	3.23	3.18	3.14	3.07	3.01	2.94	2.90	2.86	2.83	2.79	2.75	2.71
10	4.96	4.10	3.71	3.48	3.33	3.22	3.14	3.07	3.02	2.98	2.91	2.85	2.77	2.74	2.70	2.66	2.62	2.58	2.54
11	4.84	3.98	3.59	3.36	3.20	3.09	3.01	2.95	2.90	2.85	2.79	2.72	2.65	2.61	2.57	2.53	2.49	2.45	2.40
12	4.75	3.89	3.49	3.26	3.11	3.00	2.91	2.85	2.80	2.75	2.69	2.62	2.54	2.51	2.47	2.43	2.38	2.34	2.30
13	4.67	3.81	3.41	3.18	3.03	2.92	2.83	2.77	2.71	2.67	2.60	2.53	2.46	2.42	2.38	2.34	2.30	2.25	2.21
14	4.60	3.74	3.34	3.11	2.96	2.85	2.76	2.70	2.65	2.60	2.53	2.46	2.39	2.35	2.31	2.27	2.22	2.18	2.13
15	4.54	3.68	3.29	3.06	2.90	2.79	2.71	2.64	2.59	2.54	2.48	2.40	2.33	2.29	2.25	2.20	2.16	2.11	2.07
16	4.49	3.63	3.24	3.01	2.85	2.74	2.66	2.59	2.54	2.49	2.42	2.35	2.28	2.24	2.19	2.15	2.11	2.06	2.01
17	4.45	3.59	3.20	2.96	2.81	2.70	2.61	2.55	2.49	2.45	2.38	2.31	2.23	2.19	2.15	2.10	2.06	2.01	1.96
18	4.41	3.55	3.16	2.93	2.77	2.66	2.58	2.51	2.46	2.41	2.34	2.27	2.19	2.15	2.11	2.06	2.02	1.97	1.92
19	4.38	3.52	3.13	2.90	2.74	2.63	2.54	2.48	2.42	2.38	2.31	2.23	2.16	2.11	2.07	2.03	1.98	1.93	1.88
20	4.35	3.49	3.10	2.87	2.71	2.60	2.51	2.45	2.39	2.35	2.28	2.20	2.12	2.08	2.04	1.99	1.95	1.90	1.84
21	4.32	3.47	3.07	2.84	2.68	2.57	2.49	2.42	2.37	2.32	2.25	2.18	2.10	2.05	2.01	1.96	1.92	1.87	1.81
22	4.30	3.44	3.05	2.82	2.66	2.55	2.46	2.40	2.34	2.30	2.23	2.15	2.07	2.03	1.98	1.94	1.89	1.84	1.78
23	4.28	3.42	3.03	2.80	2.64	2.53	2.44	2.37	2.32	2.27	2.20	2.13	2.05	2.01	1.96	1.91	1.86	1.81	1.76
24	4.26	3.40	3.01	2.78	2.62	2.51	2.42	2.36	2.30	2.25	2.18	2.11	2.03	1.98	1.94	1.89	1.84	1.79	1.73
25	4.24	3.39	2.99	2.76	2.60	2.49	2.40	2.34	2.28	2.24	2.16	2.09	2.01	1.96	1.92	1.87	1.82	1.77	1.71
26	4.23	3.37	2.98	2.74	2.59	2.47	2.39	2.32	2.27	2.22	2.15	2.07	1.99	1.95	1.90	1.85	1.80	1.75	1.69
27	4.21	3.35	2.96	2.73	2.57	2.46	2.37	2.31	2.25	2.20	2.13	2.06	1.97	1.93	1.88	1.84	1.79	1.73	1.67
28	4.20	3.34	2.95	2.71	2.56	2.45	2.36	2.29	2.24	2.19	2.12	2.04	1.96	1.91	1.87	1.82	1.77	1.71	1.65
29	4.18	3.33	2.93	2.70	2.55	2.43	2.35	2.28	2.22	2.18	2.10	2.03	1.94	1.90	1.85	1.81	1.75	1.70	1.64
30	4.17	3.32	2.92	2.69	2.53	2.42	2.33	2.27	2.21	2.16	2.09	2.01	1.93	1.89	1.84	1.79	1.74	1.68	1.62
40	4.08	3.23	2.84	2.61	2.45	2.34	2.25	2.18	2.12	2.08	2.00	1.92	1.84	1.79	1.74	1.69	1.64	1.58	1.51
60	4.00	3.15	2.76	2.53	2.37	2.25	2.17	2.10	2.04	1.99	1.92	1.84	1.75	1.70	1.65	1.59	1.53	1.47	1.39
120	3.92	3.07	2.68	2.45	2.29	2.17	2.09	2.02	1.96	1.91	1.83	1.75	1.66	1.61	1.55	1.50	1.43	1.35	1.25
∞	3.84	3.00	2.60	2.37	2.21	2.10	2.01	1.94	1.88	1.83	1.75	1.67	1.57	1.52	1.46	1.39	1.32	1.22	1.00

Table A.5 (Contd.)

$$\alpha = 0.01 \text{ (upper 1\% fractile)}$$

v_2 \ v_1	1	2	3	4	5	6	7	8	9	10	12	15	20	24	30	40	60	120	∞
1	4052.00	4999.50	5403.00	5625.00	5764.00	5859.00	5928.00	5982.00	6022.00	6056.00	6106.00	6157.00	6209.00	6235.00	6261.00	6287.00	6313.00	6339.00	6366.00
2	98.50	99.00	99.17	99.25	99.30	99.33	99.36	99.37	99.39	99.40	99.42	99.43	99.45	99.46	99.47	99.47	99.48	99.49	99.50
3	34.12	30.82	29.46	28.71	28.24	27.91	27.67	27.49	27.35	27.23	27.05	26.87	26.69	26.60	26.50	26.41	26.32	26.22	26.13
4	21.20	18.00	16.69	15.98	15.52	15.21	14.98	14.80	14.66	14.55	14.37	14.20	14.02	13.93	13.84	13.75	13.65	13.56	13.46
5	16.26	13.27	12.06	11.39	10.97	10.67	10.46	10.29	10.16	10.05	9.89	9.72	9.55	9.47	9.38	9.29	9.20	9.11	9.02
6	13.75	10.92	9.78	9.15	8.75	8.47	8.26	8.10	7.98	7.87	7.72	7.56	7.40	7.31	7.23	7.14	7.06	6.97	6.88
7	12.25	9.55	8.45	7.85	7.46	7.19	6.99	6.84	6.72	6.62	6.47	6.31	6.16	6.07	5.99	5.91	5.82	5.74	5.65
8	11.26	8.65	7.59	7.01	6.63	6.37	6.18	6.03	5.91	5.81	5.67	5.52	5.36	5.28	5.20	5.12	5.03	4.95	4.86
9	10.56	8.02	6.99	6.42	6.06	5.80	5.61	5.47	5.35	5.26	5.11	4.96	4.81	4.73	4.65	4.57	4.48	4.40	4.31
10	10.04	7.56	6.55	5.99	5.64	5.39	5.20	5.06	4.94	4.85	4.71	4.56	4.41	4.33	4.25	4.17	4.08	4.00	3.91
11	9.65	7.21	6.22	5.67	5.32	5.07	4.89	4.74	4.63	4.54	4.40	4.25	4.10	4.02	3.94	3.86	3.78	3.69	3.60
12	9.33	6.93	5.95	5.41	5.06	4.82	4.64	4.50	4.39	4.30	4.16	4.01	3.86	3.78	3.70	3.62	3.54	3.45	3.36
13	9.07	6.70	5.74	5.21	4.86	4.62	4.44	4.30	4.19	4.10	3.96	3.82	3.66	3.59	3.51	3.43	3.34	3.25	3.17
14	8.86	6.51	5.56	5.04	4.69	4.46	4.28	4.14	4.03	3.94	3.80	3.66	3.51	3.43	3.35	3.27	3.18	3.09	3.00
15	8.68	6.36	5.42	4.89	4.56	4.32	4.14	4.00	3.89	3.80	3.67	3.52	3.37	3.29	3.21	3.13	3.05	2.96	2.87
16	8.53	6.23	5.29	4.77	4.44	4.20	4.03	3.89	3.78	3.69	3.55	3.41	3.26	3.18	3.10	3.02	2.93	2.84	2.75
17	8.40	6.11	5.18	4.67	4.34	4.10	3.93	3.79	3.68	3.59	3.46	3.31	3.16	3.08	3.00	2.92	2.83	2.75	2.65
18	8.29	6.01	5.09	4.58	4.25	4.01	3.84	3.71	3.60	3.51	3.37	3.23	3.08	3.00	2.92	2.84	2.75	2.66	2.57
19	8.18	5.93	5.01	4.50	4.17	3.94	3.77	3.63	3.52	3.43	3.30	3.15	3.00	2.92	2.84	2.76	2.67	2.58	2.49
20	8.10	5.85	4.94	4.43	4.10	3.87	3.70	3.56	3.46	3.37	3.23	3.09	2.94	2.86	2.78	2.69	2.61	2.52	2.42
21	8.02	5.78	4.87	4.37	4.04	3.81	3.64	3.51	3.40	3.31	3.17	3.03	2.88	2.80	2.72	2.64	2.55	2.46	2.36
22	7.95	5.72	7.82	4.31	3.99	3.76	3.59	3.45	3.35	3.26	3.12	2.98	2.83	2.75	2.67	2.58	2.50	2.40	2.31
23	7.88	5.66	4.76	4.26	3.94	3.71	3.54	3.41	3.30	3.21	3.07	2.93	2.78	2.70	2.62	2.54	2.45	2.35	2.26
24	7.82	5.61	7.72	4.22	3.90	3.67	3.50	3.36	3.26	3.17	3.03	2.89	2.74	2.66	2.58	2.49	2.40	2.31	2.21
25	7.77	5.57	4.68	4.18	3.85	3.63	3.46	3.32	3.22	3.13	2.99	2.85	2.70	2.62	2.54	2.45	2.36	2.27	2.17
26	7.72	5.53	4.64	4.14	3.82	3.59	3.42	3.29	3.18	3.09	2.96	2.81	2.66	2.58	2.50	2.42	2.33	2.23	2.13
27	7.68	5.49	4.60	4.11	3.78	3.56	3.39	3.26	3.15	3.06	2.93	2.78	2.63	2.55	2.47	2.38	2.29	2.20	2.10
28	7.64	5.45	4.57	4.07	3.75	3.53	3.36	3.23	3.12	3.03	2.90	2.75	2.60	2.52	2.44	2.35	2.26	2.17	2.06
29	7.60	5.42	4.54	4.04	3.73	3.50	3.33	3.20	3.09	3.00	2.87	2.73	2.57	2.49	2.41	2.33	2.23	2.14	2.03
30	7.56	5.39	4.51	4.02	3.70	3.47	3.30	3.17	3.07	2.98	2.84	2.70	2.55	2.47	2.39	2.30	2.21	2.11	2.01
40	7.31	5.18	4.31	3.83	3.51	3.29	3.12	2.99	2.89	2.80	2.66	2.52	2.37	2.29	2.20	2.11	2.02	1.92	1.80
60	7.08	4.98	4.13	3.65	3.34	3.12	2.95	2.82	2.72	2.63	2.50	2.35	2.20	2.12	2.03	1.94	1.84	1.73	1.60
120	6.85	4.79	3.95	3.48	3.17	2.96	2.79	2.66	2.56	2.47	2.34	2.19	2.03	1.95	1.86	1.76	2.66	1.53	1.38
∞	6.63	4.61	3.78	3.32	3.02	2.80	2.64	2.51	2.41	2.32	2.18	2.04	1.88	1.79	1.70	1.59	1.47	1.32	1.00

From E.S. Pearson and H.O. Hartley (1976), *Biometrika Tables for Statisticians*, Vol. I. Reproduced with the kind permission of Oxford University Press.

Table A.6 Binomial Probabilities (X is $b(X; n, p)$)

Given n and p, the table gives the binomial probability of $X = r$ successes or $P(X = r; n, p) = \binom{n}{r} p^r (1 - p)^{n-r}$, $r = 0, 1, \ldots, n$. For $p > 0.50$, $P(X = r; n, p) = P(X = n - r; n, 1 - p)$ (e.g., $P(3; 10, 0.40) = 0.2150$; $P(4; 10, 0.70) = P(6; 10, 0.30) = 0.0368$).

n	r	p 0.05	0.10	0.15	0.20	0.25	0.30	0.35	0.40	0.45	0.50
1	0	0.9500	0.9000	0.8500	0.8000	0.7500	0.7000	0.6500	0.6000	0.5500	0.5000
	1	0.0500	0.1000	0.1500	0.2000	0.2500	0.3000	0.3500	0.4000	0.4500	0.5000
2	0	0.9025	0.8100	0.7225	0.6400	0.5625	0.4900	0.4225	0.3600	0.3025	0.2500
	1	0.0950	0.1800	0.2550	0.3200	0.3750	0.4200	0.4550	0.4800	0.4950	0.5000
	2	0.0025	0.0100	0.2550	0.0400	0.0625	0.0900	0.1225	0.1600	0.2025	0.2500
3	0	0.8574	0.7290	0.6141	0.5120	0.4219	0.3430	0.2746	0.2160	0.1664	0.1250
	1	0.1354	0.2430	0.3251	0.3840	0.4219	0.4410	0.4436	0.4320	0.4084	0.3750
	2	0.0071	0.0270	0.0574	0.0960	0.1406	0.1890	0.2389	0.2880	0.3341	0.3750
	3	0.0001	0.0010	0.0034	0.0080	0.0156	0.0270	0.0429	0.0640	0.0911	0.1250
4	0	0.8145	0.6561	0.5220	0.4098	0.3164	0.2401	0.1785	0.1296	0.0915	0.0625
	1	0.1715	0.2916	0.3685	0.4096	0.4219	0.4116	0.3845	0.3456	0.2995	0.2500
	2	0.0135	0.0486	0.0975	0.1536	0.2109	0.2646	0.3105	0.3456	0.3675	0.3750
	3	0.0005	0.0036	0.0115	0.0256	0.0469	0.0756	0.1115	0.1536	0.2005	0.2500
	4	0.0000	0.0001	0.0005	0.0016	0.0039	0.0081	0.0150	0.0256	0.0410	0.0625
5	0	0.7738	0.5905	0.4437	0.3277	0.2373	0.1681	0.1160	0.0778	0.0503	0.0312
	1	0.2036	0.3280	0.3916	0.4096	0.3955	0.3602	0.3124	0.2592	0.2059	0.1562
	2	0.0214	0.0729	0.1382	0.2048	0.2637	0.3087	0.3364	0.3456	0.3369	0.3125
	3	0.0011	0.0081	0.0244	0.0512	0.0879	0.1323	0.1811	0.2304	0.2757	0.3125
	4	0.0000	0.0004	0.0022	0.0064	0.0146	0.0284	0.0488	0.0768	0.1128	0.1562
	5	0.0000	0.0000	0.0001	0.0003	0.0010	0.0024	0.0053	0.0102	0.0185	0.0312
6	0	0.7351	0.5314	0.3771	0.2621	0.1780	0.1176	0.0754	0.0467	0.0277	0.0156
	1	0.2321	0.3543	0.3993	0.3932	0.3560	0.3025	0.2437	0.1866	0.1359	0.0938
	2	0.0305	0.0984	0.1762	0.2458	0.2966	0.3241	0.3280	0.3110	0.2780	0.2344
	3	0.0021	0.0146	0.0415	0.0819	0.1318	0.1852	0.2355	0.2785	0.3032	0.3125
	4	0.0001	0.0012	0.0055	0.0154	0.0330	0.0595	0.0951	0.1382	0.1861	0.2344
	5	0.0000	0.0001	0.0004	0.0015	0.0044	0.0102	0.0205	0.0369	0.0609	0.0938
	6	0.0000	0.0000	0.0000	0.0001	0.0002	0.0007	0.0018	0.0041	0.0083	0.0156
7	0	0.6983	0.4783	0.3206	0.2097	0.1335	0.0824	0.0490	1.0280	0.0152	0.0078
	1	0.2573	0.3720	0.3960	0.3670	0.3115	0.2471	0.1848	0.1306	0.0872	0.0547
	2	0.0406	0.1240	0.2097	0.2753	0.3115	0.3177	0.2985	0.2613	0.2140	0.1641
	3	0.0036	0.0230	0.0617	0.1147	0.1730	0.2269	0.2679	0.2903	0.2918	0.2734
	4	0.0002	0.0026	0.0109	0.0287	0.0577	0.0972	0.1442	0.1935	0.2388	0.2734
	5	0.0000	0.0002	0.0012	0.0043	0.0115	0.0250	0.0466	0.0774	0.1172	0.1641
	6	0.0000	0.0000	0.0001	0.0004	0.0013	0.0036	0.0084	0.0172	0.0320	0.0547
	7	0.0000	0.0000	0.0000	0.0000	0.0001	0.0002	0.0006	0.0016	0.0037	0.0078

Table A.6 (Contd.)

n	r	0.05	0.10	0.15	0.20	0.25	0.30	0.35	0.40	0.45	0.50
							p				
8	0	0.6634	0.4305	0.2725	0.1678	0.1001	0.0576	0.0319	0.0168	0.0084	0.0039
	1	0.2793	0.3826	0.3847	0.3355	0.2760	0.1977	0.1373	0.0896	0.0548	0.0312
	2	0.0515	0.1488	0.2376	0.2936	0.3115	0.2965	0.2587	0.2090	0.1569	0.1094
	3	0.0054	0.0331	0.0839	0.1468	0.2076	0.2541	0.2786	0.2787	0.2568	0.2188
	4	0.0004	0.0046	0.0185	0.0459	0.0865	0.1361	0.1875	0.2322	0.2627	0.2734
	5	0.0000	0.0004	0.0026	0.0092	0.0231	0.0467	0.0808	0.1239	0.1719	0.2188
	6	0.0000	0.0000	0.0002	0.0011	0.0038	0.0100	0.0217	0.0413	0.0703	0.1094
	7	0.0000	0.0000	0.0000	0.0001	0.0004	0.0012	0.0033	0.0079	0.0164	0.0312
	8	0.0000	0.0000	0.0000	0.0000	0.0000	0.0001	0.0002	0.0007	0.0017	0.0039
9	0	0.6302	0.3874	0.2316	0.1342	0.0751	0.0404	0.0277	0.0101	0.0046	0.0020
	1	0.2985	0.3874	0.3679	0.3020	0.2253	0.1556	0.1004	0.0605	0.0339	0.0176
	2	0.0629	0.1722	0.7260	0.3020	0.3003	0.2668	0.2162	0.1612	0.1110	0.0703
	3	0.0077	0.0446	0.1069	0.1762	0.2336	0.2668	0.2716	0.2508	0.2119	0.1641
	4	0.0006	0.0074	0.0283	0.0661	0.1168	0.1715	0.2194	0.2508	0.2600	0.2461
	5	0.0000	0.0008	0.0050	0.0165	0.0389	0.0735	0.1181	0.1672	0.2128	0.2461
	6	0.0000	0.0001	0.0006	0.0028	0.0087	0.0210	0.0424	0.0743	0.1160	0.1641
	7	0.0000	0.0000	0.0000	0.0003	0.0012	0.0039	0.0098	0.0212	0.0407	0.0703
	8	0.0000	0.0000	0.0000	0.0000	0.0001	0.0004	0.0013	0.0035	0.0083	0.0176
	9	0.0000	0.0000	0.0000	0.0000	0.0000	0.0000	0.0001	0.0003	0.0008	0.0020
10	0	0.5987	0.3487	0.1969	0.1074	0.0563	0.0282	0.0135	0.0060	0.0025	0.0010
	1	0.3151	0.3874	0.3474	0.2684	0.1877	0.1211	0.0725	0.0403	0.0207	0.0098
	2	0.0746	0.1937	0.2759	0.3020	0.2816	0.2335	0.1757	0.1209	0.0763	0.0439
	3	0.0105	0.0574	0.1298	0.2013	0.2503	0.2668	0.2522	0.2150	0.1665	0.1172
	4	0.0010	0.0112	0.0401	0.0881	0.1460	0.2001	0.2377	0.2508	0.2384	0.2051
	5	0.0001	0.0015	0.0085	0.0284	0.0584	0.1029	0.1536	0.2007	0.2340	0.2461
	6	0.0000	0.0001	0.0012	0.0055	0.0162	0.0368	0.0689	0.1115	0.1596	0.2051
	7	0.0000	0.0000	0.0001	0.0608	0.0031	0.0090	0.0212	0.0425	0.0746	0.1172
	8	0.0000	0.0000	0.0000	0.0001	0.0004	0.0014	0.0043	0.0106	0.0229	0.0439
	9	0.0000	0.0000	0.0000	0.0000	0.0000	0.0001	0.0005	0.0016	0.0042	0.0098
	10	0.0000	0.0000	0.0000	0.0000	0.0000	0.0000	0.0000	0.0001	0.0003	0.0016
11	0	0.5688	0.3138	0.1673	0.0859	0.0422	0.0198	0.0088	0.0036	0.0014	0.0005
	1	0.3293	0.3835	0.3248	0.2362	0.1549	0.0932	0.0518	0.0266	0.0125	0.0054
	2	0.0867	0.2131	0.2866	0.2953	0.2581	0.1998	0.1395	0.0887	0.0513	0.0269
	3	0.0137	0.0710	0.1517	0.2215	0.2581	0.2568	0.2254	0.1774	0.1259	0.0806
	4	0.0014	0.0158	0.0536	0.1107	0.1721	0.2201	0.2428	0.2365	0.2060	0.1611
	5	0.0001	0.0025	0.0132	0.0388	0.0803	0.1231	0.1830	0.2207	0.2360	0.2256
	6	0.0000	0.0003	0.0023	0.0097	0.0268	0.0566	0.0985	0.1471	0.1931	0.2256
	7	0.0000	0.0000	0.0003	0.0017	0.0064	0.0173	0.0379	0.0701	0.1128	0.1611
	8	0.0000	0.0000	0.0000	0.0002	0.0011	0.0037	0.0102	0.0234	0.0462	0.0806
	9	0.0000	0.0000	0.0000	0.0000	0.0001	0.0005	0.0018	0.0052	0.0126	0.0269
	10	0.0000	0.0000	0.0000	0.0000	0.0000	0.0000	0.0002	0.0007	0.0021	0.0054
	11	0.0000	0.0000	0.0000	0.0000	0.0000	0.0000	0.0000	0.0000	0.0002	0.0005

Table A.6 (Contd.)

						p					
n	r	0.05	0.10	0.15	0.20	0.25	0.30	0.35	0.40	0.45	0.50
12	0	0.5404	0.2824	0.1422	0.0687	0.0317	0.0138	0.0057	0.0022	0.0008	0.0002
	1	0.3413	0.3766	0.3012	0.2062	0.1267	0.0712	0.0368	0.0174	0.0075	0.0029
	2	0.0988	0.2301	0.2924	0.2835	0.2323	0.1678	0.1088	0.0639	0.0339	0.0161
	3	0.0173	0.0852	0.1720	0.2362	0.2581	0.2397	0.1954	0.1419	0.0923	0.0537
	4	0.0021	0.0213	0.0683	0.1329	0.1936	0.2311	0.2367	0.2128	0.1700	0.1208
	5	0.0002	0.0038	0.0193	0.0532	0.1032	0.1585	0.2039	0.2270	0.2225	0.1934
	6	0.0000	0.0005	0.0040	0.0155	0.0401	0.0792	0.1281	0.1766	0.2124	0.2256
	7	0.0000	0.0000	0.0006	0.0033	0.0115	0.0291	0.0591	0.1009	0.1489	0.1934
	8	0.0000	0.0000	0.0001	0.0005	0.0024	0.0078	0.0199	0.0420	0.0762	0.1208
	9	0.0000	0.0000	0.0000	0.0001	0.0004	0.0015	0.0048	0.0125	0.0277	0.0537
	10	0.0000	0.0000	0.0000	0.0000	0.0000	0.0002	0.0008	0.0025	0.0068	0.0161
	11	0.0000	0.0000	0.0000	0.0000	0.0000	0.0000	0.0001	0.0003	0.0010	0.0029
	12	0.0000	0.0000	0.0000	0.0000	0.0000	0.0000	0.0000	0.0000	0.0001	0.0002
13	0	0.5133	0.2542	0.1209	0.0550	0.0238	0.0097	0.0037	0.0013	0.0004	0.0001
	1	0.3512	0.3672	0.2774	0.1787	0.1029	0.0540	0.0259	0.0113	0.0045	0.0016
	2	0.1109	0.2448	0.2937	0.2680	0.2059	0.1388	0.0836	0.0453	0.0220	0.0095
	3	0.0214	0.0997	0.1900	0.2457	0.2517	0.2181	0.1651	0.1107	0.0660	0.0349
	4	0.0028	0.0277	0.0838	0.1535	0.2097	0.2337	0.2222	0.1845	0.1350	0.0873
	5	0.0003	0.0055	0.0266	0.0691	0.1258	0.1803	0.2154	0.2214	0.1989	0.1571
	6	0.0000	0.0008	0.0063	0.0230	0.0559	0.1030	0.1546	0.1968	0.2169	0.2095
	7	0.0000	0.0001	0.0011	0.0058	0.0186	0.0442	0.0833	0.1312	0.1775	0.2095
	8	0.0000	0.0000	0.0001	0.0011	0.0047	0.0142	0.0336	0.0656	0.1089	0.1571
	9	0.0000	0.0000	0.0000	0.0001	0.0009	0.0034	0.0101	0.0243	0.0495	0.0873
	10	0.0000	0.0000	0.0000	0.0000	0.0001	0.0006	0.0022	0.0065	0.0162	0.0349
	11	0.0000	0.0000	0.0000	0.0000	0.0000	0.0001	0.0003	0.0012	0.0036	0.0095
	12	0.0000	0.0000	0.0000	0.0000	0.0000	0.0000	0.0000	0.0001	0.0005	0.0016
	13	0.0000	0.0000	0.0000	0.0000	0.0000	0.0000	0.0000	0.0000	0.0000	0.0001
14	0	0.4877	0.2288	0.1028	0.0440	0.0178	0.0068	0.0024	0.0008	0.0002	0.0001
	1	0.3593	0.3559	0.2539	0.1539	0.0832	0.0407	0.0181	0.0073	0.0027	0.0009
	2	0.1229	0.2570	0.2912	0.2501	0.1802	0.1134	0.0634	0.0317	0.0141	0.0056
	3	0.0259	0.1142	0.2056	0.2501	0.2402	0.1943	0.1366	0.0845	0.0462	0.0222
	4	0.0037	0.0349	0.0998	0.1720	0.2202	0.2290	0.2022	0.1549	0.1040	0.0611
	5	0.0004	0.0078	0.0352	0.0860	0.1468	0.1963	0.2178	0.2066	0.1701	0.1222
	6	0.0000	0.0013	0.0093	0.0322	0.0734	0.1262	0.1759	0.2066	0.2088	0.1833
	7	0.0000	0.0002	0.0019	0.0092	0.0280	0.0618	0.1082	0.1574	0.1952	0.2095
	8	0.0000	0.0000	0.0003	0.0020	0.0082	0.0232	0.0510	0.0918	0.1398	0.1833
	9	0.0000	0.0000	0.0000	0.0003	0.0018	0.0066	0.0183	0.0408	0.0762	0.1222
	10	0.0000	0.0000	0.0000	0.0000	0.0003	0.0014	0.0049	0.0136	0.0312	0.0611
	11	0.0000	0.0000	0.0000	0.0000	0.0000	0.0002	0.0010	0.0033	0.0093	0.0222
	12	0.0000	0.0000	0.0000	0.0000	0.0000	0.0000	0.0001	0.0005	0.0019	0.0056
	13	0.0000	0.0000	0.0000	0.0000	0.0000	0.0000	0.0000	0.0001	0.0002	0.0009
	14	0.0000	0.0000	0.0000	0.0000	0.0000	0.0000	0.0000	0.0000	0.0000	0.0001

Table A.6 (Contd.)

						p					
n	r	**0.05**	**0.10**	**0.15**	**0.20**	**0.25**	**0.30**	**0.35**	**0.40**	**0.45**	**0.50**
15	**0**	0.4633	0.2059	0.0874	0.0352	0.0134	0.0047	0.0016	0.0005	0.0001	0.0000
	1	0.3658	0.3432	0.2312	0.1319	0.0668	0.0305	0.0126	0.0047	0.0016	0.0005
	2	0.1348	0.2669	0.2856	0.2309	0.1559	0.0916	0.0476	0.0219	0.0090	0.0032
	3	0.0307	0.1285	0.2184	0.2501	0.2252	0.1700	0.1110	0.0634	0.0318	0.0139
	4	0.0049	0.0428	0.1156	0.1876	0.2252	0.2186	0.1792	0.1268	0.0780	0.0417
	5	0.0006	0.0105	0.0449	0.1032	0.1651	0.2061	0.2123	0.1859	0.1404	0.0916
	6	0.0000	0.0019	0.0132	0.0430	0.0917	0.1472	0.1906	0.2066	0.1914	0.1527
	7	0.0000	0.0003	0.0030	0.0138	0.0393	0.0811	0.1319	0.1771	0.2013	0.1964
	8	0.0000	0.0000	0.0005	0.0035	0.0131	0.0348	0.0710	0.1181	0.1647	0.1964
	9	0.0000	0.0000	0.0001	0.0007	0.0034	0.0116	0.0298	0.0612	0.1048	0.1527
	10	0.0000	0.0000	0.0000	0.0001	0.0007	0.0030	0.0096	0.0245	0.0515	0.0916
	11	0.0000	0.0000	0.0000	0.0000	0.0001	0.0006	0.0024	0.0074	0.0191	0.0417
	12	0.0000	0.0000	0.0000	0.0000	0.0000	0.0001	0.0004	0.0016	0.0052	0.0139
	13	0.0000	0.0000	0.0000	0.0000	0.0000	0.0000	0.0001	0.0003	0.0010	0.0032
	14	0.0000	0.0000	0.0000	0.0000	0.0000	0.0000	0.0000	0.0000	0.0001	0.0005
	15	0.0000	0.0000	0.0000	0.0000	0.0000	0.0000	0.0000	0.0000	0.0000	0.0000
16	**0**	0.4401	0.1853	0.0743	0.0281	0.0100	0.0033	0.0010	0.0003	0.0001	0.0000
	1	0.3706	0.3294	0.2097	0.1126	0.0535	0.0228	0.0087	0.0030	0.0009	0.0002
	2	0.1463	0.2745	0.2775	0.2111	0.1336	0.0732	0.0353	0.0150	0.0056	0.0018
	3	0.0359	0.1423	0.2285	0.2463	0.2079	0.1465	0.0888	0.0468	0.0215	0.0085
	4	0.0061	0.0514	0.1311	0.2001	0.2252	0.2040	0.1553	0.1014	0.0572	0.0278
	5	0.0008	0.0137	0.0555	0.1201	0.1802	0.2099	0.2008	0.1623	0.1123	0.0667
	6	0.0001	0.0028	0.0180	0.0550	0.1101	0.1649	0.1982	0.1983	0.1684	0.1222
	7	0.0000	0.0004	0.0045	0.0197	0.0524	0.1010	0.1524	0.1889	0.1969	0.1746
	8	0.0000	0.0001	0.0009	0.0055	0.0197	0.0487	0.0923	0.1417	0.1812	0.1964
	9	0.0000	0.0000	0.0001	0.0012	0.0058	0.0185	0.0442	0.0840	0.1318	0.1746
	10	0.0000	0.0000	0.0000	0.0002	0.0014	0.0056	0.0167	0.0392	0.0755	0.1222
	11	0.0000	0.0000	0.0000	0.0000	0.0002	0.0013	0.0049	0.0142	0.0337	0.0667
	12	0.0000	0.0000	0.0000	0.0000	0.0000	0.0002	0.0011	0.0040	0.0115	0.0278
	13	0.0000	0.0000	0.0000	0.0000	0.0000	0.0000	0.0002	0.0008	0.0029	0.0085
	14	0.0000	0.0000	0.0000	0.0000	0.0000	0.0000	0.0000	0.0001	0.0005	0.0018
	15	0.0000	0.0000	0.0000	0.0000	0.0000	0.0000	0.0000	0.0000	0.0001	0.0002
	16	0.0000	0.0000	0.0000	0.0000	0.0000	0.0000	0.0000	0.0000	0.0000	0.0000
17	**0**	0.4181	0.1668	0.0631	0.0225	0.0075	0.0023	0.0007	0.0002	0.0000	0.0000
	1	0.3741	0.3150	0.1893	0.0957	0.0426	0.0169	0.0060	0.0019	0.0005	0.0001
	2	0.1575	0.2800	0.2673	0.1914	0.1136	0.0581	0.0260	0.0102	0.0035	0.0010
	3	0.0415	0.1556	0.2359	0.2393	0.1893	0.1245	0.0701	0.0341	0.0144	0.0052
	4	0.0076	0.0605	0.1457	0.2093	0.2209	0.1868	0.1320	0.0796	0.0411	0.0182
	5	0.0010	0.0175	0.0668	0.1361	0.1914	0.2081	0.1849	0.1379	0.0875	0.0472
	6	0.0001	0.0039	0.0236	0.0680	0.1276	0.1784	0.1991	0.1839	0.1432	0.0944
	7	0.0000	0.0007	0.0065	0.0267	0.0668	0.1201	0.1685	0.1927	0.1841	0.1484
	8	0.0000	0.0001	0.0014	0.0084	0.0279	0.0644	0.1143	0.1606	0.1883	0.1855
	9	0.0000	0.0000	0.0003	0.0021	0.0093	0.0276	0.0611	0.1070	0.1540	0.1855
	10	0.0000	0.0000	0.0000	0.0004	0.0025	0.0095	0.0263	0.0571	0.1008	0.1484

Table A.6 (Contd.)

n	r	0.05	0.10	0.15	0.20	0.25	0.30	0.35	0.40	0.45	0.50
						p					
17	11	0.0000	0.0000	0.0000	0.0001	0.0005	0.0026	0.0090	0.0242	0.0525	0.0944
	12	0.0000	0.0000	0.0000	0.0000	0.0001	0.0006	0.0024	0.0081	0.0215	0.0472
	13	0.0000	0.0000	0.0000	0.0000	0.0000	0.0001	0.0005	0.0021	0.0068	0.0182
	14	0.0000	0.0000	0.0000	0.0000	0.0000	0.0000	0.0001	0.0004	0.0016	0.0052
	15	0.0000	0.0000	0.0000	0.0000	0.0000	0.0000	0.0000	0.0001	0.0003	0.0010
	16	0.0000	0.0000	0.0000	0.0000	0.0000	0.0000	0.0000	0.0000	0.0000	0.0001
	17	0.0000	0.0000	0.0000	0.0000	0.0000	0.0000	0.0000	0.0000	0.0000	0.0000
18	0	0.3972	0.1501	0.0536	0.0180	0.0058	0.0016	0.0004	0.0001	0.0000	0.0000
	1	0.3763	0.3002	0.1704	0.0811	0.0338	0.0126	0.0042	0.0012	0.0003	0.0001
	2	0.1683	0.2835	0.2558	0.1723	0.0958	0.0458	0.0190	0.0069	0.0022	0.0006
	3	0.0473	0.1680	0.2406	0.2297	0.1704	0.1046	0.0547	0.0246	0.0095	0.0031
	4	0.0093	0.0700	0.1592	0.2153	0.2130	0.1681	0.1104	0.0614	0.0291	0.0117
	5	0.0014	0.0218	0.0787	0.1507	0.1988	0.2017	0.1664	0.1146	0.0666	0.0327
	6	0.0002	0.0052	0.0310	0.0818	0.1436	0.1873	0.1941	0.1655	0.1181	0.0708
	7	0.0000	0.0010	0.0091	0.0350	0.0820	0.1376	0.1792	0.1892	0.1657	0.1214
	8	0.0000	0.0002	0.0022	0.0120	0.0376	0.0811	0.1327	0.1734	0.1864	0.1669
	9	0.0000	0.0000	0.0004	0.0033	0.0139	0.0386	0.0794	0.1284	0.1694	0.1855
	10	0.0000	0.0000	0.0001	0.0008	0.0042	0.0149	0.0385	0.0771	0.1248	0.1669
	11	0.0000	0.0000	0.0000	0.0001	0.0010	0.0046	0.0151	0.0374	0.0742	0.1214
	12	0.0000	0.0000	0.0000	0.0000	0.0002	0.0012	0.0047	0.0145	0.0354	0.0708
	13	0.0000	0.0000	0.0000	0.0000	0.0000	0.0002	0.0012	0.0045	0.0134	0.0327
	14	0.0000	0.0000	0.0000	0.0000	0.0000	0.0000	0.0002	0.0011	0.0039	0.0117
	15	0.0000	0.0000	0.0000	0.0000	0.0000	0.0000	0.0000	0.0002	0.0009	0.0031
	16	0.0000	0.0000	0.0000	0.0000	0.0000	0.0000	0.0000	0.0000	0.0001	0.0006
	17	0.0000	0.0000	0.0000	0.0000	0.0000	0.0000	0.0000	0.0000	0.0000	0.0001
	18	0.0000	0.0000	0.0000	0.0000	0.0000	0.0000	0.0000	0.0000	0.0000	0.0000
19	0	0.3774	0.1351	0.0456	0.0144	0.0042	0.0011	0.0003	0.0001	0.0000	0.0000
	1	0.3774	0.2852	0.1529	0.0685	0.0268	0.0093	0.0029	0.0008	0.0002	0.0000
	2	0.1787	0.2852	0.2428	0.1540	0.0803	0.0358	0.0138	0.0046	0.0013	0.0003
	3	0.0533	0.1796	0.2428	0.2182	0.1517	0.0869	0.0422	0.0175	0.0062	0.0018
	4	0.0112	0.0798	0.1714	0.2182	0.2023	0.1491	0.0909	0.0467	0.0203	0.0074
	5	0.0018	0.0266	0.0907	0.1636	0.2023	0.1916	0.1468	0.0933	0.0497	0.0222
	6	0.0002	0.0069	0.0374	0.0955	0.1574	0.1916	0.1844	0.1451	0.0949	0.0518
	7	0.0000	0.0014	0.0122	0.0443	0.0974	0.1525	0.1844	0.1797	0.1443	0.0961
	8	0.0000	0.0002	0.0032	0.0166	0.0487	0.0981	0.1489	0.1797	0.1771	0.1442
	9	0.0000	0.0000	0.0007	0.0051	0.0198	0.0514	0.0980	0.1464	0.1771	0.1762
	10	0.0000	0.0000	0.0001	0.0013	0.0066	0.0220	0.0528	0.0976	0.1449	0.1762
	11	0.0000	0.0000	0.0000	0.0003	0.0018	0.0077	0.0233	0.0532	0.0970	0.1442
	12	0.0000	0.0000	0.0000	0.0000	0.0004	0.0022	0.0083	0.0237	0.0529	0.0961

Table A.6 (Contd.)

						p					
n	r	0.05	0.10	0.15	0.20	0.25	0.30	0.35	0.40	0.45	0.50
19	13	0.0000	0.0000	0.0000	0.0000	0.0001	0.0005	0.0024	0.0085	0.0233	0.0518
	14	0.0000	0.0000	0.0000	0.0000	0.0000	0.0001	0.0006	0.0024	0.0082	0.0222
	15	0.0000	0.0000	0.0000	0.0000	0.0000	0.0000	0.0001	0.0005	0.0022	0.0074
	16	0.0000	0.0000	0.0000	0.0000	0.0000	0.0000	0.0000	0.0001	0.0005	0.0018
	17	0.0000	0.0000	0.0000	0.0000	0.0000	0.0000	0.0000	0.0000	0.0001	0.0003
	18	0.0000	0.0000	0.0000	0.0000	0.0000	0.0000	0.0000	0.0000	0.0000	0.0000
	19	0.0000	0.0000	0.0000	0.0000	0.0600	0.0000	0.0000	0.0000	0.0000	0.0000
20	0	0.3585	0.1216	0.0388	0.0115	0.0032	0.0008	0.0002	0.0000	0.0000	0.0000
	1	0.3774	0.2702	0.1368	0.0576	0.0211	0.0068	0.0020	0.0005	0.0001	0.0000
	2	0.1887	0.2852	0.2293	0.1369	0.0669	0.0278	0.0100	0.0031	0.0008	0.0002
	3	0.0596	0.1901	0.2428	0.2054	0.1339	0.0716	0.0323	0.0123	0.0040	0.0011
	4	0.0133	0.0898	0.1821	0.2182	0.1897	0.1304	0.0738	0.0350	0.0139	0.0046
	5	0.0022	0.0319	0.1028	0.1746	0.2023	0.1789	0.1272	0.0746	0.0365	0.0148
	6	0.0003	0.0089	0.0454	0.1091	0.1686	0.1916	0.1712	0.1244	0.0746	0.0370
	7	0.0000	0.0020	0.0160	0.0545	0.1124	0.1643	0.1844	0.1659	0.1221	0.0739
	8	0.0000	0.0004	0.0046	0.0222	0.0609	0.1144	0.1614	0.1797	0.1623	0.1201
	9	0.0000	0.0001	0.0011	0.0074	0.0271	0.0654	0.1158	0.1597	0.1771	0.1602
	10	0.0000	0.0000	0.0002	0.0020	0.0099	0.0308	0.0686	0.1171	0.1593	0.1762
	11	0.0000	0.0000	0.0000	0.0005	0.0030	0.0120	0.0336	0.0710	0.1185	0.1602
	12	0.0000	0.0000	0.0000	0.0001	0.0008	0.0039	0.0136	0.0355	0.0727	0.1201
	13	0.0000	0.0000	0.0000	0.0000	0.0002	0.0010	0.0045	0.0146	0.0366	0.0739
	14	0.0000	0.0000	0.0000	0.0000	0.0000	0.0002	0.0012	0.0049	0.0150	0.0370
	15	0.0000	0.0000	0.0000	0.0000	0.0000	0.0000	0.0003	0.0013	0.0049	0.0148
	16	0.0000	0.0000	0.0000	0.0000	0.0000	0.0000	0.0000	0.0003	0.0013	0.0046
	17	0.0000	0.0000	0.0000	0.0000	0.0000	0.0000	0.0000	0.0000	0.0002	0.0011
	18	0.0000	0.0000	0.0000	0.0000	0.0000	0.0000	0.0000	0.0000	0.0000	0.0002
	19	0.0000	0.0000	0.0000	0.0000	0.0000	0.0000	0.0000	0.0000	0.0000	0.0000
	20	0.0000	0.0000	0.0000	0.0000	0.0000	0.0000	0.0000	0.0000	0.0000	0.0000

Table A.7   Cumulative Distribution Function Values for the Binomial Distribution (X is $b(X; n, p)$)

$$B(x; n, p) = \sum_{i \le x} \binom{n}{i} p^i (1-p)^{n-i} = \sum_{i \le x} b(i; n, p)$$

$B(x; n, p)$ gives the probability that the random variable X assumes a value $\le x$.

				p		
n	x	**0.1**	**0.2**	**0.3**	**0.4**	**0.5**
2	**0**	0.8100	0.6400	0.4900	0.3600	0.2500
	1	0.9900	0.9600	0.9100	0.8400	0.7500
3	**0**	0.7290	0.5120	0.3430	0.2160	0.1250
	1	0.9720	0.8960	0.7840	0.6480	0.5000
	2	0.9990	0.9920	0.9730	0.9360	0.8750
4	**0**	0.6561	0.4096	0.2401	0.1296	0.0625
	1	0.9477	0.8192	0.6517	0.4752	0.3125
	2	0.9963	0.9728	0.9163	0.8208	0.6875
	3	0.9999	0.9984	0.9919	0.9744	0.9375
5	**0**	0.5905	0.3277	0.1681	0.0778	0.0312
	1	0.9185	0.7373	0.5282	0.3370	0.1875
	2	0.9914	0.9421	0.8369	0.6826	0.5000
	3	0.9995	0.9933	0.9692	0.9130	0.8125
	4	1.0000	0.9997	0.9976	0.9898	0.9688
6	**0**	0.5314	0.2621	0.1176	0.0467	0.0156
	1	0.8857	0.6554	0.4202	0.2333	0.1094
	2	0.9842	0.9011	0.7443	0.5443	0.3438
	3	0.9987	0.9830	0.9295	0.8208	0.6562
	4	0.9999	0.9984	0.9891	0.9590	0.8906
	5	1.0000	0.9999	0.9993	0.9959	0.9844
7	**0**	0.4783	0.2097	0.0824	0.0280	0.0078
	1	0.8503	0.5767	0.3294	0.1586	0.0625
	2	0.9743	0.8520	0.6471	0.4199	0.2266
	3	0.9973	0.9667	0.8740	0.7102	0.5000
	4	0.9998	0.9953	0.9712	0.9037	0.7734
	5	1.0000	0.9996	0.9962	0.9812	0.9375
	6	1.0000	1.0000	0.9998	0.9984	0.9922
8	**0**	0.4305	0.1678	0.0576	0.0168	0.0039
	1	0.8131	0.5033	0.2553	0.1064	0.0352
	2	0.9619	0.7969	0.5518	0.3154	0.1445
	3	0.9950	0.9437	0.8059	0.5941	0.3633
	4	0.9996	0.9896	0.9420	0.8263	0.6367
	5	1.0000	0.9988	0.9887	0.9502	0.8555
	6	1.0000	0.9999	0.9987	0.9915	0.9648
	7	1.0000	1.0000	0.9999	0.9993	0.9961

Table A.7   (Contd.)

				p		
n	x	**0.1**	**0.2**	**0.3**	**0.4**	**0.5**
9	**0**	0.3874	0.1342	0.0404	0.0101	0.0020
	1	0.7748	0.4362	0.1960	0.0705	0.0195
	2	0.9470	0.7382	0.4628	0.2318	0.0898
	3	0.9917	0.9144	0.7197	0.4826	0.2539
	4	0.9991	0.9804	0.9012	0.7334	0.5000
	5	0.9999	0.9969	0.9747	0.9006	0.7461
	6	1.0000	0.9997	0.9957	0.9750	0.9102
	7	1.0000	1.0000	0.9996	0.9962	0.9805
	8	1.0000	1.0000	1.0000	0.9997	0.9980
10	**0**	0.3487	0.1074	0.0282	0.0060	0.0010
	1	0.7361	0.3758	0.1493	0.0464	0.0107
	2	0.9298	0.6778	0.3828	0.1673	0.0547
	3	0.9872	0.8791	0.6496	0.3823	0.1719
	4	0.9984	0.9672	0.8497	0.6331	0.3770
	5	0.9999	0.9936	0.9527	0.8338	0.6230
	6	1.0000	0.9991	0.9894	0.9452	0.8281
	7	1.0000	0.9999	0.9984	0.9877	0.9453
	8	1.0000	1.0000	0.9999	0.9983	0.9893
	9	1.0000	1.0000	1.0000	0.9999	0.9990
11	**0**	0.3138	0.0859	0.0198	0.0036	0.0005
	1	0.6974	0.3221	0.1130	0.0302	0.0059
	2	0.9104	0.6174	0.3127	0.1189	0.0327
	3	0.9815	0.8389	0.5696	0.2963	0.1133
	4	0.9972	0.9496	0.7897	0.5328	0.2744
	5	0.9997	0.9883	0.9218	0.7535	0.5000
	6	1.0000	0.9980	0.9784	0.9006	0.7256
	7	1.0000	0.9998	0.9957	0.9707	0.8867
	8	1.0000	1.0000	0.9994	0.9941	0.9673
	9	1.0000	1.0000	1.0000	0.9993	0.9941
	10	1.0000	1.0000	1.0000	1.0000	0.9995
12	**0**	0.2824	0.0687	0.0138	0.0022	0.0002
	1	0.6590	0.2749	0.0850	0.0196	0.0032
	2	0.8891	0.5583	0.2528	0.0834	0.0193
	3	0.9744	0.7946	0.4925	0.2253	0.0730
	4	0.9957	0.9274	0.7237	0.4382	0.1938
	5	0.9995	0.9806	0.8822	0.6652	0.3872
	6	0.9999	0.9961	0.9614	0.8418	0.6128
	7	1.0000	0.9994	0.9905	0.9427	0.8062
	8	1.0000	0.9999	0.9983	0.9847	0.9270
	9	1.0000	1.0000	0.9998	0.9972	0.9807
	10	1.0000	1.0000	1.0000	0.9997	0.9968
	11	1.0000	1.0000	1.0000	1.0000	0.9998

Table A.7 (Contd.)

n	x	|	0.1	0.2	p 0.3	0.4	0.5
13	0		0.2542	0.0550	0.0097	0.0013	0.0001
	1		0.6213	0.2336	0.0637	0.0126	0.0017
	2		0.8661	0.5017	0.2025	0.0579	0.0112
	3		0.9658	0.7473	0.4206	0.1686	0.0461
	4		0.9935	0.9009	0.6543	0.3530	0.1334
	5		0.9991	0.9700	0.8346	0.5744	0.2905
	6		0.9999	0.9930	0.9376	0.7712	0.5000
	7		1.0000	0.9988	0.9818	0.9023	0.7095
	8		1.0000	0.9998	0.9960	0.9679	0.8666
	9		1.0000	1.0000	0.9993	0.9922	0.9539
	10		1.0000	1.0000	0.9999	0.9987	0.9888
	11		1.0000	1.0000	1.0000	0.9999	0.9983
	12		1.0000	1.0000	1.0000	1.0000	0.9999
14	0		0.2288	0.0440	0.0068	0.0008	0.0001
	1		0.5846	0.1979	0.0475	0.0081	0.0009
	2		0.8416	0.4481	0.1608	0.0398	0.0065
	3		0.9559	0.6982	0.3552	0.1243	0.0287
	4		0.9908	0.8702	0.5842	0.2793	0.0898
	5		0.9985	0.9561	0.7805	0.4859	0.2120
	6		0.9998	0.9884	0.9067	0.6925	0.3953
	7		1.0000	0.9976	0.9685	0.8499	0.6047
	8		1.0000	0.9996	0.9917	0.9417	0.7880
	9		1.0000	1.0000	0.9983	0.9825	0.9102
	10		1.0000	1.0000	0.9998	0.9961	0.9713
	11		1.0000	1.0000	1.0000	0.9994	0.9935
	12		1.0000	1.0000	1.0000	0.9999	0.9991
	13		1.0000	1.0000	1.0000	1.0000	0.9999
15	0		0.2059	0.0352	0.0047	0.0005	0.0000
	1		0.5490	0.1671	0.0353	0.0052	0.0005
	2		0.8159	0.3980	0.1268	0.0271	0.0037
	3		0.9444	0.6482	0.2969	0.0905	0.0176
	4		0.9873	0.8358	0.5155	0.2173	0.5920
	5		0.9978	0.9389	0.7216	0.4032	0.1509
	6		0.9997	0.9819	0.8689	0.6098	0.3036
	7		1.0000	0.9958	0.9500	0.7869	0.5000
	8		1.0000	0.9992	0.9848	0.9050	0.6964
	9		1.0000	0.9999	0.9963	0.9662	0.8491
	10		1.0000	1.0000	0.9993	0.9907	0.9408
	11		1.0000	1.0000	0.9999	0.9981	0.9824
	12		1.0000	1.0000	1.0000	0.9997	0.9963
	13		1.0000	1.0000	1.0000	1.0000	0.9995
	14		1.0000	1.0000	1.0000	1.0000	1.0000

Table A.7 (Contd.)

n	x	0.1	0.2	0.3	0.4	0.5
16	**0**	0.1853	0.0281	0.0033	0.0003	0.0000
	1	0.5147	0.1407	0.0261	0.0033	0.0003
	2	0.7892	0.3518	0.0994	0.0183	0.0021
	3	0.9316	0.5981	0.2459	0.0651	0.0106
	4	0.9830	0.7982	0.4499	0.1666	0.0384
	5	0.9967	0.9183	0.6598	0.3288	0.1051
	6	0.9995	0.9733	0.8247	0.5272	0.2272
	7	0.9999	0.9930	0.9256	0.7161	0.4018
	8	1.0000	0.9985	0.9743	0.8577	0.5982
	9	1.0000	0.9998	0.9929	0.9417	0.7728
	10	1.0000	1.0000	0.9984	0.9809	0.8949
	11	1.0000	1.0000	0.9991	0.9951	0.9616
	12	1.0000	1.0000	1.0000	0.9991	0.9894
	13	1.0000	1.0000	1.0000	0.9999	0.9979
	14	1.0000	1.0000	1.0000	1.0000	0.9997
	15	1.0000	1.0000	1.0000	1.0000	1.0000
17	**0**	0.1668	0.0225	0.0023	0.0002	0.0000
	1	0.4818	0.1182	0.0193	0.0021	0.0001
	2	0.7618	0.3096	0.0774	0.0123	0.0012
	3	0.9174	0.5489	0.2019	0.0464	0.0064
	4	0.9779	0.7582	0.3887	0.1260	0.0245
	5	0.9953	0.8943	0.5968	0.2639	0.0717
	6	0.9992	0.9623	0.7752	0.4478	0.1662
	7	0.9999	0.9891	0.8954	0.6405	0.3145
	8	1.0000	0.9974	0.9597	0.8011	0.5000
	9	1.0000	0.9995	0.9873	0.9081	0.6855
	10	1.0000	0.9999	0.9968	0.9652	0.8338
	11	1.0000	1.0000	0.9993	0.9894	0.9283
	12	1.0000	1.0000	0.9999	0.9975	0.9755
	13	1.0000	1.0000	1.0000	0.9995	0.9936
	14	1.0000	1.0000	1.0000	0.9999	0.9988
	15	1.0000	1.0000	1.0000	1.0000	0.9999
	16	1.0000	1.0000	1.0000	1.0000	1.0000
18	**0**	0.1501	0.0180	0.0016	0.0001	0.0000
	1	0.4503	0.0991	0.0142	0.0013	0.0001
	2	0.7338	0.2713	0.0600	0.0082	0.0007
	3	0.9018	0.5010	0.1646	0.0328	0.0038
	4	0.9718	0.7164	0.3327	0.0942	0.0154
	5	0.9936	0.8671	0.5344	0.2088	0.0481
	6	0.9988	0.9487	0.7217	0.3743	0.1189
	7	0.9998	0.9837	0.8593	0.5634	0.2403
	8	1.0000	0.9957	0.9404	0.7368	0.4073
	9	1.0000	0.9991	0.9790	0.8653	0.5927
	10	1.0000	0.9998	0.9939	0.9424	0.7597

Table A.7 (Contd.)

n	x	0.1	0.2	p 0.3	0.4	0.5
18	11	1.0000	1.0000	0.9986	0.9797	0.8811
	12	1.0000	1.0000	0.9997	0.9942	0.9519
	13	1.0000	1.0000	1.0000	0.9987	0.9846
	14	1.0000	1.0000	1.0000	0.9998	0.9962
	15	1.0000	1.0000	1.0000	1.0000	0.9993
	16	1.0000	1.0000	1.0000	1.0000	0.9999
	17	1.0000	1.0000	1.0000	1.0000	1.0000
19	0	0.1351	0.0144	0.0011	0.0001	0.0000
	1	0.4203	0.0829	0.0104	0.0008	0.0000
	2	0.7054	0.2369	0.0462	0.0055	0.0004
	3	0.8850	0.4551	0.1332	0.0230	0.0022
	4	0.9648	0.6733	0.2822	0.0696	0.0096
	5	0.9914	0.8369	0.4739	0.1629	0.0318
	6	0.9983	0.9324	0.6655	0.3081	0.0835
	7	0.9997	0.9767	0.8180	0.4878	0.1796
	8	1.0000	0.9933	0.9161	0.6675	0.3238
	9	1.0000	0.9984	0.9674	0.8139	0.5000
	10	1.0000	0.9997	0.9895	0.9115	0.6762
	11	1.0000	1.0000	0.9720	0.9648	0.8204
	12	1.0000	1.0000	0.9994	0.9884	0.9165
	13	1.0000	1.0000	0.9999	0.9969	0.9682
	14	1.0000	1.0000	1.0000	0.9994	0.9904
	15	1.0000	1.0000	1.0000	0.9999	0.9978
	16	1.0000	1.0000	1.0000	1.0000	0.9996
	17	1.0000	1.0000	1.0000	1.0000	1.0000
20	0	0.1216	0.0115	0.0008	0.0000	0.0000
	1	0.3917	0.0692	0.0076	0.0005	0.0000
	2	0.6769	0.2061	0.0355	0.0036	0.0002
	3	0.8670	0.4114	0.1071	0.0160	0.0013
	4	0.9568	0.6296	0.2375	0.0510	0.0059
	5	0.9887	0.8042	0.4164	0.1256	0.0207
	6	0.9976	0.9133	0.6080	0.2500	0.0577
	7	0.9996	0.9679	0.7723	0.4159	0.1316
	8	0.9999	0.9900	0.8867	0.5956	0.2517
	9	1.0000	0.9974	0.9520	0.7553	0.4119
	10	1.0000	0.9994	0.9829	0.8725	0.5881
	11	1.0000	0.9999	0.9949	0.9435	0.7483
	12	1.0000	1.0000	0.9987	0.9790	0.8684
	13	1.0000	1.0000	0.9997	0.9935	0.9423
	14	1.0000	1.0000	1.0000	0.9984	0.9793
	15	1.0000	1.0000	1.0000	0.9997	0.9941
	16	1.0000	1.0000	1.0000	1.0000	0.9987
	17	1.0000	1.0000	1.0000	1.0000	0.9998
	18	1.0000	1.0000	1.0000	1.0000	1.0000

Table A.8   Poisson Probabilities (X is $p(X; \lambda T)$)

Given λT and nonnegative integer values of k, the table gives the probability

$$p(X = k; \lambda T) = e^{-\lambda T} \frac{(\lambda T)^k}{k!} \text{ (e.g., } p(X = 2; 3.7) = 0.1692).$$

	λT									
k	**0.1**	**0.2**	**0.3**	**0.4**	**0.5**	**0.6**	**0.7**	**0.8**	**0.9**	**1.0**
0	0.9048	0.8187	0.7408	0.6703	0.6065	0.5488	0.4966	0.4493	0.4066	0.3679
1	0.0905	0.1637	0.2222	0.2681	0.3033	0.3293	0.3476	0.3595	0.3659	0.3679
2	0.0045	0.0164	0.0333	0.0536	0.0758	0.0988	0.1217	0.1438	0.1647	0.1839
3	0.0002	0.0011	0.0033	0.0072	0.0126	0.0198	0.0284	0.0383	0.0494	0.0613
4	0.0000	0.0001	0.0002	0.0007	0.0016	0.0030	0.0050	0.0077	0.0111	0.0153
5	0.0000	0.0000	0.0000	0.0001	0.0002	0.0004	0.0007	0.0012	0.0020	0.0031
6	0.0000	0.0000	0.0000	0.0000	0.0000	0.0000	0.0001	0.0002	0.0003	0.0005
7	0.0000	0.0000	0.0000	0.0000	0.0000	0.0000	0.0000	0.0000	0.0000	0.0001

	λT									
k	**1.1**	**1.2**	**1.3**	**1.4**	**1.5**	**1.6**	**1.7**	**1.8**	**1.9**	**2.0**
0	0.3329	0.3012	0.2725	0.2466	0.2231	0.2019	0.1827	0.1653	0.1496	0.1353
1	0.3662	0.3614	0.3543	0.3452	0.3347	0.3230	0.3106	0.2975	0.2842	0.2707
2	0.2014	0.2169	0.2303	0.2417	0.2510	0.2584	0.2640	0.2678	0.2700	0.2707
3	0.0738	0.0867	0.0998	0.1128	0.1255	0.1378	0.1496	0.1607	0.1710	0.1804
4	0.0203	0.0260	0.0324	0.0395	0.0471	0.0551	0.0636	0.0723	0.0812	0.0902
5	0.0045	0.0062	0.0084	0.0111	0.0141	0.0176	0.0216	0.0260	0.0309	0.0361
6	0.0008	0.0012	0.0018	0.0026	0.0035	0.0047	0.0061	0.0078	0.0098	0.0120
7	0.0001	0.0002	0.0003	0.0005	0.0008	0.0011	0.0015	0.0020	0.0027	0.0034
8	0.0000	0.0000	0.0001	0.0001	0.0001	0.0002	0.0003	0.0005	0.0006	0.0009
9	0.0000	0.0000	0.0000	0.0000	0.0000	0.0000	0.0001	0.0001	0.0001	0.0002

	λT									
k	**2.1**	**2.2**	**2.3**	**2.4**	**2.5**	**2.6**	**2.7**	**2.8**	**2.9**	**3.0**
0	0.1225	0.1108	0.1003	0.0907	0.0821	0.0743	0.0672	0.0608	0.0550	0.0498
1	0.2572	0.2438	0.2306	0.2177	0.2052	0.1931	0.1815	0.1703	0.1596	0.1494
2	0.2700	0.2681	0.2652	0.2613	0.2565	0.2510	0.2450	0.2384	0.2314	0.2240
3	0.1890	0.1966	0.2033	0.2090	0.2138	0.2176	0.2205	0.2225	0.2237	0.2240
4	0.0992	0.1082	0.1169	0.1254	0.1336	0.1414	0.1488	0.1557	0.1622	0.1680
5	0.0417	0.0476	0.0538	0.0602	0.0668	0.0735	0.0804	0.0872	0.0940	0.1008
6	0.0146	0.0174	0.0206	0.0241	0.0278	0.0319	0.0362	0.0407	0.0455	0.0504
7	0.0044	0.0055	0.0068	0.0083	0.0099	0.0118	0.0139	0.0163	0.0188	0.0216
8	0.0011	0.0015	0.0019	0.0025	0.0031	0.0038	0.0047	0.0057	0.0068	0.0081
9	0.0003	0.0004	0.0005	0.0007	0.0009	0.0011	0.0014	0.0018	0.0022	0.0027
10	0.0001	0.0001	0.0001	0.0002	0.0002	0.0003	0.0004	0.0005	0.0006	0.0008
11	0.0000	0.0000	0.0000	0.0000	0.0000	0.0001	0.0001	0.0001	0.0002	0.0002
12	0.0000	0.0000	0.0000	0.0000	0.0000	0.0000	0.0000	0.0000	0.0000	0.0001

Table A.8 (Contd.)

					λT					
k	3.1	3.2	3.3	3.4	3.5	3.6	3.7	3.8	3.9	4.0
0	0.0450	0.0408	0.0369	0.0334	0.0302	0.0273	0.0247	0.0224	0.0202	0.0183
1	0.1397	0.1304	0.1217	0.1135	0.1057	0.0984	0.0915	0.0850	0.0789	0.0733
2	0.2165	0.2087	0.2008	0.1929	0.1850	0.1771	0.1692	0.1615	0.1539	0.1465
3	0.2237	0.2226	0.2209	0.2186	0.2158	0.2125	0.2087	0.2046	0.2001	0.1954
4	0.1734	0.1781	0.1823	0.1858	0.1888	0.1912	0.1931	0.1944	0.1951	0.1954
5	0.1075	0.1140	0.1203	0.1264	0.1322	0.1377	0.1429	0.1477	0.1522	0.1563
6	0.0555	0.0608	0.0662	0.0716	0.0771	0.0826	0.0881	0.0936	0.0989	0.1042
7	0.0246	0.0278	0.0312	0.0348	0.0385	0.0425	0.0466	0.0508	0.0551	0.0595
8	0.0095	0.0111	0.0129	0.0148	0.0169	0.0191	0.0215	0.0241	0.0269	0.0298
9	0.0033	0.0040	0.0047	0.0056	0.0066	0.0076	0.0089	0.0102	0.0116	0.0132
10	0.0010	0.0013	0.0016	0.0019	0.0023	0.0028	0.0033	0.0039	0.0045	0.0053
11	0.0003	0.0004	0.0005	0.0006	0.0007	0.0009	0.0011	0.0013	0.0016	0.0019
12	0.0001	0.0001	0.0001	0.0002	0.0002	0.0003	0.0003	0.0004	0.0005	0.0006
13	0.0000	0.0000	0.0000	0.0000	0.0001	0.0001	0.0001	0.0001	0.0002	0.0002
14	0.0000	0.0000	0.0000	0.0000	0.0000	0.0000	0.0000	0.0000	0.0000	0.0001

					λT					
k	4.1	4.2	4.3	4.4	4.5	4.6	4.7	4.8	4.9	5.0
0	0.0166	0.0150	0.0136	0.0123	0.0111	0.0101	0.0091	0.0082	0.0074	0.0067
1	0.0679	0.0630	0.0583	0.0540	0.0500	0.0462	0.0427	0.0395	0.0365	0.0337
2	0.1393	0.1323	0.1254	0.1188	0.1125	0.1063	0.1005	0.0948	0.0894	0.0842
3	0.1904	0.1852	0.1798	0.1743	0.1687	0.1631	0.1574	0.1517	0.1460	0.1404
4	0.1951	0.1944	0.1933	0.1917	0.1898	0.1875	0.1849	0.1820	0.1789	0.1755
5	0.1600	0.1633	0.1662	0.1687	0.1708	0.1725	0.1738	0.1747	0.1753	0.1755
6	0.1093	0.1143	0.1191	0.1237	0.1281	0.1323	0.1362	0.1398	0.1432	0.1462
7	0.0640	0.0686	0.0732	0.0778	0.0824	0.0869	0.0914	0.0959	0.1002	0.1044
8	0.0328	0.0360	0.0393	0.0428	0.0463	0.0500	0.0537	0.0575	0.0614	0.0653
9	0.0150	0.0168	0.0188	0.0209	0.0232	0.0255	0.0280	0.0307	0.0334	0.0363
10	0.0061	0.0071	0.0081	0.0092	0.0104	0.0118	0.0132	0.0147	0.0164	0.0181
11	0.0023	0.0027	0.0032	0.0037	0.0043	0.0049	0.0056	0.0064	0.0073	0.0082
12	0.0008	0.0009	0.0011	0.0014	0.0016	0.0019	0.0022	0.0026	0.0030	0.0034
13	0.0002	0.0003	0.0004	0.0005	0.0006	0.0007	0.0008	0.0009	0.0011	0.0013
14	0.0001	0.0001	0.0001	0.0001	0.0002	0.0002	0.0003	0.0003	0.0004	0.0005
15	0.0000	0.0000	0.0000	0.0000	0.0001	0.0001	0.0001	0.0001	0.0001	0.0002

Table A.8 (Contd.)

k	5.1	5.2	5.3	5.4	5.5	5.6	5.7	5.8	5.9	6.0
					λT					
0	0.0061	0.0055	0.0050	0.0045	0.0041	0.0037	0.0033	0.0030	0.0027	0.0025
1	0.0311	0.0287	0.0265	0.0244	0.0225	0.0207	0.0191	0.0176	0.0162	0.0149
2	0.0793	0.0746	0.0701	0.0659	0.0618	0.0580	0.0544	0.0509	0.0477	0.0446
3	0.1348	0.1293	0.1239	0.1185	0.1133	0.1082	0.1033	0.0985	0.0938	0.0892
4	0.1719	0.1681	0.1641	0.1600	0.1558	0.1615	0.1472	0.1428	0.1383	0.1339
5	0.1753	0.1748	0.1740	0.1728	0.1714	0.1697	0.1678	0.1656	0.1632	0.1606
6	0.1490	0.1515	0.1537	0.1555	0.1571	0.1584	0.1594	0.1601	0.1605	0.1606
7	0.1086	0.1125	0.1163	0.1200	0.1234	0.1267	0.1298	0.1326	0.1353	0.1377
8	0.0692	0.0731	0.0771	0.0810	0.0849	0.0887	0.0925	0.0962	0.0998	0.1033
9	0.0392	0.0423	0.0454	0.0486	0.0519	0.0552	0.0586	0.0620	0.0654	0.0688
10	0.0200	0.0220	0.0241	0.0262	0.0285	0.0309	0.0334	0.0359	0.0386	0.0413
11	0.0093	0.0104	0.0116	0.0129	0.0143	0.0157	0.0173	0.0190	0.0207	0.0225
12	0.0039	0.0045	0.0051	0.0058	0.0065	0.0073	0.0082	0.0092	0.0102	0.0113
13	0.0015	0.0018	0.0021	0.0024	0.0028	0.0032	0.0036	0.0041	0.0046	0.0052
14	0.0006	0.0007	0.0008	0.0009	0.0011	0.0013	0.0015	0.0017	0.0019	0.0022
15	0.0002	0.0002	0.0003	0.0003	0.0004	0.0005	0.0006	0.0007	0.0008	0.0009
16	0.0001	0.0001	0.0001	0.0001	0.0001	0.0002	0.0002	0.0002	0.0003	0.0003
17	0.0000	0.0000	0.0000	0.0000	0.0000	0.0001	0.0001	0.0001	0.0001	0.0001

k	6.1	6.2	6.3	6.4	6.5	6.6	6.7	6.8	6.9	7.0
					λT					
0	0.0022	0.0020	0.0018	0.0017	0.0015	0.0014	0.0012	0.0011	0.0010	0.0009
1	0.0137	0.0126	0.0116	0.0106	0.0098	0.0090	0.0082	0.0076	0.0070	0.0064
2	0.0417	0.0390	0.0364	0.0340	0.0318	0.0296	0.0276	0.0258	0.0240	0.0223
3	0.0848	0.0806	0.0765	0.0726	0.0688	0.0652	0.0617	0.0584	0.0552	0.0521
4	0.1294	0.1249	0.1205	0.1162	0.1118	0.1076	0.1034	0.0992	0.0952	0.0912
5	0.1579	0.1549	0.1519	0.1487	0.1454	0.1420	0.1385	0.1349	0.1314	0.1277
6	0.1605	0.1601	0.1595	0.1586	0.1575	0.1562	0.1546	0.1529	0.1511	0.1490
7	0.1399	0.1418	0.1435	0.1450	0.1462	0.1472	0.1480	0.1486	0.1489	0.1490
8	0.1066	0.1099	0.1130	0.1160	0.1188	0.1215	0.1240	0.1263	0.1284	0.1304
9	0.0723	0.0757	0.0791	0.0825	0.0858	0.0891	0.0923	0.0954	0.0985	0.1014
10	0.0441	0.0469	0.0498	0.0528	0.0558	0.0588	0.0618	0.0649	0.0679	0.0710
11	0.0245	0.0265	0.0285	0.0307	0.0330	0.0353	0.0377	0.0401	0.0426	0.0452
12	0.0124	0.0137	0.0150	0.0164	0.0179	0.0194	0.0210	0.0227	0.0245	0.0264
13	0.0058	0.0065	0.0073	0.0081	0.0089	0.0098	0.0108	0.0119	0.0130	0.0142
14	0.0025	0.0029	0.0033	0.0037	0.0041	0.0046	0.0052	0.0058	0.0064	0.0071
15	0.0010	0.0012	0.0014	0.0016	0.0018	0.0020	0.0023	0.0026	0.0029	0.0033
16	0.0004	0.0005	0.0005	0.0006	0.0007	0.0008	0.0010	0.0011	0.0013	0.0014
17	0.0001	0.0002	0.0002	0.0002	0.0003	0.0003	0.0004	0.0004	0.0005	0.0006
18	0.0000	0.0001	0.0001	0.0001	0.0001	0.0001	0.0001	0.0002	0.0002	0.0002
19	0.0000	0.0000	0.0000	0.0000	0.0000	0.0000	0.0000	0.0001	0.0001	0.0001

Table A.8 (Contd.)

k	7.1	7.2	7.3	7.4	7.5	7.6	7.7	7.8	7.9	8.0
					λT					
0	0.0008	0.0007	0.0007	0.0006	0.0006	0.0005	0.0005	0.0004	0.0004	0.0003
1	0.0059	0.0054	0.0049	0.0045	0.0041	0.0038	0.0035	0.0032	0.0029	0.0027
2	0.0208	0.0194	0.0180	0.0167	0.0156	0.0145	0.0134	0.0125	0.0116	0.0107
3	0.0492	0.0464	0.0438	0.0413	0.0389	0.0366	0.0345	0.0324	0.0305	0.0286
4	0.0874	0.0836	0.0799	0.0764	0.0729	0.0696	0.0663	0.0632	0.0602	0.0573
5	0.1241	0.1204	0.1167	0.1130	0.1094	0.1057	0.1021	0.0986	0.0951	0.0916
6	0.1468	0.1445	0.1420	0.1394	0.1367	0.1339	0.1311	0.1282	0.1252	0.1221
7	0.1489	0.1486	0.1481	0.1474	0.1465	0.1454	0.1442	0.1428	0.1413	0.1396
8	0.1321	0.1337	0.1351	0.1363	0.1373	0.1382	0.1388	0.1392	0.1395	0.1396
9	0.1042	0.1070	0.1096	0.1121	0.1144	0.1167	0.1187	0.1207	0.1224	0.1241
10	0.0740	0.0770	0.0800	0.0829	0.0838	0.0887	0.0914	0.0941	0.0967	0.0993
11	0.4780	0.0504	0.0531	0.0558	0.0585	0.0613	0.0640	0.0667	0.0695	0.0722
12	0.0283	0.0303	0.0323	0.0344	0.0366	0.0388	0.0411	0.0434	0.0457	0.0481
13	0.0154	0.0168	0.0181	0.0196	0.0211	0.0227	0.0243	0.0260	0.0278	0.0296
14	0.0078	0.0086	0.0095	0.0104	0.0113	0.0123	0.0134	0.0145	0.0157	0.0169
15	0.0037	0.0041	0.0046	0.0051	0.0057	0.0062	0.0069	0.0075	0.0083	0.0090
16	0.0016	0.0019	0.0021	0.0024	0.0026	0.0030	0.0033	0.0037	0.0041	0.0045
17	0.0007	0.0008	0.0009	0.0010	0.0012	0.0013	0.0015	0.0017	0.0019	0.0021
18	0.0003	0.0003	0.0004	0.0004	0.0005	0.0006	0.0006	0.0007	0.0008	0.0009
19	0.0001	0.0001	0.0001	0.0002	0.0002	0.0002	0.0003	0.0003	0.0003	0.0004
20	0.0000	0.0000	0.0001	0.0001	0.0001	0.0001	0.0001	0.0001	0.0001	0.0002
21	0.0000	0.0000	0.0000	0.0000	0.0000	0.0000	0.0000	0.0000	0.0001	0.0001

k	8.1	8.2	8.3	8.4	8.5	8.6	8.7	8.8	8.9	9.0
					λT					
0	0.0003	0.0003	0.0002	0.0002	0.0002	0.0002	0.0002	0.0002	0.0001	0.0001
1	0.0025	0.0023	0.0021	0.0019	0.0017	0.0016	0.0014	0.0013	0.0012	0.0011
2	0.0100	0.0092	0.0086	0.0079	0.0074	0.0068	0.0063	0.0058	0.0054	0.0050
3	0.0269	0.0252	0.0237	0.0222	0.0208	0.0195	0.0183	0.0171	0.0160	0.0150
4	0.0544	0.0517	0.0491	0.0466	0.0443	0.0420	0.0398	0.0377	0.0357	0.0337
5	0.0882	0.0849	0.0816	0.0784	0.0752	0.0722	0.0692	0.0663	0.0635	0.0607
6	0.1191	0.1160	0.1128	0.1097	0.1066	0.1034	0.1003	0.0972	0.0941	0.0911
7	0.1378	0.1358	0.1338	0.1317	0.1294	0.1271	0.1247	0.1222	0.1197	0.1171
8	0.1395	0.1392	0.1388	0.1382	0.1375	0.1366	0.1356	0.1344	0.1332	0.1318
9	0.1256	0.1269	0.1280	0.1290	0.1299	0.1306	0.1311	0.1315	0.1317	0.1318
10	0.1017	0.1040	0.1063	0.1084	0.1104	0.1123	0.1140	0.1137	0.1172	0.1186
11	0.0749	0.0776	0.0802	0.0828	0.0853	0.0878	0.0902	0.0925	0.0948	0.0970
12	0.0505	0.0530	0.0555	0.0579	0.0604	0.0629	0.0654	0.0679	0.0703	0.0728
13	0.0315	0.0334	0.0354	0.0374	0.0395	0.0416	0.0438	0.0459	0.0481	0.0504
14	0.0182	0.0196	0.0210	0.0225	0.0240	0.0256	0.0272	0.0289	0.0306	0.0324
15	0.0098	0.0107	0.0116	0.0126	0.0136	0.0147	0.0158	0.0169	0.0182	0.0194
16	0.0050	0.0055	0.0060	0.0066	0.0072	0.0079	0.0086	0.0093	0.0101	0.0109

Table A.8   (Contd.)

k	8.1	8.2	8.3	8.4	λT 8.5	8.6	8.7	8.8	8.9	9.0
17	0.0024	0.0026	0.0029	0.0033	0.0036	0.0040	0.0044	0.0048	0.0053	0.0058
18	0.0011	0.0012	0.0014	0.0015	0.0017	0.0019	0.0021	0.0024	0.0026	0.0029
19	0.0005	0.0005	0.0006	0.0007	0.0008	0.0009	0.0010	0.0011	0.0012	0.0014
20	0.0002	0.0002	0.0002	0.0003	0.0003	0.0004	0.0004	0.0005	0.0005	0.0006
21	0.0001	0.0001	0.0001	0.0001	0.0001	0.0002	0.0002	0.0002	0.0002	0.0003
22	0.0000	0.0000	0.0000	0.0000	0.0001	0.0001	0.0001	0.0001	0.0001	0.0001

k	9.1	9.2	9.3	9.4	λT 9.5	9.6	9.7	9.8	9.9	10.0
0	0.0001	0.0001	0.0001	0.0001	0.0001	0.0001	0.0001	0.0001	0.0001	0.0000
1	0.0010	0.0009	0.0009	0.0008	0.0007	0.0007	0.0006	0.0005	0.0005	0.0005
2	0.0046	0.0043	0.0040	0.0037	0.0034	0.0031	0.0029	0.0027	0.0025	0.0023
3	0.0140	0.0131	0.0123	0.0115	0.0107	0.0100	0.0093	0.0087	0.0081	0.0076
4	0.0319	0.0302	0.0285	0.0269	0.0254	0.0240	0.0226	0.0213	0.0201	0.0189
5	0.0581	0.0555	0.0530	0.0506	0.0483	0.0460	0.0439	0.0418	0.0398	0.0378
6	0.0881	0.0851	0.0822	0.0793	0.0764	0.0736	0.0709	0.0682	0.0656	0.0631
7	0.1145	0.1118	0.1091	0.1064	0.1037	0.1010	0.0982	0.0955	0.0928	0.0901
8	0.1302	0.1286	0.1269	0.1251	0.1232	0.1212	0.1191	0.1170	0.1148	0.1126
9	0.1317	0.1315	0.1311	0.1306	0.1300	0.1293	0.1284	0.1274	0.1263	0.1251
10	0.1198	0.1210	0.1219	0.1228	0.1235	0.1241	0.1245	0.1249	0.1250	0.1251
11	0.0991	0.1012	0.1031	0.1049	0.1067	0.1083	0.1098	0.1112	0.1125	0.1137
12	0.0752	0.0776	0.0799	0.0822	0.0844	0.0866	0.0888	0.0908	0.0928	0.0948
13	0.0526	0.0549	0.0572	0.0594	0.0617	0.0640	0.0662	0.0685	0.0707	0.0729
14	0.0342	0.0361	0.0380	0.0399	0.0419	0.0439	0.0459	0.0479	0.0500	0.0521
15	0.0208	0.0221	0.0235	0.0250	0.0265	0.0281	0.0297	0.0313	0.0330	0.0347
16	0.0118	0.0127	0.0137	0.0147	0.0157	0.0168	0.0180	0.0192	0.0204	0.0217
17	0.0063	0.0069	0.0075	0.0081	0.0088	0.0095	0.0103	0.0111	0.0119	0.0128
18	0.0032	0.0035	0.0039	0.0042	0.0046	0.0051	0.0055	0.0060	0.0065	0.0071
19	0.0015	0.0017	0.0019	0.0021	0.0023	0.0026	0.0028	0.0031	0.0034	0.0037
20	0.0007	0.0008	0.0009	0.0010	0.0011	0.0012	0.0014	0.0015	0.0017	0.0019
21	0.0003	0.0003	0.0004	0.0004	0.0005	0.0006	0.0006	0.0007	0.0008	0.0009
22	0.0001	0.0001	0.0002	0.0002	0.0002	0.0002	0.0003	0.0003	0.0004	0.0004
23	0.0000	0.0001	0.0001	0.0001	0.0001	0.0001	0.0001	0.0001	0.0002	0.0002
24	0.0000	0.0000	0.0000	0.0000	0.0000	0.0000	0.0000	0.0001	0.0001	0.0001

Table A.8   (Contd.)

	λT									
k	11	12	13	14	15	16	17	18	19	20
0	0.0000	0.0000	0.0000	0.0000	0.0000	0.0000	0.0000	0.0000	0.0000	0.0000
1	0.0002	0.0001	0.0000	0.0000	0.0000	0.0000	0.0000	0.0000	0.0000	0.0000
2	0.0010	0.0004	0.0002	0.0001	0.0000	0.0000	0.0000	0.0000	0.0000	0.0000
3	0.0037	0.0018	0.0008	0.0004	0.0002	0.0001	0.0000	0.0000	0.0000	0.0000
4	0.0102	0.0053	0.0027	0.0013	0.0006	0.0003	0.0001	0.0001	0.0000	0.0000
5	0.0224	0.0127	0.0070	0.0037	0.0019	0.0010	0.0005	0.0002	0.0001	0.0001
6	0.0411	0.0265	0.0152	0.0087	0.0048	0.0026	0.0014	0.0007	0.0004	0.0002
7	0.0646	0.0437	0.0281	0.0174	0.0104	0.0060	0.0034	0.0018	0.0010	0.0005
8	0.0888	0.0665	0.0457	0.0304	0.0194	0.0120	0.0072	0.0042	0.0024	0.0013
9	0.1085	0.0874	0.0661	0.0473	0.0324	0.0213	0.0135	0.0083	0.0050	0.0029
10	0.1194	0.1048	0.0859	0.0663	0.0486	0.0341	0.0230	0.0150	0.0095	0.0058
11	0.1194	0.1144	0.1015	0.0844	0.0663	0.0496	0.0355	0.0245	0.0164	0.0106
12	0.1094	0.1144	0.1099	0.0984	0.0829	0.0661	0.0504	0.0368	0.0259	0.0176
13	0.0926	0.1056	0.1099	0.1060	0.0956	0.0814	0.0658	0.0509	0.0378	0.0271
14	0.0728	0.0905	0.1021	0.1060	0.1024	0.0930	0.0800	0.0655	0.0614	0.0387
15	0.0534	0.0724	0.0886	0.0989	0.1024	0.0992	0.0906	0.0786	0.0650	0.0516
16	0.0367	0.0543	0.0719	0.0866	0.0960	0.0992	0.0963	0.0884	0.0772	0.0646
17	0.0237	0.0383	0.0550	0.0713	0.0847	0.0934	0.0963	0.0936	0.0863	0.0760
18	0.0146	0.0256	0.0397	0.0554	0.0706	0.0830	0.0909	0.0936	0.0911	0.0844
19	0.0084	0.0161	0.0272	0.0409	0.0557	0.0699	0.0814	0.0887	0.0911	0.0888
20	0.0046	0.0097	0.0177	0.0286	0.0418	0.0559	0.0692	0.0798	0.0866	0.0888
21	0.0024	0.0055	0.0109	0.0191	0.0299	0.0426	0.0560	0.0684	0.0783	0.0846
22	0.0012	0.0030	0.0066	0.0121	0.0204	0.0310	0.0433	0.0560	0.0676	0.0769
23	0.0006	0.0016	0.0037	0.0074	0.0133	0.0216	0.0320	0.0438	0.0559	0.0669
24	0.0003	0.0008	0.0020	0.0043	0.0083	0.0144	0.0226	0.0328	0.0442	0.0557

	λT									
k	11	12	13	14	15	16	17	18	19	20
25	0.0001	0.0004	0.0010	0.0024	0.0050	0.0092	0.0154	0.0237	0.0336	0.0446
26	0.0000	0.0002	0.0005	0.0013	0.0029	0.0057	0.0101	0.0164	0.0246	0.0343
27	0.0000	0.0001	0.0002	0.0007	0.0016	0.0034	0.0063	0.0109	0.0173	0.0254
28	0.0000	0.0000	0.0001	0.0003	0.0009	0.0019	0.0038	0.0070	0.0117	0.0181
29	0.0000	0.0000	0.0001	0.0002	0.0004	0.0011	0.0023	0.0044	0.0077	0.0125
30	0.0000	0.0000	0.0000	0.0001	0.0002	0.0006	0.0013	0.0026	0.0049	0.0083
31	0.0000	0.0000	0.0000	0.0000	0.0001	0.0003	0.0007	0.0015	0.0030	0.0054
32	0.0000	0.0000	0.0000	0.0000	0.0001	0.0001	0.0004	0.0009	0.0018	0.0034
33	0.0000	0.0000	0.0000	0.0000	0.0000	0.0001	0.0002	0.0005	0.0010	0.0020
34	0.0000	0.0000	0.0000	0.0000	0.0000	0.0000	0.0001	0.0002	0.0006	0.0012
35	0.0000	0.0000	0.0000	0.0000	0.0000	0.0000	0.0000	0.0001	0.0003	0.0007
36	0.0000	0.0000	0.0000	0.0000	0.0000	0.0000	0.0000	0.0001	0.0002	0.0004
37	0.0000	0.0000	0.0000	0.0000	0.0000	0.0000	0.0000	0.0000	0.0001	0.0002
38	0.0000	0.0000	0.0000	0.0000	0.0000	0.0000	0.0000	0.0000	0.0000	0.0001
39	0.0000	0.0000	0.0000	0.0000	0.0000	0.0000	0.0000	0.0000	0.0000	0.0001

Table A.9 Fisher's $\hat{\rho}(=r)$ to ξ Transformation

The table gives values of $\xi(\hat{\rho}) = \frac{1}{2}\log_e\left(\frac{1+\hat{\rho}}{1-\hat{\rho}}\right)$, where $\xi(-\hat{\rho}) = -\xi(\hat{\rho})$ (e.g., for $\hat{\rho} = 0.76$, $\xi(0.76) = 0.996$; for $\hat{\rho} = -0.76$, $\xi(-0.76) = -0.996$).

$\hat{\rho}$	$\hat{\rho}$ (3rd decimal)					$\hat{\rho}$	$\hat{\rho}$ (3rd decimal)				
	0.000	0.002	0.004	0.006	0.008		0.000	0.002	0.004	0.006	0.008
0.00	0.0000	0.0020	0.0040	0.0060	0.0080	**0.50**	0.5493	0.5520	0.5547	0.5573	0.5600
1	0.0100	0.0120	0.0140	0.0160	0.0180	1	0.5627	0.5654	0.5682	0.5709	0.5736
2	0.0200	0.0220	0.0240	0.0260	0.0280	2	0.5763	0.5791	0.5818	0.5846	0.5874
3	0.0300	0.0320	0.0340	0.0360	0.0380	3	0.5901	0.5929	0.5957	0.5985	0.6013
4	0.0400	0.0420	0.0440	0.0460	0.0480	4	0.6042	0.6070	0.6098	0.6127	0.6155
0.05	0.0500	0.0520	0.0541	0.0561	0.0581	**0.55**	0.6194	0.6213	0.6241	0.6270	0.6299
6	0.0601	0.0621	0.0641	0.0661	0.0681	6	0.6328	0.6358	0.6387	0.6416	0.6446
7	0.0701	0.0721	0.0741	0.0761	0.0782	7	0.6475	0.6505	0.6535	0.6565	0.6595
8	0.0802	0.0822	0.0842	0.0862	0.0882	8	0.6625	0.6655	0.6685	0.6716	0.6746
9	0.0902	0.0923	0.0943	0.0963	0.0983	9	0.6777	0.6807	0.6838	0.6869	0.6900
0.10	0.1003	0.1024	0.1044	0.1064	0.1084	**0.60**	0.6931	0.6963	0.6994	0.7026	0.7057
1	0.1104	0.1125	0.1145	0.1165	0.1186	1	0.7089	0.7121	0.7153	0.7185	0.7218
2	0.1206	0.1226	0.1246	0.1267	0.1287	2	0.7250	0.7283	0.7315	0.7348	0.7381
3	0.1307	0.1328	0.1348	0.1368	0.1389	3	0.7414	0.7447	0.7481	0.7514	0.7548
4	0.1409	0.1430	0.1450	0.1471	0.1491	4	0.7582	0.7616	0.7650	0.7684	0.7718
0.15	0.1511	0.1532	0.1552	0.1573	0.1593	**0.65**	0.7753	0.7788	0.7823	0.7858	0.7893
6	0.1614	0.1634	0.1655	0.1676	0.1696	6	0.7928	0.7964	0.7999	0.8035	0.8071
7	0.1717	0.1737	0.1758	0.1779	0.1799	7	0.8107	0.8144	0.8180	0.8217	0.8254
8	0.1820	0.1841	0.1861	0.1882	0.1903	8	0.8291	0.8328	0.8366	0.8404	0.8441
9	0.1923	0.1944	0.1965	0.1986	0.2007	9	0.8480	0.8518	0.8556	0.8595	0.8634
0.20	0.2027	0.2048	0.2069	0.2090	0.2111	**0.70**	0.8673	0.8712	0.8752	0.8792	0.8832
1	0.2132	0.2153	0.2174	0.2195	0.2216	1	0.8872	0.8912	0.8953	0.8994	0.9035
2	0.2237	0.2258	0.2279	0.2300	0.2321	2	0.9076	0.9118	0.9160	0.9202	0.9245
3	0.2342	0.2363	0.2384	0.2405	0.2427	3	0.9287	0.9330	0.9373	0.9417	0.9462
4	0.2448	0.2469	0.2490	0.2512	0.2533	4	0.9505	0.9549	0.9549	0.9639	0.9684
0.25	0.2554	0.2575	0.2597	0.2618	0.2640	**0.75**	0.9730	0.9780	0.9820	0.9870	0.9910
6	0.2661	0.2683	0.2704	0.2726	0.2747	6	0.9960	1.0010	1.0060	1.0110	1.0150
7	0.2769	0.2790	0.2812	0.2833	0.2855	7	1.0200	1.0250	1.0300	1.0350	1.0400
8	0.2877	0.2899	0.2920	0.2942	0.2964	8	1.0450	1.0500	1.0560	1.0610	1.0660
9	0.2986	0.3008	0.3029	0.3051	0.3073	9	1.0710	1.0770	1.0820	1.0880	1.0930
0.30	0.3095	0.3117	0.3139	0.3161	0.3183	**0.80**	1.0990	1.1040	1.1100	1.1160	1.1210
1	0.3205	0.3228	0.3250	0.3272	0.3294	1	1.1270	1.1330	1.1390	1.1450	1.1510
2	0.3316	0.3339	0.3361	0.3383	0.3406	2	1.1570	1.1630	1.1690	1.1750	1.1820
3	0.3428	0.3451	0.3473	0.3496	0.3518	3	1.1880	1.1950	1.2010	1.2080	1.2140
4	0.3541	0.3564	0.3586	0.3609	0.3632	4	1.2210	1.2280	1.2350	1.2420	1.2490
0.35	0.3654	0.3677	0.3700	0.3723	0.3746	**0.85**	1.2560	1.2630	1.2710	1.2780	1.2860
6	0.3769	0.3792	0.3815	0.3838	0.3861	6	1.2930	1.3010	1.3090	1.3170	1.3250
7	0.3884	0.3907	0.3931	0.3954	0.3977	7	1.3330	1.3410	1.3500	1.3580	1.3670
8	0.4001	0.4024	0.4047	0.4071	0.4094	8	1.3760	1.3850	1.3940	1.4030	1.4120
9	0.4118	0.4142	0.4165	0.4189	0.4213	9	1.4220	1.4320	1.4420	1.4520	1.4620
0.40	0.4236	0.4260	0.4284	0.4308	0.4332	**0.90**	1.4720	1.4830	1.4940	1.5050	1.5160
1	0.4356	0.4380	0.4404	0.4428	0.4453	1	1.5280	1.5390	1.5510	1.5640	1.5760
2	0.4477	0.4501	0.4526	0.4550	0.4574	2	1.5890	1.6020	1.6160	1.6300	1.6440
3	0.4599	0.4624	0.4648	0.4673	0.4698	3	1.6580	1.6730	1.6890	1.7050	1.7210
4	0.4722	0.4747	0.4772	0.4797	0.4822	4	1.7380	1.7560	1.7740	1.7920	1.8120

Table A.9   (Contd.)

$\hat\rho$	$\hat\rho$ (3rd decimal)					$\hat\rho$	$\hat\rho$ (3rd decimal)				
	0.000	0.002	0.004	0.006	0.008		0.000	0.002	0.004	0.006	0.008
0.45	0.4847	0.4872	0.4897	0.4922	0.4948	**0.95**	1.8320	1.8530	1.8740	1.8970	1.9210
6	0.4973	0.4999	0.5024	0.5049	0.4075	**6**	1.9460	1.9720	2.0000	2.0290	2.0600
7	0.5101	0.5126	0.5152	0.5178	0.5204	**7**	2.0920	2.1270	2.1650	2.2050	2.2490
8	0.5230	0.5256	0.5282	0.5308	0.5334	**8**	2.2980	2.3510	2.4100	2.4770	2.5550
9	0.5361	0.5387	0.5413	0.5440	0.5466	**9**	2.6470	2.7590	2.9030	3.1060	3.4530

This table is abridged from E.S. Pearson and H.O. Hartley. (1976); *Biometrika Tables for Statisticians*, Vol. I. Reproduced with the kind permission of Oxford University Press.

Table A.10 R Distribution for the Runs Test of Randomness, $\alpha = 0.05$

Given n_1 and n_2, the table gives the lower- and upper-tail critical values r_l and r_u, respectively, such that $P(R \le r_l) + P(R \ge r_u) = \alpha/2 + \alpha/2 = 0.05$.

n_1 ＼ n_2	r_l / r_u	2	3	4	5	6	7	8	9	10	11	12	13	14	15
2	r_l											2	2	2	2
	r_u														
3	r_l				2	2	2	2	2	2	2	2	2	2	3
	r_u														
4	r_l				2	2	2	3	3	3	3	3	3	3	3
	r_u				9	9									
5	r_l				2	3	3	3	3	3	4	4	4	4	4
	r_u				9	10	10	11	11						
6	r_l		2	2	3	3	3	3	4	4	4	4	5	5	5
	r_u			9	10	11	12	12	13	13	13	13			
7	r_l		2	2	3	3	3	4	4	5	5	5	5	5	6
	r_u				11	12	13	13	14	14	14	14	15	15	15
8	r_l		2	3	3	3	4	4	5	5	5	6	6	6	6
	r_u				11	12	13	14	14	15	15	16	16	16	16
9	r_l		2	3	3	4	4	5	5	5	6	6	6	7	7
	r_u					13	14	14	15	16	16	16	17	17	18
10	r_l		2	3	3	4	5	5	5	6	6	7	7	7	7
	r_u					13	14	15	16	16	17	17	18	18	18
11	r_l		2	3	4	4	5	5	6	6	7	7	7	8	8
	r_u					13	14	15	16	17	17	18	19	19	19
12	r_l	2	2	3	4	4	5	6	6	7	7	7	8	8	8
	r_u					13	14	16	16	17	18	19	19	20	20
13	r_l	2	2	3	4	5	5	6	6	7	7	8	8	9	9
	r_u						15	16	17	18	19	19	20	20	21
14	r_l	2	2	3	4	5	5	6	7	7	8	8	9	9	9
	r_u						15	16	17	18	19	20	20	21	22
15	r_l	2	3	3	4	5	6	6	7	7	8	8	9	9	10
	r_u						15	16	18	18	19	20	21	22	22

Adapted from F.S. Swed and C. Eisenhart, "Tables for Testing Randomness of Grouping in a Sequence of Alternatives," *The Annals of Mathematical Statistics*, 14, 1943, 66–87, with kind permission from the Institute of Mathematical Statistics.

Table A.11 W^+ Distribution for the Wilcoxon Signed Rank Test

One-Tail Test. Given n and α, the table gives the (CASE I) upper-tail critical value w_u such that $P(W^+ \geq w_u) = \alpha$; or the (CASE II) lower-tail critical value w_l such that $P(w^+ \leq w_l) = \alpha$ (e.g., for $n = 17$ and $\alpha = 0.05$, the CASE I critical region is $\mathcal{R} = \{w^+ \mid w^+ \geq 112\}$; and the CASE II critical region is $\mathcal{R} = \{w^+ \mid w^+ \leq 41\}$).

Two-Tail Test. Given n and α, the table gives the (CASE III) lower- and upper-tail critical values w_l and w_u, respectively, such that $P(W^+ \leq w_l) + P(W^+ \geq w_u) = \alpha/2 + \alpha/2 = \alpha$ (e.g., for $n = 10$ and $\alpha = 0.01$, the CASE III critical region is $\mathcal{R} = \{w^+ \mid w^+ \leq 3 \text{ or } w^+ \geq 52\}$).

| | w_l | | | | w_u | | | | |
n	0.005	0.01	0.025	0.05	0.95	0.975	0.99	0.995	n
6			0	2	19	21			6
7		0	2	3	25	26	28		7
8	0	1	3	5	31	33	35	36	8
9	1	3	5	8	37	40	42	44	9
10	3	5	8	10	45	47	50	52	10
11	5	7	10	13	53	56	59	61	11
12	7	9	13	17	61	65	69	71	12
13	9	12	17	21	70	74	79	82	13
14	12	15	21	25	80	84	90	93	14
15	15	19	25	30	90	95	101	105	15
16	19	23	29	35	101	107	113	117	16
17	23	27	34	41	112	119	126	130	17
18	27	32	40	47	124	131	139	144	18
19	32	37	46	53	137	144	153	158	19
20	37	43	52	60	150	158	167	173	20
21	42	49	58	61	164	173	182	189	21
22	48	55	65	75	178	188	198	205	22
23	54	62	73	83	193	203	214	222	23
24	61	69	81	91	209	219	231	239	24
25	68	76	89	100	225	236	249	257	25

Table A.12 R_1 Distribution for the Mann-Whitney Rank-Sum Test

One-Tail Test. Given n_1, n_2, and α, the table gives the (CASE I) upper-tail critical value r_u such that $P(R_1 \geq r_u) = \alpha$; or the (CASE II) lower-tail critical value r_l such that $P(R_1 \leq r_l) = \alpha$ (e.g., for $n_1 = 8$, $n_2 = 6$, and $\alpha = 0.01$, the CASE I critical region is $\mathcal{R} = \{r_1 \mid r_1 \geq 78\}$; and the CASE II critical region is $\mathcal{R} = \{r_1 \mid r_1 \leq 42\}$.

Two-Tail Test. Given n_1, n_2, and α, the table gives the (CASE III) lower- and upper-tail critical values r_l and r_u, respectively, such that $P(R_1 \leq r_l) + P(R_1 \geq r_u) = \alpha/2 + \alpha/2 = \alpha$ (e.g., for $n_1 = n_2 = 10$ and $\alpha = 0.05$, the CASE III critical region is $\mathcal{R} = \{r_1 \mid r_1 \leq 78 \text{ or } r_1 \geq 132\}$).

$n_1 = 3$	n_2	r_l				r_u				n_2	$n_1 = 3$
		0.005	0.01	0.025	0.05	0.95	0.975	0.99	0.995		
	3	5	5	5	6	15	16	16	16	3	
	4	5	5	5	6	18	19	19	19	4	
	5	5	5	6	7	20	21	22	22	5	
	6	5	5	7	8	22	23	25	25	6	
	7	5	6	7	8	25	26	27	28	7	
	8	5	6	8	9	27	28	30	31	8	
	9	6	7	8	10	29	31	32	33	9	
	10	6	7	9	10	32	33	35	36	10	
	11	6	7	9	11	34	36	38	39	11	
	12	7	8	10	11	37	38	40	41	12	
	13	7	8	10	12	39	41	43	44	13	
	14	7	8	11	13	41	43	46	47	14	
	15	8	9	11	13	44	46	48	49	15	
	16	8	9	12	14	46	48	51	52	16	
	17	8	10	12	15	48	51	53	55	17	
	18	8	10	13	15	51	53	56	58	18	
	19	9	10	13	16	53	56	59	60	19	
	20	9	11	14	17	55	58	61	63	20	

$n_1 = 4$	n_2	r_l				r_u				n_2	$n_1 = 4$
		0.005	0.01	0.025	0.05	0.95	0.975	0.99	0.995		
	3	9	9	9	10	22	23	23	23	3	
	4	9	9	10	11	25	26	27	27	4	
	5	9	10	11	12	28	29	30	31	5	
	6	10	11	12	13	31	32	33	34	6	
	7	10	11	13	14	34	35	37	38	7	
	8	11	12	14	15	37	38	40	41	8	
	9	11	13	14	16	40	42	43	45	9	
	10	12	13	15	17	43	45	47	48	10	
	11	12	14	16	18	46	48	50	52	11	
	12	13	15	17	19	49	51	53	55	12	
	13	13	15	18	20	52	54	57	59	13	
	14	14	16	19	21	55	57	60	62	14	
	15	15	17	20	22	58	60	63	65	15	
	16	15	17	21	24	60	63	67	69	16	
	17	16	18	21	25	63	67	70	72	17	
	18	16	19	22	26	66	70	73	76	18	
	19	17	19	23	27	69	73	77	79	19	
	20	18	20	24	28	72	76	80	82	20	

Table A.12 (Contd.)

$n_1 = 5$	n_2	r_l 0.005	0.01	0.025	0.05	r_u 0.95	0.975	0.99	0.995	n_2	$n_1 = 5$
	3	14	14	15	16	29	30	31	31	3	
	4	14	15	16	17	33	34	35	36	4	
	5	15	16	17	19	36	38	39	40	5	
	6	16	17	18	20	40	42	43	44	6	
	7	16	18	20	21	44	45	47	49	7	
	8	17	19	21	23	47	49	51	53	8	
	9	18	20	22	24	51	53	55	57	9	
	10	19	21	23	26	54	57	59	61	10	
	11	20	22	24	24	27	58	61	63	11	
	12	21	23	26	28	62	64	67	69	12	
	13	22	24	27	30	65	68	71	73	13	
	14	22	25	28	31	69	72	75	78	14	
	15	23	26	29	33	72	76	79	82	15	
	16	24	27	30	34	76	80	83	86	16	
	17	25	28	32	35	80	83	87	90	17	
	18	26	29	33	37	83	87	91	94	18	
	19	27	30	34	38	87	91	95	98	19	
	20	28	31	35	40	90	95	99	102	20	

$n_1 = 6$	n_2	r_l 0.005	0.01	0.025	0.05	r_u 0.95	0.975	0.99	0.995	n_2	$n_1 = 6$
	3	20	20	22	23	37	38	40	40	3	
	4	21	22	23	24	42	42	43	44	4	
	5	22	23	24	26	26	46	48	49	5	
	6	23	24	26	28	50	52	54	55	6	
	7	24	25	27	29	55	57	59	60	7	
	8	25	27	29	31	59	61	63	65	8	
	9	26	28	31	33	63	65	68	70	9	
	10	27	29	32	35	67	70	73	75	10	
	11	28	30	34	37	71	74	78	80	11	
	12	30	32	35	38	76	79	82	84	12	
	13	31	33	37	40	80	83	87	89	13	
	14	32	34	38	42	84	88	92	94	14	
	15	33	36	40	44	88	92	96	99	15	
	16	34	37	42	46	92	96	101	104	16	
	17	36	39	43	47	97	101	105	108	17	
	18	37	40	45	49	101	105	110	113	18	
	19	38	41	46	51	105	110	115	118	19	
	20	39	43	48	53	109	114	119	123	20	

Table A.12 (Contd.)

$n_1 = 7$	n_2		r_l					r_u			n_2	$n_1 = 7$
		0.005	**0.01**	**0.025**	**0.05**	**0.95**	**0.975**	**0.99**	**0.995**		
	3	27	28	29	30	47	48	49	50	**3**	
	4	28	29	31	32	52	53	55	56	**4**	
	5	29	31	33	34	57	58	60	62	**5**	
	6	31	32	34	36	62	62	64	66	**6**	
	7	32	34	36	39	66	66	69	71	**7**	
	8	34	35	38	41	71	74	77	78	**8**	
	9	35	37	40	43	76	79	82	84	**9**	
	10	37	39	42	45	81	84	87	89	**10**	
	11	38	40	44	47	86	89	93	95	**11**	
	12	40	42	46	49	91	94	98	100	**12**	
	13	41	44	48	52	95	99	103	106	**13**	
	14	43	45	50	54	100	104	109	111	**14**	
	15	44	47	52	56	105	109	114	117	**15**	
	16	46	49	54	58	110	114	119	122	**16**	
	17	47	51	56	61	114	119	124	128	**17**	
	18	49	52	58	63	119	124	130	133	**18**	
	19	50	54	60	65	124	129	135	139	**19**	
	20	52	56	62	67	129	134	140	144	**20**	

$n_1 = 8$	n_2		r_l					r_u			n_2	$n_1 = 8$
		0.005	**0.01**	**0.025**	**0.05**	**0.95**	**0.975**	**0.99**	**0.995**		
	3	35	36	38	39	57	58	60	61	**3**	
	4	37	38	40	41	63	64	66	67	**4**	
	5	38	40	42	44	68	70	72	74	**5**	
	6	40	42	44	46	74	76	78	80	**6**	
	7	42	43	46	49	79	82	85	86	**7**	
	8	43	45	49	51	85	87	91	93	**8**	
	9	45	47	51	54	90	93	97	99	**9**	
	10	47	49	53	56	96	99	103	105	**10**	
	11	49	51	55	59	101	105	109	111	**11**	
	12	51	53	58	62	106	110	115	117	**12**	
	13	53	56	60	64	112	116	120	123	**13**	
	14	54	58	62	67	117	122	126	130	**14**	
	15	56	60	65	69	123	127	132	136	**15**	
	16	58	62	67	72	128	133	138	142	**16**	
	17	60	64	70	75	133	138	144	148	**17**	
	18	62	66	72	77	139	144	150	154	**18**	
	19	64	68	74	80	144	150	156	160	**19**	
	20	66	70	77	83	149	155	162	166	**20**	

Table A.12   (Contd.)

$n_1 = 9$				r_l				r_u					
$n_1 = 9$	n_2	0.005	0.01	0.025	0.05	0.95	0.975	0.99	0.995	n_2	$n_1 = 9$
	3	45	46	47	49	68	70	71	72	3
	4	46	48	49	51	75	77	78	80	4
	5	48	50	52	54	81	83	85	87	5
	6	50	52	55	57	87	89	92	94	6
	7	52	54	57	60	93	96	99	101	7
	8	54	56	60	63	99	102	106	108	8
	9	56	59	62	66	105	109	112	115	9
	10	58	61	65	69	111	115	119	122	10
	11	61	63	68	72	117	121	126	128	11
	12	63	66	71	75	123	127	132	135	12
	13	65	68	73	78	129	134	139	142	13
	14	67	71	76	81	135	140	145	149	14
	15	69	73	79	84	141	146	152	156	15
	16	72	76	82	87	147	152	158	162	16
	17	74	78	84	90	153	159	165	169	17
	18	76	81	87	93	159	165	171	176	18
	19	78	83	90	96	165	171	178	183	19
	20	81	85	93	99	171	177	185	189	20

$n_1 = 10$				r_l				r_u					
$n_1 = 10$	n_2	0.005	0.01	0.025	0.05	0.95	0.975	0.99	0.995	n_2	$n_1 = 10$
	3	55	56	58	59	81	82	84	85	3
	4	57	58	60	62	88	90	92	93	4
	5	59	61	63	66	94	97	99	101	5
	6	61	63	66	69	101	104	107	109	6
	7	64	66	69	72	108	111	114	116	7
	8	66	68	72	75	115	118	122	124	8
	9	68	71	75	79	121	125	129	132	9
	10	71	74	78	82	128	132	136	139	10
	11	73	77	81	86	134	139	143	147	11
	12	76	79	84	89	141	146	151	154	12
	13	79	82	88	92	148	152	158	161	13
	14	81	85	91	96	154	159	165	169	14
	15	84	88	94	99	161	166	172	176	15
	16	86	91	97	103	167	173	179	184	16
	17	89	93	100	106	174	180	187	191	17
	18	92	96	103	110	180	187	194	198	18
	19	94	99	107	113	187	193	201	206	19
	20	97	102	110	117	193	200	208	213	20

Table A.12 (Contd.)

$n_1 = 11$	n_2	r_l				r_u				n_2	$n_1 = 11$
		0.005	0.01	0.025	0.05	0.95	0.975	0.99	0.995		
	3	66	67	69	71	94	96	98	99	3	
	4	68	70	72	74	102	104	106	108	4	
	5	71	73	75	78	109	112	114	116	5	
	6	73	75	79	82	116	119	123	125	6	
	7	76	78	82	85	124	127	131	133	7	
	8	79	81	85	89	131	135	139	141	8	
	9	82	84	89	93	138	142	147	149	9	
	10	84	88	92	97	145	150	154	158	10	
	11	87	91	96	100	153	157	162	166	11	
	12	90	94	99	104	160	165	170	174	12	
	13	93	97	103	108	167	172	178	182	13	
	14	96	100	106	112	174	180	186	190	14	
	15	99	103	110	116	181	187	194	198	15	
	16	102	107	113	120	188	195	201	206	16	
	17	105	110	117	123	196	202	209	214	17	
	18	108	113	121	127	203	209	217	222	18	
	19	111	116	124	131	210	217	225	230	19	
	20	114	119	128	135	217	224	233	238	20	

$n_1 = 12$	n_2	r_l				r_u				n_2	$n_1 = 12$
		0.005	0.01	0.025	0.05	0.95	0.975	0.99	0.995		
	3	79	80	82	83	109	110	112	113	3	
	4	81	83	85	87	117	119	121	123	4	
	5	84	86	89	91	125	127	130	132	5	
	6	87	89	92	95	133	136	139	141	6	
	7	90	92	96	99	141	144	148	150	7	
	8	93	95	100	104	148	152	157	159	8	
	9	96	99	104	108	156	160	165	168	9	
	10	99	102	107	112	164	169	174	177	10	
	11	102	106	111	116	172	177	182	186	11	
	12	105	109	115	120	180	180	185	191	12	
	13	109	113	119	125	187	193	199	203	13	
	14	112	116	123	129	195	201	208	212	14	
	15	115	120	127	133	203	209	216	221	15	
	16	119	124	131	138	210	217	224	229	16	
	17	122	127	135	142	218	225	233	238	17	
	18	125	131	139	146	226	233	241	247	18	
	19	129	134	143	150	234	241	250	255	19	
	20	132	138	147	155	241	249	258	264	20	

Table A.12 (Contd.)

$n_1 = 13$		r_l				r_u					
$n_1 = 13$	n_2	0.005	0.01	0.025	0.05	0.95	0.975	0.99	0.995	n_2	$n_1 = 13$
	3	92	93	95	97	124	126	128	129	3	
	4	94	96	99	101	133	135	138	140	4	
	5	98	100	103	106	141	144	147	149	5	
	6	101	103	107	110	150	153	157	159	6	
	7	104	107	111	115	158	162	166	169	7	
	8	108	111	115	119	167	171	175	178	8	
	9	111	114	119	124	175	180	185	188	9	
	10	115	118	124	128	184	188	194	197	10	
	11	118	122	128	133	192	197	203	207	11	
	12	122	126	132	138	200	206	212	216	12	
	13	125	130	136	142	209	215	221	226	13	
	14	129	134	141	147	217	223	230	235	14	
	15	133	138	145	152	225	232	239	244	15	
	16	136	142	150	156	234	240	248	254	16	
	17	140	146	154	161	242	249	257	263	17	
	18	144	150	158	166	250	258	266	272	18	
	19	148	154	163	171	258	266	275	281	19	
	20	151	158	167	175	267	275	284	291	20	

$n_1 = 14$		r_l				r_u					
$n_1 = 14$	n_2	0.005	0.01	0.025	0.05	0.95	0.975	0.99	0.995	n_2	$n_1 = 14$
	3	106	107	110	112	140	142	145	146	3	
	4	109	111	114	116	150	152	155	157	4	
	5	112	115	118	121	159	162	165	168	5	
	6	116	118	122	126	168	172	176	178	6	
	7	120	122	127	131	177	181	186	188	7	
	8	123	127	131	136	186	191	195	199	8	
	9	127	131	136	141	195	200	205	209	9	
	10	131	135	141	146	204	209	215	219	10	
	11	135	139	145	151	213	219	225	229	11	
	12	139	143	150	156	222	228	235	239	12	
	13	143	148	155	161	231	237	244	249	13	
	14	147	152	160	166	240	246	254	259	14	
	15	151	156	164	171	249	256	264	269	15	
	16	155	161	169	176	258	265	273	279	16	
	17	159	165	174	182	266	274	283	289	17	
	18	163	170	179	187	275	283	292	299	18	
	19	168	174	183	192	284	293	302	308	19	
	20	172	178	188	197	293	302	312	318	20	

Table A.12   (Contd.)

$n_1 = 15$	n_2	r_l				r_u				n_2	$n_1 = 15$
		0.005	0.01	0.025	0.05	0.95	0.975	0.99	0.995		
	3	122	123	125	127	158	160	162	163	3	
	4	125	127	130	132	168	170	173	175	4	
	5	128	131	134	138	177	181	184	187	5	
	6	132	135	139	143	187	191	195	198	6	
	7	136	139	144	148	197	201	206	209	7	
	8	140	144	149	153	207	211	216	220	8	
	9	144	148	154	159	216	221	227	231	9	
	10	149	153	159	164	226	231	237	241	10	
	11	153	157	164	170	235	241	248	252	11	
	12	157	162	169	175	245	251	258	263	12	
	13	162	167	174	181	254	261	268	273	13	
	14	166	171	179	186	264	271	279	284	14	
	15	171	176	184	192	273	281	289	294	15	
	16	175	181	190	197	283	290	299	305	16	
	17	180	186	195	203	292	300	309	315	17	
	18	184	190	200	208	302	310	320	326	18	
	19	189	195	205	214	311	320	330	336	19	
	20	193	200	210	220	320	330	340	347	20	

$n_1 = 16$	n_2	r_l				r_u				n_2	$n_1 = 16$
		0.005	0.01	0.025	0.05	0.95	0.975	0.99	0.995		
	3	138	139	142	144	116	178	181	182	3	
	4	141	143	147	150	186	189	193	195	4	
	5	145	148	151	155	191	201	204	207	5	
	6	149	152	157	161	207	211	216	219	6	
	7	154	157	162	166	218	222	227	230	7	
	8	158	162	167	172	228	233	238	242	8	
	9	163	167	173	178	238	243	249	253	9	
	10	167	172	178	184	248	254	260	265	10	
	11	172	177	183	190	258	265	271	276	11	
	12	177	182	189	196	268	275	282	287	12	
	13	181	187	195	201	279	285	293	299	13	
	14	186	192	200	207	289	296	304	310	14	
	15	191	197	206	213	299	306	315	321	15	
	16	196	202	211	219	309	317	326	332	16	
	17	201	207	211	225	319	327	337	343	17	
	18	206	212	222	231	329	338	348	354	18	
	19	210	218	228	237	339	348	358	366	19	
	20	215	223	234	243	349	358	369	377	20	

Table A.12 (Contd.)

$n_1 = 17$	n_2			r_l				r_u			n_2	$n_1 = 17$
		0.005	0.01	0.025	0.05	0.95	0.975	0.99	0.995		
	3	155	157	159	162	195	198	200	202	3	
	4	159	161	164	168	206	210	213	215	4	
	5	163	166	170	173	218	221	225	228	5	
	6	168	171	175	179	229	233	237	240	6	
	7	172	176	181	186	239	244	249	253	7	
	8	177	181	187	192	250	255	261	265	8	
	9	182	186	192	198	261	267	273	277	9	
	10	187	191	198	204	272	278	285	289	10	
	11	192	197	204	210	283	289	296	301	11	
	12	197	202	210	217	293	300	308	313	12	
	13	202	208	216	223	304	311	319	325	13	
	14	207	213	222	230	314	322	331	337	14	
	15	213	219	228	236	325	333	342	348	15	
	16	218	224	234	242	336	344	354	360	16	
	17	223	230	240	249	346	355	365	372	17	
	18	228	235	246	255	357	366	377	384	18	
	19	234	241	252	262	367	377	388	395	19	
	20	239	246	258	268	378	388	400	407	20	

$n_1 = 18$	n_2			r_l				r_u			n_2	$n_1 = 18$
		0.005	0.01	0.025	0.05	0.95	0.975	0.99	0.995		
	3	173	175	178	180	216	218	221	223	3	
	4	177	180	183	187	227	231	234	237	4	
	5	182	185	189	193	239	243	247	250	5	
	6	187	190	195	199	251	255	260	263	6	
	7	192	195	201	206	262	267	273	276	7	
	8	197	201	207	212	274	279	285	289	8	
	9	202	207	213	219	285	291	297	302	9	
	10	208	212	219	226	296	303	310	314	10	
	11	213	218	226	232	308	314	322	327	11	
	12	218	224	232	239	319	326	334	340	12	
	13	224	230	238	246	330	338	246	352	13	
	14	229	236	245	253	341	349	358	365	14	
	15	235	241	251	259	353	361	371	377	15	
	16	241	247	257	266	364	373	383	389	16	
	17	246	253	264	273	375	384	395	402	17	
	18	252	259	270	280	386	396	407	414	18	
	19	258	265	277	287	397	407	419	426	19	
	20	263	271	283	294	408	419	431	439	20	

Table A.12 (Contd.)

$n_1 = 19$	n_2	r_l				r_u				n_2	$n_1 = 19$
		0.005	0.01	0.025	0.05	0.95	0.975	0.99	0.995		
	3	193	194	197	200	237	240	243	244	3	
	4	197	199	203	207	249	253	257	259	4	
	5	202	205	209	213	262	266	270	273	5	
	6	207	210	215	220	274	279	284	287	6	
	7	212	216	222	227	286	291	297	301	7	
	8	218	222	228	234	298	304	310	314	8	
	9	223	228	235	241	310	316	323	328	9	
	10	229	234	242	248	322	328	336	341	10	
	11	235	240	248	255	334	341	349	354	11	
	12	241	246	255	262	346	353	362	367	12	
	13	247	253	262	270	357	365	374	380	13	
	14	253	259	268	277	369	378	387	393	14	
	15	259	265	275	284	381	390	400	406	15	
	16	264	272	282	291	393	402	412	420	16	
	17	271	278	289	299	404	414	425	432	17	
	18	277	284	296	306	416	426	438	445	18	
	19	283	291	303	313	428	438	450	458	19	
	20	289	297	309	320	440	451	463	471	20	

$n_1 = 20$	n_2	r_l				r_u				n_2	$n_1 = 20$
		0.005	0.01	0.025	0.05	0.95	0.975	0.99	0.995		
	3	213	215	218	221	259	262	265	267	3	
	4	218	220	224	228	272	276	280	282	4	
	5	223	226	230	235	285	290	294	297	5	
	6	228	232	237	242	298	303	308	312	6	
	7	234	238	244	249	311	316	322	326	7	
	8	240	244	251	257	323	329	336	340	8	
	9	246	250	258	264	336	342	350	354	9	
	10	252	257	265	272	348	355	363	368	10	
	11	258	263	272	279	361	368	377	382	11	
	12	264	270	279	287	373	381	390	396	12	
	13	270	277	286	294	386	394	403	410	13	
	14	277	283	293	302	398	407	417	423	14	
	15	283	290	300	310	410	420	430	437	15	
	16	289	297	308	317	423	432	443	454	16	
	17	296	303	315	325	435	445	457	464	17	
	18	302	310	322	333	447	458	470	478	18	
	19	309	317	329	340	460	471	483	491	19	
	20	315	324	337	348	472	483	496	505	20	

Table A.13   Quantiles of the Lilliefors Test Statistic $\hat{D}_n$

The table gives the upper $100(1 - \alpha)\%$ quantile $\hat{d}_{n,1-\alpha}$ of the sampling distribution of $\hat{D}_n$ such that $P(\hat{D}_n \leq \hat{d}_{n,1-\alpha}) = 1 - \alpha$ or $P(\hat{D}_n \geq \hat{d}_{n,1-\alpha}) = \alpha$ (e.g., for $n = 15$ and $\alpha = 0.01$, the upper-tail critical region is $\mathcal{R} = \{\hat{d}_{15} | \hat{d}_{15} \geq \hat{d}_{15,0.99} = 0.257\}$).

$1 - \alpha =$	**0.80**	**0.85**	**0.90**	**0.95**	**0.99**
Sample size $n =$ **4**	0.300	0.319	0.352	0.381	0.417
5	0.285	0.299	0.315	0.337	0.405
6	0.265	0.277	0.294	0.319	0.364
7	0.247	0.258	0.276	0.300	0.348
8	0.233	0.244	0.261	0.285	0.331
9	0.223	0.233	0.249	0.271	0.311
10	0.215	0.224	0.239	0.258	0.294
11	0.206	0.217	0.230	0.249	0.284
12	0.199	0.212	0.223	0.242	0.275
13	0.190	0.202	0.214	0.234	0.268
14	0.183	0.194	0.207	0.227	0.261
15	0.177	0.187	0.201	0.220	0.257
16	0.173	0.182	0.195	0.213	0.250
17	0.169	0.177	0.189	0.206	0.245
18	0.166	0.173	0.184	0.200	0.239
19	0.163	0.169	0.179	0.195	0.235
20	0.160	0.166	0.174	0.190	0.231
25	0.142	0.147	0.158	0.173	0.200
30	0.131	0.136	0.144	0.161	0.187
Over 30	$\dfrac{0.736}{\sqrt{n}}$	$\dfrac{0.768}{\sqrt{n}}$	$\dfrac{0.805}{\sqrt{n}}$	$\dfrac{0.886}{\sqrt{n}}$	$\dfrac{1.031}{\sqrt{n}}$

Adapted from Table 1 of H.W. Lilliefors, "On the Kolmogorov-Smirnov Test for Normality with Mean and Variance Unknown," *JASA*, 62, 1967, 399–402. Reprinted with permission from *The Journal of the American Statistical Association*. Copyright [1967] by the American Statistical Association. All rights reserved.

Table A.14 Quantiles of the Kolmogorov-Smirnov Test Statistics D_n

The table gives the upper $100(1-\alpha)\%$ quantile $\hat{d}_{n,1-\alpha}$ of the sampling distribution of $\hat{D}_n$ such that $P(\hat{D}_n \leq \hat{d}_{n,1-\alpha}) = 1 - \alpha$ or $P(\hat{D}_n \geq \hat{d}_{n,1-\alpha}) = \alpha$ (e.g., for $n = 20$ and $\alpha = 0.05$, the one-tail critical region is $\mathcal{R} = \{\hat{d}_{20}|\hat{d}_{20} \geq \hat{d}_{20,0.95} = 0.265\}$; the two-tail critical region is $\mathcal{R} = \{\hat{d}_{20}|\hat{d}_{20} \geq \hat{d}_{20,0.95} = 0.294\}$).

One-Sided Test $1 - \alpha =$	0.90	0.95	0.975	0.99	0.995	$1 - \alpha =$	0.90	0.95	0.975	0.99	0.995
Two-Sided Test $1 - \alpha =$	0.80	0.90	0.95	0.98	0.99	$1 - \alpha =$	0.80	0.90	0.95	0.98	0.99
$n = 1$	0.900	0.950	0.975	0.990	0.995	$n = 21$	0.226	0.259	0.287	0.321	0.344
2	0.684	0.776	0.842	0.900	0.929	22	0.221	0.253	0.281	0.314	0.337
3	0.565	0.636	0.708	0.785	0.829	23	0.216	0.247	0.275	0.307	0.330
4	0.493	0.565	0.624	0.689	0.734	24	0.212	0.242	0.269	0.301	0.323
5	0.447	0.509	0.563	0.627	0.669	25	0.208	0.238	0.264	0.295	0.317
6	0.410	0.468	0.519	0.577	0.617	26	0.204	0.233	0.259	0.290	0.311
7	0.381	0.436	0.483	0.538	0.576	27	0.200	0.229	0.254	0.284	0.305
8	0.358	0.410	0.454	0.507	0.542	28	0.197	0.225	0.250	0.279	0.300
9	0.339	0.387	0.430	0.480	0.513	29	0.193	0.221	0.246	0.275	0.295
10	0.323	0.369	0.409	0.457	0.489	30	0.190	0.218	0.242	0.270	0.290
11	0.308	0.352	0.391	0.437	0.468	31	0.187	0.214	0.238	0.266	0.285
12	0.296	0.338	0.375	0.419	0.449	32	0.184	0.211	0.234	0.262	0.281
13	0.285	0.325	0.361	0.404	0.432	33	0.182	0.208	0.231	0.258	0.277
14	0.275	0.314	0.349	0.390	0.418	34	0.179	0.205	0.227	0.254	0.273
15	0.266	0.304	0.338	0.377	0.404	35	0.177	0.202	0.224	0.251	0.269
16	0.258	0.295	0.327	0.366	0.392	36	0.174	0.199	0.221	0.247	0.265
17	0.250	0.286	0.318	0.355	0.381	37	0.172	0.196	0.218	0.244	0.262
18	0.244	0.279	0.309	0.346	0.371	38	0.170	0.194	0.215	0.241	0.258
19	0.237	0.271	0.301	0.337	0.361	39	0.168	0.191	0.213	0.238	0.255
20	0.232	0.265	0.294	0.329	0.352	40	0.165	0.189	0.210	0.235	0.252

Adapted from L.H. Miller, "Tables of Percentage Points of Kolmogorov Statistics," *JASA*, 51, 1956, 111–121. Reprinted with permission from *The Journal of the American Statistical Association*. Copyright [1956] by the American Statistical Association. All rights reserved.

Table A.15 Quantiles of the Kolmogorov-Smirnov Test Statistic $D_{n,m}$ When $n = m$

The table gives the upper $100(1 - \alpha)$ % quantile $\hat{d}_{n,m}$ of the sampling distribution of $\hat{D}_{n,m}$ such that $P(\hat{D}_{n,m} \le \hat{d}_{n,m,1-\alpha}) = 1 - \alpha$ or $P(\hat{D}_{n,m} \ge \hat{d}_{n,m,1-\alpha}) = \alpha$ (e.g., for $n = m = 15$ and $\alpha = 0.05$, the one-tail critical region is $\mathcal{R} = \{\hat{d}_{15,15}|\hat{d}_{15,15} \ge \hat{d}_{15,15,0.95} = 0.40\}$; the two-tail critical region is $\mathcal{R} = \{\hat{d}_{15,15}|\hat{d}_{15,15} \ge \hat{d}_{15,15,0.95} = 0.467\}$).

One-Sided Test $1 - \alpha =$	0.90	0.95	0.975	0.99	0.995	$1 - \alpha =$	0.90	0.95	0.975	0.99	0.995
Two-Sided Test $1 - \alpha =$	0.80	0.90	0.95	0.98	0.99	$1 - \alpha =$	0.80	0.90	0.95	0.98	0.99
$n = 3$	2/3	2/3				$n = 20$	6/20	7/20	8/20	9/20	10/20
4	3/4	3/4	3/4			21	6/21	7/21	8/21	9/21	10/21
5	3/5	3/5	4/5	4/5	4/5	22	7/22	8/22	8/22	10/22	10/22
6	3/6	4/6	4/6	5/6	5/6	23	7/23	8/23	9/23	10/23	10/23
7	4/7	4/7	5/7	5/7	5/7	24	7/24	8/24	9/24	10/24	11/24
8	4/8	4/8	5/8	5/8	6/8	25	7/25	8/25	9/25	10/25	11/25
9	4/9	5/9	5/9	6/9	6/9	26	7/26	8/26	9/26	10/26	11/26
10	4/10	5/10	6/10	6/10	7/10	27	7/27	8/27	9/27	11/27	11/27
11	5/11	5/11	6/11	7/11	7/11	28	8/28	9/28	10/28	11/28	12/28
12	5/12	5/12	6/12	7/12	7/12	29	8/29	9/29	10/29	11/29	12/29
13	5/13	6/13	6/13	7/13	8/13	30	8/30	9/30	10/30	11/30	12/30
14	5/14	6/14	7/14	7/14	8/14	31	8/31	9/31	10/31	11/31	12/31
15	5/15	6/15	7/15	8/15	8/15	32	8/32	9/32	10/32	12/32	12/32
16	6/16	6/16	6/25	8/16	12/15	34	8/34	10/34	11/34	12/34	13/34
17	9/29	7/17	7/17	8/22	9/17	36	9/36	10/36	11/36	12/36	13/36
18	6/18	7/18	8/18	9/18	9/19	38	9/38	10/38	11/38	13/38	14/38
19	6/19	7/19	8/19	9/19	9/19	40	9/40	10/40	12/40	13/40	14/40
						Approximation for $n > 40$:	$\dfrac{1.52}{\sqrt{n}}$	$\dfrac{1.73}{\sqrt{n}}$	$\dfrac{1.92}{\sqrt{n}}$	$\dfrac{2.15}{\sqrt{n}}$	$\dfrac{2.30}{\sqrt{n}}$

Adapted from Z.W. Birnbaum and R.A. Hall, "Small Sample Distribution for Multisample Statistics of the Smirnov Type," *The Annals of Mathematical Statistics*, 31, 1960, 710–720, with kind permission from the Institute of Mathematical Statistics.

Table A.16 Quantiles of the Kolmogorov-Smirnov Test Statistic $D_{n,m}$ When $n \neq m^*$

The table gives the upper $100(1-\alpha)\%$ quantile $\hat{d}_{n,m}$ of the sampling distribution of $\hat{D}_{n,m}$ such that $P(\hat{D}_{n,m} \leq \hat{d}_{n,m,1-\alpha}) = 1 - \alpha$ or $P(\hat{D}_{n,m} \geq \hat{d}_{n,m,1-\alpha}) = \alpha$ (e.g., for $n = 6$, $m = 10$, and $\alpha = 0.05$, the one-tail critical region is $\mathcal{R} = \{\hat{d}_{6,10}|\hat{d}_{6,10} \geq \hat{d}_{6,10,0.95} = 0.567\}$; the two-tail critical region is $\mathcal{R} = \{\hat{d}_{6,10}|\hat{d}_{6,10} \geq \hat{d}_{6,10,0.95} = 0.633\}$).

One-Sided Test	$1 - \alpha =$	0.90	0.95	0.975	0.99	0.995
Two-Sided Test	$1 - \alpha =$	**0.80**	**0.90**	**0.950**	**0.98**	**0.990**
$n = 1$	$m = 9$	17/18				
	10	9/10				
$n = 2$	3	5/6				
	4	3/4				
	5	4/5	4/5			
	6	5/6	5/6			
	7	5/7	6/7			
	8	3/4	7/8	7/8		
	9	7/9	8/9	8/9		
	10	7/10	4/5	9/10		
$n = 3$	$m = 4$	3/4	3/4			
	5	2/3	4/5	4/5		
	6	2/3	2/3	5/6		
	7	2/3	5/7	6/7	6/7	
	8	5/8	3/4	3/4	7/8	
	9	2/3	2/3	7/9	8/9	8/9
	10	3/5	7/10	4/5	9/10	9/10
	12	7/12	2/3	3/4	5/6	11/12
$n = 4$	$m = 5$	3/5	3/4	4/5	4/5	
	6	7/12	2/3	3/4	5/6	5/6
	7	17/28	5/7	3/4	6/7	6/7
	8	5/8	5/8	3/4	7/8	7/8
	9	5/9	2/3	3/4	7/9	8/9
	10	11/20	13/20	7/10	4/5	4/5
	12	7/12	2/3	2/3	3/4	5/6
	16	9/16	5/8	11/16	3/4	13/16
$n = 5$	$m = 6$	3/5	2/3	2/3	5/6	5/6
	7	4/7	23/35	5/7	29/35	6/7
	8	11/20	5/8	27/40	4/5	4/5
	9	5/9	3/5	31/45	7/9	4/5
	10	1/2	3/5	7/10	7/10	4/5
	15	8/15	3/5	2/3	11/15	11/15
	20	1/2	11/20	3/5	7/10	3/4

Table A.16 (Contd.)

One-Sided Test	$1-\alpha=$	0.90	0.95	0.975	0.99	0.995
Two-Sided Test	$1-\alpha=$	**0.80**	**0.90**	**0.950**	**0.98**	**0.990**
$n=6$	$m=7$	23/42	4/7	29/42	5/7	5/6
	8	1/2	7/12	2/3	3/4	3/4
	9	1/2	5/9	2/3	13/18	7/9
	10	1/2	17/30	19/30	7/10	11/15
	12	1/2	7/12	7/12	2/3	3/4
	18	4/9	5/9	11/18	2/3	13/18
	24	11/24	1/2	7/12	5/8	2/3
$n=7$	$m=8$	27/56	33/56	5/8	41/56	3/4
	9	31/63	5/9	40/63	5/7	47/63
	10	33/70	39/70	43/70	7/10	5/7
	14	3/7	1/2	4/7	9/14	5/7
	28	3/7	13/28	15/28	17/28	9/14
$n=8$	$m=9$	4/9	13/24	5/8	2/3	3/4
	10	19/40	21/40	23/40	27/40	7/10
	12	11/24	1/2	7/12	5/8	2/3
	16	7/16	1/2	9/16	5/8	5/8
	32	13/32	7/16	1/2	9/16	19/32
$n=9$	$m=10$	7/15	1/2	26/45	2/3	31/45
	12	4/9	1/2	5/9	11/18	2/3
	15	19/45	22/45	8/15	3/5	29/45
	18	7/18	4/9	1/2	5/9	11/18
	36	13/36	5/12	17/36	19/36	5/9
$n=10$	$m=15$	2/5	7/15	1/2	17/30	19/30
	20	2/5	9/20	1/2	11/20	3/5
	40	7/20	2/5	9/20	1/2	
$n=12$	$m=15$	23/60	9/20	1/2	11/20	7/12
	16	3/8	7/16	23/48	13/24	7/12
	18	13/36	5/12	17/36	19/36	5/9
	20	11/30	5/12	7/15	31/60	17/30
$n=15$	$m=20$	7/20	2/5	13/30	29/60	31/60
$n=16$	$m=20$	27/80	31/80	17/40	19/40	41/80
Large-sample approximation		$1.07\sqrt{\dfrac{m+n}{mn}}$	$1.22\sqrt{\dfrac{m+n}{mn}}$	$1.36\sqrt{\dfrac{m+n}{mn}}$	$1.52\sqrt{\dfrac{m+n}{mn}}$	$1.63\sqrt{\dfrac{m+n}{mn}}$

*Let n be the smaller sample size and let m be the larger sample size. If this table does not cover n and m, use the large sample approximation.

Adapted from F.J. Massey, "Distribution Table for the Deviation Between Two Sample Cumulatives," *The Annals of Mathematical Statistics*, 23, 1952, 435–441. Corrections appear in Davis, L.S. (1958), Mathematical Tables and other Aids to Computation, 12, 1952, 262–263, with kind permission from the Institute of Mathematical Statistics.

Table A.17 Quantiles of the Shapiro-Wilk Test Statistic W

The table gives the $100\alpha\%$ quantile w_α of the sampling distribution of W such that $P(W \le w_\alpha) = \alpha$ (e.g., for $n = 30$ and $\alpha = 0.01$, the critical region $\mathcal{R} = \{w | w \le w_{0.01} = 0.900\}$).

n \ α	0.01	0.02	0.05	0.10	0.50	0.90	0.95	0.98	0.99
3	0.753	0.756	0.767	0.789	0.959	0.998	0.999	1.000	1.000
4	0.687	0.707	0.748	0.792	0.935	0.987	0.992	0.996	0.997
5	0.686	0.715	0.762	0.806	0.927	0.979	0.986	0.991	0.993
6	0.713	0.743	0.788	0.826	0.927	0.974	0.981	0.986	0.989
7	0.730	0.760	0.803	0.838	0.928	0.972	0.979	0.985	0.988
8	0.749	0.778	0.818	0.851	0.932	0.972	0.978	0.984	0.987
9	0.764	0.791	0.829	0.859	0.935	0.972	0.978	0.984	0.986
10	0.781	0.806	0.842	0.869	0.938	0.972	0.978	0.983	0.986
11	0.792	0.817	0.850	0.876	0.940	0.973	0.979	0.984	0.986
12	0.805	0.828	0.859	0.883	0.943	0.973	0.979	0.984	0.986
13	0.814	0.837	0.866	0.889	0.945	0.974	0.979	0.984	0.986
14	0.825	0.846	0.874	0.895	0.947	0.975	0.980	0.984	0.986
15	0.835	0.855	0.881	0.901	0.950	0.975	0.980	0.984	0.987
16	0.844	0.863	0.887	0.906	0.952	0.976	0.981	0.985	0.987
17	0.851	0.869	0.892	0.910	0.954	0.977	0.981	0.985	0.987
18	0.858	0.874	0.897	0.914	0.956	0.978	0.982	0.986	0.988
19	0.863	0.879	0.901	0.917	0.957	0.978	0.982	0.986	0.988
20	0.868	0.884	0.905	0.920	0.959	0.979	0.983	0.986	0.988
21	0.873	0.888	0.908	0.923	0.960	0.980	0.983	0.987	0.989
22	0.878	0.892	0.911	0.926	0.961	0.980	0.984	0.987	0.989
23	0.881	0.895	0.914	0.928	0.962	0.981	0.984	0.987	0.989
24	0.884	0.898	0.916	0.930	0.963	0.981	0.984	0.987	0.989
25	0.888	0.901	0.918	0.931	0.964	0.981	0.985	0.988	0.989
26	0.891	0.904	0.920	0.933	0.965	0.982	0.985	0.988	0.989
27	0.894	0.906	0.923	0.935	0.965	0.982	0.985	0.988	0.990
28	0.896	0.908	0.924	0.936	0.966	0.982	0.985	0.988	0.990
29	0.898	0.910	0.926	0.937	0.966	0.982	0.985	0.988	0.990
30	0.900	0.912	0.927	0.939	0.967	0.983	0.985	0.988	0.990
31	0.902	0.914	0.929	0.940	0.967	0.983	0.986	0.988	0.990
32	0.904	0.915	0.930	0.941	0.968	0.983	0.986	0.988	0.990
33	0.906	0.917	0.931	0.942	0.968	0.983	0.986	0.989	0.990
34	0.908	0.919	0.933	0.943	0.969	0.983	0.986	0.989	0.990
35	0.910	0.920	0.934	0.944	0.969	0.984	0.986	0.989	0.990
36	0.912	0.922	0.935	0.945	0.970	0.984	0.986	0.989	0.990
37	0.914	0.924	0.936	0.946	0.970	0.984	0.987	0.989	0.990
38	0.916	0.925	0.938	0.947	0.971	0.984	0.987	0.989	0.990
39	0.917	0.927	0.939	0.948	0.971	0.984	0.987	0.989	0.991
40	0.919	0.928	0.940	0.949	0.972	0.985	0.987	0.989	0.991
41	0.920	0.929	0.941	0.950	0.972	0.985	0.987	0.989	0.991
42	0.922	0.930	0.942	0.951	0.972	0.985	0.987	0.989	0.991
43	0.923	0.932	0.943	0.951	0.973	0.985	0.987	0.990	0.991
44	0.924	0.933	0.944	0.952	0.973	0.985	0.987	0.990	0.991
45	0.926	0.934	0.945	0.953	0.973	0.985	0.988	0.990	0.991
46	0.927	0.935	0.945	0.953	0.974	0.985	0.988	0.990	0.991
47	0.928	0.936	0.946	0.954	0.974	0.985	0.988	0.990	0.991
48	0.929	0.937	0.947	0.954	0.974	0.985	0.988	0.990	0.991
49	0.929	0.937	0.947	0.955	0.974	0.985	0.988	0.990	0.991
50	0.930	0.938	0.947	0.955	0.974	0.985	0.988	0.990	0.991

Reprinted from E.S. Pearson and H.O. Hartley (1976), *Biometrika Tables for Statisticians*, Vol. II. Reproduced with the kind permission of Oxford University Press.

Table A.18   Coefficients for the Shapiro-Wilk Test

The table gives coefficients $a_j, j = 1, \ldots, k \ (\approx n/2)$ for the Shapiro-Wilk test statistic W.

j \ n	2	3	4	5	6	7	8	9	10
1	0.7071	0.7071	0.6872	0.6646	0.6431	0.6233	0.6052	0.5888	0.5739
2		0.0000	0.1667	0.2413	0.2806	0.3031	0.3164	0.3244	0.3291
3				0.0000	0.0875	0.1401	0.1743	0.1976	0.2141
4						0.0000	0.0561	0.0947	0.1224
5								0.0000	0.0399

j \ n	11	12	13	14	15	16	17	18	19	20
1	0.5601	0.5475	0.5359	0.5251	0.5150	0.5056	0.4968	0.4886	0.4808	0.4734
2	0.3315	0.3325	0.3325	0.3318	0.3306	0.3290	0.3273	0.3253	0.3232	0.3211
3	0.2260	0.2347	0.2412	0.2460	0.2495	0.2521	0.2540	0.2553	0.2561	0.2565
4	0.1429	0.1586	0.1707	0.1802	0.1878	0.1939	0.1988	0.2027	0.2059	0.2085
5	0.0695	0.0922	0.1099	0.1240	0.1353	0.1447	0.1524	0.1587	0.1641	0.1686
6	0.0000	0.0303	0.0539	0.0727	0.0880	0.1005	0.1109	0.1197	0.1271	0.1334
7			0.0000	0.0240	0.0433	0.0593	0.0725	0.0837	0.0932	0.1013
8					0.0000	0.0196	0.0359	0.0496	0.0612	0.0711
9							0.0000	0.0163	0.0303	0.0422
10									0.0000	0.0140

j \ n	21	22	23	24	25	26	27	28	29	30
1	0.4643	0.4590	0.4542	0.4493	0.4450	0.4407	0.4366	0.4328	0.4291	0.4254
2	0.3185	0.3156	0.3126	0.3098	0.3069	0.3043	0.3018	0.2992	0.2968	0.2944
3	0.2578	0.2571	0.2563	0.2554	0.2543	0.2533	0.2522	0.2510	0.2499	0.2487
4	0.2119	0.2131	0.2139	0.2145	0.2148	0.2151	0.2152	0.2151	0.2150	0.2148
5	0.1736	0.1764	0.1787	0.1807	0.1822	0.1836	0.1848	0.1857	0.1864	0.1870
6	0.1399	0.1443	0.1480	0.1512	0.1539	0.1563	0.1584	0.1601	0.1616	0.1630
7	0.1092	0.1150	0.1201	0.1245	0.1283	0.1316	0.1346	0.1372	0.1395	0.1415
8	0.0804	0.0878	0.0941	0.0997	0.1046	0.1089	0.1128	0.1162	0.1192	0.1219
9	0.0530	0.0618	0.0696	0.0764	0.0823	0.0876	0.0923	0.0965	0.1002	0.1036
10	0.0260	0.0368	0.0459	0.0539	0.0610	0.0672	0.0728	0.0778	0.0822	0.0862
11	0.0000	0.0122	0.0228	0.0321	0.0403	0.0476	0.0540	0.0598	0.0650	0.0697
12			0.0000	0.0107	0.0200	0.0284	0.0358	0.0424	0.0483	0.0537
13					0.0000	0.0094	0.0178	0.0253	0.0320	0.0381
14							0.0000	0.0084	0.0159	0.0227
15									0.0000	0.0076

Table A.18 (Contd.)

j \ n	31	32	33	34	35	36	37	38	39	40
1	0.4220	0.4188	0.4156	0.4127	0.4096	0.4068	0.4040	0.4015	0.3989	0.3964
2	0.2921	0.2898	0.2876	0.2854	0.2834	0.2813	0.2794	0.2774	0.2755	0.2737
3	0.2475	0.2462	0.2451	0.2439	0.2427	0.2415	0.2403	0.2391	0.2380	0.2368
4	0.2145	0.2141	0.2137	0.2132	0.2127	0.2121	0.2116	0.2110	0.2104	0.2098
5	0.1874	0.1878	0.1880	0.1882	0.1883	0.1883	0.1883	0.1881	0.1880	0.1878
6	0.1641	0.1651	0.1660	0.1667	0.1673	0.1678	0.1683	0.1686	0.1689	0.1691
7	0.1433	0.1449	0.1463	0.1475	0.1487	0.1496	0.1505	0.1513	0.1520	0.1526
8	0.1243	0.1265	0.1284	0.1301	0.1317	0.1331	0.1344	0.1356	0.1366	0.1376
9	0.1066	0.1093	0.1118	0.1140	0.1160	0.1179	0.1196	0.1211	0.1225	0.1237
10	0.0899	0.0931	0.0961	0.0988	0.1013	0.1036	0.1056	0.1075	0.1092	0.1108
11	0.0739	0.0777	0.0812	0.0844	0.0873	0.0900	0.0924	0.0947	0.0967	0.0986
12	0.0585	0.0629	0.0669	0.0706	0.0739	0.0770	0.0798	0.0824	0.0848	0.0870
13	0.0435	0.0485	0.0530	0.0572	0.0610	0.0645	0.0677	0.0706	0.0733	0.0759
14	0.0289	0.0344	0.0395	0.0441	0.0484	0.0523	0.0559	0.0592	0.0622	0.0651
15	0.0144	0.0206	0.0262	0.0314	0.0361	0.0404	0.0444	0.0481	0.0515	0.0546
16	0.0000	0.0068	0.0131	0.0187	0.0239	0.0287	0.0331	0.0372	0.0409	0.0444
17			0.0000	0.0062	0.0119	0.0172	0.0220	0.0264	0.0305	0.0343
18					0.0000	0.0057	0.0110	0.0158	0.0203	0.0244
19							0.0000	0.0053	0.0101	0.0146
20									0.0000	0.0049

j \ n	41	42	43	44	45	46	47	48	49	50
1	0.3940	0.3917	0.3894	0.3872	0.3850	0.3830	0.3808	0.3789	0.3770	0.3751
2	0.2719	0.2701	0.2684	0.2667	0.2651	0.2635	0.2620	0.2604	0.2589	0.2574
3	0.2357	0.2345	0.2334	0.2323	0.2313	0.2302	0.2291	0.2281	0.2271	0.2260
4	0.2091	0.2085	0.2078	0.2072	0.2065	0.2058	0.2052	0.2045	0.2038	0.2032
5	0.1876	0.1874	0.1871	0.1868	0.1865	0.1862	0.1859	0.1855	0.1851	0.1847
6	0.1693	0.1694	0.1695	0.1695	0.1695	0.1695	0.1695	0.1693	0.1692	0.1691
7	0.1531	0.1535	0.1539	0.1542	0.1545	0.1548	0.1550	0.1551	0.1553	0.1554
8	0.1384	0.1392	0.1398	0.1405	0.1410	0.1415	0.1420	0.1423	0.1427	0.1430
9	0.1249	0.1259	0.1269	0.1278	0.1286	0.1293	0.1300	0.1306	0.1312	0.1317
10	0.1123	0.1136	0.1149	0.1160	0.1170	0.1180	0.1189	0.1197	0.1205	0.1212
11	0.1004	0.1020	0.1035	0.1049	0.1062	0.1073	0.1085	0.1095	0.1105	0.1113
12	0.0891	0.0909	0.0927	0.0943	0.0959	0.0972	0.0986	0.0998	0.1010	0.1020
13	0.0782	0.0804	0.0824	0.0842	0.0860	0.0876	0.0892	0.0906	0.0919	0.0932
14	0.0677	0.0701	0.0724	0.0745	0.0765	0.0783	0.0801	0.0817	0.0832	0.0846
15	0.0575	0.0602	0.0628	0.0651	0.0673	0.0694	0.0713	0.0731	0.0748	0.0764
16	0.0476	0.0506	0.0534	0.0560	0.0584	0.0607	0.0628	0.0648	0.0667	0.0685
17	0.0379	0.0411	0.0442	0.0471	0.0497	0.0522	0.0546	0.0568	0.0588	0.0608
18	0.0283	0.0318	0.0352	0.0383	0.0412	0.0439	0.0465	0.0489	0.0511	0.0532
19	0.0188	0.0227	0.0263	0.0296	0.0328	0.0357	0.0385	0.0411	0.0436	0.0459
20	0.0094	0.0136	0.0175	0.0211	0.0245	0.0277	0.0307	0.0335	0.0361	0.0386
21	0.0000	0.0045	0.0087	0.0126	0.0163	0.0197	0.0229	0.0259	0.0288	0.0314
22			0.0000	0.0042	0.0081	0.0118	0.0153	0.0185	0.0215	0.0244
23					0.0000	0.0039	0.0076	0.0111	0.0143	0.0174
24							0.0000	0.0037	0.0071	0.0104
25									0.0000	0.0035

Reprinted from E.S. Pearson and H.O. Hartley (1976), *Biometrika Tables for Statisticians*, Vol. II. Reproduced with the kind permission of Oxford University Press.

Table A.19 Durbin-Watson (DW) Statistic—5% Significance Points
d_L and d_U (n is the sample size and k' is the number of regressors excluding the intercept)

	$k'=1$		$k'=2$		$k'=3$		$k'=4$		$k'=5$		$k'=6$		$k'=7$		$k'=8$		$k'=9$		$k'=10$	
n	d_L	d_U	d_L	d_U	d_L	d_U	d_L	d_U	d_L	d_U	d_L	d_U	d_L	d_U	d_L	d_U	d_L	d_U	d_L	d_U
6	0.610	1.400																		
7	0.700	1.356	0.467	1.896																
8	0.763	1.332	0.559	1.777	0.368	2.287														
9	0.824	1.320	0.629	1.699	0.455	2.128	0.296	2.588												
10	0.879	1.320	0.697	1.641	0.525	2.016	0.376	2.414	0.243	2.822										
11	0.927	1.324	0.758	1.604	0.595	1.928	0.444	2.283	0.316	2.645	0.203	3.005								
12	0.971	1.331	0.812	1.579	0.658	1.864	0.512	2.177	0.379	2.506	0.268	2.832	0.171	3.149						
13	1.010	1.340	0.861	1.562	0.715	1.816	0.574	2.094	0.445	2.390	0.328	2.692	0.230	2.985	0.147	3.266				
14	1.045	1.350	0.905	1.551	0.767	1.779	0.632	2.030	0.505	2.296	0.389	2.572	0.286	2.848	0.200	3.111	0.127	3.360		
15	1.077	1.361	0.946	1.543	0.814	1.750	0.685	1.977	0.562	2.220	0.447	2.472	0.343	2.727	0.251	2.979	0.175	3.216	0.111	3.438
16	1.106	1.371	0.982	1.539	0.857	1.728	0.734	1.935	0.615	2.157	0.502	2.388	0.398	2.624	0.304	2.860	0.222	3.090	0.155	3.304
17	1.133	1.381	1.015	1.536	0.897	1.710	0.779	1.900	0.664	2.104	0.554	2.318	0.451	2.537	0.356	2.757	0.272	2.975	0.198	3.184
18	1.158	1.391	1.046	1.535	0.933	1.696	0.820	1.872	0.710	2.060	0.603	2.257	0.502	2.461	0.407	2.667	0.321	2.873	0.244	3.073
19	1.180	1.401	1.074	1.536	0.967	1.685	0.859	1.848	0.752	2.023	0.649	2.206	0.549	2.396	0.456	2.589	0.369	2.783	0.290	2.974
20	1.201	1.411	1.100	1.537	0.998	1.676	0.894	1.828	0.792	1.991	0.692	2.162	0.595	2.339	0.502	2.521	0.416	2.704	0.336	2.885
21	1.221	1.420	1.125	1.538	1.026	1.669	0.927	1.812	0.829	1.964	0.732	2.124	0.637	2.290	0.547	2.460	0.461	2.633	0.380	2.806
22	1.239	1.429	1.147	1.541	1.053	1.664	0.958	1.797	0.863	1.940	0.769	2.090	0.677	2.246	0.588	2.407	0.504	2.511	0.424	2.734
23	1.257	1.437	1.168	1.543	1.078	1.660	0.986	1.785	0.895	1.920	0.804	2.061	0.715	2.208	0.628	2.360	0.545	2.514	0.465	2.670
24	1.273	1.446	1.188	1.546	1.101	1.656	1.013	1.775	0.925	1.902	0.837	2.035	0.751	2.174	0.666	2.318	0.584	2.464	0.506	2.613
25	1.288	1.454	1.206	1.550	1.123	1.654	1.038	1.767	0.953	1.886	0.868	2.012	0.784	2.144	0.702	2.280	0.621	2.419	0.544	2.560
26	1.302	1.461	1.224	1.553	1.143	1.652	1.062	1.759	0.979	1.873	0.897	1.992	0.816	2.117	0.735	2.246	0.657	2.379	0.581	2.513
27	1.316	1.469	1.240	1.556	1.162	1.651	1.084	1.753	1.004	1.861	0.925	1.974	0.845	2.093	0.767	2.216	0.691	2.342	0.616	2.470
28	1.328	1.476	1.255	1.560	1.181	1.650	1.104	1.747	1.028	1.850	0.951	1.958	0.874	2.071	0.798	2.188	0.723	2.309	0.650	2.431
29	1.341	1.483	1.270	1.563	1.198	1.650	1.124	1.743	1.050	1.841	0.975	1.944	0.900	2.052	0.826	2.164	0.753	2.278	0.682	2.396
30	1.352	1.489	1.284	1.567	1.214	1.650	1.143	1.739	1.071	1.833	0.998	1.931	0.926	2.034	0.854	2.141	0.782	2.251	0.712	2.363
31	1.363	1.496	1.297	1.570	1.229	1.650	1.160	1.735	1.090	1.825	1.020	1.920	0.950	2.018	0.879	2.120	0.810	2.226	0.741	2.333
32	1.373	1.502	1.309	1.574	1.244	1.650	1.177	1.732	1.109	1.819	1.041	1.909	0.972	2.004	0.904	2.102	0.836	2.203	0.769	2.306
33	1.383	1.508	1.321	1.577	1.258	1.651	1.193	1.730	1.127	1.813	1.061	1.900	0.994	1.991	0.927	2.085	0.861	2.181	0.795	2.281
34	1.393	1.514	1.333	1.580	1.271	1.652	1.208	1.728	1.144	1.808	1.080	1.891	1.015	1.979	0.950	2.069	0.885	2.162	0.821	2.257
35	1.402	1.519	1.343	1.584	1.283	1.653	1.222	1.726	1.160	1.803	1.097	1.884	1.034	1.967	0.971	2.054	0.908	2.144	0.845	2.236
36	1.411	1.525	1.354	1.587	1.295	1.654	1.236	1.724	1.175	1.799	1.114	1.877	1.053	1.957	0.991	2.041	0.930	2.127	0.868	2.216
37	1.419	1.530	1.364	1.590	1.307	1.655	1.249	1.723	1.190	1.795	1.131	1.870	1.071	1.948	1.011	2.029	0.951	2.112	0.891	2.198
38	1.427	1.535	1.373	1.594	1.318	1.656	1.261	1.722	1.204	1.792	1.146	1.864	1.088	1.939	1.029	2.017	0.970	2.098	0.912	2.180
39	1.435	1.540	1.382	1.597	1.328	1.658	1.273	1.722	1.218	1.789	1.161	1.859	1.104	1.932	1.047	2.007	0.990	2.085	0.932	2.164
40	1.442	1.544	1.391	1.600	1.338	1.659	1.285	1.721	1.230	1.786	1.175	1.854	1.120	1.924	1.064	1.997	1.008	2.072	0.952	2.149
45	1.475	1.566	1.430	1.615	1.383	1.666	1.336	1.720	1.287	1.776	1.238	1.835	1.189	1.895	1.139	1.958	1.089	2.022	1.038	2.088
50	1.503	1.585	1.462	1.628	1.421	1.674	1.378	1.721	1.335	1.771	1.291	1.822	1.246	1.875	1.201	1.930	1.156	1.986	1.110	2.044
55	1.528	1.601	1.490	1.641	1.452	1.681	1.414	1.724	1.374	1.768	1.334	1.814	1.294	1.861	1.253	1.909	1.212	1.959	1.170	2.010
60	1.549	1.616	1.514	1.652	1.480	1.689	1.444	1.727	1.408	1.767	1.372	1.808	1.335	1.850	1.298	1.894	1.260	1.939	1.222	1.984
65	1.567	1.629	1.536	1.662	1.503	1.696	1.471	1.731	1.438	1.767	1.404	1.805	1.370	1.843	1.336	1.882	1.301	1.923	1.266	1.964
70	1.583	1.641	1.554	1.672	1.525	1.703	1.494	1.735	1.464	1.768	1.433	1.802	1.401	1.837	1.369	1.873	1.337	1.910	1.305	1.948
75	1.598	1.652	1.571	1.680	1.543	1.709	1.515	1.739	1.487	1.770	1.458	1.801	1.428	1.834	1.399	1.867	1.369	1.901	1.339	1.935
80	1.611	1.662	1.586	1.688	1.560	1.715	1.534	1.743	1.507	1.772	1.480	1.801	1.453	1.831	1.425	1.861	1.397	1.893	1.369	1.925
85	1.624	1.671	1.600	1.696	1.575	1.721	1.550	1.747	1.525	1.774	1.500	1.801	1.474	1.829	1.448	1.857	1.422	1.886	1.396	1.916
90	1.635	1.679	1.612	1.703	1.589	1.726	1.566	1.751	1.542	1.776	1.518	1.801	1.494	1.827	1.469	1.854	1.445	1.881	1.420	1.909
95	1.645	1.687	1.623	1.709	1.602	1.732	1.579	1.755	1.557	1.778	1.535	1.802	1.512	1.827	1.489	1.852	1.465	1.877	1.442	1.903
100	1.654	1.694	1.634	1.715	1.613	1.736	1.592	1.758	1.571	1.780	1.550	1.803	1.528	1.826	1.506	1.850	1.484	1.874	1.462	1.898
150	1.720	1.746	1.706	1.760	1.693	1.774	1.679	1.788	1.665	1.802	1.651	1.817	1.637	1.832	1.622	1.847	1.608	1.862	1.594	1.877
200	1.758	1.778	1.748	1.789	1.738	1.799	1.728	1.810	1.718	1.820	1.707	1.831	1.697	1.841	1.686	1.852	1.675	1.863	1.665	1.874

Table A.19 (Contd.)

n	k′=11 dL	dU	k′=12 dL	dU	k′=13 dL	dU	k′=14 dL	dU	k′=15 dL	dU	k′=16 dL	dU	k′=17 dL	dU	k′=18 dL	dU	k′=19 dL	dU	k′=20 dL	dU
16	0.098	3.503																		
17	0.138	3.378	0.087	3.557																
18	0.177	3.265	0.123	3.441	0.078	3.603														
19	0.220	3.159	0.160	3.335	0.111	3.496	0.070	3.642												
20	0.263	3.063	0.200	3.234	0.145	3.395	0.100	3.542	0.063	3.616										
21	0.307	2.976	0.240	3.141	0.182	3.300	0.132	3.448	0.091	3.583	0.058	3.105								
22	0.349	2.897	0.281	3.057	0.220	3.211	0.166	3.358	0.120	3.495	0.083	3.619	0.052	3.731						
23	0.391	2.826	0.322	2.979	0.259	3.128	0.202	3.272	0.153	3.409	0.110	3.535	0.076	3.650	0.048	3.753				
24	0.431	2.761	0.362	2.908	0.297	3.053	0.239	3.193	0.186	3.327	0.141	3.454	0.101	3.572	0.070	3.678	0.044	3.773		
25	0.470	2.702	0.400	2.844	0.335	2.983	0.275	3.119	0.221	3.251	0.172	3.376	0.130	3.494	0.094	3.604	0.065	3.702	0.041	3.790
26	0.508	2.649	0.438	2.784	0.373	2.919	0.312	3.051	0.256	3.179	0.205	3.303	0.160	3.420	0.120	3.531	0.087	3.632	0.060	3.724
27	0.544	2.600	0.475	2.730	0.409	2.859	0.348	2.987	0.291	3.112	0.238	3.233	0.191	3.349	0.149	3.460	0.112	3.563	0.081	3.658
28	0.578	2.555	0.510	2.680	0.445	2.805	0.383	2.928	0.325	3.050	0.271	3.168	0.222	3.283	0.178	3.392	0.138	3.495	0.104	3.592
29	0.612	2.515	0.544	2.634	0.479	2.755	0.418	2.874	0.359	2.992	0.305	3.107	0.254	3.219	0.208	3.327	0.166	3.431	0.129	3.528
30	0.643	2.477	0.577	2.592	0.512	2.708	0.451	2.823	0.392	2.937	0.337	3.050	0.286	3.160	0.238	3.266	0.195	3.368	0.156	3.465
31	0.614	2.443	0.608	2.553	0.545	2.665	0.484	2.776	0.425	2.887	0.370	2.996	0.317	3.103	0.269	3.208	0.224	3.309	0.183	3.406
32	0.703	2.411	0.638	2.517	0.576	2.625	0.515	2.733	0.457	2.840	0.401	2.946	0.349	3.050	0.299	3.153	0.253	3.252	0.211	3.348
33	0.731	2.382	0.668	2.484	0.606	2.588	0.546	2.692	0.488	2.796	0.432	2.899	0.379	3.000	0.329	3.100	0.283	3.198	0.239	3.293
34	0.758	2.355	0.695	2.454	0.634	2.554	0.575	2.654	0.518	2.754	0.462	2.854	0.409	2.954	0.359	3.051	0.312	3.147	0.267	3.240
35	0.783	2.330	0.722	2.425	0.662	2.521	0.604	2.619	0.547	2.716	0.492	2.813	0.439	2.910	0.388	3.005	0.340	3.099	0.295	3.190
36	0.808	2.306	0.748	2.398	0.689	2.492	0.631	2.586	0.575	2.680	0.520	2.774	0.467	2.868	0.417	2.961	0.369	3.053	0.323	3.142
37	0.831	2.285	0.772	2.374	0.714	2.464	0.657	2.555	0.602	2.646	0.548	2.738	0.495	2.829	0.445	2.920	0.397	3.009	0.351	3.091
38	0.854	2.265	0.796	2.351	0.739	2.438	0.683	2.526	0.628	2.614	0.575	2.703	0.522	2.792	0.472	2.880	0.424	2.968	0.318	3.054
39	0.815	2.246	0.819	2.329	0.763	2.413	0.707	2.499	0.653	2.585	0.600	2.671	0.549	2.757	0.499	2.843	0.451	2.929	0.404	3.013
40	0.896	2.228	0.840	2.309	0.785	2.391	0.731	2.473	0.678	2.557	0.626	2.641	0.575	2.724	0.525	2.808	0.477	2.892	0.430	2.974
45	0.988	2.156	0.938	2.225	0.887	2.296	0.838	2.367	0.788	2.439	0.740	2.512	0.692	2.586	0.644	2.659	0.598	2.733	0.553	2.801
50	1.064	2.103	1.019	2.163	0.973	2.225	0.927	2.287	0.882	2.350	0.836	2.414	0.792	2.479	0.747	2.544	0.703	2.610	0.660	2.675
55	1.129	2.062	1.087	2.116	1.045	2.170	1.003	2.225	0.961	2.281	0.919	2.338	0.877	2.396	0.836	2.454	0.795	2.512	0.754	2.571
60	1.184	2.031	1.145	2.079	1.106	2.127	1.068	2.177	1.029	2.227	0.990	2.278	0.951	2.330	0.913	2.382	0.874	2.434	0.836	2.487
65	1.231	2.006	1.195	2.049	1.160	2.093	1.124	2.138	1.088	2.183	1.052	2.229	1.016	2.276	0.980	2.323	0.944	2.371	0.908	2.419
70	1.272	1.986	1.239	2.026	1.206	2.066	1.172	2.106	1.139	2.148	1.105	2.189	1.072	2.232	1.038	2.275	1.005	2.318	0.971	2.362
75	1.308	1.970	1.277	2.006	1.247	2.043	1.215	2.080	1.184	2.118	1.153	2.156	1.121	2.195	1.090	2.235	1.058	2.275	1.021	2.315
80	1.340	1.957	1.311	1.991	1.283	2.024	1.253	2.059	1.224	2.093	1.195	2.129	1.165	2.165	1.136	2.201	1.106	2.238	1.076	2.275
85	1.369	1.946	1.342	1.977	1.315	2.009	1.287	2.040	1.260	2.073	1.232	2.105	1.205	2.139	1.177	2.172	1.149	2.206	1.121	2.241
90	1.395	1.937	1.369	1.966	1.344	1.995	1.318	2.025	1.292	2.055	1.266	2.085	1.240	2.116	1.213	2.148	1.187	2.179	1.160	2.211
95	1.418	1.929	1.394	1.956	1.370	1.984	1.345	2.012	1.321	2.040	1.296	2.068	1.271	2.097	1.247	2.126	1.222	2.156	1.197	2.186
100	1.439	1.923	1.416	1.948	1.393	1.974	1.371	2.000	1.347	2.026	1.324	2.053	1.301	2.080	1.277	2.108	1.229	2.135	1.229	2.164
150	1.579	1.892	1.564	1.908	1.550	1.924	1.535	1.940	1.519	1.956	1.504	1.972	1.489	1.989	1.474	2.006	1.458	2.023	1.443	2.040
200	1.654	1.885	1.643	1.896	1.632	1.908	1.621	1.919	1.610	1.931	1.599	1.943	1.588	1.955	1.576	1.967	1.565	1.979	1.554	1.991

From N.E. Savin and K.J. White, "The Durbin-Watson Test for Serial Correlation with Extreme Sample Sizes or Many Regressors," *Econometricia*, 45, 1977, 1989–1996. Corrections: R.W. Farebrother, *Econometricia*, 48, 1980, 1554. Reprinted by permission of the Econometric Society.

Table A.20 D Distribution of the von Neumann Ratio of the Mean Square Successive Difference to the Variance, $\alpha/2 = 0.05$ (or 0.01)

Table gives both the lower- and upper-tail critical values d_l and d_u, respectively, such that $P(D \leq d_l) + P(D \geq d_u) = \alpha/2 + \alpha/2 = 0.10$ (or 0.02)).

	Values of d_l		Values of d_u			Values of d_l		Values of d_u	
n	$P = 0.01$	$P = 0.05$	$P = 0.95$	$P = 0.99$	n	$P = 0.01$	$P = 0.05$	$P = 0.95$	$P = 0.99$
4	0.8341	1.0406	4.2927	4.4992	31	1.2469	1.4746	2.6587	2.8864
5	0.6724	1.0255	3.9745	4.3276	32	1.2570	1.4817	2.6473	2.8720
6	0.6738	1.0682	3.7318	4.1262	33	1.2667	1.4885	2.6365	2.8583
7	0.7163	1.0919	3.5748	3.9504	34	1.2761	1.4951	2.6262	2.8451
8	0.7575	1.1228	3.4486	3.8139	35	1.2852	1.5014	2.6163	2.8324
9	0.7974	1.1524	3.3476	3.7025					
10	0.8353	1.1803	3.2642	3.6091	36	1.2940	1.5075	2.6068	2.8202
					37	1.3025	1.5135	2.5977	2.8085
11	0.8706	1.2062	3.1938	3.5294	38	1.3108	1.5193	2.5889	2.7973
12	0.9033	1.2301	3.1335	3.4603	39	1.3188	1.5249	2.5804	2.7865
13	0.9336	1.2521	3.0812	3.3996	40	1.3266	1.5304	2.5722	2.7760
14	0.9618	1.2725	3.0352	3.3458					
15	0.9880	1.2914	2.9943	3.2977	41	1.3342	1.5357	2.5643	2.7658
					42	1.3415	1.5408	2.5567	2.7560
16	1.0124	1.3090	2.9577	3.2543	43	1.3486	1.5458	2.5494	2.7466
17	1.0352	1.3253	2.9247	3.2148	44	1.3554	1.5506	2.5424	2.7376
18	1.0566	1.3405	2.8948	3.1787	45	1.3620	1.5552	2.5357	2.7289
19	1.0766	1.3547	2.8675	3.1456					
20	1.0954	1.3680	2.8425	3.1151	46	1.3684	1.5596	2.5293	2.7205
					47	1.3745	1.5638	2.5232	2.7125
21	1.1131	1.3805	2.8195	3.0869	48	1.3802	1.5678	2.5173	2.7049
22	1.1298	1.3923	2.7982	3.0607	49	1.3856	1.5716	2.5117	2.6977
23	1.1456	1.4035	2.7784	3.0362	50	1.3907	1.5752	2.5064	2.6908
24	1.1606	1.4141	2.7599	3.0133					
25	1.1748	1.4241	2.7426	2.9919	51	1.3957	1.5787	2.5013	2.6842
					52	1.4007	1.5822	2.4963	2.6777
26	1.1883	1.4336	2.7264	2.9718	53	1.4057	1.5856	2.4914	2.6712
27	1.2012	1.4426	2.7112	2.9528	54	1.4107	1.5890	2.4866	2.6648
28	1.2135	1.4512	2.6969	2.9348	55	1.4156	1.5923	2.4819	2.6585
29	1.2252	1.4594	2.6834	2.9177					
30	1.2363	1.4672	2.6707	2.9016	56	1.4203	1.5955	2.4773	2.6524
					57	1.4249	1.5987	2.4728	2.6465
					58	1.4294	1.6019	2.4684	2.6407
					59	1.4339	1.6051	2.4640	2.6350
					60	1.4384	1.6082	2.4596	2.6294

Reproduced from B.I. Hart, "Significance Levels for the Ratio of the Mean Square Successive Difference to the Variance," *Annals of Mathematical Statistics*, 13, no. 4, 1942, 445–447, with kind permission from the Institute of Mathematical Statistics.

Solutions to Selected Exercises

2-2 A: $n = 10, \overline{X} = 8.7$, median $= 9$, two modes (7 and 10), $S = 3.59$.
B: $n = 9, \overline{X} = 5.44$, median $= 5$, mode $= 10$, $S = 4.36$.

2-3 A: $V_A = (S/\overline{X}) \times 100 = 41.26\%$.
B: $V_B = (S/\overline{X}) \times 100 = 80.15\%$. B has more variability.

2-4 $1 - \frac{1}{k^2} = 0.84$. At least 84% of the data lies within 2.5 standard deviations of the mean. $\overline{X} \pm 2.5S$ becomes $8.7 \pm 2.5(3.59)$ or $(-0.275, 17.675)$.

3-1 (a) 0.034   (f) 0.170
(b) 0.593   (g) 0.343
(c) 0   (h) No.
(d) 0.146
(e) 0.671

3-3 (a) 1;   (b) 0

3-5 (a) 0.87   (d) 0.91
(b) 0.07   (e) 0.435
(c) 0.125

3-9 0.03

3-11 0.0909

3-15 (a) Yes; No   (c) $\frac{1}{8}$

(b) $\frac{1}{2}$ (d) $\frac{3}{4}$

3-21 (a) $\frac{1}{6}$;   (b) $\frac{1}{4}$

3-23 $w_c \cdot w + m_c(1 - w)$

3-31 $P(R) = P(Q) = P(T) = \dfrac{1}{2}$ and

$P(R \cap Q) = \dfrac{1}{4} = P(R) \cdot P(Q)$

$P(R \cap T) = \dfrac{1}{4} = P(R) \cdot P(T)$

$P(Q \cap T) = \dfrac{1}{4} = P(Q) \cdot P(T)$

$P(R \cap Q \cap T) = \dfrac{1}{4} \neq P(R) \cdot P(Q) \cdot P(T) = \dfrac{1}{8}$

3-34 0.016

3-35 (a) No   (d) Yes

(b) Yes   (e) No

(c) Yes

3-36 (a) 20/100   (e) 10/40

(b) 60/100   HSG and Gender are not

(c) 70/100   independent. For instance,

(d) 40/70 $P(M \cap N) = \dfrac{20}{100} \neq P(M) \cdot P(N) = \dfrac{18}{100}$

3-37 0.056

3-38 0.008

3-39 (a) 2/15;   (b) 6/15

3-47 0.4463

3-48 0.0884

3-49 0.75

4-1 (a) $k = \dfrac{1}{3}$.

(b) $F(t) = \begin{cases} 0, & t < 0; \\ 1 - e^{-\frac{t}{3}}, & t \geq 0. \end{cases}$

(c) $1 - e^{-2}$;   (d) $e^{-\frac{1}{3}} - e^{-\frac{4}{3}}$.

4-4 (a) $F\left(\dfrac{1}{4}\right) = \dfrac{7}{16}$; (b) $1 - P(X \leq \dfrac{1}{4}) = 1 - F\left(\dfrac{1}{4}\right) = \dfrac{9}{16}$;

(c) $F\left(\dfrac{3}{4}\right) - F\left(\dfrac{1}{3}\right) = \dfrac{5}{16} - \dfrac{5}{9}$;   (d) $F'(t) = 2 - 2t = f(t)$ or

$f(x) = \begin{cases} 2 - 2x, & 0 \leq x \leq 1; \\ 0 & \text{elsewhere.} \end{cases}$

4-5 $F(X) = \begin{cases} 0, & X < 0 \\ 0.215, & 0 \le X < 1 \\ 0.648, & 1 \le X < 2 \\ 0.936, & 2 \le X < 3 \\ 1, & X \ge 3. \end{cases}$

4-6 $F(X) = \begin{cases} 0, & X < -1 \\ 0.5787, & -1 \le X < 1 \\ 0.9259, & 1 \le X < 2 \\ 0.9954, & 2 \le X < 3 \\ 1, & X \ge 3. \end{cases}$

(a) 0.4167;   (b) 0.9259;   (c) 0.4167;   (d) 0.

4-7 (a) $\dfrac{1}{2}$;   (b) $\dfrac{1}{3}$;   (c) $\dfrac{1}{6}$.

4-9 Yes (check properties 4.10a,b,c,d).

4-10 Yes (check properties 4.4a,b,g,h).

4-13

X	0	1	4
$f(X)$	1/2	1/4	1/8

4-16 f is unbounded. Yes (since $f(x) \ge 0$ for all real x; f has at most a finite number of discontinuities; and $\int\limits_{-\infty}^{+\infty} f(x)dx = 1$).

4-17 f is discontinuous. At $x = 0, 1, 4, 6$. Yes (see the requisite properties listed in Exercise 4-16).

4-18 Yes. See Example 4.2.1.

4-22 0.4375.

4-24 $1 - e^{-\theta^{1/2}}$.

4-29 No. $E(X) = \int\limits_{1}^{+\infty} dx/x = \lim_{a \to +\infty}(\ln a - \ln 1)$ does not exist.

4-33 $P(A) = 0.2591$.

4-34 24.

4-35 $E(X) = 7.5$; $E(X^2) = 58.3334$; $V(X) = 2.0834$.

4-41 0.8236.

4-43 (a) $F(\gamma_{0.25}) = \gamma_{0.25}^{0.5} = 0.25$ so that $\gamma_{0.25} = 0.0625$.

(b) $F(\gamma_{0.50}) = \gamma_{0.50}^{0.5} = 0.50$ so that $\gamma_{0.50} = 0.25$.

(c) $F(\gamma_{0.75}) = \gamma_{0.75}^{0.5} = 0.75$ so that $\gamma_{0.75} = 0.5625$.

$Q.D. = (\gamma_{0.75} - \gamma_{0.25})/2 = 0.25$.

4-44 No. $E(X) = 0.33 \neq \gamma_{0.50} = 0.25$.

4-45 $\mu_r' = 0^r(0.3) + 1^r(0.7) = 0.7, r = 1, 2, \ldots$.

4-46 (a) 277.666;   (b) −10.120;   (c) −0.651;   (d) 2.607;   (e) 5;   (f) 7.

4-51 $\mu_r' = \int_{20}^{25} \frac{1}{5} x^r dx = \dfrac{25^{r+1} - 20^{r+1}}{5(r+1)}, r = 1, 2, \ldots$.

Then: $\mu_1' = \mu = 22.50; \mu_2' = 508.3333; \mu_3' = 11{,}531.25; \mu_4' = 262{,}625$;

$\mu_2 = \mu_2' - \mu^2 = 2.08; \mu_3 = \mu_3' - 3\mu\mu_2' + 2\mu^3 = 0$;

$\mu_4 = \mu_4' - 4\mu\mu_3' + 6\mu^2\mu_2' - 3\mu^4 = 7.8125; \alpha_3 = 0$; and $\alpha_4 = 1.8058$.

4-61

X	0	1	2
$f(X)$	1/4	1/2	1/4

$$m_X(t) = \frac{1}{4} + \frac{1}{2}e^t + \frac{1}{4}e^{2t}; \mu = 1; \sigma\sqrt{1/2}.$$

4-62 $\phi_X(t) = \dfrac{1}{4} + \dfrac{1}{2}t + \dfrac{1}{4}t^2$.

$\phi_X(0) = \dfrac{1}{4} = f(0); \phi_X^{(1)}(0) = \dfrac{1}{2} = f(1); \phi_X^{(2)}(0) = \dfrac{1}{2} = 2f(2)$ so that

$f(2) = \dfrac{1}{4}$.

$\phi_X^{(1)}(1) = 1 = \mu; \phi_X^{(2)}(1) = \dfrac{1}{2} = E(X^2) - \mu$.

4-63 $m_X(t) = \dfrac{10}{10 - t}, t < 10; m_{X-\mu}(t) = \dfrac{10e^{-\mu t}}{10 - t}, t < 10$.

4-65 We know that $m_X(t) = E(e^{tx}) = \sum_{i=1}^{n} e^{tX_i} f(X_i)$ with $m_X^{(1)}(0) = E(X)$; and $\phi_X(t) = E(t^X) = \sum_{i=0}^{n} t^i f(i)$ with $\phi_X^{(1)}(1) = E(X)$. But $\phi_X(e^t) = m_X(t)$ so that $\phi_X(e^t) = \sum_{i=0}^{n} e^{ti} f(i) = \dfrac{1}{8} + \dfrac{3}{8}e^t + \dfrac{3}{8}e^{2t} + \dfrac{1}{8}e^{3t}$.

The expectation must exist near $t = 0$. At $t = 0$ $\phi_X^{(1)}(e^t) = E(X)$.

4-67 $x = w(y) = \dfrac{1}{3}y$ and $dw(y)/dy = \dfrac{1}{3} \neq 0$. Then $h(y) = \dfrac{1}{3}e^{-\frac{1}{3}y}, y \geq 0$.

5-10 $E(Z) = 1.65$.

5-11 $E(X \mid Z = 7) = 21/6$.

5-16 $P(0 \leq X \leq 100, 0 \leq Y \leq 100) = (1 - e^{-100\theta})^2$.

5-18 $\lambda = 1/2$. $P(0 < X < 1/2, 1/2 < Y < 3/4) = 1/16$.

5-19 $P(A) = 1/4$.

5-21 $F(t) = 1 - e^{-t/10} - te^{-t/10}$; $h(z) = \dfrac{1}{10}(z - 9)e^{-z/10}$, $z > 0$.

5-23 $k = [2\ln(3)]^{-1}$.

5-25 $P(1/3 < X \leq 1, Y \leq 2/3) = 32/81$.

5-27 It is the marginal probability density function of X.

5-29 $\partial^2 F/\partial s \partial t = 4e^{-2(t+s)}$.

5-33 $h(z) = ze^{-z}$.

5-34 $f(x,y) = \begin{cases} (\beta - \alpha)^{-2}, & \alpha \leq x \leq \beta \text{ and } \alpha \leq y \leq \beta; \\ 0 & \text{elsewhere.} \end{cases}$

5-35 $P(X + Y > 10) = 10e^{-10} + e^{-10}$.

5-36 (a) $k=1$; (b) $F(t,s) = ts(t + s)/2$; (c) $F\left(\dfrac{1}{2}, \dfrac{1}{2}\right) = \dfrac{1}{8}$ and

$P\left(0 \leq X \leq \dfrac{1}{2}, 0 \leq Y \leq \dfrac{1}{2}\right) = \dfrac{1}{8}$.

5-41 (a) $\dfrac{1}{3}$; (b) $\dfrac{2}{3}$; (c) $\dfrac{1}{16}$; and (d) $\dfrac{1}{18}$.

6-5 $P(X = 3) = 0.1536$; $P(X \leq 2) = 0.8208$.

6-7 (a) 0.7023; (b) 0.0014.

6-9 0.0008.

6-12 $P(X > 10) = 0.4119$.

6-14 $P(X \geq 7) = 0.1719$.

6-24 0.0147.

6-26 0.0034.

6-28 $P(X < 5) = 0.5904$; $E(X) = 5$.

6-30 $P(X < 10) = 0.009$; $E(X) = 1000$.

6-36 0.0583; 0.1654.

6-38 0.0820; 0.0915.

6-42 0.048; 0.997.

6-44 0.0067; $P(4 < X \le 10) = 0.546$.

6-46 $P(X \ge 5) = 0.042; P(X = 0) = 0.223$.

6-53 $P(X \ge 2) = 0.9241$.

6-54 0.0523.

6-55 $E(X) = 1.80; V(X) = 0.5603$.

7-6 (a) 0.8413;   (b) 0.3085;   (c) 0.3015;   and   (d) 55.25

7-8 0.9544. Yes.

7-9 66.45.

7-12 7.792.

7-13 $3,013.

7-19 All true.

7-31 $F(1000; 0.0005) = 1 - e^{-0.0005(1000)} = 1 - 0.951 = 0.049$.

8-13 (a) $P(Z_{\hat{P}} \le -1.22) = 1 - P(Z_{\hat{P}} \le +1.22) = 0.1112$;

(b) $P(Z_{\hat{P}} \ge 2.04) = 1 - P(Z_{\hat{P}} \le 2.04) = 0.0207$.

8-20 All true.

8-26 $P(\hat{P} \le 0.01) = P\left(\dfrac{\hat{P} - p}{\sigma_{\hat{P}}} \le \dfrac{0.01 - 0.03}{0.008}\right) = P\left(Z_{\hat{P}} \le -2.50\right) = 0.0062$.

8-27 $P(\overline{X} > 4.5) = P\left(\dfrac{\overline{X} - \mu}{\sigma/\sqrt{n}} > \dfrac{4.5 - 4}{1.5/5}\right) = P\left(\overline{Z} > 1.66\right) = 1 - P(\overline{Z} \le 1.66)$

$$= 1 - 0.9515 = 0.0485.$$

8-28 $P(|\overline{X} - \mu| \le 500) = P(-500 \le \overline{X} - \mu \le 500) =$

$$P\left(\dfrac{-500}{3500/\sqrt{250}} \le \dfrac{\overline{X} - \mu}{\sigma/\sqrt{n}} \le \dfrac{500}{3500/\sqrt{250}}\right) = P\left(-2.25 \le \overline{Z} \le 2.25\right) =$$

$$F(2.25; 0, 1) - F(-2.25; 0, 1) = 2F(2.25; 0, 1) - 1 = 0.9756.$$

8-29 $P(\hat{P} > 0.50) = P\left(\dfrac{\hat{P} - p}{\sigma_{\hat{P}}} > \dfrac{0.50 - 0.55}{0.035}\right) = P(Z_{\hat{P}} > -1.42) =$

$$1 - P(Z_{\hat{P}} \le -1.42) = 1 - 0.7782 = 0.9222.$$

8-31 $m'_1 = \dfrac{1}{n} \sum\limits_{i=1}^{n} x_i = \bar{x}_n = 20/6 = 3.33;$

$m'_2 = \dfrac{1}{n} \sum\limits_{i=1}^{n} x_i^2 = 104/6 = 17.33;$

$S^2 = \left(\dfrac{n}{n-1}\right) [m'_2 - (m'_1)^2] = (6/5)\, [17.33 - (3.33)^2] = 7.4893.$

$m_3 = \dfrac{1}{n} \sum\limits_{i=1}^{n} (x_i - \bar{x}_n)^3 = 78.8181/6 = 13.1364.$

$m_4 = \dfrac{1}{n} \sum\limits_{i=1}^{n} (x_i - \bar{x}_n)^4 = 545.4930/6 = 90.9155.$

9-1 $E(X) = 8; V(X) = 16; \alpha_3 = 4/\sqrt{16} = 1;$ and $\alpha_4 = 3\left(1 + \dfrac{1}{2}\right) = 4.5.$

9-3 $\chi^2_{0.90,60} \approx \dfrac{1}{2}\left(z_{0.90} + \sqrt{119}\right)^2 = \dfrac{1}{2}\,(1.28 + 10.91)^2 = 74.298;$

$\chi^2_{0.95,60} \approx \dfrac{1}{2}\left(z_{0.95} + \sqrt{119}\right)^2 = \dfrac{1}{2}\,(1.645 + 10.91)^2 = 78.814.$

9-4 $P\left(S^2 < 10\right) = P\left(\dfrac{nS^2}{\sigma^2} < \dfrac{40\,(10)}{9}\right) = P\,(Y < 44.44) = 0.95$

since $44.44 \approx \chi^2_{0.95,40} = \chi^2_{1-\alpha,v}.$

9-5 $P(S^2 < 18) = P\left(\dfrac{(n-1)S^2}{\sigma^2} < \dfrac{24(18)}{16}\right) = P(Y \leq 27) \approx 0.70;$

$P\left(12 < S^2 < 20\right) = P\left(\dfrac{24(12)}{16} < \dfrac{(n-1)\,S^2}{\sigma^2} < \dfrac{24(20)}{16}\right)$

$= P(18 < Y < 30) = P(Y < 30) - P(Y < 18) \approx 0.80 - 0.20$

$= 0.60.$

9-8 Set $\alpha = n/2, \theta = 2.$

9-16 (a) $t_{0.05,10} = 1.812;$ (b) $t_{0.025,10} = 2.228.\ V(T) = 5/4;\ \alpha_4 = 4.$

$P(T \geq 1.729) = 0.05.\ P(-1.761 \leq T \leq 1.761) = 0.90.$

9-17 $P(\bar{X} \leq 13) = P(T \leq -1.581).$ For 9 degrees of freedom, the probability lies between 0.05 and 0.10.

9-18 $P(\bar{X} \geq 19) = P(T \geq 2).$ For 15 degrees of freedom, the probability lies between 0.025 and 0.05.

9-19 $P(\bar{X} \leq 290) = P(T \leq -2.37).$ For 9 degrees of freedom, the probability is between 0.01 and 0.025.

9-21 $E\,(F) = 1.33; V(F) = 0.22; \alpha_3 = 6;\ \alpha_4$ cannot be determined if v_2 does not exceed 8; mode = 0.53.

9-22 $f_{0.975,4,8} = 5.05$; $f_{0.90,4,8} = 2.81$.

9-23 $f'_{0.99,5,4} = 15.52$. Then $f_{0.01,4,5} = 1/15.52 = 0.064$.

$f'_{0.95,7,9} = 3.29$. Then $f_{0.05,9,7} = 1/3.29 = 0.304$.

9-24 (a) $f_{0.95,10,7} = 3.64$; (b) 0.01.

10-4 No. $E(T) = \sigma^2 \left(\frac{n-1}{n^2} \right)$.

10-10 (a) $E(T_1) = \mu$ (T_1 is unbiased); $\text{MSE}(T_1, \mu) = V(T_1) = \frac{3}{8}\sigma^2$.

(b) $E(T_2) = \frac{4}{3}\mu$ (T_2 is biased); $\text{MSE}(T_2, \mu) = V(T_2) + B(T_2, \mu)^2 = \frac{6}{9}\sigma^2 + \frac{1}{9}\mu^2$.

10-12 The maximum likelihood estimate of θ is $\hat{\theta} = -n/ln(\prod_{i=1}^{n} x_i)$. The maximum likelihood estimator $\hat{T} = -n/ln(\prod_{i=1}^{n} X_i)$ is consistent.

10-44 $\hat{\theta} = -(n/\sum_{i=1}^{n} \log x_i) - 1$.

10-47 $\hat{T} = \overline{X}$.

10-49 $\hat{T}_1 = \sum_{i=1}^{n} X_i/n$; $\hat{T}_2 = \sum_{i=1}^{n} (X_i - \hat{T}_1)^2/n$. From Theorem 10.4, $S_1 = \sum_{i=1}^{n} X_i$ and $S_2 = \sum_{i=1}^{n} X_i^2$ are jointly sufficient for θ_1, θ_2 respectively.

10-55 $\hat{T} = 1/\overline{X}$.

10-62 To estimate λ, set $M'_1 = \mu'_1 = \lambda = E(X)$. Thus the method-of-moments estimator for λ is $M'_1 = \overline{X}$.

10-63 To estimate θ, set $M'_1 = \mu'_1 = \theta = E(X)$. Then the method-of-moments estimator of θ is $1/M'_1 = 1/\overline{X}$.

11-1 $\bar{x} \pm z_{\alpha/2}(s/\sqrt{n})$ becomes $4.6 \pm 1.96(2.06/\sqrt{65})$ or $(l_1, l_2) = (4.099, 5.101)$. Degree of precision is ± 0.5008.

11-3 $\bar{x} \pm t_{\alpha/2,n-1}(s/\sqrt{n})$ yields for $t_{0.025,11} = 2.201$, $81.917 \pm 2.201(13.021/\sqrt{12})$ or $(l_1, l_2) = (73.644, 90.190)$

11-7 The error bound is $\pm z_{\alpha/2}(\sigma/\sqrt{n})$, Hence set $k = z_{\alpha/2}(\sigma/\sqrt{n})$ so as to obtain $n = (z_{\alpha/2}\sigma/k)^2$.

11-9 $\hat{p} + z_{\alpha/2}\sqrt{\hat{p}(1-\hat{p})/n}$ becomes $0.15 \pm 2.58(0.0357)$ or $(l_1, l_2) = (0.0579, 0.2421)$.

11-11 $\hat{p} \pm z_{\alpha/2}\sqrt{\hat{p}(1-\hat{p})/n}$ yields $0.5714 \pm 1.96(0.0265)$ or $(l_1, l_2) = (0.0596, 0.6232)$.

11-13 $n = (1.96)^2(0.45)(0.55)/(0.03)^2 = 1057$. If p is unknown, use $p = 0.5$. Then $n = (1.96)^2(0.25)/(0.03)^2 = 1068$.

11-15 $m = 1.96(0.25)/(0.05)^2 = 196$ so that $n = 182$.

11-24 $\bar{d} = -11.4$, $s_D = 4.28$, $t_{0.025,4} = 2.776$. Then from (11.48), $(l_1, l_2) = (-16.714, -6.087)$.

11-30 $\hat{p}_x = 0.5286$, $\hat{p}_y = 0.7294$, $z_{0.025} = 1.96$. Then from (11.52), $(l_1, l_2) = (-0.3509, -0.0507)$.

11-32 $\hat{p}_x = 0.50$, $\hat{p}_y = 0.42$, $Z_{0.025} = 1.96$. Using (11.52) renders $(l_1, l_2) = (-0.0213, 0.1813)$.

11-34 $\bar{d} = -3.79$, $s_D = 5.262$, $t_{0.05,9} = 1.833$. Then from (11.48), $(l_1, l_2) = (-6.84, 0.74)$.

11-42 $(l_1, l_2) = \left(\frac{1}{30} \chi^2_{0.975,20}, \frac{1}{30} \chi^2_{0.025,22} \right) = (9.59/30, 36.78/30) = (0.320, 1.226)$.

12-1 H_0: $\mu = \mu_0 = 78$, H_1: $\mu \neq 78$. $|\bar{z}| = 2.7386 > 1.96 = z_{0.025}$. Reject H_0. p-value $= 0.0064$.

12-3 H_0: $\mu = \mu_0 = 0.9$, H_1: $\mu \neq 0.9$. $|t| = 6.5034 > 2.201 = t_{0.025,11}$. Reject H_0.

12-5 H_0: $\mu = \mu_0 = 12.97$, H_1: $\mu \neq 12.97$. $|t| = 9.2105 > 2.797 = t_{0.005,24}$. Reject H_0.

12-7 H_0: $p = p_0 = 0.5$, H_1: $p \neq 0.5$. $\left| z_{\hat{p}}^o \right| = 1.4142 < 1.96 = z_{0.025}$. Do not reject H_0. We sample from a binomial population; n is large; and the standardized observed relative frequency of a success is approximately $N(0,1)$.

12-18 H_0: $\sigma^2 = \sigma_0^2 = 230$, H_1: $\sigma^2 > 230$. $(n-1) s^2/\sigma_0^2 = 17.93 < \chi^2_{\alpha,n-1} = \chi^2_{0.05,15} = 24.996$. Do not reject H_0.

12-21 H_0: $\sigma_X^2 = \sigma_Y^2$, H_1: $\sigma_X^2 < \sigma_Y^2$. $f = s_Y^2/s_X^2 = 1.467 < f_{0.90,14,14} = 2.02$. Do not reject H_0.

12-24 Since $z_S = 0.93\sqrt{9} = 2.79 > 1.645 = z_{0.05}$, we reject H_0; i.e., there exists a statistically significant amount of agreement in the ranks at the 5% level.

12-25 Since $z_S = 0.75\sqrt{13} = 2.704 > 2.33 = z_{0.01}$, we reject H_0 — at the 1% level we have a statistically significant amount of agreement in the rankings.

12-49 Since $z_V = \sqrt{24}(0.306 - 0.35)/0.35\sqrt{0.6225} = -0.7826 > -1.645 = z_{0.05}$, we do not reject H_0.

13-1 $n_1 = 15$, $n_2 = 12$, $n_1 + n_2 = 27$ and $r = 14$. The Case III $\mathcal{R} = \{r \mid r \leq 8$ or $r \geq 20\}$. Do not reject H_0.

13-2 For this large-sample runs test, $n_1 = 17$, $n_2 = 21$, $n_1 + n_2 = 38$, and $r = 21$. With $E(R) = 19.79$, $V(R) = 9.03$, and $z_R = 0.402$, the Case III $\mathcal{R} = \{z_R \mid |z_R| \geq 1.96 \}$. Do not reject H_0.

13-4 $n = 18$, $\bar{x} = 25$, and $d = 480/514 = 1.529$. The Case I $\mathcal{R} = \{d \mid d \geq d_u = 3.1787\}$ with $P(D \geq d_u) = 0.01$. Do not reject H_0.

14-1 H_0: $p_1 = \dfrac{1}{2}, p_2 = \dfrac{1}{4}, p_3 = \dfrac{1}{4}, H_1$: at least one of these equalities does not hold. $\mathcal{R} = \left\{ u \mid u \geq \chi^2_{0.95,2} = 5.99 \right\}$.

o_i	$e_i = np_i$
$o_1 = 101$	$250 \left(\dfrac{1}{2} \right) = 125 = e_1$
$o_2 = 60$	$250 \left(\dfrac{1}{4} \right) = 62.5 = e_2$
$o_3 = 89$	$250 \left(\dfrac{1}{4} \right) = 62.5 = e_3$
250	250

$$u = \sum_{i=1}^{3} \frac{(o_i - e_i)^2}{e_i} = 15.944. \text{ Reject } H_0.$$

14-3 H_0: number of calls uniformly distributed, H_1: number of calls not uniformly distributed. $\mathcal{R} = \left\{ u \mid u \geq \chi^2_{0.95,3} = 11.34 \right\}$.

o_i	$e_i = n \left(\dfrac{1}{4} \right)$
$o_1 = 57$	$248 \left(\dfrac{1}{4} \right) = 62 = e_1$
$o_2 = 65$	$248 \left(\dfrac{1}{4} \right) = 62 = e_2$
$o_3 = 60$	$248 \left(\dfrac{1}{4} \right) = 62 = e_3$
$o_4 = 66$	$248 \left(\dfrac{1}{4} \right) = 62 = e_4$
248	248

$$u = \sum_{i=1}^{4} \frac{(o_i - e_i)^2}{e_i} = 0.871. \text{ Do not reject } H_0.$$

14-5 H_0: X is Poisson distributed with $\lambda = 0.5$, H_1: X is not Poisson distributed with $\lambda = 0.5$.

X	o_i	$p(X; \lambda = 0.5)$	$e_i = np_i$
0	82	$p_1 = 0.6065$	89.7620
1	37	$p_2 = 0.3033$	44.8884
2	20	$p_3 = 0.0758$	11.2184
3	7	$p_4 = 0.0126$	1.8648
4	2	$p_5 = 0.0016$	0.2368
	148		

Since np_4, np_5 are each less than five, combine cells 3, 4, and 5 with $p_3 = 1 - p_1 - p_2 = 0.0902$

(Cells 3, 4, and 5 combined)

X	o_i	$p(X; \lambda = 0.5)$	$e_i = np_i$
0	82	$p_1 = 0.6065$	87.7620
1	37	$p_2 = 0.3033$	44.8884
2	29	$p_3 = 0.0902$	13.3496
	148		

$$u = \sum_{i=1}^{3} \frac{(o_i - e_i)^2}{e_i} = 20.4052, \ \mathcal{R} = \left\{ u \mid u \geq \chi_{0.95,2}^2 = 5.99 \right\}. \text{ Reject } H_0.$$

14-6 H_0: X is normally distributed, H_1: X is not normally distributed. Class boundaries are:

$$95.945 - 95.965$$
$$95.965 - 95.985$$
$$95.985 - 96.005$$
$$96.005 - 96.025$$
$$96.025 - 96.045$$
$$96.045 - 96.065$$

We next determine $\bar{x} = 96.0061$ and $s = 0.0202$ (see exercise 2-13) and thus:

$$\hat{p}_1 = P(-\infty < X < 95.965) = P(-\infty < Z < -2.0347) = 0.0212,$$

$$\hat{p}_2 = P(95.965 \leq X \leq 95.985) = P(-2.0347 \leq Z \leq -1.0446) = 0.1270,$$

$$\hat{p}_3 = P(95.985 \leq X \leq 96.005) = P(-1.0446 \leq Z \leq -0.0545) = 0.3309,$$

$$\hat{p}_4 = P(96.005 \leq X \leq 96.025) = P(-0.0545 \leq Z \leq 0.9356) = 0.3484,$$

$$\hat{p}_5 = P(96.025 \leq X \leq 96.045) = P(0.9356 \leq Z \leq 1.9257) = 0.1444,$$

$$\hat{p}_6 = P(96.045 \leq X < +\infty) = P(1.9257 \leq Z < +\infty) = 0.0268.$$

Then

X	o_i	$\hat{p}_i$	$\hat{e}_i = n\hat{p}_i$
95.95–95.96	5	0.0212	6.2540
95.97–95.98	32	0.1270	37.4650
95.99–96.00	111	0.3309	97.6155
96.01–96.02	97	0.3484	102.7780
96.03–96.04	42	0.1444	42.5980
96.05–96.06	8	0.0268	7.9060
	295		

$v = \sum_{i=1}^{6} \frac{(o_i - \hat{e}_i)^2}{\hat{e}_i} = 3.2181$. With $k - r - 1 = 6 - 2 - 1 = 3, \mathcal{R} = \left\{ v | v \geq \chi^2_{0.95,3} = 7.81 \right\}$. Do not reject H_0.

14-9 Null distribution is

$$F_0(x; 10, 17) = \begin{cases} 0, & x < 10; \\ \dfrac{x - 10}{7}, & 10 \leq x \leq 17; \\ 1, & x > 17. \end{cases}$$

$H_0: F(x) = F_0(x; 10, 17), H_1: F(x) \neq F_0(x; 10, 17)$.

Ordered Realizations $x_{(i)}, i = 1, \ldots, 10$	$S_{10}(x) = i/10$, $i = 1, \ldots, 10$	$F_0(x; 10, 17)$	$\|S_{10}(x) - F_0(x; 10, 17)\|$	$\|S_{10}(x - \varepsilon) - F_0(x; 10, 17)\|$
9.9	0.10	0.0000	0.1000	0.0000
10.3	0.20	0.0429	0.1571	0.0571
11.2	0.30	0.1714	0.1286	0.0286
11.4	0.40	0.2000	0.2000	0.1000
11.8	0.50	0.2571	0.2429	0.1429
12.5	0.60	0.3571	0.2429	0.1429
13.2	0.70	0.4571	0.2429	0.1429
13.6	0.80	0.5143	0.2857	0.1857
16.3	0.90	0.9000	0.0000	0.1000
16.7	1.00	0.9571	0.0429	0.0571

$d_{10}^0 = 0.2857, \mathcal{R} = \left\{ d_{10} | d_{10} \geq d_{10, 0.95} = 0.41 \right\}$. Do not reject H_0.

14-11 $H_0: F(x) = F_0(x; 550, 100), H_1: F(x) \neq F_0(x; 550, 100)$.

Ordered Realizations $x_{(i)}, i = 1, \ldots, 10$	$S_{10}(x) = i/10$, $i = 1, \ldots, 10$	$z_{(i)} = \dfrac{x_{(i)} - 550}{100}$, $i = 1, \ldots, 10$	$F_0(z_{(i)}; 0, 1)$	$\|S_{10}(x) - F_0(z_{(i)}; 0, 1)\|$	$\|S_{10}(x - \varepsilon) - F_0(z_{(i)}; 0, 1)\|$
470	0.10	−0.8000	0.2119	0.1119	0.2119
490	0.20	−0.6000	0.2743	0.0743	0.1743
495	0.30	−0.5500	0.2912	0.0088	0.0912
500	0.40	−0.5000	0.3085	0.0915	0.0085
510	0.50	−0.4000	0.3446	0.1554	0.0554
525	0.60	−0.2500	0.4013	0.1987	0.0987
530	0.70	−0.2000	0.4207	0.2793	0.1793
550	0.80	0.0000	0.5000	0.3000	0.2000
570	0.90	0.2000	0.5793	0.3207	0.2207
580	1.00	0.3000	0.6179	0.3821	0.2821

$d_{10}^0 = 0.3821, \mathcal{R} = \left\{ d_{10} | d_{10} \geq d_{10, 0.95} = 0.41 \right\}$. Do not reject H_0.

14-13 H_0: the random sample is from a normal distribution function, H_1: not H_0. From this sample data set we obtain $\bar{x} = 140.28, s = 24.87$.

Ordered Realizations $x_{(i)}, i = 1, \ldots, 18$	$z_{(i)} = \dfrac{x_{(i)} - \bar{x}}{s}$, $i = 1, \ldots, 18$	$S_{18}(z) = i/18$, $i = 1, \ldots, 18$	$\hat{F}(z_{(i)}; 0, 1)$	$\left\| S_{18}(z) - \hat{F}(z_{(i)}; 0, 1) \right\|$	$\left\| S_{18}(z - \varepsilon) - \hat{F}(z_{(i)}; 0, 1) \right\|$
102	−1.54	0.0556	0.0618	0.0062	0.0618
105	−1.41	0.1111	0.0793	0.0318	0.0237
105	−1.41	0.1667	0.0793	0.0874	0.0318
111	−1.18	0.2222	0.1190	0.1032	0.0477
118	−0.90	0.2778	0.1841	0.0937	0.0381
121	−0.78	0.3333	0.2177	0.1156	0.0601
129	−0.45	0.3889	0.3264	0.0625	0.0069
142	0.07	0.4444	0.4721	0.0277	0.0832
148	0.31	0.5000	0.6217	0.1217	0.1773
149	0.35	0.5556	0.6368	0.0812	0.1369
150	0.39	0.6111	0.6517	0.0406	0.0961
150	0.39	0.6667	0.6517	0.0150	0.0406
152	0.47	0.7222	0.6808	0.0414	0.0141
160	0.79	0.7778	0.7852	0.0074	0.0630
162	0.87	0.8333	0.8078	0.0255	0.0300
169	1.15	0.8889	0.8749	0.0140	0.0416
174	1.36	0.9444	0.9131	0.0313	0.0242
178	1.51	1.0000	0.9345	0.0655	0.0099

$\hat{d}_{18} = 0.1773, \mathcal{R} = \left\{ \hat{d}_{18} \,\middle|\, \hat{d}_{18} \geq d_{18, 0.95} = 0.20 \right\}$. Do not reject H_0.

14-15 H_0: the distribution function is normal, H_1: not H_0.

$x_{(i)}, i = 1, \ldots, 22$		$a_j, j = 1, \ldots, 11$	$x_{(n-j+1)} - x_{(j)}$	$a_j \left(x_{(n-j+1)} - x_{(j)} \right)$
$x_{(1)} = 60$	$x_{(12)} = 120$	0.4590	$x_{(22)} - x_{(1)} = 117$	53.703
$x_{(2)} = 80$	$x_{(13)} = 120$	0.3156	$x_{(21)} - x_{(2)} = 95$	29.982
$x_{(3)} = 87$	$x_{(14)} = 140$	0.2571	$x_{(20)} - x_{(3)} = 76$	19.539
$x_{(4)} = 90$	$x_{(15)} = 145$	0.2131	$x_{(19)} - x_{(4)} = 70$	14.917
$x_{(5)} = 98$	$x_{(16)} = 150$	0.1764	$x_{(18)} - x_{(5)} = 62$	10.936
$x_{(6)} = 100$	$x_{(17)} = 155$	0.1443	$x_{(17)} - x_{(6)} = 55$	7.937
$x_{(7)} = 105$	$x_{(18)} = 160$	0.1150	$x_{(16)} - x_{(7)} = 45$	5.175
$x_{(8)} = 109$	$x_{(19)} = 160$	0.0878	$x_{(15)} - x_{(8)} = 36$	3.161
$x_{(9)} = 110$	$x_{(20)} = 163$	0.0618	$x_{(14)} - x_{(9)} = 30$	1.854
$x_{(10)} = 110$	$x_{(21)} = 175$	0.0368	$x_{(13)} - x_{(10)} = 10$	0.368
$x_{(11)} = 115$	$x_{(22)} = 177$	0.0122	$x_{(12)} - x_{(11)} = 5$	0.061
				147.633

Since $\bar{x} = 124.05$ and $\sum_{i=1}^{22} (X_i - \overline{X})^2 = 22{,}732.15$, we have $w = (147.633)^2 / 22{,}732.15 = 0.9588$. With $\mathcal{R} = \{ w \,|\, w \leq w_{0.05} = 0.911 \}$, we do not reject H_0.

14-17 X corresponds to Method A, Y corresponds to Method B. H_0: $F_X(x) = F_Y(x)$, H_1: $F_X(x) \neq F_Y(x)$ for at least one x.

$x_{(i)}, i = 1,\ldots,10$	$y_{(j)}, j = 1,\ldots,12$	$S_n(x)$	$S_m(x)$	$\left\lvert S_n(x) - S_m(x) \right\rvert$
50	50	0.100	0.083	$0.100 - 0.083 = 0.017$
55		0.200		$0.200 - 0.083 = 0.117$
58		0.300		$0.300 - 0.083 = 0.217$
62	62	0.400	0.167	$0.400 - 0.167 = 0.233$
65		0.500		$0.500 - 0.167 = 0.333$
67		0.600		$0.600 - 0.167 = 0.433$
70		0.700		$0.700 - 0.167 = 0.533$
71		0.800		$0.800 - 0.167 = 0.633$
72		0.900		$0.900 - 0.167 = 0.733$
	75		0.250	$0.900 - 0.250 = 0.650$
	78		0.333	$0.900 - 0.333 = 0.567$
	79		0.417	$0.900 - 0.417 = 0.483$
	80		0.500	$0.900 - 0.500 = 0.400$
81		1.000		$1.000 - 0.500 = 0.500$
	82		0.583	$1.000 - 0.583 = 0.417$
	83		0.667	$1.000 - 0.667 = 0.333$
	84		0.750	$1.000 - 0.750 = 0.250$
	88		0.833	$1.000 - 0.833 = 0.167$
	90		0.917	$1.000 - 0.917 = 0.083$
	91		1.000	$1.000 - 1.000 = 0$

$$d_{10,12} = 0.733, \quad \mathcal{R} = \left\{ d_{10,12} \,\middle|\, d_{10,12} \geq d_{10,12,0.95} = 1.36\sqrt{\tfrac{12+10}{120}} = 1.36(0.4282) = 0.5823 \right\}. \text{ Reject } H_0.$$

15-1 H_0: preference and sex are independent,
H_1: not H_0.

	Mini-Van	SUV	
Male	10 (=a)	30 (=b)	40 (=e)
Female	20 (=c)	20 (=d)	40 (=f)
	30 (=g)	50 (=h)	80 (=n)

From (15.10.1)

$$u'' = \frac{80\,(|200 - 600| - 40)^2}{40(40)(30)(50)} = 4.32.$$

For 1 degree of freedom, $0.025 < p$-value < 0.05.

15-3 H_0: $p_{ij} = p_{i\cdot}p_{\cdot j}$, H_1: $p_{ij} \neq p_{i\cdot}p_{\cdot j}$ for at least one cell (i,j), $i = 1,\ldots,4; j = 1,\ldots,3$.

	Poor	Good	Excellent	
1 day	10 (5.92)	15 (14.21)	2 (6.86)	27 $(=n_{1\cdot})$
2 days	7 (5.70)	10 (13.68)	9 (6.61)	26 $(=n_{2\cdot})$
3 days	5 (6.14)	15 (14.73)	8 (7.12)	28 $(=n_{3\cdot})$
4 days	3 (7.23)	20 (17.36)	10 (8.39)	33 $(=n_{4\cdot})$
	25 $(=n_{\cdot 1})$	60 $(=n_{\cdot 2})$	29 $(=n_{\cdot 3})$	114 $(= n)$

From (15.9.2)

$$u = \frac{(10 - 5.92)^2}{5.92} + \cdots + \frac{(10 - 8.39)^2}{8.39} = 11.93.$$

For degrees of freedom $= (r - 1)(c - 1) = 6$, $0.05 < p$-value < 0.10.

15-5 H_0: the four samples come from the same population, H_1: not H_0.

	In Favor	Neutral	Opposed	
Clerical	27 (19)	33 (33.5)	15 (22.5)	75 $(=n_1)$
Sanitation	13 (19)	41 (33.5)	21 (22.5)	75 $(=n_2)$
Transp.	25 (19)	10 (33.5)	40 (22.5)	75 $(=n_3)$
Bldg.&Grnds.	11 (19)	50 (33.5)	14 (22.5)	75 $(=n_4)$
	76	134	90	300 $(=n)$

From (15.18),

$$u = \frac{(27 - 19)^2}{19} + \cdots + \frac{(14 - 22.5)^2}{22.5} = 56.22.$$

For degrees of freedom $= (r - 1)(c - 1) = 6$, $\mathcal{R} = \left\{ u \,\middle|\, u \geq \chi^2_{0.99,6} = 16.81 \right\}$. Reject H_0.

15-7 $H_0: p_1 = \cdots = p_5$, H_1: the p_i's are not all equal, $i = 1, \ldots, 5$.

	In Favor	Opposed	
Officers	37	13	$n_1 = 50$
Hunters	10	40	$n_2 = 50$
Teachers	25	25	$n_3 = 50$
Clergy	40	10	$n_4 = 50$
Shopkeepers	30	20	$n_5 = 50$
	142	108	250 (=n)

$\hat{p} = \dfrac{142}{250} = 0.568$,

$1 - \hat{p} = 0.432$

$e_{i1} = 28.4$, $e_{i2} = 21.6$, $i = 1, \ldots, 5$.
From (15.14),

$$u = \frac{(37 - 28.4)^2}{28.4} + \cdots + \frac{(20 - 21.6)^2}{21.6} = 51.71.$$

$\mathcal{R} = \left\{ u \,\middle|\, u \geq \chi^2_{0.90,4} = 7.78 \right\}$. Reject H_0.

15-8 $\hat{\psi} = \dfrac{n_{11}n_{22}}{n_{12}n_{21}} = 0.33$, $\hat{\sigma}(\ln \hat{\psi}) = (0.233)^{1/2} = 0.4827$.

$\ln \hat{\psi} \pm 1.96\hat{\sigma}(\ln \hat{\psi})$ becomes $(l, u) = (-2.0548, -0.1626)$. Then $(e^l, e^u) = (0.1281, 0.8499)$.

$$\varphi^2 = \frac{4.32}{80(2 - 1)} = 0.054.$$

15-11

	Attribute B			
	10	15	2	27 $(=n_{1.})$
	7	10	9	26 $(=n_{2.})$
Attribute A	5	15	8	28 $(=n_{3.})$
	3	20	10	33 $(=n_{4.})$
	25$(=n_{.1})$	60$(=n_{.2})$	29$(=n_{.3})$	114 $(=n)$

$$\hat{\lambda}_{A/B} = \frac{10 + 20 + 10 - 33}{114 - 33} = 0.0864,$$

$$\hat{\sigma}(\hat{\lambda}_{A/B}) = \left[\frac{(114 - 40)(40 + 33 - 2(20))}{(114 - 33)^3} \right]^{1/2} = 0.0678,$$

$\hat{\lambda}_{A/B} \pm 2.58 \hat{\sigma}(\hat{\lambda}_{A/B})$ becomes $(-0.0885, 0.2613)$.

$$\hat{\lambda}_{B/A} = \frac{15 + 10 + 15 + 20 - 60}{114 - 60} = 0,$$

$$\hat{\sigma}(\hat{\lambda}_{B/A}) = \left[\frac{(114 - 60)(60 + 60 - 2(20))}{(114 - 60)^3} \right]^{1/2} = 0.1656,$$

$\hat{\lambda}_{B/A} \pm 2.58 \hat{\sigma}(\hat{\lambda}_{B/A})$ becomes $(-0.4273, 0.4273)$.

15-13 H_0: the proportion of pans exhibiting food sticking is the same for No-Stick and margarine,
H_1: not H_0.

		No-Stick		
		Sticking	Not Sticking	
Margarine	Sticking	12	20 $(=n_{12})$	32
	Not Sticking	15 $(=n_{21})$	13	28
		27	33	60 $(=n)$

From (15.32),

$$\chi^2 = \frac{(|20 - 15| - 1)^2}{20 + 15} = 0.4571.$$

$\mathcal{R} = \left\{ \chi^2 \,\middle|\, \chi^2 \geq \chi^2_{0.95,1} = 3.84 \right\}$. Do not reject H_0.

16-1 $\overline{X} = 9.2, \overline{Y} = 8.8, \sum_{i=1}^{10} x_i^2 = 171.6, \sum_{i=1}^{10} y_i^2 = 145.6, \sum_{i=1}^{10} x_i y_i = 142.4.$

a. $\hat{\beta}_1 = 0.8298, \hat{\beta}_0 = 1.1655. \hat{Y} = 1.1655 + 0.8298X.$

b. From (16.7), $s_\varepsilon^2 = 3.4289$.

c. $s_\varepsilon = 1.8517$.

d. From (16.8), $s_{\hat{\beta}_0} = 1.4262, s_{\hat{\beta}_1} = 0.1414$.

16-3 $\hat{\eta} = 0.8298(9.2/8.8) = 0.8675$. Inelastic case. A 1% increase in X is associated with a 0.87% increase in Y on the average.

References and Suggested Reading

Agresti, A. (1984). *Analysis of Ordinal Categorical Data*. New York: Wiley.

Agresti, A. (1990). *Categorical Data Analysis*. New York: Wiley.

Agresti, A. (1996). *An Introduction to Categorical Data Analysis*. New York: Wiley.

Bowman, K. O., Shenton, L. R. (1975). Omnibus Test Contours for Departures from Normality Based on $\sqrt{b_1}$ and b_2. *Biometrika*, 12, 243–252.

Brillouin, L. (1962). *Science and Information Theory*. New York: Academic Press.

Canavos, G. C. (1984). *Applied Probability and Statistical Methods*. Boston: Little, Brown and Co.

Casella, G., Berger, R. L. (1990). *Statistical Inference*. Belmont, California: Duxbury.

Cochrane, D., Orcutt, G. H. (1949). Application of Least Squares Regressions to Relationships Containing Autocorrelated Error Terms. *Journal of the American Statistical Assn.*, 44, 32–61.

Conover, W. J. (1999). *Practical Nonparametric Statistics*, 3rd ed. New York: Wiley.

Dudley, R. M. (1999). *Uniform Central Limit Theorems*. New York: Cambridge University Press.

Durbin, J., Watson, G. S. (1951). Testing for Serial Correlation in Least Squares Regression. *Biometrika*, 37, 409–428.

Feller, W. (1968). *An Introduction to Probability Theory and Its Applications*, Volume I. New York: Wiley.

Feller, W. (1968). *An Introduction to Probability Theory and Its Applications*, Volume II. New York: Wiley.

Freeman, H. (1963). *Introduction to Statistical Inference*. Reading, MA: Addison-Wesley.

Freund, J. E. (1971). *Mathematical Statistics*, 2nd ed. New Jersey: Prentice-Hall, Inc.

Goldfeld, S. M., Quandt, R. E. (1965). Some Tests for Homoscedasticity. *Journal of the American Statistical Assn.*, 60, 539–547.

Hogg, R. V., Graig, A. T. (1970). *Introduction to Mathematical Statistics*, 3rd ed. New York: The MacMillan Co.

Hogg, R. V., Tanis, E. A. (1997). *Probability and Statistical Inference*, 5th ed. New Jersey: Prentice-Hall, Inc.

Hollander, M., Wolfe, D. A. (1999). *Nonparametric Statistical Methods*, 2nd ed. New York: Wiley.

Johnston, J. (1984). *Econometric Methods*, 3rd ed. New York: McGraw Hill Book Co.

Kleinbaum, D. G., Kupper, L. L., Muller, K. E. (1998). *Applied Regression Analysis and Other Multivariable Methods*. Boston: PWS-Kent.

Kmenta, Jan (1986). *Elements of Econometrics*, 2nd ed. New York: MacMillan.

Kolmogorov, A. N. (1956). *Foundations of the Theory of Probability*, 2nd ed. New York: Chelsea Publishing Co.

Larsen, R. J., Marx, M. L. (1981). *An Introduction to Mathematical Statistics and Its Applications*. New Jersey: Prentice-Hall, Inc.

Lehmann, E. L. (1986). *Testing Statistical Hypotheses*, 2nd ed. New York: Springer-Verlag.

Lehmann, E. L. (1999). *Elements of Large-Sample Theory*. New York: Springer-Verlag.

Lehmann, E. L., Casella, G. (1998). *Theory of Point Estimation*, 2nd ed. New York: Springer-Verlag.

Lohr, S. L. (1999). *Sampling: Design and Analysis*. Belmont, California: Duxbury.

Maddala, G. S. (1992). *Introduction to Econometrics*, 2nd ed. New York: MacMillan Publishing Co.

Mendenhall, W., Scheaffer, R. L., Wackerly, D. D. (1986). *Mathematical Statistics with Applications*, 3rd ed. Massachusetts: Duxbury.

Mood, A. M., Graybill, F. A., Boes, D. C. (1974). *Introduction to the Theory of Statistics*, 3rd ed. New York: McGraw-Hill Book Co.

Mukhopadhyay, N. (2000). *Probability and Statistical Inference*. New York: Marcel Dekker, Inc.

Myers, R. H. (1990). *Classical and Modern Regression with Applications*, 2nd ed. Boston: PWS-Kent.

Neter, J., Wasserman, W., Kutner, M. H. (1989). *Applied Linear Regression Models*, 2nd ed. Homewood, IL: Irwin.

Pearson, K., ed. (1956). *Tables of the Incomplete Beta Function.* New York: Cambridge University Press.

Pearson, K., ed. (1956). *Tables of the Incomplete Gamma Function.* New York: Cambridge University Press.

Pielou, E. C. (1966). The Measurement of Diversity in Different Types of Biological Collections. *J. Theoret. Biol.*, 13, 131–144.

Prais, S. J., Winsten, C. B. (1954). *Trend Estimators and Serial Correlation,* Discussion Paper no. 383. Chicago: Cowles Commission.

Pratt, J. W., Raiffa, H., Schlaiter, R. (1995). *Introduction to Statistical Decision Theory.* Cambridge, Massachusetts: MIT Press.

Scheaffer, R. L., Mendenhall, W., Ott, L. (1986). *Elementary Survey Sampling,* 3rd ed. Massachusetts: Duxbury.

Sheskin, D. J. (2000). *Handbook of Parametric and Nonparametric Statistical Procedures,* 2nd ed. Boca Raton, Florida: Chapman & Hall.

Theil, H., Nagar, A. L. (1961). Testing the Independence of Regression Disturbances. *Journal of the American Statistical Assn.*, 56, 93–806.

Winkler, R. L., Hays, W. L. (1975). *Statistics: Probability Inference, and Decision,* 2nd ed. New York: Holt, Rinehart and Winston.

Index

Printed and bound by CPI Group (UK) Ltd, Croydon, CR0 4YY

08/05/2025

01864788-0001